ENCYCLOPAEDIA
OF
FUNGI

ENCYCLOPAEDIA OF FUNGI

Vol. 1

Edited by

A.M. Bagulia

ANMOL PUBLICATIONS PVT. LTD.

NEW DELHI - 110 002 (INDIA)

ANMOL PUBLICATIONS PVT. LTD.
H.O.: 4374/4B, Ansari Road, Darya Ganj,
New Delhi-110 002 (India)
Ph.: 23278000, 23261597
B.O.: No. 1015, Ist Main Road, BSK IIIrd Stage
IIIrd Phase, IIIrd Block,
Bangalore - 560 085 (India)
Visit us at: www.anmolpublications.com

Encyclopaedia of Fungi

First Published, 2008

ISBN 978-81-261-3526-4 (Set)

PRINTED IN INDIA

Printed at Mehra Offset Press, Delhi.

Contents

Preface

Fungi are a kingdom of eukaryotic organisms. The fungi are heterotrophic organisms characterised by a chintinous cell wall, and in majority of species, filamentous growth as multicellular hyphae forming a mycelium; while some fungal species grow as single cells. Sexual and asexual reproduction is via pores, often produced on specialised structures or in fruiting bodies. For example, yeasts, moulds, and mushrooms are fungi.

Fungi are found all over the world, and grow in a wide range of habitats, including desert. Most fungi grow in terrestrial environments, but several species occur only in aquatic habitats. Fungi along with bacteria are the primary decomposers of organic matter in almost all terrestrial ecosystems, worldwide. As per estimates there are 1.5 million fungal species of which around 70,000 have been marked. Most fungi grow as thread-like filaments called hyphae, which form mycelium, while others grow as single cells. Human use of fungi for preparation or preservation and other purpose is extensive: yeasts are required for fermentation of beer and bread, some other fungal species are used in the production of soy sauce and tempeh.

Mushroom farming and gathering is a large industry in many countries. Many fungi are producers of antibiotics, including β-lactam antibiotics, such as a pencillin and cephalosporin.

Some of the most well-known types of fungi are edible or poisonous mushrooms. Many species are commercially raised but others are grown in forests. *Agaricus bisporus*, sold as button mushrooms, when small and Protobello mushroom when grownup, are the most commonly eaten species, used in salads, soups, and many other dishes.

Many fungi have important symbiotic relationships with organisms from most, if not all kingdoms. The mycorrhiza symbiosis is most ancient and its existence dates back to 400 million years ago.

This eminent work covers all perfect and realistic features of fungi in an interesting manner. Expectedly, this work would create new waves in the discipline and would prove to be useful for all—students, scholars and general readers.

Editor

1

Introduction

Modes of Existence

The common characteristic of all living organisms is that they possess metabolic power, i.e., they continually use up and renew the materials of which their bodies are composed; and thus, to sustain life, they need constant and unfailing supplies of food as fuel for their vital energies. These foods, which are essentially the same in both plants and animals, and which consist of carbohydrates, proteins, fats, etc., with mineral salts and water as vital adjuncts, can only be constructed by "green" or chlorophyll-bearing plants.

Thus all other forms of life, including such widely divergent organisms as fungi and animals, are directly or indirectly dependent on green plants for their supplies of these necessary foods. In the case of the fungi, this complete dependence has done much to shape their destinies. Because of it, they themselves can never be pioneers, but must remain—to continue the simile—the camp followers of the plant world; for until other organic life has first occupied a location, no fungus can there find subsistence. Further, this very dependence has served to evolve a variety not only of form, but also of habit, probably greater than that of any other form of life—a variety so great that concise description would fail in the task of depicting it.

We have, in this matter, likened the fungi to animals, and if we turn for a moment to the latter, we find that, in feeding, certain creatures are herbivorous or "vegetarians"; while others are carnivorous or "meat-eaters." Of the latter, again, some are predatory, others are "carrion-feeders." The members of each class select that type of food to which, in course of time, they have become accustomed or adapted, and normally feed only upon it. In so doing, they are not exercising individual choice so much as following natural instinct.

Fungi, of course, do not ingest their food, but absorb it through the cell walls of the mycelium, after it has been converted, by means of the ferments or enzymes which they secrete, into a suitable soluble form: the starch into sugar or glucose; the albumen into other nitrogenous compounds, and so forth. But, in this matter of "choice" of foods, we find among them a state of affairs somewhat similar to that existing among animals.

A fungus spore, under suitable weather conditions, will germinate almost anywhere, but unless the resultant germ tube almost immediately encounters a hospitable medium, i.e., one capable of furnishing the accustomed type of food, it soon perishes, for, unlike the animals, it naturally cannot roam in search of its food.

This factor of natural selection plays an important part in the lives of fungi, since it tends to correlate a fungus with its host substance, and, through the latter, to fix its habits and habitat. Moreover, this factor proves a safeguard in preventing these plants from completely overrunning the earth.

Pursuing our theme, we find, excluding for the moment certain rather exceptional cases, that the fungi from the humus ever present in the soil, and arising from dead grass and decaying plant debris generally. The composition of this humus naturally varies with its original source, and as, within limits, the food requirements of individual species of fungi also vary, it is found that meadow fungi, like the Common Mushroom, never occur in woods; while woodland species seldom invade pastures. Again, in the woods, as the tree formation changes, so do the kinds of fungi encountered.

The fungus flora of a beech wood differs materially from that of an oak wood, while a plantation of conifers supplies yet another range of species. Another factor which has a marked effect on the prevalence of individual species, is the composition of the soil, whether chalk, sand or clay. It will be readily seen that the combination of these characters tends to segregate these fungi into a number of communities, more or less well defined; and thus the chaotic profusion in which they appear to occur is more orderly than might at first be supposed.

Wood-destroying Saprophytes: A further large class of fungi confine their attention to wood—old fence rails and posts, fallen logs, tree stumps, and the like. The mycelia of such fungi secrete ferments which break down the woody tissues of these materials, and in this way not only liberate quantities of nutrient substance for their own use, but in course of time reduce the timber to a powdery mass which mingles with the soil, and thus becomes available for the sustenance of other plants.

This process, however laudable when applied to useless tree stumps, is naturally not appreciated when exercised on useful timbers (the woodwork of a house, railway ties, etc.), and among this class of fungi are to be found the agents of many of the wood rots.

Dung Saprophytes: The dung of certain animals affords an excellent medium for the growth of many fungi, both large and minute. Many of these result from spores deposited from the air, but, in other cases, the spores pass through the animal with its food. In the latter case, not only are the spores unharmed, but their germinative powers are frequently increased by the action of the digestive juices of the animal.

Other Saprophytes: Although the former classes cover most of the larger fungi, some consideration should be given to a large number of microscopic fungi, which live saprophytically on foods, clothes, feathers and various other materials, when conditions are suitable.

Among those quite familiar to the housewife are the blue and green moulds (certain of the *Aspergillaceae)* often found growing on the surface of jams, preserved fruits, potted meats and other foods, and arising from the development of spores deposited thereon from the air. The germination of these airborne spores results in a film of mycelium, carrying enormous numbers of bluish spores, to which the colour of the mass is due. Of the same type are the black moulds (certain of the *Mucorineae)* of bread, cake, cheese, etc.

This list might be extended almost indefinitely, for, as we have said, there is no form of organic material which does not, under suitable conditions, provide the necessary food materials for the existence of some fungi. Sufficient examples have, however, already been given, to illustrate the versatility of saprophytic fungi as a whole, and the variety of the substances on which they can take up their abode.

Parasitic Fungi

Parasitism (Gr. *parasites:* one who sups at another's table) may be defined as that property which causes an organism to attach itself to another, and to nourish itself at its host's expense. When, as is usually the case, this dual strain is brought to bear on resources which were only designed to cater for one, the results can hardly fail to be detrimental to the host.

Indeed, so serious may be the check to its vital processes that a state of disease is provoked. It is this power of causing disease, both in animals and plants, which forms the heaviest item in the indictment which can be made against the parasitic fungi.

At the same time, this parasitic habit, though admittedly

degenerate, is one which often entails considerable adaptation to the vicissitudes of the host, and it is this necessity for adaptation which has caused the parasitic fungi to exhibit in most cases a degree of specialisation, and a variability of life methods, greater than that shown by the saprophytic kinds.

While not necessarily characteristic of them, parasitism is found chiefly among the more minute groups of fungi, and since these provide a complete and an intricate study in themselves, we do not propose to deal with other than general principles.

The power of attacking, which obviously precedes their existence on living organisms, is manifested by parasitic fungi in varying degrees; but in all cases it appears to be due primarily to a definite, if unwilling, attraction exercised by the host, rather than to any innate pugnacity on the part of the fungus. This attraction is furnished by certain chemical substances, present in the tissues of the host, which react on the germ tube of a fungus spore, and supply this organ with the necessary impulse to attack.

Conversely, when these substances are inimical to the fungus, no stimulus is received, and the germinating spore soon dies from lack of nourishment. Thus it is the composition of the host which is the potential deciding factor in regulating the number and the kinds of parasitic fungi which can subsist upon it. The only hosts available are members of either the animal or the vegetable kingdoms, and these form convenient headings for their further consideration.

Fungi Parasitic on Animals: Of the thousands of known fungus parasites, those which are capable of subsisting on the tissues of living animals—the term is used in its widest sense—are in the vast minority. The explanation of this relative immunity is found in the operation of several factors, of which the chief are:

(i) The alkaline reaction of animal tissues—fungi generally prefer acid or neutral media,

(ii) Their comparative freedom from air cavities in which the fungus mycelium can develop readily,

(iii) The body temperature, which, in the higher animals, is greater than that at which most fungi develop,

(iv) The natural combative powers of the white corpuscules of the blood.

In view of these facts, it is correct to assume that of the fungus parasites of animals, the larger proportion is supported by the lower members of the animal world, insects being especially vulnerable. Even so, there is hardly a class of animal, from the smallest unicellular organisms of our ponds, right up to man himself, which does not contain potential hosts for these fungi. Lest, however, the reader should imagine himself in the light of a peripatetic mushroom bed, let us hasten to add that in almost every case these fungi are microscopic growths.

Fungi Parasitic on Plants: The vast proportion of parasitic fungi have linked their destinies with those of other plants, and it is here that we find the most remarkable examples of adaptation to a parasitic existence.

The versatility exhibited by many individuals is really extraordinary, for not only do they change their complete forms two or three times during the course of their life, but in the same time even utilise two distinct host plants. Some idea of the completeness of this apparent change of identity may be gathered from the fact that the three stages of the Wheat Rust *(Puccinia graminis)* were considered by early investigators to be three distinct species of fungi.

Even the mycelium, which is usually the most stable organ, participates in this versatility. In the simplest forms (the ectoparasites), it remains external, merely ramifying on the surface of the plant, usually the leaves, in the form of a loose web, and taking up nourishment, either by means of small suckers *(haustoria)*, which are thrust beneath the surface of the plant, or by direct osmotic absorption through the epidermis, which in this case naturally remains unbroken.

In the more advanced section of parasites (the endoparasites), the mycelium itself lives wholly within the tissues of the plant, where it has not only greater facilities for feeding, but also greater opportunities for harm. Even so, it is not in the interest of the fungus to kill its host plant—although this too frequently is the result—and many species, apparently in an attempt to obviate this dilemma, have in the past confined their attention chiefly to one organ of the host plant, and on this organ alone have they now the power to live.

Leaf Parasites: Of the various parts of a plant, the leaves are the organs most open to attack. Their tender tissues; the breathing pores (stomata) which communicate with the outer air, and which, therefore, give convenient ingress to marauding fungi; the numerous moist air cavities; and the proximity of the starch-making apparatus (the chlorophyll): all assist in rendering the leaves ideal breeding places for parasitic fungi. Thus it is they who have to support the most extensive, and, for the same reason, the most destructive group of parasites. In this group are the various rusts, mildews, leaf-curls, leaf-spots, etc., and although their actual presence usually passes unnoticed, the damage which they cause is only too evident. The various coloured spots, scabs, etc., on the surfaces of leaves, as well as the wilting and yellowing of these organs, are due mostly to the action of these fungi.

Stem Parasites: The conditions existing in the leaves are in part reproduced in the younger shoots, which have, therefore, to accommodate the same or a similar band of parasites. As the stem gets older, the tissues firmer, and a tough outer rind or cuticle is formed, the difficulties of entrance on the part of the fungus are increased, and, therefore, the liability to attack is lessened. Finally, when the stem becomes truly woody, it can only provide a satisfactory medium for those fungi which develop wood-destroying ferments.

Root Parasites: The roots of a plant do not provide conditions so suitable to fungus attack, yet many parasites which prefer the moisture conditions provided by the soil make their homes in these organs, where they often cause such damage as to prevent their proper functioning.

Flower and Fruit Parasites: A sufficiently large number of fungi infest the flowers and especially the fruits of plants. The latter are particularly susceptible, for not only are they, in their later stages, often gorged with foods intended as an attractive present to those animals and birds on which they rely for the distribution of their seeds—foods which prove equally acceptable to many fungi—but there is the additional advantage that resting spores or dormant mycelium may be carried with the seeds, the fungi being thus enabled to repeat their life cycles on the new plants.

Effects of Parasitic Fungi on their Hosts: Many parasitic fungi have little or no effect on their host plants, but others have tremendous disease-causing propensities. With regard to the latter, speculation may arise as to how this damage is brought about. The mycelium of the invading fungus may nourish itself by drawing on the foods which the plant is producing for its own future use, in which case the ill-effects on the host are not pronounced; or, it may destroy and absorb the complete structures of the host in its immediate vicinity. In the latter event, the amount of injury caused depends upon the activities of the fungus.

In some cases the area occupied by the fungus may be small, the result being merely the production of unsightly, decayed patches, which reduce by this amount the utility of the organs on which they occur. In other cases the damage is very considerable, extending frequently to areas widely separated from the immediate seat of the fungus.

This extended field is due to toxic substances liberated in the vessels of the plants and carried along them to more distant parts; or, to interference with the normal flow of foods to these parts. The irritation set up by the fungus frequently causes reaction on the part of the host, which reaction usually takes the form of a greatly-increased production of tissue, resulting in the appearance of fungus galls, "witches, brooms," etc. In short, parasitic fungi have, in most cases, a distinctly injurious effect upon their hosts, and this injury is evidenced by certain external symptoms, the chief of which are given below.

Rot, due to the breaking down of tissue, is probably the most common symptom of disease to be noted in any plant part. It has numerous forms, and may be of a wet, dry, soft, or hard character, and may be accompanied by an offensive odour, or be odourless.

Blight is a symptom often exhibited by leaves, roots, stems, flowers or fruits. It indicates complete death of the part affected, may be local or general, and may or may not be followed by rot.

Wilt of stems or leaves is caused usually, either by a plugging, or other interruption, of the sap-carrying vessels, or by some injury to the absorbing root system. Spots on leaves, bark or fruits, may be caused by local blight, rot of tissue, imperfect colouration, or by local overdevelopment or underdevelopment of colour. Scab in fruit, leaves, or bark, is due to the growth of the causal parasite upon or near the surface, or to thickening of the outer layer or layers of tissue caused by the irritation of the parasite.

Mildew upon any plant surface consists of a whitish powdery fungus growth. Burn is a general term applied to all cases where the leaves turn red or brown, especially if the edge first shows the symptom. Smut consists of a mass of spores, usually black and powdery. It is most common in the ovaries and other floral parts of various grasses.

Rust should properly be applied to the diseases caused by the rust fungi (the *Uredineae).* These are accompanied by rust-coloured streaks due to the appearance of the spore clusters. Yellowing of leaves, due to diminution in green colouring matter, is a common symptom of disease.

Chlorosis is also a lack of proper green, the normal green tissues becoming bleached. Canker is a roughening and splitting of the bark. The Effects of the Host upon the Fungus. When such powerful influences are at work, it might well be assumed that any effect which the host plant has upon the fungus would be negligible; yet such an assumption would be far from the truth.

Certainly, when once infection has taken place, the fungus is usually able to proceed without much interference from its victim. But, for all that, plants have had a very considerable effect in controlling the life methods of the parasitic fungi which prey upon them.

We have already seen that the attractive power of the host tends to restrict the fungi which can exist upon it, but in addition to this, a factor controlled by the seasons is introduced. The dependence of these fungi on plants which themselves have a restricted season of growth forces the fungi to experience, and, therefore, to make provision for, the periods during which the host plant is unavailable. These "close" seasons they have learned to meet in a number of ways, by means of sclerotia, resting spores, dormant mycelium, etc. all of which entail a degree of hibernation.

A more active manner of bridging the gap is found among certain of the rusts *(Uredineae)*, where the fungus abruptly changes its character just before the hiatus, and passes on to another host plant which *is* available in the interim, or at least in part of it. This mode of existence is known as heteroecism (Gr. *heteros*—different and *oikos* = a house) and is well exemplified in the Black Rust of Wheat *(Puccinia graminis)*. Here the spring form of the fungus occurs on the barberry; the summer and autumn stages on wheat or similar grasses; while the winter or resting stage, when neither host is available, is passed on the dead wheat straw, etc., or upon the surface of the soil.

This alternation of host, however, is not absolutely necessary to the existence of the fungus, for, in the absence of the alternative host, this stage of the life cycle can be omitted.

For example, the black rust of wheat has been prevalent in Australia for almost a century, yet no barberry bushes occur there. The exact advantage of heteroecism to the fungus is in this case not obvious, while in certain other instances it is almost as obscure.

Intermediate Forms: Since parasitism is an acquired character, it would be natural to suppose that, between the two extremes of true saprophytism and true parasitism, would occur a number of intermediate forms in various transition stages of development. Of such forms two main types may be distinguished: the *hemi-saprophytes,* which are capable of completing their life cycles as saprophytes, but which, under certain conditions, have the faculty of attacking living tissues; and, the *hemi-parasites,* which usually feed on living organisms, but which can, if necessary, pass part of their lives as saprophytes.

Hemi-saprophytes: Of this class of fungi, the most important section comprises species known as "wound-parasites," of which, again, the majority are ravagers of trees. These fungi are unable to attack uninjured, healthy trees, but if access can be gained by means of a wound—a broken branch, an abrasion caused by browsing animals, or an insect puncture—the mycelium lives first on the dead and rotting tissue at the seat of the injury, and, having gained a firm footing at this point, is then able to pass on to the healthy tissue, and may, in this way, soon involve the whole tree. Among the larger wound parasites of note are various polypores, such as *Fomes igniarius, F. fomentarius, F. applanatus, Polyporus betulinus, P. squamosus,* etc. various species of *Hydnum, Stereum,* and many others.

Hemi-parasites: These fungi, while normally acting as parasites, are able at need to exist saprophytically, although when reduced to the latter existence, they are seldom able to complete their full development. Of this type are the smuts, which, in the absence of a living host, are able to live for a considerable time upon the soil. Many of the ripe fruit rots, which exist saprophytically on ripe fruits, commenced their life as parasites upon the living tissues of the unripe fruits. In both of these groups the parasitism displayed is frequently of an exceedingly weak order, in that the plant parts selected for attack are seldom in their full vigour.

Symbiotic Unions

The living together in intimate association of two organisms does not lead always to a benefit for one at the expense of the other. In some cases a partnership is formed, in which each participant undertakes certain work for their mutual benefit. Such unions are termed *symbiosis* (Gr. *syn* = together: *bios*—life) and that they probably arose as abortive attempts at parasitism, does not alter the fact that the ultimate result has, in almost every case, been so advantageous that the dual organism is stronger, and better able to battle for existence, than would be either organism independently.

Symbiotic unions are found in all sections of life. Among the fungi the best-known examples of this phenomenon are:

1. The lichens, which are unions between fungi and algae; and
2. *Mycorrhiza*, unions between fungi and the higher plants.

Lichens: In the lichens, those foliaceous incrustations commonly seen on rocks and on the bark of trees, the component organisms are usually one of the green algae such as are found on flower pots, or as scums on ponds, and a member of the Ascomycetes; the dual entity being quite unlike either of the participant organisms.

In the economy of these dual plants, to the alga, since it contains chlorophyll, is assigned the task of starch-making for itself and its partner; while the filaments of the latter not only supply a protective coat, but also absorb the water and mineral salts necessary for both. It is interesting to note that, while this partnership is of joint benefit, the fungus is apparently the predominant organism, in that, in most cases, it alone is capable of producing reproductive bodies.

Mycorrhiza: During the past eighty years, numerous cases of symbiosis between fungi of various kinds and the higher plants have come to light. The fungi are usually found associated with the roots of the plant, and, therefore, are termed mycorrhiza ("fungus-roots"). Of these, probably the most universal examples occur in the case of forest trees growing in soil rich in humus. In these, parts of the mycelium of the fungus appear as compact sheaths around the root tips, the remainder acting as the absorbent medium. In this way, nutrient materials obtained from the humus by the fungus can be utilised by the tree, and the reliance which the latter places on this mode of nutrition is illustrated by the loss of its root hairs which would normally perform this function. In a somewhat similar manner, many bog-loving plants employ mycorrhiza for obtaining part of their food. In other instances, notably with the orchids and the heaths, the mycelium dwells actually within the root-cells, which supply the necessary organic food, and in return are enabled to absorb waste material resulting from the death of parts of the fungus.

It is of interest to note that mycorrhiza are most abundant in plants which inhabit localities rich in humus, and may be entirely absent from other plants of the same species growing in places relatively devoid of this material. That in all cases the plants so infected not only suffer no inconvenience, but rather gain by the presence of the fungus, is shown by the fact that infected plants are, on the whole, more vigorous and more fertile than uninfected ones. This is especially the case with orchids, where the presence of the appropriate fungus is absolutely necessary for the successful raising of seedlings. It should be noted, however, that the fungus can be artificially replaced by the presence of certain carbohydrates in solution.

Conclusion: This brief sketch of the life methods of the fungi is, of necessity, extremely curtailed, and for this reason much of interest has had to be omitted. It is hoped, however, that, short as it is, it will serve to illustrate some of the vagaries of these peculiar plants, and the manner in which they live and have their being.

2

Study of Fungi

Mycology (from the Greek meaning "fungus") is the study of fungi, their genetic and biochemical properties, their taxonomy, and their use to humans as a source for tinder, medicinals (e.g., penicillin), food (e.g., beer, wine, cheese, edible mushrooms) and entheogens, as well as their dangers, such as poisoning or infection. From mycology arose the field of phytopathology, the study of plant diseases, and the two disciplines remain closely related. A biologist who studies mycology is called a mycologist.

Historically, mycology was a branch of botany (though fungi are evolutionarily more closely related to animals than plants this was not recognised until recently). Pioneer *mycologists* included Elias Magnus Fries, Christian Hendrik Persoon, Anton de Bary and Lewis David von Schweinitz. The British Mycological Society was founded in 1896.

Today, the most comprehensively studied and understood fungi are the yeasts and eukaryotic model organisms *Saccharomyces cerevisiae* and *Schizosaccharomyces pombe*.

Many fungi produce toxins, antibiotics, and other secondary

metabolites. For example, the cosmopolitan (worldwide) genus *Fusarium* and their toxins associated with fatal outbreaks of alimentary toxic aleukia in humans were extensively studied by Abraham Joffe.

Fungi are fundamental for life on earth in their roles as symbionts, e.g. in the form of mycorrhizae, insect symbionts and lichens as well as their potency in breaking down complex organic biomolecules such as wood as well as xenobiotics, a critical step in the global carbon cycle.

Fungi and other organisms traditionally recognised as fungi (such as oomycetes and slime moulds) often are economically and socially important as some cause diseases of animals (such as histoplasmosis) as well as plants (such as Dutch elm disease and Rice blast).

Field meetings to find interesting species of fungi are known as 'forays', after the first such meeting organised by the Woolhope Naturalists' Field Club in 1868 and entitled "a foray among the funguses."

Auricularia Auricula-judae

Auricularia auricula-judae (syn. *Auricularia auricula, Hirneola auricula-judae*) is commonly known as the Jew's ear fungus or Judas's ear fungus, the name from which it derives the "judae" in its scientific name, or as the jelly ear fungus. This jelly fungus is conspicuously ear shaped, ranging from purple to dark brown or black in colour with a rubbery texture, and most often found on dead elder trees but also on elms. It was said that Judas hanged himself on an elder tree, which is the origin of the name.

Edible Mushroom

There are thousands of regularly harvested edible mushrooms in the world, in addition to numerous other potentially edible species. Some species are highly prized because they are not or cannot be cultivated and must be harvested from natural settings. Check safety rules under mushroom hunting, however, before assuming any wild mushroom is edible.

Even if a mushroom is edible, it doesn't mean it will taste good. Some non-poisonous mushrooms like Bitter bolete, have an extremely bad taste.

The pharaohs of Egypt enjoyed mushrooms so much, that they decreed mushrooms could only be eaten by royalty and that no commoner could even touch them, thus giving the royal family the entire available supply.

In some parts of Eurasia, especially in Russia and Nordic countries, mushrooms are an important part of the diet. Several mushrooms are especially tasty and many are rich on nutrients. Mushrooms are also easily preserved, and historically have provided additional nutrition over winter.

Many pre-historic and a few modern cultures around the world used psychedelic mushrooms for ritualistic purposes. Before 10,000 BCE while people were still hunting and gathering as a part of every day life, women did the gathering.

Women were said to be blessed with the ability to see in the dim light so they were successful in foraging for mushrooms and fungi amongst other things.

Mushroom cultivation reached the United States in the late 1800s with imported spores from Mexico. Some species such as death cap are extremely poisonous and have been deliberately used as instruments of assassination.

Current Culinary Use

A fraction of the many fungi consumed by humans are currently cultivated and sold commercially. Pioneers such as Paul Stamets are introducing more into cultivation.

Commercial cultivation is important ecologically, as there have been concerns of depletion of larger fungi such as chanterelles in Europe, possibly because the group has grown so popular yet remains a challenge to cultivate.

Commercially Cultivated Fungi

Agaricus bisporus, also known as champignon and the button mushroom, the most extensively cultivated mushroom in the world, accounting for 38 per cent of the world production of cultivated mushrooms. This species also includes the portobello and crimini mushrooms.

Agaricus Campestris

Auricularia polytricha or *Auricularia auricula-judae* (Tree ear fungus), two closely related species of jelly fungi that are commonly used in Chinese cuisine. Flammulina velutipes, the "winter mushroom", also known as *enokitake* in Japan.

Gyromitra esculenta is cultivated by the Finns. This mushroom is deadly poisonous if eaten raw, but highly regarded when correctly prepared.

Hypsizygus tessulatus (also *Hypsizygus marmoreus*), called *shimeji* in Japanese, it is a common variety of mushroom available in most markets in Japan.

Lentinus edodes, also known as shiitake, oak mushroom. *Lentinus edodes* is largely produced in Japan, China and South Korea. *Lentinus edodes* accounts for 10 per cent of world production of cultivated mushrooms. Common in Japan, China, Australia and North America.

Pleurotus species, (The oyster mushroom and king trumpet mushroom). *Pleurotus* mushrooms are the second most important mushrooms in production in the world, 25 per cent of total world production of cultivated mushrooms. *Pleurotus* mushrooms are worldwide, China is the major producer. Several species can be grown on carbonaceous matter such as straw or newspaper. In the wild they are usually found growing on wood.

- *Pleurotus cornucopiae*
- *Pleurotus eryngii*
- *Pleurotus ostreatus* (Oyster mushroom)

Sparassis crispa—recent developments have led to this being cultivated in California.

Tremella fuciformis (Snow fungus), another type of jelly fungus that is commonly used in Chinese cuisine.

Tuber species, (the truffle), Truffles belong to the ascomycete grouping of fungi. The truffle fruitbodies develop underground in mycorrhizal association with certain trees e.g. oak, poplar, beech, and hazel.

Being difficult to find, trained pigs or dogs are often used to sniff them out for easy harvesting.

- *Tuber aestivum* (Summer or St. Jean truffle)
- *Tuber brumale*
- *Tuber magnatum* (Piemont white truffle)
- *Tuber melanosporum* (Perigord truffle)

Volvariella volvacea (the "Paddy straw mushroom"). *Volvariella* mushrooms account for 16 per cent of total production of cultivated mushrooms in the world.

Other Edible Fungi

Wild fungi commonly picked and consumed include:

- *Amanita caesarea*
- *Armillaria mellea*
- *Boletus badius*

Boletus edulis or edible Boletus, native to Europe, known in Italian as Fungo Porcino (plural 'porcini') (Pig mushroom), in German as Steinpilz (Stone mushroom). It also known as the king bolete, *cep*, is renowned for its delicious flavour. It is sought after worldwide, and can be found in a variety of culinary dishes.

Boletus Elegans

Cantharellus cibarius (The chanterelle): The yellow chanterelle is one of the best and most easily recognisable mushrooms, and can be found in Asia, Europe, North America and Australia.

Caution must be used, as there are several types of very poisonous (although not usually lethal) lookalikes.

- *Chroogomphus rutilus*
- *Calvatia gigantea* (Giant Puffball)

Coprinus comatus, the Shaggy mane. Must be cooked as soon as possible after harvesting or the caps will deliquesce and turn to ink.

Only the fresh young caps and stems are edible as the mature caps will turn black and unappetising.

- *Cortinarius variecolour*
- *Craterellus cornucopioides*
- *Fistulina hepatica*

Grifola frondosa, known in Japan as *maitake* (also "hen of the woods" or "sheep's head"); a large, hearty mushroom commonly found on or near stumps and bases of oak trees, and believed to have medicinal properties.

Hericium erinaceus, a tooth fungus; also called "lion's mane mushroom."

- *Hydnum coralloides*
- *Hydnum repandum*
- *Hygrophorus chrysodon*
- *Lactarius deliciosus*
- *Lactarius salmonicolour*
- *Lactarius volemus*

Laetiporous sulphureus (Sulphur shelf), also known by names such as the "chicken mushroom", "chicken fungus", sulphur shelf is a distinct bracket fungus popular among mushroom hunters.

Lepiota Procera

Morchella species, (The morel), morels belong to the ascomycete grouping of fungi. They are usually found in open scrub, woodland or open ground in late spring. When collecting this fungus, care must be taken to distinguish it from the poisonous false morel, Gyromitra esculenta.

- *Morchella conica var. deliciosa*
- *Morchella esculenta var. rotunda*
- *Polyporus squamosus*
- *Polyporus sulphureus*
- *Rhizopogon luteolus*

Sparassis crispa. Also known as "cauliflower mushroom".

- *Suillus bovinus*
- *Suillus luteus*

Tricholoma matsutake (The Matsutake), a mushroom highly prized in Japanese cuisine.

Tricholoma Terreum

Conditionally Edible Species: There are treatments to reduce or eliminate the toxicity of certain species to the point that they can be edible. This can be done even with fungi that are widely regarded as toxic.

Amanita muscaria, edible if parboiled to leach out toxins. Fresh mushrooms are hallucinogenic and may cause seizures or coma due to the presence of ibotenic acid.

Coprinopsis atramentaria, edible without special preparation however consumption with alcohol is toxic due to the presence of coprine.

Gyromitra esculenta, edible if parboiled. Fresh mushrooms are hemolytic- and hepato-toxins due to the presence of gyromitrin.

Agaricus Augustus

Agaricus augustus, also known as the Prince is a mushroom of the genus *Agaricus*. This mushroom is a good edible. It has the smell and taste of almonds due to the presence of benzaldehyde and benzyl alcohol.

Description

Cap: 5-30 cm across, dry, and has many and brown coloured scales covering the entire cap surface. The flesh is thick, white, bruises yellow in age.

Gills: Close and cream coloured at first, turning dark brown in age.

Spores: The spores are 9×6 cm and coloured chocolate brown.

Stipe: Has a skirt-like ring and is 12 to 40 cm long, 2-6 cm thick, long, white, and is smooth above the ring and scaly below:

Taste: Good.

Odour: Smells like anise.

The yellow-brown scales on the cap and anise odour differentiate this mushroom from others which resemble it.

Habitat

This mushroom is widely distributed and is found in the woods and in gardens. Though this mushroom is saprophytic, it is often found under conifers.

Agaricus Bisporus

Agaricus bisporus is a gilled fungus which naturally occurs in Europe and North America, though now occurs much more widely. It is known by many names, including table mushroom, cultivated mushroom, white mushroom, common mushroom, button mushroom (when sold, collected or eaten in young unopened form), Portobello mushroom or Crimini mushroom. It is known as the champignon de Paris in France, and is one of the most widely cultivated mushrooms in the world.

Some grocery stores in the Western world sell this mushroom in canned and fresh preparations. An agaric, its gills are often left on in preparations. It can be found cooked on pizzas and casseroles, stuffed mushrooms, raw on salads, and in various forms in a variety of dishes.

Note that while this specific mushroom is sometimes called simply champignon in the English-speaking world, this word means "fungus" in general in French, including all mushrooms, toadstools and even fungal infections.

Agaricus Subrufescens

Agaricus subrufescens (formerly known as; *Agaricus blazei*) is a species of mushroom, sometimes known as himematsutake and by a number of other names.

This Agaricus is a choice edible, with a somewhat sweet taste and fragrance of almonds. The almond flavour is due to the presence of benzaldehyde, benzyl alcohol, benzonitrile, and methyl benzoate. This mushroom is also well known for its purported medicinal properties.

Agaricus subrufescens was first described by the American mycologist William Alphonso Murrill in 1947.

It was discovered again in Brazil during the 1970s, and thought to be a new species, *Agaricus blazei*. It was soon marketed for its purported medicinal properties under various names, including:

- Cogumelo do Sol (mushroom of the sun)
- Cogumelo de Deus (mushroom of god)
- Cogumelo de Vida (mushroom of life)
- Himematsutake (in the country of Japan)
- Royal Sun Agaricus
- Mandelpilz

Almond Mushroom

In 2002, Didukh and Wasser discovered the species was identical to *Agaricus brasiliensis*; this was confirmed in 2005, when Kerrigan used DNA sequencing as proof. Because the latter is the oldest name, it is traditionally considered the scientifically, historically correct name.

Commercial Use

Because of its high beta glucan content – higher than both Reishi and Shiitake mushrooms – *Agaricus subrufescens* is used in oncological therapy, mainly in Japan and California.

It has been commercially cultivated in Asia and South America since 1993. Because of this valuable polysaccharide, and lack of supply, *Agaricus blazei* used to be relatively expensive, until it was successfully artificially cultivated in mushroom farms. China (Maucua) and Brazil are major exporters.

Agaricus blazei mushroom specifically assists in the production of interferon and interleukin, which are potent in fighting off cancer cell metastasis, especially cancer of the uterus. It also reduces blood glucose, blood pressure, cholesterol levels and the effects of arteriosclerosis.

ABM is known to contain three different beta-glucans–beta-(1-3)-D-glucan, beta-(1-4)-a -D-glucan, and beta-(1-6)-D-glucan.

Amanita Caesarea

Amanita caesarea (Caesar's Mushroom), is an edible mushroom in family Amanitaceae, native to southern Europe. The common name comes from its being a favourite of the Roman emperors, which took the name Caesar (originally a family name) as a title. Other common names include Royal Amanita, Aminite de Cesar, Oronge, or Kaiserling in German. Because this mushroom closely resembles the fly agaric, many people do believe that the Caesar's mushroom is actually the edible version of this poisonous mushroom.

This mushroom has a tawny cap with a yellow stem and gills. It may be seated in a cup-like volva (remnant of universal veil) and have the remains of a partial veil hanging from the stipe. The base of the stipe is thicker than the top. The spores are white. This mushroom favours oak woodland, sometimes mixed with conifers.

It has also been classified as *A. umbonata*. The relationship of the similar North American species *A. hemibapha* and *A. jacksonii* to *A. caesarea* is not clear. Moreover, the edibility of North American species is also unclear.

Amanita Fulva

The tawny grisette (*Amanita fulva*), or less poetically the orange-brown ringless amanita, is a basidiomycete mushroom of the genus *Amanita*.

A. fulva is one of the grisettes that are part of the genus *Amanita*. The species in this group have slightly different colours and habitats, but look similar otherwise.

The cap is tan with a clearly lined margin. The gills are white and free from the stem. The stem is white, relatively long and thin, often hollow, and without a ring. The sack-like volva is patchy white with brown. The mushroom grows near various kinds of trees, often birch. The tawny grisette is found in conifer, birch, beech and oak woodlands in Europe.

Similar Species occur in north America Historically, both the tawny grisette and the grisette (*Amanita vaginata*) were placed in the genus *Amanitopsis* due to their lack of a ring, unlike other *Amanita* species. However this distinction is now seen as insufficient to warrant a separate genus.

Edibility

Amanita fulva is edible and has a pleasant taste, but it can cause stomach problems when consumed raw. Because of the white, free-standing gills, *A. fulva* can be confused with poisonous *Amanita* species, and therefore some authors discourage picking it for the table. The other grisettes are likewise edible.

Amanita Lanei

Amanita lanei (= *Amanita calyptrata* and *Amanita calyptroderma*), also called coccoli or coccora, is a white-spored mushroom that fruits naturally in the coastal forests of the western United States during the fall and winter.

Description

This mushroom is recognised by its large, orange-brown cap partially covered by a thick patch of universal veil, its white gills and spores, its cream-coloured stalk adorned with a partial veil, and by the presence of a large, saclike volva at the base of the stalk.

The spores of this species do not change colour when placed in a solution of Melzer's reagent, and thus are termed "inamyloid". This characteristic in combination with the absence of a bulb at the base of the stalk place this mushroom in the Section *Vaginatae*.

Classification

Rodham E. Tulloss suggests that there may in fact be several distinct species currently grouped under the name *Amanita lanei.*

Distribution and Habitat

This mushroom forms mycorrhizae with Madrone (*Arbutus menziesii)* in the southern part of its range (Central California northwards to Washington).

However, in the northern part of its range (Washington to southern Canada), its preferred host is Douglas Fir *(Pseudotsuga menziesii*).

Edibility

Experienced mushroom hunters regard this mushroom as a good edible species, but caution must be exercised when collecting *A. lanei* for the table, since it can be confused with other species in the genus *Amanita*.

This genus contains some of the deadliest mushrooms in the world, most notably *A. phalloides, A. ocreata* and *A. virosa*.

Amanita Velosa

Amanita velosa (also known as the springtime amanita or bittersweet orange ringless amanita) is an edible species of agaric found in California, as well as Oregon and Baja California.

It is part of *Amanita* section *Vaginatae*, and like other species in this group, it is characterised by its lack of an annulus, striate pileus margin, thick universal veil remnants comprising the veil, volva, and pileus patches, inamyloid spores, and lack of characteristic *Amanita* toxins such as amatoxins and ibotenic acid. It is distinguished from other species in section *Vaginatae* by its lack of any kind of umbo on its pileus, its short pileus striae, and its distinct pale orange to pale salmon colouration when young. Its colouration can become more brownish with age and entirely white specimens are occasionally seen as well. Like many other *Amanita*, the gills are white, but occasionally have a distinct pinkish or orangish tint. In older specimens, the odour can become pungent and fishy.

Habitat and Range

Amanita velosa is a late-season mushroom in its range of occurrence, being found from midwinter into spring, up until the end of the California rainy season. Its favoured habitat is the ecotone between oak (particularly coast live oak) woodlands and open grassland, living in an ectomycorrhizal relationship with young oak trees.

Although this species is primarily known from the coastal regions of California, Oregon, and Baja California, it is also reported to have been found in association with aspen and conifers in the Sierra Nevada, and there is also one report of this species being found growing with spruce in the Great Smoky Mountains National Park, in the eastern United States.

Edibility and Identification

It is considered to be an outstanding edible species with a distinctively sweet or nutty flavour. However, great caution must be exercised to properly identify it, as the deadly *Amanita ocreata* (and occasionally *Amanita phalloides*) is also found in the same habitat at the same time of year, and can often be found in close proximity to each other. *A. ocreata* and *A. phalloides* have thin universal veil remnants, a sac-like volva, an annulus, a non-striate pileus margin, and a pileus that is a different colour than *A. velosa*.

These differences can fade as the fruiting body ages, making it important to collect only specimens that have all of their identifying characteristics intact. Some authorities go so far as to say no *Amanita* should be considered for eating.

This species is used often in Asian cooking. In Chinese, it is called ("wood ear" or "tree ear") or ("black wood ear"), and in Japanese it is called *kikurage* ("tree-jellyfish"). *Auricularia polytricha* (cloud ear fungus), a closely related species, is also used in Asian cooking.

Snow fungus, another edible fungus which is white in colour, is a separate species, *Tremella fuciformis*.

Birch Bracket

Birch Bracket (*Piptoporus betulinus*—also known as Razor Strop) is one of the most common polyporous bracket fungi and, as the name suggests, grows almost exclusively on Birch trees. The brackets burst out from the bark of the tree, and these fruiting bodies can last for more than a year.

Technically, it is an edible mushroom, with a strong, pleasant "mushroomy" odour but a bitter taste. It is said to have medicinal properties, and the velvety cut surface of the fruiting body were used as a strop for finishing the finest of edges on razors. Dried specimens have also been used as tinder, and this fungus was carried by "Otzi the Iceman"—the 5,000 year old mummy found in the Tyrol.

It is a necrotrophic parasite on weakened Birches, and will cause brown rot and eventually death, being one of the most common fungi visible on dead Birches. It is likely that the birch bracket fungus becomes established in small wounds and broken branches and may lie dormant for years, compartmentalised into a small area by the tree's own defence mechanisms, until something occurs to weaken the tree. Fire, drought and suppression by other trees are common causes of such stress.

In most infections, there is only one fungal individual present, but occasionally several individuals may be isolated from a single tree, and in these cases it is possible that the birch bracket fungus entered after something else killed the tree. These fungal "individuals" can sometimes be seen if a slice of brown-rotted birch wood is incubated in a plastic bag for several days.

This allows the white mycelium of the fungus to grow out of the surface of the wood. If more than one individual dikaryon is present, lines of intraspecific antagonism form as the two individual mycelia interact and repel each other.

The fruiting bodies (basidiocarps) are pale, with a smooth greyish-brown top surface, with the underside a creamy white and with hundreds of pores that contain the spores. The fruiting body has a rubbery texture, becoming corky with age. Wood decayed by the fungus, and cultures of its mycelium, often smell distinctly of green apples.

P. betulinus has a bipolar mating system where monokaryons or germinating spores can only mate and form a fertile dikaryon with an individual that possesses a different mating-type factor.

There are at least 33 different mating-type factors within the British population of this fungus. These factors are all variants or alleles of a single gene, as opposed to the tetrapolar mating system of some other basidiomycete species, which involves two genes.

The geographic distribution of *Piptoporus betulinus* appears to be restricted to the Northern hemisphere. There is some doubt about the ability of isolates from the European continent, North America and the British Isles to interbreed.

Blewit

Blewit refers to two closely-related species of edible agarics in the genus *Clitocybe*, the wood blewit (*Clitocybe nuda*) and the field blewit (*Clitocybe saeva*). Both species are treated by some authorities as belonging to the genus *Lepista*.

Classification

Both species have been treated by many authorities as belonging to the *Clitocybe* segregate genus *Lepista*. Recent molecular research suggests the genus *Lepista* is nested within *Clitocybe*.

Edibility

Both wood blewits and field blewits are generally regarded as edible, but they are known to cause allergic reactions in sensitive individuals. This is particularly likely if the mushroom is consumed raw, though allergic reactions are known even from cooked blewits. Wood blewits contain the sugar trehalose, which is edible for most people.

Field blewits are often infested with fly larvae and don't store very well; they should therefore be used soon after picking.

Wood Blewit

The wood blewit (*Clitocybe nuda = Lepista nuda = Tricholoma nudum*), is an edible mushroom, found in both coniferous and deciduous woodlands. It is especially common where there is a large amount of duff built up on the forest floor. It is a fairly distinctive mushroom which is widely eaten, though there is some caution about edibility. Nevertheless it has been cultivated in Britain, Holland and France.

Description: This mushroom can range from lilac to purple-pink. Some North American specimens are duller and tend towards tan, but usually have purplish tones on the stem and gills. The gills are attached to the short, stout stem. Mature specimens have a darker colour and flatter cap; younger ones are lighter with more convex caps. Wood blewits have a very distinctive odour, which has been likened by one author to that of frozen orange juice.

Wood blewits can be confused with certain purple *Cortinarius* species, many of which may be poisonous. Wood blewits can be easily distinguished by their odour, as well as by their spore print.

Wood blewits have a light (white to pale pink) spore print; *Cortinarius* species produce a rusty brown spore print after several hours on white paper.

Distribution and Habitat: The wood blewit is found in Europe and North America and is becoming more common in Australia, where it appears to have been introduced. It is a saprotrophic species, growing on decaying leaf litter.

Field Blewit: The field blewit (*Clitocybe saeva* = *Lepista saeva*) is a mushroom similar to the wood blewit. It is commonly found growing in open plains.

Description: The convex cap is light cream-brown with a smooth texture. The ringless stem is thin and has light purple-blue colour, hence an alternative name Blue-Leg.

Distribution and Habitat: It prefers to grow in grasslands and dirt areas across Europe.

Boletaceae

Boletaceae is a family of mushrooms, primarily characterised by holding their spores in small pores on the underside of the mushroom, instead of gills, as are found in agarics. Nearly as widely distributed as agarics, they include the Cep or King Bolete (*Boletus edulis*), much sought after by mushroom hunters. As a whole, the typical members of the family are commonly known as boletes.

Boletes are a relatively safe group of mushrooms for human consumption, as none are known to be deadly to adults, and they are the most sought after fungi for mushroom hunting. They are especially suitable for novice mushroom hunters, since there is little danger of confusing them with deadly mushrooms, like various *Amanita* agarics, which are the most poisonous mushrooms in the world. They are easily distinguished from agarics, and easily recognised for colour, pores and thick stems and caps.

Genera

Devil's bolete (Boletus satanas): Even after recent changes in taxonomy that have moved many members out of the Boletaceae, it remains a large family with many genera. Currently, 27 genera are recognised in Boletaceae.

Afroboletus, Aureoboletus, Austroboletus, Boletellus, Boletochaete, Boletus, Buchwaldoboletus, Chalciporus, Chamonixia, Fistulinella, Gastroboletus, Gastroleccinum, Gastrotylopilus, Leccinum, Paxillogaster, Phylloboletellus, Phylloporus, Pulverboletus, Royoungia, Setogyroporus, Singeromyces, Sinoboletus, Strobilomyces, Tubosaeta, Tylopilus, Veloporphyrellus, Xanthoconium

Many other genera formerly part of this family have been moved into other, smaller families as work with molecular phylogeny shows that they are more distantly related, even if physically similar. Representative of this adjustment is the move of the slimy-capped genus *Suillus* to Suillaceae.

Distribution

Boletes are found worldwide, on every continent except Antarctica. Well-known and well-described in the temperate latitudes in the northern hemisphere, newer research has shown significant diversity in tropical and southern hemisphere regions as well. E. J. H. Corner found evidence of at least 60 species on the island of Singapore alone. In 1972, he described 140 species from the Malay Peninsula and Borneo and estimated there were an equal number again to be found.

Similar statements about the biodiverse richness of Australian Boletaceae have also been made.

Edibility

Many of the boletes are considered to be true culinary delicacies, especially King Bolete (cep, porcini) (*Boletus edulis*); the Scandinavian cuisine praises boletes. In the Finnish cuisine, the King Bolete is universally considered to be the tastiest culinary mushroom, comparable only to false morel. A large number of boletes are delicious or at least edible.

Poisonous or otherwise inedible species do exist, however, such as the gastrointestinal irritant Devil's bolete (*Boletus satanas*), the unpalatable Bitter bolete (*Tylopilus felleus*) with a taste compared to gall bladder juice, and some orange-capped species of *Leccinum*. As the bitter bolete resembles somewhat the king bolete, it can produce literally a bitter disappointment to the mushroom hunter.

The rule of thumb is that bitter bolete has pink pores and brownish stem, while cep has whitish pale grey pores and stem. They also grow on different places and terrain. Peppery bolete (*Chalciporus piperatus*) has extremely strong taste, and has been used as ersatz for pepper.

Finnish cuisine uses boletes for various soups, sauces, casseroles and hotpots. They are sometimes also used as pizza filling, not unlike champignons, shiitake or portobellos.

Two of the best common edible boletes however are the Bay bolete (*Boletus badius*), whose pores bruise blue-green, and the orange birch bolete, which is a *Leccinum* with an orange cap and which bruises a bluish grey.

Boletus Aereus

Boletus aereus is a bolete, an edible mushroom frequently consumed in the Basque Country and Italy. Known as ontto beltza in Basque, porcino nero in Italian, tete de negre (negro's head) in French and Queen bolete in California, it is well known in southern Europe for its culinary qualities that make it the king of boletus, even more appreciated than *Boletus edulis*.

Description: The height of this boletus is 15-20 cm, broad at maturity; specimens of 40 cm have been found in some cases. The cap is dark brown and the ringless stalk is 6-10 cm high, usually shorter than the cap diameter. The pores are greyish white when young, then becoming yellow.

Distribution and Habitat: *Boletus aereus* is found mainly in central and southern Europe, being rare in colder climes such as England, and also in Western North America, where it is also uncommon. Mushrooms are found in summer and autumn.

Boletus Badius

The Bay bolete, *Boletus* (formerly *Xerocomus*) *badius*, is a common edible pored mushroom from Europe and North America. Often considered a poor relation of the Cep (*Boletus edulis*), it is nevertheless highly regarded by some authors such as Antonio Carluccio.

Description: It derives its common name from the bay or chestnut coloured cap, which is almost spherical in young specimens before broadening and flattening out to 18 cm. The pores are cream to pale yellow, but stain blue when bruised or cut. The flesh is white and stains pale blue, the stem is up to 12.5 cm and similar in colour to the cap but paler. It appears less affected by maggots than other boletes.

Distribution and Habitat: The Bay Bolete is common in mixed woodlands in Europe and North America from Eastern Canada west to Minnesota and south to North Carolina in autumn.

Boletus Edulis

Porcini (*Boletus edulis,* the Latin name) is a highly regarded edible mushroom. It has a number of English names, including cep (from its Catalan name *cep* or its French name cepe), king bolete and penny bun. A common term in current use is porcini (from the plural of its Italian name porcino).

The scientific name, *Boletus*, from the Latin stem *bolet-*, meaning "superior mushroom" and *edulis*, meaning edible, explains the species' culinary qualities. This mushroom has a distinct aroma reminiscent of fermented dough. It has a higher water content than other edible mushrooms.

Other Names

The mushroom is known as *Ontto txuri* or "the blond" in Basque, *hoib pravy* (Czech), *cep* (Catalan) *varganya* (Hungarian), *vrganj* (Croatian), *borowik szlachetny* (Polish), *baravykas* (Lithuanian), *belyj grib* "white mushroom" or *borovik* (Russian), *dubak* or *hrib smrekovy* (Slovakian), *jureek* or *jesenski goban* (Slovenian), *hrib* or *manatarca* (Romanian), *manatarka* (Bulgarian), *herkkutatti* or "delicious bolete" (Finnish), *harilik kivipuravik* or "common bolete" (Estonian), *Karljohan* after King Charles XIV John) or *stensopp* (Swedish), *steinsopp* (Norwegian), *Karl Johan* or *spiselig rorhat* (Danish), *eekhoorntjesbrood* (Dutch), *vrganj* (Serbian), or "beef liver mushroom".

Description: The cap of this mushroom is 7–30 cm broad at maturity, and mostly reddish-brown fading to white in areas near the margin; the colour continues to darken as it matures. The stalk is 8–25 cm in height, and up to 7 cm thick—rather large in comparison to the cap. The pores, which do not stain when bruised, are white in youth, fading to yellow and then to brown with age. Fully mature specimens can weigh about 2 pounds (1 kg). However, the most appreciated by gourmets are the young small porcini, as the large ones often harbour maggots (insect larvae), and they become slimy and less tasty with age.

Distribution and Habitat: *Boletus edulis* can be found most commonly in Europe and western North America. The mushroom can grow singly or in clusters. Its habitat consists of areas dominated by oak, pine, spruce, and fir trees. Not limited to these locations, the King Bolete is also found in hardwood forests containing oaks. It fruits from summer to autumn, following sustained rainfall. This mushroom can be found during the Fall in Syria and Lebanon where it grows in large clusters on decaying oak tree stumps.

It is well known to grow in the Borgotaro area of Parma, Italy, and has PGI status there.

Boletus edulis has been reported from the vicinity of Christchurch in New Zealand, where it is likely to have been somehow introduced. In South Africa it has been growing plentifully in pine forests around the country, for more than 50 years. It is not indigenous to South Africa, and is believed to have been brought here with the import of pine trees.

As the name implies, *Boletus edulis* is edible, and most people consider it superior in flavour and in texture. It is described as nutty and slightly meaty, with a smooth, creamy texture. Porcini are eaten and enjoyed raw, sauteed with butter, ground into pasta, in risotto, in soups, and in many other dishes.

They are widely exported and sold in dried form, reaching countries where they do not occur naturally, such as Australia.

Boletus edulis, along with *Boletus badius* and other boletes can be dried by stringing them separately on twine and hanging close to the ceiling of a kitchen. Alternatively they can be cleaned, but they should not be washed, and then placed in a wicker basket or bamboo steamer on top of a boiler or hot water tank. Once dry, they are best kept in an airtight jar. Drying them in the oven is inadvisable as it can result in them being cooked and spoiling. Considering that in the UK, boletes cost GB£2.50 for 25 grams dried, it is important to preserve them without damage.

Related Species

There are many fungi belonging to the genus, some edible, and some poisonous. In France, in addition to *Boletus edulis* (or *cepe de Bordeaux*), the most popular are

Tete de negre (negro's head; *Boletus aereus*), much rarer than the *Boletus edulis*, is by far the most appreciated by gourmets, as well as the most expensive. Usually smaller than the *Boletus edulis*, it is also distinctively darker in colour.

Cepe des pins (pine tree cep; *Boletus pinophilus* or *Boletus pinicola*) grows among pine trees. Its pores are characteristically bright yellow. It is less appreciated by gourmets than the two other kinds of porcini, but remains a mushroom ranking above most others.

In parts of Colorado and New Mexico (and possibly elsewhere) there is a species *Boletus barrowsii*, named after its discoverer Chuck Barrows. It is mycorrhizal with Ponderosa pine and hence tends to grow in areas where there is less rainfall. Some find it as good as if not better than *Boletus edulis*.

Boletus Reticulatus

Boletus reticulatus (formerly known as *Boletus aestivalis*), or Summer cep is an edible mushroom in the cepe (porcini) family. These are mushrooms whose tubed hymenium can be easily separated from the flesh of the hat, with a relatively thick central foot and compact flesh. They have a round cap which becomes convex as they age. They are easily distinguished by the browny-red colour of the cap and by the stalk with a white pattern on a reddish background.

Description: The summer cep is a mushroom with a bulbous cap (5 to 20 centimetres), with tawny brown flesh. In dry weather, they often have crevices giving the appearance of a fine net, with tubes on the end of which pores become brownish with age.

They have a central stipe (13-16 cm) of pale brown, which has a strongly marked reticulated pattern. The flesh is white and thick and remains firm if yellowish as the mushroom ages, and is often attacked by insect larvae. Its odour is pleasant.

The Summer cep is found in woods in France, after hot and humid weather, from the start of summer until the end of autumn. It is particularly common in the south and west of the country.

Crimini Mushroom

Although sometimes described a sub-variety of the portobello mushroom, the Crimini or Cremini mushroom is actually an immature portobello. Marketers have begun to refer to Crimini mushrooms as baby Portobellos. Left to grow another 48 to 72 hours, a Crimini mushroom will more than quadruple in size, taking on the large-capped Portobello shape. They are more delicate in texture but still have the meaty Portobello flavour.

Mushroom Hunting

Mushroom hunting can be a satisfying hobby. However, only expert mushroom hunters should look for common mushrooms in the wild. The common mushroom can easily be confused with young specimens of the destroying angel (*Amanita virosa*). The resemblance is significant enough to have caused fatal mushroom poisonings. Some of the differences are:

Upon slicing a picked mushroom in half, the Destroying angel is completely white, while common mushrooms have pink or brown gills and greyish flesh.

The destroying angel grows on mossy woods and lives symbioticaly with spruce. Various common mushrooms grow on open ground.

The destroying angel has white/cream coloured gills and spores, gills that are attached to the cap, but not the stalk. Common mushrooms have brown gills.

The base of the destroying angel stalk closest to the ground has a little cup which is a leftover piece of the veil that covers the mushroom during the button stage of growth. The common mushroom does not.

Some destroying angels have a ring or skirt on the stalk which is another remnant of the veil. The common mushroom, on the other hand, does not.

Calocybe Gambosa

St. Georges mushroom (*Calocybe gambosa*) is an edible mushroom that grows mainly in woodlands. Deriving its common name from when it first appears in the UK, namely on St. George's Day. It is considered a delicacy, especially when fried in butter. It was previously considered a part of the large genus *Tricholoma* and is still seen as *T. gambosum* in older texts.

Idenification

The cap has a smooth texture and has ridges on it. The cap, stem and flesh can be from white-creamy coloured to bright yellow. The flesh can be peeled. The spore print is white to pinkish white.

Habitat

It is common in woodlands in Europe, often in areas rich in limestone. It is common on the Swedish islands Oland and Gotland, both situated in the Baltic sea.

Cantharellus

Cantharellus is a genus with many delicious and popular edible mushrooms. It is a mycorrhizal edible fungus, meaning it forms symbiotic associations with plants, making it very challenging to cultivate. Caution must be used when identifying chanterelles for consumption; lookalikes, such as the Jack-O-Lantern *(Omphalotus olearius)*, can make a person very ill. Still, the yellow chanterelle is one of the most recognised edible mushrooms and can be found in Asia, Europe, North America and Australia.

Some species of *Cantharellus*, such as the yellowfoot chanterelle, have been re-examined and moved to the closely related genus *Craterellus*.

The genus *Cantharellus* contains many species known generally as chanterelles, though for the most part the name refers to the most famous species *C. cibarius*.

C. cibarius: The best known species of this genus is the Yellow chanterelle, which is orange or yellow, meaty and funnel-shaped. It has forking gills on the underside, running all the way down its stalk, which tapers down seamlessly from the cap. It has a fruity smell and a peppery taste, and is considered an excellent food mushroom.

C. subalbidus: In California and the Pacific Northwest of USA there is also the White chanterelle, which looks like the yellow except for its off-white colour. It is more fragile and found in lesser numbers than the yellow chanterelle, but can otherwise be treated as its yellow cousin.

C. formosus: The Pacific golden chanterelle *(C. formosus)* has recently been recognised as a separate species from the yellow chanterelle. It forms a mycorrhizal association with the Douglas-fir and Sitka spruce forests of the Pacific Northwest. This chanterelle has been designated Oregon's state mushroom, due to its economic value and abundance.

Use in food

Chanterelles in general go well with eggs, curry, chicken, pork and veal, can be used as toppings on pizzas, be stewed, marinated, fried in butter, or used as filling for stuffed crepes. Of course, these are just examples; chanterelles are versatile and can be added as an ingredient to most dishes.

Preparation and Storage

Since the mushrooms hold a lot of water, a good way of preparing them is to "dry saute" them: after cleaning, the mushrooms are sliced and put in a pan over medium heat. When covered in the water they've released, they are removed from the heat and frozen in their own water. Alternatively, the water can be used in sauces or simply discarded.

Chanterelles can also be pickled in brine. Salted water is brought to the boil, and pickling spices such as peppercorns, mustard seeds, and thyme are added. The mushrooms are then cooked in this solution for 5-10 minutes before being transferred to sterilised bottles along with some of the liquid. Sliced garlic and dill can be added to the bottles for extra flavour. The remaining liquid forms an excellent stock for making soup. When pickled in this way, chanterelles can last from six to twelve months.

Another storage technique is drying. The chanterelles are dried in open air and then stored without a lid. A few hours before final preparation the mushrooms are put in water which they absorb and return to practically their original size. They can then be used as fresh.

Fresh chanterelles can generally be stored up to ten days in a refrigerator.

Similar Species

The False chanterelle (*Hygrophoropsis aurantiaca*) has finer, more orange gills and a darker cap. It is edible, but typically a culinary disappointment. The very similar Jack O'Lantern mushroom (*Omphalotus olearius*) and its sister species (*Omphalotus olivascens*) are very poisonous, though not lethal. They have true gills (unlike chanterelles) which are thinner, have distinct crowns, and generally do not reach up to the edge. Additionally, the Jack-O-Lantern mushroom is bioluminescent.

Chalciporus Piperatus

The Peppery bolete, *Chalciporus* (formerly *Boletus*) *piperatus*, is a small pored mushroom of the Boletaceae family found in mixed woodland in Europe.

Edible but very peppery, Antonio Carluccio recommends only using it to add a peppery flavour to other mushrooms. It has been used as a peppery condiment in many countries.

Description: One of the smaller boletes, the cap is 3-6cm across and orange-fawn coloured. The pores are orange and the flesh lemon-yellow. The pore print is cinnamon.

Distribution and Habitat: Found naturally in coniferous and beech and oak woodland in Europe in autumn, *Chalciporus piperatus* has also been able to spread into native forest in northeastern Tasmania, having been found growing with the native Myrtle beech (*Nothofagus cunninghamii*).

Classification

The cultivated mushroom is a member of the large genus *Agaricus,* which has numerous members which are edible, tasty and collected worldwide. The next best-known is the commonly collected wild mushroom *A. campestris,* known in North America as the meadow mushroom or field mushroom in England and Australia. This can be found throughout much of the United States and Europe.

Culinary Use

Common mushrooms are fairly rich in vitamins and minerals. The mushroom contains an especially high amount of vitamin B and potassium. Raw mushrooms are naturally cholesterol, fat, and sodium free. The mushrooms also have very low energy levels — five medium-sized common mushrooms added together only have 20 calories.

Common mushrooms have a unique flavour that can be matched by few other mushrooms. No specific flavour can be defined; most people describe the mushroom as "plain", but other people say that the common mushroom tastes slightly sweet or "meaty".

Like potatoes and apples, table mushrooms "rust" quickly when exposed to air. When sliced and exposed to air for ten minutes or more, the mushrooms quickly soften, turn a brownish colour, and lose their original flavour.

The cultivated *Agaricus bisporus* common mushroom originated in Fránce. Today's commercial variety of the common mushroom was originally a light brown colour. In 1926, a Pennsylvanian mushroom farmer found a clump of common mushrooms with white caps in his mushroom bed. As was done with the navel orange and Red Delicious apple, cultures were grown from the mutant individuals, and most of the cream-coloured store mushrooms we see today are products of this chance natural mutation.

In most supermarkets, common mushrooms are marketed as "table mushrooms" and are often packed in small quantities. Mushrooms may be sold sliced or whole.

Portobello Mushroom

The 'Portobello mushroom' is a large brown strain of the same fungus, left to mature and take on a broader, more open shape before picking. Portobello mushrooms are distinguished by their large size, thick cap and stem, and a distinctive musky smell. Because of their size and the thickness of their fleshy caps, these mushrooms can be cooked in a range of different ways, including grilling and frying.

Clavicorona Pyxidata

Clavicorona pyxidata is a coral fungus that is commonly called "crown coral" or "crown-tipped coral fungus". Its most identifying feature is the crown shape extending from the tips of its branches.

Clavicorona pyxidata can be observed in the state of Wisconsin, most commonly found in the months of May through October, residing on decaying logs. This fungus is usually found in concentrated groups and large quantities. In order to produce a larger amount of spores, the coral fungus's basidia are covered with basidiospores, which increase the overall surface area of the organism and enable it to reproduce in larger quantity.

Each different kind of coral fungus has upright branches, much resembling the branches of ocean-inhabiting coral. There are many different species of coral fungus, and although they may look alike, scientists have found that they are actually not closely related. *Clavicorona* fungi are easily distinguished from the rest of the genera of coral fungi because they grow directly from wood. Also, their tips have crown-like structures at the end, something that is absent in most other coral fungi. Clavicorona pyxidata is very pale in colour when compared to the other fungi of the Clavicorona genera. These fungi are also edible when cooked. They have a very spicy taste when put into the mouth, but the flavour is not distinguishable when cooked and added to various other foods.

Cloud Ear Fungus

Cloud ear fungus (*Auricularia polytricha*, syn. *Hirneola polytricha*) is an edible jelly fungus. It is grey-brown in colour and often used in Asian cooking. In Chinese, it is known as"cloud ear", "hairy wood ear", or "wood ear" or "tree ear", and in Japanese it is called *kikurage* ("tree jellyfish"). It is also known as black fungus, black Chinese fungus, wood fungus, ear fungus, or tree ear fungus, an allusion to its rubbery ear-shaped growth.

The fungus grows in frilly masses on dead wood. It is a dark brown colour but somewhat translucent. It is usually sold dried and needs to be soaked before use. While almost tasteless, it is prized for its slightly crunchy texture and supposed medicinal properties, including its newly discovered anti-coagulant properties. Of note, the slight crunchiness persists despite most cooking processes.

Auricularia auricula-judae, a closely related species, is also used in Asian cooking.

Coprinus Comatus

Coprinus comatus, the shaggy ink cap, lawyer's wig, or shaggy mane, is a common fungus often seen growing on lawns, along gravel roads and waste areas. The young fruiting bodies first appear as white cylinders emerging from the ground, then the bell-shaped caps open out. The caps are white, and covered with scales - this is the origin of the common names of the fungus. The gills beneath the cap are white, then pink, then turn black and secrete a black liquid filled with spores (hence the "ink cap" name). Its specific name derives from *coma*, or "hair", hence *comatus*, "haired" or "shaggy".

When young it is an excellent edible mushroom provided that it is eaten soon after being collected (it keeps very badly because of the autodigestion of its gills and cap). The species is cultivated in China as food.

Coprinus comatus is the type species for the genus *Coprinus*. This genus was formerly considered to be a large one with well over 100 species. However, molecular analysis of DNA sequences showed that the former species belonged in 2 families, the Agaricaceae and the Psathyrellaceae. *Coprinus comatus* is the best known of the true *Coprinus*. Adding to its unusual features, a recent study has found the shaggy ink cap kills nematode species *Panagrellus redivivus* and *Meloidogyne arenaria*.

Description

The shaggy ink cap is easily recognisable from its cap which initially covers almost the whole of its stem. Entirely white to begin with, the cap becomes shaggy as separating scales develop. The gills change rapidly from white to pink, then to black. It is deliquescent. Microscopically it lacks pleurocystidia.

Distribution and Habitat

It grows in groups in places which are often unexpected, such as green areas in towns. It occurs widely in grasslands and meadows in Europe and North America. It appears to have been introduced to Australia and New Zealand.

Corn Smut

Corn smut is a disease of maize caused by the pathogenic plant fungus *Ustilago maydis*. *U. maydis* causes smut disease on maize (*Zea mays*) and teosinte (*Euchlena mexicana*). Although it can infect any part of the plant it usually enters the ovaries and replaces the normal kernels of the cobs with large distorted tumours analogous to mushrooms. These tumours, or "galls", are made up of much-enlarged cells of the infected plant, fungal threads, and blue-black spores. The spores give the cob a burned, scorched appearance. In fact, the name *Ustilago* comes from the Latin word *ustilare* (to burn).

Considered a pest in most of the United States, smut feeds off the corn plant and decreases the yield. Usually smut-infected crops are destroyed. However, in Mexico corn smut is called *huitlacoche* (sometimes spelled *cuitlacoche*), an Aztec word reportedly meaning *raven's excrement*. It is considered a delicacy, even being preserved and sold for a higher price than corn. For culinary use, the galls are harvested while still immature—fully mature galls are dry and almost entirely spore-filled. The immature galls, gathered two to three weeks after an ear of corn is infected, still retain moisture and, when cooked, have a flavour described as mushroom-like, sweet, savoury, woody, and earthy. Flavour compounds include sotolon and vanillin, as well as the sugar glucose.

The fungus has had difficulty entering into the American and European diets as most farmers see it as blight, despite attempts by government and high profile chefs. In the mid-1990s and due to demand created by high-end restaurants, Pennsylvania and Florida farms were allowed by the United States Department of Agriculture (USDA) to intentionally infect corn with *huitlacoche*. Most observers consider the programme to have had little impact, although the initiative is still in progress. Regardless, the cursory show of interest is significant because the USDA has spent a considerable amount of time and money trying to eradicate *huitlacoche* in the United States. Moreover, in 1989 the James Beard Foundation held a high-profile *huitlacoche* dinner. This dinner famously tried to get Americans to eat more of it by renaming it the Mexican truffle.

Huitlacoche grows best during times of drought in a 78°F to 93°F (25°C-34°C) temperature range. Aztecs purposely inoculated corn with the spores by scratching their corn plants at the soil level with a knife—thereby allowing the water-borne spores easy entrance into the plant.

Life Cycle

When grown in the lab on very simple media, it behaves like baker's yeast, forming single cells going by the name sporidia. These cells multiply by budding off daughter cells. When two compatible sporidia meet on the surface of the plant, they will switch to a different mode of growth. First, they'll send out conjugation tubes to find each other, after which they fuse and make a hypha to enter the maize plant. Hyphae growing in the plant are dikaryotic; they possess two haploid nuclei per hyphal compartment. In contrast to sporida, the dikaryotic phase of *U. maydis* requires infection of the plant in order to grow and differentiate and cannot be maintained in the laboratory.

Proliferation of the fungus inside the plant leads to disease symptoms as chlorosis, anthocyanin formation, reduced growth and the appearance of tumours harbouring the developing teliospores (Banuett, 1995; Christensen, 1963).

Mature spores are released from the tumours and spread by rain and wind. Under appropriate conditions a probasidium is formed in which meiosis occurs. Resulting haploid nuclei migrate into elongated single cells. These cells detach from the probasidium; these are the sporidia, completing the life cycle.

Uses

Native Americans of the American Southwest, including the Zuni tribe, have used corn smut to induce labour. It has similar medicinal effects to ergot, but weaker, due to the presence of the chemical *ustilagine*.

Model System

The yeast-like growth is a big advantage for research, although its relevance in nature is unknown. The fungus is exceptionally well-suited for genetic modification. This allows researchers to study the interaction between fungus and host with relative ease. The availability of the entire genome is another advantage of this fungus as model system.

Ustilago maydis is not only used to study plant disease. It has also been used to study recombination and DNA repair as well. That the function of the breast-cancer gene BRCA2 is now known is largely due to work with *Ustilago maydis*.

Availability

Huitlacoche can be bought as a canned good in some markets and over the net. Some farmers markets and organic growers are endeavouring to bring fresh huitlacoche to their customers and local food service trade. The Troy Community Farm of Madison, Wisconsin started a project in 2003 to grow corn and produce huitlacoche to make available in the Madison area and now sell it directly to the public from their farm stand once a week.

Craterellus

Craterellus is a genus of generally edible fungi similar to the closely related chanterelles, with some species recently reassigned to this genus. They are distinguished by their lack of gill-like structures on the underside of their caps.

Black chanterelle, also black trumpet, horn of plenty, or trumpet of the dead, are the common names for the edible mushroom *Craterellus cornucopioides*. The Cornucopia, in Greek mythology, referred to the magnificent horn of the goat (or goat of the nymph) Amalthea, that filled itself with whatever meat or drink its owner requested. It has become the symbol of plenty.

The mushroom is dark coloured, almost black and looks rather unattractive, but has a very good flavour. It is hard to find because of its dark colour, which easily blends in with the leaf litter on the forest floor.

The yellowfoot or funnel chanterelle (*Craterellus tubaeformis*) is a yellowish-brown and trumpet-shaped chanterelle found in great numbers late in the mushroom season, thus earning the common name winter mushroom. The cap is convex and sometimes hollow down the middle, and because of this it is also known as funnel chanterelle. The gills are widely separated, and of lighter colour than the cap. It grows on moss or rotten wood, and is an excellent food mushroom, especially fried or in soups.

Both species are gathered commercially and, unlike *Cantharellus,* can be easily preserved by drying.

Dryland Fish

Dryland fish is a type of morel mushroom that grows near river banks in the Appalachian mountains in Tennessee and Kentucky; they also are known as hickory chickens. They are generally found in 2 varieties: black (mostly found in mid to late April) and the larger yellow variety (found in early to mid-May), which is more highly prized for its taste. They are usually prepared by coating the mushrooms in corn meal, then placing them in a frying pan with hot grease and cooking thoroughly. They sprout in early spring and are soon gone, after roughly a month and a week.

Enokitake

Enokitake are long, thin white mushrooms used in Asian cuisines, particularly those of Japan, Korea, China, and Vietnam. These mushrooms are cultivars of *Flammulina velutipes* also called golden needle mushroom. Wild forms differing in colour, texture, and sliminess are called winter mushrooms, velvet foot, or velvet stem among other names.

The mushroom is available fresh or canned, the fresh mushroom being preferred by many. They are traditionally used for soups, but can also be used for salads and other dishes. They have a fruity flavour and a crisp texture. The mushroom can be refrigerated for about one week.

The mushroom naturally grows on the stumps of the Chinese Hackberry tree, called *enoki* in Japanese, but also on some other trees as for example mulberry and persimmon trees. There is a significant difference in appearance between the wild and the cultivated mushrooms.

Cultivated mushrooms are not exposed to light resulting in a white colour, whereas wild mushrooms usually have a dark brown colour. The cultivated mushrooms are also grown to produce long thin stems, whereas wild mushrooms produce a much shorter and thicker stem.

The variety available in the supermarket is always cultivated. The mushroom is cultivated in a plastic bottle or a vinyl bag for 30 days at 15°C and 70 per cent humidity, on a substrate of saw dust or corn cobs, and a number of additional ingredients.

Afterwards, the mushroom is grown for another 30 days in a slightly cooler but more humid environment. The growth is constricted to force the mushroom to grow long and thin. The mushroom available in the supermarket often still shows the impression of the bottle around the base of the mushroom.

The mushroom is very easy to cultivate, and has been cultivated in Japan for over 300 years, initially on wood, and later in the bottles as described above. Home cultivation kits are also available.

In North America, a second species, *Flammulina populicola* has been cultivated and may be available in kits.

Fistulina Hepatica

The beefsteak fungus (*Fistulina hepatica*) is an unusual polyporous bracket fungus that is most commonly seen in Britain, but can be found in North America, Australia, and the rest of Europe.

It is sometimes called the *beefsteak polypore* or the *ox tongue*. As its name suggests, it looks remarkably similar to a slab of raw meat. It has been used as a meat substitute in the past, and can still be found in some French markets. It has a sour, slightly acidic taste.

The shape resembles a large tongue, and it is rough-surfaced with a reddish-brown colour. The spores are released from minute pores on the creamy-white underside of the fruit body. A younger beefsteak fungus is a pinkish-red colour, and it darkens with age. It bleeds a dull red juice when cut, and the cut flesh further resembles meat.

The species is fairly common, and can often be found on oaks and sweet chestnut, from August to the end of Autumn, on either living or dead wood. In Australia, they can be found growing from wounds on *Eucalyptus* trees.

3

Structure and Organisation

Mechanism of Fungi

Dispersal is one of the fundamental problems that has faced the Fungi and much of their beauty of structure is associated with the solution of this problem. In fungi, the vegetative part is usually hidden away as a feeding mycelium in the nutrient substratum and the conspicuous part is an apparatus concerned with the production and liberation of spores.

A spore to students of archegoniate plants (Bryophyta and Pteridophyta) is definable as a unicellular asexual reproductive body formed within a sporangium following meiosis. However, the mycologist uses the term in a much wider sense.

To him a spore is any simple reproductive unit, often unicellular but frequently multicellular, and usually with some food reserve. He recognises many different kinds of spore: zoospores, sporangiospores, chlamydospores, zygospores, oospores, ascospores, uredospores, basidiospores, teleutospores, brand spores, and various types of conidia such as cnidiospores, aleuriospores, phialospores and blastospores. With the possible exception of ascospores and basidiospores none of these can be considered as homologous with the spores of archegoniate plants. However, all these kinds of spore agree in being microscopic, normally devoid of vacuoles and, more often than not, unicellular.

Not all spores are units of dispersal, some are merely resting-spores that tide over an unfavourable season, usually winter. This is true of most zygospores and oospores and of the teleutospores of rusts. However, certain spores, such as the brand spores of some smuts (e.g. *Ustilago avenae,* the loose smut of oat), are concerned both with hibernation and with dispersal.

The size and shape of fungal spores vary considerably. Many species have very small spores. At the other extreme are the oval ascospores of certain lichens: *Pertusaria velata* with a unicellular spore and *Varicellaria microsticta* with an even larger two-celled spore. Spores of this size are just visible to the naked eye and are of much the same size as the smallest seeds of flowering plants.

It is interesting to note that those phanerogams, mostly orchids and saprophytes, which produce very small seeds would seem to be reversing the course of evolution. Their seeds have become so simplified that they are almost spores and, correlated with the decrease in size and differentiation, the number produced has greatly increased.

Except for the naked zoospores of water moulds, spores have cell-walls usually quite thin, but becoming very thick in some species, especially in those forming subterranean fruit-bodies (e.g. *Elaphomjces granulatus).*

The spore-wall is mostly transparent and colourless, but in a number of fungi it is coloured and quite frequently black. Sometimes the colour of a spore is not entirely due to the pigment in the wall. Thus the red uredospores of *Puccinia graminis* owe their colour to a combination of a brown wall colour with an orange pigment in the cytoplasm, although in this species mutant strains are known where one or the other pigment is suppressed, giving in one variety greyish-brown uredospores and in the other orange ones. In most fungi the spore-wall is smooth, but it may be wrinkled, pitted, warted, spiny, or alveolate.

An outstanding feature of most species is the enormous spore production. Students of green plants and of animals are very familiar with the principle of the necessary wastage of reproductive units, but in Fungi this wastage seems to reach its extreme expression.

On the average not more than one or two spores from each individual fungus succeed in their reproductive function, since each species is more or less in equilibrium with its environment and its numbers, though they may fluctuate from year to year, show no steady increase.

Many estimates of spore output have been made, but here only a few examples are selected to give a clear picture of the gigantic scale of spore production in a wide range of types. A big specimen of the giant puffball *(Calvatia gigantea)* has been estimated to contain 7,000,000,000,000 spores.

The large bracket fungus *Ganoderma applanatum* may liberate 30,000,000,000 spores a day, apparently maintaining this output for the whole six months (May to October) of its spore-fall period. The small apothecium of the cup-fungus *Sclerotinia sclerotiorum* has been shown to produce 30,000,000 ascospores.

A perithecial stroma of the flask-fungus *Daldinia concentrica* may discharge over 100,000,000 spores each day. In the stinking smut of wheat due to *Tilletia caries* a single diseased grain may contain over 12,000,000 brand spores. A colony of blue mould *(Penicilliwn* sp.) 2-5 cm. in diameter may bear 400,000,000 conidia.

However, production of spores is not always on this gigantic scate and, especially in aquatic fungi, spore output may be quite low. For example, in the Chytridiales the zoospores formed by each individual fungus are often less than 100. In *Rhizophydium planktonicum*, a parasite of the diatom *Asterionella* so abundant in freshwater plankton, the number of spores liberated from the single sporangium which each individual forms is only ten to twenty, yet this fungus manages to survive and is probably very common in lakes and reservoirs.

In some land fungi, the spore output is also quite small. Thus in the coprophilous species *Dasyobolus immersus* and *Sordaria fimiseda* the total spore production from each fruit-body is not great.

As with many seed plants, so with the fungi there seems to have been a struggle between two tendencies: the tendency to produce an enormous number of very small units readily dispersed but each with a relatively small food reserve and thus with a relatively poor chance of success, and the tendency to concentrate on a smaller number of larger units not, perhaps, so readily dispersed but with a greater chance of succeeding.

An important attribute of a spore from the point of view of dispersal is its viability, and spores vary greatly in this respect. The fact that a spore may be transported a long distance is of no significance if at the end of its journey it is incapable of germination.

The bearing of viability on effective dispersal is well illustrated by a study of the epidemjology of certain rusts. On the whole, uredospores are very resistant, and because they can survive for a long time in the air—resisting both desiccation and light—they can carry rust infection in a single step to a distance of a hundred miles and more. In contrast the smaller, thin-walled basidiospores (sporidia) are short-lived and are seldom capable of causing infection at a distance of more than a few miles from a source of infection.

The vast majority of fungi reproduce by spores, but some have reproductive units which cannot be classified as spores by any stretching of definitions. In the minute agaric *Omphalia flavida*, for example, which causes a leaf-spot disease of coffee and many other plants in the New World, fruit-bodies are formed which liberate basidiospores in the normal way, but in addition reproduction occurs by means of multicellular macroscopic gemmae. Each of these appears to be homologous with a pileus.

In *Sclerotium coffeicola*, another coffee parasite, reproduction and dispersal are entirely by gemmae which seem to be of the same nature as those of *Omphalia flavida*. Again, the sclerotium of ergot *(Claviceps purpurea)*, like the host seed that it replaces, becomes detached from the host, and may play a part in dispersal although the spread of this fungus, under natural conditions, is mainly brought about by conidia and ascospores.

However, the commonest type of this kind of vegetative reproduction occurs in lichens in which reproduction by soredia is so frequent. In many lichens (e.g. *Cladonia* spp.) there occurs on the surface of thallus a dry, greyish powder each grain of which is a single soredium consisting of a tangled ball of fungal hyphae wrapped around a few algal cells. This powder is easily blown away by the wind. Many species rely for reproduction and dispersal entirely on soredia.

Further, somewhat as bracken spreads over the hillsides by the growth of its underground rhizome, so a few fungi are dispersed locally by the growth of an underground rhizomorph. The most outstanding example is *Armillaria mellea* the honey fungus. Basidiospores, carried perhaps from a considerable distance, establish the fungus as a saprophyte on a dead stump in a forest, but from there rhizomorphs grow out through the soil and attack living trees many yards from the base of operations.

Finally, a number of specialised parasitic fungi are dispersed by means of a resting mycelium carried by the reproductive unit, usually the seed, of the host plant. The most outstanding examples are the endophytes so commonly present in rye-grass *(Lolium perenne)* and its allies. These do not apparently sporulate in nature, but depend, with complete success, for dispersal on a mycelial infection of the seeds.

It is often clear that the rate of spread of a fungus is conditioned by the efficiency of its dispersal apparatus. This can, perhaps, be illustrated most readily by a consideration of the two well-known potato diseases, blight due to *Phytophthora infestans* and wart disease caused by *Synchytrium endobioticum*. Blighted potato leaves bear, on their under surfaces, numerous conidiophores of *P. infestans* projecting through the stomata and bearing an abundant supply of the small conidia (sporangia) which are easily detached and scattered by the wind.

This, combined with the speed of completion of the life-cycle, explains the epidemic spread of the disease when the weather is suitable. There are no practical steps that can be taken to control *dispersal* and the farmer can check the disease only by preventing infection, usually by spraying the foliage with Bordeaux Mixture before the spores of the pathogen become prevalent in the air.

On the other hand, in *Synchytrium endobioticum*, which attacks the underground parts, mainly the tubers, there are no airborne spores. The short-lived zoospores discharged into the soil water are probably responsible for dispersal over a very short distance, at most a few feet, and the relatively large black resting-spores, liberated into the soil when the warted tubers decay, are not normally dispersed by the wind.

The disease may be spread by the transport of 'seed' potatoes carrying resting-spores from one area to another, by contaminated soil attached to the wheels of agricultural machines, and, more naturally but probably very occasionally, in infected soil on the feet of rooks and other birds. Thus the dispersal of the fungus can largely be prevented and its spread has been successfully limited by the measures adopted by the Ministry of Agriculture. The disease is notifiable, and susceptible varieties may not be grown in contaminated regions. Nor may tubers from an infected zone be transported to other regions and used as 'seed'.

In many-fungi more than one kind of spore is produced

in the course of the life-cycle, the rusts being, perhaps, the extreme example, with often five different spores (basidiospores, pycnidiospores, aecidiospores, uredospores, and teleutospores) in the cycle. Very often, where more than one kind of spore is formed they have different roles in dispersal. Thus in *Endothia parasitica*, causing the bark disease of American chestnut-trees, the ascospores, violently discharged into the air and readily" distributed by the wind, are responsible for dispersal from tree to tree and from plantation to plantation, whilst the pycnidiospores, exuded in slimy masses from the pycnidia and washed down by the rain, are responsible only for the local spread of the disease on the trunk of the infected tree.

Spore Liberation

In a great many terrestrial fungi the spores are violently discharged and are thus set free from the parent tissue, but in others liberation is passive so far as the fungus is concerned, and it is the energy of wind, rain, and animals that is responsible for the liberation of the spores and also, frequently, for their full dispersal.

Passive Liberation

In the vast array of conidial apparatus exhibited by Hyphomycetes it has been suggested that one of the most fundamental biological distinctions is between 'dry-spore' and 'slime-spore' types. Mostly, where the spores are dry they can readily be blown off their sporophores and carried away by the wind, whilst slime-spores are involved in a sticky liquid so that they cannot normally be dislodged by wind and appear to be dispersed, at least in the first instance, by water; or by insects. A consideration of dispersion has led to much the same distinction and the concept is clearly of general application in Fungi, so that most examples of passive spore liberation can be classed either as dry-spore or as slime-spore types, the former being the more numerous.

Among the larger fungi the most outstanding dry-spore forms are puffballs and their allies *(Lycoperddh* spp., *Geaster* spp., and *Scleroderma* spp.). An interesting but rather isolated example is the small agaric *Nyctalis asterophora,* which lives as a parasite on larger toadstools, especially *Russula nigricans.* In the parasite basidiospore formation on the greatly reduced gills is completely, or almost completely, suppressed and the whole upper surface of the cap is converted into a fawn-coloured, dry, powdery mass of stellate chlamydospores, easily blown away, which are the main units of dispersal. It is of interest to note that in the other common species, *Nyctatis parasitica,* the chlamydospores do not become powdery, and It is riot at all clear how they are dispersed. The candle snuff fungus *(Xylana hypoxylori)* is another common dry-spore fungus in which the white tips of the erect stroma are covered by minute powdery conidia readily scattered by the wind.

Amongst the microfungi many of the common saprophytes have dry spores, for example *Rhizopus yigrican), Penicillium* spp., *Aspergillus" spp., Botrjtis cinerea, Clado sporium herbarum, Monilia (Neurospom)' sitophila, Alternaria* spp., and many more. Anyone who has worked in a mycological or bacteriological laboratory knows to his cost that it is these dry-spore moulds which from time to time cause trouble as contaminants in cultures of other microorganisms. However, mankind must be grateful for one such contaminant, for it was a spore of *Penicillium notatum* drifting by chance into a culture of bacteria that led Fleming in 1929 to his discovery of penicillin.

Amongst obligate parasitic fungi that spread with epidemic speed under suitable conditions, dry-spore types prevail, occurring, for example, in rusts (uredosporestage), in downy mildews and their allies (e.g. *Phytophthora infestans),* and in powdery mildews (oidial stage).

Although in most dry-spore fungi the spores are easily blown away by wind, sometimes their attachment to the sporophore is so firm that even the strongest winds fail to dislodge them. An outstanding example of this is the conidial

(Fusicladiuni) stage of the apple scab fungus *(Ventuna inaequalis)* in which wind alone does not seem to detach the spores, but in the presence of water they separate easily from their conidiophores and dispersal is, apparently, brought about in the main by wind-driven rain. Even in highly developed dry-spore types, rain may play a part in spore liberation.

Thus, in the puff-balls *(Lycoperdon* spp.) and earth-stars *(Geasterspp.),* wind blowing across the apical pore of the ripe fruit-body may suck out spore-laden air from within, but liberation of spores is brought about even more effectively by heavy rain, for when a large drop strikes the papery, waterproof peridium the fruit-body is momentarily compressed and a visible cloud of spores is expelled. In the larger fungi there are few slime-spore forms, the most outstanding being the stink-horns (Phallaceae), but amongst micro-fungi numerous examples occur.

Many soil fungi are of this type for example: *Tnchoderma viride, Fusarium* spp., *Pullulana pullylaps.* Amongst plant pathogens the *Graphium* stage of *Opostoma ulmi* (the cause of Dutch elm disease), *Colletotrictfum lindemuthianum* (causing anthracnose of dwarf and runner beans), and the conidial stage of ergot of rye *(Claviceps purpured)* are familiar example. Most pycnidial types have spores embedded in slime which exudes through the ostiole as in the imperfect stage of *Endothia parasitica* (American chestnut bark disease) and in the pycnidial *pi*age of rusts.

Active Liberation

In some fungi the spores occur in or on a turgid cell which finally bursts, discharging the spores into the air. This type of active liberation is especially characteristic of the ascus of Ascomycetes. There are, however, many examples, widely scattered in any scheme of classification, in which the ascus is not explosive. Some would regard all such Ascomycetes as degenerate in this respect, but others would consider some genera primitive (e.g. *Endomyces, Monascus, Gymnoascus)* whilst agreeing that certain types with non-explosive asci (e.g. Tuberales, *Chaetomium* spp., *Ophiostoma ulmi)* are clearly degenerate in their ascus dehiscence. However, in the great majority of Ascomycetes, including lichens, the ascus is explosive and spore discharge is violent.

The typical ascus just before explosion is a turgid cell with the general properties of a thin-walled, living plant cell. The thin, stretched, elastic cell-wall is freely permeable to water and to substances in true solution. However, the thin lining layer of enucleate protoplasm has the properties of a more or less semi-permeable membrane.

These permeability relationships are indicated by the normal plasmolysis that occurs in an ascus if it is immersed in a hypertonic solution. In *Ascobolus stercorarius* it has been estimated that the osmotic pressure of the ascus sap is in the neighbourhood of 10-13 atmospheres. The considerable stretching of the ascus that usually occurs as it ripens may be caused by an increase in the osmotic pressure.

The ascospores occur suspended in the sap with the topmost spore at the apical end of the vacuole and, in some species at least *(Aleuria repanda)* the spores are maintained in this position by a fine bridle of protoplasm. When the ascus bursts it does so in a definite manner: by a hinged lid in most operculate Discomycetes (Pezizales), by the separation of an apical cap *(Dasyobolus immersus* and in *Sordaria curvula),* by the development of a minute apical pore as in Helotiales and Clavicipitales, or by an apical slit as in *Sphaerotheca.*

The ascospores in different species vary very considerably in size, shape, and in the distance to which they are discharged. This is in striking contrast to Basidiomycetes in which the size and shape of the basidiospores are subject to little variation from species to species and they are discharged to distances varying within narrow limits. Ascospores range in size from the huge spores of the lichen *Varicellaria microsticta* to the minute ones of *Diatrype disciformis,* and in shape from the spherical spores *of Boudiera* spp. to the long thread-like ones of *Cordyceps militaris*.

Again, the distances of violent spore discharge range from 0.03 cm. in *Cordyceps militaris* to 60 cm. for *Dasyobolus immersus*. On the whole there is a correlation between the diameter of the spore-projectile and the distance to which it is shot.

In nearly all Ascomycetes complex sporophores or fruit-bodies occur, and in the larger forms a single fruit-body may produce millions of asci. It is easy to recognise four fairly distinct types of organisation in connection with spore liberation:

1. In the *Discomycete* type, the spore-producing surface, consisting of asci intermixed with parallel paraphyses, is more or less exposed, most often as a lining to a shallow cup-shaped apothecium. The extensive exposed hymenium allows opportunities for 'puffing'—the simultaneous bursting of numerous asci.
2. In the *Pyrenomycete* type, the asci are contained in a small flask-shaped structure (perithecium) which opens to the outside by a minute ostiole. Before each ascus can discharge the spores, its tip must reach the ostiole, and the canal of the neck is usually so narrow that normally only one ascus can emerge at a time.
3. In the *Erysiphales* type, the fruit-body is a cleistocarp. This is rather like a perithecium but is completely closed; there is no ostiole. In this type the swelling asci must first burst the wall of the cleistocarp before they can emerge and discharge their spores.
4. In the *Myriangium* type, though the hymenium, is exposed in a structure like a small apothecium, the spherical asci are embedded in a plectenchymatous tissue and are free to discharge only when this gradually undergoes gelatinisation. Species of *Myriangium* and allied genera are nearly all fungi of warmer regions and are little known to British mycologists.

The first three types will now be considered in some detail.

Discomycetes

Discomycetes range from minute forms 0.2 mm. in diameter (e.g, *Saccobolus* spp.) to the large morels (*Morchella* spp.). Many are more or less cup-shaped with the hymenium lining the inside of the cup as in the genera *Pezizr Aleuria, Dasyscypha, Bulgaria, Helotium, Mollisia;* in others there is a more or less flat horizontal surface (e.g. *Rhizina*); again, in some genera the hymenium is fully exposed on a stalked, club-like fruit-body as *Geoglossum* and *Mitrula;* in the morels there is a stout, sterile stalk and the extensive hymenium is greatly convoluted, forming a series of broad alveoli or pits.

The types of fruit-body architecture that have evolved seem to have depended on two main characteristics of the ascus: first, that its spores are usually discharged to a distance of 0-5-4-0 cm.; second, that the ascus is positively phototropic. A third point of importance is that the ascomycete hymenium is not apparently harmed by rain, in contrast to the hymenomycete hymenium which is completely disorganised by a film of water.

Since the ascospores are normally discharged to a distance of more than a centimetre, the spores, if shot vertically into the air, stand a reasonable chance of being blown away without falling back on to the parent hymenium. Consequently, the horizontal upward-facing type of hymenium is quite consistent with efficient dispersal and is, indeed, very frequently developed.

The upward-facing hymenium hardly ever occurs in Hymenomycetes (toadstools, polypores, and their allies) where the basidiospores are rarely shot from the basidia to a distance of more than 0-02 cm. and where the hymenium is so susceptible to injury by water. On the other hand, the polypore or toadstool type of fruit-body, in which the vertical hymenial surfaces are closely opposed, could not have been developed in Discomycetes without a very considerable decrease in the violence of ascus discharge.

The phototropism of the ascus, originally recognised by

Zopf in *Ascobolus,* is probably a very common feature of the elongated type of ascus. It has been reported in numerous genera, all belonging to the 'operculate' series, but probably it is of general occurrence. A hymenium lining a deep narrow cup, as in *Sarcoscypha protracta,* or the alveoli of such an elaborate fruit-body as that of *Morchella* spp., can function efficiently only because of this phototropism, for, as Buller has pointed out, if the asci were simply arranged at right angles to the general surface of the hymenium, a high proportion of discharged spores would be wasted by being flung against an opposing hymenial surface within range and directly in the line of fire of the asci.

Spore Discharge in the **Ascobolus** *type of Apothecium*

Apart from its size *Ascobolus stercorarius* is fairly typical of a great many Discomycetes. This species is exceedingly common, growing in great numbers on horse-dung.

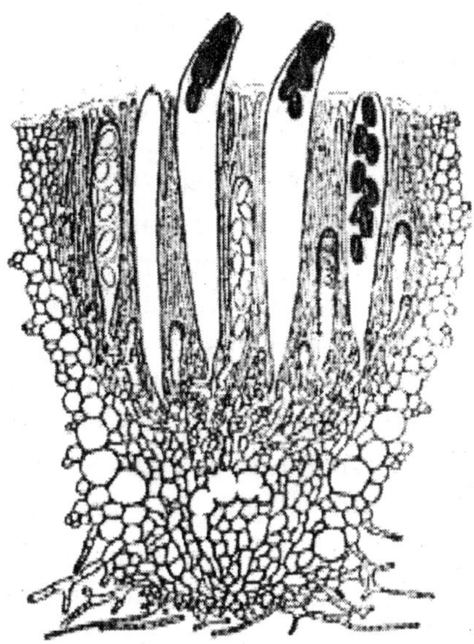

Fig. Ascobolus stercorarius. Vertical section through apothecium.

The projecting asci are pointing towards the incident light.

In *Ascobolus stercorarius,* as in nearly all operculate

Discomycetes, when the ascus bursts a small apical hinged lid is fluni; back, the stretched ascus-wall contract longitudinally and transversely, and instantly the spores and some of the ascus sap is squirted into the air to a distance of several centimetres.

It has been pointed out that the phototropism of the asci in such a species as *Sarcoscypha protracta* permits the escape of spores that would otherwise be is, thrown against an opposite hymenial surface and so wasted.

However, if there be a biological advantage in the phototropism of the asci in *Ascobolus,* it cannot be of this nature. Nevertheless, growing on an irregular surface such as the dung pellets of horse or rabbit, phototropism of the asci does help to ensure that the spores are thrown towards an open space away from the dung. This is clearly of some importance since successful dispersal in coprophilous fungi depends on the spores reaching the surrounding herbage.

Spore Discharge in Peziza Type

In *Aleuria vesiculosa* the eight ascospores form a single series in the ascus and are attached to its apex and to one another by a fine granular thread. When the ascus bursts a hinged lid flies back and the spores are squirted out to a height of 1-2 cm., the eight spores from one ascus being separated from one another in the air and not held together in a single mass. Buller demonstrated this in a dark room by passing a strong beam of light above an apothecium from which spores were being liberated. The eight spores on discharge appeared 1-2 cm. above the hymenium as eight little motes spread out vertically in the beam of light.

In all these cup-fungi, as in *Ascobolus,* the eight spores appear to be discharged simultaneously from the ascus. However, it is likely that discharge in reality is successive, with an exceedingly minute fraction of a second elapsing between the liberation of sister spores, the interval being so small that to the eye discharge appears to be simultaneous. This theory is supported by the fact that the diameter of the ascus mouth is usually less than that of the spore, so that it is difficult to see how more than one could escape at a time. With the discharge of each spore the pressure in the ascus presumably falls, successive spores receive decreasing momenta, and are, therefore, shot to decreasing distances so that they become spread out vertically.

The daily puff of spores in *Ascobolus* has already been considered, but the phenomenon of puffing is especially associated with the larger cup-fungi. During a period of quiescence thousands of asci reach a highly stretched condition of unstable equilibrium, so that the least shock —for instance a jerk, a sudden increase in temperature or illumination, or a change in humidity—is sufficient to cause all these asci to burst simultaneously, throwing into the air a visible cloud of spores that drift away like smoke.

If a cup-fungus is picked and at once placed near the ear, puffing may occur and is then audible as a hissing sound. Once an apothecium has puffed it cannot, as a rule, be induced to do so again for a time, presumably because all the ripe asci burst at each puff. There is no doubt that puffing does occur under natural conditions, but it is not clear if it represents the usual method of spore discharge or whether, under field conditions, spore liberation may be more continuous, with only a few asci bursting at a time.

Falck, in the course of his investigations of spore dispersal in Discomycetes, recognised two types. In the first, the *radiosensitive,* puffing does not occur, but spore discharge is greatly stimulated by incident radiation. *Morchella, GyroTnitra, Helvetia,* and *Geoglossum* are of this type. In the second, the *tactiosensitive,* puffing occurs in response to mild shocks, as when an apothecium is touched or blown upon. To this type most of the larger cup-fungi belong.

In *Aleuria vesiculosa* all the elements of the hymenium are phototropic, but this seems to be exceptional and usually response to light is limited to the projecting parts of the, ripe asci, as in *Ascobolus*. This is so, for example, in *Ciliaria scutellata* and *Melastiza miniata*. In *Sarcoscypha protracta* there is no *evident* curvature even in the projecting parts of the ripe asci. Curvature does, however, occur but is limited to the extreme tip, and the only result of this reaction to light is that the lid is displaced slightly towards the more strongly illuminated side of the apex, but this displacement is sufficient to direct the issuing ascus-jet towards the incident light.

Sometimes, as in *Aleuria repanda*, the whole apothecium turns towards light, due to growth movement of the stipe, and in addition the individual asci are positively phototropic.

Discharge in Morchella Type

The morels (*Morchella* spp.) have the most elaborate kind of apothecium. There is a stout, sterile stalk and the spore-producing layer is spread over a surface consisting of a number of alveoli or shallow pits. The hymenium of asci and paraphyses is continuous except for the exposed crests of the major ridges, separating neighbouring pits, which are sterile. The asci are phototropic and thus the escape of spores from the alveoli is assured.

Morchella has been regarded as the discomycete equivalent of *Boletus* in Hymenomycetes. Both have the hymenium lining pores, the striking difference in the size of these pores being associated with differences in the range of the ascus and the basidium. In *Morchella* it is the phototropism of the asci that determines the escape of the spores from the pits, which are, in the main, horizontal; in *Boletus* it is the positive geotropism of the downward-facing hymenial tubes that is so important, allowing the spores discharged into the middle of the tubes to fall freely until they reach the air below the pileus.

Successive Discharge of Ascospores in Geoglossum

In the examples considered above, all the spores of the ascus are liberated simultaneously or almost so. In sharp contrast are species of *Geoglossum* and *Trichoglossum,* in which the spores are shot from the ascus in definite succession.

The apothecium of *Trichoglossum hirsutum* or *Geoglossum ophioglossoides,* species quite often seen on lawns, is black, club-shaped, and 5-10 cm. high, with the hymenium covering the upper half of the club. The hymenium has the usual organisation, but the ascus contains a sheaf of eight long, septate ascospores. When ripe it dehisces by the development of a minute apical pore. As soon as this is formed one of the spores is squeezed into it, completely stoppering the ascus. At first this spore is squeezed very slowly out of the ascus by the hydrostatic pressure within until about half of it is projecting. Then it rapidly gathers velocity and is finally discharged with such speed that its flight cannot be followed by the eye until, slowed down by friction with the air, it flashes into view about 0.5 cm. from the hymenium. Immediately following the discharge of one spore another takes its place, blocking the pore before the ascus can undergo any shrinkage. In its turn this spore is shot away a few seconds after the first, and so on until the whole complement of eight has been discharged. As soon as the last spore has been ejected the ascus suddenly contracts, since there is no longer a spore to act as a temporary stopper.

Pyrenomycetes

In Pyrenomycetes, the largest division of Ascomycetes, the asci are contained within a small flask-shaped structure (perithecium) rarely more than a millimetre in diameter. This opens by a narrow canal traversing the neck which may be very short (e.g. *Sordaria* spp.) or quite long (e.g. *Ceratostomella* spp.). Lining this canal are upward-projecting filaments (periphyses). For successful spore liberation the tip of the ascus must project beyond the mouth (ostiole) of the perithecium before it explodes. Owing to the narrowness of the neck-canal only one ascus can, as a rule, reach the ostiole at a time, so that simultaneous bursting of asci, so characteristic of Discomycetes, cannot occur in Pyrenomycetes. In an active perithecium no gas-phase is present. Sometimes the asci, paraphyses, and periphyses are so tightly packed that there are no spaces between them, but if spaces do occur they are filled with slime.

Amongst Pyrenomycetes several more or less distinct types of spore discharge can be recognised and will now be considered in turn.

Sordaria Type

For watching spore discharge the small Pyrenomycetes that appear with such regularity on the dung of herbivores are especially favourable. This is particularly true of the two very common and closely similar species *Sordaria curvula* and *S. tetraspora*. In most Pyrenomycetes the wall of the perithecium is quite opaque, but in these two species it is semi-transparent, so by focusing an optical section the behaviour of the living asci can be seen in a specimen mounted in water.

Most of the asci occur crowded in the lower half of the perithecium attached to a basal cushion of tissue, but the upper half is occupied by the greatly swollen upper parts of those asci that are fully mature. Between the perithecium wall and the mass of asci are very thin-walled, much-inflated cells of the paraphyses.

The top of the perithecial cavity tapers and finally merges into a short, narrow canal leading to the outside. One of the mature asci in the upper region of the perithecium is usually slightly in advance of the others, and if this is watched it may be seen to elongate, in part probably due to actual growth and in part to the pressure of surrounding turgid cells.

Finally the tip of the ascus reaches the ostiole and protrudes very slightly beyond it. It then explodes, discharging a projectile, consisting of the ascus cap, eight spores bound together by ropes of mucilage, and some ascus sap, to a height of about 20 cm. On discharge the elastic wall of the ascus, which is enormously stretched but still attached to the interior of the perithecium, retracts into the perithecium, where it soon disintegrates. As soon as one ascus has discharged, another begins to elongate up the neck, and so the process goes on in orderly succession.

One interesting feature of *Sordaria,* also found in a number of allied genera, is that the neck of the perithecium is positively phototropic. The essential features of the *Sordaria* type are, first, that the ascus remains attached to the basal cushion inside the perithecium, and second, that it reaches the outside by elongation involving the gradual stretching of the whole ascus wall.

Sporormia Type

The distinctive feature of this type is that the asci has a double wall. This is a fairly common type frequently noted and figured since it was first described by Prings-heim in *Pleosbgra mrtoicola.* A common example is *Sporormia intermedia,* a frequent member of the horse-dung flora. The asci occur attached to a basal cushion within the perithecium, but each has a jack-in-the-box construction. The ascus wall consists of an outer relatively thick and fairly rigid membrane and an inner thin, extensible one.

Shortly before discharge, the outer membrane is ruptured apically and, the wall pressure being now reduced, the ascus is free to absorb water with the result that it elongates very rapidly up the neck-canal of the perithecium and soon projects through the ostiole. The ascus then bursts, squirting its spores simultaneously into the air.

Leptosphaeria acuta, a very common saprophyte on the dead stalks of nettle, agrees in general with *Sporormia,* but the spores (as in *Geoglossum)* are discharged in succession through an apical pore.

In the double-wall type of asci apparently constant feature is the enormous swelling of the thinner wall that occurs immediately after discharge, due, no doubt, to the absorption of water. This swelling was clearly illustrated by Pringsheim when he described this kind of discharge nearly a hundred years ago.

Cordyceps Type

Spore discharge in *Cordyceps militaris* is rather similar to that in *Leptosphaeria acuta,* but there are some interesting points of difference. *C. militaris* is a common parasite of the larvae and pupae of Lepidoptera. From the dead, mummified pupa in the soil, orange, club-shaped stromata grow above the ground. The upper parts of these stromata bear numerous reddish-orange perithecia, partly projecting as conical pimples.

Spore discharge is easy to see in this fungus and the spores are so long that the process can be watched. In the Clavicipitales generally, the ascus long, narrow, and cylindrical wall consists of a single layer.

The ascus is characteristically thickened and on dehiscence is pierced by a narrow, tubular pore. The spores are very long, and extremely narrow, forming a sheaf of eight, and, as generally happens with long, narrow spores, they are discharged in obvious succession. On watching an active perithecium under the high power of a miscroscope, the tip of an ascus suddenly appears at the ostiole and elongates rapidly. Then suddenly an ascospore flashes into sight and quickly sinks out of view.

Then, a second or two later, another appears, and so on. After each spore is discharged the ascus may retract by a few microns and may or may not elongate again before the next spore is discharged. Between the liberation of sister; spores some liquid contents are seen to ooze out of the ascus and flow down its outside. As soon as all eight spores are discharged the empty ascus collapses. Then, after an interval varying from half a minute to ten minutes, another ascus protrudes and discharges its spores.

The most curious feature of discharge in *Cordyceps* is that the spore leaves the ascus at such speed that nothing is to be seen of it until it flashes into sight about a quarter of a millimetre from the ostiole; at no stage is it seen protruding from the ascus and stoppering it as in *Geoglossum* or as in *Leptosphaeria acuta*. It is difficult to understand how the long, thread-like spore can reach sufficient speed within the ascus so as to leave it without being visible to an observer.

In *Epichloe typhina* another member of the Clavicipitales, spore discharge is much as in *Cordyceps militaris*, but the interval between the liberation of successive spores is very small and usually the escape of all eight spores occupies less than a second. Sometimes a spore follows its predecessor so rapidly that its apex sticks to the base of the first and a long filament consisting of several spores end to end may then be discharged in one piece.

Detached-ascus Type

This type is found particularly amongst long-necked species of Pyrenomycetes. It has been described in *Endothia parasitica*, *Gnomonia rubi* and *Ceratostomella ampullasca*. In this last species the Derithecium is immersed in rotting oak wood with only the tip of the neck projecting. Within the peritherm the asci arise from a basal cushion but they break free when mature. The flask thus becomes filled with many thousands of small asci, mostly detached. These absorb water and the perithecium which is relieved by the asci passing put in single file through the narrow neck-canal. In *Ceratostomella* the neck is opaque and the actual movement of the asci in the canal cannot be observed, but in *Gnomonia jrubi* due to the transparency of the neck, this movement can be seen.

In *Ceratostomella* there is a narrow, slit-like ostiole, and when an ascus reaches it pushes the slit wider open and protrudes until the spore-containing part is exposed. At this stage the lower part of the ascus is still held by the lips of the ostiole. The ascus then bursts, scattering its spores. Immediately afterwards the empty ascus is pushed out by the one below, and so on. This may be a very quick-firing mechanism with only a few seconds elapsing between successive ejections.

This detached-ascus type represents a mechanism of discharge which allows rapid spore liberation from a long-necked perithecium. It is clear that in optical longitudinal that had each ascus to elongate from the basal cushion up a neck-canal 1-2 mm. in length, spore discharge would certainly be slow.

Non-explosive Types

In a number of Pyrenomycetes the asci are no longer explosive. Their walls deliquesce within the perithecium and the liberated spores form a mass internuxed with mucilage. This absorbs water, swells, andoozes out through the ostiole, like toothpaste out of a tube, often forming a tendril-like thread composed of innumerable spores. This type of spore liberation is characteristic of the large genus *Chaetomium* and occurs also in some species of *Ceratostomella*.

In *Daldmia concentrica*, which grows so abundantly on dead ash-trees, violent spore discharge is the rule, but at times the normal mechanism fails and spore-tendrils are produced. They are not, however, formed in quite the same manner as in *Chaetomium*. An ascus elongates and nearly reaches the ostiole, but instead of bursting explosively its contents merely ooze out. The next ascus behaves likewise and piles its eight spores just behind the first eight, pushing them forward a trifle. As the process goes on the spore-tendril grows longer and longer. Spore-tendrils or spore-horns are also a feature of the pycnidial stage of many Ascomycetes.

Thus in *Diatrype stigma*, so common on beech bark, the pycnidia exude orange tendrils several centimetres in length and it is these that are the most conspicuous feature of the fungus at this stage of its life-history.

Spore Output by Perithecia: The perithecium of Pyrenomycetes is always a small structure and is not subject to the great range of size-variation seen in the apothecia of

Discomycetes. This is very understandable, for in Pyrenomycetes discharge is normally limited to one ascus at a time and, therefore, increase beyond a certain size is of no value. Perithecia of different species do, however, vary considerably in their spore-liberating capacity which gives the rate under favourable conditions of spore discharge from perithecia of a number of species.

Conditions Affecting Spore Discharge in Pyrenomycetes

The principal conditions that affect spore discharge in the Pyrenomycetes are the supply of water, temperature and light. Since violent spore discharge involves the bursting of turgid cells, it is clear that a supply of water is essential if it is to go on. The majority of Pyrenomycetes have hard, resistant perithecial walls or have their perithecia immersed in hard stromatal tissue.

Most can be air-dried without injury, and this is true even of some fleshy forms (e.g. *Nectria cinnabarina).* They are essentially drought-resisting xerophytes which, however, discharge their spores only during wet periods when the fungal tissue is fully turgid. This has, for example, been clearly demonstrated for *Endothia parasitical* the cause of American chestnut blight, in which the ascospores are ejected into the air only during rain and so long as the bark remains wet. From March to October discharge occurs during each spell of rain. Observations on the apple scab fungus *(Venturia inaequalis),* which discharges its ascospores from perithecia developed on the old fallen leaves in spring, show a close correlation between periods of rainfall and periods of spore discharge. In some species (e.g. *Hypoxylon pruinatum]* it has been shown that a saturated atmosphere is not sufficient to initiate spore discharge, but that the perithecial stroma must first be wetted before discharge will begin.

Daldinia concentrica is a special kind of xerophyte able to continue spore discharge during dry periods by virtue of a considerable reserve of water in the stromatal tissue. This fungus is very common on dead ash branches and is to be found on other trees that have been scorched by fire. The fungus is usually somewhat hemispherical and often about the size of half an apple.

The surface layers of the stroma are extremely hard and resistant, and no doubt the nature of this tissue helps in reducing loss of water by evaporation. The very numerous perithecia occur immersed in this peripheral tissue. The bulk of the stroma is composed of tissue showing the curious concentric zoning from which the specific epithet is derived.

A single stroma lasts only for one season, and the perithecia ripen and begin to shed their spores in May and may continue to do so until September. During this period discharge is not dependent upon rain but on the water reserve of the stroma. If an active stroma is brought into a dry room and left fully exposed without any extraneous water-supply, spore discharge will continue for weeks.

Even when placed in a desiccator with the air kept bone dry by anhydrous calcium chloride, discharge may go on for many days. It is only when evaporation has led to the reduction of the density of the stroma that spore discharge finally ceases. In its water relations *Daldinia* is rather like a cactus. It is a succulent xerophyte.

Another interesting example of a stromatal pyrenomycete that is not directly dependent on rain for its spore discharge is *Epichloe typhina,* which occurs commonly on a large number of grasses, especially *Dactylis glomerata, Holcus mollis,* and *Agrostis tennis.* The orange perithecial stroma is produced in July and August as a conspicuous crust around the grass-shoot just above a node. Immersed in the stromatal tissue are numerous perithecia from which long, thread-like spores are violently discharged.

If a stroma is held in the hand with sunlight falling on it from just the right direction, the escaping spores can be seen with the unaided eye. Spore discharge is not dependent on external water, and if a stalk of grass bearing a stroma is brought into a dry room, discharge continues for days, provided that the cut end of the grass stalk is in water or, in other words, that the transpiration stream of the host is maintained. If this stream is interrupted, spore liberation from the fungus soon ceases. In this species the water for spore discharge is clearly being derived from the living tissue of the host.

Concerning the effect of temperature on spore discharge little information is available. Below 5° C discharge is probably arrested in most species.

Above 5° C and below 35° C a rise in temperature tends to increase the rate of spore liberation. The result of a change from the lower to the higher temperature is a sudden increase in the rate of spore output which, however, rapidly falls off.

A study of the daily march of spore ejection has demonstrated in most species examined a periodicity governed, in all probability, by the periodic changes in illumination that occur as night follows day.

Some species are essentially nocturnal in the spore liberation (e.g. *Daldinia concentrica* and *Hypoxylon fuscuni)* whilst others are diurnal *(Nectria cinnabarina* and *Sordaria curvula*. The whole question of the cause of this periodicity is in need of careful experimental study.

Spore Liberation from the Cleistocarps of Powdery Mildews and from the Ascus of Protomyces

In Erysiphales (powdery mildews) the cleistocarp (cleistothecium) is like a small perithecium, but without an ostiole. It is essentially a hibernating structure which remains dormant throughout the winter, but shows the renewed activity in spring that eventually leads to the discharge of the ascospores. Within the firm wall of the cleistocarp one or several oval ascioccur.

Sphaerotheca morsuvae is an example of the first type. The single ascus within the cleistocarp swells in the spring, and this ieads to the rupture of the cleistocarp wall by a slit through which the ascus protrudes as it continues to swell. Finally the ascus bursts and the ascospores are squirted out.

In *Podosphaera leucotricha* double discharge occurs. This has been observed by Woodward and is best described in his own words:

> The perithecium is a true cleistocarp, the ascus being ejected through an irregular rupture of the wall. The opening formed gradually widens owing to the expansion of the ascus following the absorption of water.. .. The perithecial wall is elastic and responds to the pressure exerted from within until the opening, although small in diameter, can no longer retain the ascus which squeezes through the orifice and is violently ejected. The force of ejection is considerably increased by the snapping together of the 'jaws' of the orifice, due to the elasticity of the perithecial case. The ascus is often thrown several centimetres in the air. On reaching water or in a damp atmosphere, it continues to swell and within one or two minutes explodes, scattering the ascospores in all directions.

Spore discharge in *Protomyces* may be considered at this point because it resembles, somewhat, discharge in *Sphaerotheca*. *Protomyces* is one of those unfortunate genera which have failed to settle neatly into any scheme of classification. Some have placed it in Phycomycetes as an archimycete type; others have tried to push it into Ustilaginales; others, in despair, have made for it a special group of fungi, Protomycetes; but now most mycologists are resigned to including it amongst the 'lower' Ascomycetes.

Species of *Protomyces* occur as obligate parasites on herbaceous plants, but are of no economic importance. *P. macrosporus* on *Aegopodium, P. packydermus* on *Taraxacum,* and *P. inundatus* on *Apium nodiflorum* are common species. They produce small warts on the stalks and veins of the leaves and on the stems. Each wart is the result of a single infection and, in section of a wart, the fungus is seen as an intercellular mycelium of limited extent with many much-enlarged intercalary chlamydospores each with a two-layered wall.

These are resting spores in *P. macrosporus* and *P. pachydermus*, but in *P. inundatus* they germinate at once. When fully developed, the warts on *Apium* split open exposing the parasite. Each exposed 'chlamydospore' may be regarded as an ascus confined within an outer rigid cell-wall. As the cell absorbs water the ascus expands, bursts through the outer rigid cell-wall and lies free as a spherical cell containing a large number of minute spores clumped together. The ascus then bursts by a slit and the spores are thrown into the air to a distance of about 2 cm.

Spore Discharge by Water-squirting in Phycomycetes

There are apparently few examples of fungi discharging their spores by a water-squirting mechanism outside Ascomycetes, although possibly discharge of basidiospores, in Basidiomycetes may turn out to be of this nature. In Phycomycetes there are, however, three rather isolated but interesting examples: *Pilobolus, Basidiobolus,* and *Empusa.*

Pilobolus

Species *of Pilobolus,* a genus of Mucorales, develop with great regularity and in abundance if freshly deposited horse-dung is kept in the light under a bell-jar, and may also frequently be seen on dung in the field, especially under the humid conditions of early autumn. The commonest species are *P. kleinii* and *P. longipes.* There is a diurnal periodicity, conditioned by light, in the development of the crops of ripe sporangiophores which usually mature and discharge their sporangia round about noon.

The structure of the asexual apparatus in *Pilobolus* is fundamentally like that of *Mucor,* but differs in one notable feature, the presence of cross-walls separating the sporangiophore from the feeding mycelium.

The sporangium is bounded by a wall which is relatively thick and blackened in its upper regions, but thin and transparent below. Within are numerous spores occupying the upper part of the sporangium, and below them is a ring of mucilage encircling the columella.

Shortly before the sporangium is discharged its wall ruptures along a transverse line of weakness in the transparent region so that the sporangium gapes open there, exposing the mucilage.

The sporangiophore is a single cell composed of a basal bulb that is more or less buried in the substratum, a straight stalk part, and a Sub-sporangial bulb the apex of which bulges into the sporangium as a substantial columella. There is a thin lining layer of protoplasm, but where the Sub-sporangial bulb meets the stalk this layer is thicker and more conspicuous, being coloured orange by innumerable oil drops containing a carotinoid pigment.

Within the protoplasmic layer of the sporangiophore is a large vacuole filled with cell-sap having an osmotic pressure of some 5 atmospheres. The ripe sporangiophore is a highly turgid cell covered with numerous minute droplets of fluid presumably exuded under the influence of the hydrostatic pressure within.

There is a circular transverse line of weakness in the sporangiophore just below the junction of the sub-sporangial bulb with the columella. Finally, rupture occurs along this line and the stretched sporangiophore wall contracts, squirting out a drop of sap that carries the sporangium, including the columella, to a height of 50-200 cm. The sporangium wall, like the back of a duck, cannot be wetted.

The ring of mucilage below the spores is, however, very readily wettable, as is also the inner surface of the columella, with the result that the sporangium projects partially from the drop of sap with which it is associated. When the projectile strikes an object, although the sporangium may be momentarily submerged by the accompanying drop of sap, it bobs up to the surface of the drop with the unwettable wall outwards. As the drop dries, the ring of mucilage comes in contact with the surface of the object and finally dries to a hard cement, so that the sporangium is firmly stuck with the spores completely covered by the black sporangium wall.

It is interesting to notice in passing the wide variety of spore-liberation types in Mucorales—a very natural order of Fungi. *Pilobolus* is the only genus with violent spore discharge. In most species of *Mucor* the sporangiospores are slime spores. Thus in *M. hiemalis* and M. *racemosus* when the sporangium is ripe its wall dissolves (it is said to be diffluent) and becomes converted into a sporangial drop.

When this dries the spores are firmly cemented to the columella and cannot be dislodged by wind. Just how they are dispersed is not at all clear— perhaps it is by rain splash. The same type of sporangial drop is found in the beautiful *Phycomyces blakesleeanus*—a fungus of great value to the botanist and to be found in every botanical laboratory, but little known in nature. *Phycomyces* has xerophytic features. If it is grown on nutrient agar in an open petri dish freely exposed in a desiccator containing anhydrous calcium chloride, sporangiophores are produced as readily as in damp air and grow to the same considerable length (10-15 cm.), and at their ends the usual slimy masses of spores are produced.

This unique xerophytism of *Phycomyces* is due in part to the chitinised wall of the sporangiophore and in part to the freedom with which water streams into the sporangiophore, making good any loss by evaporation. *Pilaria* is a curious genus with a sporangium structure exactly like that of *Pilobolus,* but with a sporangiophore of the *Mucor* type. *Pilaria anomala* is very common on horse-dung, developing just before the *Pilobolus* crop. There is a straight and, at first, short sporangiophore terminated by the sporangium, which splits, as *in Pilobolus,* by a transverse line of weakness, exposing the mucilage within. Then the phototropic sporangiophore elongates enormously and very rapidly, often carrying the sporangium on to the surrounding herbage, to which it becomes firmly cemented like a discharged sporangium of *Pilobolus.*

Rhizopus nigricans, black bread-mould, is a typical dry-spore fungus. When the ripe sporangium is exposed to dry air its wall cracks into numerous small pieces and exposes, on the collapsed columella, a mass of dry, powdery spores which are easily blown away.

A very similar example is *Sporodinia grandis,* common on decaying agarics in autumn. Here the sporangiophore is dichotomously branched. Each sporangium has an excessively thin wall. This breaks down completely at maturity, exposing the dry, powdery spores.

Cunninghamella elegans is another example of a highly successful dry-spore mechanism. In this species, swollen heads of conidia are produced instead of sporangia. These conidia are finely attached and easily blown away. *Thamnidium elegans* is of special interest. It produces a long sporangiophore with a large terminal sporangium of the *Mucor* type which finally becomes a sporangial drop, but in addition short but much-branched laterals from the sporangiophore bear minute sporangioles, each containing a small number of spores, usually four. These sporangioles when ripe readily break off and can be dispersed by wind.

Basidiobolus

Basidiobolus ranarum occurs with great regularity on the excrement of frogs, the conidiophores projecting into the air looking like minute specimens of *Pilobolus*. The conidiophore arises from a single cell of the septate mycelium or directly from a discharged conidium. There is a straight stalk and a sub-conidial bulb, the tip of which projects as a minute columella into the spore.

The conidiophore is a highly turgid cell which finally bursts along a transverse line of weakness near the base of the bulb, not near the top as in *Pilobolus,* and the elastic part of the stretched wall is above this line, not below it. Thus, on bursting, the stretched elastic wall of the upper part of the bulb contracts, squirts sap backwards, and flies off on the recoil carrying the conidium with it. It is a minute water rocket. During the flight through the air the two parts of the projectile usually separate so that, although the conidium is shot to a distance of 1-2 cm., the accompanying portion of the conidiophore (the 'basidium') is rarely thrown farther than 0-5 cm.

Empusa Muscae

House flies are frequently seen in the late summer attacked by *Empusa muscae* (fly cholera)). Just before death a doomed fly, plugged with the fungus, usually attaches itself to some object, often a pane of glass, by its sucker-like proboscis. Then in a few hours thousands of conidiophores protrude between the segments of the abdomen so that it appears to have several white transverse stripes. A conidium is borne terminally on each conidiophore which is simple and somewhat inflated. The conidiophore bursts, just as in *Pilobolus*, shooting the conidium to a distance of 1-1-5 cm. The result of the bursting of thousands of conidiophores is that the dead fly becomes surrounded by discharged spores forming a white halo 4-5 cm. in diameter.

Discharge due to rounding-off of turgid cells The shape of a turgid cell is often modified by adhesion to other cells. Thus each polyhedral cell of the cortical parenchyma of a stem or root tends to become spherical, but is prevented from so doing by the presence of its neighbours. The individual cells of the tissue are under conditions of strain. In Fungi, there are a number of scattered examples in which sudden change of shape by turgid cells in a state of strain results in violent spore discharge.

Discharge in Entomophthorales

Entomophthorales is a small order, mainly of fungi parasitic on insects, and one of its most constant features is that the terminal conidium is shot from its conidiophore. Two examples *(Basidiobolus* and *Empusa)* where discharge is due to the bursting of the conidiophore have already been considered, but in species of *Entomophthora* the conidia are discharged just as far, but by a mechanism involving the sudden rounding-off of turgid cells.

Entomophthora coronata (Conidiobolus villosus) which grows very readily as a saprophyte on ordinary laboratory media, but in addition parasitises termites and aphids, is a useful species for seeing this type of discharge. The terminal conidium is separated from its simple conidiophore by a cross-wall which bulges into the spore as a hemispherical columella. The cross-wall is a double structure, one layer belonging to the spore-wall, the other to the apex of the conidiophore.

Suddenly the in curved part of the spore-wall bulges outwards so that the spore springs off from the conidiophore. The walls of the spore and of the conidiophore, except where they are in contact in the columella zone, are two-ply structures, the outer layer of the conidiophore wall being continuous with the outer layer of the spore-wall. As a preliminary to discharge, rupture of this outer layer occurs along the circular line where the spore joins its conidiophore. Probably as soon as this rupture occurs the spore is shot away.

In *Entomophthora spkaerosperma* a parasite of the firefly, discharge occurs just as in *E. coronata.*

Downy Mildews

Weston has reported spore discharge very closely resembling the *Entomophthora* type in *Sclerospora philipinensis,* one of the downy mildews (Peronosporales) which attacks maize in the Philippines. The mycelium of the fungus is in the intercellular spaces of the leaf, and the branched conidiophores project in tufts through the stomates, being produced only during the night when a film of dew is deposited on the surface of the leaf.

Although the lower parts of the branched conidiophores are immersed in dew, the tips bearing the spores project into the air. From each tip a single conidium is formed, and the small surface of contact between the two is flat. Both the spore and the conidiophore are turgid cells, so there is a tendency for each to round off in this flattened region—a tendency counteracted by the adhesion of the two walls to one another. It seems that the rounding-off of the spore and the conidiophore tip occurs suddenly and in consequence the spore is shot away to a distance of a millimetre or two. All the spores of a single conidiophore are, apparently, discharged simultaneously. It is only in *Sclerospora* that violent discharge of this type has been reported, but another type of mechanism—a hygroscopic mechanism— has been described in other genera of downy mildews.

Aecidiospore Liberation in Rusts

In rust-fungi at the base of each aecidium or 'cluster-cup' is a closely set palisade of basal cells each cutting off an ever-growing chain of cells with aecidiospores and separating-cells alternating. The separating-cells soon break down, leaving the spores stuck together by blobs of sticky material resulting from their disintegration. In the aecidium the spores are tightly packed and each assumes a polyhedral form due to the pressure of its neighbours. Under damp conditions the spores absorb water, and with increasing turgor each tends more strongly towards the spherical form. Spores in the outermost layer of aecidiospore mass round off suddenly and in doing so spring out either singly or in small groups. These spores are followed by more and so the aecidium tends to empty, but the supply is renewed from the base.

Sphaerobolus

In *Sphaerobolus stellatus,* classified in Gasteromycetes, the sudden release of a tissue tension, which does not involve the bursting of any cells, leads to the discharge of a macroscopic spore-mass to a considerable distance.

It seems quite clear that Gasteromycetes is a polyphyletic

group of fungi, derived, probably, from hymenomycete ancestors. However, if these fungi ever possessed the hymenomycete equipment of spore liberation, it has now been lost and other methods have been developed.

The Gasteromycetes can fairly be regarded as an assorted collection of experiments in dispersal. Thus the spores of the stink-horns and their allies (Phallaceae) are spread by insects, the puff-balls and earth-stars (Lycoperdaceae) have evolved dry-spore wind-dispersal mechanisms, the subterranean forms (Hymenogastraceae) apparently rely on rodents, the bird's-nest fungi *(Crucibulum* and *Cyathus)* have a highly specialised rain-splash mechanism, and in *Sphaerqbolus* a unique type of violent discharge has been developed.

Sphaerobolus stellatus is a common saprophyte on old, rotten wood, and the tiny fruit-bodies, each about 2 mm. in diameter, are commonly to be seen in autumn, usually crowded together in large numbers. The opened fruit-body is spherical and anchored to its substratum by fine hyphae. In section it is found to consist of a central spore-containing region (gleba) surrounded by a wall (peridium) differentiated into six histologically distinct layers.

When the fruit-body is mature, the peridium splits along a number of lines radiating from the apex. In this way from four to eight little teeth of peridial tissue are delimited which bend outwards, exposing the spherical gleba. At the same time the peridium becomes separated into two little cups, fitting one inside the other, each with a toothed margin and attached together only at the points of the teeth. The outer cup is composed of the three outer layers of the peridium, the inner of two of the three remaining layers.

It seems quite clear that Gasteromycetes is a polyphyletic group of fungi, derived, probably, from hymenomycete ancestors. However, if these fungi ever possessed the hymenomycete equipment of spore liberation, it has now been lost and other methods have been developed.

The Gasteromycetes can fairly be regarded as an assorted collection of experiments in dispersal.

Thus the spores of the stink-horns and their allies (Phallaceae) are spread by insects, the puff-balls and earth-stars (Lycoperdaceae) have evolved dry-spore wind-dispersal mechanisms, the subterranean forms (Hymenogastraceae) apparently rely on rodents, the bird's-nest fungi *(Crucibulum* and *Cyathus)* have a highly specialised rain-splash mechanism, and in *Sphaerqbolus* a unique type of violent discharge has been developed.

Sphaerobolus stellatus is a common saprophyte on old, rotten wood, and the tiny fruit-bodies, each about 2 mm. in diameter, are commonly to be seen in autumn, usually crowded together in large numbers. The opened fruit-body is spherical and anchored to its substratum by fine hyphae.

In section it is found to consist of a central spore-containing region (gleba) surrounded by a wall (peridium) differentiated into six histologically distinct layers.

When the fruit-body is mature, the peridium splits along a number of lines radiating from the apex. In this way from four to eight little teeth of peridial tissue are delimited which bend outwards, exposing the spherical gleba. At the same time the peridium becomes separated into two little cups, fitting one inside the other, each with a toothed margin and attached together only at the points of the teeth.

The outer cup is composed of the three outer layers of the peridium, the inner of two of the three remaining layers. The sixth layer, which was in immediate contact with the gleba, has at cells of the palisade layer in contact with the lubricating fluid absorb water, so that the area of the inner surface of the cup tends to increase, but where they are in contact with the layer of interwoven hyphae such increase is impeded.

In this way strains are set up which are suddenly and violently released by the inner cup turning inside out, thereby catapulting the gleba to a distance of up to 550 cm.

Ballistospores

The name 'ballistospore' has been given to the type of spore which is perched asymmetrically on a fine sterigma and in which the secretion of a drop of liquid by a small projection (hilum) of the spore immediately precedes spore discharge.

Basidiospores of Hymenomycetes and Gelatinous

Basidiomycetes are ballistospores, and so also are the sporidia (basidiospores) of rusts (Uredinales) and the conidia of certain smuts (e.g. *Tilletia caries* and *Entyloma* spp.) and of shadow yeasts (Sporobolomycetaceae).

The basidiospore of *Puccinia malvacearum* (hollyhock rust) may be considered as a typical ballistospore. In this rust, when the two-celled teleutospore is placed in a damp atmosphere, a short hypha is produced from each cell. This curved hypha is the young basidium.

It be comes divided into four cells from each of which, on the convex side of the curve, a short sterigma grows out and swells at its tip to form a spore. From its first appearance as a minute enlargement of the tip of the sterigma it grows to full size in about forty minutes, and ten minutes later is shot away. Near its attachment to the sterigma the spore has a small blunt, beak-like projection—the hilum. From this, just before discharge, a drop of fluid appears, grows steadily to a certain definite size, and then at once the spore is shot away, carrying the drop with it. No liquid is to be seen at the end of the vacated sterigma, which exhibits no observable shrinkage.

Ballistospores show little range in size and shape as compared with ascospores, and the distance of discharge is small also and varies within narrow limits.

The mechanism responsible for the violent discharge of ballistospores still remains in doubt, but there are three current theories worthy of mention. One theory is that discharge is due to the bursting of a turgid cell, as in *Pilobolus*. Now the unicellular basidium of Hymenomycetes bears four ballistospores which are discharged in succession, with a short interval between the liberation of sister spores. If each spore is discharged by the bursting of its sterigma, it is difficult to see how, after one sterigma has burst at its apex, projecting its spore, the end becomes sealed to allow maintenance of turgor in the basidium so that the next spore in succession can be shot away.

Further, there is no noticeable and immediate shortening of the sterigma following discharge, nor is any fluid normally to be seen at its apex. Certainly, if the sterigma bursts it would be expected that escape of basidial content, in the form of a small drop, would occur before the sterigma could have time to become sealed again. A second theory is that the surface energy of the drop excreted at the hilum is used in some way to effect discharge, but there is no definite evidence to support this view and its only merit is that it brings the exuded drop of the ballistospore into the picture. A third theory is that discharge may be of the type already recorded for *Entomophthora* and *Sclerospora*.

In recent years, this old theory has been advocated again by Prince, who made a careful study of basidiospore liberation in *Gymnosporangiwn nidus-avis*. The basidiospore of this rust is a little unusual because the prominent beak on the spore is not the hilum, but is where the spore is actually attached to the sterigma. The hilum itself cannot be detected in the liberated spore, and in the attached one can be located only when drop-exudation occurs. Prince found that where the mature spore joins the sterigma there is a clearly defined cross-wall which is flat. It is interesting to note that Buller, in his observations on the development and liberation of basidiospores, was unable to satisfy himself that such a cross-wall existed.

However, *Gymnosporangium idusavis* appears to be exceptionally suitable material for seeing the nature of the junction between the spore and its sterigma, and there can be no doubt of the correctness of Prince's observation. Further, Prince found that immediately following discharge both the end of the sterigma and the beak of the spore showed convex surfaces, not flat ones. He suggests that it is sudden rounding-off at the flat surface of contact that is responsible for violent discharge, just as in *Sclerospora*.

Fruit-bodies of Hymenomycetes in Relation to Ballistospore Discharge

In Hymenomycetes the fruit-body, usually large and conspicuous, bears extensive exposed hymenia in which millions of basidia occur interspersed with sterile cells. Each basidium is a single turgid cell which, like the ascus, is a special kind of sporangium, but instead of the spores being produced internally they are formed externally and are characteristic ballistospores. Usually there are four spores grouped around the apex of the basidium. Each is borne asymmetrically on a very fine sterigma with the hilum pointing inwards.

It has been seen that the chief factors governing the form of the apothecium in Discomycetes are that the spores are usually squirted to a distance of several centimetres, that the asci are positively phototropic, and that the hymenium is not damaged by water. In the same way the architecture of the hymenomycete fruit-body seems to have been governed by certain simple facts, namely that the spores are discharged to a distance of less than half a millimetre, that the hymenial surfaces can be orientated by geotropic movements, and that the hymenium is spoilt, at least temporarily, by water.

Since the distance of spore discharge is so small the upward-facing hymenium is inefficient as the sticky spores would tend to fall back almost immediately on to the parent surface and so be wasted. Further, such an hymenium would be continually subject to injury by rain. Thus, hymenial surfaces tend to be displayed in fruit-bodies either in a vertical or a downward-facing position between the horizontal and the vertical, and where the vertical position is assumed there is usually, except in the Clavariaceae, an adequate protection from rain.

Escape of Spores from Agaric Fruit-body

Russula ochroleuca, one of the commonest of woodland agarics, may be taken as a fairly typical example of a toadstool fruit-body. The hymenium covering the gills consists of basidia intermixed with sterile cells of two kinds: paraphyses that are difficult to distinguish from young basidia from which sterigmata have not yet sprouted, and conspicuous cystidia with pointed ends and granular contents.

In the majority of toadstools cystidia are absent and the hymenium is composed solely of basidia and paraphyses. In any small area of the hymenium basidia are to be seen at all stages of development. Those that have discharged all four spores are still turgid cells and look very much like young basidia that have just formed their sterigmata but in which spore-formation has not yet begun. However, the exhausted basidia are clearly differentiated from young ones by having little remaining protoplasm, and they soon collapse and disappear.

The pileus in *Russula ochroleuca* is supported on a central stalk (stipe) which, as in most toadstools, is a fairly rigid structure so that the sporophore shows no tendency to sway in the wind. The importance of the stipe depends partly on its rigidity and partly on the fact that it provides a space between the pileus and the ground, thus affording an opportunity for the falling spores to be swept away by air currents.

In connection with the function of the stipe it is interesting to note how size and form seem to be related in the agaric fruit-body. The weight of the cap, roughly proportional to its volume, acts on stipe and lie strain is distributed over the area of the cross-section, (if it is assumed that organisms are economical with their tissue, it might be expected that the stipe would tend to be of a thickness just sufficient to support the pileus with the rigidity necessary for the proper functioning of the sporophore. Now in solid figures, that vary in size but not in form, the volume varies as the cube of the linear dimensions, but two-dimensional features vary as the square. Thus the volume and, therefore, the weight of the pileus varies as the cube of its diameter, but the cross-section of the stipe varies as the square of its width.

On these considerations it would be expected that the agaric sporophore would not remain constant in form with change of size, but that, as compared with a median type, the smaller fruit-bodies would tend to have relatively slender stipes and the larger ones relatively thick ones. Again, the height of an agaric is mainly due to the length of the stipe, which provides a space between the under side of the cap and the ground, so that spores in their downward course have a good chance of being blown away before they reach the ground in the immediate vicinity of the fruit-body. With the development of larger and smaller types of agarics it might be expected that the general form would tend to be preserved.

However, in the larger types the interval between cap and substratum would tend to become greater than necessary and it might be expected that mutants having shorter stalks, and thereby economising tissue, would tend to survive in the struggle for existence.

On the other hand, in the smaller types, if the form remained unaltered, the interval between the pileus and the ground might be reduced too far for efficient dispersal and mutant types with longer stalks would tend to be selected. Thus we might expect that the larger sporophores would tend to be relatively short and the smaller ones relatively tall.

Both these expectations are fulfilled, and it is found that in fact small agarics tend to be relatively tall and have relatively thin stalks and large agarics tend to be relatively short and have relatively thick stalks as compared with those of medium size.

In a fruit-body such as that of *Russula ochroleuca* the hymenial surfaces are brought into a vertical position by growth movements in response to external stimuli. In some agarics (e.g. *Coprinus cinereus)* the stipe in the young fruit-body is positively piotolropic. The fine adjustment by positive geotropism of the hymenium-bearing structures is not limited to the gills of agarics, but is also found in the tubes of *Boletus* spp. and in the spines of *Hydnum* spp.

Although in fleshy toadstools gravity plays such an important part in the final adjustment of the fruit-body, it has *no formative* effect. Thus if a tube-slant of nutrient agar is inoculated with *Collybia velutipes,* a species that fruits exceptionally well in culture, and if this tube is rotated continuously for several weeks about a horizontal axis on a klinostat, thus eliminating any unilateral action of gravity, normal fruit-bodies are produced.

In toadstools, Buller has shown that the spores are shot from the horizontal basidia to a distance of 0-01-0-02 cm. into the spaces between the gills. They then fall and, on emerging into the open air below the pileus, can readily be dispersed by quite gentle winds or even convection currents. If the sticky spores fall on to the hymenium surface they remain stuck there. For this reason, the vertical position of the gills is essential if wastage of this nature is to be avoided. In addition, for successful spore liberation the gills must be farther apart than the distance to which the spores are shot and there is, in fact, usually a considerable margin of safety.

Spore Liberation in Coprinus

The organisation of the fruit-body in the genus *Coprinus* (ink-cap fungi) is especially interesting and has been thoroughly investigated by Buller. Probably the *Coprinus* apparatus represents the most advanced type amongst toadstools. *Coprinus atramentarius* may be taken as a beautiful example. Most ink-caps are more or less coprophilous, but *C. atramentarius,* a relatively large and very common species, grows on buried wood and is often to be seen at the base of old gate-posts. In most agarics basidia on any small area of the hymenium are at very different stages of development, but in the genus *Coprinus*, in a small area such as appears in the high-power field of the microscope when the gill is seen in surface view, all the basidia are roughly at the same stage. A wave of development passes from the free margin of each gill upwards towards the pileus tissue.

Young basidiospores are colourless, but change as they ripen through pink to dark brown or almost black. It is the spores that give the colour to the gill. So in a fruit-body from which spores are escaping each gill is blackish towards its free margin and grades through pink to white as the pileus tissue is approached. As the spores are shot away a very narrow zone near the gill margin becomes white due to the loss of its spores, and this spore-depleted area rapidly undergoes a process of autodigestion resulting in its liquefaction.

The fluid formed in this way turns black and under highly humid conditions may flow down the inclined margins of the gills and accumulate as inky drops at the edge of the pileus. However, if conditions are drier the liquid produced by autodigestion evaporates and no inky drops are formed. Most of the spores shot from the basidia escape, but a certain number drift into the inky liquid and get stuck.

In *Coprinus atramentarius* the stipe is negatively geotropic and this reaction brings the gills roughly into a vertical plane. But the gills themselves are not geotropic. However, their failure to respond to gravity does not militate against the efficiency of spore liberation, since, due to the removal of exhausted hymenium by autodigestion, the spores have only a fraction of a millimetre to fall between gill surfaces before emerging into the free air below the cap.

The behaviour of the cystidia is especially interesting. These undergo autodigestion somewhat before the rest of the gill and disappear just before spore discharge in their vicinity begins. In this way they offer no obstacle to the escape of the falling spores. In *C. atramentarius* the basidia are dimorphic—there are longer and shorter ones. This arrangement allows closer packing in the hymenium than would otherwise be possible. Further, the longer basidia mature and discharge their spores slightly earlier than the shorter ones, so that the ballistospores of the shorter basidia can be discharged without colliding with spores on the longer basidia.

Spore Liberation from an Annual Bracket Polypore

Polyporus betulinus, a birch parasite which, having killed the host, produces its fruit-bodies. Gravity has profound formative influence on the sporophore. The young appears as a small, undifferentiated spherical knob 2-3 cm. across. If this is on the main vertical trunk, it then grows out horizontally (diageotropism) to form a firm, more or less semi-circular structure, with a radius of 15-20 cm.

If however, the original spherical primordium is on the under side of an approximately horizontal branch, the fruit-body develops a roughly circular form with a central attachment to the tree Fruit-bodies do not normally arise on the upper side of a branch, but if a dead tree bearing primordia is felled, those on the recumbent trunk may continue their development. A primordium thus exposed on the upper surface of a fallen trunk grows out on one side only, more or less at right angles to the pull of gravity.

These pores are at first very shallow. They grow by means of an active zone around the mouth of each pore, so that gradually they become longer. The direction of growth is conditioned by gravity so that the tubes produced are orientated precisely in the vertical direction. This gradual geotropic growth of the tubes is in contrast to the state of affairs in a toadstool, where gravity is not concerned in the development of the hymenial surfaces but only in their final orientation. In bracket polypores geotropic growth achieves the desirable result of vertical hymenial surfaces, but if these, once formed, are slightly displaced from the vertical, there is no mechanism of readjustment in the pores. This is in striking contrast to the *Boletus* type of polypore.

In *Trametes gibbosa,* so common as a saprophyte on beech stumps, the morphogenesis of the sporophore is similar to that of *Polyporus betulinus.* Here the fruit-bodies arising on the more or less vertical surface of the trunk are of the bracket form, whilst those on the transversely cut surface of the stump are radially symmetrical with a very broad central attachment. The pores are normally formed on the under surface, probably in response to gravity, but the mechanism appears not infrequently to fail, since shallow pores bearing an active hymenium are occasionally to be found on the upward-facing surfaces.

In *Polyporus betulinus* the hymenial tubes almost from their first inception are lined with active hymenium and spore discharge takes place during the whole period of several months during which the pores are elongating.

Apart from the rather striking difference in the developmental picture of the sporophore, the general story of spore liberation is the same as in agarics. The spores discharged into the tubes fall vertically and on emerging from the under side of the bracket are carried away by wind.

Perennial Bracket Polypores

In the genera *Femes,* with white spores, and *Ganoderma,* with brown, certain species produce large woody brackets which are perennial. As an example the common and worldwide species *Ganoderma applanatum* may be considered. It is frequently found on the trunks of broad-leaved trees, especially beech. As in *Polyporus betulinus,* gravity has a determining formative effect on fruit-body development. Because of its extremely hard and rigid structure, and because of the broad and firm attachment to the tree trunk, the hymenial tubes formed by geotropic growth are very accurately orientated and there is little likelihood of even the slightest subsequent displacement from the vertical. Apparently correlated with this rigidity of construction the hymenial tubes are exceedingly narrow, having a diameter (0-15 mm.) probably little more than the distance to which the spores are discharged. Unfortunately there are no measurements of this distance for the ballistospores of *Ganoderma applanatum*. The distance apart of the gills in *Coprinus atramentarius* is almost the same as the diameter of the tubes in *Ganoderma applanatum* but in the former the spores have only a fraction of a millimetre to fall between hymenial surfaces before emerging into the open air, whereas in the latter a spore may have to fall several centimetres before escaping from the tubes.

In *Ganoderma applanatum*, and also in *Fomes fomentarius*, not only the fruit-bodies but also the hymenial surfaces themselves are perennial. A typical specimen of G. *applanatum* collected in July 1949. Six annual layers of hymenium tubes are seen. In the 1949 layer active hymenium was found throughout the length of each tube with the exception of the youngest half millimetre; there basidia were absent because, no doubt, of its recent formation.

In the 1948 and 1947 layers more than 90 per cent, of the tubes were lined with active hymenium throughout their length and the density of basidia in the hymenium appeared just as great as in that of the current year. In the 1946 layer, however, only about 50 per cent, of the tubes had a functioning hymenium; the remainder had no basidia and many of these tubes were plugged with white mycelium. Finally, in the 1945 and 1944 layers no tubes contained basidia and most were plugged. When a single tube was followed upwards from its orifice it was found normally to be continuous and to be lined with active hymenium in the lower three or four annual sections, but to be devoid of basidia higher up.

If a sporophore is cut vertically with a saw it is impossible to distinguish me boundaries separating the annual layers of tubes. If, however, a deep incision is made in the sterile upper tissue and the fruit-body is then broken by bending, the vertical surface of tubes thus exposed usually has a step-like appearance and the annual layers of tubes can easily be seen. Some structural peculiarity associated with the cessation of growth during the winter is no doubt responsible for this tendency for one annual stratum of tubes to break away from that above and below, in spite of the fact that they are not separated by layers of sterile tissue.

Although for most of the hymenial tubes sealing over does not occur before the production of the next annual layer, this does happen along a semi-circular band forming the circumference of the under surface of the sporophore. The hymenial tubes thus sealed have only their one year of activity, and from the under side of the semi-circular strip of sterile tissue new tubes are formed in the following growing season. Thus the greater part of each annual stratum of hymenial tubes is derived from the continued growth of tubes of the previous year, but a small part consists of new tubes.

The account of the organisation of the sporophore given above agrees with the observations of Faull on *Fames fomentarius* but is at variance with Buller's account *of Ganoderma applanatum*. He figured a sporophore in which each annual layer of tubes was completely sealed off by sterile tissue before the production of the next, so that the tubes could function only for a single season. This would seem to be what happens in a variety of *Ganoderma applanatum* sometimes separated as a distinct species, *G. vegetus*. It also is a regular feature *of Femes ulmarius*, a common perennial bracket polypore on elm trunks.

Certain types of sporophore have been considered illustrating the principles involved in the architecture of large structures concerned with the production and liberation of ballistospores, but in addition there are other types. Thus, in *Clavaria* and related genera the fruit-body consists either of a simple club (e.g. *Clavaria pistillaris* and *C. inaequalis)* or of a branched system with the branches orientated vertically by negative geotropism and with the hymenium covering the branches except towards the base of the sporophore; in *Hydnum* (e.g. *H. repandum]* the general form is that of a toadstool, but the hymenium covers positively geotropic spines on the under side of the cap; and in *Stereum* and other members of

Thelephoraceae the hymenium is more or less smooth, but in the development of the sporophore gravity determines the position of the hymenium so that it is vertical, horizontal and downward-facing, or at some intermediate position, but it does not face upwards.

Although in some gelatinous Basidiomycetes (e.g. *Auricularia mesenterica, Hirneola auriculajudae* and *Tremelledori)* the position of the hymenium is apparently determined during development by gravity as in *Stereum,* in others (e.g. *Tremella mesenterica* and *Dacryomyces deliquescent)* the spore-producing layer may face in any direction and gravity appears to have nothing to do with its development. Critical experimental work on morphogenesis in the larger fungi is much to be desired.

External Conditions and Discharge of Basidiospores

Information concerning the effect of external conditions on spore discharge in Hymenomycetes is not extensive and there are few quantitative data. The main external factors that might be expected to be important are temperature, water supply, and light.

Temperature: Buller has shown that spore discharge in *Lenzites betulina* can continue over a range of temperature of 0°-29° C., but at the extremes of this range discharge is clearly much less than at intermediate temperatures. Again, it has been shown that chilling an active fruit-body of *Femes fomentarius* leads to cessation of spore discharge for some days even after a return to normal temperatures.

Humidity and Water-supply: The interpretation of experiments on the effect of humidity on fungi is difficult because, whatever may be the level of general humidity maintained in an experimental chamber, the actual humidity of the air in immediate contact with the surface of damp fungal tissue is likely to be close to saturation-point. So long as the water content of the hymenial tissue is kept up, spore discharge can occur irrespective of the general humidity of the surrounding air. However, dry air leads to rapid evaporation from the spore-producing surface and, unless there are available reserves of water to replenish the hymenium, loss of turgor in the basidia soon leads to a cessation of spore discharge. The great majority of the large fleshy fungi are mesophytes that can function only under damp conditions.

They rapidly lose water and cease to liberate spores in dry air and have practically no power of recovery when free water is again available. These fungi tend to be produced at periods in the year when the air and soil are relatively moist and when there is a freedom from hard frosts. In their ecological studies on the occurrence of grassland agarics Wilkins and Patrick have demonstrated two maxima of fruit-body production—a minor one in May and the major one in autumn.

The gelatinous and leathery lignicolous basidiomycetes, on the other hand, are in the main drought-enduring xerophytes. Under dry conditions water is lost and spore discharge ceases, but with the return of wet conditions water is rapidly absorbed and spore discharge soon starts again. In this connection *Schizophyllum commune,* a small bracket agaric commonly seen on fallen beech trunks, has been carefully studied and it has been shown in tests, started by Buller and now carried on by Bisby, that fruit-bodies can remain stored in a dry state for twenty-five years and then, a few hours after wetting, freely liberate viable spores.

However, in nature, in *Schizophyllum* and in most other leathery bracket fungi, a fruit-body lasts for only a single season. The drought-enduring type of xerophytism of these lignicolous Basidiomycetes is comparable with the xerophytism of the Pyrenomycetes of wood and bark.

The water-relations of spore liberation in the perennial polypores call for special mention. In *Ganoderma applanatum* spore discharge is continuous day and night from May to October, and even after long periods of summer drought of six to eight weeks' duration, ferruginous clouds of spores can be seen escaping from the under side of a sporophore. However, when a sporophore is detached from its tree trunk, brought into a dry laboratory, and carefully arranged so that its hymenial tubes are vertical, spore discharge continues only for a day or two. Water rapidly evaporates both from the under side and from the firm crust of the upper surface, and no doubt this leads to the turgor of the hymenial surfaces falling below the critical level for spore discharge.

Thus the continuance of discharge from the fungus on the tree cannot be due to water reserve in the fruit-body, but is probably due to maintenance of turgor by the translocation of water from the tree trunk making good the considerable loss of water in transpiration by the sporophore. The water reserve in the wood is not only the water present as such in the tree trunk, but also the water combined in the carbohydrates of the woody tissues which are broken down by the fungal mycelium, with the incidental production of water.

It has been calculated that the decay by a wood-destroying fungus of the cellulose in 1 cubic metre of wood might yield 139 litres of water. Buchwald records a striking instance which shows how little these perennial polyphores depend on atmospheric humidity and rain. A large specimen of *Fomes fomentarius* was found in January on poplar, and, the host being unusual, a section of the trunk bearing the sporophore was cut from the tree and brought into a Copenhagen museum, where it was exhibited uncovered on a table.

The following spring the fruit-body began to liberate spores in dense white clouds. Spore discharge continued for many weeks. The present writer has brought a healthy fruit-body of *Fomes fomentarius,* detached from its substratum, into the laboratory in May, but spore liberation, though at first very vigorous, continued only for two days, so that evidently there is insufficient water reserve in the sporophore to maintain discharge. Here again it seems that the reserve of water in the wood is being used by the fungus and translocated into the fruit-body, thus maintaining the turgor necessary for spore discharge.

It is interesting to compare the xerophytism of *Ganoderma applanatum* and *Fomes fomentarius* with that of the pyrenomycete *Daldinia concentrica*. In all spore discharge occurs during the summer when the air is driest. In all discharge goes on irrespective of rain or humidity. But in *Daldinia* there is a water reserve in the stroma which is gradually used up, whilst in the perennial polyphores the reserve is in the wood.

Light: In general, light appears to exert little direct infiuence on spore liberation in Hymenomycetes. Most workers have reported discharge as occurring day and night throughout the whole spore-fall period in agarics and bracket fungi. Recently, however, *Corticium filamentosa* which forms its basidia on the leaves of *Hevea*, has been shown to be nocturnal in its spore discharge. Again, in small toadstools with a spore-fall period of only a few hours the production of fruit-bodies and, therefore, the liberation of spores may be periodic. Thus, in *Coprinus curtis* sporophores mature and shed their spores in the morning, those of *C. plicatilis* from midday to early afternoon, and those of *C. fimetarius* from midnight to dawn.

Discharge due to Hygroscopic Movements

Liberation of spores from branched conidiophores by the hygroscopic movements of the main axis was described long ago by de Bary. He said:

> Where filiform sporophores rise free into the air, a further mechanical arrangement is found which greatly assists the shedding and scattering of the abscised spores. It may be readily observed in the Hyphomycetes, in *Peronospora, Phytophthora infestans,* and in the gonidiophores of *Peziza fuckeliana*. The hyphae of these fungi are cylindrical in the moist and turgescent state, but collapse when dry and especially when the spores are ripe into and flat ribbon-like form, and the drier they are the more strongly do they become twisted round their own longitudinal axis. They are so highly hygroscopic that the slightest change in the.

humidity of the surrounding air, such for instance as may be caused by the breath of the observer, at once produces changes in their turgescence and torsion; the latter give a twirling motion to the extremity of the gonidiophore, and the ripe spores are thereby thrown off in every direction.

More recently it has been described and figured for *Peronospora tabacina* by Pinckard, whose account agrees closely with that of de Bary. However, in downy mildews and their like the ripe spores are very finely attached and easily blown off their conidiophores by wind, so that the biological importance of hygroscopic discharge seems very uncertain.

Spores in Air

Growth of moulds on damp bread, moist leather, and decaying vegetable matter must have been familiar to man from early times, but the recognition of these growths as fungi came only with the use of the compound microscope. Robert Hooke, the first of the great microscopists, seems to have been the earliest to suggest (1665) that moulds are 'nothing else but several kinds of small and variously figur'd Mushrooms', but even when recognised as fungi they were still generally regarded merely as by-products of organic decay without the power of reproduction.

Clear evidence that these organisms develop from 'seeds' was given by Micheli (1729) who not only observed the microscopic 'seeds', but also showed by experiment that they could reproduce the species. The 'seeds' were, of course, spores, but the academic distinction between seeds and spores was made much later by Hedwig in 1789. Spallanzani, arguing from well-conceived controlled experiments, threw additional doubt on the generally accepted view of spontaneous generation and pointed out that the apparently spontaneous development of moulds was probably due to the omnipresence of their spores. He wrote: 'The seeds of mould may be disseminated in such abundance as to enter into the composition of all animal and vegetable substance. It is the readiest method of accounting for the extraordinary abundance and universal existence of mould'.

Later (1865) Pasteur proved, in so far as a negative can be proved, that neither fungi nor bacteria are generated spontaneously, but that their reproductive units normally occur in air, so that following brief exposure any suitable sterilised nutrient develops colonies of bacteria, yeasts, and moulds.

The spore content of the air has been extensively studied in modern times. It is a matter of prime concern to the mycologist studying the epidemiology of plant diseases, and recently, on the medical side, the importance of fungal spores in the air has been emphasised in connection with inhalant allergy.

Before discussing the main facts about the occurrence and distribution of fungal spores in the air, it is necessary to examine the principal experimental methods that have been used.

The student of spores in the air, whether he is concerned with spores as allergens or with the spread of plant pathogens, is interested not only in what kinds are present but also in their concentration. Consequently the best methods would seem to be those in which the number of spores per unit volume is directly determined. As early as 1887 Frankland, developing a method used by Pasteur and attempted determinations of this nature. A known volume of air was aspirated through a tube stuffed with gun cotton.

This was later dissolved in an ether-alcohol mixture and the spores in the liquid were counted. A modern form of this method, developed in connection with the analysis of particle size in fogs and aerial dusts, is the cascade impactor. The whole outfit with its wind-vane attachment is suspended by its centre of gravity in such a way that it can turn freely so that the plane of the first greased slide faces the wind. Air is sucked through the apparatus at a definite rate. It is found that the speed of intake should be approximately the same as the velocity of the wind, for if not the lines of flow of the air-stream entering the impactor are considerably distorted, leading to important errors. The larger particles are deposited on the first slide, smaller ones on the second, and so on.

The range of particle size trapped on each slide depends on the shape and density of the particles and on the rate of inflow of air into the apparatus. Clearly most fungal spores would be impacted on the first two slides. After the apparatus has been in operation for the required time, the slides are removed and examined microscopically and the spores on each are counted. Thus, knowing the volume of air that has been sucked into the apparatus, the concentration of each recognisable kind of spore in the air can be calculated.

The gravity slide method has been widely used, especially by allergists, in estimating the pollen-grain and spore content of air. In this a horizontal slide is exposed, protected from rain but otherwise so placed that air has free access to it. After a period of exposure a defined area of the slide is examined microscopically, the spores identified, if possible, and counted. The merit of this method is its extreme simplicity.

However, since those studying spores in the air are concerned primarily with the number present in unit volume, the gravity-slide method is justified only if there is a reasonable correlation between spore deposition and spore concentration. No such correlation has been found to exist. Thus Buller and Lowef in an investigation of the microorganisms present in the air of Winnipeg throughout the year found that the ratio: microorganisms deposited per sq. foot per min.

In perfectly still air, no doubt, the number of spores reaching a horizontal surface would be proportional to their concentration in the air, but normal air is not still but is of ever-changing turbulence, and most of the spores coming to rest on a surface do so by impaction rather than by steady fall under the influence of gravity. Thus the gravity-slide method can be considered to give only rough quantitative information about the spore content of the air. Some workers have used vertical greased slides mounted on a wind-vane so that the wind may always impinge directly on the surface of the glass.

This was the method employed long ago by Cunningham in his aeroscope. It has been used in this country for the study of atmospheric pollen and in connection with the epidemiology of rust in Canada. Here again the catch can bear little relation to the number of spores in unit volume of air, especially as the volume of air impinging on the slide must vary with the velocity of the wind.

When spores are caught on a slide and examined microscopically, the fundamental difficulty of the method is identification. Rarely are specific determinations possible, and the genus, family, order or even class may be impossible to ascertain when the spores, as are so many fungal spores, are small, unicellular, and hyaline. However, in some genera the spores are large and unmistakable (e.g. in *Helminthosporium* and *Alternaria).* The uredospores of rusts and the spore-chains of *Cladosporium* are also easy to identify.

Some workers use petri dishes containing sterile nutrient agar as spore-traps. When sampling is done at or near ground level, the sterile agar surface is briefly exposed in a horizontal position, but when exposures are made from aeroplanes the dish is exposed vertically and facing forwards so that spores are impacted on the agar.

Following exposure the dishes are incubated and after several days an attempt is made to identify the fungal colonies that develop. An outstanding objection is that obligate parasites, especially rusts and downy and powdery mildews, do not figure in counts made by this method. Further, many slow-growing saprophytes are swamped by more vigorous species before they can form visible colonies. Again, the results obtained depend on the nature of the nutrient medium, since one type of medium may favour the growth of certain species more than others.

It is noteworthy that none of the usual methods normally reveals the presence of the basidiospores of the larger fungi in the air. They grow too slowly to be recorded in the exposed petri dish method, and they are not sufficiently characteristic to be identified microscopically when caught on a greased slide; yet at certain times of the year spores of the larger fleshy fungi presumably form a significant part of the aerial plankton.

Those who have used the exposed petri dish method in investigating the spore population of the air have usually found that the most abundant species are dry-spore moulds such as species of *Cladosporium* (especially *C. herbarum), Altcrnaria, Penicillinm, Botrytis (B. cinerea), Sporotrichum,* and *Epicoccum.* But in addition a few colonies of slime-spore fungi such as *Mucor* spp., *Phoma* spp., and *Fusarium* spp. usually make their appearance on exposed plates.

This is understandable, for although in slime-spore species the spores are not liberated directly into the air, they may, after a primary dispersal by rain or some other agency, become associated with dry soil or other particles and these may become suspended as dust in the air. However, the preponderance of dry-spore types in the air is unmistakable.

In an extensive investigation carried out by Home at two separate localities the ratio of dry-spore colonies to slime-spore ones was 24:1 at Swanley and 28:1 at East Mailing. Investigators who have examined exposed sticky slides have been particularly struck by the abundance of spores *of Alternaria* spp. at certain times of the year in some regions, and great attention has been given by allergists to spores of this genus because there is clear evidence that people may be allergic to them and because they are so easy to identify. A thorough study has been made of the seasonal fall of *Alternaria* spores at over fifty stations scattered throughout the United States. The fall is heaviest in the great agricultural areas, especially in the zone of the Mississippi valley, and comparatively slight in the western states. *Alternaria* occurs as a very common saprophyte on old straw and its local abundance depends on the presence of this substratum in sufficient amount. Again, as is to be expected, the frequency of the spores in the air varies from season to season.

Thus in South Wales the spores are abundant in the air during summer months but relatively rare at other times. Sometimes the fall of *Alternaria* spores may be on a remarkable scale. An extreme example is the great spore-shower recorded in America between 6 and 8 October 1937. Spores rained down over the whole eastern area of the United States and at some places the deposition on horizontal slides reached 800-1,200 per sq. cm, per day.

Seasonal fluctuations in the spore content of the air have been noted by all workers who have pursued their researches over the course of a year. Home, for example, exposed petri dishes containing nutrient agar on stands 10 feet from ground level at Swanley. For each determination ten dishes were used, exposed for three minutes and examined for fungal colonies after several days' incubation.

Four distinct sites were studied, three in an orchard and the other on arable land. During the summer and autumn spores were relatively abundant in the air, but during the winter there were very few. Here again the seasonal variations of spores in the air doubtless correspond with the seasonal development of the fungi, although day-to-day fluctuations in counts, which may be very considerable, are probably connected with current meteorological conditions influencing spore liberation, dispersal, and deposition.

Since fungi with airborne spores are land organisms, one would expect the air over the sea to become progressively freer from spores on receding from the land, and this is clearly indicated by data obtained by exposing petri dishes at masthead level on a ship at increasing distances from the shore. Over the sea in mid-ocean the air may be almost free from fungal spores, but not from bacteria, since marine species are continually replenishing the bacterial population of the air in spray from the crests of breaking waves.

The question of the distribution of spores in the upper air first came into prominence in 1876. Plant pathologists were anxious to know if via Bleuredospores were to be found there in sufficient abundance to spread infections to distant regions. The pioneer work on this problem was carried out under the leadership of Stakman, who has contributed so much to the study of rust epidemiology in America. Sticky slides were exposed from an aeroplane operating at varying heights.

In June and July, over the wheatlands of the Mississippi valley at heights up to 11,000 feet, spores were found to be relatively abundant, those of *Alternaria* being by far the commonest, with uredospores of cereal rusts next in frequency, followed by the conidia of *Helminthosporium*. All these could be recognised with reasonable certainty, but there were many besides which could not be identified.

Many of the spores were viable. Thus *Alternaria* conidia caught at altitudes (up to 10,500 feet and uredospores trapped at 7,000 feet germinated readily, indicating their ability to withstand the conditions of desiccation, insolation, and temperature obtaining at these altitudes. Since 1923 many workers, usually with the aid of aeroplanes, have studied the spore content of the air far above ground level, and their results in the main confirm Stakman's conclusions without adding any important facts. Mention may be made, however, of the trapping of spores belonging to several genera of Fungi Imperfecti above 36,000 feet by a sampling apparatus released from the stratosphere balloon *Explorer II.*

In recent years, plant pathologists have given considerable attention to the question of the dispersal of airborne spores around a centre of liberation in an attempt to define the size of the danger zone around a locus of infection. An important contribution to this subject was made by Stepanov. In the attack on pines by the heteroecious rust *Cronartium ribicola* it has been found that in trees around rusted currant bushes the percentage of infected needles declines so rapidly with distance that beyond 60 yards from the bushes the pines are infected to a negligible degree. Potato blight may present a similar picture. In parts of the United States the first infection of potato fields by *Phytophthora infestans* is due to conidia developing on shoots derived from diseased tubers on old potato-refuse piles. A study of primary infection in a potato field adjoining such a pile showed clearly that the number of blight lesions per 100 plants decreased rapidly with distances. Many similar examples could be cited.

In general it is clear that even if spores are liberated at a point by the thousand millign, the number falling on a small area a mile away may be almost negligible. This, perhaps, helps us to understand the enormous spore production of so many fungi. However, although the general picture is clear, more exact information is needed about dispersal around a centre using more satisfactory methods of determining the actual spore concentration in the air at increasing distances.

The theoretical basis of spore dispersal in the air has been considered by a number of workers. Some consider that the wind velocity and the rate of fall of spores in still air are the essential features determining dispersal.

This rate of fall varies from 0-05 cm. sec. for the minute spores of *Lycoperdon pyriforme* to 2-8 cm. sec. for the large elongated conidia of *Helminthosporium sativum*, but for most fungal spores it is of the order of 1 cm. sec. It has been calculated that if a spore of *Ustilago zeae* fell by its own weight after attaining a height of 1 mile in the air and if the wind blew at 20 m.p.h., it would be borne to a distance of about 2,500 miles in the nine days of its descent to earth.

A number of similar calculations have been made, but they seem to be of little or no significance since an all-important property of the air—its turbulence—is ignored. It is quite clear that individual spores can be carried for hundreds or even thousands of miles by strong winds, but it is doubtful if the rate of fall in still air has anything to do with the distance of transport. Further, what is of most concern to the student of plant pathology is not the occasional transport of individual spores to very great distances, but rather the normal distance of effective dispersal.

Since the wind is so capricious in its behaviour, it may well be doubted if any quantitative generalisations can be made about dispersal, but Gregory has pointed out that 'the problem is statistical because, whilst the destination of a single spore in a wind eddy is inscrutable, the average distribution of vast numbers of spores over a uniform area and over a period of time offers hope of rational treatment'. He applies the modern meteorological concepts of eddy diffusion to the problem and develops a somewhat elaborate formula to describe dispersion around a point source, but the great advance of his formula over those of others is that the constants used are not arbitrary, but have a defined physical significance. Further, reviewing the experimental results of many workers, Gregory finds a reasonable correspondence between theory and fact.

A rough picture of dispersal by wind and eddy-diffusion can be obtained by watching the smoke-trail from a factory chimney. It tends to be conical with the apex at the chimney's mouth and the smoke cloud is visibly turbulent. At the periphery eddies, are continually mixing the smoke cloud with smoke-free air, thereby increasing the width of the cloud and diluting it. In connection with this conical model of dispersal experiments by Wilson and Baker are of interest. As a convenient experimental material they used spores of the club-moss *Lycopodium*. Large numbers of spores were liberated over a period of fifteen minutes at a point 7 feet above the ground and trapped to leeward on vertical greased slides exposed 5, 10, and 15 feet away.

In considering gradients of spore concentration on receding from a source attention has been focused on horizontal gradients, but there are vertical gradients as well. Vertical distribution has been studied by collecting spores on sticky surfaces exposed from aeroplanes operating at different heights. Stakman and his colleagues found in a general way that the greater the altitude the fewer the spores, and this has been borne out by later workers. Two examples may be given. The rust *Gymnosporangium juniperi-virginianae* attacks cedars, forming galls bearing the teleutospore stage.

Following rain, basidiospores are discharged in enormous numbers. From an aeroplane flying at about 85 m.p.h. over a heavily infected area of red cedar, petri dishes containing agar were exposed for a standard time. These dishes were later examined and the trapped spores counted. Again, Graigie exposed greased slides in Manitoba above wheat fields heavily infected with stein rust.

Each slide was exposed from an aeroplane for ten minutes and the number of uredospores caught on one square inch was counted. These vertical gradients of spore concentration are of considerable interest since they help to explain the efficacy of certain physical barriers to the spread of airborne disease. Thus a mountain chain may protect an area of wheat from uredospores originating in a neighbouring country, since the vast majority of the spores, being in the lower levels of the air, are precipitated on to the mountain-sides.

The principal fact emerging from a study of wind dispersal from a centre is that it is essentially local, and plant pathologists find that the practical safety zones around centres of infection are remarkably small. However, the spread of cereal rusts calls for special attention, since in certain regions long-distance infection on a considerable scale may frequently occur.

During the past quarter of a century the epidemiology of black stem rust of wheat caused by *Puccinia graminis tritici* has been extensively studied on the North American continent where there is a more or less continuous belt of wheat cultivation spreading northwards from Mexico through the Mississippi valley to western Canada. Infection of the wheat in early summer may be by aecidiospores from local barberry bushes, by uredospores that have survived the winter, or by airborne uredospores coming from far away. In west Canada barberry is rare, the coldness of the winter normally prevents the survival of uredospores, and it seems, therefore, that the early primary infection of the wheat is largely due to uredospores blown from the faraway warmer south, where the fungus develops earlier.

In connection with the epidemiology of black stem rust Craigie has made a long-term study of uredospores in the air at a number of stations in western Canada. The technique used was to expose at each place for one or two days a vertical sticky glass slide mounted on a vane so that it continually faced the wind. It was found that viable spores occurred in the air of Manitoba for a week or two each year before any infections could be seen in the field.

Sometimes the vaselined slides showed sudden but temporary increases in uredospore deposition associated with 'spore showers'. Such a shower was usually followed, after a necessary incubation period of a week or two, by the development of stem rust on the wheat in the neighbourhood. It is of interest to consider an unusually early and rather spectacular example that occurred in 1929. The actual shower occurred between 16 and 18 June, and from 12 to 19 June a south wind prevailed.

The evidence suggested that the uredospores came from a region of the Mississippi valley at least 500 miles away. Another equally remarkable instance is cited by Stakman and Christensen. Early in June 1925 black stem rust was general in the wheat-growing region south of lat. 40°, but not north of this parallel. In the first week of June, however, strong winds blew from the south and by 10 June rust had appeared in nearly all fields of susceptible wheat up to 600 miles north of lat. 40°, reaching almost to the Canadian border. Thus early in the rust season, apparently, a great shower of uredospores spread infection almost simultaneously over an area of nearly a quarter of a million square miles.

As well as the infection of northern regions in early summer by uredospores from the south, a reverse infection may occur in autumn.

Uredospores often fail to survive the hot summer in the south (e.g. in Texas), but the young winter wheat may be infected in early autumn by uredospores blown southwards from rusted areas hundreds of miles to the north.

The long-distance dispersal of rust spores in America and in the other great wheat regions of the world raises the question of whether it is different in kind from the local dispersal around a centre.

It is difficult to answer this question with confidence, but it should be noted that the uredospores certainly do not come from a 'point' source, but perhaps from an area of rusted wheat of several thousands of square miles, and a rusted crop may develop an astronomical number of spores. On an acre of slightly rusted wheat there may be 10,000,000,000,000 uredospores and several times this number on an acre of heavily infected grain.

Indeed, clouds of "rust dust" can often be seen rising into the air when heavily infected wheat is being harvested; and machines, men and soil are often red with countless numbers of spores.

The long-distance dispersal of the uredospores of black rust in America raises the question of the possible intercontinental spread of plant diseases by airborne spores. For example: are viable spores airborne from America to Europe and vice versa?

This is still an open question, although the available evidence suggests that the Atlantic acts as an effective barrier to the spread of airborne plant diseases. More information about the spore content of the air over mid-ocean would be extremely interesting in this connection.

The complete dispersal story of an airborne fungus involves not only the liberation if its spores and their actual passage through the air, but also their deposition.

This matter has recently been considered in some detail by Gregory. Two processes can be recognised: (1) *impaction* on a surface due to the momentum of a wind-driven spore, and (2) *sedimentation* under the influence of gravity.

A study of impaction on sticky vertical cylinders of *Lycopodium* spores blown through a wind-tunnel showed that the number of spores caught per unit area increased greatly with decrease in the diameter of the cylinders.

Similar results were obtained outdoors on trapping airborne fungal spores above a field of grass. A consideration of impaction in relation to spore size led to the conclusion that, although larger spores may frequently be impacted on a stem or leaf, the smaller ones can rarely be brought to rest on such surfaces by this means.

They may come to rest on a stem or leaf by sedimentation when the air is still or be deposited with falling rain. In connection with the greater efficiency of impaction of larger spores, Gregory drew attention to the interesting fact that most airborne leaf parasites have relatively large spores (e.g. rust uredospores, conidia of downy and powdery mildews, spores of *Helminthosporiurri).*

4

Qualitative Features

Characteristics of Fungi

One is loath to shatter these charming fantasies, but truth must prevail, for despite the queer properties with which they have been invested, and the romantic tales woven around them in the past, fungi are, after all, only commonplace natural objects, holding a definite position in the plant world. Since, however, they differ so materially from the popular conception of a plant, it is interesting to consider, by comparison with their associates, what these differences really are.

Mode of Nutrition of Fungi: A smaller group of plants, of which the fungi are the chief representatives, are, however, destitute of chlorophyll, and are thus incapable of manufacturing their food in the manner described. They are, therefore, compelled to absorb it in the form of the complex organic compounds already elaborated by other plants, either *directly* from these plants, or *indirectly,* from animals which have fed on them. This somewhat immoral habit, of living on the energies of their neighbours, renders it unnecessary for them to perform many of the duties of the flowering plants, and this has caused their evolution to develop on simpler lines. These facts explain why the fungi are so primitive in growth, when compared with the higher and infinitely more complex development exhibited, for example, by an oak tree.

Evolution of the Fungi: These being thus so diverse from the flowering plants, one may well enquire how they originated; and on this point there is a conflict of opinion. The oldest and most widely-accepted theory is that they have degenerated from the algae, or "seaweeds," and, by the gradual adoption of their present mode of life, have lost the need for chlorophyll, and, therefore, the chlorophyll itself. The more modern line of thought, and one that is probably nearer the truth, is that the fungi have evolved directly from the *protists,* that indefinite and indefinable group of primitive forms of life, which exist on the borderline of the animal and vegetable kingdoms. In the latter case, they naturally never possessed chlorophyll. Whichever view is taken, there is little doubt that the early ancestors of the fungi emerged from the primeval slime, and, gradually adapting themselves, by various biological changes, to their new surroundings, have developed in course of time into the multitude of diverse types existing today.

Extent of the Fungi: It is this multiplicity of forms, many of whose life-cycles are extremely complicated, and of which some have never been elucidated, together with the vast number of individuals concerned, that renders the intensive study of the fungi as a whole the work of a lifetime. It is thus obvious that, in the limited space here available, it is only possible to discuss the botanical aspects of this huge subject very broadly, and even then our remarks must be considered as referring more particularly to that sufficiently numerous class commonly designated—the larger British fungi. Of these, the most familiar is of course the Common Mushroom *(Psalliota campestris)* and since this species is typical of the vast majority of the kinds under immediate consideration, it will simplify matters if first we trace out its life history and structure before attempting any comprehensive description of the fungi in bulk.

Life History of the Common Mushroom: If, then, a brick of mushroom spawn, which acts as the basis for the cultivation of this esculent, is obtained from a nurseryman, and is carefully examined, it will be found to consist of a compressed mass of compost, permeated in all directions by a vast network of minute interwoven branching whitish threads, in appearance not unlike a loose weft of wadding. This system, which is known as the *mycelium* constitutes the vegetative portion of the mushroom plant, and is the part concerned with growth and nutrition. The function of this organ is to absorb, from the organic materials of the mushroom bed, the food requisite for its own development, and for the eventual production of the fruiting bodies. It is in this respect analogous in a sense to the root, stem, and leaves of a flowering plant.

Production of Fruiting Bodies: When the spawn is placed in a mushroom bed, this dormant mycelium springs into life and "runs" or spreads in all directions, gradually permeating the compost, as it did the brick.

Having reached a suitable stage of development, the threads of the mycelium near the surface of the bed bunch themselves together, here and there, into small knots, which enlarge, take shape, and eventually burst through the surface of the bed in the form of "button" mushrooms. The "buttons" continue to expand, resulting in the mature specimens with which everyone is familiar.

These mushrooms constitute the reproductive portion of the fungus, and may be termed the fruiting-bodies. Their function is to produce and liberate the spores, the microscopic mushroom "seeds." If we examine one of these fruiting-bodies more closely, the mechanism for accomplishing this task becomes obvious. A mushroom consists of an umbrella-like *cap*, supported on a stout *stem*, and bearing on its lower surface, radiating outward like the spokes of a wheel, a number of vertical knife-like plates, known as the *gills* which are covered with the spore-bearing surface or *hymenium*. During the "button" stage, these delicate gills are protected from injury, partly by the convexity of the cap, and partly by a veil-like membrane, which joins the edge of the cap to the stem. In expanding, in order to liberate the spores, the cap tears this membrane, some remnants of which remain attached to the edge of the cap, while the major portion is left on the stem in the form of a *ring*.

The gills are thus completely exposed and the spores have access to the air. The latter bodies are liberated from the gills in vast numbers, and are so impalpable that they are carried away on the lightest puff of air, and scattered far and wide. On reaching a suitable substratum—a thing which obviously seldom occurs—the spore germinates, and produces a new mycelium, the life-cycle being thus complete.

This, then, is briefly the story of the mushroom, and what is true of it is equally true, with but varying details, of almost all of the larger species of fungi.

It is seen that, fundamentally, a fungus plant consists of two distinct parts; the vegetative system or mycelium, and the reproductive system or fruiting-body. The mycelium may, and frequently does, live for years, without producing any fruiting-bodies; while, on the other hand, the latter can only arise from a mycelium.

These basic facts having been clearly grasped, we may pass on to the consideration of these two systems, vegetative and reproductive respectively, as they appear in the larger fungi as a whole.

Spawn or Mycelium

The spawn or mycelium of a fungus, as has already been seen in the case of the Common Mushroom, is necessary to its nutrition and is, therefore, a constant feature in the life history of the plant. True, in some of the microscopic forms it may be so rudimentary as to be almost indistinguishable, and even in the larger fungi, it is, owing to its subterranean mode of growth, a not very obvious feature.

Nevertheless, it is always present. In most places where undisturbed organic matter abounds, a careful search will usually reveal the mycelium of one fungus or another; while in its more usual haunts—the woods and forests—every fragment of leaf-mould, every hedge-bottom, every decaying stump or log, will be found to be literally teeming with it.

Mode of Development of Mycelium

The mycelium usually arises from a spore in the following manner. On germination the spore throws out one or more minute germ-tubes, which elongate by apical growth and branch into a number of filaments or *hyphae,* these continuing to ramify until a typical mycelium is formed.

With those fungi which inhabit decaying wood and the like, the mycelium is naturally confined by the limits of the host. In terrestrial species, however, especially where no obstacles impede growth, the general tendency is for the mycelium to spread in all directions, so that it gradually assumes the form of a large subterranean rosette, of ever-increasing size.

The rosette thus becomes a ring of mycelium, whose outer filaments are continually pushing forward into areas of fresh nutriment, and whose inner filaments are equally continuously dying of starvation.

Since the mycelium requires oxygen for its growth, it remains near the surface of the soil, and the chemical and physical changes which it produces, react on the local vegetation and cause the appearance on the surface of a bare ring corresponding to the ring of growing mycelium. This bare ring is the "fairy ring" of mythology.

Ritzema Bos has very aptly compared the formation of a fairy ring, to the circular flame which results from dropping a lighted match in dry grass, where, on a still day, the flame spreads regularly outward. A fairy ring is usually bordered by vegetation of a more vigorous growth, while that within the area of the ring is often quite rank.

Shantz and Piemeisel account for these three zones in the following manner: 1.(Outer)—Acceleration of growth of vegetation, owing to the chemical action of the mycelium on the organic contents of the soil. 2. (The actual ring)—Death of vegetation, owing to the drought brought about by the physical action of the mycelium. 3. (Inner)—Stimulation of the vegetation, owing to the decay of fungus mycelium and fruiting-bodies, and the subsequent chemical changes.

The advance of the mycelium in fairy rings is, of course, extremely slow, and naturally varies with the season. In the case of the Champignon *(Marasmius oreades)*, Bayliss found the rate to vary from three to thirteen and a half inches a year. Thus, comparing the size of rings with the average annual increase, it seems probable that their age often exceeds a century.

In Colorado, Shantz and Piemeisel calculated that certain rings were 250 and 420 years old respectively, whilst certain fragmentary rings would appear to be about 600 years old. The mycelia of some hundred species of the larger fungi are known to grow in this annular formation, and to this number might probably be added many more, did not their natural habitats possess obstructions of various kinds which interfere with this tendency.

In the case of those fungi which do not appear to grow in rings, it is probable that the mycelium ramifies either in a mass insufficiently dense to cause surface reaction, or in a more erratic fashion. It should be also noted that many rings grow in woods where there is no grass, and where the reactive effect is therefore missing.

Function of Mycelium—Storage Bodies: As we have already indicated, the primary work of the mycelium is to absorb food, of which it requires little for its normal growth, but naturally much more at the fruiting period.

Some fungi have learned to provide for this heavy demand by storing large supplies of nutriment, which can be rapidly drawn on for the production of fruiting-bodies, and which will also tide them over adverse periods, when growing conditions are unfavourable. The storage organs used for this purpose consist of densely-compacted mycelium, and take various forms in the different species in which they occur.

Some, known as *sclerotia,* appear usually as dark-coloured, rounded or elongated tubercles, varying in size from that of a pin's head in some species to that of a man's fist, or even his head, in others. Examples of these may be found in the "ergot of rye" *(Claviceps purpurea),* or the "black fellows' bread" *(Mylifta australis = Polyporus mylittae).* Other kinds, called from their root-like appearance, *rhizomorphs,* consist of thick and tough cord-like strands, as in the "dry rot fungus" *(Merulius lacrymans),* the honey-tuft *(Armillaria melted),* etc. In the latter cases, besides acting as storage organs, they facilitate the wide distribution of the fruiting-bodies, by reason of the distances over which they frequently extend.

Except, however, when such special adjuncts as those mentioned are present, the mycelia of the vast bulk of the larger fungi are similar, both in character and in general appearance, to that of the Common Mushroom; indeed, so great is this similarity that in spite of variations of colour and habitat, it is usually impossible to establish the identity of a fungus from this organ alone; and hence the characters necessary for this purpose must be sought in the fruiting-bodies.

Fruiting Bodies or Sporophores

The fruiting-bodies—the "mushrooms," "toadstools," "puffballs," etc.—comprise the reproductive system of the fungus. They are the structures by whose means the subterranean mycelium is enabled to produce and liberate its reproductive bodies—the spores—above the surface of the leaf-mould, decaying wood, or other material in which the plant is living, so giving them access to the air. Essentially, these structures consist of complex, compact masses of mycelium, bearing in, or upon some portion of them, the spore-producing layer or *hymenium,* on which the spores are generated, in one of the two following manners. In the Common Mushroom, the spores are budded off and set free from the tips of minute horns, which are borne, in fours, on the extremities of club-shaped cells, called *basidia;* while in the Morel, the spores are produced, usually eight together, within cylindrical sac-like cells, known as *asci* and are liberated at maturity by the rupture of these sacs.

These characters, which are naturally microscopic ones, provide the division between the two great primary groups of the larger fungi, viz.: the *Basidiomycetes* and the *Ascomycetes.* Whichever mode of production is adopted, the spores are liberated in prodigious numbers, a fact which can be accounted for by the peculiar conditions requisite for their germination. Fungus spores are microscopically small and are in most cases unable to withstand circumstances adverse to them.

Unless, therefore, they encounter conditions of substratum, weather, etc., suitable for their germination, they die, and, since all these conditions are seldom contemporaneously propitious, it follows that for every spore which produces a mature fungus plant, myriads do not. It is to counteract this state of affairs that a beneficent providence has arranged this prodigality of spore production.

Spore Dispersal: These two factors, their minute size, and the numbers in which they are liberated, render the distribution of fungus spores an extremely simple, and in most cases a very haphazard proceeding; yet it is undoubtedly to these factors that the cosmopolitan character of the fungi is due.

The distances over which such light and invisible bodies can be carried by the wind, and by insects, birds, animals and so on, is of course a matter of pure conjecture. But it must be very great, and is certainly far in excess of the distances covered by the larger and less portable seeds of the flowering plants. This is supported by the fact that there are hundreds of species of fungi which are ubiquitous throughout the world, at any rate where climatic conditions are suitable, whereas very much more limited habitats are found in the case of many flowering plants which climatically are equally favoured.

As would be natural to suppose, the vast majority of fungi have their spores disseminated by the wind. In most of these species we find the same tremendous degree of spore production, a fact which presupposes a large spore-bearing surface. This is obtained with, at the same time, an economy of structure, by having this surface spread over corrugations, projections or convolutions of various kinds, such as the gills of the Agarics, the tubes of the Polypores, the teeth of the Hydnums, etc. These projections are usually on the lower surface of the fruiting-body, so that the spores as they reach maturity may drop free and be carried away. In some species, notably the puffballs, the spore-bearing surface occupies the greater part of the interior of the "plant," and, at maturity and under suitable atmospheric conditions, the spores emerge from an apical vent in puffs, often visible as small clouds resembling smoke.

The members of another group of fungi have specialised in the employment of insects for the distribution of their spores, by producing the latter in sugary secretions on which these insects feed. A notable example of these is the Stinkhorn, the spores of which are embedded in a green viscid mucus of powerful carrion-like odour, having a great attraction for flies. So greedily do they devour it that they become almost comatose. Eventually the spores, which are quite unharmed in the bodies of the insects, are deposited, often at a great distance from the parent fungus, and, moreover, in a medium suitable for their germination.

In the case of spore dispersal by animals, it must be admitted that the results undoubtedly obtained are largely fortuitous. Even so, they must take a prominent position, for slugs are proverbial mycophagists, and after feeding they carry away great numbers of spores with them in much the same way as do insects; while squirrels, rabbits, and other rodents, which feed at times on fungi, also play their part in this work. Man, however, is a much more powerful agent of distribution. His various commercial transactions; the transhipment of grain and other commodities; the introduction of plants, bulbs, seeds, etc., from other countries, all lead to the distribution of millions of fungus spores over distances which no natural means could so readily traverse. The most serious aspect of this unwilling transportation is the introduction of fungus diseases from an infested country to another which was formerly immune.

There is, lastly, a class which is by no means the least interesting, those fungi which, by various expulsive methods, actually hurl their spores to a distance, greater or less, according to the efficiency of the apparatus employed, and the degree of assistance rendered by the wind.

Most of these fungi are minute and thus hardly come within our purview. The ingenuity displayed, however, renders them worthy of at least passing reference, and we may consider two examples here: In the case of *Empusa Muscae,* the fungus parasitic on flies, which causes the death of such multitudes of these insects, the spores, at maturity, are shot off from the tips of their hyphae like the bullet from a rifle.

These spores can often be seen in the form of a whitish halo, round a fly which has died on a window pane. With *Sphaerobolus,* a minute puffball, the inner spore case is thrown, as from a catapult, a yard or more into the air by the bursting and sudden inversion of the outer coats of the fruiting body.

The varying methods of production, liberation and dispersal of spores are naturally expressed in a varying structure of the fruiting body, and thus this organ provides a series of characters, the permutations and combinations of which serve as the basis for the classification of the fungi.

This scheme of classification, when viewed in tabular form, gives a graphic summary of the main groups of the larger fungi, and the relationship they bear to each other. It must, however, be borne in mind that essential characters are not necessarily the most obvious ones, and, therefore, the following tabulation should be read in conjunction with the fuller descriptions given at the pages indicated.

It should further be remarked that the English names we have given must not be considered as strictly correct, or even as "popular" names, although some undoubtedly *are* so, but merely as a rough indication of some prominent characteristic. They are frequently an approximate rendering of the botanical ones. To the strict botanist, such names are objectionable, in certain aspects rightly so.

Classification of Larger British Fungi

Class I: Basidiomycetes: Spores borne on the exterior of more or less club-shaped *basidia,* usually in fours.

Order I:

Hymenomycetes: Spore-bearing surface *(hymenium)* superficial, exposed either from the first, or at a very early stage of development.

Family 1: Agaricaceae (The Agarics or Gill Fungi): Spore-bearing surface covering radiating plates or gills.

Family 2: Polyporaceae (The Polypores or Pore Fungi): Spore-bearing surface lining the interiors of tubes or tube-like recticulations, the mouths of which appear as pores.

Family 3: Hydnaceae (The Toothed Fungi): Spore-bearing surface spread over spines, teeth or papillae.

Family 4: Thelephoraceae (The Leathery Fungi): Spore-bearing surface covering a firm, smooth or corrugate surface.

Family 5: Clavariaceae (The Fairy Clubs): Spore-bearing surface smooth, covering erect club-shaped or variously-branched fruiting bodies.

Family 6: Tremellinaceae (The Jew's Ears, etc.): Spore-bearing surface spread over smooth or irregularly-folded gelatinous surface.

Order II:

Gasteromycetes: Spore-bearing surface produced in "tissue" (the *glebd)* within the more or less globular body of the plant (the *peridium).*

Family 1: Phalloidaceae (The Carrion Fungi): Peridium globular, with a middle gelatinous layer. At maturity the foetid mucilaginous gleba, containing the spores, is raised above the ruptured peridium by the elongation of a stalked or latticed receptacle.

Family 2: Lycoperdaceae (The Puffballs): Peridium usually globular, of two or more layers; dehiscing at maturity by an apical vent or by the rupture of the walls. Spores when ripe a powdery mass intermixed with a web of threads *(capillitium).*

Family 3: Sclerodermaceae (The Earth Balls): Peridium usually globular, thick, of a single layer, dehiscing irregularly; capillitium absent or rudimentary.

Family 4: Nidulariaceae (The Bird's Nest Fungi): Peridium bell- or cup-shaped, containing usually several egg-like spore chambers *(peridiola),* each sometimes attached by a thread *(funuulus),* but free at maturity.

Family 5: Hymenogastraceae (The False Truffles): Subterranean; peridium globular or nodular, indehiscent. Spores released by the decay of the plant.

Class II: Ascomycetes: Spores produced *within* cylindrical, spherical or club-shaped *asci,* usually in eights.

Order I:

Discomycetes: Spore-bearing surface exposed on a flat, cup-shaped or convex receptacle. (The Cup Fungi, Morels, etc.

Order II:

Pyrenomycetes: Spore-bearing surface lining minute flask-shaped bodies *(feritkecia)*; of which, in the larger species, many hundreds are borne embedded in a fleshy, corky, or carbonaceous receptacle (the *stroma).*

Order III:

Tuberaceae (The True Truffles) Subterranean: Spore-bearing surface lining chambers in a globular or nodular peridium, the decay of which releases the spores.

A short account of each of these families is given below:

Basidiomycetes

The Agarics or Gill Fungi *(Aaricaceae),* comprising the mushrooms and toadstools proper, form the most numerous class of the larger fungi, and are represented in Britain by over fifteen hundred species, contained in fifty-eight genera. Their common distinguishing feature is the presence, on the fruiting body, of vertical knife-like plates, the gills, which carry the spore-bearing layers.

The fruiting body takes most frequently the parasol-like form of a convex cap, supported on a stem, and is well exemplified by the Common Mushroom *(Psalliota campestris);* or, less frequently, appears as a shelf-like or bracket-shaped organ, as in the Oyster fungus *(Pleurotus ostreatus)*; the gills being carried on the lower surface, radiating outward from the stem, or in the stemless species, from the point of support.

In *Amanitopsis,* the entire fruiting body when young is enclosed in a veil-like covering, called the "volva," which ruptures during development and remains as a sheath surrounding the base of the stem; particles of it being sometimes carried upward on the cap, on which they remain as loose patches. In others, such as the Common Mushroom, only the gills are enclosed, and in these the ruptured veil is left on the stem, as a collar or "ring," while fragments are often found hanging to the edge of the cap. In *Amanita,* both of these characters are present.

In the classification of the Agarics, the primary character selected is that of the colour of the spore mass. If the cap of a mature specimen is placed, gills downwards on a sheet of paper, and left undisturbed for a few hours, the spores are deposited in such vast numbers that they form on the paper a complete plan of the gills. This plan will be in one of five colours, viz.: (1) White; (2) Pink; (3) Ochre, cinnamon or rust; (4) Purple; or (5) Black.

This somewhat laborious process is not, however, always essential, since the colour of the spores is usually the same as that of the gills, although some species with dark coloured gills produce white spores. Lest this colour character be considered to constitute an extremely artificial means of classification, it may be pointed out that not only is it very stable, but it possibly indicates the degree of evolution of the individual.

The mode of attachment of the gills is another character of importance; whether they are completely free from the stem *(free)*; partially attached to the stem *(adnexed)*; attached for their whole width *(adnate)*; attached by means of a small tooth or sinus *(sinuate)*; or, not only attached to the stem, but running for a considerable distance down it *(decurrent)*.

Other characters which are of service in distinguishing the individual genera and species of this huge group are:

(1) The consistency of the plant itself, whether soft and soon decaying; or, membranous, leathery or corky, and therefore more persistent.

(2) The general shape and colour of the cap, and further, whether its surface is viscid or dry, smooth, warted, or scaly, etc.

(3) The general character of the stem, and whether furnished with a ring and/or volva.

(4) The size of the plant, its habitat, etc.

Some genera have characteristics which separate them from all others, and in this connection reference may be made to the species of *Lactarius* which exude a copious milky juice when broken, and to those of *Schizophyllum* in which the edges of the gills are split longitudinally.

The habitats selected by the Agarics are numerous and varied. The majority, it is true, prefer the moist, shady, and relatively warm situations afforded by woods; a number inhabiting the trees themselves. But even so, a sufficiently large number are found in other localities, and pastures, heaths, farmyards, gardens, dunghills, old gate-posts and fences, old thatch, dead leaves and fircones, are but a few of the sites which demand their share of attention. In short, it is hardly an exaggeration to say that some species of the Agarics will be found in any situation which is capable of supporting fungus life. The preferences of individual species become, therefore, a special factor in their detailed study.

The Polypores (Polyporaceae): This group comprises some two hundred and fifty British species, contained in eleven genera. In superficial appearance many of them are not unlike some of the Agarics, but the gills of the latter are here replaced by a mass of compacted tubes, of which the lower surface appears punctured by a multitude of pores, round or angular, occasionally sinuate or torn. In some genera, as *Boletus*, the plant is terrestrial, and has an umbrella-shaped cap, supported on a stout central stem. In others, such as *Merulius* and *Poria*, the fruiting body is flat and crust-like; the mouths of the pores, which in some species are mere shallow depressions, appearing on the free surface. In the vast majority, however, are those forms, like *Polyporus*, *Polystictus*, *Fomes* and *Fistulina*, which appear as bracket-shaped, or hoof-like masses, adhering to trees, both living and dead. In the Polypores, the fruiting body shows every variation of consistency, from the soft and juicy, through the leathery varieties, to those really woody. The latter are perennial, and their annual layers of growth may be easily traced.

On account of their arboreal habits, many of these species are very destructive in woods and forests. *Merulius lacrymans* is the agent of "dry rot" in houses.

Toothed Fungi (Hydnaceae): In this group, the spore-bearing surface is spread over spines, teeth, tubercles, papillae, or wart-like folds, which take the place of the gills in the Agarics and the tubes in the Polypores. The Hydnaceae are represented in Britain by nine genera, containing in all about one hundred species. Of these, a few have a central stem, and conform more or less to the orthodox parasol-like shape; but most of them resemble hedgehog-like masses, or else fleshy or leathery crusts of various shapes, bearing the spine-like process on the lower, or the free, surface.

In the largest and best known genus, *Hydnum,* of which there are some fifty species, the spines are large and almost awl-like. From this, they vary in shape and prominence, in the different genera, down to *Phlebia,* which has radiate veins or wrinkles. The consistency of the fruiting body shows also every variation and may be gelatinous, fleshy, membranous, wax-like or leathery. Many of them are terrestrial and are found in woods or on heaths; others inhabit decaying logs, while some are known to attack living trees. Most of the fleshy kinds are edible; none is considered poisonous.

The Leathery Fungi *(Thelephoraceae)* comprise a heterogeneous collection of fungi, numbering about one hundred and fifty species, contained in eleven genera, whose identity, in the absence of a microscope, is often difficult to determine. In general they have a smooth, flat or slightly undulating hymenium, spread over the exterior of an irregular cup-like body, the free surface of a prostrate crust, or the lower surface of a bracket-shaped body. In most cases their consistency is membranous or leathery, though occasionally waxy or even gelatinous. The members of the order have no particular points of interest except that all the species of *Graterellus* are edible, C. *cornucopioides* being formerly sold in Covent Garden market, while of the other genera, many species are destructive pests. In this connection, note may be made of *Stereum purpureum,* which is the cause of "silver leaf" disease of plum, apple, and similar fruit trees, and *Coniophora cerebella,* which is one of the fungi causing dry rot in buildings. The latter is particularly dangerous in that it paves the way for *Merulius lacrymans.*

Fairy Clubs (Clavariaceae): This group contains some of the most graceful fungi known. Most of them are small in stature, delicate in appearance and often bright in colour. Of the sixty or so British species, comprised in six genera, forty alone are contained in the genus *Glavaria,* in which the fruiting bodies occur as small erect tufts of simple or much branched clubs; many of the latter having an elaborate antlered or coral-like appearance.

In *Sparassis,* the fruiting body is so much divided that it resembles a cauliflower. In *Calocera,* which has a superficial resemblance to Clavaria, the clubs are gelatinous, becoming horny when dry.

Most of the Club fungi are found among grass in meadows or woods, or upon lawns. A few inhabit decaying stumps, dead leaves, etc. All of the larger kinds are edible: none is known to be poisonous.

Jew's Ears, etc. (Tremellinaceae): In these fungi, of which there are some three dozen British species, contained in eleven genera, the fruiting bodies are of gelatinous consistency, becoming rigid and horny in dry weather, and regaining their original condition when moistened. In shape they are very variable and extremely irregular. In *Tremella* the largest genus, the fruiting bodies are often brightly coloured and appear as irregular, lobed or brain-like masses. In *Exidia, Hirneola* and *Auricularia* they are irregularly cup-shaped, or convoluted like a human ear. Others appear as small variously-shaped lumps of jelly. Most of these fungi inhabit decaying logs, the stumps of trees, old fence posts, etc.

The Carrion Fungi *(Phalloidaceae)* constitute a family whose affinity with the puffballs would not be suspected by the casual observer. When young, the complete fungus is enclosed in a gelatinous "egg," which eventually bursts open and permits the erection of a stem-like or latticed receptacle, which bears the spores, embedded in an evil-smelling mucus, the last being responsible for making the presence of these fungi known long before they are seen. There occur in Britain five genera, containing seven species, of which three are introductions from abroad.

The Puffballs *(Lycoperdaceae)* are one of the most familiar groups of fungi, numbering some forty British species, of six genera. The fruiting bodies are usually more or less spherical in shape and are provided with a cuticle of two layers.

In *Ly-coperdon* and *Bovista,* the outer coat flakes away in spine-like tufts or warts during development; while in *Geaster* it is permanent, becoming split and reflexed like a conventional star. At maturity, the inner coat opens by a apical vent and releases the spores in smoke-like puffs; or the top may collapse completely, exhibiting the powdery spores enclosed in a brownish wadding-like mass.

In the young state the flesh is white, and in this condition is in most cases edible. Most of the puffballs grow in pastures, though a number occur on the ground in woods, and one species on decaying stumps.

The Earth Balls *(Sclerodermaceae)* are close allies of the puffballs, from which they differ chiefly in their thick, hard, leathery rind, and by the absence of a capillitium. The flesh when young is generally more or less white, becoming darker with age; while the peridium opens at maturity by splitting irregularly, releasing a mass of dark violet spores. There are two British genera, *Scleroderma* and *Polysaccum,* containing respectively five and one species, the last being extremely rare. All are terrestrial and appear to favour a sandy soil.

Birds' Nest Fungi (Nidulariaceae): This group contains a number of charming little fungi, which, despite their affinity, would hardly be recognised as akin to the puffballs. At maturity, most of them appear as miniature, fanciful birds nests, containing minute eggs, well shown in mature specimens, x 3- genera occurring in Britain are *Cyathus, Nidularia* and *Crucibuhtm,* containing respectively two, three and one species. The remaining genus, *Sphaerobolus,* has only one egg, which at maturity is hurled away as from a catapult.

False Truffles (Hymenogastraceae): This family is remarkable for the underground habit of its members.

The fruiting bodies, which are usually just below the surface, and are found near trees, appear as globose or irregular tubers, somewhat resembling a small warted potato. There are some twelve British species contained in half that number of genera. They are seldom found except by those who are making deliberate search for them.

Ascomycetes

This class contains a tremendous number of fungi, whose characteristics are too varied to consider in detail. Most of them are minute and, therefore, require the aid of a microscope for their determination and study. So distinctive in appearance are most of them, and so unlike the ordinary conception of a toadstool, that a popular description will almost certainly fail to conjure up the plant we have in mind.

Discomycetes

Cup Fungi (Pezizae): As their name suggests, these fungi have fruiting bodies more or less cup shaped in form, though often shallow and distorted, or even disc-like; the hymenium being carried on the inner or the upper surface.

They are mostly soft and fleshy in onsistency and brilliant in colour. In the largest and best-known genus, *Peziza,* which contains some dozen British species, the cups vary from bright scarlet, through various shades of orange to lemon yellow, and from scarlet to brown. They are thus very striking and very beautiful objects, and attract attention wherever they occur. Most of them grow on the ground or on decaying stumps, though some of the smaller inhabit dead sticks, herbaceous stems, etc.

Morels and Helvells (Helvelleae): The members of this group are close allies of the preceding one. They differ, however, in having a distinctly-stalked fruiting body, with a variously-shaped cap. In *Morchella*, it is irregularly conical and is a mass of ridges and hollows, deeply folded and pitted like a sponge. In *Gyromitra*, the cap is approximately spherical and convoluted like a brain. In *Helvella,* it is grotesquely saddle-shaped, with irregular drooping lobes. In other genera, this organ is smooth, and is spatula-like, conical, etc. Most of these fungi are terrestrial in woods.

Pyrenomycetes

In this order are a great number of fungi which are of little interest to folk other than those engaged in their actual study. Most of them are small, and even the larger ones are far from attractive. The best known is probably the Candle Snuff fungus, *Xylaria hypoxylon,* a small, branched, fungus common on old stumps.

True Truffles (Tuberactae)

In superficial appearance, these fungi have almost all of the attributes of the *Hymenogastraceae;* from which they differ, however, in the spores being produced in *asci.* Like them, these plants are subterranean and grow near, and often associated with, the roots of trees. The principal genus is *Tuber,* which comprises the edible truffles, a subject dealt with elsewhere.

In this brief survey of the classes of the larger fungi we have not attempted to go very deeply into the matter, for to do so would be belying the sentiment which we expressed in our preface. It is, however, desirable that the reader should understand something of the numerous and varied kinds of fungi which exist, even if only to bring realisation that these larger kinds comprise considerably more than the class which he has perhaps formerly lumped together, under the comprehensive term, "toadstool".

5

The Pathogens

Fungal Pathogens

Fungal infections have recently emerged as a growing threat to human health, especially in persons whose immune systems are compromised in some way. For example, fungi are associated with complex disease entities in complex medical patients, e.g. cryptococcosis in AIDS patients, or aspergillosis in bone marrow or organ transplant patients.

Clinicians are particularly concerned that the increasing use of anti-fungal drugs will lead to drug-resistant fungi, especially in settings such as hospitals where nosocomial (hospital-acquired) infections are a growing problem. Recent studies have documented resistance of *Candida* species to fluconazole and other azole and triazole drugs, used widely to treat patients with systemic fungal diseases. In addition, primary or inherent resistance limits the activity or currently available anti-fungal drugs for fungi such as *Aspergillus* and other emerging moulds.

Fungi present an especially complex challenge to researchers, in part because pathogenicity is often associated with certain morphological forms or a certain part of the life cycle of a fungal species. For example, pathogens such as *Histoplasma capsulatum*, *Coccidioides immitis* or *Sporothrix schenckii* convert from one morphological form to another in the host tissue before they propagate to cause disease.

Cryptococcus neoformans causes infection only in the asexual form of its life cycle (yeast cells). The molecular biological progress made in *Cryptococcus neoformans* over the past five years attests to the importance of collaborative research between medical and molecular researchers. Researchers are identifying a series of antigenic peptides from fungal pathogens that can generate immune responses, which may assist in developing an anti-fungal vaccine.

Fusarium

Fusarium is a large genus of filamentous fungi widely distributed in soil and in association with plants. Most species are harmless saprobes and are relatively abundant members of the soil microbial community. Some species produce mycotoxins in cereal crops that can affect human and animal health if it enters the food chain. The main toxins produced by these *Fusarium* are fuminosins and trichothecenes.

The genus includes a number of economically important plant pathogenic species. The genome of the wheat and maize pathogen, *Fusarium graminearum*, has been sequenced. In addition, some species may cause a range of opportunistic infections in humans. In humans with normal immune systems, fusarial infections may occur in the nails (onychomycosis) and in the cornea (keratomycosis or mycotic keratitis).

In humans whose immune systems are weakened in a particular way (neutropenia, i.e., very low count of the white blood cell type called neutrophils), aggressive fusarial infections penetrating the entire body and bloodstream (disseminated infections) may be caused by members of the *Fusarium solani* complex, *Fusarium oxysporum*, *Fusarium verticillioides*, *Fusarium proliferatum* and rarely other fusarial species. The neutropenia in such cases is almost always the result of chemotherapy against certain kinds of leukaemia or else heavy use of immunosuppressive drugs in problematic cases of major organ transplant surgery.

Biological Warfare

Mass casualties occurred in the Soviet Union in the 1930s and 1940s when *Fusarium*-contaminated wheat flour was baked into bread, causing alimentary toxic aleukia with a 60 per cent mortality rate. Symptoms began with abdominal pain, diarrhoea, vomiting, and prostration. Within days fever, chills, myalgias and bone marrow depression with granulocytopenia and secondary sepsis. Further symptoms included pharyngeal or laryngeal ulceration and diffuse bleeding into the skin (petechiae and ecchymoses), melena, bloody diarrhoea, hematuria, hematemesis, epistaxis, vaginal bleeding, Pancytopenia and gastrointestinal ulceration. *Fusarium sporotrichoides* contamination was found in affected grain in 1932, spurring research for medical purposes and for use in biological warfare. The active ingredient was found to be trichothecene T-2 mycotoxin, and was produced in quantity and weaponised prior to the passage of the Biological Weapons Convention in 1972. The Soviets were accused of using the agent, dubbed *"yellow rain"*, to cause 6,300 deaths in Laos, Kampuchea, and Afghanistan between 1975 and 1981.

Toxic Effects of Mycotoxins in Humans

In experimental animals, trichothecenes are 40 times more toxic when inhaled than when given orally.

Mycotoxicoses are diseases caused by mycotoxins, i.e. secondary metabolites of moulds. Although they occur more frequently in areas with a hot and humid climate, favourable for the growth of moulds, they can also be found in temperate zones. Exposure to mycotoxins is mostly by ingestion, but also occurs by the dermal and inhalation routes. Mycotoxicoses often remain unrecognised by medical professionals, except when large numbers of people are involved. Epidemiological, clinical and histological findings (when available) in outbreaks of mycotoxicoses resulting from exposure to aflatoxins, ergot, trichothecenes, ochratoxins, 3-nitropropionic acid, zearalenone and fumonisins are discussed.

Mycotoxins are secondary metabolites of moulds that exert toxic effects on animals and humans. The toxic effect of mycotoxins on animal and human health is referred to as mycotoxicosis, the severity of which depends on the toxicity of the mycotoxin, the extent of exposure, age and nutritional status of the individual and possible synergistic effects of other chemicals to which the individual is exposed.

The chemical structures of mycotoxins vary considerably, but they are all relatively low molecular mass organic compounds. The untoward effect of moulds and fungi was known already in ancient times. In the seventh and eighth centuries BC the festival "Robigalia" was established to honour the god Robigus, who had to be propitiated in order to protect grain and trees. It was celebrated on 25 April because that was the most likely time for crops to be attacked by rust or mildew.

In the Middle Ages, outbreaks of ergotism caused by ergot alkaloids from *Claviceps purpurea* reached epidemic proportions, mutilating and killing thousands of people in Europe. Ergotism was also known as ignis sacer (sacred fire) or St. Anthony's fire, because at the time it was thought that a pilgrimage to the shrine of St. Anthony would bring relief from the intense burning sensation experienced.

The victims of ergotism were exposed to lysergic acid diethylamide (LSD), a hallucinogen, produced during the baking of bread made with ergot-contaminated wheat, as well as to other ergot toxins and hallucinogens, as well as belladonna alkaloids from mandragora apple, which was used to treat ergotism. While ergotism no longer has such important implications for public health, recent reports indicate that outbreaks of human mycotoxicoses are still possible. Some mycotoxicoses have disappeared owing to more rigorous hygiene measures.

For example, citreoviridin-related malignant acute cardiac beriberi ("yellow rice disease" or shoshin-kakke disease in Japanese) has not been reported for several decades, following the exclusion of mouldy rice from the markets. Citreoviridin is a metabolic product of Penicillium citreonigrum, which grows readily on rice during storage after harvest, especially in the colder regions of Japan. Another mycotoxicosis not seen for decades is alimentary toxic aleukia, common in the 1930s and 1940s in the USSR. This disease was caused by trichothecenes produced by Fusarium strains on unharvested grain.

General interest in mycotoxins rose in 1960 when a feed-related mycotoxicosis called turkey X disease, which was later proved to be caused by aflatoxins, appeared in farm animals in England. Subsequently it was found that aflatoxins are hepatocarcinogens in animals and humans, and this stimulated research on mycotoxins. There is a long history of the use of certain moulds in the production of cheese and salami and in the fermentation of beer and wine.

Moulds are also used in the production of drugs (antibiotics). The classification of mould metabolites as antibiotics or mycotoxins is based on their toxicity or beneficial effect in treating diseases. Some mould metabolites that were initially considered to be antibiotics (e.g. citrinin) were subsequently found to be highly toxic, and are currently classified as toxins. Ergot alkaloids are still used, inter alia, in the treatment of parkinsonism, as prolactin inhibitors, in cerebrovascular insufficiency, migraine treatment, venous insufficiency, thrombosis and embolisms, for the stimulation of cerebral and peripheral metabolism, in uterine stimulation, as a dopaminergic agonist.

The toxic effects of mycotoxins (e.g. ochratoxins, fumonisins, zearalenone, etc.) are mostly known from veterinary practice.

Mycotoxicoses, which can occur in both industrialised and developing countries, arise when environmental, social and economic conditions combine with meteorological conditions (humidity, temperature) which favour the growth of moulds.

Involvement of mycotoxins in disease causation should be considered in instances when a disease appears in several persons, with no obvious connection to a known etiological agent, such as microorganisms. Given current trade patterns, mycotoxicoses resulting from contaminated food, locally grown or imported, could occur in developing and developed countries alike. Strict control of food and feed and appropriate public health measures are therefore of considerable importance in reducing the risks to human and animal health.

This review covers only the human aspects of the untoward effects of mycotoxins. However, owing to the frequent non-specific effects of mycotoxin involvement, the results of animal experiments are useful for understanding possible effects on humans. Since review articles and books are available dealing with specific topics such as the chemistry, analytical procedures, metabolism, and economic aspects of mycotoxins, these aspects of mycotoxin toxicology are not presented here. Mycotoxicoses are usually insufficiently treated in medical textbooks and are not covered in curricula of many medical schools. The aim of this article is to summarise current understanding of the clinical aspects mainly of mycotoxicoses in humans, and to stress the importance of this class of naturally occurring toxins.

Ergot is the common name of the sclerotia of fungal species within the genus Claviceps, which produce ergot alkaloids. The sclerotium is the dark-coloured, hard fungal mass that replaces the seed or kernel of a plant following infestation. Ergot alkaloids are also secondary metabolites of some strains of Penicillium, Aspergillus and Rhizopus spp.

The ca. 40 ergot alkaloids isolated from Claviceps sclerotia can be divided into three groups:

- Derivatives of lysergic acid (e.g. ergotamine and ergocristine);

- Derivatives of isolysergic acid (e.g. ergotaminine);
- Derivatives of dimethylergoline (clavines, e.g. agroclavine).

The source of the ergot strongly influences the type of alkaloids present, as well as the clinical picture of ergotism. Claviceps purpurea produces ergotamine-ergocristine alkaloids, which cause the gangrenous form of ergotism because of their vasoconstrictive activity.

The initial symptoms are oedema of the legs, with severe pains. Paraesthesias are followed by gangrene at the tendons, with painless demarcation. The last-recorded outbreak of gangrenous ergotism occurred in Ethiopia in 1977-78; 140 persons were affected and the mortality was high (34 per cent).

The other type of ergotism, a convulsive form related to intoxication with clavine alkaloids from Claviceps fusiformis was last seen during 1975 in India when 78 persons were affected. It was characterised by gastrointestinal symptoms (nausea, vomiting and giddiness) followed by effects on the central nervous system (drowsiness, prolonged sleepiness, twitching, convulsions, blindness and paralysis). The onset of symptoms occurred 1-48 hours following exposure; there were no fatalities.

Ergotism is extremely rare today, primarily because the normal grain cleaning and milling processes remove most of the ergot so that only very low levels of alkaloids remain in the resultant flours. In addition, the alkaloids that are the causative agents of ergotism are relatively labile and are usually destroyed during baking and cooking.

Aflatoxins

Aflatoxins occur in nuts, cereals and rice under conditions of high humidity and temperature and present a risk to human health that is insufficiently recognised. The two major Aspergillus species that produce aflatoxins are A. flavus, which produces only B aflatoxins, and A. parasiticus, which produces both B and G aflatoxins. Aflatoxins [M.sub.1] and [M.sub.2] are oxidative metabolic products of aflatoxins [B.sub.1] and [B.sub.2] produced by animals following ingestion, and so appear in milk (both animal and human), urine and faeces. Aflatoxicol is a reductive metabolite of aflatoxin [B.sub.1].

Aflatoxins are acutely toxic, immunosuppressive, mutagenic, teratogenic and carcinogenic compounds. The main target organ for toxicity and carcinogenicity is the liver. The evaluation of epidemiological and laboratory results carried out in 1987 by the International Agency for Research on Cancer (IARC) found that there is sufficient evidence in humans for the carcinogenicity of naturally occurring mixtures of aflatoxins, which are therefore classified as carcinogens, except for aflatoxin [M.sub.1], which is possibly carcinogenic to humans.

Several outbreaks of aflatoxicosis have occurred in tropical countries, mostly among adults in rural populations with a poor level of nutrition for whom maize is the staple food. The clinical picture presented by cases indicated acute toxic liver injury, which was confirmed by morphological changes in liver autopsy specimens that were indicative of toxic hepatitis. Mortality rates in the acute phase were 10-60 per cent. At the end of one year, surviving patients had no jaundice, and most of them had recovered clinically.

A case of attempted suicide with purified aflatoxin [B.sub.1] is reported to have occurred in 1966 in the USA. A young woman ingested a total of 5.5 mg of aflatoxin [B.sub.1] over 2 days and, 6 months later, a total of 35 mg over 2 weeks. Following the first exposure, she was admitted to hospital with a transient, non-pruritic, nausea and headache; the second time she reported nausea only. On both occasions, physical, radiological and laboratory examinations were normal and liver biopsies appeared normal by light microscopy. A follow-up examination 14 years later did not reveal any signs or symptoms of disease or lesions. These findings suggest that the hepatotoxicity of aflatoxin [B.sub.1] may be lower in well nourished persons than in experimental animals or that the latent period for turnout formation may exceed 14 years.

Aflatoxins have been detected in the blood of pregnant women, in neo-natal umbilical cord blood, and in breast milk in African countries, with significant seasonal variations. Levels of aflatoxins detected in some umbilical cord bloods at birth are among the highest levels ever recorded in human tissue and fluids.

Aflatoxins have been suggested as an etiological factor in encephalopathy and fatty degeneration of viscera, similar to Reye syndrome, which is common in countries with a hot and humid climate. The clinical picture includes enlarged, pale, fatty liver and kidneys and severe cerebral oedema. Aflatoxins have been found in blood during the acute phase of the disease, and in the liver of affected children. However, use of aspirin or phenothiazines is also suspected to be involved in the etiology.

In tropical countries, clinically recognisable jaundice is frequent during the neo-natal period. In a large investigation undertaken on 327 babies with jaundice and 80 matching controls in Nigeria, it was found that the occurrence of glucose-6-phosphate dehydrogenase (G6PD) deficiency together with the presence of aflatoxins in the serum are significant risk factors for the development of neo-natal jaundice.

The geographical and seasonal prevalences of aflatoxins in food and of kwashiorkor show a remarkable similarity. In several tropical countries, aflatoxins have been found more frequently and in higher concentration in liver specimens from children with kwashiorkor than in controls.

Clinical investigation of aflatoxin elimination in children with kwashiorkor and marasmic kwashiorkor, who were fed an aflatoxin-free diet, proved that aflatoxins in these children are slowly eliminated. In several studies, aflatoxicol was found in the serum, liver, urine and stools of children with kwashiorkor and marasmic kwashiorkor, in contrast to marasmic and control children where this metabolite was not found. It is not clear whether this difference is causally related to kwashiorkor or is a consequence of the disease.

In recent studies, aflatoxins were found in the brain and lungs of children who had died from kwashiorkor and in control children who had died from various other diseases. It was suggested that the presence of aflatoxins in the brains of control children might be due to metabolic imbalance or to a failure in the excretory mechanisms of children with conditions such as measles (which in 25 per cent of cases precedes kwashiorkor), renal failure, pyloric stenosis, gastroenteritis. Aflatoxins in the lungs were found in all children diagnosed to have pneumonia, irrespective of the presence of kwashiorkor. This could be due to a reduced clearing ability of the lungs in pulmonary diseases or to exposure via the respiratory route. In the Philippines, a study of the relationship between the presence of aflatoxin in the serum and urine of children and the outcome of acute lower respiratory infection failed to prove a correlation.

However, aflatoxin [B.sub.1] was found in the lungs of one textile and two agricultural workers who died from pulmonary interstitial fibrosis. These individuals were probably occupationally exposed to aflatoxin [B.sub.1] via the respiratory route. Aflatoxin [B.sub.1] was also detected in the lung tissue of a chemical engineer who had worked for 3 months on a method for sterilising Brazilian peanut meal contaminated with Aspergillus flavus, and who died of alveolar cell carcinoma.

In the United Kingdom, it was found that intravenous heroin users can be exposed to aflatoxin [B.sub.1] from samples of heroin on sale. Through intravenous administration, aflatoxin [B.sub.1] bypasses the detoxifying mechanisms of the liver, which results in direct systemic exposure. In the United Kingdom and the Netherlands, analysis of 121 urine samples obtained from heroin addicts revealed a higher proportion of samples contaminated with aflatoxins [B.sub.1], [B.sub.2], [M.sub.1] and [M.sub.2] and aflatoxicol (20 per cent) than those from normal adult volunteers (2 per cent). In addition, aflatoxin [B.sub.1] was found at much lower concentrations in the latter group.

3-Nitropropionic acid 3-Nitropropionic acid (3-NPA) is a secondary metabolite of Arthrinium sp., considered to cause a form of acute food-poisoning called "mouldy sugarcane poisoning". The problem occurred during winter (February and March) in 13 provinces of northern China as a consequence of ingesting sugarcane that had been stored for at least two months and which was infested with Arthrinium sp. In the period 1972-88, a total of 884 persons were involved in outbreaks, with 88 (10 per cent) fatalities.

The main epidemiological feature is the small number of persons in one outbreak (one to five persons), with the victims being mostly children and young people. The incubation period is generally 2-3 hours following the ingestion of mouldy sugarcane, and the main clinical symptoms are vomiting, dystonia, staring to one side, convulsions, carpopedal spasm and coma. Delayed dystonia develops in 10-50 per cent of patients as a consequence of bilateral symmetric necrosis of the basal ganglia. The development of delayed symptoms can be predicted by abnormality in the basal ganglia on cranial computed tomography (Et) scans. In adults, 3-NPA causes gastrointestinal symptoms; signs of severe encephalopathy are not common.

Ochratoxins

Ochratoxins are secondary metabolites of Aspergillus and Penicillium strains, found on cereals, coffee and bread, as well as on all kinds of food commodities of animal origin in many countries. The most frequent is ochratoxin A, which is also the most toxic. It has been shown to be nephrotoxic, immunosuppressive, carcinogenic and teratogenic in all experimental animals tested so far.

Acute renal failure in one person, possibly caused by inhalation of ochratoxin A in a granary which had been closed for 2 years, was reported in Italy. The symptoms developed after 24 hours of transitory epigastric tension, respiratory distress, and retrosternal burning. Acute tubular necrosis was found on biopsy, but the blood was not analysed for ochratoxin A. The presence of the mycotoxin in wheat from the granary was proved qualitatively by thin-layer chromatography.

Owing to the similarity of morphological and functional kidney lesions in ochratoxin A-induced porcine nephropathy and endemic nephropathy, this mycotoxin has been proposed as the causative agent of endemic nephropathy, although the evidence for this is not substantial. This fatal renal disease occurs among rural populations in Croatia, Bosnia and Herzegovina, Yugoslavia, Bulgaria, and Romania, where it has been estimated that about 20,000 people are either suffering from or are suspected to have the disease. There is no acute phase of the illness; the first signs and symptoms of the disease are not specific and include fatigue, headache, loss of body weight and pale skin. A mild low-molecular-mass proteinuria without hypertension but with either aplastic or normochromic anaemia gradually develops over several years. The main features of endemic nephropathy are bilateral, primarily chronic lesions of the renal cortex (tubular degeneration, interstitial fibrosis and hyalinisation of the glomeruli). In the advanced stage of the disease, the size and weight of kidneys are remarkably reduced, with diffuse cortical fibrosis, usually without signs of inflammation.

Ochratoxin A is found more frequently and in higher concentrations in the blood of inhabitants from endemic regions than control regions. Many samples of locally produced food and feed collected in the endemic area contained ochratoxin A. It should be emphasised that the grain analysed had been kept for many months in the inadequate food stores of individual families.

In Tunisia, ochratoxin A has been detected in high concentrations in the blood and food of patients with kidney impairment of unknown etiology. It has also been found in several countries, both in food and feed and in humans.

In endemic regions of Croatia, Bulgaria and Yugoslavia, the incidence of otherwise rare urothelial tumours of the pelvis and ureter is 50, 90 and 100 times greater, respectively, than in non-endemic regions. It has been suggested that ochratoxin A may be the causal agent for both endemic nephropathy and urothelial tumours. IARC classified ochratoxin A as a compound possibly carcinogenic to humans.

Trichothecenes

Trichothecenes are mycotoxins produced mostly by members of the Fusarium genus, although other genera (e.g. Trichoderma, Trichothecium, Myrothecium and Stachybotrys) are also known to produce these compounds. To date, 148 trichothecenes have been isolated, but only a few have been found to contaminate food and feed. The most frequent contaminants are deoxynivalenol (DON), also known as vomitoxin, nivalenol (NIV), diacetoxyscirpenol (DAS), while T-2 toxin is rarer.

Common manifestations of trichothecene toxicity are depression of immune responses and nausea, sometimes vomiting. The first recognised trichothecene mycotoxicosis was alimentary toxic aleukia in the USSR in 1932; the mortality rate was 60 per cent. In regions where the disease occurred, 540 per cent of grain samples cultured showed the presence of Fusarium sporotrichoides, while in those regions where the disease was absent this fungus was found in only 2-8 per cent of samples. The severity of mycotoxicosis was related to the duration of consumption of toxic grain. Such severe trichothecene mycotoxicoses, the consequence of continuous ingestion of toxins, have not been recorded since this outbreak.

In several cases, trichothecene mycotoxicosis was caused by a single ingestion of bread containing toxic flour or rice. In experimental animals, trichothecenes are 40 times more toxic when inhaled than when given orally. Trichothecenes were found in air samples collected during the drying and milling process on farms, in the ventilation systems of private houses and office buildings, and on the walls of houses with high humidity. There are some reports showing trichothecene involvement in the development of "sick building syndrome". The symptoms of airborne toxicosis disappeared when the buildings and ventilation systems were thoroughly cleaned.

Zearalenone (previously known as F-2) is produced mainly by Fusarium graminearum and related species, principally in wheat and maize but also in sorghum, barley and compounded feeds. Zearalenone and its derivatives produce estrogenic effects in various animal species (infertility, vulval oedema, vaginal prolapse and mammary hypertrophy in females and feminisation of males—atrophy of testes and enlargement of mammary glands).

In Puerto Rico, zearalenone was found in the blood of children with precocious sexual development exposed to contaminated food. Zearalenone was also found together with other Fusarium mycotoxins in "scabby grain toxicosis" in China, but the significance of this finding is not clear.

Fumonisins

Fumonisins are mycotoxins produced throughout the world by Fusarium moniliforme and related species when they grow in maize. Fumonisins [B.sub.1] and [B.sub.2] are of toxicological significance, while the others (B.sub.3], [B.sub.4], [A.sub.1] and [A.sub.2]) occur in very low concentrations and are less toxic.

In India a single outbreak of acute foodborne disease possibly caused by fumonisin [B.sub.1] has been reported. In the 27 villages involved, the individuals affected were from the poorest social strata, who had consumed maize and sorghum harvested and left in the fields during unseasonable rains. The main features of the disease were transient abdominal pain, borborygmus and diarrhoea, which began half an hour to one hour following consumption of unleavened bread prepared from mouldy sorghum or mouldy maize. Patients recovered fully when the exposure ceased and there were no fatalities. Fumonisin [B.sub.1] was found in much higher concentrations in the maize and sorghum from the affected households than from controls.

Fumonisin [B.sub.1] was found more frequently and in much higher concentrations in maize in regions of Transkei, China and northeast Italy with a higher incidence of oesophageal cancer than other regions. It was postulated that the high incidence of oesophageal cancer was related to the presence of this mycotoxin in maize, which is a staple food in these regions. The incidence and concentration of aflatoxin [B.sub.1], deoxynivalenol and fumonisins [B.sub.1], [B.sub.2] and [B.sub.3] were recently determined in maize samples from an area of China (Haimen) with a high incidence of primary liver cancer and from an area with a low incidence (Penlai). Aflatoxin [B.sub.1] was found in low concentrations in almost all maize samples from both these areas, but the incidence and concentration of deoxynivalenol and fumonisins were much higher in the samples from the area where the incidence of primary liver cancer was high. The authors put forward the hypothesis that fumonisins, which have known cancer-promoting activity in rat liver, and deoxynivalenol promote the initial lesion caused by aflatoxin [B.sub.1]. An IARC working group classified the toxins from F. moniliforme as possibly carcinogenic to humans.

Other Unidentified Mycotoxins

The impact of other mycotoxins on human health was reported in persons occupationally exposed to large amounts of different mycotoxin-producing fungi (farmers, workers in silos, etc.). In such cases, exposure to spores via the respirator), tract seems to be of considerable importance.

In Norway an extensive epidemiological study was undertaken between 1967 and 1991 on 192, 417 births to test the hypotheses that perinatal death was associated with parental exposure to pesticides, Toxoplasma gondii infection from sheep or pigs, or mycotoxins found in grain. The proportion of late-term abortions (gestational age 16-27 weeks) was higher among farmers. The risk associated with grain farming was higher after the harvest, in seasons with a poor quality harvest and in pregnancies with multiple fetuses, which suggests that mycotoxins in grain induce labour at an early stage of pregnancy.

Pulmonary mycotoxicosis has been reported in ten persons exposed to large quantities of fungal hyphae and spores during the cleaning of silos. The clinical picture developed several hours afterwards, with burning eyes, throat and chest, irritating cough and fever. There was no wheezing, cyanosis or other sign of bronchospasm. In five patients, chest X-rays revealed reticular and fine nodular features compatible with interstitial pneumonitis.

Histological study of a lung biopsy from one patient showed a multifocal acute process, with primary involvement of terminal bronchioles containing numbers of various spores. Cultures from lung biopsy material revealed at least five fungal species, including one Fusarium and one Penicillium. However, blood samples were not checked for the presence of mycotoxins. In contrast with the findings in patients with farmer's lung disease, these patients did not develop positive serological reactions to thermophilic actinomycetes or to extracts of fungi obtained from hay or silage. The patients were followed for periods of 1-10 years; they continued their work, avoiding massive re-exposure to fungal dust, and during the observation period there were no further incidents.

Acute mycotoxicoses can cause serious and sometimes fatal diseases. The possibility of mycotoxin intoxication should be considered when an acute disease occurs in several persons when there is no evidence of infection with a known etiological agent, and no improvement in the clinical picture following treatment. Most of the outbreaks of mycotoxicoses described are a consequence of the ingestion of food that is contaminated with mycotoxins. The strict control of food quality, in both industrialised and developing countries, is therefore necessary to avoid such outbreaks.

Trichothecene Mycotoxins (T2) Signs and Symptoms

Exposure causes skin pain, pruritus, redness, vesicles, necrosis and sloughing of epidermis. Effects on the airway include nose and throat pain, nasal discharge, itching and sneezing, cough, dyspnea, wheezing, chest pain and haemoptysis. Toxin also produces effects after ingestion or eye contact. Severe poisoning results in prostration, weakness, ataxia, collapse, shock, and death. Diagnosis: Should be suspected if an aerosol attack occurs in the form of "yellow rain" with droplets of yellow fluid contaminating clothes and the environment. Confirmation requires testing of blood, tissue and environmental samples.

Treatment: There is no specific antidote. Superactivated charcoal should be given orally if swallowed. Prophylaxis: The only defence is to wear a protective mask and clothing during an attack. No specific immunotherapy or chemotherapy is available for use in the field.

Decontamination: The outer uniform should be removed and exposed skin should be decontaminated with soap and water. Eye exposure should be treated with copious saline irrigation. Once decontamination is complete, isolation is not required.

Overview: The trichothecene mycotoxins are low molecular weight (250-500 daltons) non-volatile compounds produced by filamentous fungi (moulds) of the genera Fusarium, Myrotecium, Trichoderma, Stachybotrys and others. The structures of approximately 150 trichothecene derivatives have been described in the literature. These substances are relatively insoluble in water but are highly soluble in ethanol, methanol and propylene glycol. The trichothecenes are extremely stable to heat and ultraviolet light inactivation. Heating to 500°F for

30 minutes is required for inactivation, while brief exposure to NaOH destroys toxic activity. The potential for use as a BW toxin was demonstrated to the Russian military shortly after World War II when flour contaminated with species of Fusarium was baked into bread that was ingested by civilians.

Some developed a protracted lethal illness called alimentary toxic aleukia (ATA) characterised by initial symptoms of abdominal pain, diarrhoea, vomiting, prostration, and within days fever, chills, myalgias and bone marrow depression with granulocytopenia and secondary sepsis.

Survival beyond this point allowed the development of painful pharyngeal/laryngeal ulceration and diffuse bleeding into the skin (petechiae and ecchymoses), melena, bloody diarrhoea, hematuria, hematemesis, epistaxis and vaginal bleeding. Pancytopenia, and gastrointestinal ulceration and erosion were secondary to the ability of these toxins to profoundly arrest bone marrow and mucosal protein synthesis and cell cycle progression through DNA replication.

History and Significance Mycotoxins allegedly have been used in aerosol form ("yellow rain") to produce lethal and non-lethal casualties in Laos (1975-81), Kampuchea (1979-81), and Afghanistan (1979-81). It has been estimated that there were more than 6,300 deaths in Laos, 1,000 in Kampuchea, and 3,042 in Afghanistan.

The alleged victims were usually unarmed civilians or guerrilla forces. These groups were not protected with masks and chemical protective clothing and had little or no capability of destroying the attacking enemy aircraft. These attacks were alleged to have occurred in remote jungle areas which made confirmation of attacks and recovery of agent extremely difficult. Much controversy has centred about the veracity of eyewitness and victim accounts, but there is enough evidence to make agent use in these areas highly probable.

Clinical Features T2 and other mycotoxins may enter the body through the skin and aerodigestive epithelium. They are fast acting potent inhibitors of protein and nucleic acid synthesis. Their main effects are on rapidly proliferating tissues such as the bone marrow, skin, mucosal epithelia, and germ cells. In a successful BW attack with trichothecene toxin (T2), the toxin(s) will adhere to and penetrate skin, be inhaled, and swallowed. Clothing will be contaminated and serve as a reservoir for further toxin exposure.

Early symptoms beginning within minutes of exposure include burning skin pain, redness, tenderness, blistering, and progression to skin necrosis with leathery blackening and sloughing of large areas of skin in lethal cases. Nasal contact is manifested by nasal itching and pain, sneezing, epistaxis and rhinorrhea; pulmonary/tracheobronchial toxicity by dyspnea, wheezing, and cough; and mouth and throat exposure by pain and blood tinged saliva and sputum.

Anorexia, nausea, vomiting and watery or bloody diarrhoea with abdominal crampy pain occurs with gastrointestinal toxicity. Eye pain, tearing, redness, foreign body sensation and blurred vision may follow entry of toxin into the eyes. Skin symptoms occur in minutes to hours and eye symptoms in minutes. Systemic toxicity is manifested by weakness, prostration, dizziness, ataxia, and loss of coordination. Tachycardia, hypothermia, and hypotension follow in fatal cases. Death may occur in minutes, hours or days. The commonest symptoms were vomiting, diarrhoea, skin involvement with burning pain, redness and pruritus, rash or blisters, bleeding, and dyspnea.

Diagnosis: Rapid onset of symptoms in minutes to hours supports a diagnosis of a chemical or toxin attack. Mustard agents must be considered but they have an odour, are visible, and can be rapidly detected by a field available chemical test. Symptoms from mustard toxicity are also delayed for several hours after which mustard can cause skin, eye and respiratory symptoms. Staphylococcal enterotoxin B delivered by an aerosol attack can cause fever, cough, dyspnea and wheezing but does not involve the skin and eyes. Nausea, vomiting, and diarrhoea may follow swallowing of inhaled toxin. Ricin inhalation can cause severe respiratory distress, cough, nausea and arthralgias. Swallowed agent can cause vomiting, diarrhoea, and gastrointestinal bleeding, but it spares the skin, nose and eyes. Specific diagnosis of T-2 mycotoxins in the form of a rapid diagnostic test is not presently available in the field. Removal of blood, tissue from fatal cases, and environmental samples for testing using a gas liquid chromatography-mass spectrometry technique will confirm the toxic exposure. This system can detect as little as 0.1-1.0 ppb of T-2. This degree of sensitivity is capable of measuring T-2 levels in the plasma of toxin victims.

Medical Management: Use of a chemical protective mask and clothing prior to and during a mycotoxin aerosol attack will prevent illness. If a soldier is unprotected during an attack the outer uniform should be discarded within 4 hours and decontaminated by exposure to 5 per cent hypochlorite for 6-10 hours. The skin should be thoroughly washed with soap and uncontaminated water if available. The M291 skin decontamination kit should also be used to remove skin adherent T-2. Superactive charcoal can absorb swallowed T-2 and should be administered to victims of an unprotected aerosol attack. The eyes should be irrigated with normal saline or water to remove toxin. No specific antidote or therapeutic regimen is currently field available. All therapy is symptomatic and supportive.

Prophylaxis

Physical protection of the skin and airway are the only proven effective methods of protection during an attack. Immunological (vaccines) and chemoprotective pretreatments are being studied in animal models, but are not available for field use by the warfighter.

Staphylococcal Enterotoxin B: Summary Signs and Symptoms

From 3-12 hours after aerosol exposure, sudden onset of fever, chills, headache, myalgia, and non-productive cough.

Some patients may develop shortness of breath and retrosternal chest pain. Fever may last 2 to 5 days, and cough may persist for up to 4 weeks. Patients may also present with nausea, vomiting, and diarrhoea if they swallow toxin. Higher exposure can lead to septic shock and death.

Diagnosis: Diagnosis is clinical. Patients present with a febrile respiratory syndrome without CXR abnormalities. Large numbers of soldiers presenting with typical symptoms and signs of SEB pulmonary exposure would suggest an intentional attack with this toxin.

Treatment: Treatment is limited to supportive care. Artificial ventilation might be needed for very severe cases, and attention to fluid management is important. Prophylaxis: Use of protective mask. There is currently no human vaccine available to prevent SEB intoxication. Decontamination: Hypochlorite (0.5 per cent for 10-15 minutes) and/or soap and water. Destroy any food that may have been contaminated.

Overview: Staphylococcus aureus produces a number of exotoxins, one of which is Staphylococcal enterotoxin B, or SEB. Such toxins are referred to as exotoxins since they are excreted from the organism; however, they normally exert their effects on the intestines and thereby are called enterotoxins. SEB is one of the pyrogenic toxins that commonly causes food poisoning in humans after the toxin is produced in improperly handled foodstuffs and subsequently ingested.

SEB has a very broad spectrum of biological activity. This toxin causes a markedly different clinical syndrome when inhaled than it characteristically produces when ingested. Significant morbidity is produced in individuals who are exposed to SEB by either portal of entry to the body.

History and Significance: SEB has caused countless endemic cases of food poisoning. Often these cases have been clustered, due to common source exposure in a setting such as a church picnic or passengers eating the same toxin-contaminated food on an airliner. Although this toxin would not be likely to produce significant mortality on the battlefield, it could render up to 80 per cent or more of exposed personnel clinically ill and unable to perform their mission for a fairly prolonged period of time. Therefore, even though SEB is not generally thought of as a lethal agent, it may incapacitate soldiers for up to two weeks, making it an extremely important toxin to consider.

Toxin Characteristics: Staphylococcal enterotoxins are extracellular products produced by coagulase-positive staphylococci. They are produced in culture media and also in foods when there is overgrowth of the staph organisms. At least five antigenically distinct enterotoxins have been identified, SEB being one of them. These toxins are heat stable. SEB causes symptoms when inhaled at very low doses in humans: a dose of several logs lower than the lethal dose by the inhaled route would be sufficient to incapacitate 50 per cent of those soldiers so exposed. This toxin could also be used (theoretically) in a special forces or terrorist mode to sabotage food or low volume water supplies.

Mechanism of Toxicity: Staphylococcal enterotoxins produce a variety of toxic effects. Inhalation of SEB can induce extensive pathophysiological changes to include widespread systemic damage and even septic shock.

Many of the effects of staphylococcal enterotoxins are mediated by interactions with the host's own immune system. The mechanisms of toxicity are complex, but are related to toxin binding directly to the major histocompatibility complex that subsequently stimulates the proliferation of large numbers of T cell lymphocytes.

Because these exotoxins are extremely potent activators of T cells, they are commonly referred to as bacterial superantigens. These superantigens stimulate the production and secretion of various cytokines, such as tumour necrosis factor, interferon– interleukin-1 and interleukin-2, from immune system cells. Released cytokines are thought to mediate many of the toxic effects of SEB.

Clinical Features: Relevant battlefield exposures to SEB are projected to cause primarily clinical illness and incapacitation. However, higher exposure levels can lead to septic shock and death. Intoxication with SEB begins 3 to 12 hours after inhalation of the toxin. Victims may experience the sudden onset of fever, headache, chills, myalgias, and a non-productive cough. More severe cases may develop dyspnea and retrosternal chest pain.

Nausea, vomiting, and diarrhoea will also occur in many patients due to inadvertently swallowed toxin, and fluid losses can be marked. The fever may last up to five days and range from 103 to 106°F, with variable degrees of chills and prostration. The cough may persist up to four weeks, and patients may not be able to return to duty for two weeks. Physical examination in patients with SEB intoxication is often unremarkable. Conjunctival injection may be present, and postural hypotension may develop due to fluid losses. Chest examination is unremarkable except in the unusual case where pulmonary edema develops. The chest X-ray is also generally normal, but in severe cases increased interstitial markings, atelectasis, and possibly overt pulmonary edema or an ARDS picture may develop.

Diagnosis: As is the case with botulinum toxins, intoxication due to SEB inhalation is a clinical and epidemiologic diagnosis. Because the symptoms of SEB intoxication may be similar to several respiratory pathogens such as influenza, adenovirus, and mycoplasma, the diagnosis may initially be unclear. All of these might present with fever, non-productive cough, myalgia, and headache. SEB attack would cause cases to present in large numbers over a very short period of time, probably within a single 24 hour period. Naturally occurring pneumonias or influenza would involve patients presenting over a more prolonged interval of time. Naturally occurring staphylococcal food poisoning cases would not present with pulmonary symptoms. SEB intoxication tends to progress rapidly to a fairly stable clinical state, whereas pulmonary anthrax, tularemia pneumonia, or pneumonic plague would all progress if left untreated. Tularemia and plague, as well as Q fever, would be associated with infiltrates on chest radiographs.

Nerve agent intoxication would cause fasciculations and copious secretions, and mustard would cause skin lesions in addition to pulmonary findings; SEB inhalation would not be characterised by these findings. The dyspnea associated with botulinum intoxication is associated with obvious signs of muscular paralysis, bulbar palsies, lack of fever, and a dry pulmonary tree due to cholinergic blockade; respiratory difficulties occur late rather than early as with SEB inhalation. Laboratory findings are not very helpful in the diagnosis of SEB intoxication.

A non-specific neutrophilic leukocytosis and an elevated erythrocyte sedimentation rate may be seen, but these abnormalities are present in many illnesses. Toxin is very difficult to detect in the serum by the time symptoms occur; however, a serum specimen should be drawn as early as possible after exposure. Data from rabbit studies clearly show that SEB in the serum is transient; however, it accumulates in the urine and can be detected for several hours post exposure. Therefore, urine samples should be obtained and tested for SEB. High SEB concentrations inhibit kidney function. Because most patients will develop a significant antibody response to the toxin, acute and convalescent serum should be drawn which may be helpful retrospectively in the diagnosis.

Medical Management: Currently, therapy is limited to supportive care. Close attention to oxygenation and hydration are important, and in severe cases with pulmonary edema, ventilation with positive end expiratory pressure and diuretics might be necessary. Acetaminophen for fever, and cough suppressants may make the patient more comfortable. The value of steroids is unknown. Most patients would be expected to do quite well after the initial acute phase of their illness, but most would generally be unfit for duty for one to two weeks.

Prophylaxis

Although there is currently no human vaccine for immunisation against SEB intoxication, several vaccine candidates are in development. Preliminary animal studies have been encouraging and a vaccine candidate is nearing transition to advanced development and safety and immunogenicity testing in man. Experimentally, passive immunotherapy can reduce mortality, but only when given within 4-8 hours after inhaling SEB.

Moist conditions involving drywall, wood, carpeting, or paper material are the focal proliferation medium in the indoor environment. Since Americans spend 75 to 90 per cent of their time indoors, they are exposed to moulds that may grow indoors.

Moulds enter the indoor environment through doorways, windows, heating and ventilation systems, and air conditioning systems, given the appropriate circumstances. Spores in the air deposit on people, animals, clothing, shoes, and bags, turning them into common and potential carriers of moulds into the indoor environments.

Indoor environments that contain excessive moisture such as leakage from roofs, walls, plant pots, or pet urine cause proliferation and development of moulds. The most common moulds which are found indoor are Cladosporium, Penicillium, and Aspergillus. In order to proliferate moulds need nutrients which are commonly present in building environments such as cellular substrates in paper, paper products, cardboard, ceiling tiles, wood, wood products, drywall, carpet, fabric, insulation materials, wallpaper, paints, and dusts. Some of the indoor moulds have the potential to produce extremely toxic materials called mycotoxins. Those moulds which have the potential of producing toxic materials include Fusarium and Stachybotrys, among others. Depending on the quantities produced and consumed, mycotoxins can cause acute or chronic toxicity in animals and humans. Home dampness with resulting mould growth may be associated with several medical conditions (one or sometimes all) including immediate hypersensitivity reaction, hypersensitivity pneumonia, or what has been described as "humidifier fever".

Onset of asthma, recent onset sinusitis, and/or recent onset skin rashes. Several studies have shown a clear correlation and association between the occurrence of moulds in the inside air environment, dampness in the indoor environment, and the symptomatology of the skin, and respiratory tract, especially in children. This has been summarised in an interesting study published in the American Journal of Epidemiology by Robert E. Dales. Since the symptoms in this study were comparable to the symptoms described with humidifier fever and mycotoxicosis, the authors suggested a common pathogenic and etiological mechanism.

The role of indoor moulds, especially the most toxic one–Stachybotrys, has been shown recently in a scientific paper published in the journal Pediatrics. The authors described a child with pulmonary haemorrhaging where Stachybotrys was isolated from the lung.

Indeed, epidemiological data to support the connection between mould exposure and lung haemorrhage was published in the scientific literature from Cleveland, Ohio, which was later examined by the Centre for Disease Control. The scientific data clearly demonstrates a high spore count of Stachybotrys in 9 out of 10 of the houses where these infants lived, and 5 infants had recurrence of the bleeding of the lungs on re-entry to their homes, implicating that the fungus is a potential agent in the pathogenesis of infantile pulmonary haemorrhage. The study by Okan Alidemir, *et al.*, shows the isolation of Stachybotrys atra from the BAL fluid of a child with pulmonary haemorrhage, thus connecting the epidemiological data and the historical data in this case report with objective findings of Stachybotrys from the lung fluids. In the scientific paper entitled "Stachybotrys: Mycotoxin Producing Fungus of Increasing Toxicological Importance", the investigators concluded "Current data on the toxicology of mycotoxins produced by Stachybotrys demonstrates that this group of mycotoxins is capable of producing immunosuppression and inflammatory insults to the gastrointestinal and pulmonary system".

While it is an ideal situation to have "statistical firmness", in medicine the clinician established a diagnosis and causation based on known and accepted factors where statistical firmness is not a prerequisite. The causal clinical association between allergic reaction to the sinuses in the form of rhinitis, sinusitis, or asthma and indoor air mould exposure has been very well documented in the scientific literature in an early review by Susan Gravesen.

That indoor moisture and moulds represents a public health issue is described in the scientific paper by Hodgson. These authors report an outbreak of disease associated with exposure to these moulds in 2 buildings in Florida. The specific buildings were a new court house and office building which were constructed between 1986 and 1989. Within weeks after moving in patients described mucous membrane irritation, fatigue, headaches, and chest tightness. Moisture problems such as window and roof leaks have been described as starting in 1987 and persisting through 1992.

Utilising epidemiological methodology the investigators concluded that this outbreak represents a likely human response to inhaled fungal toxins in indoor air environments. What to do when you suspect moulds as a cause for symptomatology. First and most importantly is to see a doctor who specialises in the fields of internal medicine, occupational medicine and toxicology with the understanding of building-related illnesses and toxic moulds. The doctor will have to rule out other diseases, perform laboratory studies, and provide an opinion as to whether these symptoms can be and have been described with moulds. Upon determination that these symptoms may be related to mould exposure, you should have an industrial hygienists go and inspect your residence or alternatively office/

work place (depending on where the suspected mould resides) to do a careful investigation of any water damages, and air counts both inside and outside at several locations for moulds and spores.

A well-trained industrial hygienist will not only take air counts but also will go under and behind the walls and/or carpeting where the water damage is anticipated to be in order to further evaluate for mould spores and mould growth.

Once moulds are discovered, depending on the damage that occurred, either expert remediation (with appropriate protective devices and removal of the inhabitants from the area) or at times destruction and rebuilding of the damaged house or building area is necessary.

The most common symptoms of mould related diseases are those of cough, asthma, atypical asthma, nasal congestion, sinusitis/rhinitis, skin rashes, and generalised fatigue.

On many occasions the patients presented with neurological symptoms such as headaches, reduced concentration ability, and memory loss. The patients may present with only one symptom (such as sinusitis) or a combination of symptoms.

Nachman Brautbar, M.D. is board certified in internal medicine, forensic medicine, and nephrology with a specialisation in toxicology. Dr. Brautbar is a medical professor at USC, school of medicine, has published over 230 journal articles, manuscripts, abstracts and books.

Thielaviopsis

Thielaviopsis is a small genus of fungi in the order Microascales. The genus includes several important agricultural pathogens. The most widespread is *T. basicola*, the causal agent in several root rot diseases of economically important crop species including cotton and a variety of vegetables. In cotton, *Thielaviopsis* root rot, also known as black root rot causes necrosis of the roots and stunting of the crop plants.

Verticillium

Verticillium is a genus of fungi in the division Ascomycota. The majority of *Verticillium* species are saprobes and are abundant in soils worldwide. A small number of species, cause wilt diseases in economically important plant species such as cotton, tomatoes, potatoes, eggplants, peppers, ornamental woody plants, as well as plants in natural vegetation communities.

Symptoms are superficially similar to *Fusarium* wilts. Crop rotation, the use of resistant crop varieties and deep ploughing, may be useful in controlling *Verticillium* wilt.

Selected Species

- *Verticillium lecanii Viegas:* This fungus species was first described in 1861. It has a worldwide distribution found on insects. In horticulture and agriculture it is sometimes used as an entomopathogen (infects insects) for controlling insect pests like mealybugs and mites and aphids. It's a bio-pesticide/bio-insecticide that is sprayed on insects while they are feeding off of desirable plants. The insects are infected when they come into contact with the sticky fungal spores which then grow and invade the body, thus pathogenesised, the insects internal organs are consumed, leading to the insects death.
- *Verticillium alboatrum Reinke and Berthold:* Causes Verticillium Wilt or Maple Wilt. First identified from potatoes in Germany in 1870. This species attacks over 300 different cultivated plants and can persist as a saprophytic soil organism for more that 15 years. When infecting ornamental trees like Maples, Elms, Aspen, Ash, Beech, Catalpa Oak plus many more the first symptoms are midsummer wilting on one side of a tree or branch. The sapwood has greenish or brownish streaks, and the infection can take a few years to progress to the rest of the tree or move rapidly. The

fungi universally move up the xylem vessels. In fruit trees, the infection is know as 'Black Heart' and is common in apricots and sometimes effects almonds, peaches, plums and avocado. This Fungus also effects herbaceous ornamentals and vegetables like Chrysanthemum, mints, Lychnis, tomatoes, eggplants, okra, rhubarb. Causing wilting and death. Identification can be made by looking for one-celled conidia, hyaline round to ellipsoid which are formed at the tips of whorled branches. They are easily separated from the tips.

Magnaporthe Grisea

Magnaporthe grisea, also commonly know as rice blast fungus, is a plant-pathogenic fungus that causes an important disease affecting rice. It can also infect a number of other agriculturally important cereals including wheat, rye, barley, and pearl millet causing diseases called blast disease or blight disease. M. grisea causes economically significant crop losses annually, each year it is estimated to destroy enough rice to feed more than 60 million people. The fungus is known to occur in 85 countries worldwide. It is sometimes thought of as a model organism in the study of phytopathogenic fungi.

Biology and Pathology

M. grisea is an ascomycete fungus. It is an extremely effective plant pathogen as it can reproduce both sexually and asexually to produce specialised infectious structures known as appressoria that infect aerial tissues and hyphae that can infect root tissues.

The asexual life cycle begins when the hyphae of the fungus undergo sporulation to produce fruiting structures called conidia which contain many spores. When these spores land on leaves and other aerial tissues of susceptible plants they germinate, developing the appressorium. The appressorium penetrates the plant cell by producing a penetration peg.

Pressure in the appressorium increases and the structure explodes, forcing the penetration peg through the cell wall and into the cell. The fungus can then grow hyphae within the leaf and form lesions. Once established in the host plant the fungal hyphae can undergo asexual sporulation again. Sexual reproduction occurs when two strains of opposite mating types meet and form a perithecium in which ascospores develop. Once released, ascospores can develop appressoria and infect host cells. Spores are transmitted between plants by the wind.

In 2004, it was shown that in addition to infecting plants through the leaf, *M. grisea* can also infect the plant roots. The mode of root infection is the same as most root infecting fungi; it grows long hyphae that form an infection pad to gain entry to the root's interior. Once embedded in the root the fungus can produce resting structures. The blast fungus can also invade the plant's vascular system, growing inside the xylem and phloem and blocking the transport of nutrients and water from the roots, and produce lesions on aerial plant parts.

The fungus can kill or damage the plant in a number of ways. Aerial infection in seedlings is usually lethal because the young plants are unable to photosynthesise. Aerial infection in mature plants does not kill them, but reduced photosynthesis due to lesions on the leaf and use of photosynthate by the fungus greatly reduces yield. Infection of root and vascular tissues has the potential to kill the plant by cutting off the supply of water and nutrients to the root.

In 2005, the complete genome of *M. grisea* was sequenced. The organism is predicted to have over 11,000 genes. It is expected that the genome will reveal the mechanisms of fungal pathogen-plant interaction, in both aerial and root infection.

Distribution

Rice blast was probably first recorded as *rice fever disease* in China in 1637. It was later described as *imochi-byo*in Japan in 1704, and as *brusone* in Italy in 1828. The fungus is currently reported to be present in at least 85 countries. In 1996, rice blast was found in rice in California, and has since been found in grasses on golf courses in the midwestern United States.

Strains of the fungus can infect domesticated grasses such as barley, wheat, rye, pearl millet, and turf grasses in addition to rice. Thus, even when crops are burned to destroy fungal infection, grass weeds can act as a disease reservoir. The disease may be called different names depending on the crop infected; In rice it is called rice blast, in wheat it is called wheat blast, in rye it is called rye blast and so on.

Control

M. grisea outbreaks are controlled through the application of expensive and potentially hazardous fungicides. Among the current blasticides are probenazole, tricyclazole, pyroquilon and phthalide. Infected crops are also burned in some areas.

With plant diseases, control is a relative term defining disease progress and subsequent loss of desirable plant product (rice grain here). The magnitude of control varies from very little to complete. With rice blast, efficacy of fungicides and plant genetic resistance is determined by the interaction of environmental and cultural conditions.

The fungus has been able to develop resistance to both chemical treatments and genetic resistance developed by plant breeders in some types of rice. It is thought that the fungus can achieve this by genetic change through mutation. Researchers hope that by having the full genome sequence of the fungus the development of effective control methods will be possible. For example, a mutant *M. grisea* unable to produce a penetration peg has been shown to lack the ability to infect rice.

Biological Weapon

M. grisea spores were prepared as an anti-plant biological weapon independently by the United States and the USSR during World War II. There are some anecdotal reports that US officials had considered using rice blast agents to destroy

Japan's rice crop during the closing months of World War II, forcing surrender by starving the Japanese people. There are concerns that *M. Grisea* may be used as a biological weapon by a terrorist organisation.

Soybean Rust

Soybean rust, also known as Asian soybean rust, is a disease that affects soybeans and other legumes. It is caused by two types of fungi, Phakopsora pachyrhizi and Phakopsora meibomiae. P. meibomiae is the weaker pathogen of the two and generally does not cause widespread problems. The disease has been reported across Asia, Australia, A'frica, South America and the United States.

Habitat

Soybean rust thrives on green, growing plants, and requires a climate which provides high moisture and moderate heat. It is unable to survive the cold winters of northern habitats.

Propagation

Soybean rust is spread by wind-borne spores, which are released in cycles of seven days to two weeks. It is commonly believed that the disease was carried from Colombia to the United States by Hurricane Ivan.

Treatment

When properly applied, fungicides can be used preventively or to cure infected plants which, if left untreated, will eventually die. Recent research from Washington State University indicates that the herbicide Glyphosate may be effective in dealing with the fungus. Rust-resistant varieties of soybeans are currently in development by both public universities and private industry.

Oomycetes

The oomycetes are fungal-like organisms that until recently used to be mistaken for fungi. They include some of the most destructive plant pathogens including the genus *Phytopthora* which includes the casual agents of potato blight and sudden oak death.

Despite not being closely related to the fungi, the oomycetes have developed very similar infection strategies and so many plant pathologists group them with fungal pathogens.

Significant oomycete plant pathogens

- *Pythium* spp.
- *Phytophthora* spp.; including the causal agent of the Irish Potato Famine

Barley Diseases

This article focuses on minor diseases of barley. These are diseases of barley that are rare, of minor economic significance, and/or have not been researched very extensively.

Anthracnose

Anthracnose is a fungal diseased caused by *Colletotricum graminicola*. It is diseases of many grasses and cereals, but is rarely of economic importance.

In barley, it has been found worldwide and is more prevalent in crops grown on sandy soils with low fertility.

Ascochyta Leaf Blight

Asochyta leaf spot is a fungal diseased caused by *Ascochyta hordei*. It is only been found in Japan, Korea and the United States (Virginia, West Virginia, Washington).

Symptoms

Lesions are round, oval or irregular, 3-11 mm in diameter, sometimes with rings, and occur primarily on lower leaves. There are rarely more than a few lesions on an individual plant.

Silver Leaf

Silver leaf or *Chondrostereum purpureum* is a fungal disease of trees. It attacks most species of the rose family Rosaceae, particularly the genus Prunus. The disease is progressive and often fatal. The common name is taken from the progressive silvering of leaves on affected branches. It is spread by airborne spores landing on freshly exposed sapwood. For this reason cherries and plums are pruned in summer, when spores are least likely to be present.

False Loose Smut

False loose smut is a fungal disease of barley caused by *Ustilago nigra*. This fungus is very similar to *U. nuda*, the cause of loose smut, and was first distinguished from it in 1932.

Symptoms: The disease is not apparent until heading, at which time, smutted heads emerge slightly earlier than healthy heads. At first, each smutted head is covered by a delicate, paperlike, grayish membrane. These membranes break shortly after the smutted heads have emerged and expose a dark brown to black, powdery mass of spores. This spores are easily dislodged, leaving only the bare rachis.

Disease Cycle: The disease cycle of *Ustilago nigra* is similar to that of *U. hordei*, the cause of covered smut of barley. The teliospores survive on the surface or in the soil. In some cases, the teliospores that are deposited under the hull, may germinate immediately. The mycelium then grows into the lower layers of the seed and then remains dormant until seed germination.

Infection of seedling occurs between germination and emergence. Infection can occur from seed-borne teliospores or by teliospores residing in the soil. Relatively dry soil at temperatures of 15-21 C, are most favourable for infection. The invading mycelium becomes established within the growing point. As the plant enters the boot stage, the mycelium grows rapidly into the floral tissue which is converted to masses of black teliospores. Teliospores are disseminated by wind or during combining. The teliospores may remain viable for several spores.

Management: The incidence of false loose smut can be reduced by using clean seed, treated seed and resistant cultivars.

Rhynchosporium Secalis

Rhynchosporium secalis is the causal agent of barley scald.

Morphology: No sexual stage is known. The mycelium is hyaline to light gray and develops sparsely as a compact stroma under the cuticle of the host plant. Condia (2-4 x 12-20 μm) are borne sessilely on cells of the fertile strom.

They are hyaline, 1-septate, and cylindric to ovate, mostly with a short apical beak. Microconida have been reported, but their function is unknown. They are exuded from flasklike mycelial branches.

Host Species

- *Agropyron dasystachyum,*
- *A. desertorum,*
- *A. elmeri,*
- *A. intermedium,*
- *A. repens,*
- *A. riparium,*
- *A. scabriglume,*
- *A. semi-costatum,*
- *A. subsecundum,*
- *A. trachycaulum,*
- *A. trachycaulum var. trachycaulum,*
- *A. trachycaulum var. unilaterale*
- *Agrostis alba,*
- *A. stolonifera,*

- *A. tenuis*
- *Alopecurus geniculatus,*
- *A. pratensis*
- *Bouteloua gracilis,*
- *B. hirsuta*
- *Bromus aleutensis,*
- *B. carinatus,*
- *B. ciliatus,*
- *B. frondosus,*
- *B. inermis,*
- *B. pumpellianus,*
- *B. secalinus,*
- *B. vulgaris*
- *Calamagrostis arundinacea,*
- *C. epigeios*
- *Chrysopogon gryllus*
- *Critesion murinum*
- *Cynodon dactylon*
- *Dactylis glomerata*
- *Danthonia sp.*
- *Deschampsia caespitosa*
- *Elymus angustus,*
- *E. canadensis,*
- *E. chinensis,*
- *E. condensatus,*
- *E. glaucus,*
- *E. innovatus,*
- *E. junceus,*

- *E. triticoides,*
- *E. virginicus*
- *Elytrigia repens*
- *Festuca pratensis,*
- *F. rubra*
- *Hordeum aegiceras,*
- *H. brachyantherum,*
- *H. distichon,*
- *H. hexastichon,*
- *H. jubatum,*
- *H. leporinum,*
- *H. murinum,*
- *H. vulgare,*
- *H. vulgare var. nudum,*
- *H. vulgare var. trifurcatum*
- *Lolium multiflorum,*
- *L. perenne,*
- *L. rigidum*
- *Microlaena stipoides*
- *Panicum sp.*
- *Phalaris arundinacea*
- *Phleum pratense*
- *Poa annua,*
- *P. eminens,*
- *P. pratensis*
- *Quercus chrysolepis*
- *Roegneria sp.*
- *Secale cereale,*

- *S. montanum*
- *Triticosecale sp.*
- *Stem rust (barley)*

Stem rust is a fungal diseases of barley caused by *Puccinia graminis* f.sp. *tritici* and *Puccinia graminis* f.sp. *secalis*.

Symptoms: Leaf rust is characterised by large pustules that are oval to elongated and are often surrounded by a characteristic torn margin. The red rust or summer spore stage (uredinia) is evident on stems, leaf sheaths, glumes, awns, and leaves as elongate pustules containing reddish brown spore masses which fall away easily. The black rust or autumn spore stage (telia) looks like the red stage except for the black colour, but do not dislodge.

Disease Cycle: Stem rust survives the summer mainly on volunteer wheat, barley, triticale and some grasses including common wheat grass and barley grass. Barberry species (*Berberis vulgaris. B. canadensis, B. fendleri*) serve as an alternate host for *Puccinia graminis*.

Spores are spread by wind to susceptible hosts. High humidity and heavy dew favour development of the disease which develops rapidly at temperatures near 20°C. Disease development is markedly reduced by temperatures below 15°C and above 40°C.

Chondrostereum Purpureum

Chondrostereum purpureum is a plant pathogen which causes the Silver leaf disease of fruit trees; plums are especially vulnerable. In the past the name *Stereum purpureum* Pers. was widely used for this fungus, but according to modern taxonomy it is only distantly related to *Stereum,* actually belonging to order Polyporales whereas *Stereum* is in order *Russulales*.

Characteristics

After starting as just a crust on the wood, the fruiting structure develops undulating intergrowing brackets up to about 3cm broad, which have a tough rubbery texture. The edges and fertile lower surfaces show a fairly vivid violet colour while the fungus is growing, and the upper surfaces have a grey aspect (sometimes with zonation) and are covered with whitish hairs.

After a week or two the fructification dries out, becomes brittle, and turns a drab brown or beige. Infected wood can be recognised because it is stained a darker tint.

The spores are rounded cylinders approximately 5-8μm x 3-4μm in size. The hyphal structure is monomitic with clamp connections.

It is often found on old stumps and dead wood, but can also be a serious parasite of living trees. As well as plum trees it attacks many other broad-leafed species (other *Prunus*, apple, pear, willow, poplar, maple, hornbeam, plane, oak, elm, lilac, and many others. Occasionally it also infects conifers (fir, spruce, *Thuya*, ...). Geographically it is roughly speaking just as widespread as its hosts - it is common in woods, orchards and tree plantations in temperate climates.

A Biological Control Agent

C. purpureum is commercially available as a method of combating forest "weed" trees such as red alder, aspens, and others. The fungus is applied directly to the weed trees in a nutrient paste which can be stored and handled conveniently. According to a report of the Canadian Pest Management Agency, the use of this control method will only have a limited impact on non-target trees since the fungus spores are ubiquitious anyway and healthy trees are resistant to attack.

Tapesia Yallundae

Tapesia yallundae is the causal agent for a variety of cereal and forage grass diseases. The anamorph of *T. yallundae* is the W-type strain of *Pseudocercosporella herpotrichoides*. The R-type strain of *Pseudocercosporella herpotrichoides* is now known as *Tapesia acuformis*.

Morphology: Produces two types of mycelium–one vegetative, yellow-brown, linear, and branching, the other dark and stromalike. Condiophores are simple or sparingly branched. Conidia (1.5-3.5 x 37-70 μm) are hyaline, curved, and mostly five- to seven-celled. Sclerotia or sclerotialike stromatic mycelium, at first white to yellow-brown but later dark brown, may also be found on the lesions of infected plants.

Black apothecia, 0.2 to 0.5 mm of diameter, form at the base of host culms. Apothecium contain cylindric to fusoid asci, of 35 to 38 μm x 5.9 to 7.4 μm. Ascospores are hyaline, fusoid, 0-1 septate, with a rounded end and an average size of 8.9 μm (7.4 to 10.3) x 2.07 μm (1.95 to 2.34).

Growth Media

Tapesia yallundae can be grown on a moist, sterile wheat and barley straw, oat kernels, and a variety of simple agar media, preferably supplemented with wheat extract. Sporulation in vitro tends to originate from loose sporodochai. Young colonies on potato-dextrose agar are gray, compact and mounded.

Molecular Characterisation

Polymerase chain reaction (PCR) combined with restriction enzyme digestion of an amplified ribosomal DNA fragment, are now used to characterise *T. yallundae* isolates. Novartis produces a polymerase chain reaction (PCR) diagnostic tool that provides cereal growers with an efficient means for checking the progression of eyespot disease in cereals. The tool reveals the presence and extent of disease before symptoms develop and can identify the different eyespot strains including the W-type (*Tapesia yallundae'*) and R-type (*Tapesia acuformis*).

Host Species

Aegilops cylindrica, Aegilops ovata, Aegilops sp., *Aegilops triuncialis, Agropyron cristatum, Agropyron dasystachyum, Agropyron inerme, Agropyron repens, Agropyron riparium, Agropyron* sp., *Avena fatua, Avena sativa, Avena* sp., *Balsamorhiza* sp., *Bromus carinatus, Bromus inermis, Bromus japonicus, Bromus sterilis, Bromus tectorum, Delphinium* sp., *Festuca idahoensis, Hordeum distichon, Hordeum vulgare, Koeleria cristata, Lithospermum ruderale, Lomatium triternatum, Poa sandbergii, Poa secunda, Secale cereale, Sitanion hystrix, Trisetum aestivum, Triticum aestivum, Triticum dicoccum, Triticum durum, Triticum monococcum, Triticum* sp., *Triticum spelta, Triticum vulgare.*

Pyrenophora Graminea

Pyrenophora graminea is the causal agent of barley stripe.

Identification

Asexual Stage: Pycnidia are rarely observed in nature. They are 70-176 µm in diameter, globose to pear-shaped, and develop superficially or partly submerged. The wall is thin and fragile and is yellow to brown, with a short ostiole. Pycnidiospores are 1.4-3.2 x 1.0-1.6 µm, spherical or ellipsoidal, hyaline, and non-septate.

Sexual Stage: Perithecia are rare in nature, they occur in barley straw in the autumn. The perithecia are 576-728 x 442-572 µm. They are superficial to partly submerged and are elongate, with rigid setae on the surface. Acsi are club-shaped or cylindrical, clearly bitunicate, and rounded at the apex, with a short stalk at the base. Ascospores are 43-61 x 16-28 µm, light yellow-brown, ellipsoidal, and rounded at both ends, with transverse septa and one, occasionally two, septum in the median cells but never in the terminal cells.

Conidia are borne laterally and terminally on conidiophores, which usually occur in clusters of three to five. The conidia are straight with rounded ends and measure 11-24 x 30-100 µm. They are subhyaline to yellow-brown and have up to seven transverse septa.

In culture, mycelium is gray to olivaceous and is often sterile. Conidia may be formed when infected barley pieces as placed on water agar and incubated under diurnal light conditions followed by a period of chilling.

Glomerella Graminicola

Glomerella graminicola (anamorph *Colletotrichum graminicola*) is the causal agent of anthracnose on cereals and grasses.

Identification

Stromata

- 70-300 µm in diameter;
- Bear prominent, dark, septate spines (setae) up to 100 xxm long.

Condia

- Developing at the base of the spines;
- Hyaline to pale yellow, unicellular, sickle-shaped, falcate to fusiform, tapered towards both ends;
- 3-5 x 19-29 µm.

Phialides

- Unicellular, hylanine and cylindrical,
- 4-8 x 8-20 µm.

Growth on PDA

- Gray and feltlike;
- Conida and appressoria are numerous when culture are well aerated, and sclerotia sometimes occur.
- Appressoria are diagnostic: they are tawny brown, irregular-shaped in edge, prominent, and terminal on thickened hyphae.

Leaf Rust (Barley)

Leaf rust is a fungal disease of barley caused by *Puccinia hordei*. It is also as brown rust and it is the most important rust disease on barley.

Symptoms: Pustules of leaf rust are small and circular and produce a mass of orange-brown powdery spores. They appear on the leaf sheaths and predominantly on the upper leaf surfaces. Heavily infected leaves did prematurely.

Disease Cycle: In many areas, leaf rust requires green volunteer barley to survive between cropping seasons. However, in areas where the alternate host, the weed, Star of Bethlehem (*Ornithogalum umbellatum*) occurs, teliospores are able to infect the plant. They then can undergo a sexual reproduction cycle and produce aceiospores which can·then re-infect barley.

Leaf rust develops rapidly between 15 and 22°C when moisture is not limiting. Following infection, new pustules and spores may be produced within 8 days. Urediniospores are wind-borne and can be spread long distances.

Teliospore development occurs later in the season either within uredial sori or within separate telial sori. The teliospores are thick-walled and can persist between crops. Upon germination, they a promycelium (basidium) with four haploid sporidia (basidiospores). The basidiospores can infect both barley and the alternate host.

In western Canada, where the rust cannot overwinter, and the alternate host is not present, the fungus may be reintroduced annually as the rust moves northward from the southern United States. In eastern Canada, the alternate host is present, however, infection of barley by aeciospores has not been reported.

Crop Losses: Leaf rust of barley is considered a relatively minor disease in the United States, however, sporadic outbreaks have occurred in the southeastern and mid-western regions of the country.

Pathotypes and Host Resistance: Most of the barley cultivars grown in the United States are susceptible to *Puccinia hordei*. Sixteen major resistance genes (*Rph*1 to *Rph*16) have been identified, but only three (*Rph*3, 7 and 9) have been deployed in commercial cutlivars worldwide. In the United States, the *Rph7* gene effectively controlled the disease for over twenty years. However, in 1993, pathotypes with virulence to the *Rph7* resistance gene were identified in Virginia, California and Pennsylvania.

Cochliobolus Sativus: Cochliobolus sativus is the teleomorph (sexual stage) of *Bipolaris sorokiniana* (anamorph) which is the causal agent of a wide variety of cereal diseases. The pathogen can infect and cause disease on the root, leaf and stem, and head tissue. C.sativus is extremely rare in nature and thus it is the asexual or anamorphic stage which causes infections. The two most common diseases caused by B. sorokiniana are spot blotch and common root rot, mainly on wheat and barley crops.

Identification

The mycelium of B.sorokiniana is usually deep olive-brown. New cultures produce abundant simple conidiophores, which may be single or clustered and measure 6-10 x 110-220 µm with septations. Conidia develop laterally from pores beneath each conidiophore setpum. Conidia are olive-brown and ovate to oblong, with rounded ends and a prominent basal scar. They measure 15-28 x 40-120 µm and are 3- to 10- septate. Some may be slightly curved. Their walls are smooth and noticeably thickened at the septa.

The sexual state (C. sativus), when formed in culture, is in the form of black, globose pseudothecia 300-400 µm in diameter, with erect beaks 50-200 µm long. Asci are clavate and measure 20-35 x 150-250 µm. Ascospores are hyaline, uniformly filamentous, and spirally flexed within asci. They measure 5-10 x 200-250 µm and are 4- to 10-septate.

Host Species

- *Agropyron cristatum,*
- *Allium sp.,*
- *Alopecurus pratensis,*
- *Aneurolepidium chinense,*
- *Avena sativa,*

- *Bromus inermis,*
- *B. marginatus,*
- *B. willdenowii,*
- *Calluna vulgaris,*
- *Chloris gayana,*
- *Cicer arietinum,*
- *Clinelymus dahuricus,*
- *C. sibiricus,*
- *Cynodon dactylon,*
- *C. transvaalensis,*
- *Dactylis glomerata,*
- *Echinochloa crus-galli,*
- *Elymus junceus,*
- *Festuca sp.,*
- *Guzmania sp.,*
- *Hordeum brevisubulatum,*
- *H. distichon,*
- *H. sativum var. hexastichon,*
- *H. vulgare,*
- *H. vulgare var. hexastichon*
- *Lablab purpureus,*
- *Linum usitatissimum,*
- *Lolium multiflorum,*
- *Pennisetum typhoides,*
- *Roegneria semicostata,*
- *Saccharum sp.,*
- *Secale cereale,*
- *Setaria italica,*
- *Sorghum sp.*
- *Taraxacum kok-saghyz,*
- *Trisetum aestivum,*
- *Triticum aestivum,*

- *T. secale,*
- *T. turgidum subsp. durum,*
- *T. vulgare"*
- *Zea mays,*

Fusarium Culmorum

Fusarium culmorum is the causal agent of seedling blight, foot rot, ear blight, stalk rot, and other diseases of cereals, grasses, and a wide variety of monocots and dicots.

Identification

Growth on PDA: Colonies grow rapidly. Aerial mycelium is whitish to yellow, tan or pale orange, but become brown to dark brown to red-brown with age.

Under alternating conditions of light and temperature, rings of spore masses may be formed by some isolates.

Microconidia

- Absent

Macroconidia

- *Abundance:* Usually abundant.
- *Sporodochia:* Orange to brown colour and relatively common
- *Morphology:* Thick and bluntly pointed at their apex. Conspicuously wider above the centre of the spore. The dorsal side is somewhat curved, but the ventral side is almost straight. Distinguishing character from *F. sambucinum* is the broader macroconidia.
- *Size:* Range from 4 to 7 µm wide and from 25 to 50 µm long. Number of septa: Usually 3- or 5-septate
- *Development:* Develop singly from phialides (5 x 15-20 µm). They are loose at first and are later aligned in sporodochia.

Chlamydospores

- *Abundance/Speed of Formation:* Usually abundant and form relatively quickly, requiring 3-5 weeks on CLA.
- *Location:* In hyphae and macroconidia. Those found in the macroconidia persist longer than those found in the hyphae under field conditions.
- *Morphology:* Thick-walled, globose.
- *Appearance:* Found singly, in clumps or chains.

Dutch Elm Disease

Dutch elm disease is a fungal disease of elm trees which is spread by the elm bark beetle. Although believed to be originally native to Asia, it has been accidentally introduced into America and Europe, where it has devastated native populations of elms which had not had the opportunity to evolve resistance to the disease. The name *Dutch elm disease* refers to the identification of the disease in the 1920s in the Netherlands; the disease is not specific to the Dutch Elm hybrid.

The causative agents are microfungi. Three strains are now recognised, *Ophiostoma ulmi,* which afflicted Europe in 1910, reaching North America on imported timber in 1928, *Ophiostoma himalulmi,* a strain endemic to the western Himalaya, and *Ophiostoma novoulmi,* the extremely virulent strain that has devastated Europe's elms since the late 1960s. *O. novoulmi* is now considered to be a hybrid between *O. ulmi* and *O. himalulmi.* The new strain was widely believed to have originated in China, but a comprehensive survey there in 1986 found no trace of it, although elm bark beetles were very common.

The disease is spread by two species of bark beetles (Family: Curculionidae, Subfamily: Scolytinae): the native elm bark beetle, *Hylurgopinus rufipes* , and the European elm bark beetle, *Scolytus multistriatus*. Both act as vectors for infection. The tree reacts to the presence of the fungus by plugging its own cambial tissue in an attempt to block the fungus from spreading further. As the area around cambium (the vascular tissue) is crucial for delivering nutrients and water to the rest of the plant, these plugs or tyloses prevent them from travelling up the trunk of the tree, eventually killing it. The first symptom of infection is usually an upper branch of the tree with leaves starting to wither and yellow in summer, months before the normal autumnal leaf shedding. This progressively spreads to the rest of the tree, with further dieback of branches. Eventually, the roots die, starved of nutrients from the leaves.

Often, not all the roots die: the roots may put up small suckers. These may grow up for some years into small elm trees, but after a decade or so the new trunks become large enough to support the bark beetles, and with their inevitable arrival the fungus returns, and the new tree dies.

Europe

Dutch elm disease was first noticed in Europe in 1910, and spread slowly, reaching Britain in 1927. This first strain was a relatively mild one, which only killed a small proportion of elms, more often just killing scattered branches, and had largely died-out by 1940. It was isolated in Holland in 1921 by Marie Beatrice Schwarz, a pioneering Dutch phytopathologist, and this discovery would lend the disease its name.

In about 1967, a new, far more virulent strain arrived in Britain on a shipment of Rock Elm logs from North America, and this strain proved both highly contagious and lethal to all of the European native elms; more than 25 million trees died in the UK alone. By 1990-2000, very few mature elms were left in Britain or much of northern Europe.

One of the most distinctive English countryside trees, the English Elm *U. procera* is particularly susceptible. Thirty years after the epidemic, these magnificent trees, which often grew to > 45 m high, are long gone. The species still survives in hedgerows, as the roots are not killed and send up root sprouts ("suckers"). These suckers rarely reach more than 5 m tall before succumbing to a new attack of the fungus. However, established hedges kept low by clipping have remained apparently healthy throughout the nearly 40 years since the onset of the disease in the UK.

The largest concentration of mature elm trees remaining in Britain is found in Brighton, where 15,000 elms still stand (2005 figures). Their survival is due to a concerted effort by local authorities to identify and remove infected sections of trees as soon as they show signs of the disease to save the tree and prevent it spreading.

United States

The disease was first reported in the United States in 1928, with the beetles believed to have arrived in a shipment of logs from the Netherlands destined for the Ohio furniture industry. The disease spread slowly from New England westward and southward, almost completely destroying the famous Elms in the 'Elm City' of New Haven, reaching the Detroit area in 1950[3], the Chicago area by 1960, and Minneapolis by 1970.

Canada

Dutch elm disease reached Eastern Canada during the Second World War, and spread to Manitoba in 1975 and Saskatchewan in 1981. In Toronto, Ontario, as much as 80 per cent of the elm trees have been lost to Dutch elm disease, and many more have fallen victim to the disease in Ottawa and Montreal and other cities during the 1970s and 1980s. Alberta and British Columbia are the only provinces that are currently free of Dutch elm disease, although an elm tree in southeastern Alberta was found diseased in 1998 and was immediately destroyed before the disease could spread any further. Thus, this was an isolated case. Today, Alberta has the largest number of elms unaffected by Dutch elm disease in the world. Aggressive measures are being taken to prevent the spread of the disease into Alberta as well as further progression of the disease in other parts of Canada. The City of Edmonton has banned elm pruning from March 31 to October 1, since fresh pruning wounds will attract the beetles during the warmer months.

Treatment

The first fungicide used for preventive treatment of Dutch elm disease was Lignasan BLP (carbendazim phosphate), which was introduced in the 1970s. This had to be injected into the base of the tree using specialised equipment, and was never especially effective. It is still sold under the name "Elm Fungicide". Arbotect (thiabendazole hypophosphite) became available some years later, and it has been proven effective. Arbotect must be injected every 2 to 3 years to provide ongoing control; the disease generally cannot be eradicated once a tree is infected.

Alamo (propiconazole) has become available more recently and shows some promise, though several university studies show it to be less effective than Arbotect treatments. Alamo is primarily recommended for treatment of Oak Wilt.

Treatment of diseased trees is costly and at best will prolong the life of the tree, perhaps by as many as five or ten years. It is usually only justified when a tree has unusual symbolic value or occupies a particularly important place in the landscape.

Resistant Trees

Research to select resistant cultivars and varieties began in the Netherlands in 1928, and in the USA since the disease became endemic there. Initial efforts in the Netherlands involved crossing varieties of *U. minor* and *U. glabra*, but later included the Himalayan or Kashmir Elm *U. wallichiana* as a source of anti-fungal genes. Early efforts in the USA involved the hybridisation of the Chinese Elm with the American Elm, and produced a resistant tree that lacked the beauty, traditional shape, and landscape value of the American Elm. Few were planted.

Three major groups of resistant cultivars are commercially available now:

- The Princeton Elm, a cultivar selected in 1922 by Princeton Nurseries for its landscape value. By happy coincidence, this cultivar was revealed to be highly resistant in inoculation studies carried out by USDA in the early 1990s, to Dutch elm disease. Because mature trees planted in the 1920s still remain, the properties of the mature plant are well known.

- The Liberty Elm, a set of five cultivars produced through selection over several generations starting in the 1970s. Marketed as a single variety, nurseries selling the "Liberty Elm" actually distribute the five cultivars at random. Two of the cultivars are covered by patents.
- The Valley Forge elm, and some related cultivars, have demonstrated resistance to Dutch elm disease approximately equal to that of the Princeton elm cultivar, in controlled USDA tests.

Even resistant cultivars can become infected, particularly if the tree is under stress from drought and other environmental conditions, and if the disease pressure is high. With the exception of the Princeton Elm, no trees have yet been grown to maturity. The oldest liberty elm was planted in about 1980, and the trees cannot be said to be mature until they have reached an age of sixty years.

The Dutch research programme ended in 1992, after raising two complex hybrids, later released as Columella and found to be actually immune to the disease when inoculated with unnaturally high doses of the fungus. In Italy, research is continuing at the Istituto per la Protezione delle Piante, Florence, to produce a wide range of disease-resistant trees using a variety of Asiatic species crossed with the early Dutch hybrid Plantyn as a safeguard against any future mutation of the disease. Two trees with very high levels of resistance, San Zanobi and Plinio, were released in 2003. Both feature the Siberian Elm *U. pumila* as the male parent.

The Red Elm *U. rubra* is less susceptible to Dutch elm disease than many elms, but this quality seems to have somehow largely evaded the attention of the resistance programme.

In 2001, English Elm was genetically engineered to resist disease in experiments at Abertay University, Dundee, by transferring anti-fungal genes into the elm genome using minute DNA-coated ball bearings. However, there are no plans to release the trees into the countryside.

Trees in the genus *Zelkova,* closely related to elms, are also planted as resistant substitutes for susceptible elms. *Zelkova serrata,* the Japanese Zelkova, the most commonly planted *Zelkova* tree, is similar to the American Elm in size and the vase-shaped crown.

Prion

A prion — short for proteinaceous infectious particle that lacks nucleic acid (by analogy to virion) — is a type of infectious agent composed only of protein. They cause a number of diseases in a variety of animals, including bovine spongiform encephalopathy (BSE, also known as mad cow disease) in cattle and Creutzfeldt-Jakob disease in humans. All prion diseases affect the structure of the brain or other neural tissue, and all are untreatable and fatal.

Prions are believed to infect and propagate by refolding abnormally into a structure which is able to convert normal molecules of the protein into the abnormally structured form. However, the term in itself does not preclude other mechanisms of transmission.

All known prions induce the formation of an amyloid fold, in which the protein polymerises into a fibre with a core consisting of tightly packed beta sheets. Other mechanisms may exist in yet undiscovered infectious protein particles. This altered structure renders them quite resistant to denaturation by chemical and physical agents, although infectivity can be reduced by these treatments, making disposal and containment of these particles difficult.

Proteins showing prion behaviour are also found in some fungi. Some fungal prions may not be associated with any disease; it is unknown whether these prions represent an evolutionary advantage for their hosts.

Discovery: Radiation biologist Tikvah Alper and physicist J.S. Griffith developed the theory in the 1960s that some TSEs are caused by an infectious agent made solely of protein. This theory was developed to explain the discovery that the mysterious infectious agent causing the diseases scrapie and Creutzfeldt-Jakob Disease resisted ultraviolet radiation (which breaks down nucleic acids - present in viruses and all living things) yet responded to agents that disrupt proteins.

Stanley B. Prusiner of the University of California, San Francisco announced in 1982 that his team had purified infectious material and that the infectious agent consisted mainly of a specific protein, although they had not managed to satisfactory isolate the protein until two years after making his announcement.

Prusiner coined the word "prion" as a name for the infectious agent, by combining the first two syllables of the words *proteinaceous* and *infectious*. While the infectious agent was named a prion, the specific protein that the prion was made of was named PrP, an abbreviation for "protease-resistant protein". Prusiner was awarded the Nobel Prize in Physiology or Medicine in 1997 for his research into prions.

Prions cause neurodegenerative disease by aggregating extracellularly within the central nervous system to form plaques known as amyloids, which disrupt the normal tissue structure. This disruption is characterised by holes in the tissue with resultant spongy architecture due to the loss of neurons. Other histological changes include astrogliosis and the absence of an inflammatory reaction. While the incubation period for prion diseases is generally quite long, once symptoms appear the disease progresses rapidly, leading to brain damage and death. Neurodegenerative symptoms can include convulsions, dementia, ataxia (balance and coordination dysfunction), and behavioural or personality changes.

All known prion diseases, collectively called *transmissible spongiform encephalopathies* (TSEs), are untreatable and fatal. However, a vaccine has been developed in mice that may provide insight into providing a vaccine in humans to resist prion infections. Additionally, in 2006 scientists announced that they had genetically engineered cattle lacking a necessary gene for prion production–thus theoretically making them immune to BSE, building on research indicating that mice lacking normally-occurring prion protein are resistant to infection by scrapie prion protein.

Prions are able to affect a variety of different species, however the prions involved are somewhat species-specific: they are similar but not identical. However overlap may occur; the human prion disease *variant Creutzfeldt-Jakob disease* is believed to be caused by a prion which typically infects cattle and is transmitted through infected meat.

Metal ion interactions with prion proteins might be relevant to the progression of prion-mediated disease, based on epidemiological studies of clusters of prion disease in locales with low soil concentrations of copper.

The following diseases are believed to be caused by prions.

- In animals:
  - Scrapie in sheep.
  - Bovine spongiform encephalopathy (BSE) in cattle (known as *mad cow disease*).
  - Transmissible mink encephalopathy (TME) in mink.
  - Chronic wasting disease (CWD) in elk and mule deer.
  - Feline spongiform encephalopathy in cats.
  - Exotic ungulate encephalopathy (EUE) in nyala, oryx and greater kudu.
- In humans:
  - Creutzfeldt-Jakob disease (CJD) and its varieties: iatrogenic Creutzfeldt-Jakob disease (iCJD), variant

Creutzfeldt-Jakob disease (vCJD), familial Creutzfeldt-Jakob disease (fCJD), and sporadic Creutzfeldt-Jakob disease (sCJD).

— Gerstmann-Straussler-Scheinker syndrome (GSS).

— Fatal familial insomnia (fFI).

— Sporadic fatal insomnia (sFI).

— Kuru.

— Alpers syndrome.

Isoforms: The protein that prions are made of is found throughout the body, even in healthy people and animals. However, the prion protein found in infectious material has a different structure and is resistant to proteases, the enzymes in the body that can normally break down proteins. The normal form of the protein is called PrP^C, while the infectious form is called PrP^{Sc} — the C refers to 'cellular' or 'common' PrP, while the Sc refers to 'scrapie', a prion disease occurring in sheep. While PrP^C is structurally well-defined, PrP^{Sc} is certainly polydisperse and defined at a relatively poor level. PrP can be induced to fold into other more-or-less well-defined isoforms in vitro, and their relationship to the form(s) that are pathogenic in vivo is not yet clear.

PrP^C: PrP^C is a normal protein found on the membranes of cells. Several topological forms of it exist; one cell surface form anchored via glycolipid and two transmembrane forms, however its function has not been fully resolved. PrP^C is readily digested by proteinase K and can be liberated from the cell surface by the enzyme *phosphatidyl inositol-specific phospholipase C*, which cleaves the *phosphatidyl inositol* glycolipid anchor. A typical yeast prion protein contains a core region (domain) with many repeats of the amino acids glutamine and asparagine. Normal yeast prion domains are flexible and lack a defined structure.

PrP^{Sc}: The infectious isoform of PrP^C, known as PrP^{Sc}, is able to catalyse the formation of other normal PrP^C proteins into the infectious isoform by changing their conformation.

Although the exact 3D structure of PrP^{Sc} is not known, there is increased α-sheet content in the diseased form of the molecule, replacing normal areas of α-helix.

Aggregations of these abnormal isoforms forms a highly structured amyloid fibre. The end of the fibre acts as a template for the free protein molecules, causing the fibre to grow. Small differences in the amino acid sequence of prion-forming regions lead to distinct structural features on the surface of prion fibres. As a result, only free protein molecules that are identical in amino acid sequence to the prion protein can be recruited into the growing fibre. The mammalian prion proteins do not resemble the prion proteins of yeast in their amino acid sequence, however, they are still known as PrP^C and PrP^{Sc} and share basic structural features.

Fungal Prions: Fungal prions have been investigated, leading to a deeper understanding of disease-forming mammalian prions.

Prion-like proteins are found naturally in some plants and non-mammalian animals. Some of these are not associated with any disease state and may possibly even have a useful role. [1]. Because of this, scientists reasoned that such proteins could give some sort of evolutionary advantage to their host. This was suggested to be the case in a species of fungus, *Podospora anserina*.

Genetically compatible colonies of this fungus can merge together and share cellular contents such as nutrients and cytoplasm. A natural system of protective "incompatibility" proteins exists to prevent promiscuous sharing between unrelated colonies. One such protein, called HET-S, adopts a prion-like form in order to function properly. The prion form of HET-S spreads rapidly throughout the cellular network of a colony and can convert the non-prion form of the protein to a prion state after compatible colonies have merged. However, when an incompatible colony tries to merge with a prion-containing colony, the prion causes the "invader" cells to die, ensuring that only related colonies obtain the benefit of sharing resources.

Sup35p and Ure2p: In 1965, Brian Cox, a geneticist working with the yeast *Saccharomyces cerevisiae*, described a genetic trait (termed PSI+) with an unusual pattern of inheritance. Despite many years of effort, Cox could not identify a conventional mutation that was responsible for the PSI+ trait. In 1994, yeast geneticist Reed Wickner correctly hypothesised that PSI+ as well as another mysterious heritable trait, URE3, resulted from prion forms of certain normal cellular proteins. It was soon noticed that heat shock proteins (which help other proteins fold properly) were intimately tied to the inheritance and transmission of PSI+ and many other yeast prions. Since then, researchers have unravelled how the proteins that code for PSI+ and URE3 can convert between prion and non-prion forms, as well as the consequences of having intracellular prions. When exposed to certain adverse conditions, PSI+ cells actually fare better than their prion-free siblings; this finding suggests that, in some proteins, the ability to adopt a prion form may result from positive evolutionary selection. It has been speculated that the ability to convert between prion infected and prion-free forms enables yeast to quickly and reversibly adapt in variable environments. Nevertheless, Wickner maintains that URE3 and PSI+ are diseases.

Classification

Fungal Prions

Protein	*Natural Host*	*Normal Function Phenotype*	*Prion State*	*Prion*
Ure2p	Saccharomyces cerevisiae	Nitrogen catabolite repressor	[URE3]	Growth on poor nitrogen sources
Sup35p	Saccharomyces cerevisiae	Translation termination factor	[PSI+]	Increased levels of nonsense suppression
Rnq1p	Saccharomyces cerevisiae	Protein template factor	[RNQ+]	Promotes aggregation of other prions
HET-S	Podospora anserina	Regulates heterokaryon incompatibility	[Het-s]	Heterokaryon formation between incompatible strains

As of 2003, the following proteins in Saccharomyces cerevisiae had been identified or postulated as prions:

- Sup35p, forming the [PSI+] element;
- Ure2p, forming the [URE3] element;
- Rnq1p, forming the [RNQ+] element (also known as [PIN+]);
- A fifth prion protein, forming the [ISP+] element remains to be identified.

Spot Blotch (Barley)

Spot blotch is a disease of barley caused by *Cochliobolus sativus*. The disease is found everywhere that barley is grown, but only significant yield losses in warm, humid climates.

Symptoms: Infections appear as dark, chocolate-coloured blotches. The spots merge, eventually forming irregular dead patches on the leaves. Leaf spots may be surrounded by a zone of yellow leaf tissue of varying width. Spot may also appear on the leaf sheaths, necks and heads of the plant. Heavily infected leaves dry up completely, and infections on the flag leaf during kernel filling are the most serious.

Disease Cycle: The fungus overwinters in barley straw and stubble, in the soil, or on the seed. Spores are produced on barley debris in the spring and are dispersed by wind and rain. Barley seedlings may become infected from incolum on the seed or in the soil. Temperatures above 20°C and moist humid conditions within the crop canopy favour spot blotch development. Conducive weather conditions may favour successive production of new spores and lesions leading to rapid disease development during the growing season.

Yield Loss

The disease may be particularly damaging in the Upper Midwest of the United States. Yield losses of 10-30 per cent may be occur when weather conditions are conducive to disease development.

Management: The disease is managed used resistant varieties, clean seed, seed treatments, foliar fungicide and rotation to non-cereal crops.

Scald (Barley)

Scald is common disease of barley in temperate regions. It is caused by the fungus Rhynchosporium secalis and can cause significant yield losses in cooler, wet seasons.

Symptoms: Scald is a foliar disease of barley affecting the leaves and sheaths of the plant, however, lesions may also occur on coleoptiles, glumes, floral bracts and awns.

Initial sympotms are oval, water-soaked, greyish-green spots, 1.0-1.5 cm long. As the disease develops, the centres of the lesions dry and bleach, becoming light gray, tan, or white with a dark brown margin. The lesions are not delimited by the leaf veins and often coalesce.

Disease Cycle: The fungus can infected and survive in barley seed. It exists as mycelium in the pericarp and hull of infected seeds. Infection of the coleoptile occurs as it emerges from the embryo. Optimal infections occurs at soil temperatures of 16°C. At soil temperatures of 22°C or higher, very little infection occurs.

In spring cropping systems, the fungus overwinters on the crop debris and stubble of previous diseased barley crops. The fungus produces abundant conidia on wet lesions during cool, damp weather after the leaf tissue has become necrotic. Conidia, spread by wind and splashing rain, infect young leaves of spring-planted grain. Optimum temperatures for sporulation and infection range from 10-18°C. Hot, dry weather reduces the rate of disease development.

Geographical Distribution and Economic Importance: The disease is economically important barley disease in Europe, North America and Australia. It is been reported from South America, Africa, the Middle east, Japan and Korea. Yield losses as high as 35-40 per cent have been reported, however, losses of 1-10 per cent are more common. Yield loss is primarily due to reduced kernel weight, but both kernels per head and number of heads per plant may also been affected.

In the wetter areas of the United Kingdom, scald is the most damaging disease of barley, affecting both spring and winter crops. In the Victoria area of Australia, scald is widespread in barley crops in most seasons, but its severity varies greatly from crop to crop and between seasons. In Victoria, scald causes annual average yield losses of 10-20 per cent, with individual losses as high as 45 per cent in susceptible varieties. In Canada, yield losses in Alberta have been calculated at 2.4 per cent, although losses in particular fields may exceed 25 per cent. Losses are due to a decrease in photosynthetic area on the flag and second leaves resulting in reduced seed weight.

Management: Management of the disease involves the use of clean and/or treated seeds, resistant cultivars, crop rotation, residue management, and foliar fungicides.

Resistant Cultivars

Cultivars with scald resistance are available in all major barley growing areas. The level of resistance occurs along a spectrum and the degree of resistance that is required to effectively control the disease will depend on the region where it is grown, cropping practices that reduce initial disease inoculum, wetter conditions and fungal pathotypes.

Fungicides

Foliar fungicides can be used to effectively control disease development. Product selection, application rates and timing, depend upon numerous factors. The availability of fungicides as a management tool depends on whether the product has been evaluated and registered for use in a specific country or region.

Foliar Fungicides: Azoxystrobin, bromuconazole, cyproconazole, epoxiconazole, fluquinconazole, flusilazole, propiconazole, prochloraz, pyraclostrobin, tebuconazole.

Seed Treatment Fungicides: triadimenol

1. *EPPO:* Europe
2. *Canada:* Alberta
3. *US:* Oregon

Fungicide Resistance

In the United Kingdom, there are issues concerning MBC fungicides and some of the triazole fungicides (e.g. flusilazole and epoxiconazole). To date there are no issues with strobilruin fungicides (QoI fungicides), but the situation is being monitored closely. There are currently no issues concerning chlorothalonil, cyprodinil or the morpholines.

Linear Immunoglobulin

Linear IgA dermatosis is a very rare autoimmune blistering disorder, which manifests via blistering of the skin and the mucous membranes. An autoimmune disorder is a malfunction of the body's immune system causing the body to attack its own tissues.

Linear immunoglobulin A (IgA) dermatosis is also referred to as Linear IgA Bullous Disease (LABD or LAD). The childhood version is called Linear IgA Bullous Disease of Childhood (LABDC), and is immunologically identical to the adult disease.

Drug-induced cases resolve once the causative drug is withdrawn. The most common causative agent for LAD is Vancomycin. Seventeen case reports have implicated vancomycin in LAD. Of all reported causative drugs, it is the best-documented agent in the literature. Other potential triggers and their respective number of published case reports (listed in parentheses) include the following:

- Amiodarone (2)
- Ampicillin sodium (1)
- Captopril (2)
- Cefamandole nafate (1)

- Cyclosporine (1)
- Depot sulphonamide (1)
- Diclofenac (3)
- Glibenclamide (1)
- Interferon gamma and interleukin (2)
- Iodine contrast agent (1)
- Lithium carbonate (1)
- Penicillin sodium (2)
- Phenytoin sodium (2)
- Somatostatin (1)
- Sulphamethoxazole/trimethoprim (1)
- Sulphisoxazole (1)
- Topical sodium hypochlorite (1)
- Vigabatrin (1)

Symptoms

Onset may be slow and insidious but is more usually sudden. Symptoms vary from mild to severe pruritis (itching), accompanied by a burning sensation, and blisters. Contact from such articles as clothes, labels, shoes, waistbands and spectacles can cause blistering. Surgery and medical procedures also excite blistering, for example, use of airway apparatus during anaesthesia, skin biopsies, dental work or pap smears. As with most autoimmune diseases, patients report tiredness, with some patients also reporting fevers and digestive problems.

Presentation

Clinically, LAD has a heterogeneous (mixed) presentation, the lesions comprising urticated plaques, papules and vesicles and blisters. Blisters can arise from normal skin and from urticated plaques. Blisters appear in two distinct ways, either in a linear presentation like a string of beads, or in a ring. Once the first blister has scabbed over, a second ring of blisters will form around it, forming the "crown of jewels" effect. Blisters can vary in size from small (2mm in diameter) to much larger (up to about 5 cm). They can be tense and full of clear fluid, or reddish lumps.

LAD affects all types of skin including mucous membranes. Mucosal involvement is common, causing ulcers and erosions and can involve the mouth, gums, (thereby damaging teeth), pharynx, larynx, nose, rectum, vagina, oesophagus and eyes.

Lesions in mucous membranes heal with scarring and therefore pose considerable morbidity, as scarring in this type of tissue continues to grow, and can cause serious problems if in the throat or gullet, and if in the eyes, blindness. Cutaneous lesions normally heal without scarring.

Hoarseness is an indication of throat involvement; stuffiness, crustiness and bleeding an indication of nasal involvement. Involvement of the eyes is different and far more serious, in that changes, such as fine scarring, may be present in the absence of ocular complaints and can lead to blindness. The progression of ocular LAD can prove to be rapid and is indistinguishable from Cicatricial Pemphigoid.

Similar bullous diseases are Cicatricial Pemphigoid, Bullous Pemphigoid, Pemphigus and Dermatitis Herpetiformis.

Diagnosis

LAD is diagnosed by immunofluorescence shone onto a salt split skin biopsy. The distinguishing feature of the biopsy is a linear deposit of IgA in the Base Membrane Zone (BMZ).

Duration

The mean duration of idiopathic LAD of childhood is 3.9 years, ranging from 2.1-7.9 years, and remission has been reported to occur in approximately 64 per cent of these cases within 2 years.

LAD in adults can commence at any age from post puberty to the ninth decade, most commonly occurring after the age of 60 years. It can last for anywhere between 1-15years, and the remission rate is much less than in children, around 48 per cent. The disease tends to wax and wane in severity and is totally unpredictable.

Autoimmune Disease

The immune system is the body's secret army, protecting the body from invasion from bacteria, viruses, fungi, etc. All the cells in the immune system come from the same basic stem cells in the bone marrow, and they in turn produce white blood cells called lymphocytes. These fall into two categories: B lymphocytes and T lymphocytes. It is the B lymphocytes that are involved in this disease.

B Lymphocytes are made in the bone marrow and produce antibodies, also called immunoglobulins. These antibodies are large proteins, not cells. Invading bacteria and virus cells have proteins on their surfaces (as do all cells) called antigens, and antibodies can recognise intruders and latch onto them and destroy them.

There are five groups of antibodies: IgG, IgA, IgM, IgE and IgD, with IgG and IgM being most involved in fighting infection. It is the IgA antibody in LAD that is abnormal.

Every line of defence is heavily reliant on good communication between its commanders and officers, and the immune system is no exception. It has chemical messengers known as cytokines, which tell the immune cells when they need to attack and when they may leave.

When the immune system malfunctions and attacks its own body tissues, it produces abnormal antibodies, called autoantibodies. This response is called an autoimmune reaction, which results in inflammation and tissue damage. In the case of LAD, the IgA autoantibody attacks a normal protein (found in all human skin), which is in the lower layer of the epidermis, (the outer layer of skin) known as the BMZ. These proteins are important in maintaining the integrity of skin, and when the autoantibody binds with this protein it causes the cells to come apart (acantholysis) leading to fluid accumulation between layers of skin and forms blisters, which are the manifestation of the disease. The target antigen for LAD is as yet unidentified.

The exact mechanisms causing these changes are not fully understood, but scientists are working to understand the events that precede the production of the antibody, and what causes the IgA antibody to be abnormal. The cells that control antibody production, (B lymphocytes–a type of white blood cell) may malfunction and produce this IgA antibody abnormally (now referred to as an IgA autoantibody).

It is accepted that individuals having this disease possess a genetic make-up that renders them susceptible to the disease, and it is theorised that the inflammation initiated by bacteria, viruses, toxins, some drugs and other environmental agents may play a role in provoking or triggering the autoimmune response, in someone who is genetically predisposed to develop such a disorder.

Treatment

Linear IGA disease cannot be cured or made to "disappear" with treatment. Treatment is a drug regime designed to suppress the immune system.

Specialised Treatment is Required for all Aspects of this Disease: Dermatologists who specialise in bullous diseases, pathology blood reports to monitor drug side effects, dentists, endocrinologists and, in particular, ophthalmologists, all of whom monitor the patient closely and at regular intervals. A general practitioner is usually the coordinator of the treatment regime, referring the patient to various specialists and monitoring side effects of the drug regime.

The goals of treatment are control of the autoimmune reaction by balancing the suppression of the immune system, reducing the symptoms of the autoimmune process to self-tissues, while maintaining the body's normal immune response to infection.

The mainstay of the treatment regime is corticosteroids and dapsone or sulphasalazine. If this is not satisfactory, corticosteroids are added to a long-term immunosuppressant and, sometimes, an antibiotic.

Corticosteroids such as prednisolone or prednisone are used as short term immunosuppressants and anti-inflammatories, at the start of treatment, or if symptoms become worse. Once the long-term immunosuppressants take effect, corticosteroid doses can be reduced. It is seen as desirable to reduce the prednisone as it has more toxic side effects than the long term immunosuppressants.

Dapsone is usually combined with the corticosteroids. Sulphasalazine is sometimes used as an alternative to dapsone.

Some long-term immunosuppressant drugs used in the treatment of LAD are Dapsone, Cyclophosphamide, Cyclosporine, Azathioprine, or Methotrexate. These are usually administered in conjunction with prednisone. An additional antibiotic may also be prescribed. The antibiotics are used for their anti-inflammatory effects.

Intravenous immunoglobulins may also be used in treatment.

Drugs used to control the autoimmune reaction (immunosuppressants) interfere with the body's ability to fight disease, the drugs suppressing not only the autoimmune reaction but also the body's ability to defend itself from microorganisms that cause infection and cancer cells. All of these drugs have side effects, which in themselves may require additional treatment, such as calcium supplementation to reduce bone thinning, Omeprazole (Acimax) to prevent possible stomach ulcers and/or bleeding from corticosteroids, and Vitamin E 800mg to reduce dapsone-induced headache. Anti-fungal medication may also be necessary as immunosuppressed patients are vulnerable to fungal infections.

Topical steroids are usually prescribed to relieve itching and to dry up sores left by blisters. Wet dressing treatment using topical steroids may be used to control bad outbreaks of lesions. This treatment increases absorption of the steroid cream to lower levels of the skin and helps prevent blister formation.

Patients on these drug regimes require close monitoring via regular blood tests, blood pressure monitoring and weight monitoring.

Neotyphodium

Neotyphodium is a form genus containing species of endophytic fungi. These endophytes are asexual, seed-borne symbionts of cool-season grasses, and grow intercellularly throughout the aerial tissues of their hosts, including shoot apical meristems, leaf sheaths and blades, inflorescences, seeds and embryos.

Taxonomic Considerations

Neotyphodium species are closely related to teleomorphic species of the genus *Epichloe*, from which many have evolved by processes involving interspecific hybridisation. Molecular phylogenetic evidence demonstrates that asexual *Neotyphodium* species are derived either from individual *Epichloe* species, or more commonly, from hybrids with at least two ancestral *Epichloe* species. Hence, the form genus *Neotyphodium* is very closely associated with the teleomorphic genus *Epichloe*. In keeping with the code of botanical nomenclature, the form genus refers to the asexual spore or vegetative state, and the teleomorphic genus refers to the sexual state.

Life Cycle

The taxonomic dichotomy is especially interesting in this group of symbionts, because vegetative propagation of fungal mycelium occurs by vertical transmission, i.e. fungal growth into newly developing host tillers (=individual grass plants). Importantly, all *Neotyphodium* and many *Epichloe* species infect new grass plants solely by growing into the seeds of their grass hosts, and infecting the growing seedling. Manifestation of the sexual state–which only occurs in *Epichloe* fungi — causes choke disease, a condition in which grass inflorescences are engulfed by rapid fungal outgrowth forming a stroma. The fungal stroma suppresses host seed production and culminates in the ejection of meiospores (ascospores) that mediate horizontal (contagious) transmission of the fungus to new plants.

Effects on the Grass Plant and on Herbivores

It has been proposed that vertically transmitted symbionts should evolve to be mutualists since their reproductive fitness is intimately tied to that of their hosts. In fact, some positive effects of *Neotyphodium* species on their host plants include increased growth, drought tolerance, and herbivore and pathogen resistance.

Resistance against herbivores has been attributed to endophyte-produced alkaloids. Although grass-endophyte symbioses have been widely recognised to be mutualistic in many wild and cultivated grasses, the interactions can be highly variable and sometimes antagonistic, especially under nutrient-poor conditions in the soil.

Neotyphodium Alkaloids

Ergoline alkaloids (which are ergot alkaloids, named after the ergot fungus, *Claviceps purpurea*, a close relative of the *Neotyphodium/Epichloe* endophytes) are characterised by a ring system derived from 4-prenyl tryptophan. Among the most abundant ergot alkaloids in endophyte-symbiotic grasses is ergovaline, comprising an ergoline moiety attached to a bicyclic tripeptide containing the amino acids, L-proline, L-alanine, and L-valine. Another group of endophyte alkaloids are the indole-diterpenoids, such as lolitrem B. Both the ergoline and indole-diterpenoid alkaloids have biological activity against mammalian herbivores, and also activity against some insects. Peramine is a pyrrolopyrazine alkaloid thought to be biosynthesised from the guanidinium-group-containing amino acid, L-arginine, and pyrrolidine-5-carboxylate, a precursor of L-proline and is an insect-feeding deterrent.

The loline alkaloids are 1-aminopyrrolizidines with an oxygen atom linking bridgehead carbons 2 and 7, and are biosynthesised from the amino acids, L-proline and L-homoserine. The lolines have insecticidal and insect-deterrent activities comparable to nicotine. Many, but not all, *Neotyphodium* species produce up to three classes of these alkaloids.

Species

- *Neotyphodium aotearoae*
- *Neotyphodium australiense*
- *Neotyphodium chilense*
- *Neotyphodium chisosum*
- *Neotyphodium coenophialum*
- *Neotyphodium gansuense*
- *Neotyphodium huerfanum*
- *Neotyphodium lolii*
- *Neotyphodium melicicola*
- *Neotyphodium occultans*
- *Neotyphodium siegelii*
- *Neotyphodium starrii*
- *Neotyphodium tembladerae*
- *Neotyphodium typhinum*
- *Neotyphodium uncinatum*

Aspergillus Niger

Aspergillus niger is a fungus and one of the most common species of the genus *Aspergillus*. It causes a disease called black mould on certain fruits and vegetables such as grapes, onions, and peanuts, and is a common contaminant of food.

It is ubiquitous in soil and is commonly reported from indoor environments, where its black colonies can be confused with those of *Stachybotrys* (species of which have also been called "black mould"). Some strains of *A. niger* have been reported to produce potent mycotoxins called ochratoxins, but other sources disagree, claiming the latter report is based upon misidentification of the fungal species. Recent evidence suggests some true *A. niger* strains do produce ochratoxin A.

Taxonomy

A. niger is included in *Aspergillus* subgenus *Circumdati,* section *Nigri.* The section *Nigri* includes 15 related black-spored species that may be confused with *A. niger,* including *A. tubingensis, A. foetidus, A. carbonarius,* and *A. awamori.* A number of morphologically similar species were recently described by Samson *et al.*

Pathogenicity

Plant Disease: *A. niger* causes black mould of onions. Infection of onion seedlings by *A. niger* can become systemic, manifesting only when conditions are conducive. *A. niger* causes a common post-harvest disease of onions, in which the black conidia can be observed between the scales of the bulb. The fungus also causes disease in peanuts and in grapes. The spore comes to common trees such as maple.

Human and Animal Disease: *A. niger* is less likely to cause human disease than some other *Aspergillus* species, but if large amounts of spores are inhaled, a serious lung disease, aspergillosis can occur. Aspergillosis is particularly frequent among horticultural workers who inhale peat dust, which can be rich in *Aspergillus* spores. Less commonly, it has been found on the walls of ancient Egyptian tombs and can be inhaled when the area is disturbed. *A. niger* is one of the most common causes of otomycosis (fungal ear infections), which can cause pain, temporary hearing loss and, in severe cases, damage to the ear canal and tympanic membrane.

Industrial Uses

A. niger is cultured for the industrial production of many substances. Various strains of *A. niger* are used in the industrial preparation of citric acid (E330) and gluconic acid (E574) and have been assessed as acceptable for daily intake by the World Health Organisation.

Many useful enzymes are produced using industrial fermentation of *A. niger*. For example, *A. niger* glucoamylase is used in the production of high fructose corn syrup, and pectinases are used in cider and wine clarification. α -galactosidase, an enzyme that breaks down certain complex sugars, is a component of Beano and other medications which the manufacturers claim can decrease flatulence. Another use for *A. niger* within the biotechnology industry is in the production of magnetic isotope-containing variants of biological macromolecules for NMR analysis.

Other Uses

In 2006, it was reported that a secreted RNase produced by *A. niger* called *actibind* has anti-angiogenic and anti-carcinogenic characteristics.

A. niger is also cultured for the extraction of the enzyme glucose oxidase (GO). This enzyme is used in the design of glucose biosensors, due to its high affinity for a-D-glucose.

Genetics

The genomes of two different *A. niger* strains have been fully sequenced.

Common Root Rot (Barley)

Common root rot is a fungal disease of barley caused by *Cochliobolus sativus, Fusarium culmorum* and *F. graminearum.*

Symptoms: Initial symptoms appear as small brown spots on the crown roots, the lower sheath, and the subcrown internode. This spots will elongate and coalesce and can cause an extensive brown discolouration at the crown, and the entire length of the subcrown internode. Healthy root tissue is white to slightly cream-coloured.

Disease Cycle: In western Canada, infections occur mainly from soilborne spores, whereas in eastern Canada, seed-borne inoculum is more important. The spores can survive for many years, and germinate in the presence of susceptible plants. At infected plants reach maturity, abundant sporulation occurs.

Management: Disease inoculum is reduced by long rotations with non-host crops. Shallow seeding and optimum fertilization levels, especially of phosphate and potassium, reduce disease severity. Resistance cultivars are also available.

Powdery Mildew (Barley)

Powdery mildew is a fungal disease of barley caused by *Blumeria graminis* sp. *hordei*. The disease has a worldwide distribution and is most damaging in cool, wet climates. The host range of the form species *hordei* is restricted to barley and other *Hordeum species*.

Symptoms: At first, powdery mildew can be observed as small greyish patches of fluffy fungal growth (mycelium) on the upper surface of the lower leaves. These spots resemble small cushions of white powder. Leaf tissue on the opposite side of an infected leaf turns pale green to yellow. The fungus only infects the epidermal layer and can be easily scrapped off with a fingernail. Infections can also occur on the leaf sheaths and ears. Leaves remain green and active for some time following infection, then gradually become chlorotic and die off. As the disease progresses, the mycelium often becomes dotted with minute black points (cleistothecia), which are the sexual fruiting bodies of the fungus.

Disease Cycle

The fungus ovewinters as cleistothecia on straw, and in milder climates, also as mycelium and conidia on stubble and straw or volunteer barley and certain grasses. Windborne ascospores or conidia are the primary inoculum and can be dispersed over considerable distances. Infection by conidia requires high humidity, but not free water on the leaf surface. Sporulation and spore dispersal are favoured by drier conditions. Thus the disease does well under alternating wet and dry conditions. Production of conidia declines markedly as the colony ages. Cleistothecia develop on older leaves as the plant matures. Low temperatures, together with the wetting of the cleistothecia for at least 72 hours, induces the maturations of the ascospores. Ascospores are released following rains, but are relatively sparse in comparison to the condia.

Phycomycosis

Phycomycosis is an uncommon condition of the gastrointestinal tract and skin most commonly found in dogs and horses. The condition is caused by a variety of moulds and fungi, and individual forms include pythiosis, zygomycosis, and lagenidiosis. Pythiosis is the most common type and is caused by *Pythium*, a type of water mould. Zygomycosis can also be caused by two types of zygomycetes, *Entomophthorales* (such as *Basidiobolus* and *Conidiobolus*) and *Mucorales* (such as *Mucor, Mortierella, Absidia, Rhizopus, Rhizomucor*, and *Saksenaea*). The latter type of zygomycosis is also referred to as mucormycosis. Lagenidiosis is caused by a *Lagenidium* species, which like *Pythium* is a water mould. Since both pythiosis and lagenidiosis are caused by organisms from the class Oomycetes, they are sometimes collectively referred to as oomycosis.

Pythiosis

Pythiosis is caused by *Pythium insidiosum* and occurs most commonly in dogs and horses, but is also found in cats, cattle, and humans. In the United States it is most commonly found in the Gulf states, especially Louisiana, but has been found in midwest and eastern states. It is also found in South East Asia, eastern Australia, New Zealand, and South America. Pythiosis occurs in areas with mild winters due to the organism surviving in standing water that does not reach freezing temperatures. *Pythium* occupies swamps in late summer and infects dogs who drink water containing it. The disease is typically found in young, large breed dogs.

It is suspected that pythiosis is caused by invasion of the organism into wounds, either in the skin or in the gastrointestinal tract. The disease grows slowly in the stomach and small intestine, eventually forming large lumps of granulation tissue. It can also invade surrounding lymph nodes. Symptoms include vomiting, diarrhoea, depression, weight loss, and a mass in the abdomen. Pythiosis of the skin in dogs is very rare, and appears as ulcerated lumps. Primary infection can also occur in the bones and lungs.

In horses, subcutaneous pythiosis is the most common form and infection occurs through a wound while standing in water containing the pathogen. The disease is also known as leeches, swamp cancer, and bursatti. Lesions are most commonly found on the lower limbs, abdomen, chest, and genitals. They are granulomatous and itchy, and may be ulcerated or fistulated. The lesions often contain yellow, firm masses of dead tissue known as kunkers. It is possible with chronic infection for the disease to spread to underlying bone.

In humans it can cause arteritis, keratitis, and periorbital cellulitis. In cats pythioisis is almost always confined to the skin as hairless and edematous lesions. It is usually found on the limbs, perineum, and at the base of the tail. Lesions may also develop in the nasopharynx.

Pythium insidiosum is different from other members of the genus in that human and horse hair, skin, and decaying animal tissue are chemoattractants for its zoospores, in addition to decaying plant tissue.

Zygomycosis

Zygomycosis usually is a disease of the skin, but can also occur in the sinuses or gastrointestinal tract. In humans it is most prevalent in immunocompromised patients (HIV/AIDS, the elderly, SCID, etc.) and patients in acidosis (diabetes, burns), particularly after barrier injury to the skin or mucus membranes. Zygomycosis caused by *Mucorales* causes a rapidly progressing disease of sudden onset in sick or immunocompromised animals. *Entomophthorales* cause chronic, local infections in otherwise healthy animals. The important species that cause entomophthoromycosis are *Conidiobolus coronatus, C. incongruous,* and *Basidiobolus ranarum*. *Conidiobolus* infections of the upper respiratory system have been reported in humans, sheep, horses, and dogs, and *Basidiobolus* has been reported less commonly in humans and dogs. Horses are one of the most common domestic animals to be affected by entomophthoromycosis. *C. coronatus* causes lesions in the nasal and oral mucosa of horses that may cause nasal discharge or difficulty breathing. *B. ranarum* causes large circular nodules on the upper body and neck of horses. *Entomophthorales* is found in soil and decaying plant matter, and specifically *Basidiobolus* can be contracted from insects and the faeces of reptiles or amphibians.

Zygomycosis of the sinuses can extend from the sinuses into the orbit and the cranial vault, leading to rhinocerebral mucormycosis.

Lagenidiosis

The best known species of *Lagenidium* is *Lagenidium giganteum,* a parasite of *mosquito* larvae used in biological control of mosquitoes. A different species causes disease exclusively in dogs. Lagenidiosis is found in the southeastern United States in lakes and ponds. It causes progressive skin and subcutaneous lesions in the legs, groin, trunk, and near the tail. The lesions are firm nodules or ulcerated regions with draining tracts. Regional lymph nodes are usually swollen. Spread of the disease to distant lymph nodes, large blood vessels, and the lungs may occur. An aneurysm of a great vessel can rupture and cause sudden death.

Diagnosis and Treatment

Diagnosis is through biopsy, although an ELISA test has been developed for *Pythium insidiosum* in animals. Treatment is very difficult and includes surgery when possible. Post-operative recurrence is common. Anti-fungal drugs show only limited effect on the disease, but itraconazole and terbinafine hydrochloride are often used for two to three months following surgery. Humans with *Basidiobolus* infections have been treated with amphotericin B and potassium iodide. For pythiosis and lagenidiosis, a new drug targeting water moulds called caspofungin is available, but it is very expensive. Immunotherapy has been used successfully in humans and horses with pythiosis. The prognosis for any type of phycomycosis is poor.

Chagas Disease

Chagas disease (also called *American trypanosomiasis*) is a human tropical parasitic disease which occurs in the Americas, particularly in South America. Its pathogenic agent is a flagellate protozoan named *Trypanosoma cruzi*, which is transmitted to humans and other mammals mostly by haematophagous assassin bugs of the subfamily Triatominae (Family Reduviidae).

Those insects are known by numerous common names varying by country, including benchuca, vinchuca, kissing bug, chipo and barbeiro. The most common insect species belong to the genera *Triatoma*, *Rhodnius*, and *Panstrongylus*. Other forms of transmission are possible, though, such as ingestion of food contaminated with parasites, blood transfusion and fetal transmission.

Trypanosoma cruzi is a member of the same genus as the infectious agent of African sleeping sickness, but its clinical manifestations, geographical distribution, life cycle and insect vectors are quite different.

The disease was named after the Brazilian physician and infectologist Carlos Chagas, who first described it in 1909 but, the disease was not seen as a major public health problem in humans until the 1960s (the outbreak of Chagas disease in Brazil in the 1920s went widely ignored). He discovered that the intestines of Triatomidae harboured a flagellate protozoan, a new species of the *Trypanosoma* genus, and was able to prove experimentally that it could be transmitted to marmoset monkeys that were bitten by the infected bug.

Chagas named the pathogenic parasite that causes the disease *Trypanosoma cruzi* and later that year as *Schizotrypanum cruzi*, both honouring Oswaldo Cruz, the noted Brazilian physician and epidemiologist who fought successfully epidemics of yellow fever, smallpox, and bubonic plague in Rio de Janeiro and other cities in the beginning of the 20th century.

Chagas' work is unique in the history of medicine because he was the only researcher so far to describe completely a new infectious disease: its pathogen, vector, host, clinical manifestations, and epidemiology. Nevertheless, he at least believed falsely until 1925, that the main infection route is by the bite of the insect and not by its faeces, as was proposed by his colleague Emile Brumpt 1915 and assured by Silveira Dias 1932, Cardoso 1938 and Brumpt himself 1939. Chagas was also the first to unknowingly discover and illustrate the parasitic fungal genus *Pneumocystis*, later to infamously be linked to PCP (Pneumocystis pneumonia in AIDS victims). Confusion between the two pathogens' life-cycles led him to briefly recognise his genus *Schizotrypanum*, but following the description of *Pneumocystis* by others as an independent genus, Chagas returned to the use of the name *Trypanosoma cruzi*.

On another historical point of view, it has been hypothesised that Charles Darwin might have suffered from this disease as a result of a bite of the so-called Great Black Bug of the Pampas (vinchuca). The episode was reported by Darwin in his diaries of the Voyage of the Beagle as occurring in March 1835 to the east of the Andes near Mendoza. Darwin was young and in general good health though six months previously he had been ill for a month near Valparaiso, but in 1837, almost a year after he returned to England, he began to suffer intermittently from a strange group of symptoms, becoming incapacitated for much of the rest of his life. Attempts to test Darwin's remains at the Westminster Abbey by using modern PCR techniques were met with a refusal by the Abbey's curator.

Epidemiology and Geographical Distribution

Chagas disease currently affects 16-18 million people, with some 100 million (25 per cent of the Latin American population) at risk of acquiring the disease, killing around 50,000 people annually. Chronic Chagas disease remains a major health problem in many Latin American countries, despite the effectiveness of hygienic and preventive measures, such as eliminating the transmitting insects, which have reduced to zero new infections in at least two countries of the region. With increased population movements, however, the possibility of transmission by blood transfusion has become more substantial in the United States. Also, *T. cruzi* has already been found infecting wild opossums and raccoons as far north as the state of North Carolina.

The disease is distributed in the Americas, ranging from the southern United States to southern Argentina, mostly in poor, rural areas of Central and South America.

The disease is almost exclusively found in rural areas, where the Triatominae can breed and feed on the natural reservoirs (the most common ones being opossums and armadillos) of *T.cruzi*.

Depending on the special local interactions of the vectors and their hosts, other infected humans, domestic animals like cats, dogs, guinea pigs and wild animals like rodents, monkeys, ground squirrels (*Spermophilus beecheyi*) and many others could also serve as important parasite reservoirs. Though Triatominae bugs feed on birds, these seem to be immune against infection and therefore are not considered to be a *T. cruzi* reservoir; but there remain suspicions of them being a feeding resource for the vectors near human habitations.

The triatomine insects are known popularly in the different countries as *vinchuca, barbeiro* (the barber), *chipo* and other names, so called because it sucks the blood at night by biting the face of its victims. The insects, who develop a predominantly domiciliary and anthropophilic behaviour once they have infested a house, usually hide during the day in crevices and gaps in the walls and roofs of poorly constructed homes.

More rarely, better constructed houses may harbour the insect vector, because of the use of rough materials for making roofs, such as bamboo and thatch. A mosquito net, wrapped under the mattress, will provide protection in these situations, when the adult insect might sail down from above, but one of the five nymphal stages (instars) could crawl up from the floor.

Even when the colonies of insects are eradicated from a house and surrounding domestic animal shelters, they can arrive again (e.g., by flying) from plants or animals that are part of the ancient, natural sylvatic infection cycle. This can happen especially in zones with mixed open savanna, clumps of trees, etc. interspersed by human habitation.

Dense vegetation, like in tropical rain forests, and urban habitats, are not ideal for the establishment of the human transmission cycle. However, in regions where the sylvatic habitat and its fauna are thinned out by economical exploitation and human habitation, such as in newly deforested, piassava palm (*Leopoldinia piassaba*) culture areas, and some parts of the Amazon region, this may occur, when the insects are searching for new prey.

Clinical Manifestations

The human disease occurs in two stages: the acute stage shortly after the infection, and the chronic stage that may develop over 10 years.

In the acute phase, a local skin nodule called a *chagoma* can appear at the site of inoculation. When the inoculation site is the conjunctival mucous membranes, the patient may develop unilateral periorbital edema, conjunctivitis, and pre-auricular lymphadenitis. This constellation of symptoms is referred to as Romana's sign. The acute phase is usually asymptomatic, but may present symptoms of fever, anorexia, lymphadenopathy, mild hepatosplenomegaly, and myocarditis. Some acute cases (10 to 20 per cent) resolve over a period of 2 to 3 months into an asymptomatic chronic stage, only to reappear after several years.

The symptomatic chronic stage may not occur for years or even decades after initial infection. The disease affects the nervous system, digestive system and heart. Chronic infections result in various neurological disorders, including dementia, damage to the heart muscle (cardiomyopathy, the most serious manifestation), and sometimes dilation of the digestive tract (megacolon and megaesophagus), as well as weight loss. Swallowing difficulties may be the first symptom of digestive disturbances and may lead to malnutrition. After several years of an asymptomatic period, 27 per cent of those infected develop cardiac damage, 6 per cent develop digestive damage, and 3 per cent present peripheral nervous involvement. Left untreated, Chagas disease can be fatal, in most cases due to the cardiomyopathy component.

Infection Cycle

An infected triatomine insect vector feeds on blood and releases trypomastigotes in its faeces near the site of the bite wound. The victim, by scratching the site of the bite, causes trypomastigotes to enter the host through the wound, or through intact mucosal membranes, such as the conjunctiva. Then, inside the host, the trypomastigotes invade cells, where they differentiate into intracellular amastigotes.

The amastigotes multiply by binary fission and differentiate into trypomastigotes, then are released into the circulation as bloodstream trypomastigotes. These trypomastigotes infect cells from a variety of biological tissues and transform into intracellular amastigotes in new infection sites. Clinical manifestations and cell death at the target tissues can occur because of this infective cycle. For example, it has been shown by Austrian-Brazilian pathologist Dr. Fritz Koberle in the 1950s at the Medical School of the University of Sao Paulo at Ribeirao Preto, Brazil, that intracellular amastigotes destroy the intramural neurons of the autonomic nervous system in the intestine and heart, leading to megaintestine and heart aneurysms, respectively.

The bloodstream trypomastigotes do not replicate (unlike the African trypanosomes). Replication resumes only when the parasites enter another cell or are ingested by another vector. The "kissing" bug becomes infected by feeding on human or animal blood that contains circulating parasites. Moreover the bugs might be able to spread the infection to each other through their cannibalistic predatory behaviour. The ingested trypomastigotes transform into epimastigotes in the vector's midgut. The parasites multiply and differentiate in the midgut and differentiate into infective metacyclic trypomastigotes in the hindgut.

Trypanosoma cruzi can also be transmitted through blood transfusions, organ transplantation, transplacentally, breast milk and in laboratory accidents. According to the World Health Organisation, the infection rate in Latin American blood banks varies between 3 per cent and 53 per cent, a figure higher than of HIV infection and hepatitis B and C.

Children can also acquire Chagas Disease while still in the womb. Chagas disease accounts for approximately 13 per cent of stillborn deaths in parts of Brazil. It is recommended that pregnant women be tested for the disease.

Alternative Infection Mechanism

Researchers suspected since 1991 that the transmission of the trypanosome by the oral route might be possible, due to a number of micro-epidemics restricted to particular times and places (such as a farm or a family dwelling), particularly in non-endemic areas such as the Amazonia (17 such episodes recorded between 1968 and 1997). In 1991, farm workers in the state of Paraiba, Brazil, were apparently infected by contamination of food with opossum faeces; and in 1997, in Macapa, state of Amapa, 17 members of two families were probably infected by drinking assai palm fruit juice contaminated with crushed triatomine vector insects. In the beginning of 2005, a new outbreak with 27 cases was detected in Amapa.

In March 2005, a new startling outbreak was recorded in the state of Santa Catarina, Brazil, that seemed to confirm this alternative mechanism of transmission. Several people in Santa Catarina who had ingested sugar cane juice ("garapa", in Portuguese) by a roadside kiosk acquired Chagas disease. As of March 30, 2005, 49 cases had been confirmed in Santa Catarina, including 6 deaths. The hypothesised mechanism, so far, is that trypanosome-bearing insects were crushed into the raw preparation. The health authorities of Santa Catarina have estimated that 60,000 people might have had contact with the contaminated food in Santa Catarina and urged everyone in this situation to submit to blood tests. They have prohibited the sale of sugarcane juice in the state until the situation is rectified.

The unusual severity of the disease outbreak has been blamed on a hypothetical higher parasite load achieved by the oral route of infection. Brazilian researchers at the Instituto Oswaldo Cruz, Rio de Janeiro, were able to infect mice via a gastrointestinal tube with trypanosome-infected oral preparations.

Laboratory Diagnosis

Demonstration of the causal agent is the diagnostic procedure in acute Chagas disease. It almost always yields positive results, and can be achieved by:

- *Microscopic examination:* (a) of fresh anti-coagulated

blood, or its buffy coat, for motile parasites; and (b) of thin and thick blood smears stained with Giemsa, for visualisation of parasites; it can be confused with the 50 per cent longer *Trypanosoma rangeli*, which has not shown any pathogenicity in humans yet.

- *Isolation of the agent by:* (a) inoculation into mice; (b) culture in specialised media (e.g., NNN, LIT); and (c) xenodiagnosis, where uninfected Reduviidae bugs are fed on the patient's blood, and their gut contents examined for parasites 4 weeks later.
- Various Immunodiagnostic tests; (also trying to distinguish strains (zymodemes) of *T.cruzi* with divergent pathogenicities).
  - — Complement fixation.
  - — Indirect haemagglutination.
  - — IFA, Indirect fluorescent assay.
  - — RIA, Radio-immunoassay.
  - — ELISA, Enzyme-Linked Immunosorbent Assay.
- Diagnosis based on Molecular Biology techniques.
  - — PCR, Polymerase chain reaction.

Prognosis

An index for classification of patients who have Chagas disease was published in the August 24, 2006 edition of the *New England Journal of Medicine*. Based on over 500 patients, this index includes clinical aspects, X-ray findings, EKG, echocardiography and Holter.

Chagas risk factors

Risk factor	*Points*
NYHA class III or IV	5
Cardiomegaly	5
Wall motion abnormalities	3
Non-sustained ventricular tachycardia	3

Low voltage on ECG	2
Male sex	2

Total points	***Risk of death in 10 years***
0–6	10 %
7–11	40%
12–20	85%

Treatment

Medication for Chagas disease is usually only effective when given during the acute stage of infection. The drugs of choice are azole or nitroderivatives such as benznidazole or nifurtimox (under an Investigational New Drug protocol from the CDC Drug Service), but resistance to these drugs has already been reported.

Furthermore, these agents are very toxic and have many adverse effects, and cannot be taken without medical supervision. The anti-fungal agent Amphotericin B has been proposed as a second-line drug, but cost and this drug's relatively high toxicity have limited its use. Moreover, 10-year study of chronic administration of drugs in Brazil has revealed that current chemotherapy does not totally remove parasitemia. Thus, the decision about whether to use anti-parasitic therapy should be individualised in consultation with an expert.

In the chronic stage, treatment involves managing the clinical manifestations of the disease, e.g., drugs and heart pacemaker for chronic heart failure and heart arryhthmias; surgery for megaintestine, etc., but the disease per se is not curable in this phase.

Chronic heart disease caused by Chagas is now a common reason for heart transplantation surgery. Until recently, however, Chagas disease was considered a contraindication for the procedure, since the heart damage could recur as the parasite was expected to seize the opportunity provided by the immunosuppression that follows surgery. The research that changed the indication of the transplant procedure for Chagas disease patients was conducted by Dr. Adib Jatene's group at the Heart Institute of the University of Sao Paulo, in Sao Paulo, Brazil. The research noted that survival rates in Chagas patients can be significantly improved by using lower dosages of the immunosuppressant drug cyclosporin.

Recently, direct stem cell therapy of the heart muscle using bone marrow cell transplantation has been shown to dramatically reduce risks of heart failure in Chagas patients. Patients have also been shown to benefit from the strict prevention of reinfection, though the reason for this is not yet clearly understood.

Some examples for the struggle for advances:

- Use of oxidosqualene cyclase inhibitors and cysteine protease inhibitors has been found to cure experimental infections in animals.
- Dermaseptins from frog species *Phyllomedusa oreades* and *P. distincta*. Anti-*Trypanosoma cruzi* activity without cytotoxicity to mammalian cells.
- Design of inhibitors to enzymes involved in trypanothione metabolism, which is unique to the kinetoplastid group of parasites.
- The sesquiterpene lactone dehydroleucodine (DhL) affects the growth of cultured epimastigotes of *Trypanosoma cruzi*
- The genome of Trypanosoma cruzi has been sequenced. Proteins that are produced by the disease but not by humans have been identified as possible drug targets to defeat the disease.

Prevention

A reasonably effective vaccine was developed in Ribeirao Preto in the 1970s, using cellular and subcellular fractions of the parasite, but it was found economically unfeasible. More recently, the potential of DNA vaccines for immunotherapy of acute and chronic Chagas disease is being tested by several research groups. Prevention is centred on fighting the vector (*Triatoma*) by using sprays and paints containing insecticides (synthetic pyrethroids), and improving housing and sanitary conditions in the rural area.

For urban dwellers, spending vacations and camping out in the wilderness or sleeping at hostels or mud houses in endemic areas can be dangerous, a mosquito net is recommended. If the traveller intends to travel to the area of prevalence, he/she should get information on endemic rural areas for Chagas disease in traveller advisories, such as the CDC.

In most countries where Chagas disease is endemic, testing of blood donors is already mandatory, since this can be an important route of transmission. In the past, donated blood was mixed with 0.25 g/L of gentian violet successfully to test for parasites.

With all these measures, some landmarks were achieved in the fight against Chagas disease in Latin America: a reduction by 72 per cent of the incidence of human infection in children and young adults in the countries of the Initiative of the Southern Cone, and at least two countries (Uruguay, in 1997, and Chile, in 1999), were certified free of vectorial and transfusional transmission. In Brazil, with the largest population at risk, 10 out of the 12 endemic states were also certified free.

Some stepstones of vector control:

- A yeast trap has been tested for monitoring infestations of certain species of the bugs: "Performance of yeast-baited traps with *Triatoma sordida, Triatoma brasiliensis, Triatoma pseudomaculata*, and *Panstrongylus megistus* in laboratory assays."
- Promising results were gained with the treatment of vector habitats with the fungus *Beauveria bassiana*, "Activity of oil-formulated *Beauveria bassiana* against *Triatoma sordida* in peridomestic areas in Central Brazil."
- Targeting the symbionts of Triatominae.

Chaga Mushroom

The Chaga mushroom (*Inonotus obliquus*), also known as tinder mushroom, is a fungus in *Hymenochaetaceae* family. It is a parasitic fungus of the Birch and other trees. The sterile conk is irregularly formed and has the appearance of burnt charcoal. It causes the tree's death in 5-7 years, and is known as "birch cancer" in Russia, but like many medicinal mushrooms contains many of its own anti-tumour compounds.

Medicinal Use

The anti-mutagenic action of the molecules found in the white part of birch bark where chaga feeds inhibit free-radical oxidation; and, induce the production of interferons, which help induce DNA repair. The substances, contained in white part of birch bark contribute to the decrease of hypoxia and to increase of the stability of organism to the oxygen deficiency, being anti-hypoxant correcting the metabolism of cells. The anti-cancer properties of betulin or betulinic acid, a chemical isolated from birch trees, is now being studied for use as a chemotherapeutic agent. Chaga contains large amounts of betulinic acid in a form that can be ingested orally, and it also contains the full spectrum of immune-stimulating phytochemicals found in other medicinal mushrooms such as maitake mushroom and shiitake mushroom. Since the 16th century, Chaga Mushroom has been used in folk medicine and the botanical medicine of the Eastern European countries as a remedy for cancer and gastritis, ulcers, tuberculosis (TB) of the bones. In 1958, scientific studies in Finland and Russia found this mushroom provided an epochal effect in breast cancer, liver cancer, uterine cancer, and gastric cancer, as well as in hypertension and diabetes. Herbalist David Winston maintains that it is the strongest anti-cancer medicinal mushroom.

Research

In 1998, there was a study in Poland that demonstrated Chaga's inhibiting effects on tumour growth. Noda *et al.* found that betulin seems to work highly selectively on tumour cells because the interior pH of tumour tissues is generally lower than that of normal tissues, and betulinic acid is only active at those lower levels. Fulda *et al.* found in 1997 that once inside the cells, betulinic acid induces apoptosis (programmed cell death) in the tumours. In 2005, there was a study done at Department of Medical Nutrition in South Korea. The Chaga Mushroom was evaluated for their potential for protecting against oxidative damage to DNA in human lymphocytes. The study found that the polyphenolic extract protected these cells against hydrogen peroxide-induced oxidative stress. Another study that year found the endopolysaccharide of Chaga produced indirect anti-cancer effects via immunostimulation.

The mycelial endopolysaccharide of I. obliquus was identified as a candidate for use as an immune response modifier and indicate that the anti-cancer effect of endopolysaccharide is not directly tumorcidal but rather is immuno-stimulating. It has also been demonstrated as anti-inflammatory. Saitoh Akiko published on the anti-mutagenic effects of Chaga in 1996, and Mizuno *et al.* published on the anti-tumour and hypoglycemic activities of the polysaccharides from the sclerotia and mycelia of Chaga.

Chaga grows in birch forests of Russia, Korea, Eastern Europe, Northern areas of the United States and in the North Carolina mountains.

Morgellons

"Morgellons" or "Morgellons disease" is a controversial medical condition, described as a multiple-symptom syndrome, characterised by skin lesions, a sensation that insects are crawling on or under the skin, reporting that fibres and granules are coming out of the skin, and fatigue. Reported mental symptoms may include short-term memory loss and impaired thinking, described as brain fog. The cause is unknown, and no diagnostic criteria have been established.

Dissatisfied with doctors diagnosing her two-year-old son's rash as eczema, Mary Leitao, of McMurray, Pennsylvania, coined the name Morgellons, naming it after a condition in Sir Thomas Browne's 1690 monograph *A Letter to a Friend,* wherein he describes "that endemial distemper of children in Languedoc, called the morgellons, wherein they critically break out with harsh hairs on their backs."

A 1935 paper by British doctor C.E. Kellett identifies the name *morgellons* with the Provencal term *masclous,* or "little flies". Researchers from the Morgellons Research Foundation consider Morgellons a distinct and novel disease. However, the contemporary syndrome is not confined to children, and the reported fibres do not appear to be hair.

The Morgellons Research Foundation states that Morgellons has been reported in every American state, and 15 other countries. Most of the US cases are in California, Texas, and Florida.

Morgellons shares characteristics with a number of recognised medical conditions, including attention-deficit disorder, chronic fatigue syndrome, Lyme disease, obsessive-compulsive disorder, but especially with delusional parasitosis, a mental illness involving false beliefs about infestation by parasites.

The Journal of the American Academy of Dermatology printed two letters, and a commentary on them, in its November 2006 issue. Their consensus was that dermatologists should agree with patients who have self-diagnosed "Morgellons disease", but treat them for delusional parasitosis. Morgellons has been successfully treated with the anti-psychotic medication Pimozide, which has been used to treat delusional parasitosis.

A Nature Medicine article reported, "Most dermatologists deny the disease exists, saying the people who claim to suffer from it have either common skin illnesses or psychological disorders such as delusional parasitosis, in which people become irrationally convinced that they harbour parasites."

Nanette Orman, an adjunct clinical professor from Stanford University, however, believes it may be related to Lyme disease.

The Centres for Disease Control and Prevention (CDC) has scheduled an investigation, but postponed it six times since June 2006.

Morgellons Research Foundation

Mary Leitao has a Bachelor of Science degree in biology from the University of Massachusetts at Boston, and has worked as a chemist. She formed the Morgellons Research Foundation (MRF) in 2002, and is its Executive Director.

In May 2006, the Morgellons Research Foundation was featured in a number of local TV news segments coordinated by the MRF's director of communications. This resulted in a significant rise in the public awareness of the term Morgellons. In response, the Los Angeles County Department of Health services issued a statement that: No credible medical or public health association has verified the existence or diagnosis of "Morgellons Disease." The current description of the disease is vague and covers many conditions. Until there is a credible, national standard for the diagnosis of this condition, there is no basis for making it a reportable disease.

Three of the eight MRF board members, including former chairman Charles Holman, medical director Greg Smith, and treasurer Judy Smith, resigned in August 2006, disagreeing with Leitao over the treatment of charitable contributions to the organisation. Dr. Randy Wymore, an Oklahoma State University assistant professor of pharmacology and MRF's former director of research, has also resigned from his position at the organisation, though his research programme continues.

Symptoms and Diagnosis

According to an opinion piece published by the MRF in the American Journal of Clinical Dermatology, Morgellons symptoms include skin lesions which can be anything from minor to disfiguring in their appearance, sensations of crawling and stinging on and under the skin, and the appearance of fibres and granules coming out of the skin. In addition, "according to statistics from the Morgellons Research Foundation (MRF), the majority (95 per cent) of affected patients also report symptoms of disabling fatigue and self-described "brain fog" or problems with attention. Patients also report a high incidence (50 per cent) of fibromyalgia, joint pain, and sleep disorders. Other symptoms include hair loss, decline in vision, neurological disorders and disintegration of teeth in the absence of cavities or gingivitis. Most patients are unable to continue working, and those who are able to continue working report that they do not function optimally." The authors "declared that they have no conflicts of interest related to the contents of this article." In fact, the paper was co-authored by the founder of the Morgellons Research Foundation and other MRF members involved in the treatment of patients exhibiting these symptoms.

There is no agreed-upon differential diagnosis, because Morgellons is not an accepted medical condition. The symptoms described by patients with Morgellons may result in the diagnosis of one or more of these conditions:

- *Atopic Dermatitis or Eczema:* A common skin condition with various causes including stress.
- *Cellulitis:* A skin infection.
- *Compulsive skin picking:* Obsessive picking at one's own skin.
- Contact dermatitis, an allergic rash.
- *Delusional Parasitosis:* a tactile hallucination of insects, snakes, or other vermin crawling over the skin.
- *Drug side effects:* from use, overuse, or withdrawal.
- *Folliculitis:* An infection of the hair follicles.
- Compulsive Methamphetamine smoking.
- *Impetigo:* A rash caused by bacterial infection.
- *Lice:* an infestation of parasitic insects.

- *Nematodes:* helminth or worm infestation.
- *Neurodermatitis:* Eczema or other skin condition exacerbated by scratching.
- *Scabies:* an infestation of the mite *Sarcoptes scabei.*
- *Seborrhea:* A condition due to over-active sebaceous glands.
- *Tinea:* A fungal infection of the skin or hair.

If a specific complaint is not identified, a doctor might diagnose a medically unexplained physical symptoms (MUPS) syndrome, such as fibromyalgia or chronic fatigue syndrome.

Delusional Parasitosis

Delusional parasitosis is one of the most common differential diagnoses that is used for these patients, and it is the primary source of controversy. The symptoms of delusional parasitosis are very similar to those presented by a Morgellons sufferer who rejects conventional diagnosis of their symptoms, or who presents a belief in the existence of an organism that cannot be observed except by the patient.

In her article *Delusory Parasitosis,* entomologist Nancy C. Hinkle outlines thirteen indications attributed to delusional parasitosis, many of which are shared by patients claiming to have Morgellons. These indications include:

- The presentation of physical evidence such as skin scrapings and debris.
- Obsessive cleaning and use of disinfectants and insecticides.
- Rejection of the possibility of psychological or other explanations.
- Emotional trauma, desperation, social isolation.
- Having seen numerous physicians, to no avail.

The belief that fibres are emerging from their skin is still present in these patients. This belief is generally regarded by doctors as either delusional or a result of simply mistaking fibres from clothing (lint) as fibres emerging from the body. One doctor, declining to be named for a news article, indicated that he treats patients simply by placing a cast over the affected area of skin, protecting it from the patient's scratching, which results in healing in a matter of weeks. This lends weight to the theory that perceived skin abnormalities interpreted as the manifestation of Morgellons are a side-effect of habitual scratching.

Patients who suffer from delusional parasitosis and who also believe they have Morgellons will often move from doctor to doctor in search of one who will offer them the diagnosis they are seeking. As a result, some doctors have adopted the use of the term "Morgellons disease" as "a rapport-enhancing term for delusions of parasitosis".

In the letter cited, Dr. Jenny Murase wrote, "because the term "Morgellons disease" does not have the word "delusions" embedded in the term, it is a useful way to communicate with patients regarding their disease. As a case in point, I have established a close relationship with the patient described above by referring to her delusions of parasitosis as Morgellons disease."

Many entomologists have joined dermatologists in treating Morgellons as a disorder, rather than a disease. Davis, says:

> "If symptoms of itching and crawling sensations in the skin persist and no evidence of parasites can be found, then a syndrome called Delusional Parasitosis must be considered. A variety of causes have been suggested for these sensations, including parasitism by Collembola and Strepsiptera or the presence of organisms called Morgellons. "Morgellons" is a term used to describe what are purported to be fibre-like parasites of the skin, but after decades of detailed study there is no evidence of an unknown organism fitting this description. There is also no evidence that Collembola or Strepsiptera are biologically capable of parasitising

humans. However, there are quite a number of physiological, hormonal and neurological syndromes that will cause these symptoms."

In February 2005, a report on a Fox affiliate in Duval County, Florida, prompted the Duval County Health Department (DCHD) to investigate. Their report concluded:

"An in-depth literature search was initiated and a bibliography is attached for reference; it was determined after extensive reviewing of these articles that Morgellons Disease is synonymous with delusional parasitosis (CDC, 1999). DCHD Epidemiology consulted a pediatric dermatologist within the health department for his professional opinion. It was concluded that this is a psychological condition that has been mentioned in literature for hundreds of years. Additionally, further investigation would be very costly to perform skin scrapings and laboratory testing."

The DCHD report noted there was a significant spike in reported cases after the news report, and said, "Calls began in February 2005 and DCHD continues to receive sporadic calls. Although there appears to be a cluster of reports around March 3,2005, this is attributed to the airing of Fox News' coverage of the illness and is not a true cluster of disease."

Morgellons Theories and Research

To date, studies from the medical community indicate that Morgellons and delusional parasitosis are the same condition referred to by different names. No formal clinical studies have yet demonstrated an etiology of Morgellons, though one study has demonstrated that patients can be cured using Pimozide, reinforcing the contention that the condition is synonymous with delusional parasitosis, which is treated with the same drug.

Stephen Stone, president of the American Academy of Dermatology, has stated: "There really is no scientific basis at this point to believe that this is real". Only one paper treating Morgellons as a disease has been published in a medical journal;

the paper was published as an opinion piece (not peer reviewed), co-authored by the founder of the Morgellons Research Foundation and other MRF-supported researchers. The Morgellons Research Foundation has published its own Case Definition for physicians, and MRF-supported researchers at Oklahoma State University and the State University of New York at Albany are undertaking ongoing clinical research into Morgellons.

The Pittsburgh Post-Gazette, reporting on Ms. Leitao's plight, noted that Dr. Randy Wymore, Oklahoma State University assistant professor of pharmacology and physiology, and former MRF director, recruited two Oklahoma State faculty physicians who tweezed fibres from beneath the skin of some Morgellons patients.

The samples were sent to the Tulsa Police Department's forensic laboratory. The police checked the samples against carpet and clothing fibres and other materials, and conducted chemical analyses and other tests, and found no matches against any fibre in their databases. However, the fibres taken from the Morgellons patients matched each other. There are apparently no published records of the composition of the fibres from this case or other Morgellons cases.

George Schwartz of Santa Fe, New Mexico initially believed the cause may be the bacterium *Stenotrophomonas maltophilia*, and has claimed success in treating patients with antibiotics that target this waterborne bacterium. He now believes it is a parasitic infection. In his booklet "Lisa's Disease, A Fibre Disease", he describes four stages to this condition. Stage four places body organs in jeopardy.

Schwartz also treated patients with anti-worm medication and diatomaceous earth. He says it is "a modern day plague which silently grows within the host and after weeks may explode into cavernous, thread- bearing skin lesions, and can extrude eggs and larvae" and is "a highly contagious, worldwide epidemic which will soon reach a critical mass".

He has developed a treatment plan for the early stages. In August, 2006 he published a new book describing this disease and treatment plans. Currently he is barred from practising medicine, due to a narcotics violation.

On the fringes of Morgellons research, there is a contention that the condition might be the first known outbreak of a nanotech parasite–a man-made molecular machine designed to be spread pathologically through as yet undetermined methods.

CDC Investigation

The Centres for Disease Control and Prevention are launching a study of Morgellons disease: "We're going into this with an open mind," said Dan Rutz, spokesman for the CDC Morgellons task force that first met in June 2006.

"The 12-person CDC task force includes two pathologists, a toxicologist, an ethicist, a mental health expert and specialists in infectious, parasitic, environmental and chronic disease. Among other tasks, the group is developing a case definition of Morgellons."

Lick Granuloma

A lick granuloma, also known as acral lick dermatitis, is a skin disorder in dogs resulting from an urge to lick the lower portion of the leg. The lesion from the incessant licking is a thickened, firm, oval plaque.

Causes

The cause of a lick granuloma is most often psychogenic and is considered to be a form of canine obsessive-compulsive disorder, but other causes include bacterial or fungal infections, demodectic mange, trauma causing nerve damage, allergies, or joint disease. Large breed dogs appear to be predisposed.

Psychogenic causes include boredom or separation anxiety. Lick granulomas are especially seen in large active dogs left alone for long periods of time. The condition becomes a vicious cycle–erosion of the skin from licking leads to pain and itching, which leads to more licking. One theory is that excessive licking causes endorphin release, causing an addiction to licking. The lick granuloma often becomes infected with bacteria.

Commonly Affected Breeds

- Dobermann
- Great Dane
- Labrador Retriever
- Irish Setter
- Golden Retriever
- German Shepherd Dog

Treatment

Treatment of the primary cause, if known, is essential. In psychogenic cases, dealing with psychological factors is most important. Factors should be identified such as being left alone all day, being confined, and changes in the household. Correction of these causes may include increased walks, avoiding confinement, and more interaction in the home.

Drugs may be used until behaviour modification has had time to take effect. Anti-depressants are most commonly used, including doxepin, amitriptyline, fluoxetine, and clomipramine. If the psychological factors are not corrected, the dog will usually relapse after the drugs are discontinued. Endorphin blockers such as naltrexone can be used to reduce addiction to licking, or endorphin substitutes such as hydrocodone may decrease the urge to lick.

The lesion should also be treated. Licking can be prevented by the use of Elizabethan collars, bandages, or anti-licking ointments (which are bad tasting). Topical medications such as corticosteroids or DMSO may be effective if used early. Small lesions may be injected with triamcinolone or methylprednisolone. Oral antibiotics are used to control infection. Surgery may be performed to remove whole lesions, but there is risk of continued self-mutilation to the area afterwards. Other potential treatments include cryosurgery, laser, radiation therapy, and acupuncture.

Overall, lick granulomas are very difficult to treat, with control only being achieved in about 65 per cent of cases.

Lingzhi

Lingzhi is the name for one form of the mushroom *Ganoderma lucidum*. This fungal species has a worldwide distribution in both tropical and temperate geographical regions, including North and South America, Africa, Europe, and Asia, growing as a parasite or saprophyte on a wide variety of trees.

Ganoderma lucidum enjoys special veneration in Asia, where it has been used in traditional Chinese medicine as a herbal medicine for more than 4,000 years, making it one of the oldest mushrooms known to have been used in medicine. Similar ganodermas have been found growing in the Amazon, according to mycologist Christopher Hobbs.

The word *lingzhi*, in Chinese, means "herb of spiritual potency" and has also been described as "mushroom of immortality". Because of its presumed health benefits and apparent absence of side-effects, it has attained a reputation in the East as the ultimate herbal substance. Lingzhi has now been added to the American Herbal Pharmacopoeia and Therapeutic Compendium.

Varieties

Ganoderma lucidum generally occurs in two growth forms, one, found in North America, is sessile and rather large with only a small or no stalk, while the other is smaller and has a long, narrow stalk, and is found mainly in the tropics. However, many growth forms exist that are intermediate to the two types, or even exhibit very unusual morphologies, raising the possibility that they are separate species. Environmental conditions also play a substantial role in the different morphological characteristics lingzhi can exhibit. For example, elevated carbon dioxide levels result in stem elongation in lingzhi. Other forms show "antlers', without a cap and these may be affected by carbon dioxide levels as well.

According to *The Chinese Herbal Materia Medica*, lingzhi may be classified into six categories according to their shapes and colours, each of which is believed to nourish a different part of the body.

1. *Red:* heart
2. *Purple:* joints
3. *Green:* liver
4. *White:* lungs and skin
5. *Yellow:* spleen
6. *Black:* kidneys and brain

Morphology and Constituents

Lingzhi is a polypod mushroom that is soft (when fresh), corky, and flat, with a conspicuous red-varnished, kidney-shaped cap and, depending on specimen age, white to dull brown pores underneath. It lacks gills on its underside and releases its spores through fine pores, leading to its morphological classification as a polypod.

Ganoderma lucidum is the only known source of a group of triterpenes, known as ganoderic acids, which have a molecular structure similar to steroid hormones. It is a source of biologically active polysaccharides with presumed medicinal properties, and it also contains:

- Ergosterol
- Coumarin
- Mannitol
- Lactones
- Alkaloids

- Unsaturated fatty acids
- Vitamins and minerals.

Unlike many other mushrooms, which have up to 90 per cent water content, fresh Lingzhi only contains about 75 per cent water.

Source

In nature, Lingzhi grows at the base and stumps of deciduous trees, especially maple (National Audubon Society; Field guide to Mushrooms, 1993). Only two or three out of 10,000 such aged trees will have Lingzhi growth, and therefore its wild form is generally rare. Today, Lingzhi is effectively cultivated both indoors under sterile conditions and outdoors on either logs or woodchip beds.

Lingzhi can be found for sale in many Asian markets as well as Western health shops. Extracts of 'lingzhi,' which may also be called 'reishi' are also available.

In general, a hot water extract is best at concentrating the polysaccharides in lingzhi and alcohol extracts are best at concentrating the triterpenoids in lingzhi but an extract can also be made with a blend of both extracts. There are multiple species of lingzhi, scientifically known to be within the Ganoderma lucidum species complex and mycologists are still researching the differences between species within this complex of species.

Medicinal Uses

Lingzhi may possess some anti-tumour, immunomodulatory and immunotherapeutic activities, supported by some studies on polysaccharides, terpenes, and other bioactive compounds isolated from fruiting bodies and mycelia of this fungus.

However, the efficacy of these compounds in the treatment of cancer has not yet been shown in clinical trials. Moreover, as with any herb, variation between preparations and potential negative side effects cannot be ruled out. It is understood as adaptogenic, anti-allergenic and anti-hypertensive due to the presence of triterpenes. Apart from these properties, lingzhi has been found to be anti-inflammatory, anti-viral, anti-parasitic, anti-fungal, anti-diabetic, anti-hypotensive, and hepatoprotective. It has also been found to inhibit platelet aggregation, and to lower blood pressure, cholesterol and blood sugar. Because of these properties, lingzhi has been regarded as blood pressure stabiliser, anti-oxidant, analgesic, a kidney and nerve tonic. It has been used in bronchitis prevention and in cardiovascular treatment, and in the treatment of high triglycerides, high blood pressure, hepatitis, allergies, chemotherapy support, HIV support, and even for fatigue and altitude sickness.

Some peer-reviewed studies indicate that ganoderic acid has some protective effects against liver injury by viruses and other toxic agents in mice, suggesting a potential benefit of this compound in the treatment of liver diseases in humans.

Although the experiences in fighting cancer are more inconsistent, the extract has been claimed to be effective in regressing tumours. The results depend on the type of cancer and the severity of the condition. It is usually recommended that it be used in combination with other prescribed medical treatments and as part of a fu zheng formula with a variety of supporting herbs. The Ganoderma extract has been employed to help substantially reduce or eliminate the side-effects of radio- and chemotherapies if it is taken before, during and after the treatments. It has been found clinically to reduce side-effects like hair loss, nausea, vomiting, stomatitis, sore throat, loss of appetite and insomnia.

Preparation

Because mushrooms contain chitin which locks up medicinal components, preparations of lingzhi are unlikely to be medicinally active unless there has been a prolonged water extraction at low temperatures (below 110°F). Simply tincturing the mushroom in ethanol or powdering it and encapsulating it makes preparations that are essentially inert and may account for some of the inconsistency in research results. Additionally, mushrooms traditionally incorporate or transform constituents from their host trees and mycelial fractions grown in sawdust or other substrate may differ appreciably from the whole fungus.

Lingzhi is traditionally prepared by simmering in water. Thinly sliced or pulverised lingzhi (either fresh or dried) is added to a pot of boiling water, the water is then brought to a simmer, and the pot is covered; the lingzhi is then simmered for two hours. The resulting liquid should be fairly bitter in taste, with the more active red lingzhi more bitter than the black. The process may be repeated. Alternatively, it can be used as an ingredient in a formula decoction or used to make an extract (in liquid, capsule, or powder form). The more active red forms of lingzhi are far too bitter to be consumed in a soup, as long cooked shitaki mushrooms might be.

Side Effects

There are no known side effects on its use.

Historical Chinese Records

The "Shen Nong's Herbal Classic", a 2000-year old medicinal Chinese book, considered today as the oldest book on oriental herbal medicine classifies 365 species of roots, grass, woods, furs, animals and stones into three categories of herbal medicine:

- The first category, called "superior", includes herbs effective for multiple diseases and are mostly responsible for maintaining and restoring the body balance. They have almost no unfavourable side-effects.
- The second category comprises tonics and boosters, for which their consumption must not be prolonged.
- The third category must be taken, usually in small doses, and for the treatment of specific ailments only.

Lingzhi ranked number one of the superior medicines, and was therefore the most exalted medicine in ancient times.

Modern Scientific Studies

Numerous studies of lingzhi, mainly in China, Korea, Japan and the United States, have shown its effectiveness in the treatment of a very wide range of diseases and symptoms. But the studies have not given any explanation of exactly how lingzhi has so many diverse effects, because none of the known active components taken alone have produced results as powerful as the intake of lingzhi itself, suggesting synergy is important. For example, reports of lingzhi's effect on stamina, appetite, and other human conditions are largely anecdotal and haven't been studied scientifically. It is perhaps more comprehensible at this time to explain lingzhi's "miraculous powers" from the traditional Chinese medicine point of view.

In the West, scientists have traditionally separated and classified each disease meticulously, and have specialised in each of them to such a degree that it seems as if each disease is autonomous and standing alone. Oriental medicine, resulting from knowledge accumulated through 4,000 years of human observation, asserts that health can be maintained by sustaining the proper balance within the body and that diseases can be cured by restoring this balance through nutrition, including medicinal herbs, exercise and mental peace. Traditional oriental medicine believes that a disease is but the mere tip of an iceberg, the result of the underlying imbalance of the body which must be restored.

Observations have shown that lingzhi generally has only slight side effects and can be consumed in high doses, in parallel with other medications. Its main properties are adaptogenic which mean that it is non-toxic, it works in a generalised manner on the hypothalamic-pituitary-adrenal axis and the neuroendocrine system. Its actions are alterative, enhance the immune system and lessen nervous tension. These properties are conducive to normalising and balancing the body (homoeostasis and allostasis), and as a result, lingzhi is able to help the body cure a multitude of disease states from within.

Lingzhi has been found to strengthen the respiratory system and to have a healing effect on the lungs, and is particularly beneficial for individuals with asthma, cough and other respiratory complaints.

At least one population study conducted in the 1970s confirms this claim. When more than 2,000 Chinese with chronic bronchitis took lingzhi syrup, 60 to 90 per cent felt better within two weeks and reported an improved appetite, according to an article entitled, *Medicinal Mushrooms*, written by Christopher Hobbs, and published in *Herbs for Health, Jan/ February 97.*

In Japan, after daily injections in mice with cancer it was reported that tumours in 50 per cent of the animals had completely regressed within 10 days. (Ikekawa *et al.*, 1968; Japanese Journal of Cancer Research; 59: 155-157) The host-dependent anti-tumour activity has been subsequently confirmed to be from the polysaccharide fractions of Ganoderma by Sasaki *et al.* [1]Multiple similar studies subsequently confirms this observation and anti-tumour efficacy of Ganoderma has been demonstrated from various species, at different stages of growth and using different solvents for extraction and different routes of administration. Anti-tumour activity has been demonstrated *in vitro* as well as in syngeneic tumour systems in animals. However, no human trials of Ganoderma against cancer in peer reviewed journals nor any controlled clinical trials in humans have yet been conducted or published.

There has been research showing lingzhi an effective supplement during chemotherapy or radiotherapy to reduce side-effects such as fatigue, loss of appetite, hair loss, bone marrow suppression and risk of infection. Ganodermas was shown effective against fatigue, hair loss, and bone marrow suppression. There is similar clinical evidence for other glucan BRMs applied in the setting of cancer chemotherapy or radiotherapy lending further support to the supplementation of Ganoderma in combination with cytotoxic cancer therapies.

The recommended dose should be in the range of five to ten grams of fruiting body or equivalent per day.

In an animal model, Ganoderma has been demonstrated to effectively prevent cancer metastasis, and these results are comparable to those of Lentinan from shitaki mushrooms. While only anecdotal or clinical data exists indicating ganoderma supplementation may enhance survival of *human* cancer patients, this survival advantage has been demonstrated for a number of comparable glucan BRMs like lentinan. Lentinan use in advanced gastric cancer demonstrated a significant life span prolongation advantage at 1, 2, 3 and 4 years in a randomised control trial. Lentinan is however injected. More appropriate for comparison to Ganoderma is perhaps PSK or PSP, which are orally administered. Mitomi *et al.* found significantly improved survival and disease-free survival (P=0.013) in colourectal cancer given PSK supplementation over three years when compared to control in a multicentre randomised controlled trials.

Meningitis

Meningitis is the inflammation of the protective membranes covering the central nervous system, known collectively as the meninges. It may develop due to a variety of causes, including infective agents, physical injury, cancer, or certain drugs. Meningitis is a serious condition owing to the proximity of the location to the brain and spinal cord. The potential for serious damage to motor control, thought processes, or even death warrants prompt medical attention.

Causes

Most cases of meningitis are caused by microorganisms, such as viruses, bacteria, fungi, or parasites, that spread into the blood and into the cerebrospinal fluid (CSF). Non-infectious causes include cancers, systemic lupus erythematosus and certain drugs. Although the most common cause of meningitis is viral, bacterial meningitis—the second most frequent cause—

can be serious and life-threatening. Anyone suspected of having meningitis should have prompt medical evaluation.

Epidemiology

Meningitis can affect anyone in any age group, from the newborn to the elderly.

Age group	*Causes*
Neonates	Streptococci Group B, E. Coli, Listeria
Infants	Neisseria meningococci, Haemophilus influenzae, Streptococci pneumonia
Children	Neisseria meningococci, Streptococci pneumonia
Adults	Streptococci pneumonia, Neisseria meningococci, Mycobacteria, Cryptococci

African Meningitis Belt

The "Meningitis Belt" is an area in sub-Saharan Africa which stretches from Senegal in the west to Ethiopia in the east in which large epidemics of meningococcal meningitis occur (this largely coincides with the Sahel region). It contains an estimated total population of 300 million people. The largest epidemic outbreak was in 1996, when over 250,000 cases occurred and 25,000 people died as a consequence of the disease.

Clinical Symptoms

Meningitis usually presents with one or more of the following symptoms.

- High fever, sometimes with chills
- Severe headache
- Nausea or vomiting
- Light sensitivity (photophobia)
- Sound sensitivity
- Neurological signs such as drowsiness or confusion
- Twitching
- Sleepiness

- Opisthotonus
- Irritability
- Sore throat
- Delirium (particularly in children)
- Seizures (occurs in about 20 to 40 per cent of patients).
- Nuchal rigidity (stiff neck) (occurs in less than 50 per cent of cases, but if seen, it is considered pathognomon).

One large prospective study demonstrated that the classic triad of nuchal rigidity, fever, and mental status change was present in only 44 per cent of confirmed cases of community acquired bacterial meningitis. Headache was the most common reported symptom (87 per cent) followed by neck stiffness (83 per cent).

Nuchal rigidity is typically assessed with the patient lying supine, and both hips and knees flexed. If pain is elicited when the knees are passively extended (Kernig's sign), this indicates nuchal rigidity and meningitis. In infants, forward flexion of the neck may cause involuntary knee and hip flexion (Brudzinski's sign). Although commonly tested, the sensitivity and specificity of Kernig's and Brudzinski's tests are uncertain.

Neck stiffness may prevent the head from bending forwards. If the child can touch the chest with the chin by bending the head forwards the physical sign of neck stiffness is probably not present. Neck stiffness in a child with a fever is a medical emergency.

In "meningococcal" meningitis (i.e. meningitis caused by the bacteria Neisseria meningitidis), a rapidly-spreading petechial rash is typical, and may precede other symptoms. The rash consists of numerous small, irregular purple or red spots on the trunk, lower extremities, mucous membranes, conjunctiva, and occasionally on the palms of hands and soles of feet.

Symptoms in infants under 12 months include high fever, fretfulness, irritability–particularly when handled, difficulty awakening, drowsiness, difficulty feeding, and/or a stiff neck, or bulging fontanelle (soft spot on top of head).

In a recent validation study published in the Journal of the American Medical Association (JAMA), a Bacterial Meningitis Score in children over the age of 2 months with at least 1 risk factor (positive CSF Gram Stain, CSF absolute neutrophil count e" 1000 cell/µL, CSF protein e" 80 mg/dL, peripheral blood absolute neutrophil count e" 10,000 cell/µL, history of seizure before or at presentation time) has a sensitivity of 100 per cent (95 per cent CI, 96.9 per cent-100 per cent), specificity of 63.5 per cent (95 per cent CI, 61.4 per cent-65.6 per cent), and negative predictive value of 100 per cent (95 per cent CI, 99.8 per cent-100 per cent) in predicting bacterial meningitis based on data collected from 20 academic medical centres as part of the Pediatric Emergency Medicine Collaborative Research Committee of the American Academy of Pediatrics.

Diagnosis

Laboratory Tests: If meningitis is suspected based on clinical examination, the patient should be given antibiotics. The next step is to perform laboratory tests on the blood and cerebrospinal fluid (CSF).

CSF is obtained by means of a lumbar puncture (LP). However, if the patient is at risk for elevated intracranial pressure, a lumbar puncture may be contraindicated because of the possibility of fatal brain herniation.

In such cases a CT or MRI scan should be performed prior to the lumbar puncture in order to make sure there are no large masses compressing the brain. Patients at risk for elevated intracranial pressure include those with recent head trauma, the immunocompromised, those with a known CNS neoplasm, or focal neurologic deficits such as papilledema or altered consciousness. Otherwise, the CT or MRI should be performed after the LP, with MRI preferred over CT due to its superiority in demonstrating areas of cerebral edema, ischemia, and meningeal inflammation.

The opening pressure is noted during the LP and the CSF sent for examination of white blood cell, red blood cell, glucose, and protein, and may have different tests performed, such as gram staining, culture, and possibly latex agglutination test, limulus lysates, or polymerase chain reaction (PCR) for bacterial or viral DNA.

If the patient is immunocompromised, the doctor may also consider testing the CSF for toxoplasmosis, Epstein-Barr virus, cytomegalovirus, JC virus and fungal infection.

CSF Analysis in Bacterial Meningitis

An autopsy demonstrating signs of pneumococcal meningitis. The forceps (centre) are retracting the dura mater (white).

Underneath the dura mater are the leptomeninges, which are edematous and have multiple small haemorrhagic foci (red).

- *Opening pressure:* > 180 mm H_2O.
- *White Blood Cells:* 10-10,000/uL with neutrophil predominance.
- *Glucose:* < 40 mg/dL.
- *CSF glucose to serum glucose ratio:* < 0.4.
- *Protein:* > 4.5 mg/dL.
- *Gram stain:* positive in >60%.
- *Culture:* positive in >80%.
- *Latex agglutination:* may be positive in meningitis due to *Streptococcus pneumoniae, Neisseria meningitidis, Haemophilus influenzae, Escherichia coli,* Group B Streptococci.
- Limulus lysates: positive in Gram-negative meningitis.

Cultures are often negative if CSF is taken after the administration of antibiotics. In these patients, PCR can be helpful in arriving at a diagnosis. It has been suggested that CSF cortisol measurement may be helpful.

CSF Analysis in Viral Meningitis

- Appearance is clear
- Will contain lymphocytes (lymphocytosis)
- Normal level of protein
- Normal or increased level of glucose

Treatment

Bacterial Meningitis: Bacterial meningitis is a medical emergency and has a high mortality rate if untreated. All suspected cases, however mild, need emergency medical attention.

Empiric antibiotics must be started immediately, even before the results of the lumbar puncture and CSF analysis are known. Antibiotics started within 4 hours of lumbar puncture will not significantly affect lab results.

The choice of antibiotic depends on local advice. In most of the developed world, the most common organisms involved are *Streptococcus pneumoniae* and *Neisseria meningitidis*: first line treatment in the UK is a third-generation cephalosporin (such as ceftriaxone or cefotaxime).

In those under 3 years of age, over 50 years of age, or immunocompromised, ampicillin should be added to cover *Listeria monocytogenes*. In the US and other countries with high levels of penicillin resistance, the first line choice of antibiotics is vancomycin and a carbapenem (such as meropenem). In sub-Saharan Africa, oily chloramphenicol or ceftriaxone are often used because only a single dose is needed in most cases.

Staphylococci and gram-negative bacilli are common infective agents in patients who have just had a neurosurgical procedure. Again, the choice of antibiotic depends on local patterns of infection: cefotaxime and ceftriaxone remain good choices in many situations, but ceftazidime is used when *Pseudomonas aeruginosa* is a problem, and intraventricular vancomycin is used for those patients with intraventricular shunts because of high rates of staphylococcal infection.

In patients with intracerebral prosthetic material (metal plates, electrodes or implants, etc.) then sometimes chloramphenicol is the only antibiotic that will adequately cover infection by *Staphylococcus aureus* (cephalosporins and carbapenems are inadequate under these circumstances).

Once the results of the CSF analysis are known along with the Gram-stain and culture, empiric therapy may be switched to therapy targeted to the specific causative organisms. Because antibiotic-resistance is a prevalent problem, information from drug susceptibility testing should also be gathered.

- *Neisseria meningitidis* (Meningococcus) can usually be treated with a 7-day course of IV antibiotics:
  - *Penicillin-sensitive:* Penicillin G or ampicillin.
  - *Penicillin-resistant:* Ceftriaxone or cefotaxime.
  - *Prophylaxis for close contacts (contact with oral secretions):* Rifampin 600 mg bid for 2 days *(adults)* or 10 mg/kg bid *(children)*. Rifampin is not recommended in pregnancy and as such, these patients should be treated with single doses of ciprofloxacin, azithromycin, or ceftriaxone.
- *Streptococcus pneumoniae* (Pneumococcus) can usually be treated with a 2-week course of IV antibiotics:
  - *Penicillin-sensitive:* Penicillin G.
  - *Penicillin-intermediate:* Ceftriaxone or cefotaxime.
  - *Penicillin-resistant:* Ceftriaxone or cefotaxime + vancomycin.
- *Listeria monocytogenes* is treated with a 3-week course of IV ampicillin + gentamicin.
- *Gram negative bacilli:* Ceftriaxone or cefotaxime.
- *Pseudomonas aeruginosa:* Ceftazidime.
- *Staphylococcus aureus.*
  - *Methicillin-sensitive:* Nafcillin.
  - *Methicillin-resistant:* Vancomycin.

- *Streptococcus agalactiae:* Penicillin G or ampicillin.
- *Haemophilus influenzae:* Ceftriaxone or cefotaxime.

Viral Meningitis

Unlike bacteria, viruses cannot be killed by antibiotics. Patients with very mild viral meningitis may only have to spend a few hours in a hospital, while those who have a more serious infection may be hospitalised for many more days for supportive care.

Patients with mild cases, which often cause only flu-like symptoms, may be treated with fluids, bed rest (preferably in a quiet, dark room), and analgesics for pain and fever. The physician may prescribe anti-convulsants such as phenytoin to prevent seizures and corticosteroids to reduce brain inflammation. If inflammation is severe, pain medicine and sedatives may be prescribed to make the patient more comfortable.

Fungal Meningitis

This form of meningitis is rare in healthy people, but is a higher risk in those who have AIDS. Anti-fungals to combat the infection are usually administered, as well as fluids and medicine to control pain and fever.

Complications

There are several potential disabilities resulting from damage to the nervous system. These include seizures (with its concommitant brain damage), sensorineural hearing loss, hydrocephalus, and cerebral palsy.

A common complication is loss of hearing as a result from damages to the hair cells in the cochlea, making them unable to transmit fluid-borne sound vibrations into electrical signalling to the auditory cortex.

Hearing may be restored with the use of a cochlear implant, with better prognosis for the patient if they receive it early. As time passes from the meningitis until the implantation, the liquid in the cochlea may become progressively stiffer, first like jelly and then already within two months it may become hard like bone, making the implantation impossible to perform. But if the implant is received quickly enough, the resulting artificial hearing may be very good. A one-year-old child completely losing all hearing and receiving implants within 4 weeks can get an artificial hearing that is good enough to understand spoken language as well as, or even better than average children of the same age.

Vaccination

All current vaccines target only bacterial meningitis.

Vaccinations against *Haemophilus influenzae* (Hib) have decreased early childhood meningitis significantly.

Vaccines against type A and C *Neisseria meningitidis,* the kind that causes most disease in preschool children and teenagers in the United States, have also been around for a while. Type A is also prevalent in sub-Sahara Africa and W135 outbreaks have affected those on the Hajj pilgrimage to Mecca.

A vaccine called *MeNZB* for a specific strain of type B Neisseria meningitidis prevalent in New Zealand has completed trials and is being given to many people in the country under the age of 20. There is also a vaccine, MenBVac, for the specific strain of type B meningoccocal disease prevalent in Norway, and another specific vaccine for the strain prevalent in Cuba.

Pneumovax (also known as Prevenar) against *Streptococcus pneumoniae* is recommended for all people 65 years of age or older and for all newborns starting at 6 weeks - 2 months, according to American Association of Pediatrics (AAP) recommendations.

History

Meningitis may have been described in the Middle Ages, but it was first accurately identified by the Swiss Vieusseux (a scientific-literary association) during an outbreak in Geneva, Switzerland in 1805.

In the 19th century, meningitis was a scourge of the Japanese imperial family, playing the largest role in the horrendous prematurity death rate the family endured. In the mid-1800s, only the Emperor Komei and two of his siblings reached maturity out of fifteen total children surviving birth.

Komei's son, the Emperor Meiji, was one of two survivors out of Komei's six children, including an elder brother of Meiji who would have taken the throne had he lived to maturity. Five of Meiji's 15 children survived, including only his third son, Emperor Taisho, who was feeble-minded, perhaps as a result of having contracted meningitis himself. By Emperor Hirohito's generation the family was receiving modern medical attention. As the focal point of tradition in Japan, during the Tokugawa Shogunate the family was denied modern "Dutch" medical treatment then in use among the upper caste; despite extensive modernisation during the Meiji Restoration the Emperor insisted on traditional medical care for his children.

Colony Collapse Disorder

Colony Collapse Disorder (or CCD) is a poorly understood phenomenon involving the abrupt disappearance of the worker bees in a beehive or Western honey bee colony. CCD has alternatively been referenced as Vanishing Bee Syndrome (VBS), but this name is not in common use.

Apparently, CCD was originally found only in colonies of the Western honey bee in North America, but European beekeepers have recently claimed to be observing a similar phenomenon in Poland, Greece, Italy, Portugal and Spain, with initial reports coming in from Switzerland and Germany, albeit to a lesser degree.

The cause (or causes) of the syndrome is not yet well understood and even the existence of this disorder remains disputed. Theories include environmental change-related stresses, malnutrition, unknown pathogens (i.e., disease), mites, pesticides such as neo-nicotinoids, emissions from cellular phones or other man-made devices, and genetically modified (GM) crops. That the disappearances have only been reported from a subset of the commercial beekeepers in affected areas (i.e., not feral colonies or organic beekeepers), suggests to some that beekeeping practices can be a primary factor.

From 1971 to 2006 approximately half of the US honey bee colonies have vanished, but this decline includes the cumulative losses from all factors such as urbanisation, pesticide use, tracheal and *Varroa* mites and commercial beekeepers retiring and going out of business, and has been somewhat gradual. Late in the year 2006 and in early 2007, however, the rate of attrition was alleged to have reached new proportions, and the term "Colony Collapse Disorder" was proposed to describe this sudden rash of disappearances.

Limited occurrences resembling CCD have been documented as early as 1896, and this set of symptoms has in the past several decades been given many different names (disappearing disease, spring dwindle, May disease, autumn collapse, and fall dwindle disease).

Most recently, a similar phenomenon in the winter of 2004/2005 occurred, and was attributed to Varroa mites (the "Vampire Mite" scare), though this was never ultimately confirmed. In none of the past appearances of this syndrome has anyone been able to determine its cause(s). Upon recognition that the syndrome does not seem to be seasonally-restricted, and that it may not be a "disease" in the standard sense—that there may not be a specific causative agent—the syndrome was renamed.

Symptoms

A colony which has collapsed from CCD is generally characterised by all of these conditions occurring simultaneously:

- Complete absence of adult bees in colonies, with little or no build-up of dead bees in or around the colonies.

- Presence of capped brood in colonies. Bees normally will not abandon a hive until the capped brood have all hatched.
- Presence of food stores, both honey and bee pollen:
    i. Which is not immediately robbed by other bees.
    ii. When attacked by hive pests such as wax moth and small hive beetle, the attack is noticeably delayed.

Precursor symptoms that may arise before the final colony collapse are:

- Insufficient workforce to maintain the brood that is present.
- Workforce seems to be made up of young adult bees.
- Queen is uncharacteristically evident outside the hive.
- The colony members are reluctant to consume provided feed, such as sugar syrup and protein supplement.

Possible Causes and Research

While the exact mechanisms of CCD are unknown, malnutrition, pesticides, pathogens, immunodeficiencies, mites, fungus, genetically modified (GM) crops, beekeeping practices (such as the use of antibiotics, or long-distance transportation of beehives) and electromagnetic radiation have all been proposed as causative agents. Whether any single factor is responsible, or a combination of factors (acting independently in different areas affected by CCD, or acting in tandem), is still unknown. It is likewise still uncertain whether CCD is a genuinely new phenomenon, as opposed to a known phenomenon that previously only had a minor impact.

At present, the primary source of information, and presumed "lead" group investigating the phenomenon, is the Colony Collapse Disorder Working Group, based primarily at Penn State University. Their preliminary report pointed out some patterns, but drew no strong conclusions.

Poor Nutrition or Malnutrition

One of the patterns reported by the aforementioned group at Penn State was that all producers in a preliminary survey noted a period of "extraordinary stress" affecting the colonies in question prior to the die-off, most commonly involving poor nutrition and/or drought.

To date, this is the only factor that all of the reported cases of CCD have in common; accordingly, there is at least some significant possibility that this phenomenon is correlated to nutritional stress, and may not manifest in healthy, well-nourished colonies.

Some researchers have attributed the syndrome to the practice of feeding high fructose corn syrup (HFCS) to supplement winter stores. The variability of HFCS may be relevant to the apparent inconsistencies of results. European commentators have suggested a possible connection with HFCS produced from genetically modified corn. If this were the sole factor involved, however, this should also lead to the exclusive appearance of CCD in wintering colonies being fed HFCS, but many reports of CCD occur in other contexts, with beekeepers who do not use HFCS.

Pesticides

One of the more common general hypotheses concerns pesticides (or, more specifically, insecticides), though several studies have found no common environmental factors between unrelated outbreaks studied.

It is particularly difficult to evaluate pesticide contributions to CCD for several reasons. First, the variety of pesticides in use in the different areas reporting CCD makes it difficult to test for all possible pesticides simultaneously. Second, many commercial beekeeping operations are mobile, transporting hives over large geographic distances over the course of a season, potentially exposing the colonies to different pesticides at each location. Third, the bees themselves place pollen and honey into long-term storage, effectively, meaning that there may be a delay of anywhere from days to months before contaminated provisions are fed to the colony, negating any attempts to associate the appearance of symptoms with the actual time at which exposure to pesticides occurred.

Pesticides used on bee forage are far more likely to enter the colony via the pollen stores rather than via nectar (because pollen is carried externally on the bees, while nectar is carried internally, and may kill the bee if too toxic), though not all potentially lethal chemicals, either natural or man-made, affect the adult bees–many primarily affect the brood, but brood die-off does not appear to be happening in CCD.

Most significantly, brood are not fed honey, and adult bees consume relatively little pollen; accordingly, the pattern in CCD suggests that if contaminants or toxins from the environment are responsible, it is most likely to be via the honey, as it is the adults that are dying (or leaving), not the brood.

One recently published view is that bees are falling victim to new varieties of nicotine-based pesticides; beekeepers in Canada are also losing their bees and are blaming neo-nicotinoid pesticides. To date, most of the evaluation of possible roles of pesticides in CCD have relied on the use of surveys submitted by beekeepers, but it seems likely that direct testing of samples from affected colonies will be needed, especially given the possible role of systemic insecticides such as the neo-nicotinoid imidacloprid (which are applied to the soil and taken up into the plant's tissues, including pollen and nectar), which may be applied to a crop when the beekeeper is not present.

The known effects of imidacloprid on insects, including honey bees, are consistent with the symptoms of CCD; for example, the effects of imidacloprid on termites include apparent failure of the immune system, and disorientation. In Europe the interaction of the phenomenon of "dying bees" with imidacloprid, has been discussed for quite some time now. It was a study from the "Comite Scientifique et

Technique (CST)" which was in the centre of discussion recently, which led to a partial ban of imidacloprid in France, primarily due to concern over potential effects on honey bees. Consequently when fipronil, a phenylpyrazole insecticide and in Europe mainly labelled "Regent", was used as a replacement, it was also found to be toxic to bees, and banned partially in France in 2004. Five other insecticides based on "fipronil" were also "accused" of killing bees.

However, the scientific committees of the European Union still are of the opinion "that the available monitoring studies were mainly performed in France and EU-member-states should consider the relevance of these studies for the circumstances in their country." Recently the British non-governmental 'Institute for Science in Society' published a short review of the scientific literature on the dying of honey bees and neo-nicotinoids.

In 2005, a team of scientists led by the National Institute of Beekeeping in Bologna, Italy, found that pollen obtained from seeds dressed with imidacloprid contains significant levels of the insecticide, and suggested that the polluted pollen might cause honey bee colony death. Analysis of maize and sunflower crops originating from seeds dressed with imidacloprid suggest that large amounts of the insecticide will be carried back to honey bee colonies.

Sub-lethal doses of imidacloprid in sucrose solution have also been documented to affect homing and foraging activity of honeybees. Imidacloprid in sucrose solution fed to bees in the laboratory impaired their communication for a few hours. Sub-lethal doses of imidacloprid in laboratory and field experiment decreased flight activity and olfactory discrimination, and olfactory learning performance was impaired. However, no detailed studies of toxicity or pesticide residue in remaining honey or pollen in CCD-affected colonies have been published so far, so, despite the similarity in symptoms, no connection of neo-nicotinoids to CCD has yet been confirmed.

Antibiotics and Miticides

Most beekeepers affected by CCD report that they use antibiotics and miticides in their colonies, though the lack of uniformity as to which particular chemicals are used makes it seem unlikely that any single such chemical is involved. However, it is possible that not all such chemicals in use have been tested for possible effects on honey bees, and could therefore potentially be contributing to the CCD phenomenon. Some reports indicate that "organic" beekeepers (who do not use antibiotics or miticides) are not affected by CCD, despite proximity to non-organic beekeepers that have been affected.

Pathogens and Immunodeficiency

Some researchers have commented that the pathway of propagation functions in the manner of a contagious disease; however, there is some sentiment that the disorder may involve an immunosuppressive mechanism, potentially linked to the aforementioned "stress" leading to a weakened immune system. Specifically, according to researchers at Penn State: "The magnitude of detected infectious agents in the adult bees suggests some type of immunosuppression." These researchers have further suggested a connection between *Varroa destructor* mite infestation and CCD, suggesting that a combination of these bee mites, deformed wing virus (which the mites transmit) and bacteria work together to suppress immunity and may be one cause of CCD. This research group is reported to be focusing on a search for possible viral, bacterial, or fungal pathogens which may be involved.

When a colony is dying, for whatever cause, and there are other healthy colonies nearby (as is typical in a bee yard), those healthy colonies often enter the dying colony and rob its provisions for their own use. If the dying colony's provisions were contaminated (by natural or man-made toxins), the resulting pattern (of healthy colonies becoming sick when in proximity to a dying colony) might suggest to an observer that a contagious disease is involved. However, it is typical in CCD cases that provisions of dying colonies are not being robbed, suggesting that at least this particular mechanism (toxins being spread via robbing, thereby mimicking a disease) is not involved in CCD.

Some have suggested that the syndrome may be an inability by beekeepers to correctly identify known diseases such as European foulbrood or the microsporidian fungus *Nosema*. The testing and diagnosis of samples from affected colonies (already performed) makes this highly unlikely, as the symptoms are fairly well-known and differ from what is classified as CCD. A high rate of *Nosema* infection was reported in samples of bees from Pennsylvania, but this pattern was not reported from samples elsewhere.

A recent paper, published by the Journal of Invertebrate Pathology reports that when hives of European honey bees were infected with *Nosema ceranae*, a recently-described microsporidian fungus, the colonies were wiped out within eight days.

Various areas in Europe have reported this fungus, but no direct link to CCD has yet been established. Highly preliminary evidence of *N. ceranae* was recently reported in a few hives in the Merced Valley area of California (USA). "Tests of genetic material taken from a "collapsed colony" in Merced County point to a once-rare microbe that previously affected only Asian bees but might have evolved into a strain lethal to those in Europe and the United States."

The researcher did not, however, believe this was conclusive evidence of a link to CCD; "We don't want to give anybody the impression that this thing has been solved." A USDA bee scientist has similarly stated, "while the parasite nosema ceranae may be a factor, it cannot be the sole cause. The fungus has been seen before, sometimes in colonies that were healthy." Likewise, a Washington State beekeeper familiar with *N. ceranae* in his own hives discounts it as being the cause of CCD. The primary antibiotic used against *Nosema* is Fumagillin, which has been used in a German research project to reduce the microsporidian's impact, and is mentioned as a possible remedy by the CCDWG.

According to a 2007 article, *Varroa* mites remain the world's most destructive honey bee killer due in part to the viruses they carry. As such, *Varroa* have been considered as a possible cause of CCD, though not all dying colonies contain these mites.

Genetically Modified Crops

Potential effects of gathering pollen and nectar from Genetically Modified (GM) crops that produce *Bacillus thuringiensis* (Bt) toxin have not been thoroughly investigated. Corn (maize), the major such crop, is not a preferred plant for honey bees, although beekeepers who keep bees near corn fields state that "corn is an excellent source of pollen when in tassel". Cotton, the second important Bt crop, is highly subject to bee visitation for nectar (pollen is only consumed if there is no other pollen available), but there is no credible evidence of toxicity of GM cotton, other than that from insecticides used during bloom.

The primary effects of Bt on insects is in the larval stage. Thus the studies on Bt-toxins and effects on honey bees originally concentrated more on larvae and their development. However, as pollen is an important part of bee bread, which is also food for adult bees, some beekeepers think that adult bees may be more affected by ingredients of pollen, because adult bees are something like a filter for larvae. And as the CCD phenomenon involves the disappearance of the adult bees, some think there could be a direct connection despite the absence of symptoms in the larvae, and despite any evidence that the bees experiencing CCD have ever been exposed to GM crops.

Most of the short summaries of US risk assessment studies on Bt in relation to honey bees are published on the United States Environmental Protection Agency (EPA) homepage for

Bio-pesticides Registration Action Documents; in particular, there is a document concerning the environmental effects of Bacillus thuringiensis as plant incorporated protectant. Literature references of studies, which are in the public domain, are included.

For Bt cotton, there are written some paragraphs in a fact sheet with the title "Bacillus thuringiensis Cry2Ab2 protein and the Genetic Material Necessary for Its Production in Cotton". These tests were usually made according to "Honey bee testing Tier I". Such tests seem to have a rather short duration time. ("Control and treated bees should be observed for at least 30 days after dosing.") The fact sheets for plant-incorporated protectants—retrievable and not retrievable—are listed in a special EPA homepage. The original studies on the effects of Bt pollen on honeybees do not seem to be in the public domain.

In 2005 Bt maize, which is commercially planted in the US since 1996, accounted for 35 per cent (10.64 million ha) of total US maize plantings. GM insect resistant Bt cotton has also been grown commercially in the US since 1996 and by 2005, was planted on 52 per cent (2.8 million ha) of total cotton plantings. According to David Hackenberg, former president of the American Beekeeping Federation and leading the public information concerning CCD as a beekeeper, "beekeepers that have been most affected so far have been close to corn, cotton, soybeans, canola, sunflowers, apples, vine crops and pumpkins", though Hackenberg personally attributes CCD to pesticides. Thus, some of the commercially grown Bt plants seem to be included in gaps of pollination management. However, similar massive bee die-offs have been recorded for decades prior to the introduction of these crops, and also occur in areas in Europe and Canada where there are no GM crops grown at all.

In 2004, the knowledge of GMO authorisation agencies was mainly based on a comprehensive review of the scientific literature published in *Bee World* which examined the effects of various commercialised and uncommercialised transgenes on honey bees. The review concludes that "evidence available so far shows that none of the GM plants currently commercially available have significant impacts on honey bee health." However, in 2005 a new publication in the Journal Apidologie suggested that (though in the treatment with CRY1Ab- enriched feed no significant differences in bee mortality were found at different treatment stages) foraging activity of bees fed with CRY1Ab may decline continuously through the treatment stages without any recovery between treatments.

The European Union GMO Panel of the European Food Safety Authority (EFSA) did not share the view by the authors "that the above results were mainly CRY1Ab dependent." The Panel was of the opinion that "negative effects on bees are likely not directly associated with exposure to the CRY1Ab protein because of the design of the experiment and lack of simultaneous controls or replication.

Research conducted in Germany suggests that exposure to maize pollen containing genes for Bt production may weaken the adult bees' defence against *Nosema,* though in the absence of such an infection, there were no detectable effects: "When the trial was repeated the colonies were treated prophylactically with antibiotics to prevent re-infection.

This indicates that healthy bee colonies are not impaired in any way by the toxin in any of the tested vital functions of colony size, foraging activity, brood care activity or development, even when exposed to extreme levels of Bt maize pollen over a period of six weeks." However, if "the bee colonies happened to be infested with parasites (microsporidia), this infestation led to a reduction in the number of bees and subsequently to reduced broods. This effect was significantly more marked in the Bt-fed colonies." It has further been suggested that "genetically modified corn may have altered the surface of the bee's intestines, sufficiently weakening the bees to allow the parasites to gain entry—or perhaps it was the other way around" though it was also noted "Of course, the concentration of the toxin was ten times higher in the experiments than in normal Bt corn pollen. In addition, the bee feed was administered over a relatively lengthy six-week period." Other more recent studies have failed to show any adverse effects of Bt pollen on healthy bee colonies.

The preliminary report of the Colony Collapse Disorder Working Group concerning "Fall Dwindle Disease" indicated that "all PA samples were found to have *Nosema* spores in their rectal contents. The sting gland of many examined bees was obviously scarred with distinct black "marks"; this type of pin-point melanisation or darkening is indicative of an immune response to some sort of pathogen." If the bees in Pennsylvania were gathering Bt-toxin-containing corn pollen, it could potentially have interacted with *Nosema* and thus contributed to CCD in those colonies; however, there is no evidence that these colonies were gathering corn pollen at any point prior to their deaths, nor has it been reported that colonies afflicted by CCD elsewhere had been collecting corn pollen.

The vast majority of the colonies reported to be dying from CCD occur in locations where GM corn is not grown (at least in the United States; also, 5 of the 10 states with the greatest amount of corn production–Illinois, Indiana, Kansas, Missouri, and Nebraska–have had no reported cases of CCD), nor have bees from other areas outside of Pennsylvania been reported to be significantly infected by *Nosema*.

In 2006, the "Committee on Status and Trends of Pollinators" of the United States National Research Council published a report on the "Status of Pollinators in North America". It suggested that GMO, besides other factors, might contribute to pollinator decline because, according to one scientific review of "the small literature on this topic,...in some cases, there are negative but sublethal effects attributable to consumption of transgenic pollens." The report goes on to say that, "These effects varied with the identity of the transgene and the amount of its expression, but in no case have any effects of transgenic crops on honey bee populations been documented."

The Sierra Club Genetic Engineering Committee recently published a letter to Senator Thomas Harkin with the title "GE and bee Colony Collapse Disorder—science needed!" They are of the opinion that "highly respected scientists believe that exposure to genetically engineered crops and their plant-produced pesticides merit serious consideration as either the cause or a contributory factor to the development and spread of CCD." Nine literature references which might support this theory are cited.

On March 28, 2007, the "Mid-Atlantic Apiculture Research and Extension Consortium" published a new "Summary of Research on the Non-Target Effects of Bt Corn Pollen on Honeybees", which states that according to "a field study...(soon to be published in the bee journal Apidologie) there is no evidence thus far of any lethal or sub-lethal effects of the currently used Bt proteins on honey bees", and, specifically regarding the possible causal connections between Bt pollen and CCD, stated "While this possibility has not been ruled out, the weight of evidence reported here argues strongly that the current use of Bt crops is not associated with CCD".

Saprophytic Activity of the Root-infecting Fungi

The parasitic activity of the root-infecting fungi has been closely studied by many investigators; their saprophytic existence under natural conditions in the soil has received comparatively little attention, though the earlier literature abounds in assertions as to the capacity of this or that fungus for indefinite survival as a saprophyte in the soil; such statements have usually been deemed self-evident, though sometimes they have been supported by reference to the capacity of the fungus for maintained saprophytic growth on sterile culture media or in sterilised soil. With the realisation of the part played by competition of organisms in the microbiology of the soil, such assertions have become appreciably more rare, but the need for experimental investigation is still very great.

Two types of saprophytism can be distinguished amongst the root-infecting fungi:

(i) Active saprophytic colonisation of dead plant tissues lying on or in the soil,

(ii) Limited saprophytic survival in tissues originally invaded as parasite.

The first type of behaviour is typical of the primitive parasites of the *soil inhabitant* class, and the second of the more specialised parasites of the *soil invader* class.

Active Saprophytism: Root-infecting fungi behaving as active saprophytes will be considered, for convenience, under the following three heads:

(i) Fungi making a macroscopically-visible saprophytic spread through the soil,

(ii) Fungi making a microscopically-visible saprophytic spread through the soil,

(iii) Fungi making no extensive saprophytic spread through the soil.

Fungi Making Macroscopically-visible Saprophytic Spread Through the Soil: The species of *Rosellinia* may be selected as furnishing the best known examples of this class. These fungi are especially important in West Indian plantations; their biology and control has been described in detail by Nowell (1923), from whose book the following general description is taken: "In the West Indies the fungi concerned are mainly species of *Rosellinia*. Other forms occur but are comparatively rare. In cultivations of an open nature where sun and wind have access to the soil, and specially in those of arable crops, there is a rapid dissipation of the decaying material necessary for the development of the fungus and the diseases as a rule soon disappear.

They can persist, however, in cultivations such as that of cacao in which the conditions, especially where shade trees are abundant, approach to those of the forest in respect of shade and humidity. They also occur in windbreaks and hedges of certain susceptible trees and shrubs in wet or sheltered districts. The causative fungi infect shaded soil rich in decaying vegetable matter and spread slowly through it, destroying every plant with which they come into contact.

A description of a particular occurrence of *Rosellinia arcuata* in Ceylon, which reveals the observer as well as the observed, has been contributed by Fetch (1928): "Almost my first experience of *Rosellinia arcuata* was its occurrence in a Panax hedge round my bungalow garden at Peradeniya. A mango tree which stood on the adjoining estate had been felled and the log left lying with one end near the hedge.

The *Rosellinia* began to develop among the dead leaves which had accumulated alongside the log, and then spread to the hedge, where it ran along the surface of the ground under and among the dead leaves in the hedge bottom. As the mycelium reached each of the closely-planted Panax stems, it enveloped it at the base. The cortex of the Panax stem rapidly became soft and rotten, and in a very short time the plant was dead. In a few weeks the hedge had been killed for a length of about 6 feet. It was then considered that sufficient information had been obtained concerning the method of progress of the fungus.

Fungi Making Microscopically-visible Saprophytic Spread Through the Soil: Free mycelial spread of *Rhizoctonia solani* through the soil can easily be demonstrated under the dissecting microscope; it is necessary only to place an agar inoculum disc of this fungus on the surface of moist unsterilised soil in a covered dish, and examine again in a few days' time. Blair (1943) has studied the spread of this fungus through the soil, using the Rossi-Cholodny slide technique. Growth of the fungus in 4 types of soil was most rapid at the lowest soil moisture content tested, 30 per cent saturation, and was improved by forced aeration of the soil. Growth was depressed by addition to the soil of such fresh organic material as ground wheat straw or dried grass. Blair showed that this depression was due chiefly to the competitive effect of other microorganisms better able to develop upon such substrates; under sterile conditions, the organic supplements failed to depress the growth of *R. solani*. Pre-decomposed materials depressed the growth of the fungus less than the fresh materials.

In pure culture experiments on cellulose substrates, Blair demonstrated that the cellulose decomposing capacity of *R. solani* was weak by comparison with that of typical strong cellulose decomposers. He further analysed this depressing effect of the cellulose decomposing microflora upon the growth of *R. solani* by demonstrating that (i) the effect was much diminished by addition of excess nitrogen to the soil, and was therefore due in part to competition for available nitrogen between the cellulose decomposers and *R. solani* (ii) the effect was also greatly reduced by removal of the respiratory carbon dioxide produced by the cellulose decomposers, through absorption over strong alkali in closed vessels. In other experiments, Blair demonstrated that whereas growth of *R. solani* through untreated soils was most rapid at pH 7, in soils with fresh organic amendments the optimum pH value for growth was approximately 8; this he attributed to the action of alkaline soil as a carbon dioxide acceptor (Garrett, 1936).

Although such experiments with Rossi-Cholodny slides demonstrated an effect of soil conditions upon the rate of growth of *R. solani* through the soil, they offered no clue as to how far such growth was supported by the 6-mm. disc of agar inoculum, and how far by nourishment from the soil itself. In an attempt to answer this question, Blair carried out experiments on the growth of the fungus through glass tubes filled with three different types of moist soil and with quartz sand, respectively; the tubes were inoculated at one end with an agar inoculum disc of the same diameter. Whereas growth of the fungus ceased at a distance of some 5 cm. from the inoculum in the tubes of sand, it extended for some 20 cm. beyond the inoculum in the three types of soil, during 21 days. In another experiment, Blair removed the agar inoculum disc from some tubes 2 days after the start of the experiment, and inoculated other tubes not with agar discs but with mycelial strands picked off the vertical glass walls of petri dishes containing agar cultures of the fungus.

Removal of the agar inoculum discs after 2 days, or the substitution of mycelial strands as inoculum, greatly diminished the growth of *R. solani* in the tubes of sand; in the three soils, however, removal of the agar discs reduced growth by only a negligible amount. Growth of the fungus from the mycelial strand inoculum had almost caught up with that in the undisturbed agar disc controls after 23 days in one type of soil, but in the other two soils growth was only about half that in the agar disc control tubes after 23 days. Blair suggested that a certain *food-potential* of the inoculum might be necessary to initiate saprophytic growth of *R. solani* through the soil by analogy with the need of most root-infecting fungi for a *food-base* before they can establish root infection.

The saprophytic persistence of *R. solani* in the soil has been correlated by Elmer (1942) in Kansas with summer weather conditions. Elmer states: "The commercial potato crop in eastern Kansas matures and usually is harvested in early June. Throughout the rest of the summer the fields usually remain fallow.

Summer temperature in this area is too high for the production of *Rhizoctonia* sclerotia as is disclosed by the fact that they are rarely found here on tubers of the early summer crop. Thereafter the fungus must consequently exist saprophytically as soilborne mycelium.

In this state its persistence depends on the presence of sufficient soil moisture to prevent death from desiccation. The springtime prevalence of soilborne *Rhizoctonia* in these fallow fields indicates to what extent this fungus is able to persist in such field soils during the preceding summer." Variation in the prevalence of *R. solani* stem canker on potato crops established from clean seed was therefore attributed by Elmer to variation in saprophytic survival of *R. solani* in the soil, following potato harvest of the previous July. An annual survey of the incidence of stem canker in potato plots established from clean seed in some 15 scattered localities of eastern Kansas was made by ELMER during the 13-year period 1928-1940. A high percentage infection in any year could be correlated with the occurrence of soil conditions suitable for saprophytic survival of *R. solani* mycelium in the soil during July and August of the preceding year, *viz.* high rainfall, relatively low temperature, and absence of drying winds. Percentage infection was low, on the other hand, following a year in which July-August rainfall had been low, and hot drying winds had been frequent, leading to a drying-out of the surface soil. Variation in the incidence of stem canker from year to year was evidently not due to the effect of the current season's weather on development of infection, because plants established from infected tubers showed a high percentage of infection throughout the whole period of the survey.

Survival of *R. solani* on a host plant was not adversely affected by dryness of the soil in the summer months; if the land was occupied in August by a late-planted potato crop, or by some other crop susceptible to *R. solani,* then the potato crop of the following year showed considerable stem canker, regardless of weather conditions in the preceding July and August. From the results of these various experiments, Elmer concludes: "The prevalence of soilborne *Rhizoctonia* in eastern Kansas can be approximately predicted from the relative soil moisture content of the preceding summer. In regions like eastern Kansas, where soilborne *Rhizoctonia* may periodically be practically eradicated because of environment, effective seed-potato treatments are valuable not only in preventing infection of the current season's crop but also in preventing reinfestation of the soil with this fungus."

Fungi Making no Extensive Saprophytic Spread through the Soil: A well-established example of this type of fungus is provided by *Fusarium cithnorum,* the active saprophytism of which was shown by Sadasivan (1939). Both natural, untreated wheat straw and straw sterilised in the autoclave were buried in 6 different types of soil for periods not exceeding 4 months; at intervals, samples of straw were removed from the soil, washed, and plated out after a light surface sterilisation with mercuric chloride.

High percentages of *F. culmorum* colonies were obtained from the straws at many of the samplings; F. *culmorum* and *Penicillium* spp. were the dominant organisms forming colonies from the straws on the agar plates. Sadasivan further demonstrated that the isolates of *F. culmorum* thus obtained possessed parasitic ability comparable to that of isolates derived from diseased plants, when tested against wheat seedlings.

Sadasivan's discovery was confirmed and extended by Walker (1941), who demonstrated the saprophytic development of *F. culmorum* in straws buried in samples of soil taken from a number of fields at monthly intervals throughout the year. There appeared to be little correlation between percentage colonies of *F. culmorum* appearing on the isolation plates, and season or cropping history of individual fields.

The relative numbers of colonies of *F. culmorum, Penicillium* spp. and other fungi obtained by Walkek were, however, greatly influenced by the surface sterilising agent employed and by the time of sterilisation; this suggested that a number of fungi were present in each straw, but that a "selection", which determined colony production on the agar plate, occurred during the surface sterilising and plating-out. Similar conclusions concerning the saprophytic ability of F. *culmorum* were reached by Weise (1939) in his study of the foot rot of asparagus due to this fungus in Saxony. F. *culmorum* was invariably found to be present in samples of soil taken from the Weinbohla district, and was found living as a saprophyte both in compost heaps and in decomposing straw.

Saprophytic Survival in Tissues Invaded as Parasite: Various more or less casual observations on the survival of root-infecting fungi in diseased crop residues have been reported. The longest survival periods have been recorded for mycelium in infected tree roots buried deeply in the soil; the most famous instance is provided by a much quoted piece of *Mesua ferrea* root infected by *Fames noxius* which, according to Tunstall (1926), must have lain buried in the soil at the Tocklai Experiment Station in North East India for at least 14 years. Cooley (1942) successfully isolated *Xylaria malt* from undisturbed roots of apple trees, affected by black root rot, at periods of 4 years and more after death of the host; he quoted a survival period of 16 years for another infection by the same fungus, as recorded by Schneideriian (1936).

Taubenhaus and Ezekiel (1930&) have drawn attention to the necessity for distinguishing between real and apparent saprophytic survival of a root-infecting fungus. These authors confirmed the earlier observation of Taubenhaus and Killough (1923) that the roots of cotton plants might live on underground throughout the Texas winter after the tops of the plants had been killed by frost.

The root rot fungus, *Phymatotrichum omnivorum*, not only survived but also spread along such living, overwintering roots; in one instance, Taubenhaus and Ezekiel observed a spread of as much as 12 feet along a row of overwintering roots under dead tops. The viability of *P. omnivorum* on living overwintering roots was demonstrated by isolation of the fungus in pure culture, by its ability to put out fresh mycelial strands when the root was placed in a moist chamber, and by its rapid infection of cotton and turnip roots. No isolations were secured, on the other hand, from dead decaying roots, no strands were put out by the fungus, and no infection of healthy roots could be obtained by the use of such dead decaying roots as inoculum.

Taubenhaus and Ezekiel therefore concluded that, unless sclerotia were formed on the infected root, the fungus could not survive the winter on dead, decaying roots, but only on living overwintered ones. In confirmation of this claim, Taubenhaus and Ezekiel cited an experiment in which they had successfully inoculated healthy cotton plants with tap roots of cotton plants killed by root rot only 1 day to 2 weeks earlier, but not with those of plants killed from 3 to 9 weeks earlier. Neal and Maclean (1931) reported that mycelial strands of *P. omnivorum* recovered at various depths from the soil of infected cotton fields were viable as late as December, but that no viable strands could be found in January, February or March. More recently, Ezekiel (1940) has shown that topping or girdling infected cotton plants at the advancing margins of root rot spots in midsummer may greatly shorten the survival period of *P. omnivorum* on the roots.

From the practical point of view, however, interest in the saprophytic survival of *P. omnivorum* in dead infected root tissues has declined since the original announcement by King and Loomis (1929) of their discovery of a sclerotial stage of this fungus in pure culture.

The survival of *Ophiobolus graminis*, causing the take-all disease of cereals, in infected plant tissues seems to have received greater attention than that of any other root-infecting fungus. The influence of soil conditions upon the survival period of *O. graminis* was examined by Garrett (1938b, 1940), using a glass tumbler technique; at regular intervals, the viability of the fungus in 100 or more individual pieces of infected straw from each of the different soil series was ascertained by means of a wheat seedling test. From results thus obtained, Garrett was able to correlate the rapidity of disappearance of the fungus with general microbiological activity of the soil; loss of viability was most rapid under conditions of medium to high temperature, suitable moisture content, and good soil aeration. Viability of *O. graminis* was preserved, on the other hand, at a low temperature (2°-3°C) and in air-dry soil, and was actually maintained for longer in a waterlogged soil than in one held at a moisture content of *50 per cent* saturation. Loss of viability was particularly hastened by the addition of fresh energy materials, poor or entirely lacking in nitrogen, such as glucose, starch, or rye-grass meal, to the soil; this effect diminished as the nitrogen content of the material increased, so that rape dust with a nitrogen content of 5.8 per cent was found to exert no significant influence upon the survival of *O. graminis* in the straws, whilst dried blood with a nitrogen content of 13 per cent actually increased the survival period.

Ammonium carbonate was found also to promote survival. Garrett attempted to explain these results by the hypothesis that the "resting" mycelium of *O. graminis* in the infected wheat straw was subject to decomposition by the other microorganisms of the soil; he suggested that the mycelium was attacked as a source of nitrogen by the microorganisms decomposing the matrical straw (nitrogen content only 0.4 per cent). The addition of nitrogen-poor organic materials to the surrounding soil would enhance the nitrogen scarcity and increase the number of microorganisms engaged in decomposition, whereas the addition of nitrogen-rich materials would provide an alternative and more readily available source of nitrogen than the mycelium of *O. graminis* for straw decomposition.

In a second paper, Garrett (1940) reported survival of the fungus in almost 100 per cent of infected straws buried for 8 months in a garden soil rich in humus and high in nitrogen content. Reconciliation of this observation with the hypothesis that the "resting" mycelium of the fungus was a passive subject for decomposition by the other soil microorganisms was difficult, and led to further direct microscopical observation of the fungus in the infected straws.

A slow but apparently continuous branching and development of the mycelium of *O. graminis* was found to occur for weeks or even months after the straws had been buried in the soil. At time of removal of the infected straws from the pure-culture flasks, the mycelium in the tissues was mainly of the hyaline type, corresponding to the "microhyphae" of Fellows (1928), but, after the straws had been buried in the soil, the dark mycelium, or "macrohyphae" of Fellows, developed. The development of dark mycelium was noticeably greater, and also more sustained, in the straws buried in soils well supplied with soluble nitrogen. Garrett therefore concluded that survival of *O. graminis* depended upon the continued slow development of the mycelium inside the infected straw; a plentiful supply of soluble nitrogen from the soil prolonged-this development, and thereby extended the life of the fungus. Some typical results obtained by Garrett, which illustrate this important effect of nitrogen in prolonging the survival of *O. graminis* in the infected straw.

The survival of *O. graminis* in the infected plant tissues thus seems to depend especially upon its ability to form fresh hyphae. This, in turn, waits upon an adequate supply of nutrients; lack of nitrogen usually seems to be the limiting factor for *O. graminis,* as well as for other soil microorganisms engaged in decomposition of the straw tissues.

Now Chang (1939) has shown that decomposition of composted wheat straw by a general soil flora proceeds substantially further in the presence of an adequate supply of phosphate, in addition to the usual supplement of nitrogen. Recent unpublished evidence suggests that the survival of *O. graminis* may be slightly prolonged by supplementary phosphate, if the straws are buried in a medium completely lacking this nutrient.

The gap between the conditions obtaining in such laboratory tumblers and those in soil *in situ* in the field has been bridged by the experiments of Fellows (1941), who followed the disappearance of *O. graminis in* naturally infected soil taken straight from the field and incubated under the eight possible combinations of high arid low temperature, loose and firm packing, and moist and air-dry conditions.

The low-temperature samples were stored in a deep cave and the high-temperature ones were kept in the glasshouse. Looseness was maintained by occasional stirring, and a firm packing achieved by thorough tamping of the soil whilst moist. The moist soils were maintained at approximately *60 per cent* of saturation, whereas the others were allowed to air-dry. Four experiments were made, and periods of incubation varied from 230 to 777 days; at the end of the incubation period, all soil samples were brought to favourable tilth and moisture content, and a test crop of wheat grown in them to maturity, in the glasshouse. The degree of infection on the plants at harvest was expressed by means of a "severity rating". In almost every comparison provided by this series of extensive factorial experiments, the fungus disappeared more quickly in a warm soil than in a cold one, and more quickly in a loose soil than in a compact soil. The influence of moisture was less straightforward; under cool conditions survival appeared to be rather better in a moist than in a dry soil, whereas under warm conditions the reverse was the case. In all four series of experiments, without exception, disappearance of the fungus was most rapid in the warm, moist, loose soil, *i.e.* under just those conditions of high temperature, adequate moisture content and good aeration that were postulated by Garrett (1938) as most conducive to rapid loss of viability.

Poisonous Fungi

Galen affirms, that they are all very cold and moist, and therefore to approach unto a venomous and murthering facultie, and ingender a clammy, pituitous, and cold nutriment if they be eaten. To conclude, few of them are good to be eaten, and most of them do suffocate and strangle the eater.

It is common knowledge that many of the fungi are extremely poisonous, and this aspect, more than any other, has firmly impressed itself upon the public mind. Indeed, so strong is popular prejudice on the point, that it has invested with the most deadly attributes, a great number of perfectly innocuous species. Yet, on the other hand, many folk, while pronouncing anathema upon unmistakeable species of great esculence, are often only too willing to risk their lives by experimenting with anything that "looks like" a mushroom. When, therefore, the layman proposes to indulge in a pleasant repast which includes fungi, the question, whether it is quite safe, or whether it may be followed with undesirable, or even fatal, consequences, depends frequently upon the insecure foundations of the ludicrously ineffective" rules and as these are usually only sketchily applied, it is not surprising that the historians of all times, and in all countries, have had to record numbers of distressing fatalities from this cause. Among the great ones of the earth, whose lives have been so sacrificed, are Pope Clement VII, the Emperor Jovian, Charles VI of France, Berronill of Naples, and the widow of Tsar Alexis; while those of less exalted rank who have paid the same dread penalty are legion. It, therefore, behoves us to examine a little more closely these unconscious assassins, and the lethal weapons which they unwittingly employ.

Early Classical Cases of Fungus Poisoning: The earliest records of poisoning by fungi are those of the Greeks, and date back some centuries before the Christian Era. Thus, according to Eparchides, when Euripides (480-406 BC) was on a visit to Icarus, a woman with her daughter and two full-grown sons perished in one day through eating poisonous fungi; while, about the same time, Hippocrates, the father of medicine, who was born about 460 BC, instances the poisoning of the daughter of Pausanias by a raw fungus, a case which, fortunately, did not terminate fatally. In the second century BC, experience of these accidents had become more extensive, and definite if somewhat empiric remedies came into vogue.

Primitive Remedies: It was about this time that Nicander of Colophon, the celebrated Greek poet, physician and grammarian, described his universal prescription for fungus poisoning in the following terms:

> "Let not the evil ferment of the earth which often causes swellings in the belly or strictures in the throat distress a man; for when it has grown up under the viper's deep hollow track, it gives forth the poison and hard breathing of its mouth; an evil ferment is that; men generally call the ferment by the name of fungus, but different kinds are distinguished by different names; but do thou take the many-coated heads of the cabbage,

or cut round the twisting stems of the rue, or take the efflorescence which has accumulated on old corroded copper, or steep the ashes of clematis in vinegar, then bruise the roots of pyrethrum, adding a sprinkling of lye or soda, and the leaf of cress which grows in gardens, with the medic plant and pungent mustard, and burn wine-lees or the dung of the domestic fowl into ashes; then, putting your right finger in your throat to make you sick, vomit forth the baneful pest."

It is not recorded what success usually attended this treatment, but the remedy would appear to be almost as bad as the complaint.

Supposed Origins of the Poisons in Fungi: Later classical writers attempted to explain how the poisonous principles of fungi originated. Pliny (23-79 AD) whose opinions are obviously influenced by those of Nicander, says "noxious kinds must be entirely condemned; for if there be near them a hobnail *(caligaris clavus)* or a bit of rusty iron, or a piece of rotten cloth, forthwith the plant, as it grows, elaborates the foreign juice and flavour into poison; who could discern the different kinds, except country-folk and those who gather them? Moreover they imbibe other noxious qualities besides; if, for instance, the hole of a venomous serpent be near, and the serpent breathe upon them as they open, from their natural affinity with poisonous substances, they are readily disposed to imbibe such poison. Therefore, "he naively counsels, "it will be well to exercise care in gathering them until the serpents retire into their holes." *A* little later, Dioscorides, the renowned Greek physician, records that "fungi have a twofold difference, for they are either good for food or poisonous; their poisonous nature depends on various causes, for such fungi grow either amongst rusty nails or rotten rags, or near serpents' holes, or on trees producing noxious fruits." According to the same writer, even the good kinds "if partaken of too freely, are injurious, being indigestible, causing stricture and cholera," and he recommended the taking of an emetic after every meal in which fungi figured. Dioscorides thus appears to have had grave doubts as to the wholesomeness of the fungi, and he cites as a common remedy, when poisonous fungi had been eaten, a dose of nitre exhibited in vinegar and water.

Distrust of the Fungi: Other writers, as well they might from the numerous fatalities that occurred, shared the doubts of Dioscorides. In this connection, the opinions of Seneca, Galen and others, have already been cited before dealing with the fungi as foods, from which aspect the consideration of the poisonous fungi is, of course, inseparable. We may also refer to an Epigram written by Martial (43-104 AD) upon the death of Macrinus, who died from eating poisonous fungi, in which Epigram, to point the moral, the poet makes two punning allusions to the fungi: "*Defungi fungis* homines, Macrine, negabas; *Boleti leti* causa fueri tui."

"You used to deny, Macrinus, that men could be killed by fungi; yet Boleti destroyed yourself."

Survival of Classical Beliefs: The ingenious hypotheses advanced by the classical writers as to the origins of the fungus poisons, as with those involved in their general practice of medicine, appear to have held ground for centuries, and even at so late a period as the year 1560, Matthiolus includes two serpents in his general illustration of fungi.

Real Origin of the Fungus Poisons: Modern research has, however, revealed the fact that the poisonous substances contained in certain fungi are, as in the case of other poisonous plants, not absorbed from without, but manufactured and liberated in the tissues of the fungus, during the normal process of metabolism.

They consist of various alkaloids, glucosides, toxalbumins, etc., and are probably waste products somewhat akin to those produced in, and excreted from, the human body. It *has* been suggested that they serve in some measure as a protection against marauders, but this would appear to be fortuitous; since they are harmless to the slug, one of the chief ravagers of fungi. Moreover, this hypothesis does not explain why certain fungi should require this protection, while other closely-allied species do not.

The Toxicology of the Fungi has not been investigated, either from the chemical or the medical standpoints, to the extent which the importance of the subject warrants; but from the work which has been done, it is evident that not only do the compositions of these toxins often vary in the different genera, but even in allied species of the same genus. Again, most genera containing poisonous species include also harmless ones. The matter is further complicated by the fact that individual fungi sometimes contain two separate poisons: some even more. Thus, in the present state of our knowledge, the whole subject is exceedingly baffling.

Classification of Fungus Poisons: This seeming chaos has been reduced to some state of order by Roch, a leading Swiss toxicologist, who has classified those fungi containing specific poisons, according to the effect which they have upon the human system, and since this aspect is the most vital one, we have used his scheme in tabulating our remarks on the individual species. The main groups into which poisonous fungi, according to their toxic effects, may be divided, are as follows:

1. Fungi containing poisons which, after a long period of incubation, cause the degeneration of the cells of the organism, particularly those of the nervous system and the glandular parenchymatous tissues (the liver, etc.).

   The three most poisonous fungi known are of this type. They are characterised by toxins of a very virulent description, rendered more deadly by their delayed action on the human system. This delay allows them to become completely absorbed into the blood before the obnoxious qualities of the fungus are realised, and remedial measures can be taken.

The most lethal of these three species is *Amanita phalloides* a fungus whose harmless appearance and not unpleasant taste and smell have, in spite of the fact that the two have little in common, repeatedly led careless collectors to mistake it for the Common Mushroom. Such unhappy folk have cause, if only for a short time, to regret the error, for this fungus is capable of causing an enormous amount of suffering. After a meal in which it plays a prominent part, the unwary eater experiences little or no discomfort for some ten hours. Then he is seized with sudden abdominal cramp-like pains, accompanied by nausea, vomiting, extreme thirst and cholera-like diarrhoea, which includes the passing of blood and mucus. These and other more distressing symptoms may abate and recur for many hours, during which there is a gradual but serious loss of weight, and considerable exhaustion. The symptoms may then slowly pass away and a tardy recovery is followed by a very protracted period of convalescence. In most cases, however, the symptoms continue with greater or less intensity, and on the third to the eighth day the unhappy sufferer is released from his agony by coma and death.

The actual poison *(Amanita toxiri)* which is responsible for the calamity is an alcohol-soluble substance whose exact composition has not been ascertained, and to which no antidote is known. Another poison *(Amanita haemolysin)*, a glucoside present in the raw plant, is capable, when injected hypodermically, of destroying the red corpuscles of the blood; but, since it is decomposed by a heat less than that necessary for cooking, and also, usually, by the gastric juices, it apparently plays no part in the actual poisoning.

The two other species in this group are *Amanita virosa* and *A. verna*. These closely resemble *A. phalloides* in appearance and probably are equally deadly in effect.

This inglorious trio is responsible for fully ninety per cent, of the deaths from fungus poisoning.

2. Fungi containing poisons which, while often capable of causing serious gastric disturbances, act chiefly by exciting and then paralysing the central nervous system.

   The poisonous principles of these fungi belong to that class of toxins known as *deliriants*, which, acting on the nerve centres, produce dilation of the pupils of the eyes, spectral illusions, incoordination and delirium, and occasionally, though rarely, tetanoid spasms and paralysis.

   These symptoms are frequently accompanied by such gastric symptoms as vomiting, excessive thirst, etc. The most notorious member of this group, and one which almost advertises its baneful properties, is the beautiful Fly Agaric *(Amanita musearia)*.

   The bright scarlet or orange-red cap of this species, studded with white warts, should prevent anyone, however careless, from mistaking it for a mushroom; but if this were not sufficient, its bitter acrid taste ought to discourage the most venturesome. Nevertheless, it has frequently been eaten, more especially in those countries where the allied *Amanita Caesarea* occurs, and for which species it has often been mistaken. Although its aftereffects are neither so painful nor so dangerous as those of the preceding group, yet many deaths have resulted from it.

   The poisonous principles of this species, and, therefore, their effect on the human system, appear, however, to vary in different countries, for among many of the native tribes of northeastern Siberia and Kamchatka, it has long been employed as a stimulant, the effect being similar to that of hashish. This use has been observed by a number of travellers in these regions, notably Krasheninnikoff (1733-1744), Langsdorf (1803-1806), and Kennan (1865-67); and in more recent times by Lansdell

(1882) and Vanderlip (1903). The effects of the fungus are so remarkable that, rather than appear to exaggerate, we will leave these writers to tell their own tales, in the course of which the reader may trace the inspiration of "The Purple Pileus" of H. G. Wells. According to Langsdorf, as translated by Greville, "this variety of *Amanita muscana* is used by the inhabitants of the northeastern parts of Asia in the same manner as wine, brandy, arrack, opium, etc. by other nations. Such fungi are found most plentifully about Wischna, Kamchatka, and Wilkowa Derecona, and are very abundant in some seasons, and scarce in others. They are collected in the hottest months, and hung up by a string in the air to dry; some dry of themselves on the ground, and are said to be far more narcotic than those artificially preserved. Small deep-coloured specimens, thickly covered with warts, are also said to be more powerful than those of a larger size and paler colour. The usual mode of taking the Fungus is, to roll it up like a bolus, and swallow it without chewing, which, the Kamchatkadales say, would disorder the stomach. It is sometimes eaten fresh in soups and sauces, and then loses much of its intoxicating property; when steeped in the juice of the berries of *Vaccinium uliginosum* its effects are those of strong wine. One large, or two small Fungi, are a common dose to produce a pleasant intoxication for a whole day, particularly if water be drunk after it, which augments the narcotic principle. The desired effect comes on from one to two hours after taking the fungus. Giddiness and drunkenness result in the same manner as from wine or spirits; cheerful emotions of the mind are first produced; the countenance becomes flushed; involuntary words and actions follow, and sometimes at last an entire loss of consciousness. It renders some remarkably active, and proves highly stimulant to muscular exertion: by too large a dose, violent spasmodic effects are produced.

So very exciting to the nervous system in many individuals is this Fungus, that the effects are often very ludicrous. If a person under its influence wishes to step over a straw or small stick, he takes a stride or a jump sufficient to clear the trunk of a tree; a talkative person cannot keep silence or secrets; and one fond of music is perpetually singing."

Vanderlip, who observed the customary use of this fungus among the Koraks of Kamchatka, confirms Langdorf's remarks on its effects, and adds:

"Curiously enough, after recovering from one of these debauches, they claim that all the antics performed were by command of the mushroom. The use of it is not unattended with danger, for unless a man is well looked after he is likely to destroy himself. The Koraks sometimes take this drug in order to work themselves up to the point of murdering an enemy. Three or four of the mushrooms is a moderate dose, but when one wants to get the full effect, one takes ten or twelve."

Kennan, who also visited the Koraks, and who describes them as "a quiet inoffensive hospitable tribe of semi-barbarians, notable only for honesty, general amiability and comical reverence for legally-constituted authority" remarks of the fungus:

"Its habitual use, however, completely shatters the nervous system, and its sale by Russian traders to the natives has consequently been made a penal offence by Russian law. In spite of all prohibitions, the trade is still secretly carried on, and I have seen twenty dollars worth of furs bought with a single fungus. The Koraks would gather it for themselves but it requires the shelter of timber for its growth and is not to be found on the barren steppes over which they wander; so that they are obliged for the most part to buy it, at enormous prices, from the Russian traders. ... As the supply of these toadstools is by no means equal to the demand, Korak ingenuity has been greatly exercised in the endeavour to economise the precious stimulant, and make it go as far as possible. Sometimes in the course of human events, it becomes imperatively necessary that a whole band should get drunk together and they have only one toadstool to do it with. For a description of the manner in which this band gets drunk collectively and individually upon one fungus, and keeps drunk for a week, the curious reader is referred to Goldsmith's Citizen of the World/ Letter 32. It is but just to say, that this horrible practice is almost confined to the settled Koraks of Penzhinsk Gulf—the lowest, most degraded portion of the whole tribe. It may prevail to a limited extent among the wandering natives but I have never heard of more than one such instance outside of the Penzhinsk Gulf settlements."

Whether the consumption of such large, yet non-fatal, doses of an admittedly poisonous fungus is due to a less virulent form, or to a hardier people inured through generations to its use, is difficult to say. Certainly, the species is sufficiently poisonous in more temperate countries. The active principles of this species are a number of toxins, of which the best known is the alkaloid, muscarine. Among other poisonous fungi which belong to this class are *Amanita pantherina Stropharia coronaria,* etc.

3. Fungi containing irritant principles which cause gastroenteritis by direct action on the mucous membrane of the digestive system.

   It is this group to which the vast bulk of poisonous fungi would appear to belong. Their active principles are probably of varying constitution, but their action on the human system is materially the same. Owing to the solubility of these acrid principles, the fungi produce symptoms soon after they are eaten, and these usually appear as griping pains in the stomach, a quickening and weakening of the pulse, depression, lethargy and vertigo, accompanied by vomiting and diarrhoea. These symptoms usually abate soon after the fungus is removed from the system, and recovery is in most cases rapid. Only in rare instances, where the patient is of weak constitution, or where a large quantity of the fungus has been eaten, does death occur. To this group belong the acrid species of *Russula, Lactarius* and *Boletus,* some species of *Amanita, Entoloma lividum,* and several others.

4. Fungi containing haemolytic or blood-destroying principles.

   Curiously enough, this group is composed almost wholly of fungi which are generally considered to be edible, namely, the morels *(Morchella sp.)*; the helvells *(Helvella sp.)*; *Gyromitra esculenta, Amanita rubescens,* etc. The explanation of this seeming paradox is found in the unstable nature of these *haemolysins,* which, while present in the raw fungus, are partially removed by the preliminary preparation of washing, while the remainder are destroyed in cooking. It follows, therefore, that the rare cases of poisoning which have been caused by eating these fungi, are probably always due to insufficient cooking. In such cases, the poisonous substances act by attacking and partially destroying the red corpuscles of the blood, the process being accompanied by symptoms of uneasiness, vertigo, vomiting and the passing of blood, and sometimes by convulsions and profound sleep.

5. Fungi containing substances which excite the muscular system, especially the smooth muscle fibres (the uterus, vessels, etc.

   These poisonous fungi (ergot of rye, etc.) are mostly very minute, and are only eaten inadvertently as contaminants of the more usual foods. In this connection reference has already been made to them. In some cases, the toxins are employed in medicine, and this aspect is dealt with elsewhere.

Doubtful Fungi: In addition to these five classes of fungi, in which the presence of specific poisons has already been proved, there is also a certain number of species whose wholesomeness is open to grave doubt. True, many of them have been eaten by some persons without the slightest ill-effects; while with others the reverse has been the case.

The question of digestive idiosyncracies must not be overlooked, for, as with lobsters and various other foods, certain people become ill through eating quite wholesome fungi, which do not at all affect normal individuals. This personal peculiarity has at times brought harmless species under unjust suspicion. Again, the eating of fungi normally quite innocuous may have harmful effects in certain diseases, e.g., parenchymatous nephritis. Kakowski cites *Boletus edulis* in this connection. There are also the questions of the freshness of the specimens, their mode of preparation and of cooking, and the quantity eaten. The variable effects of these fungi do not indicate whether they should be regarded as definitely edible or equally definitely poisonous. Therefore, until further information is available, it is best to regard them all with suspicion. These species are too numerous to tabulate, and lest this be considered a serious omission, it may be pointed out that it is better for the reader to learn what he *may* eat, rather than what he *may not.*

Treatment: No account of the poisonous fungi could be considered complete without some remarks on the treatment to be adopted in cases of fungus poisoning. Yet, so obviously is this a matter for the medical man, that we hesitate in making any suggestions which might be construed as an attempt to usurp his powers. On the other hand, the severity of an attack, and its after effects, can be considerably lessened by prompt treatment, and since so many of these cases occur in country districts, where a certain delay in obtaining a doctor is often inevitable, it is desirable that everyone who is in the habit of collecting and eating fungi, should be capable of applying first-aid remedies. Therefore, in giving the following facts, we are animated rather with the desire to lighten the physician's labour than to trespass on his preserves.

When any one, after eating fungi of any description, is seized with symptoms akin to those described, no time should be lost in summoning a doctor, who should be informed of the nature of the case which he is required to attend, so that he may bring with him appropriate appliances. In the interval, preliminary treatment may be commenced. The general aims of the treatment are:

(i) To remove from the system the remains of the fungi eaten.

(ii) To bring about the elimination or exhaustion of the toxins already absorbed by the blood.

(iii) To counteract pain and all other distressing symptoms.

(iv) To guard against collapse, and generally to keep a watch on the action of the heart.

Of these items, the attendant will probably only be required to act on the first. Indeed, so long as valuable time is not lost in so doing, as much as possible should be left to the doctor.

(i) *Expulsion of the Fungus:* In cases of poisoning, the usual procedure is to resort immediately to emetics. With fungus poisoning, these may not only be useless, but actually harmful, for the first symptoms are usually violent sickness, during which practically the whole of the fungus in the stomach is rejected. Therefore, there is no point in increasing the irritation to this organ, by deliberately causing further retching. In some cases, however, the patient is unable to vomit, when a tablespoonful of mustard in half a tumblerful of warm water should be given. Apomorphine, tartar emetic or zinc sulphate may alternatively be used. Since, however, each is, in itself, a powerful poison, any of these should only be employed under medical direction. On no account should salt and water be administered, as unless this acts at once, it merely accentuates the solubility of the poisons.

If possible, the stomach should afterwards be washed out by means of a stomach tube, or, failing this, copious draughts of warm water should be given.

After this treatment, any of the fungus not accounted for has naturally passed into the intestines, and therefore the next step should be to administer a purgative, preferably of an alkaline nature, such as:

sulphate of soda (Glauber's salt) and sulphate of magnesium (Epsom salts). Each should be used in a tumblerful of warm water. Should there be severe abdominal griping, it is better to substitute for the above, one tablespoonful of castor oil in a little milk, while a further soothing drink is two tablespoonfuls of olive oil, beaten up with the yolk of an egg, and four or five tablespoonfuls of milk or warm water, to which, *under the doctor's instructions,* may be added 20 drops of laudanum.

(ii) *The Elimination of the Toxin:* The natural elimination or exhaustion of the poison which has already been absorbed into the blood, can only be assisted or expedited by subcutaneous injections of atropine, and other medical means only possible to a doctor.

(iii) *The Counteraction of the Various Distressing Symptoms:* Which almost always accompany an attack of fungus poisoning should likewise be left in the hands of the doctor. In case of delirium, sedatives such as potassium bromide are necessary; lethargy and depression can be counteracted by strong tea or coffee, dry friction, etc.; pain, by the use of various opium compounds; while prolonged and excessive vomiting can be combated with soda water (aerated water) or by giving the patient small pieces of ice to suck.

(iv) *Effect on the Heart:* Many fungi have a very marked effect on the heart's action, and resort may have to be made to stimulants.

Utilisation of Poisonous Fungi as Foods: Having already emphasised the serious danger of eating fungi without knowing precisely what they are, any consideration of the gastronomic use of fungi which are unquestionably poisonous may at first appear not only a folly, but almost a madness. However, there's method in it, and since undoubtedly many poisonous kinds are so used, the practice being furthermore of very ancient standing, it is interesting to discuss it. The crux of the matter is, that these poisons are soluble, and may first in various ways be extracted, when the remainder of the fungus may be consumed without harm. At the same time, he would be a very foolish person who would recommend experimenting with any of them, when there are so many harmless edible species available, and nothing is further from our thoughts. The consideration of the matter, however, can do no harm, and for the novice it will be safer to stop at that.

Ancient Use of Poisonous Fungi: Dealing first with the historical side of the matter, Celsus, who lived about the beginning of the Christian era, tells us that poisonous fungi "can be rendered serviceable by a mode of cooking them; for if they have been boiled in oil or with the young twig of a pear tree, they become free from any bad quality," while Pliny (23-79 AD) says that fungi "are rendered more safe still if they are cooked with meat or with pear stalks; indeed, it is good to eat pears immediately after fungi. Vinegar being of a nature contrary to them neutralises their dangerous qualities." That the ancients were substantially correct with regard to the efficacy of vinegar has since been amply demonstrated by the experiments of Paulet and Gerard.

Experiments of Paulet and Gerard: In the year 1793, Paulet stated that if poisonous fungi are cut up into small pieces, and soaked in water containing either salt, vinegar or alcohol, the poisons are extracted, so that they become innocuous to animals fed on them. About half a century later, it was demonstrated by the experiments of Gerard, that even the deadly *Amanita phalloides* and *A. muscaria*, when so treated, are rendered equally harmless to man.

In 1851, Gerard communicated to the Conseil de Salubrite of Paris, an account of various experiments he had made on the preparation of poisonous fungi, so as to render them quite safe for consumption, and stated that, as a result of these, he and his family were in the habit of eating, without any ill-effects, all kinds of deadly varieties, to the extent even of 75 kilogrammes within one month.

To investigate these claims, a commission of three was delegated by the Conseil, and in the presence of these gentlemen, Gerard, after preparation according to his method, consumed 500 grammes (a little more than i Ib.) of *Amanita muscaria*, and 70 grammes (about 2-J oz.) of *A. phalloides*, without harm, to the complete persuasion of the commission, two of whose members even had the temerity to taste some small pieces themselves, no doubt with some inward trepidation, with which one can entirely sympathise. The commission reported, however, that there was no doubt as to the success of the demonstration.

Similar Uses among Peasants, etc.: The practice of eating poisonous fungi, after the poisons had been extracted, was current in parts of France long before Gerard, in this notable manner, brought it to public notice, and it was similarly practised in other countries; while, from the evidence of Pliny, it no doubt dates back to remote antiquity. Buller suggests that "the secret may have been known only to groups of peasants in various localities. That it was not known universally appears to be shown by the numerous references in classical literature to cases of fungus poisoning. In Russia and Germany, many different species, edible and poisonous alike, have long been largely gathered for winter consumption, being preserved in vinegar, which extracts the toxic alkaloids contained in many of those used, thus preventing the ill-effects that would otherwise result. Some interesting details of a similar practice among the French peasants round Serignan are given by Fabre, who says:

"Often, when rambling in the neighbouring woods, I inspect the baskets of the mushroom-pickers, who are delighted for me to look. I see things fit to make mycological experts stand aghast. I often find the purple boletus *[B. purpureus]*, which is classed among the dangerous varieties. I made the remark one day. The man carrying the basket stared at me in astonishment:

"That a poison ! The wolf's bread he said, patting the plump boletus with his hand. ' What an idea ! Its beef-marrow, sir, regular beef-marrow"

"He smiled at my apprehensions and went away with a poor opinion of my knowledge in the matter of mushrooms".

"In the baskets aforesaid, I find the ringed agaric *(Armillaria mellea)*, which is stigmatised as *valde venenatus* by Persoon, an expert on the subject. It is even the mushroom most frequently made use of, because of its being so plentiful, especially at the foot of the mulberry-trees. I find the Satanic boletus *[B. Satanas]*, that dangerous tempter; the belted milk-mushroom *(Lactarius zonarius)*, whose burning flavour rivals the pepper of its woolly kinsman; the smooth-headed amanita *(Amanita leiocephala]*, a magnificent white dome rising out of an ample volva and fringed at the edges with flowery relics resembling flakes of casein. Its poisonous smell and soapy aftertaste should lead to suspicion of this ivory dome; but nobody seems to mind them."

"How, with such careless picking, are accidents avoided? In my village and for a long way round, the rule is to blanch the mushrooms, that is to say, to bring them to the boil in water with a little salt in it. A few rinsings in cold water conclude the treatment. They are then prepared in whatever manner one pleases. In this way, what might at first be dangerous becomes harmless, because the preliminary boiling and rinsing have removed the noxious elements."

Fabre says further that during the thirty years that he lived at Serignan, he never heard of one case of mushroom poisoning, even the mildest, in the village. This may well have been the case, if the villagers confined their attentions to the species indicated, for, as we now know, not one of them is poisonous.

In the light of this evidence, it is not surprising to find it quite a usual practice with certain of the Negro women of the United States to eat *Amanita muscaria* with impunity, after a suitable preparation involving extraction by both salt and vinegar, thus carrying out very thoroughly the recommendations of Gerard, although the worthy ladies have probably never even heard of him.

Some interesting details of this custom are given by Coville. It appears that most of the coloured women of the Washington markets wisely regard the species in question with dread, and its use as a food is confined to a few, one of whom "recited in detail how she was in the habit of cooking it. She prepared the stem by scraping, the cap by removing the gills and peeling the upper surface.

Thus dressed, the mushrooms were first boiled in salt and water, and afterwards steeped in vinegar. They were then washed in clear water, cooked in gravy like ordinary mushrooms, and served with beef steak. This is an exceedingly interesting operation from the fact that although its author was wholly ignorant of the chemistry of mushroom poisons, she had, nevertheless, been employing a process for the removal of these poisons which was scientifically correct.

The gills, according to various pharmacological researches, are the chief seat of the poisonous principles, and their removal at once takes away a large part of the poison.

The salt and water would remove phallin or any other toxalbumin the mushroom contained, and although the presence of phallin or any of this class of poisons has not been demonstrated in *Amanita muscaria* there is a strong suspicion that it may occur in slight amount.

The vinegar, secondly, removes the alkaloid poison, muscarine, and the mushroom after the two treatments is free from poisons. This process is cited, not to recommend its wider use, but as a matter of general interest.

The writer's recommendation is that a mushroom containing such a deadly poison should not be used for food in any form, particularly at a season when excellent non-poisonous species may be had in abundance, with which opinion we entirely agree, and there we may leave the matter.

Bibliography

Aberdeen, J. E. C.: *Introduction to the Mushrooms: Toadstools and Larger Fungi of Queensland,* Queensland Naturalists Club, Brisbane, 1979.

Ahmadjian, V. and Paracer, *Symbiosis: An Introduction to Biological Associations,* University Press, New England, 1986.

Ainsworth, G. C.: *The Fungi: An Advanced Treatise,* Academic Press, New York, 1973.

——————— : *Ainsworth and Bisby's Dictionary of the Fungi,* Commonwealth Mycol, Kew, 1971.

——————— : *Introduction to the History of Mycology,* Cambridge University Press, Cambridge, 1976.

Ajello, L. Georg L.K. Kaplan W. and Kaufman L.: *Laboratory Manual for Medical Mycology,* US Publ., New York 1963.

Amaranthus, M. P. and Luoma D. L.: *Diversity of Ectomycorrhizalfungi in Forest Ecosystems,* International Symposium on Microbiology—Santos, Brazil, 1997.

Ames, L. M.: *A Monograph of the Chaetomiaceae,* MacMillan, New York, 1963.

Ammirati, Joseph F. James A.: *Poisonous Mushrooms of the Northern United States and Canada,* University of Minnesota Press, Minneapolis, 1985.

Arnold, Katya and Swope Sam: *Katya's Book of Mushrooms,* Henry Holt and Co., New York, 1997.

Arora, David and Bowers J.: *Mushrooms of the World Coloring Book*, Dover Publications, New York, 1984.

Arora, David: *All that the Rain Promises*, Ten Speed Press, Berkeley, 1991.

——————— : *Mushrooms Demystified*, Ten Speed Press, Berkeley, 1986.

Arx, J. A.: *The Genera of Fungi Sporulating in Pure Culture*, J. Cramer, Vaduz, 1981.

Barnes, R. S. K.: *Diversity of Living Organisms*, Blackwell Publishers, New Jersey, 2002.

Barnett, J. A. and R. J. Pankhurst: *A New Key to the Yeasts*, North Holland Publ. Co., Amsterdam, 1974.

Barnett, H. L. and Hunter B. B.: *Illustrated Genera of Imperfect Fungi*, MacMillan, New York, 1987.

Barron, G. L.: *The Genera of Hyphomycetes from Soil*, Williams and Wilkins, Baltimore, 1968.

Barron, George: *Mushrooms of Northeast North America*, Lone Pine Publishing, Renton, 1999.

Bell, A.: *Dung Fungi: An Illustrated Guide to Coprophilous Fungi in New Zealand*, Victoria University Press, New Zealand, 1983.

Beneke, E. S.: *Medical Mycology Laboratory Manual*, Burgess Publ. Co., Minneapolis, 1958.

Benjamin, D. R.: *Mushrooms: Poisons and Panaceas*, WH Freeman and Company, New York, 1995.

Bessette, Alan and Sundberg, Walter J.: *Mushrooms: A Quick Reference Guide to Mushrooms of North America*, MacMillan Publishing Co., New York, 1987.

——————— : *Mushrooms of North America in Color: A Field Guide Companion to Seldom-Illustrated Fungi*, Syracuse University Press, Syracuse, 1995.

Bessette, Alan E. Roody, William C.: *North American Boletes*, Syracuse University Press, Syracuse, 2000.

G

H

I

O

P

R

S

T

U

V

W

X

Y

Z

□□□

ENCYCLOPAEDIA
OF
FUNGI

ENCYCLOPAEDIA
OF
FUNGI

Vol. 2

Edited by
A.M. Bagulia

ANMOL PUBLICATIONS PVT. LTD.
NEW DELHI - 110 002 (INDIA)

ANMOL PUBLICATIONS PVT. LTD.
H.O.: 4374/4B, Ansari Road, Darya Ganj,
New Delhi-110 002 (India)
Ph.: 23278000, 23261597
B.O.: No. 1015, Ist Main Road, BSK IIIrd Stage
IIIrd Phase, IIIrd Block,
Bangalore - 560 085 (India)
Visit us at: www.anmolpublications.com

Encyclopaedia of Fungi

First Published, 2008

ISBN 978-81-261-3526-4 (Set)

PRINTED IN INDIA

Printed at Mehra Offset Press, Delhi.

Contents

Preface

Fungi are a kingdom of eukaryotic organisms. The fungi are heterotrophic organisms characterised by a chintinous cell wall, and in majority of species, filamentous growth as multicellular hyphae forming a mycelium; while some fungal species grow as single cells. Sexual and asexual reproduction is via pores, often produced on specialised structures or in fruiting bodies. For example, yeasts, moulds, and mushrooms are fungi.

Fungi are found all over the world, and grow in a wide range of habitats, including desert. Most fungi grow in terrestrial environments, but several species occur only in aquatic habitats. Fungi along with bacteria are the primary decomposers of organic matter in almost all terrestrial ecosystems, worldwide. As per estimates there are 1.5 million fungal species of which around 70,000 have been marked. Most fungi grow as thread-like filaments called hyphae, which form mycelium, while others grow as single cells. Human use of fungi for preparation or preservation and other purpose is extensive: yeasts are required for fermentation of beer and bread, some other fungal species are used in the production of soy sauce and tempeh.

Mushroom farming and gathering is a large industry in many countries. Many fungi are producers of antibiotics, including β-lactam antibiotics, such as a pencillin and cephalosporin.

Some of the most well-known types of fungi are edible or poisonous mushrooms. Many species are commercially raised but others are grown in forests. *Agaricus bisporus*, sold as button mushrooms, when small and Protobello mushroom when grownup, are the most commonly eaten species, used in salads, soups, and many other dishes.

Many fungi have important symbiotic relationships with organisms from most, if not all kingdoms. The mycorrhiza symbiosis is most ancient and its existence dates back to 400 million years ago.

This eminent work covers all perfect and realistic features of fungi in an interesting manner. Expectedly, this work would create new waves in the discipline and would prove to be useful for all—students, scholars and general readers.

Editor

1

The Phenomena

Curious Phenomena Exhibited by Fungi

"The giant puffball *Lycoperdon Bovista]*... increases from the size of a pea to that of a melon in a single night"—1861, H. Macmillan, *Footnotes from the Page of Nature*, 199.

"Matthiolus mentions mushrooms which weighed thirty pounds each, and were as yellow as gold; Fer. Imperatus tells us, he saw some which weighed above a hundred pounds a-piece; and to add no more, the Journal des Sqavans furnishes us with an account of some growing on the frontiers of Hungary, which made a full cart-load"—1819, REES, *Cyclopaedia*, Vol. 24, under Mushroom".

Although popular antipathy to the fungi is certainly due in no small measure to the poisonous qualities which many possess, there are other characters which, to the layman, appear strange and inexplicable, and these have doubtless assisted in maintaining these plants in the invidious position which they hold in popular estimation. The uncanny mode of appearance; the rapidity of growth of some species, often inordinately magnified in general belief; the huge sizes frequently attained; the power, enormous for such frail bodies, which many are capable of exerting; the chameleon-like tendency, which certain species show, to change colour when broken; their phosphorescence; and the often unpleasant mode of dissolution—such characters are enough in themselves to make the casual wayfarer a little chary of closer investigation. Yet, when any of these seemingly magical properties is examined more closely, one can see how truly it is governed by natural causes, and, incidentally, how erroneous are many popular conceptions concerning fungi.

Rapidity of Growth: Among the phenomena which are usually considered to be characteristic of fungi, that which has most firmly impressed itself upon the public mind is that they grow at an enormously rapid rate. One is continually encountering the statement that mushrooms grow up in a night, while puffballs of stupendous size are said to appear suddenly in places where there was nothing before; and so on. We need adduce no further examples—this belief is proverbial—and although for reasons which we shall shortly enumerate there is a good deal of truth in the statement, yet in another sense, it must to some extent be discounted.

In the first place, it should be remembered that the fruiting bodies of fungi are produced for a specific purpose—that of generating spores—a process which, in most cases, must be accomplished rapidly at a time when climatic conditions are suitable. It is, therefore, only natural that these plants should prepare for this eventuality by constructing organs, which can be gorged with water, and rapidly distended and erected at this appropriate time. From this point of view, it is not surprising that their growth should be comparatively rapid.

Further, it should be pointed out that in what is usually regarded as the "growth" of fungi, at least two different factors are operating. There is in the first place the true growth, which we may regard as the formation of new tissue; and, secondly, there is an apparent growth, which consists of the rapid distension of tissue already formed, but previously unexpanded. The two senses are not quite the same. The apparent growth of unexpanded tissue may be likened to the regular expansion of a deflated tube on pumping in air, or of a dry sponge which has become full of water.

Leaving for a moment the question of the rapidity of "true" growth and the reasons for it, we may consider briefly one effect of the "apparent" growth, which is undoubtedly that responsible for the popular belief that mushrooms grow in a night. In the egg stage, and as an example we may take that of the Dog Stinkhorn *(Mutinus caninus)*, the substance of the stalk is much compressed; and when the spores are formed, the hymenial part of the fungus perforates and bursts through the peridium, and is carried upwards by a very rapid elongation of the stalk.

The rate of elongation in the case of an ally, the Common Stinkhorn *(Phallus impudicus)* has been recorded by Lees as about six or eight inches in two or three hours. With both fungi the egg itself is often an inconspicuous object, partly buried in the ground and sometimes partly or almost entirely covered by grass or fallen leaves; while on the other hand, the expanded fungus is very prominent, in which the egg is shown with the fungus emerging therefrom. There is thus a very good reason for thinking that it has "grown" in a night, and the same remarks apply in much the same degree to the Common Mushroom. Extended preparation has, of course, been made on the part of the plant for such growth; but without a knowledge, on the part of the observer, of the causes which have led up to it, the result is naturally somewhat startling.

It must, at the same time, be realised that the true growth of fungi is much more rapid than in the case of most other plants. On consideration, the reasons for this are quite clear. Fungi occur usually in places where unlimited supplies of food material are found; this material is already in a state fit for immediate utilisation; the vast majority of fungi contain very little solid matter, but are composed very largely of water, which can be absorbed very quickly; while finally, fungi have in general a large number of growing points, instead of the very limited and localised growing layers found in most other plants.

These factors, as a whole, combine to produce in many of the fungi that "rapidity of growth" which has always been a source of amazement to the casual observer, and even when all considerations are appreciated, one can still marvel at the enormous size of many of the fruiting bodies, and the relative speed at which they are constructed, when compared with the total quantity of mycelium from which they arise.

A statement by Macmillan, concerning the rate of development of the Giant Puffball, which well illustrates popular belief in the matter, but which can scarcely be correct. Sowerby cites the instance of a specimen of *Polyporus squamosus* (Frontispiece), as having attained a weight of 34 Ib in three weeks from the time of its first appearance, and this is probably by no means unusual for this and allied species in our own country.

Sizes Attained: Very closely related to the question of rapidity of growth is that of the size which, under favourable conditions, may be attained. In the case of the larger polypores, a weight of thirty pounds is quite possible, and under exceptional conditions, those, for example, of the tropics, there would appear no reason why a much greater development should not occur.

When, however, we read of some on the frontier of Hungary, which made a full cart-load we are moved to enquire as to the size of the cart. Bulliard certainly mentions puffballs of nine feet in circumference, and some of the tropical fungi are said to have been mistaken for sleeping lions, but it would not be a very large cart for which either would constitute a "load" within the meaning of the term.

To consider some examples which have been recorded with a little more exactitude, we note a field mushroom from

Bewdley, 30 inches in circumference; while of the same year and county, a Mr. Cuthbert Bede records that "field mushrooms would seem to have been very large this year. In my own meadow I gathered several of great size, two of which measured respectively 33 and 36 inches in circumference."

In this present year of 1924, large examples have also been found. From Ingoldisthorpe, Norfolk, is recorded one measuring 151 in. across and 47 in. round. This is, however, by no means a record for this species, and an example four feet in circumference and weighing five pounds was found by a farmer in the woods near Chateau Roux, in France.

It is, however, extremely doubtful if most of the large mushrooms recorded are really Field Mushrooms *(Psalliota arvensis)*. More probably are they *P. villaticus*, a much larger fungus. The record for abnormal size appears to be held by a specimen of the Common Mushroom grown in 1846 in the quarries at Vitry, near Paris, and presented by the growers to King Louis Philippe. The cap of this mushroom had a circumference of a little over 43 inches, and the stalk one of 18 inches, while the weight was over 5 pounds.

A good-sized puffball *(Lycoperdon Eovistd)* which was found in a meadow at Goldington, near Bedford, about the end of July, 1894. When found, it was 34 inches in circumference and weighed over 9 Ib. When, three days later, it came into the possession of our friend, Mr. E. M. Langley, M.A., it was 333- inches round and weighed just over 6 Ib.; while today, thirty years later, it is very small and shrivelled, and weighs 10 oz. This of course is by no means unusual, and is indeed insignificant, when compared with an American example of the same species which has been described. This measured 5 ft. 4 in. in its greatest diameter, by 4 ft. 6 in. in its least, and was 9 in. high. A simple calculation will show that this specimen was approximately 15-feet in circumference, which is certainly phenomenal.

Forces Exerted by Growing Fungi: There have, from time to time, been recorded instances of paving stones and similar heavy bodies being forced upward from their positions by the levitational action of fungi growing under them, and great astonishment is often expressed that such fragile organisms should be capable of exerting such tremendous forces. Some cases may, with advantage, be referred to.

In the town of Basingstoke, in the year 1830, a large paving-stone measuring 22 by 21 in., and weighing 83 lb., which had been cemented down, was dislodged and elevated an inch and a half out of its bed by the growth of two mushrooms beneath it, and other stones becoming deranged for a similar reason and in varying degrees, much perturbation was naturally occasioned to the paving contractor, who had but recently completed the work, "for it seemed doubtful whether the whole town of Basingstoke might not want re-paving during the term of his contract," but whether, as Cooke asserts, the whole town was repaved, we are not able to say, and are certainly doubtful whether such a heroic expedient could have been really necessary.

Cooke records a similar incident, which came under his own notice, "of a large kitchen hearthstone, which was forced up from its bed by an undergrowing fungus, and had to be relaid two or three times, until at last it reposed in peace, the old bed having been removed to a depth of six inches, and a new foundation laid." Many similar cases, most of which have been due to *Coprinus atramentarius* have occurred, for example, at Cheltenham in 1840, at Ely in 1901, and elsewhere. Another remarkable instance of a fungus growth is related by Sir Joseph Banks. A cask of wine, which had been left undisturbed for three years in a cellar, was at the end of that time found firmly fixed against the roof of the same, having been carried thereto in the embrace of an enormous growth of fungus, evidently developed from the leakage of the wine from the cask, which was empty.

From the superficial evidence, such occurrences are certainly puzzling. If, however, the matter be carefully considered, it becomes fairly obvious that the water absorbed by the thousands, possibly millions, of mycelial threads, during the simultaneous fruiting, enables the vessels to act as minute hydraulic rams, the combined effort resulting in a slow but inexorable movement. It is clear, also, that no greater pressure can be generated than the individual vessels are able to stand. Even so, the results are sufficiently astonishing.

Colour Changes Exhibited by Fungi: Among the many queer characteristics of the fungi, this is not the least curious. That fungi, which change colour when broken, are unwholesome, has already been cited, as one of the empirical and quite valueless "rules" by which poisonous species may be detected. Such a colour change is well shown in many species of Boletus and in representatives of various other groups, the flesh of which, when broken and exposed to the air, changes from white or yellow to a deep indigo blue. According to Bourquelot and Bertrand, the change is due to the presence in the fungus of an oxidising ferment, which they call tyrinosase, and such fungi apparently also contain chromogenous substances which when thus oxidised are converted into coloured products. An analogous change of colour is seen when the flesh of an apple is exposed to the air for a short time.

Luminosity of Fungi

Another somewhat eerie manifestation occurring in some fungi is their luminosity, of which many examples have been recorded. In a description of a forest in Sumatra, Forbes says: "The stem of every tree blinked with a pale greenish-white light which undulated also across the surface of the ground like moonlight coming and going behind the clouds, from a minute thread-like fungus invisible in the day time to the unassisted eye; and here and there thick dumpy mushrooms displayed a sharp, clear dome of light, whose intensity never varied or changed till the break of day; long phosphorescent caterpillars and centipedes crawled out of every corner, leaving a trail of light behind them, while fireflies darted about above like a lower firmament." A remarkable account is given by Berkeley, which shows how striking this phenomenon may be. A quantity of wood, having been bought in an adjoining parish, had to be dragged up a very steep hill to its destination. Part of the consignment consisted of a log of larch or spruce, it is not quite certain which, about twenty-four feet long and a foot in diameter.

The same night, some young people, having occasion to go up this hill, were astonished to find the roadway strewn with luminous particles, which, on investigation, were found to consist of small pieces of bark and wood. Following the trail, they came to a very strong blaze of white light, proceeding from the log in question. On closer investigation, it was found that the whole of the inside of the bark was covered by a white byssoid mycelium, having an extremely strong smell, but unfortunately so damaged that its true form could not be determined. It was from the mycelium that the light was proceeding, and the brightness was most intense in places where the spawn had penetrated more deeply, where, indeed, the roughest treatment, if, for example, one attempted to rub off the luminous matter, only caused it to glow more brightly. If wrapped up in five folds of paper, the light shone as strongly through the paper as if the specimen had been fully exposed, and if placed in the pocket, the latter on opening again was a perfect blaze of light. When these tests were made, the luminosity had been going on for three days. Fabre considers the mechanism of the luminosity to be an expression of the respiration of the cells of the fungus, and says:

> "Let us begin by questioning the olive-tree agaric or luminous mushroom *(Pleurotus phosphoreus)*, a magnificent mushroom coloured jujube-red.... A more remarkable feature distinguishes it from all the other European mushrooms; it is phosphorescent. On the lower surface and there only, it sheds a soft, white gleam, similar to that of the Glowworm. It lights up to celebrate its nuptials and the emission of its spores.

There is no question of chemists' phosphorus here. This is a slow combustion, a sort of more active respiration than usual. The luminous emission is extinguished in the unbreathable gases, nitrogen and carbonic acid; it continues in aerated water, it ceases in water deprived of its air by boiling. It is exceedingly faint, however, so much so that it is not perceptible except in the deepest darkness. At night and even by day, if the eyes have been prepared for it by a preliminary wait in the darkness of a cellar, this agaric is a wonderful sight, looking indeed like a piece of the full moon."

The most interesting aspect of this luminosity is its reason, for Nature does not work blindly, although to mankind its object may sometimes be obscure. To seek such a parallel from the animal world as is afforded by the glowworm, it would appear not unreasonable to suppose that luminosity is sometimes a device to attract insects and so ensure spore dissemination, and it must obviously be the night-flying insects which are so attracted, in much the same manner as the brilliant tints of flowers serve to allure those which fly by day.

That any luminous object possesses an irresistible attraction for night-flying insects is proverbial, and since flies and beetles use certain luminous fungi in which to lay their eggs, the conjecture is a not unlikely one. This explanation, however, cannot always hold good, for in certain cases, e.g., that of *Armillaria mellea*, it is only the young mycelium that is luminous, and not the fruiting bodies. Why some fungi should possess this property, and others not, is as inscrutable as is the erratic manner in which poisons occur in fungi. It should be noted that fungi are only luminous while growing, and the emission ceases when the plant is dead.

Fungi Responsible for the Rain of Blood, and Allied Phenomena

There have been recorded, from time to time, instances of the falling of coloured rain, sometimes blood-red, which the superstitious have naturally regarded as a warning of the direst calamities; sometimes of other colours, giving rains of ink, or of sulphur, according to the prevailing hue; each alike being probably vastly disquieting to the uninitiated. It would appear that various causes may at different times be responsible for these effects, which were closely studied by Ehrenberg, and may be sometimes due to soot, sometimes to pollen of different trees, and in other cases to the presence of fungi.

The red rain is generally considered to be caused by a fungus of the lowest order, and in the opinion of Berkeley "it is a state of some mould analogous to the gelatinous spots of various colours which appear on meat or paste when in a state of incipient decomposition. It increases with immense rapidity, and is easily propagated, a very singular circumstance being that from particular spots multitudes of lesser spots extend in a straight line, as if blood had been spurted out from a wounded artery." The appearance of blood-coloured spots on various articles of food was very common in the hot days of July, 1853, particularly in France, and when at Rouen at this time, Montagne investigated many such cases, on both animal and vegetable food, which became so spotted in a few hours. Food cooked in the evening was found the next day to be covered with this growth. Montagne found the red substance to be the same as that previously investigated by Ehrenberg, in Berlin, and he was able to propagate it with ease on rice paste. It is by this means that the supposed appearance, in medieval times, of stains of blood upon the Host, which was then naturally regarded as miraculous, has been explained.

Curious Growth of Fungi

It is a not uncommon thing to find two mushrooms growing with their caps joined together at the edge in a Siamese-twin manner. An even more curious example of the blewitt *(Tricholoma personatum)*, shown to the authors, had three individuals growing one above, and out of, the other, in the form of a pagoda. In another interesting case which was brought to our notice, the growth of a large Parasol *(Lepiota procera)* had carried upwards on its cap a small quantity of humus and mycelium, out of which was growing a complete but smaller specimen of the same species. A large number of monstrosities have no doubt been observed.

Rapidity of Decay in Fungi

This is naturally intimately related to that of rapidity of growth. When compared with that of most of the flowering plants, the reproductive portion of most fungi not only grows to maturity, but, its purpose accomplished, decays also so rapidly that it would appear to have been given a dose of that new accelerator, the effects of which one of our authors has so humorously described.

Thus, in the space of some forty-eight hours, many species of *Coprinus* will erect their caps, fructify, and deliquesce into a collapsed, black and semi-liquid mass, popularly regarded as "a horrible putrescence." When this process is considered by an enquiring mind, however, one loses such a viewpoint and, instead, can have nothing but admiration for another of Nature's devices by which spore dissemination is facilitated. In the first place, these fungi are auto-digestive, and contain a ferment or oxidising enzyme which gives the dark colour to the liquid produced.

The fact that fungi are composed very largely of water naturally simplifies such a process, which, by the removal of those portions of the gills which have already discharged their spores, provides the necessary space for the shedding of others in their turn, this being further facilitated by the gradual opening and reflection of the cap as the process continues.

It must be presupposed that fructification only takes place when external conditions are favourable, and thus, by the speeding-up of the whole process of fructification, it is sought to ensure that the spores will be liberated while such conditions continue. The change in the appearance of the fungus during this process is often phenomenal; indeed, the neophyte might be forgiven for refusing to credit the identity of the old, with the young form of the same species.

Many other curious phenomena are exhibited by fungi, but it has here been our object, in the first place, to illustrate some of those characteristics of the fungi which have often excited popular speculation, and, secondly, to give some explanation as to the causes for the results observed. In spite of the view which holds *omne ignotum pro magnifico,* we think that the reasons often afford as much additional cause for wonder as the original phenomena, in which belief we hope for the support of the reader.

Dormancy of the Root-infecting Fungi

Specialisation of parasitism in a root-infecting fungus brings with it the hazard of local extinction when the infected host plant dies; the greater the degree of specialisation, the more remote is the chance that the parasite will be able to reach another susceptible host before the death of the first one.

This hazard has been substantially reduced through the development of resting spores and sclerotia by many of the root-infecting fungi. Whereas long-lived resting spores are typically produced by members of the Phycomycetes, sclerotia are especially characteristic of the Ascomycetes and Fungi Imperfecti.

By means of these organs, root-infecting fungi are enabled to survive for a period of years in the soil in the absence of a suitable host plant; a small proportion of such resting spores or sclerotia eventually re-establish the parasitic phase of the fungus, provided that sooner or later contact is made with the roots of a susceptible host.

Resting Spores: The thick-walled resting spores of the Phycomycetes, resulting from the process of sexual reproduction, are responsible for long survival of the organisms in soil carrying no susceptible host plants.

Thus Schaffnit (1922) reported that potatoes contracted wart disease, due to *Synchytrium endobioticum,* on a plot that had been followed and kept free of weeds for no less than 10 years.

A four-year rotation is frequently quite inadequate to control the clubroot disease of crucifers, due to *Plasmodiophora brassicae;* on Agdell field at the Rothamsted Experimental Station in England, which has been under a 4-course rotation of swedes with three non-cruciferous crops since 1848, a severe outbreak of clubroot first attracted attention in 1920.

Gibbs (1939) has recorded survival of the resting spores of *P. brassicae* for 5 years in New Zealand, and Fedorintchik (1935) in Russia obtained 27 *per cent* infection in cabbage seedlings potted up in soil taken from a field not sown to crucifers for 7 years.

Geacii (1936) found *Aphanomyces euteiches,* the cause of pea root rot, to survive for at least 2 years in pots of infected soil kept free of weeds.

Jones and Drechsler (1925) secured field evidence that *A. euteiches* might survive under apparently non-susceptible crops for a period of 6 years, which was extended to 10 years by the survey of Linford and Vaughan (1925); later, however, Linford (1927) attributed these long survival periods to mild infections of the parasite in hosts other than pea.

There is nothing inherently improbable in such long survival of resting spores in the soil, especially in view of the long survival of certain types of weed seed under suitable conditions.

Nevertheless, when considering evidence from length of crop rotation, it is unwise to exclude the possibility that weeds susceptible to the parasite may have established themselves at some time or other on the land, though it seems very unlikely that cruciferous weeds could have been responsible for the high percentage of infection reported by Fedorintchik (1935), quoted above.

Scanty information is available concerning possible variation in survival of such resting spores under different soil conditions; such data as have been published have been derived from the field rather than from the laboratory.

Fedorintchik (1935) could detect no beneficial effect of ploughing fields two or three times a year in hastening the disappearance of the clubroot organism from the soil in Russia, and Gibbs (1939) concluded that survival was influenced by the length but not by the type of rotation and management employed by farmers in New Zealand.

In contrast to these field studies by Fedorintchik and Gibbs, in which cruciferous host plants were used to detect viable spores of *Plasmodiophora brassicae* in the soil, Bremer (1923) developed a direct microscopical technique for the same purpose.

Sections of infected cabbage roots containing spores of *P. brassicae* were buried in the soil, and viability of the spores was determined at intervals by plasmolysis in salt solution, followed by deplasmolysis in water.

Employing this technique, Bremer (1924) reported that a high proportion of spores germinated even in the soil freshly treated with lime, although germination was more rapid and more extensive in the acid, untreated soil.

He found that injury to the spores through treatment with mercuric chloride, or through exposure to a temperature of 45°C, was least severe in the alkaline soil at pH 8.0, in which the dormancy of the spores was presumably more profound.

As a result of this work, Bremer concluded that, in the absence of cruciferous host plants, *P. brassicae* was likely to survive for longest under those soil conditions, such as alkaline reaction, least favourable for spore germination.

This important conclusion is likely to apply to the survival not only of other resting spores, but also of sclerotia, and even of mycelium in infected host tissue as well.

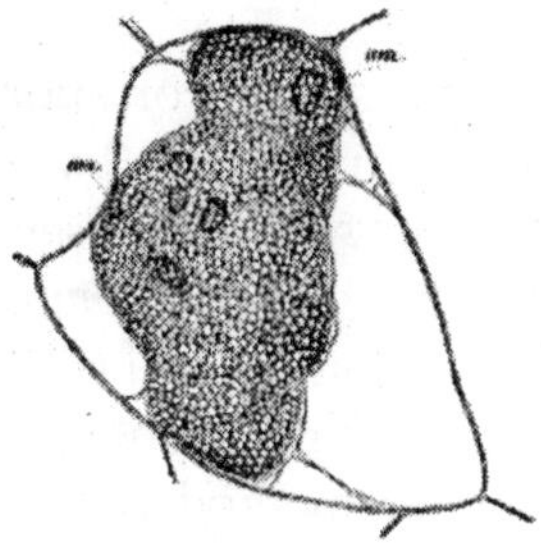

Formation of resting spores of Plasmodiophora brassicae in cells of host. (After M. Woronin).

A direct microscopical method was also developed by Glynne (1926) for studying survival of the resting bodies (winter sporangia) of *Synchytrium endobioticum*, the fungus causing wart disease in potatoes. The sporangia were mounted in 1-2 per cent aqueous acid fuchsin, and the thick brown sporangial walls were broken by pressure of the cover slip, so that the cell contents were expelled into the stain.

The contents of living sporangia slowly assumed a pale pink colour, which gradually deepened; the contents of dead sporangia rapidly stained deep crimson.

Whereas Bremer got over the difficulty of recovering the spores of *P. brassicae* from the soil by using sections of infected host tissue as a mount, Glynne solved the problem in another way, by extraction of the spores from soil with chloroform, which was found not to affect their viability.

The complete method was apparently never employed by Glynne to study survival of *S. endobioticum* under different soil conditions, as her published experiments were concerned chiefly with determination of lethal temperatures *in vitro*.

Both the Bremer and Glynne techniques, however, merit further consideration by investigators; although the resting bodies of these organisms survive for long periods in the soil, there must be a considerable reduction in numbers during the first few months, or even weeks, which could well be studied by such techniques.

Long-lived resting spores, or sporangia, of this type are characterised, as populations, by marked physiological heterogeneity. *Under no single set of soil conditions are all resting spores in a population likely to germinate.* Such physiological heterogeneity amongst populations of resting spores complicates the agricultural problem of eliminating them by cultivation or by other methods.

Other things being equal, the greater the physiological heterogeneity in a population of resting spores, the longer is the probable survival period of the parasite. For this reason, the more rapid and diverse the fluctuations in the soil environment, the more extensively will resting spores germinate (and disappear from the soil in the absence of host plants).

Factors especially likely to encourage germination are fluctuating soil temperature, moisture content, aeration and reaction; germination may also be stimulated by volatile substances emanating from plant residues buried in the soil, and from the roots of non-host as well as of host plants (NOBLE, 1924; Platz *et al*, 1927).

Some information is available concerning the survival of spores of the cereal smut fungi in the soil; these spores are rather different from the thick-walled resting bodies of the Phycomycetes, and are less effective in securing prolonged survival of the fungus in soil, though fully effective in seed transmission of the organism. Thus in many parts of the world, soil infection by *Titletia tritici*, causing bunt of wheat, is negligible, and satisfactory control is secured by seed treatment with a fungicide.

In the Pacific Northwest of the US, Fedorintchik USA, however, considerable losses from bunt were reported by Hungerford (1922) to follow contamination of fallows ready for autumn drilling of winter wheat, by clouds of bunt spores released during the threshing of crops on nearby fields. In a small plot experiment in which the soil was artificially inoculated with bunt spores, Hungerford found the percentage of bunted heads to decrease from a maximum of 35 on the plot sown immediately after soil inoculation to a minimum of 4.5 on the plot sown a month later.

In similar small-scale plot experiments, Hanna and Popp (1934) found that percentage of bunted heads in spring wheat varied according to the type of inoculum mixed with the surface soil in the previous autumn; sifted bunt spores gave *11 per cent* of bunted ears, bunt balls gave 29 per cent, and whole bunted heads gave 45 per cent. Spore germination in the spring was found to be higher in spores taken from bunted heads overwintering on the surface of the ground than in those taken from buried overwintered heads.

Bokzini (1935) reported that spores of *Ustilagoszcae,* the maize smut, survived better on the surface of soil than when buried in the soil; the buried spores survived better in air-dry than in moist soil. Similar experiments on survival amongst spores of *Urocystis tritici,* causing flag smut of wheat, were carried out in Missouri by Griffiths (1924).

Diseased leaves were ground up, mixed with 2 parts of soil and placed in 2-inch flowerpots, of which half were buried 1 inch below the soil surface, and the other half 5 inches below. Samples of pots were taken at approximately fortnightly intervals from the beginning of January, and tested for viability of *U. tritici* by the planting of wheat seedlings, which were held in the glasshouse. The percentage of infected wheat plants thus obtained decreased markedly at successive samplings; survival of the organism was more prolonged in the pots buried more deeply.

Sclerotia: Unlike resting spores, sclerotia are multicellular bodies, in which there is typically, though not always, differentiation of an outer skin or rind; this rind is assumed to exercise a protective function. The diameter usually varies from 0.5 to 2 mm., though a diameter of 1 cm. or more is sometimes encountered.

The sclerotia of root-infecting fungi germinate to produce

new vegetative mycelium; the minimum effective size of the sclerotium is determined by the volume of the food-base necessary for establishment by the fungus of root infection in the host. The most economical size of sclerotium for a root-infecting fungus may, therefore, be dinned as the minimum effective size. The sclerotia of flower and foliage-infecting fungi, on the other hand, usually germinate to produce fruiting bodies, from which air-borne spores are ejected.

The most economical size for sclerotia of this type is not necessarily the minimum effective size; in general, the sclerotia of the flower and foliage-infecting fungi are notably larger than those of the root-infecting fungi.

Even the ergots of *Claviceps purpurea* are larger than typical sclerotia of most root-infecting fungi; the large sclerotia formed by species of *Sclcrotinia* and *Botryis* are familiar to most mycologists. Brown and Butler (1936) found sclerotia of *Sclerotinia sclerotiorum* to attain a length of 3 cm. on infected lettuce plants.

Sclerotia of Phymatotrichum Omnivorum: These have received more attention than those of any other root-infecting fungus. The sclerotia of *P. omnivorum* were discovered by King and Loomis (1929) in laboratory cultures; they were first found in the field (in Texas) by Neal (1929). This discovery of true sclerotia in *P. omnivorum* accounted for the longevity of the fungus under non-susceptible crops; this would otherwise have been difficult to reconcile with the demonstration by Taùbenhaus and Ezekiel (19306) and by Neal and Maclean (1931) that the vegetative strands of *P. omnivorum* do not survive for more than a few weeks on dead roots.

Thus Ratliffe (1934), Rogers (1937), and Rea (1939) have concluded that satisfactory control of cotton root rot is not given by rotations of less than 4 years' duration, or, in other words, by a break of less than 3 years under non-susceptible crops. The sclerotia usually vary in diameter from 1 to 2 mm., though a diameter of 1 cm. may be attained (Rogers, 1937), and they are formed in chains on the vegetative strands.

The histology of the sclerotia has been described by King and Loomis (1929): "Microscopic examination of sections of the sclerotia showed that they were formed by the division and growth of the cells of portions of the large strands. This division of the cells apparently took place in the large central cells of the strands as well as in the smaller cells which surrounded the larger, as no continuous series of large cells of similar appearance extended through the sclerotia. Interior of the sclerotia is made up of closely packed, colourless, thin-walled cells of large and small size intermixed and varying in shape from roughly rounded to elongate oval and having no definite arrangement. From some of these surface cells, acicular hyphae, or setae, similar to those of the strands themselves, protrude at right angles, giving the sclerotia a somewhat bristly aspect under the microscope. The process of germination has recently been described in detail by Presley (1939), and appears to be distinctly unusual. The new hyphae do not arise as branches or elongations of the resting cells, but appear to be formed actually inside the cells, whose contents apparently go through a process of disorganisation and reorganisation that results in the formation of new hyphae, which eventually break through the old walls of the "mother cells".

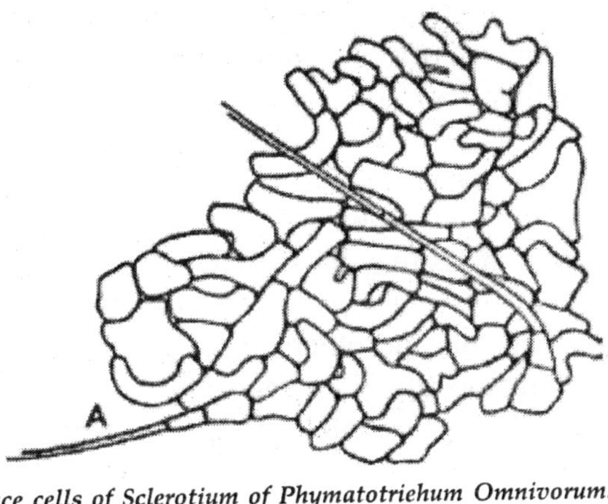

Surface cells of Sclerotium of Phymatotriehum Omnivorum. Acicular hypha at A. (After C. J. King and H. F. Loomis).

Noteworthy studies on this aspect of the cotton root rot problem are those by King and Hope (1932) and by McNamara and Hooton (1933). King and Hope found sclerotia abundant under cotton at depths between 12 and 24 inches, being most numerous between 16 and 20 inches. Under alfalfa they were abundant between 6 and 30 inches, and most numerous between 12 and 18 inches. Under a Chinese elm, sclerotia were found at levels ranging from 6 to 48 inches.

McNamara and Hooton made numerous excavations around the first cotton plants to die of root rot after 3 or 4 years of clean fallow or non-susceptible crops, on the assumption that such plants represented primary infections. Most of these primary infections were successfully traced either to sclerotia, or to infected plant material.

McNamara\ *et al.* (1934) later reported that sclerotia were not invariably the source of primary infections, many of which were traced to "persistent strands". Such strands were described as reddish to dark brown in colour, in contrast to the buff or light brown colour of the strands on the still living host plant; the persistent strands were also smooth, having lost the mantle of acicular hyphae that gives the active strands their characteristically fuzzy appearance.

Such persistent strands were sometimes found upon the decaying remains of old cotton roots; at other times they occupied empty root channels. Primary infections in cotton after 2 years of oats were traced to such persistent strands, which were also found interlacing a colony of sclerotia in a plot that had been under fallow for no less than 5 years. In the latter instance, the strands may have arisen through germination of the sclerotia at any time during the 5-year period; the sclerotia associated with this and other primary centres of infection in the first cotton crop after the 5 years fallow may also have been of secondary origin, through a process of "sclerotial budding", such as was described by King *et al.* (1931).

At first sight, these observations of McNamaka *et al.* (1934) upon the persistent strands may seem difficult to reconcile with those of Taubenhaus and Ezekiel (1930) and Neal and

Maclean (1931) on the very limited survival of vegetative strands upon plants just dead from root rot. The persistent strands, however, are probably more akin to sclerotia than to the ordinary vegetative strands; the sclerotia themselves, moreover, are merely a specialised development of the vegetative strands, upon which they arise.

Development either of sclerotia or of the persistent type of strand may be associated with a decline in nutritional level of the mycelium after a period of abundance; thus King and Loomis (1929) first observed formation of sclerotia on strands that had grown out from infected cotton roots into the surrounding moist sand in which the roots were buried. McNamara and Hooton (1933) reported, moreover, that they did not find sclerotia on the tap roots or larger lateral roots of cotton plants, but out in the soil where only small rootlets occurred. It is pertinent to recall at this point that the studies of Taubenhaus and Ezekiel (1930b) were carried out with the tap roots and larger lateral roots of infected plants.

McNamara and Hooton (1933) stressed the fact that in their excavations of primary centres of infection, they found no sclerotia in plots planted each year to cotton, whereas they were abundant under cotton following non-susceptible crops or fallow; they concluded, therefore, that the sclerotia "are connected primarily with the saprophytic phase of the fungus and serve as a resting stage". Rogers (1937), employing a highly effective method for separating sclerotia of *P. omnivorum* from the soil (Rogers, 1936), obtained, on the other hand, higher numbers of sclerotia under continuous cotton than under cotton or any other crop in a 4-year rotation of cotton with 3 non-susceptible crops.

Rogers' data do, however, tend to confirm the claim made by McNamara and Hooton that formation of sclerotia occurs after parasitic activity has ceased; the highest sclerotial population, and the greatest percentage viability of sclerotia in the population, was found by Rogers to occur not at the end of the cotton year, but at the end of the first year under a non-susceptible crop. In a laboratory study, Rogers (1939) found that in Houston black clay soil (saturation capacity 70 per cent) sclerotia were produced at moisture contents ranging from 21 to 43 per cent saturation, with an optimum moisture content at 36 per cent saturation. The temperature range for production of sclerotia was found to be 11°-37°C, with an optimum at 27°C.

Longevity of sclerotia of *P. omnivorum* in the soil. The sclerotia of *P. omnivorum,* like the strands, are very sensitive to desiccation; King *et al.* (1931) found them unable to survive drying for 1.4 hours in the air of the open laboratory. Other sclerotia kept under distilled water gave 81 per cent germination after 92 days, and 20 per cent germination after 121 days.

The susceptibility of the sclerotia to desiccation has been confirmed by Taubeniiaus and Ezekiel (1936) and by King and Eaton (1934) in experiments on survival and germination of sclerotia in soil at different moisture contents. Taubenhaus and Ezekiel studied loss of viability in sclerotia kept in Houston clay soil (saturation capacity 70 per cent moisture) maintained at different moisture contents in small, stoppered glass vials; 25 sclerotia were placed in the soil of each vial.

Germination was tested 9 days after setting up the experiment and subsequently at yearly intervals; 50 sclerotia were apparently selected from the 75 sclerotia originally placed in 3 vials. No sclerotia from the air-dry soil germinated at the first test, 9 days after setting up the experiments, or at any subsequent test. None of the sclerotia buried in the soils maintained at 14 per cent, 71 per cent or 86 per cent saturation was viable at the end of the first year; some of the sclerotia buried in the soils at intermediate moisture contents, *viz. 29 per cent,* 43 per cent and 57 per cent saturation, survived the whole test period of 5 years, viability at the final test ranging from 10 to 16 per cent. Ezekiel (1940) has since reported that mean germination was 10 per cent after 6 years, and 8 per cent after 7 years, but that no germination was obtained after 9 years or more.

Now survival of sclerotia depends upon two contingencies (i) that the sclerotia do not germinate and dissipate themselves as mycelium (ii) that they retain their capacity for germination. Since Taubenhaus and Ezekiel apparently selected 50 out of 75 sclerotia for their germination test, it is not possible to estimate, how many of the sclerotia were lost through germination or decay, but the proportion evidently did not exceed one third. The conditions of reduced aeration in the stoppered vials might have been expected to reduces spontaneous germination of the sclerotia; conditions of reduced aeration might also be deduced from the more rapid loss of viability in sclerotia kept in the two wettest soils.

In the experiments of King and Eaton (1934), on the other hand, disappearance of sclerotia through germination was much more frequent. Their experiment was carried out in a coarser-textured soil, Gila fine sand (saturation capacity 27 per cent moisture), in open containers; aeration conditions were obviously much better than in the stoppered vials used by Taubenhaus and Ezekiel.

The test was set up in May and the sclerotia were buried in the soil inside cheesecloth bags, for easier recovery, but in spite of this precaution King and Eaton remark: "Most of the cheesecloth bags had become badly decomposed by December, and many of the sclerotia having germinated in place, it became increasingly difficult to recover the desired numbers for making the monthly viability tests.

Survival in air-dry soil was better than that in the experiment of Taubenhaus and Ezekiel; 15 per cent of sclerotia were viable after 2 months, though none after 3 months. In the soil held at 19 per cent saturation, 33 per cent sclerotia were viable after 2 months, but none after 3. There was not much difference in rate of loss of germinability amongst sclerotia recovered from the three wettest soils, held at 37, 93, and 100 per cent saturation; germination varied from 10 to 18 per cent at the 12 months test.

A comparison of the results obtained by King and Eaton

with those of Taubenhaus and Ezekiel thus indicates the great importance of spontaneous germination as a factor reducing survival of sclerotia. The factors affecting survival of *P. omnivorum* sclerotia have been analysed with considerable ingenuity by Mitchell *et al.* (1941).

These authors found in laboratory experiments that the percentage of sclerotia recoverable from the soil was greatly reduced by the addition of 3 per cent ground organic materials, such as farmyard manure, cotton seed meal, cotton roots, sorghum, alfalfa or straw to the soil; these reductions in numbers of sclerotia occurred chiefly in the first two weeks' incubation in the organic-amended soils, and hence coincided with the period of most active decomposition of the organic material. Since observations made on Rossi-Cholodny slides suggested that this rapid disappearance of sclerotia was due chiefly to germination, an experiment was planned in which the recovery of germinable sclerotia was compared with that of non-germinable sclerotia. Two lots of sclerotia were available, *viz.* young sclerotia (80 per cent germinable) and old sclerotia (50 per cent germinable). Half of the sclerotia in each batch were killed by heating to 85°C for 5 minutes. Ten sclerotia were buried in each container at known locations on glass slides, so as to facilitate recovery.

Mitchell *et al.* concluded that germination was one of the factors contributing to the disappearance of the scierotia in the soils treated with fresh organic material. The living sclerotia disappeared much more rapidly than the killed ones, and the young ones (80 per cent germinable) more quickly than the old ones (50 per cent germinable).

Even the killed sclerotia disappeared more rapidly from the organic-amended than from the untreated soil, however; Mitchell *et al.* comment upon this as follows: "It is not known why heat-killed sclerotia are more rapidly eliminated from organic-amended soil during the early period of incubation. It is possible that a mass-inoculation effect with more numerous and perhaps more diverse soil microorganisms is obtained;

it is also possible that the sclerotial substance presents a source of supply of one or more elements required during the initial decomposition stages of the added organic material by the soil microflora concerned."

Clark (1942), who was one of the co-workers in Mitchell *et al.* later reported further experiments of this type, but on a more extensive scale, with the sclerotia of *P. omnivorum.* In a similar but larger experiment comparing rates of disappearance of viable *versus* heat-killed sclerotia, conditions were more favourable for decay of the dead sclerotia than for germination of the living ones; the influence of microbiological activity in accelerating disappearance of both living and dead sclerotia was demonstrated by a comparison between rates of disappearance of surface-sterilised sclerotia buried in sterile soil, on the one hand, and of untreated sclerotia in unsterilised soil, on the other.

The effect of organic materials in accelerating rate of disappearance of viable sclerotia was once again demonstrated, and was shown to increase with rising soil temperature, and with increase in moisture content of the soil. The effect also varied with soil reaction, being greater in an acid-treated soil of *pH* 3.0 and in a limed soil of pH 7.9 than in the untreated soil, or in the same soil receiving smaller doses of acid; the effect was more pronounced in sclerotia buried at some depth in the soil than in those buried near the surface.

These and other results obtained by Mitchell *et al.* and Clakk help to elucidate the controlling effect of organic manures upon cotton root rot, which is especially pronounced when such manures are applied some months in advance of sowing time. Mitchell *et al.* also reported a field experiment on land heavily infected with *P. omnivorum.* The usual late autumn ploughing in preparation for spring sowing of cotton was compared with deep rotary tillage carried out in early autumn; of the three rotary-tilled areas, the first received nothing, the second farmyard manure, and the third sorghum fodder. In August of the following year, excavations were made under the cotton crop to determine the relative abundance of sclerotia in the different areas. Sclerotia were not found in the upper 12 inches of soil in any of the areas that had been rotary tilled, though they were present below the 12 inch level; sclerotia were abundant, however, between the 6 and 12 inch levels in the control area that had received the usual late autumn ploughing. Working with another sclerotiurn-forming fungus, *Sclerotium rolfsii*, Leach and Davey (1938) obtained good correlation between numbers of viable sclerotia and incidence of root rot in the sugar-beet crop.

They found approximately 80 per cent of the sclerotia to occur in the upper 6 inches of soil, and less than 2 *per cent* of the sclerotia below 12 inches; soil sampling for sclerotia to a depth of 8 inches was accordingly considered adequate. An immediate practical use was found for this method of sampling for the sclerotia of S.*rolfsii*, by means of which probable losses from root rot in the next beet crop could be predicted; if probable loss exceeded an agreed level, the grower could be advised to substitute another crop for beet.

The figures of Leach and Davey showed a comparatively rapid decline in the population of viable sclerotia under a crop non-susceptible to this fungus; numbers were commonly reduced to one third or one quarter of the original population in an interval of 6 months, and in one instance numbers were reduced from 4810 in December to 150 (viable sclerotia per sq. foot of soil to a depth of 8 inches) in the following July.

The formation of sclerotia by *Rhizoctonia solani* in pure culture under different conditions of nutrition, temperature, atmospheric humidity, reaction of the medium, etc. has been studied by Tyner and Sanford (1935); the optimum temperature for sclerotial production was found to be 18°-21°C which is lower than that for *rate* of mycelial advance.

Formation of scelerotia was favoured by high atmospheric humidity, but was not greatly reduced even by a humidity as low as 31 *per cent*. In respect of their toleration of desiccation, the mature sclerotia of *R. solani* afford an interesting contrast to those of *Phymatotrichum omnivorum*. Thus Palo (1926) found them to survive for longer in air-dry than in moist soil, and Gadd and Bertus (1928) recorded survival of sclerotia stored dry in a corked tube in a Ceylon laboratory for 6 years.

Control of Root Disease in Field Crops

The problems of root disease control in field crops are perhaps more interesting, because more diverse, than those involved in the culture of either plantation or glasshouse crops. In plantation cropping, the grower is usually concerned with a single crop, to the needs of which everything can be subordinated, as in a factory for the mass production of a single article.

In commercial glasshouse practice, again, the grower is often engaged in production of only a single major crop, or perhaps two. In field cropping, on the other hand, the farmer is generally concerned not only with a rotation of three, four, or more crops, but also with the maintenance of stock as well. Any measures that are suggested for control of a soilborne disease affecting a single crop have, therefore, to be considered in relation to their effect not only upon that particular crop, but also upon the other crops in the rotation and upon the maintenance of stock as well.

Needless to say, the farmer will adopt those control measures that involve minimum expense and interference with his established routine; control of soilborne diseases must, therefore, be sought rather through improvement of general farming policy than through expensive special operations. Above all, the effectiveness of crop rotation in keeping the great majority of soilborne diseases and pests within bounds must be held responsible for the stillbirth of many ingenious schemes for root disease control. Again, whereas the soil and aerial environment of the crop can be directly controlled in glasshouse practice, in the field control of the crop's environment must usually be indirectly sought. Control measures employed against soilborne diseases of field crops will be classified and discussed under the following heads:

(i) Crop rotation

(ii) Plant sanitation

(iii) Disease control under the growing crop.

Crop Rotation: Crop rotation is the oldest and most effective method of root disease control in field crops; its origins are a matter for the agricultural historian, but its mechanism still presents to the plant pathologist some interesting problems, the existence of which has only recently come to be realised.

The effectiveness of crop rotation depends upon specialisation of parasitism in the soilborne parasites, amongst which may be included not only fungi and a few plant-pathogenic bacteria, but also plant-attacking nematodes as well; the very primitive parasites that are not limited in this way by their host range are limited in other ways, and so do not seriously impair the effectiveness of crop rotation for root disease control.

The efficiency of crop rotation as a root disease control measure may be defined as varying inversely with the length of rotation necessary to obtain adequate control of the particular disease. Adequate control does not imply complete elimination, but simply reduction of the parasite to a low level, at which it causes commercially negligible reduction in yield of the host crop.

Rotation is most effective in control of root diseases caused by those highly specialised root-infecting fungi that have only a limited power of saprophytic survival in infected host tissue, and do not form resting spores or sclerotia. It is much less effective against fungi producing resting spores or sclerotia at conclusion of their period of parasitic activity; highly specialised parasites of this type present the most difficult problem of all. The efficiency of rotation for root disease control also decreases as the saprophytic capacity of the parasite increases; the requisite length of rotation is inevitably prolonged by ability of such a facultative saprophyte to compete with obligate saprophytes for colonisation of virgin crop residues. Fortunately, however, this faculty of active saprophytism is possessed by such fungi at the expense of their parasitic capacity, so that these fungi are not so difficult to control as the more highly specialised parasites forming resting spores or sclerotia.

Elimination of Fungi Surviving Only as Mycelium Inside Infected Host Tissues: *Ophiobolus graminis,* causing the take-all disease of cereals, is a fungus with a strictly limited power of saprophytic survival, in tissues invaded as a parasite, from an examination of all available published evidence, Garrett (1942) has concluded that a one-year break under any non-cereal crop except pasture will give adequate control, though occasionally the disease may cause serious trouble in a two-course rotation such as that of sugar-beet and barley.

Outbreaks of take-all in a cereal crop grown in two-course rotation are probably limited to light-textured, alkaline soils exceptionally favourable to underground spread of the fungus along the roots of the host crop; in such soils, the small amount of inoculum remaining after the one-year break is much more dangerous than in heavier or more acid soils, in which the fungus spreads more slowly along the roots. As Garrett has pointed out, adequate control of take-all, and not complete elimination of *Ophiobolus graminis* from the soil, is the object of crop rotation; complete elimination of this fungus from the soil is not achieved even by a two-year break under fallow or a non-susceptible crop.

Another fungus that seems to have a strictly limited term of survival in tissues invaded as a parasite is *Verticillium alboatrum.* Thus in field experiments on control of the potato wilt due to this fungus, Mckay (1926) found adequate control to be given by a three-year rotation of potatoes with grain and clover (4 per cent infected' tubers), though not by a two-year rotation of potatoes with grain *(22 per cent* infected tubers); under continuous potatoes, there were 34 per cent infected tubers. In another experiment, the percentages of infected tubers in potato crops grown under rotations of 1-4 years duration were 25, 21, 1 and 0.3, respectively. A rather similar range of infection was obtained by Zeller (1936) in rotations designed for control of blue stripe wilt, due to *V'. alboatrum*, in black raspberries. The percentages of infected plants in raspberries grown under rotations of 1-4 years were 38, 21, 2.5 and 1.0, respectively.

McKay remarked upon the discrepancy between his cultural tests for recovery of *V. alboatrum* from infected potato haulm buried in the soil, and the results of his rotation experiments; the fungus was isolated from the infected haulms in May, after they had lain in the soil over winter, but not in May of the following year. Zeller buried a number of 10-inch sections of infected raspberry canes 6 inches deep in the soil at the end of October, 1927, and periodically attempted to isolate the fungus from each of 15 cane sections recovered from the soil at successive samplings.

The proportion of successful isolations varied as follows: February 1928 15/15, May 1928. 12/15, October 1928 6/15, and February 1929 0/15. Had Mckay and Zeller used the host plant instead of the agar plate as the ideal test for survival of *V. alboatrum* (admittedly a difficult proposition with this fungus), the discrepancy between these tests and the results of their rotation experiments would probably have been diminished.

Elimination of Fungi Surviving as Resting Spores: The resting spores of most cereal smut fungi have a relatively short life in the soil; adequate control of the kernel smuts is usually provided by ,seed disinfection, as soil contamination generally occurs only through liberation of spore clouds during threshing. Such soil contamination by the spores of *Tillctia tritici* was reported by Hungerford (1922) to cause losses of as much as 85 per cent through bunt in the Pacific Northwest of the USA; losses of this order through soil infection by bunt spores are fortunately rare.

Disappearance of free and unprotected smut spores from the soil seems to be very rapid, whether through germination or through death and decay; spores protected by enclosing host tissues have a better chance of survival. Whereas crop rotation is usually unnecessary for control of soil infection by the kernel smuts, it may be essential for control of the leaf smuts, because infected crop residues are turned into the soil; the spores of such leaf smuts enjoy the protection of enclosing host tissues.

Thus *Urocystis tritici*, causing flag smut of wheat, was found by Griffiths (1924) to survive for a year in a small plot originally heavily contaminated by the digging-in of infected wheat straw; 9/377 (2.4 per cent) test wheat plants became infected. A very similar result was later obtained by Tisdale *et al.* in (1927). Nevertheless, under ordinary farm practice, a one-year break is usually sufficient to keep the flag smut disease under control.

Another parasite that persists in the soil only in the form of resting spores, but has a greater power of prolonged survival, is *Plasmadiophora brassicae,* causing clubroot of crucifers. Although Fedorintchik (1935) has reported that this organism survived in a heavily infected field for 7 years and then caused 27 per cent infection of potted cabbage seedlings, this is probably an extreme case, and rotations of shorter duration than this period usually give adequate control of the disease in practice. Four-year rotations, in which cruciferous crops are kept off the land for three years, are sometimes adequate, but a five-year rotation is usually found to be the minimum that can safely be employed.

A shorter rotation than this may appear to control the disease on land that carries only a small population of *P. brassicae,* but may result in a gradual accumulation of this parasite in the soil; this is often overlooked until a fairly serious outbreak of clubroot has occurred. According to Fedorintchik (1935), a high concentration of spores in the soil is required before severe clubbing can occur. As with the take-all disease of cereals, the length of rotation required to keep clubroot under control can probably be correlated with suitability of the soil for parasitic activity; in alkaline soil, a much higher concentration of the organism is required for successful establishment of infection than in acid soil.

Another soilborne parasite perpetuated by means of resting spores is *Aphanomyces euteiches,* which causes root rot of peas. Geach (1936) found this fungus to survive for at least two years in pots of soil kept fallow and free of weeds. Field evidence at first suggested a much longer survival period; out of 9 fields that had not carried diseased peas for 10 or more years, three were found by Linford and Vaughan (1925) to be suffering from widespread and severe root rot. Later, however, Linford (1927), commenting upon the wide host range of *A. euteiches,* suggested that the fungus might be carried over these long periods in between two pea crops as a weak parasite on the roots of apparently resistant crops, and on those of weeds; he also reported a serious incidence of this disease in some crops sown on virgin land.

Elimination of Fungi Surviving as Sclerotia: Cotton root rot in the USA, due to *Phymatotrichum omnivorum,* furnishes the most outstanding example of an important root disease in which the length of rotation necessary for control has been correlated with longevity of the sclerotia in the soil. Ratliffe (1934), Rogers (1937) and Rea (1939) agreed that a two-year rotation exercised no check upon the disease at all, that a three-year rotation reduced it somewhat, but that for satisfactory control a four-year rotation had to be followed. Even a four-year rotation by no means prevented the appearance of root rot, but by reducing the number of plants affected, and especially by delaying the development of the disease until later in the cotton season, the four-year rotation markedly reduced losses in yield due to root rot.

Elimination of Facultative Saprophytes: Next may be considered those diseases caused by organisms with greater saprophytic potentialities. Little precise information is to be found concerning the saprophytic life of *Actinomyces scabies* in the soil, but the relatively long rotation required for control of potato scab suggests that the organism may survive as a saprophyte in the soil, independently of infected potato tissue. Goss and Afanasiev (1938) have discussed the incidence of scab in rotation experiments carried out under irrigation in Western Nebraska; two-year rotations failed to reduce the disease at all (and indeed it was worse in such rotations than under continuous potatoes), a three-year rotation appreciably reduced it, and four to six-year rotations gave satisfactory results, eliminating the severe form of scab, and giving the highest proportion of sound, unmarked tubers.

Another disease of the potato in which free saprophytic survival of the causal fungus may complicate control by rotation is stem canker and tuber scurf, due to *Rhizoctonia solani*. Blair (1943) has shown by laboratory experiments that this fungus can grow through untreated field soil as a saprophyte, though its activity is depressed by addition of 1 per cent or more of ground dried green manure, apparently through the competition of other microorganisms better fitted to attack the organic material.

Although the superficial tuber scurf form of this disease is caused by the formation of sclerotia, these bodies are often little more than mycelial plates, nor do they show any tissue differentiation; the part that they play in determining survival of *R. solani* in the field cannot as yet be estimated. Sanford (1939) remarked that, in some areas of Alberta, a year's summer fallow gave no appreciable control of the tuber diseases due to *R. solani* and *A. scabies*. Nevertheless, Goss and Afanasiev (1938) found a marked reduction in incidence of tuber scurf under two-year rotations in Western Nebraska, as compared with that under continuous potatoes, and a four-year rotation reduced it to a negligible minimum. Blodgett (1939) reported incidence of tuber scurf in New York State to be markedly reduced by a two-year rotation, as compared with continuous potatoes, but it did not seem to be further diminished by a five-year rotation. Goss and Afanastev, and Blodgett, observed incidence of tuber scurf to be substantially reduced both by farmyard manure and by a green manure cover crop; it is tempting to correlate this observation with the results obtained by Blair, but it must be noted that Blodgett also found a reduction in incidence of tuber scurf to follow the application of an artificial fertilizer.

The foot rot of cereals due to *Fusarium,* is well known to be less amenable to control by crop rotation than is the take-all disease; Sanford (1939) has commented upon this fact: "For example, one asks why one year of summer fallow produces such a marked effect in reducing the damage of *O. graminis* in the wheat crop following it, but gives much less control of *H. sativum* and *F. culmorum,* no control of *Pythium* sp., and in certain areas at least, no appreciable control of the potato diseases caused by *A. scabies* and *R. solani* An explanation of this difference in behaviour is perhaps afforded on the one hand by the findings of Garrett (1938-1940) concerning the limited power of survival of *O. graminis,* in tissues invaded as a parasite, and on the other by the demonstration by Sadasivan (1939) and Walker (1941) that F. *culmorum* can actually invade fresh wheat straw buried in the soil, as a free-living saprophyte. The foot rot of cereals due to *F. culmorum* is of little cosmopolitan importance, as compared with the take-all disease, but in Canada conditions seem to favour the parasitism of this fungus upon wheat; in combination with *Helminthosporium sativum* it causes serious losses. In view of the widespread distribution of *F. culmorum* in the wheat-growing soils of other countries, its failure to act at all commonly as a serious parasite of wheat elsewhere than in Canada requires further investigation.

For certain other root-infecting fungi, apparent survival periods of such length have been reported that, if the complete absence of host plants in the interim period can be confirmed, the parasites must be regarded also as saprophytes of *soil* inhabitant status. For example, Barker (1923) reported that it was frequently impossible to grow flax varieties susceptible to wilt *(Fusarium llni)* more than once every 10-12 years on the same land, and Meliius *et al.* (1926) found the cabbage yellows organism, *Fusarium conglutinans,* to last for at least 11 years to the soil; a survival period of 14 years under grass had previously been reported by Jones and Gilman (1915) for the same fungus, which was responsible for a severe outbreak of yellows in the first cabbage crop planted on the ploughed-up soil. It is known, however, that *Fusarium lini* can be seed-borne, so that this may explain the occurrences cited by Barker; Baylts (1940), reporting survival periods of 8 and 11 years in the absence of flax crops in New Zealand, took note of this possibility of seed transmission by observing that: "The seed came frôm crops in which no wilt was discernible, and the great majority of crops raised from it showed no sign of the disease".

Before such long periods of survival by a parasite in between one susceptible host crop and the next can be accepted as evidence of its capacity for indefinite saprophytic survival in the soil, it is necessary to exclude possibility of seed transmission. Again, Linford's (1927) suggestion, made in connection with the long apparent survival periods of *Aphanomyces euteiches,* that a fungus may survive as a weak parasite on the roots of apparently immune crops *(i.e.* showing no disease symptoms above ground), or on those of weeds, deserves wider consideration. The need for experimental investigation to supplement such field records as those just cited is thus very apparent.

Other effects of crop rotation upon survival and activity of root-injecting fungi. So far, only the obvious and well-established effect of crop rotation upon the numbers of a parasitic organism in the soil has been considered; the numbers of the parasite increase with the frequency of host crops. Little attention has yet been paid to the relative effect of different non-host plants upon the survival of a parasite.

Data have recently been provided by Bose (1938) in India which suggest that the incidence of root-disease in a crop may be profoundly affected by the nature and sequence of preceding crops in the rotation, even though none of these crops can act as host plants to the particular parasite concerned. Bose demonstrated that incidence of wilt, due to *Fusarium vasinfectum*, in pigeon pea was significantly less after a crop of tobacco than after one of linseed or fallow.

The design of the experiments failed to permit discrimination between the effect of the living roots of the preceding crops and that of the crop residues. This result was consistently obtained in experiments carried out in three consecutive seasons, and appeared both on naturally-infected plots, and on plots in which the natural inoculum had been artificially reinforced.

In pot experiments, Hildebrand and West (1941) found that the growth and turning-under of several consecutive crops of soybeans markedly reduced the incidence of Ontario root rot (the etiology of which is still incompletely understood) in strawberries; the growth of red clover cover crops, on the other hand, produced no such beneficial effect.

West and Hildebrand (1941) discovered that this beneficial effect of soybeans in reducing root rot could be simulated by addition of glucose to the soil. Later Richardson (1942), working in the same laboratory, reported that the growth and turning-under of four successive crops of soybeans in pots reduced the incidence of root rot, due to species of *Pythium*, *Helminthosporium*, and *Fusarium*, in corn seedlings; a succession of red clover cover crops, on the other hand, produced no comparable improvement in the health of the roots of the following corn crop.

A more prolonged survival of the take-all fungus in infected wheat stubble in fallow soil than in soil under crop has been reported by Garrett (1943). Eight weeks after burial of the pieces of infected straw in the soil, 68 per cent of those from the fallow soil contained viable mycelium of *Ophiobolus graminis*, but only 18, 17 and *4-per cent* of those from soil under trefoil, mustard and oats, respectively. Garrett attributed the more rapid disappearance of *O. graminis* from soils under crop to competition between the take-all fungus and plant roots for the limited supply of available nitrogen. He suggested that this observation might explain the success of a system followed by certain English farmers for continuous barley growing.

Barley is susceptible to the take-all disease, and the second or third barley crop on the same land is sometimes severely attacked. But barley undersown with trefoil (which makes a luxuriant growth in late summer and autumn after the barley has been cut, and is then ploughed in as a preparation for the next barley crop) had not thus far been observed to suffer from take-all.

To account for this observation, Garrett supposed that the active growth of the legume after harvest kept the available nitrogen content of the soil at a very low level, with resulting detriment to the longevity of O. *graminis* most of which was unable to survive until sowing of the new barley crop in the spring. The effects of crop growth upon the soil are many and diverse; one crop differs from another not only in its effect upon the microbiological equilibrium of the soil, as Hildebramd and West (1941) and West and Hildebrand (1941) have so extensively demonstrated, but also in its demands upon reserves of plant nutrients.

A particular crop may reduce reserves of available nitrogen, phosphate and potash in equivalent proportions, or it may absorb relatively greater quantities of one particular nutrient.

This effect of crop growth, unless subsequent adjustment be made by addition of artificial fertilizers to the soil, must sometimes modify the resistance of the following crop to certain root-infecting parasites.

Thus Vanterpool (1930-40), for example, demonstrated that the browning root rot of cereals in Saskatchewan, due to *Pythium* spp., was worse after fallow than after a wheat crop; he attributed this unwelcome effect of fallow to an increase in the nitrogen/phosphate ratio of the soil, which apparently lowered the resistance of wheat plants to this disease.

Condition of a Fungus in Substratum

Life Phases of Fungi: Fungi like other plants show two phases in their life period. When a fungus begins growth on a given substratum its first activities are directed to the procuring (absorbing) of food and the building up of the fungous body, while the formation of specialised structures for the production and dissemination of new individuals follows as soon as the fungous body has obtained a certain size, vigour of development or age. The condition of a fungus in or on the substratum may be then either in one of the *vegetative stages* or the purely vegetative structures may have given rise to specialised *reproductive* stages or structures.

The object on or in which a fungus establishes itself may be designated as its *substratum.* This may be the humus of gardens or fields, the dead remains or products of animals or plants, an artificial medium, another living plant or animal or portions of such living organisms—in fact, any material which will furnish the necessary food.

Host and Parasite: The plant pathologist is concerned primarily with fungi for which the substratum is a living plant or some organ or part of a living plant. The substratum may then be designated as the *host* (also *suscept)* and the fungus as the *parasite* which preys upon it. The fungus is simply carrying out its normal physiological processes— taking food from its host, growing and finally producing its reproductive structures. In this development it produces more or less serious disturbances in the life of its host, or disease in slight, pronounced or severe form is the result.

Vegetative Stages or Structures

Hyphae and Mycelia: The typical fungous body consists of a delicate, branched, tubular or filamentous structure, the *mycelium,* microscopic in size or barely visible to the unaided eye. These mycelial threads or *hyphae* may form an interlacing tangle, or a loose, woolly mass or they may be densely interwoven or even compacted into solid bodies. The hyphae of a given fungus are either *septate* or *non-septate,* that is, divided by cross-walls or partitions at intervals or entirely devoid of cross-walls, and so distinctly tubular in character.

Certain groups of fungi produce only non-septate hypha?, while other groups exhibit only septate hyphae of characteristic form and size. In working with fungi it is therefore, of importance to determine first the septate or non-septate character of the hyphae.

While the typical fungous body is of the nature described, certain fungi may consist of only a single globular cell no larger than individual cells of a hypha. These simple fungous bodies may represent primitive forms that have not yet organised a more complex plant body, or they may represent degenerates from more high specialised ancestors.

A parasitic mycelium may be either *external or internal,* that is, it may develop on the surface of its host, or it may grow within the tissues of its sooty moulds. Internal mycelia are confined to the intercellular spaces of the host tissue or penetrate into the interior of the host cells.

These also may be clear or hyaline or smoky to dark brown in colour. As they ramify through the host tissue, they are apparent to the unaided eye only when aggregated into dense masses. The presence of hyphae within the invaded tissues of blighting leaves or rotting fruits can only be demonstrated by microscopic examinations.

When a mycelium begins to spread from a point or focus in which it is established, it shows a tendency to grow radially in all directions, and thus occupy an ever widening zone. This tendency is very frequently interfered with by the character of the sub-stratum on which or in which the mycelium is developing. This peculiarity is manifested by both saprophytic and parasitic forms. Fungus "fairy rings" are the fruiting bodies developed on or near the periphery of an advancing mycelium, which is concealed within the soil. Many fungous leaf spots are distinctly circular, due to the manner of growth of the internal mycelium, while rotting areas on fruits show a circular surface outline for the same reason. Fungous invasions of the bark of woody hosts generally develop slightly elongated or somewhat elliptical lesions or affected regions, due to the fact that the lengthwise advance of the mycelium is more rapid than the transverse growth.

Mycelial Plates or Fans: In woody hosts and in some herbaceous hosts the mycelium becomes dense and compact and forms whitish pockets or radially or longitudinally elongated white plates or bands which are very evident when the invaded structures are cut open or broken.

Such mycelial plates are especially characteristic of several rots of coniferous trees, such as the pines and spruces, where they occur within the disintegrating wood. The serious chestnut blight of the eastern United States is caused by a fungus which makes its greatest development in the inner layers of the bark or in the cambium, where very characteristic tawny sheets of mycelium spread out in fan-like forms.

Mycelial Strands or Rhizomorphs: The hyphae of a mycelium are sometimes aggregated in the form of cord-like or thread-like strands, which are in reality fungous cables. Such mycelial strands vary from slender threads to others that are equal in diameter to good-sized strings. The strands generally branch more or less and may frequently fuse to form a network. In some forms they are white and very conspicuous, while in other cases they are tawny or even dark brown.

The honey agaric which is responsible for the so-called mushroom root rot of numerous hosts, especially fruit and forest trees, is frequently called the shoestring fungus, because of its brown root-like strands which mingle with the roots of the host or run over their surface.

Mycelium from these strands penetrates the bark and wood of roots or crown and causes the fatal disintegration or rotting of the tissues. On account of their root-like character these rhizomorphs frequently pass unobserved, but to one who becomes familiar with their appearance they offer a certain means of determining the presence of root rot in suspected cases.

While these fungous strands may store up some food, they are not primarily storage organs, but serve rather to bring the fungus to new hosts or to new parts of the same host and thus widen or enlarge the area occupied. In this way, the fungus may spread from a single diseased tree to adjacent trees by a natural growth. It is undoubtedly true that in orchard, practice fragments of rhizomorphs may be torn loose in cultivating and carried away to other parts of the field.

Sclerotia or Storage Organs: Some fungi have developed the habit of producing dense compacted aggregates of hyphae which become filled with food materials in the form of oil and other compounds. These structures vary in size from those that are hardly visible to the unaided eye to others that are as large as a cantaloupe or even larger, and are called *sclerotia,* the term being derived from the Greek work meaning "hard." Sclerotia are generally more or less rounded, elongated, cylindrical, globular or ellipsoidal masses, but they are sometimes more or less flattened and irregular in form. In many cases they are dark coloured, either on the surface or throughout the entire aggregate of cells. In parasitic fungi, they are formed either upon the surface of host parts, concealed between rotting leaves or within internal cavities. Sclerotial formation is not peculiar to parasitic fungi, but some saprophytes, like the carrion fungi and pore fungi, produce underground storage organs of considerable size. The so-called "Tuckahoe Indian Bread" of the southeastern states is a large sclerotium of the basidiomycete, *Poria cocos.*

A few examples of sclerotial formation by parasitic fungi which show interesting adaptations will be noted. In many regions potato tubers which show "dirt that will not wash off" are of very common occurrence.

These brown or black, hard masses which are the size of a pin head or larger (maximum 1 inch) are the sclerotia of a

Rhizoctonia, a fungus which is responsible for the Rhizoctonia disease of potatoes and many other hosts.

Alfalfa, clover, beans and some other non-leguminous hosts when suffering from a disease known as wilt produce on the crown or roots, or within the pith cavity of the basal portions of the affected stems, rounded or elongated, seed-like bodies, black without but whitish within, which are the storage organs or sclerotia of the wilt fungi *(Sclerotinia spp.).*

The ergots of rye and other grasses are elongated, cylindrical, black or purple, horny sclerotia, which take the place of normal grains.

Some fungi that are parasitic on insects show a very similar development. The fungus gains an entrance into the body of the insect, and in the caterpillar fungus fills up the entire larval host structures. The larva is killed and the "fungus caterpillar" or cast takes its place.

Sclerotia are, in reality, storage organs filled with the special kinds of reserve food peculiar to the species of fungus by which they are formed.

Although they are vegetative structures, they serve essentially the same purpose in the life of many fungi as spores, for they are able to withstand adverse conditions which would prove fatal to ordinary mycelia.

In the sclerotial condition fungi are able to endure extreme desiccation, or long periods of, high temperatures or the rigours of winter. Attached to host parts, such as seed or propagating stock, or mingled with the soil or seed, sclerotia are frequently very effective means of the dissemination of parasitic fungi.

There is great variation in the behaviour of sclerotia. Those of Rhizoctonia reproduce the fungus by the production of new vegetative hyphae when favourable conditions for growth are offered; the sclerotia of wilt and other Sclerotinias after passing through a winter period give rise to apothecial fruits, while the ergots of rye and other grasses develop new structures, which in their turn produce perithecia.

Reproductive Stages or Structures

The common reproductive structures of fungi are very small bodies of microscopic size, the *spores,* which are either cut off from hyphae or produced by specialised structures known as *spore fruits.*

Some fungi have retained a very simple method of propagation which does not involve the formation of specialised spores. The production of new cells, by a process of budding as in the yeast plant, essentially similar to the parent cell, is a primitive type of reproduction, which is retained by some of our parasitic fungi, notably some smuts, at certain stages in their life cycle.

Spores

General Characters: A spore is generally one to several specialised cells which serve the purpose of disseminating and reproducing the fungus. A single spore under proper conditions of temperature, moisture and host relations may start the development of a new fungous body, a *mycelium.* From the standpoint of function or the part which they play, spores are the "seeds" of the fungus, serving the same purpose for fungi as the true seeds do for flowering or seed plants.

Spores vary greatly in size and form and may be either clear or dark, varying from slightly smoky to almost black. The micron or micromillimetre is the standard unit for spore measurement, the smallest forms being 1/x or less in diameter, while in a few cases they may reach 1 millimetre or slightly more. In form they may be globular, ellipsoid, ovate, cylindric, filamentous, club shaped, star shaped, etc.

Sexual and Asexual Spores: Spores may be formed as the result of a breeding act, that is, the union of two separate and distinct cells or elements (gametes) which represent male and female. Spores of this type may be classed as *sexual,* while those which are formed direct from hyphae without the intervention of a breeding act are *asexual,* or without sex. It is undoubtedly true that asexual methods of spore formation were the most primitive, while sexuality was a later acquirement.

Kinds of Spores: Considerable variation is shown in the structure and origin of spores and special names are used for the different kinds. *Chlamydospores* are formed by the direct transformation of certain cells of a hypha, without the production of specialised spore-bearing branches.

Only scattered cells may be organised as spores (Mucor species) or the entire hypha may be used up in spore production (Smuts). *Swarm spores* or *zoospores* are naked protoplasmic masses provided with delicate vibratile filaments or thread-like processes, the *cilia,* by means of which they are able to swim about.

It is this power of locomotion which has suggested the name zoospore, or animal spore. The cilia, or organs of locomotion, are variously arranged, different species or groups of fungi showing one, two or a tuft of polar cilia, or they may be numerous and distributed over the entire surface of the spore. Certain swarm spores show a very characteristic kidney shape, with two cilia originating at the suture. All other spores are provided with a définite cell wall or membrane, and in many forms it is variously modified or thickened.

Conidia is a general term applied to spores which are pinched off or cut off from the ends of special spore-bearing hyphae, known as *conidiophores* or conidia bearers. Such spores may accumulate in clusters or in chains, and when in chains the terminal spore of the series may be the oldest in one case, while the basal spore is the oldest in others. There are many different kinds of conidia.

Ascospores are produced within a sac-like or club-shaped structure, the *ascus.* In a young ascus there are two nuclei which soon fuse to form a single nucleus. In a typical ascus this nucleus soon divides into two, each dividing again to produce four, while the number is again increased by division, the final number being eight. Each one of these nuclei organises a spore by surrounding itself and some of the adjacent ascus contents with a protecting cell wall. The result is a definite number of ascospores enclosed within the parent cell. There may be a deviation from the typical number of spores in some cases, one, two, four or sixteen being the common exceptions. Where there is an odd number of spores in an ascus it is due to the degeneration of one or more of the nuclei.

The production of ascospores is characteristic of one great group of fungi, the ascomycetes or sac fungi. In certain of the higher fungi, like toadstools, shelf fungi, puffballs and related forms, spores are produced on structures very similar to asci in form, but called *basidia.*

The spores, the *basidiospores,* instead of being produced within the structure, are developed on the tips of either two or four (rarely more) slender terminal projections (sterigmata). The number of spores is always definite and limited, a single spore being produced on each sterigma. In certain forms the basidia differ from the typical form by being divided lengthwise into four cells, each of which bears a spore, or they may be filamentous and divided transversely into four cells, each of which forms a spore. This later condition prevails in the true rust fungi, the basidium being designated as the *promycelium* and the spores as *sporidia.*

A spore formed by the union of two equal and similar gametes is called a *zygospore.* This is well illustrated in certain species of black moulds. In bringing about this union or *conjugation,* adjacent hyphae produce lateral branches, which develop in pairs, separate off a terminal multinucleate gamete from each, which fuse and form a thick-walled spore. In this case, the two gametes which unite are equal in size and similar in behaviour, so that male and female cannot be recognised. These gametes are, therefore, designated as positive and negative. Two positive gametes or two negative gametes would not have affinity for each other and could not unite to form a spore.

When the gametes which unite are of unequal size, the resulting spore is called an *oospore* and the union is designated as *fertilization*. The large gamete, the female cell, is passive, while the small gamete, the male cell, is active.

This type of spore formation is well illustrated in the white rust of Crucifers. In the formation of both zygospores and oospores nuclei of positive and negative gametes, or male and female gametes, fuse as the final act in conjugation or fertilization. Both zygospores and oospores are thick-walled resting spores which may carry the fungus over an unfavourable period.

The special types of spores peculiar to the true rust fungi will be reserved for consideration in the treatment of that group.

Spore Fruits

General Characters: In the simpler forms of spore production, the spores are cut off from the ends of specialised, free aerial branches or conidiophores, or organised within the interior of specialised cells called *sporangia*, these spore-bearing branches being produced direct from the mycelium.

With further development, there is the organisation of definite complex aggregates of spore-bearing hyphae, frequently surrounded by more or less supporting and protecting tissue. It is these complex aggregates which can with real propriety be designated as *spore fruits*, since they are highly specialised structures adapted to spore production, protection and dissemination. Aerial conidiophores and sporangia represent a more primitive condition before the organisation of the more complex spore fruits.

Kinds of Spore Fruits: *Aerial conidiophores* are special branches of the mycelium set apart for spore production. They may be simple or variously branched, single or in groups or tufts and extend upward from the sub-stratum so as to facilitate the separation and setting free of the spores. This type of spore production is very characteristic of the summer or conidial stage of powdery mildews, certain downy mildews and many forms of imperfect fungi.

The aerial conidiophores may arise singly from vegetative hyphae either on or within the sub-stratum. In the case of leaf parasites, it is not uncommon for one or more to emerge to the exterior through a stomatal opening, thus forming conidial tufts or fascicles. Each spore-bearing branch may give rise to a single spore only or an indefinite number may be produced, which become grouped in chains or clumps or are detached as soon as mature.

When chains of spores are formed, either the proximal one or the distal one may be the youngest. A conidiospore may be cut off from the end of the conidiophore by a cross-septum, or it may develop as a bud-like outgrowth from the conidiophore or from another spore.

In some of the imperfect fungi dense fascicles of erect conidiophores may be grouped together, thus producing a fruiting body known as a *coremium,* composed of a sterile stalk of parallel hyphae and a terminal head of fertile or spore-bearing branches. This is the common or normal type of fruiting body for some fungi, while in others it appears to be induced by environmental conditions.

Sporangia are generally borne on aerial hyphae which bear the same relation to the substratum as conidiophores. The special cell which is destined to form spores is separated off from the free end of the hypha by a cross-wall and organises within its interior an indefinite number of spores, which are later set free by the breaking of the sporangial wall or through a special opening. In strictly aquatic fungi, the sporangia form swimming or *swarm spores* adapted for life in the water, while terrestrial species may form only non-motile spores adapted to air or wind dissemination. Some terrestrial species very closely related to aquatic ancestors have retained the swarm-spore habit—for example, the white rusts and some of the downy mildews.

In certain fungi, the spores are grouped in small or large masses or clusters to which the name son, which really means "little heaps," is applied. The whitish spore dots or pustules of the white rusts, and spore dots or pustules of the red or black rust stages of the true rusts, are designated as sori, while the term is also applied to the much larger, black, powdery spore masses of the smut fungi. A single sorus is a group of either spores alone or spore-bearing hyphae or stalk cells with spores which may sometimes be mingled with sterile filaments.

Sori may be naked or covered, that is, they may remain covered with some of the host tissue, or the spore masses may be exposed to the surface so as to appear granular or powdery. The red powder which is abundant on the leaves and stems of badly rusted wheat or oats or other cereals originates from numerous sori which break out through the host tissue. The smut dust which comes from a threshing machine working on heavily smutted grain is due to the breaking up of the smut masses or sori, the dust particles being either single spores or groups of spores.

Many of the true rust fungi also produce small, flask-shaped fruiting bodies, the *pycnia* or *spermogonia* embedded in the host tissue, generally leaves. These appear on the surface as minute specks or pimples barely visible to the unaided eye. The inner wall is lined with numerous hyphal branches which separate off minute, bacteria-like bodies known as *pycniospores* or *spermatia*.

Certain species of the true rusts form another type of fruiting body which appears soon after the *pycnia* and generally in connection with them. In the most typical form of these *cecia* or *cecidia*, there is a mass of closely packed, yellow- or orange-coloured spores borne within a cup-like structure composed of specialised but sterile fungous cells. The surrounding membrane may be toothed, lacerate or cut into long lobes, irregular or entirely absent. The secia are generally produced in groups or clusters, which with the cup-like form has suggested the name 'cluster cup' which is frequently applied to this spore stage. The *ceciospores,* when mature, readily rattle out of the little cups and are borne away by air or wind.

The *pycnidium* (pycnid for short) is a type of fungous fruit that is peculiar to certain imperfect fungi or may occur in the life cycle of some sac fungi. In typical form it is a more or less globular structure, embedded in the sub-stratum and opening out to the surface by a pore known as the *ostiole.* The wall of the pycnid generally consists of one to several layers of fungous cells with simple or branched conidiophores lining its inner surface. Spores are produced in large numbers from these conidiophores and either accumulate within the pycnid or are forced out through the ostiole.

These *pycnospores* frequently accumulate in sticky masses over the pycnid or are forced out in long coils or tendrils. These tendrils or *"spore horns"* vary from those of microscopic size to others an inch or more in length. These spore threads are frequently hard and horny when dry, but when soaked by rains become soft and permit the separation and washing away of the spores by solution of the mucilaginous matrix in which they are embedded.

Pycnidia are generally minute structures, barely visible to the naked eye as pale, smoky, black or even coloured bodies immersed in the substratum or resting on its surface. They may be embedded in an aggregated mass of fungous tissue known as a *stroma.*

The *acervulus* (plural acervuli) is a type of spore fruit that is peculiar to the Melanconiales, a group of imperfect fungi, and appears in the life cycle of some sac fungi. In typical form it consists of a saucer-shaped, depressed structure which bears conidiophores over its exposed surface and sets the spores free by rupture of the superficial cell layers of the host. Acervuli are comparable to pycnids in size and in the manner of spore production and liberation. The basal matrix which bears the conidiophores, and is comparable to the pycnidial wall, may be well developed and very evident, or it may be almost lacking.

Certain other conidial fruits, the *sporodochia* (singular, sporodochium) are very similar to the acervuli, except that there is a development of a pronounced stroma-like cushion of fungous tissue, which breaks through the host tissue and bears conidiophores over most of its exposed surface. Sporodochia may also be seated in a mass or tangle of mycelium, which has become superficial. The spores may accumulate in sticky masses or they may be dry and powdery.

In some of the more primitive sac fungi, the asci or spore sacs are arranged side by side in an extensive layer over the surface of host parts without being collected in definite fruiting bodies (leaf-curl fungi). A closed ascus or ascigerous fruit, the *cleistothecium* (also called a perithecium) is characteristic of the powdery mildews. It is developed immediately following the union of male and female gametes, the product of the union growing at once into a new structure rather than organising a definite resting spore.

The ascus fruits are borne in the superficial mycelium and appear as minute, globular or slightly depressed black bodies easily visible to the unaided eye. A single fruit consists of a surrounding envelope of brown, sterile fungous cells which completely encloses one or more asci. Sterile threads or hyphae known as *appendages* extend out from the cleistothecium wall and exhibit characteristic forms in different generic types. The ascospores are not discharged until the warm spring rains, when the wall of the spore fruit ruptures and exposes the asci which burst and forcibly eject the ascospores.

The typical *perithecium* (plural perithecia) resembles a pycnidium in the form, size and character of its wall and exhibits a similar opening or ostiole. It is then a more or less flask-shaped body of fungous tissue enclosing a group of elongated or club-shaped asci which arise from a basal aggregate of cells. Mingled with the asci there may be sterile slender filaments, the *paraphyses,* but these structures are not always present. Perithecia may be borne single or in closely associated groups, either immersed in host tissue or superficial, or they may be aggregated in compact masses of fungous tissue, or *stromata*. Some forms show no definite perithecial walls but appear as cavities in an undifferentiated mass of parenchyma-like tissue.

Ascospores are sometimes set free from the perithecium in sticky or gelatinous masses, but in most of the parasitic species they are forcibly expelled. In a typical case, an ascus elongates and pushes its tip through the ostiole. A strong hydrostatic pressure is developed within the ascus, and this finally ruptures the ascus wall or forces an opening. With the sudden release of pressure the charge of ascospores, eight in number, is shot out into the outside air. The old ascus wall then collapses and another ascus is pushed up to the ostiole and the process of spore discharge is repeated. This expulsion then continues until all of the asci have liberated their spores.

The *apothecium* is a third type of ascus fruit and in its typical form consists of a disk-like, saucer-shaped or cup-like body, seated on or in the substratum or raised on a long or short stalk or stipe. The exposed flat or concave surface consists of an extended layer of cylindrical asci, closely packed side by side (the hymenium) and generally mingled with more slender sterile hyphae, the paraphyses.

The wall of the disk or cup is made up of supporting fungous tissue, filamentous in character or of closely packed cells (pseudoparenchyma). Apothecia vary from minute structures barely visible to the unaided eye to others several inches in length or diameter, and are usually more or less fleshy in character, which is in contrast to the firmer texture of perithecia.

Some of the saprophytic fungi produce greatly modified apothecia, as illustrated by morels, saddle fungi and their allies. In these forms the cup has been practically turned inside out, and in the morels shows a greatly convoluted hymenial surface, giving much larger space for the display of asci, while in the saddle fungi the enlarged cup is folded so as to suggest the common name. In the truffles the cup remains closed and develops underground, never coming to the surface to discharge its spores. In the great majority of apothecia, the ascospores are set free by explosion of the asci. In these, however, there is a simultaneous discharge of numerous asci, with a resultant cloud of spores generally visible to the naked eye. This phenomenon is spoken of as the "puffing of spores," and serves to separate the spores from the fruiting body and to facilitate effective air or wind dissemination.

In some of the more primitive basidium fungi, there is no organisation of a definite fruiting body, the basidia being produced from the general mycelium which develops on or within the substratum. In certain parasitic gall-producing forms (Exobasidiales) the basidia originate from an internal mycelium and are produced over the surface of the enlarged and modified structures, the relation being very similar to that of the asci to the mycelium in the leaf-curl fungi.

The production of spores gives a whitish, powdery appearance to the basidial surface. In some of the mould-like basidium fungi, the basidia are borne in irregular clusters or groups over the surface of a superficial mould-like growth. This is the condition which prevails in *Corticium vagum*, the cause of the Rhizoctonia disease of potatoes and other hosts when the basidium stage appears as a whitish, powdery, mould-like covering on the surface of the stem just above ground level.

In the great majority of basidium fungi, the basidia are grouped and supported by rather complex structures which may be designated as *compound sporophores*. These sporophores are of varying texture and may be fleshy, leathery, horny or woody. Some are small but they are generally large in comparison with other types of fungous fruits, the greatest size being reached in some of the giant puffballs and monstrous bracket or shelf fungi. The prime purpose of the sporophore is to afford a surface for the support and display, and in some cases for the protection, of the basidial layers, and the surface is increased by various structural modifications. The simplest sporophores are simply prostrate layers of fungous tissue with a smooth, plain basidial surface. Other forms may be simple clubs or compound clubs and more or less coralloid, while others are of the toadstool form or still others bracket-like or shelf-like in character.

Puffballs represent a still different modification of the basidium fruit, with basidia developed entirely within the closed fruit. The "smoking" of puffballs when squeezed is due to the liberation of innumerable numbers of basidiospores. In the different forms of sporophores the basidial surface may be increased by wrinkles, ridges, tooth-like projections, thin plates or lamellae packed closely together, a honeycomb-like structure or minute parallel tubes which appear as pores on the under surface of the sporophore.

But very few of the club or coral fungi or the puffballs furnish parasitic species, but there are numerous parasitic species which illustrate the other types of sporophores. The basidium fungi are the principal wood-destroying forms and may produce either root or trunk rots.

2

Different Phases

Division Mycota

Fungus (pi., fungi) is a Latin word which means mushrooms. In usage meaning of the word has been expanded to include thallus-like, non-green plants such as the moulds, yeasts and other similar organisms related to mushrooms. The fungi thus are a large group of simple, thallus-like plants which lack chlorophyll. The older botanists employed the term 'fungi' in a wider sense to include bacteria, slime moulds as well as true fungi (Eumycetes). In a narrow sense, the fungi include only the Eumycetes (true fungi). Hence the fungi are now defined as *non-green, nucleated thallophytes.* Familiar examples of true fungi (Eumycetes) are the yeasts, moulds, morels, toadstools, mushrooms, polypores, puffballs, rusts and smuts. In all they number from 50,000 to 100,000 species.

The branch of botany that deals with fungi is called mycology (Gr. *mykes* means mushroom and *logos* means study of). Etymologically mycology is the study of mushrooms. It deals with life histories, relationships and evolutionary tendencies of fungi. The scientist who is concerned with fungi is called a mycologist. He studies structure, reproduction, physiology and taxonomy of fungi.

Habitat: The fungi grow in diverse habitats. In fact they are found in almost every available habitat on earth where organic material (living or dead) is present. They are thus universal in their distribution. Great many of them are terrestrial. They occur in soil which abounds in dead, decaying, organic materials. The terrestrial fungi thrive best in humus soil. They are considered more advanced.

They produce non-motile reproductive cells which are dispersed passively by wind, water or animals. Some fungi attack living organisms. They live in tissues of plants and animals. Some fungi are aquatic. The aquatic fungi are considered primitive. They live on decaying organic matter and living organisms found in fresh water and produce flagellate (motile) reproductive cells which swim to new localities. Many fungi grow on our foodstuffs such as bread, jams, pickles, fruits and vegetables. Some fungi are found in drinking water. Thus the fungi contaminate both our food and drinking water. They are present all the time in air that we breathe. Majority prefer to grow in darkness and dim light in moist habitats.

Habits or Modes of Life: The fungi are plants of very varied habits. They lack chlorophyll and like animals are unable to manufacture their own food from carbon dioxide and water. In their mode of nutrition, the fungi thus differ from all green plants. They obtain food ready made from an external source. All fungi are, therefore, heterotrophs. In their mode of nutrition they are heterotrophic.

However, like all other plants, they cannot ingest solid food but absorb it directly through cell membranes either by living as saprophytes *(saprobes)* or parasites. Thus according to their mode of nutrition the fungi are classified into two categories, the saprophytes or saprobes and the parasites. The saprophytes grow where dead organic matter abounds in the substratum. This mode of life is called saprophytic (saprobic). The parasites live in or on the living bodies of other organisms (plants and animals) and obtain food from them. This mode of life is called parasitic. The best examples of parasitic fungi are the *rusts* and *smuts*. The organism on which the parasite feeds is called the host or suscept. The presence of the parasite may cause an abnormal condition of the host which is called disease. The parasitic fungi are harmful organisms. The thallus of a parasitic fungus may grow on the external surface of the host but usually it is hidden from view. The former are called the ectoparasites (Vine moulds) and the latter endoparasites *(Pythiumde baryanum, Ustilago* and *rusts)*. In the endoparasites the unseen but vitally active thallus grows in the tissues of host plant.

There is, however, not always a sharp distinction between parasites and saprophytes. Among the parasites one can distinguish the following three degrees of parasitism:

1. *Obligate Parasites:* They can grow only upon suitable living host tissues. The best examples of obligate parasites are the *downy* and *powdery mildews*. So far all attempts to grow them on dead artificial media have failed.
2. *Facultative Saprophytes*: Normally they live as parasites and attain their best development as such. However, they are able to grow apparently indefinitely as saprophytes under emergent circumstances. To this category belong *Taphrina deformans,* the *leaf curl fungi* and some *smuts. Phytophtbora infestans* is a parasite on potato. It causes *potato blight diseases*. In the laboratory it can be grown as a saprophyte on cultures of oat meal agar.
3. *Facultative parasites:* Normally they are saprophytes in their habit but may now and then become parasites *Botrytis cinerea* is good examples of this type. They live as saprophytes in the soil. When suitable host plants are sown in such a soil, they attack them and start living as parasites.

The fungi which live strictly as saprophytes are called the obligate saprophytes. They are incapable of infecting plants or animals. The common examples of obligate saprophytes are *Mucor* many field toadstools and most of the species of *Penioillium*. According to the damage done to the host plant, the parasitic fungi may be divided into two sorts, destructive parasites and balanced parasites. The destructive parasites cause the worst diseases of our field crops, garden crops and forest trees. They kill the host calls or tissues or produce substances toxic to the host. The balanced parasites are so adapted in their demands upon the host that the latter is able to supply the needs of the parasites and continues to live itself.

In parasitism advantages are all on one side, *i.e.* to the parasite. The disadvantages to the host may vary from a trifling inconvenience to complete destruction. There is, However, another type of association in which the fungus enters into a partnership with another organism. In this case, the advantages are not all on one side. Both the partners are mutually benefited. This type of association is known as symbiosis. The best example of symbiosis is furnished by lichens. The thallus of a lichen consists of a fungus living symbiotically with an alga. In this partnership, each member profits by the presence of the other.

Another example of symbiosis is furnished by mycorrhiza. In this case, a fungus establishes a relationship with the roots of certain higher plants particularly the forest trees. For example, in *Pinua* the fungal hyphae form a weft around the pine roots. The root hair are absent in this region of the root. It is suggested that the fungal hyphae do the work of the root hairs. They absorb water and minerals in solution from the soil and transmit them to the roots of the tree. The tree provides food to the fungus. This type of mutually beneficial association between a fungus and roots of a higher plants is called mycorrhiza. It is ectotrophic when the fungal hyphae live on the surface and endotrophic when they penetrate the root and grow inside the root tissues. The life cycle of a fungus consists of two phases, the somatic or vegetative phase and the reproductive phase.

Somatic or Vegetative Phase

1. *Thallus:* The vegetative phase of a fungus is a thallus. It may be unicellular or filamentous.

   (a) *Unicellular Thallus:* In some of the lower fungi such as the chytrids the thallus is more or less a spherical, single-celled structure. At the time of reproduction it becomes a reproductive unit. The latter produces the asexual or sexual cells. Such fungi are called holocarpic. In them the vegetative and reproductive stages do not occur together in the same thallus. *Plasmodiophora* has a vegetative phase consisting of a naked, multinucleate, amoeboid mass of protoplasm. It is termed plasmodium. The protoplast of the diploid plasmodium cleaves to from the resting spores. The yeasts, which are related to the filamentous forms, also have a unicellular thallus. In the unicellular holocarpic forms *(Synchylrium,)*, the mycelium is absent.

   (b) *Filamentous Thallus:* The vast majority of the fungi have a filamentous thallus. It originates through the germination of a spore. The spore germinates as it lands on a suitable substratum where other conditions of life are also favourable. In some species the spore, on germination, produces only a short, tubular structure of limited growth. It constitutes the thallus and is technically called a hypha. The spores of most of the fungi, however, give rise to a fluffy thallus consisting of a cottony mass of fine, branched filaments. These long, fine filaments are called the hyphae (sing, hypha). Some of these hyphae, at a certain stage of maturity, extend into the air and bear the reproductive bodies. The rest continue the normal activities. Such fungi are called eucarpic. Collectively the hyphae comprise the vegetative body (thallus) of a fungus which is called the mycelium. The hypha is thus a structural unit of the mycelium. The medium upon which the mycelium grows is known as substratum. The mycelium is the food procuring structure in the life cycle. It carries on the general activities of a plant cell such as absorption, digestion, respiration, excretion and growth but not photosynthesis. The hyphae constituting the mycelium branch and spread in all directions within or over the substratum to form a loose and ramifying network. The aerial hyphae in some fungi become coloured. Black, orange, yellow, red, blue and brown are the usual tints. The colour is usually confined to the hyphal wall. Even when the pigments are present in the protoplasmic contents they do not form an integral part of the living matter. The pigments play no role in the physiology of the fungus. The growth in length takes place at the tips of the hyphae. It is thus termed as apical.

Kinds of Mycelia: The fungus mycelium, in fact, is a multinucleate structure lacking complete internal cell boundaries.

(i) *Aseptate Mycelium:* In the algal fungi (Phycomycetes) the mycelium in the vegetative phase usually lacks internal partitions of any kind. The hyphae are thus multinucleate and aseptate. The mycelium is a continuous mass. It grows terminally by the apical elongation of the hyphae accompanied by increase in the number of nuclei by nuclear divisions. The aseptate, multinucleate mycelium is called coenocytic. The septa in the aseptate, Septate mycelium with unimicleate cells, coenocytic mycelium, however, are formed only to cut off reproductive structures, or to seal off a damaged portion. Such septa are solid plates. They lack pores.

(ii) *Septate Mycelium:* The members of other classes of fungi (Ascomycetes and Basidiomycetes) develop internal cross walls called the septa which divide the hyphae into segments. The septa appear at regular intervals behind the hyphal tip. The segments may be uninucleate or multinucleate. In a septate mycelium the septa between the cells are transverse. Longitudinal or oblique septa are rare. The formation of septum is always preceded by the division of nucleus. If the cell has more than one nuclei there is simultaneous division of all the nuclei before wall formation. It must, however, be borne in mind that the septa in the septate mycelium of fungi are incomplete. Bach has a central pore and rarely.

It is evident, therefore, that the distinction between coenocytes and septate hyphae is not so profound as it was formerly thought to be. The presence of septa gives mechanical support to the hyphae. Complete partitions do not occur in the vegetative phase of fungi. In some cases, the septa possess more than one pore and rarely none at all.

Origin of Septa: The septa in all the fungi (septate Phycomycetes, Ascomycetes, Basidiomycetes and Imperfect Fungi) have a similar origin. They arise at regular intervals behind the hyphal tip. The septum originates at the periphery on the inside of tubular hyphal wall as a ring (annular) or rim of wall material. The annular growth grows slowly inwards towards the centre (centripetally) like an iris diaphragm increasing in width and decreasing the diameter of pore. A cross wall or septum is finally formed which is seldom complete. The centripetal growth of the septum in fungi is rapid. According to Buller (1933) the aseptate *Rhizopus* takes 20-25 minutes to completely seal off a damaged portion. The complex dolipore septum *in Rhizoctonia* (Basidiomycetes) is completed in 10 minutes, *Oiboria* takes only 6 minutes.

Septal Pore: Usually a small hole or a pore remains in the centre of the septum. The perforated septum gives rigidity to the tubular hypha maintaining, at the same time, the protoplasmic continuity from cell to cell. In all the septate fungi with the exception of the Basidiomycetes, the septum has a simple central pore in the middle of the cross wall. The septum in some may be slightly swollen where the pore is formed. In the Basidiomycetes, there is further elaboration of the septum to produce a more complex pore named a *dolipore* (Latin *dolium* meaning a large jar).

The rim of the central pore, in this case, is swollen and thickened to form a barrel-shaped structure with open ends which are guarded by cap like covers. The latter in section appear as a round bracket or parentheses. The septal pore cap is, therefore, called the parenthesome. The complete septal pore ordinarily permits the movement of cytoplasm, mitochondria and nuclei from cell to cell but under certain conditions the parenthesome may close the passage. The dolipore parenthesome is the hall mark of the Basidiomycetes. With the exception of slime moulds the fungal cell usually consists of a strong, rigid cell wall enclosing the protoplast.

(a) *The Cell Wall:* The composition of cell wall is variable among the different groups of fungi or between the different species of the same group. The chief components appear to be cellulose, fungous cellulose (a form of chitin) or a combination of both.

In the majority of fungi the wall lacks cellulose but contains a form of chitin known as the fungous cellulose which is strictly not identical with insect chitin.

Electron microscope studies reveal that cellulose and chitin occur as elongated microfibrillar units. Precise analysis of the cell wall of *Phytophthora* and *Pythium* by Bartnicki-Garcia (1966), Mitchell and Sabar (1968) have revealed that cellulose is a minority component or even lacks altogether. On the other hand glucans predominates in their walls.

Yeast cell based on an electron micrograph. Thus the Oomycetes may be said to have cellulose in their cell walls but it may not be the predominant material.

The clear evidence of such an association is, however

lacking. Burnet (1968) is of the opinion that insoluble glucan forms the predominant structural material of the wall of Ascomycetes and Basidiomycetes. In addition chitin may as well be present in appreciable amounts. In the yeasts and a few other Hemi-ascomycetideae chitin is absent. Their walls are mainly composed of microfibrils of mannan and glucan (Hawker, 1966).

It is doubtful whether this substance is chemically the same as the lignin of higher plants. It is obvious that our present knowledge of the chemical composition of the cell wall of fungi is incomplete. Like the cellulose wall, the chitin wall of most fungi is permeable both to water and substances in true solution.

The living substance of the cell within the cell wall is the protoplast. It lacks the chloroplasts but is differentiated into the other usual cell parts such as plasma or cell membrane vacuolated cytoplasm, cell organelles and one or more nuclei.

Cell Membrane: It is a delicate, extremely thin, living membrane which closely invests the protoplast. The cell or plasma membrane is pressed against the cell or hyphal wall except for occasional imaginations, in some regions. The Pagination is either m the form of an infolded convoluted pocket or a pouch enclosing granular or vesicular material Moore and Mcalear (1961) named it Actually the plasma membrane is the surface layer of altered to reform special functions. It is differentially typical tripartite structure.

The presence of endoplasmic reticulum in the fungal cytoplasm has been demonstrated by the use of electron microscope. It is composed of a system of membranes or microtubular structures usually beset with small granules which by some scientists are likened to the ribosomes. In many fungi, the endoplasmic reticulum is highly vesicular. Usually it is loose and more irregular than in the cells of green plants.

Mitochondria: The cytoplasm contains small, usually spherical bodies known as the mitochondria. Each mitochondrion is enveloped by a double membrane. The inner membrane is infolded to form the cristae which are in the form of parallel flat plates or irregular tubules.

The cristae contain the same fluid that fills the space between the two membranes. The mitochondria function as the power house of the cell. There is no fundamental difference between the mitochondria of fungi and those of green plants. However, Hawker (1965) holds that the cristae of fungal mitochondria are fewer, flatter and more irregular than those of the green plants.

Golgi Apparatus (Dictyosomes): With the exception of Oomycetes there is less certainty of the occurrence of structures similar to those of the golgi apparatus (dictyosomes) in fungi.

Moore and Muhlethaler (1963) reported a golgi apparatus consisting of three flattened sacs surrounded by many bubble-like structures in *Saccharomyces* cells.

Vacuoles: The cytoplasm of young hyphae or fungal cells and hyphal tips lacks vacuoles. They appear further back or in the old cells. With age they enlarge and show a tendency to coalesce and ultimately reduce the cytoplasm to a thin lining layer immediately within the cell wall.

Inclusions: The cytoplasm contains various kinds of inclusions. Examples of stored foods are lipid globules, granules of glycogen, oils and the carbohydrate trehalose, proteinaceous material and volutin. The glycogen may occur in vacuoles. There are no starch grains. Of the pigments, the fungi lack chlorophyll carotenoids are often conspicuous by their presence and may occur throughout the cytoplasm or concentrated in the lipid granules or distributed in the cell walls. The cytoplasm, in addition, secretes several kinds of ferments, enzymes and organic acids.

Nucleus: The cytoplasm contains one or more globose or ellipsoid nuclei. They cannot be seen without special techniques. Structurally the nucleus consists of a central, dense body with a clear area around it and the whole surrounded by a definite nuclear membrane. The central body takes heavy iron haematoxylin stain and is usually Feulgen-negative. In electron micrographs it appears as an amorphous or granular mass. Mycologists usually designate it as the nucleolus.

Bakerspigel (1960) stated that it contains RNA. Under the electron microscope the nuclear membrane is seen to consist of inner and outer layers of electron dense material and the middle one of electron transparent substance. The nuclear membrane has pores. At certain points the nuclear membrane is continuous with the endoplasmic reticulum.

Structure of the Flagellum: Forming the core of the flagellum is a central or axial filament known as the axoneme. The axoneme is surrounded by a double cytoplasmic sheath or membrane which is an extension of the plasma membrane of the motile cell. In a cross-section, the axoneme is seen to be composed of eleven fibrils. Of these two are central and nine peripheral. The two central single fibrils and form an elastic axial thread which is sometimes enclosed by a subsidiary sheath of its own. Of the nine peripheral fibrils, each consists of two thin fibrils. The nine peripheral doublet fibrils surround the central pair of singlet fibrils.

Kinds of Flagella: The fungal flagella are reported to be of the following three kinds:

(i) *Whiplash Flagellum with an End Piece:* It has a smooth surface and narrows abruptly towards the tip to form a distinct end piece. The end piece is a thin extension of variable length. It is flexible. According to Manton, the end piece is the naked portion of the axoneme. It is composed of eleven fibrils. Koch (1956) and Kole (1957) stated that the end piece consists of the two central fibrils only.

(ii) *Whiplash Flagellum with a Blunt Tip:* The blunt whiplash flagellum has smooth surface but lacks the narrow tip (end piece).

Tinsel Flagellum: It does not obviously end in a narrow tip like the blunt whiplash type flagellum but bears fine, lateral hair-like appendages known as the flimmer hairs or mastigoneme.

The flimmer hairs on the tinsel flagellum arise from the axoneme. The tinsel flagellum is also known by other names such as pantonematic, flimmer or flimmergeisel flagellum. The other names of the whiplash flagellum are acronematic or peitchgeisal flagellum.

Flagellation in the Lower Fungi: The number, kind and position of flagella on the motile cell constitutes its flagellation. Flagellation plays an important role in the taxonomy of the Lower Fungi (Phycomycetes). It is constant in each class of the Lower Fungi and differs from class to class. Thus it forms an important feature in classifying the Lower Fungi into classes. The Higher Fungi and Zygomycetes lack flagella.

The motile cell in the class chytridiomycetes of the Lower Fungi is furnished with a single whiplash flagellum with a pointed tip. It is inserted at the posterior end (opisthooont). The representatives of the class Hypho-chytridiomycetes possess a single flagellum of tinsel type.

The flagella are inserted at the anterior end. Both of them are of whiplash type but one of them is several times longer than the other (helerkont). The longer whiplash flagellum has a sharply pointed tip. The shorter whiplash flagellum has a blunt tip. Biflagellate motile cells are also characteristic of the class Oomycetes. In the motile cells of this class with a pyriform shape, the two flagella are inserted at the anterior end whereas in the kidney-shaped motile cells, the flagella arise from the depression. One of these is of tinsel type and the other whiplash with a sharply pointed tip, In motion, the tinsel flagellum is directed forward and the whiplash is directed backward.

Fungal Tissues: The fungus mycelium normally, as mentioned above, is a mass of hyphae interwoven loosely to form a network. In some fungi, the whole mycelium or parts thereof undergo various modifications. The hyphae become altered in shape and become closely packed together to form a false tissue. Such a false tissue is called plectenchyma, all fungal tissues come under this general term. Plectenchyma is of two types, prosenchyma and pseudoparenchyma.

Pseudoparenehyma: In the fructifications of higher fungi the hyphae become woven and intertwined into a compact mass. The walls of the hyphae in the mass get fused and they lose their individuality. It bears a striking superficial resemblance to the parenchyma tissue of the higher plants and is called pseudoparenchyma.

Nutrition in Fungi: It is thc process of procuring nourishment or nutrient and is the principal function of the vegetative phase (mycelium). The supply of nutrients is essential for every living organism because the nutrients serve as structural units and also as energy sources for the building up of protoplasm which is the basis of life.

Nutritional Requirements: The fungi utilise both organic compounds and inorganic materials as the source of their nutrient supply. In other words organic and inorganic compounds constitute their food. No fungus is able to make any increase in its dry weight in the absence of organic food materials, why ? Lacking chlorophyll the fungi are unable to photosynthesise or use carbon dioxide to build up organic food materials.

They are thus heterotrophic for carbon (organic) food compounds which they in their natural habitats obtain by living as saprophytes or parasites from dead or living plants, animals or microorganisms or their wastes.

The carbohydrates are needed for building up the body and also as a source of energy. In a typical fungus 50 per cent of the dry weight is carbon. Of the carbohydrate source of carbon most fungi use simple sugars. Glucose, for instance, is suitable for almost all fungi. Next in preference are the fructose. Less commonly used are the hexose sugars and some pentoses. Xylose among the pentoses has been reported to be superior to glucose for some fungi. Mannitol is equivalent to glucose for many fungi. Maltose which occurs in nature as a by-product of starch hydrolysis is utilised by many fungi. Sucrose is also a good source of carbon for some.

From among the polysaccharides starch and cellulose are utilised by a fewer fungi which can synthesise the appropriate hydrolytic enzymes. Some fungi are able to make good growth on fats as the only source of carbon. Organic acids are generally poorer sources of carbon for most fungi. Basidiomycetes include most of the lignin-utilising fungi. Proteins, lipids, some organic acids and higher alcohols are utilised by some fungi as a sole energy source. Growth, however, is always better on a substance containing a suitable carbohydrate.

Besides carbon fungi require nitrogen. To obtain nitrogen they utilise both organic and inorganic materials as the source. The chief organic sources of nitrogen are protein, peptide or an amino acid. Certain groups of fungi show specialisations in respect of certain nitrogen sources. For example, the Saprolegniaceae and Blastocladiales include a number of species which grow only with organic nitrogen such as amino acid. In nature, fungi decompose proteins and other materials to obtain their supply of nitrogen. In pure cultures amino acids, peptides, or peptones, gelatin, casein and eggs albumin can serve as sources of organic nitrogen for building up protoplasm. Urea is also considered as a utilisable nitrogen source for some fungi.

Many fungi, however, obtain nitrogen from inorganic sources. A number of fungi are known which use both nitrate and ammonium salts. Bobbin (1939) and Lindberg (1944) reported that *Absidia* sp., *Mucor hiemalis, Lenzites trabea* and *Marasmius* sp. use ammonia but are incapable of utilising nitrate salts. Fewer fungi are able to utilise nitrate salts.

Nitrites can be toxic. Organic sources of nitrogen can also serve as sources of carbon. There is not much evidence to support nitrogen fixation in saprophytic fungi. Metcalfe and

Chayen (1954) reported that soil inhabiting *Rhodotorula* and yeast-like *Pullularia pullans* fix atmospheric nitrogen. It is certain, however, that nitrogen fixation is not a widespread ability in fungi.

Hydrogen and oxygen are supplied in the form of water which is the major constituent of fungus mycelium forming about 85-90 per cent of the entire weight.

The chief axaong the inorganic nutrients which the fungi require in fairly large amounts for their mineral nutrition are sulphur, phosphorus, potassium and magnesium. These macronutrients the fungi obtain from simple inorganic salts or sources such as sulphates for sulphur, and phosphates for phosphorus. These must be supplied in any culture medium. Calcium is not known to be needed by the fungi in general. Some, however, require it as a micronutrient. Some fungi are reported to require only minute traces of iron, zinc, copper, manganese, boron, and cobalt, and molybdenum. These trace elements or micronutrients are considered essential for growth. The form in which the major and the minor metallic element requirements are utilised is the anion.

The fungi, like all other organisms, require minute amounts of specific, relatively complex organic compounds for growth. These are the vitamins or growth factors. Many fungi synthesise their own supply of appropriate growth factors from a simple nutrient solution of defined composition. Such fungi are thus autotrophic for vitamins and are called araxo-aiatotrophic. They need no exogenous supply. There are others which depend in whole or in part on an external source, because they are unable to synthesise one or more of the essential growth substances. The fungi heterotrophic for their needs of growth factors are termed auxotrophic. There are marked differences between the vitamin demands of the different species of a genus or even the strains of a single species. The important fungal vitamins, which may function in enzyme systems include thiamine (B), biotin, pyredoxine (B_6) and riboflavin (B_2). A few fungi also need nicotinic acid and pantothenic acid. The vast majority, however, require thiamine. The growth factors are catalytic in their actions. To sum up, the basic nutritional needs of fungi are:

A suitable organic compound as a source of carbon and energy.

A suitable source of nitrogen. Inorganic ions of sulphur, phosphorus, potassium and magnesium in significant amounts. Inorganic ions of iron, zinc, copper, manganese and molybdenum only in minute traces. Certain vitamins or organic growth factors in trace amounts.

Besides the nutritional requirements listed above the growth of fungi is influenced by certain habitat factors such as temperature, oxygen supply, moisture, *pH* value and by-products of metabolism.

Modes of Nutrition: The fungi lack chlorophyll. They are, therefore, unable to synthesise carbohydrate food from inorganic materials and get it readymade from sources external to themselves. These heteromorphs according to their method of obtaining food are divided into two categories, namely, the saprophytes or saprobes and parasites. The former absorb the required organic nutrients directly through the cell membrane from the substratum which abounds in dead organic matter of both animal and plant origin. The saprophytes cannot ingest solid food. Yeast and moulds are the common examples of saprophytic fungi. The parasite lives in or on the living body of a plant or animal and absorbs organic molecules as nutrients through the cell walls from the tissues of the host. Rusts and smuts are the common parasites.

Ectophytic Hypha

In saprophytic fungi, the hyphae (Mucor mucedo) or rhizoidal hyphae (Ehizopus stolonifer) come in intimate contact with the substratum. They secrete enzymes which convert insoluble complex food materials in the substratum to soluble ones. The latter are then absorbed by direct diffusion either through the hyphal walls of the hyphae that penetrate the substratum or by the rhizoidal hyphae.

The mycelium of the parasites is rarely ectophytic but frequently it grows inside the host. The hyphae either ramify in the intercellular space between the host cells or penetrate into the host cells. The former are called intercellular hyphae.

The secretion from the haustorium upon the plasma membrane makes it permeable to solutes contained in the sap cavity. They diffuse out and are then absorbed by the haustorium. The parasitic fungi do not produce haustoria in artificial cultures. The haustoria are also not produced by fungi which live as parasites on animals.

The fungi, as mentioned above, are unable to synthesise sugars from carbon dioxide and water. They, however, can synthesise from soluble sugars the more complex carbohydrates which are the chief components of their cell walls. They are also able to synthesise proteins and eventually protoplasm if supplied with carbohydrates and simple nitrogen compounds such as ammonium salts. Besides ammonium salts they can absorb and utilise many complex but soluble organic nitrogenous compounds.

Reproductive Phase

When the mycelium has reached a certain stage of maturity and has accumulated reserve food, it starts reproducing. Reproduction involves the production of new individuals resembling the parent or parents. In fungi reproduction is of three kinds, vegetative, asexual and sexual. Many mycologists, however, include all methods of reproduction (vegetative and spore formation) which do not involve union of nuclei or sex cells or sex organs under asexual or somatic reproduction.

From the general account of the fungi given in the preceding pages it is evident that they are sharply set off from the algae in general, in the absence of chlorophyll, mode of nutrition and accumulation of carbohydrate reserve as glycogen instead of starch.

In spite of the differences between the algae and fungi, the two are similar in their methods of reproduction, unicellular nature of sex organs, thallus-like plant body, lack of tissue differentiation of any kind and no embryo formation after gametic union.

Classification of Fungi

The classification of fungi is still in a state of flux. A stable or ideal scheme is yet to be proposed. Many systems have been proposed from time to time by mycologists.

The latest among them are of Bessey (1950), Sparrow (1958), Ainsworth. (1961), Alexopoulos (1962), Gaumann (1964) and Hawker (1967). They are all suitable and mostly based on the features noted below:

(a) Nature of somatic phase whether unicellular or a mycelium, if latter septate or aseptate.

(b) Kinds of asexual spores (mitospores) sporangiospores or conidia; if former motile or non-motile; number, form and arrangement of flagella in the motile spores.

(c) Kinds of sporangia.

(d) Nature of the life cycle, whether asexual, haplontic, diplontic, haplodiplontic or haploid dikaryotic.

(e) Presence or absence of the perfect or sexual stage.

Algae and Fungi

The algae possess chlorophyll which in some of them is marked by other pigments such as brown, red, etc. With the help of chlorophyll, the algae are able to synthesise their own carbohydrate food and thus are called autotrophic or autophytic in their mode of nutrition.

1. The fungi lack chlorophyll.
2. Lacking chlorophyll, the fungi are unable to synthesise

their own organic food. They get it ready made from external sources either by living as parasites or saprophytes. For this reason the fungi are referred to as heterotrophic or heterophytic in their mode of nutrition.

3. The fungus thallus which is termed the mycelium is not formed by the division and differentiation of cells but is composed of a false tissue called pseudoparenchyma. The latter is an interwoven mass of fine, tubular structures called the hyphae. Thus the unit of structure of a fungal thallus or mycelium is a hypha and not a cell.
4. The cell or hyphal wall generally consists of fungus cellulose often called chitin.
5. The reserve food invariably accumulates in the form of glycogen and not starch.
6. Fungi generally grow in darkness and dim light.
7. The fungi grow in a wide variety of habitats; as parasites in the tissues of plants and animals causing disease; as saprophytes in decaying organic matter in soil or water.
8. There is gradual and progressive simplification and ultimate elimination of the sexual apparatus from the lower to higher forms.

The algal thallus is composed of true parenchyma cells. The unit of structure of an algal thallus thus is a cell. The cell wall in algae typically consists of cellulose. The food reserve is chiefly in the form of starch. The algae usually grow in habitats where light is available.

The algae grow in water or damp soil and sometimes as epiphytes rarely as endophytes.

The sexual apparatus increases in complexity from the simple to the higher forms.

The modern mycologists are unanimous on the exclusion of bacteria from the fungi on the basis of lack of a true nucleus in the bacterial cell. There is a difference of opinion as regards the slime moulds. Many mycologists favour their inclusion among the fungi.

The Committee on the International Rules of Botanical Nomenclature has made the following recommendations:

(a) The names of divisions of fungi should end in mycota.

(b) The subdivisions should end in mycotina.

(c) The names of classes should have a suffix mycetes.

(d) The subclasses should end with a suffix mycetideae.

(e) The names of orders should end in ales.

(f) The families should end with a suffix aceae.

Division Mycota

The mycologists place division Mycota in kingdom Plantae. This division includes non-green, nucleated thallophytes (fungi) which are characterised by the following features:

(a) Obtain nutrition either by living as saprophytes or as parasites.

(b) Somatic phase may be filamentous, unicellular or a plasmodium.

(c) Presence of a true nucleus with a nuclear membrane and nucleoli in the cell.

(d) Reserve food accumulates in the form of glycogen and not starch, In addition there are oil drops.

(e) Reproduce both by asexual and sexual methods.

The division Mycota includes two subdivisions, Myxomycotina and Eumycotina.

L Sub-Division Myxomycotina. This subdivision includes the slime moulds which possess the following distinctive features:

(i) The thallus is a non-green, multinucleate mass of protoplasm called the plasmodium.

(ii) The plasmodium lacks a definite cell wall, and thus is amoeboid in shape.

(iii) It is free living, diploid and holocarpic.

(iv) The propagative units (spores) are differentiated by meiosis.

All the free-living slime moulds are placed in the class Myxomycetes.

Subdivision Eumycotina

All the other fungi except the slime moulds are included in this subdivision. It is a heterogeneous group comprising about 80,000 known species.

They are distinguished as below:

(i) With the exception of a few unicellular forms the somatic phase is typically a filamentous structure called the mycelium.

(ii) The unit of structure of the thallus or mycelium is not a cell but a hypha.

(iii) The cell or hyphal wall is usually composed of fungus cellulose (chitin) or a mixture of cellulose and chitin or rarely of cellulose.

(iv) There is a definite nucleus with a nuclear membrane and nucleoli in the cell.

(v) Excepting a few all are eucarpic.

(vi) The septa between the cells when present have each a central pore.

(vii) The growth in length in the filamentous forms is apical.

(viii) Reproduce both by asexual and sexual methods.

(ix) Asexual reproduction or sporulation takes place by naitospores which may be motile or non-motile.

(x) There is gradual and progressive simplification and ultimate elimination of the sexual apparatus from the lower to the higher forms.

The older mycologists divide the subdivision Eumycotina, which includes the true fungi, into four classes, namely, Phycomycetes, Ascomycetes, Basidiomycetes and Deuteromycetes.

This scheme of classification of true fungi is still in vogue in many countries. The modern mycologists, however, consider that the class Phycomycetes, which includes all the lower fungi, is not a natural group of closely related forms. Alexopoulos (1962), therefore, took a logical step and split the lower fungi included in the class Phycomycetes into six classes.

An outline scheme of classification of Eumycotina (true fungi) followed in this edition of the text, is set out below. It is customary to divide the true fungi into two main groups, the lower and the higher fungi.

The Lower Fungi (Phycomycetes): They have comparatively a simple thallus which in some is unicellular and in others filamentous (mycelium). The septa usually remain suppressed in the actively growing vegetative mycelium. They reproduce asexually by sporangiospores (generally motile, rarely non-motile), rarely by conidia. The early mycologists placed the lower fungi in a single class Phycomycetes which is split in this text into six classes on the basis of presence or absence of motile cells in the life cycle and the number, form and position of flagella on motile cells.

Motile Cells

1. *Class Chytridiomycetes:* Included in this class are the lower fungi in which motile cells have a single flagellum of whiplash type inserted at the posterior end (episthocont). The members of this class are called the chytrids.
2. *Glass Hyphochytridiomycetes:* The motile cells (zoo-spores) possess a single flagellum of tinsel type which is inserted at the anterior end.
3. *Glass Plasmodioplioromycetes:* The motile cells (zoo-spores)

are biflagellate. Both the flagella are of whiplash type but one of these is longer than the other. The longer one has a sharply pointed end and the shorter one has a blunt end. The club root organisms are the examples of this class.

4. *Class Oomycetes:* The motile cells are biflagellate. The two flagella are usually of nearly equal length. One of these points forwards and the other trails behind. The former is of tinsel type and latter of whiplash type. The common representatives of this class are the water moulds, downy mildews and blights.

Motile Cells Lacking

5. *Class Zygomycetes:* Motile cells are absent. Asexual reproduction takes place by sporangiospores which are encapsulated and wind disseminated. The common examples are the pin moulds and the bread moulds.
6. *Class Trichomycetes:* The motile cells are Jacking. The asexual spores are produced exogenously at the tips of special hyphae (conidiophores) and are called the conidia.
7. *The Higher Fungi:* The somatic phase consists mostly of a mycelium which is usually septate. The septa have each a central pore and thus are incomplete. No motile cells are produced in the life cycle. The higher fungi comprise the following three classes:
8. *Class Ascoxnyeetes:* The characteristic spores of the sexual or perfect stage are endogenous in origin and are called the ascospores. They are produced within special, sac-like structures called the asci. Plasmogamy takes place by the fusion of gametangia or somatic hyphae. Somatic phase is usually a septate mycelium which multiplies asexually by conidia, rarely it is unicellular (as in the yeasts) and is propagated by budding or rarely by fission.
9. *Class Basidiomycetes:* The characteristic spores of the

sexual or perfect stage are exogenous in origin and are called the basidiospores. They are borne externally usually on club-shaped structure called basidia. The somatic phase consist of a simple or complex mycelium with or without clamp connections.

10. *Form-class Deuteromycetes or Fungi Imperfecta:* Included in this class are the higher fungi in which the sexual or perfect stage is unknown. The somatic phase consists of a septate mycelium which multiplies asexually by conidia. The conidia are produced on condiophores which may occur isolated or may be aggregated to form complex structures such as synnema, pycnidia and acervuli.

Origin and Phylogeny of Fungi: The origin and evolutionary relationships of fungi are not definitely known. They are a matter of speculation and problem for future research for mycologists. Concerning the origin of fungi, first and the traditional hypothesis regards algae as the ancestral stock. Some mycologists consider protozoa as the ancestral stock.

The supporters of the algal hypothesis hold that the fungi are degenerate algae which have lost their chlorophyll and thus have adapted themselves to a heterotrophic mode of nutrition. It is not difficult to believe it because certain flagellates under different conditions develop chlorophyll and prepare their own food or lose chlorophyll and live as saprobes.

The proponents of algal hypothesis are divided into two camps. Some suggest a monophyletic origin. They hold that with the exception of slime moulds and some simplest fungi all others, namely, Phycomycetes, Ascomycetes and Basidiomycetes represent one main evolutionary line of fungi evolved from an algal ancestor.

To view that all fungi have a common ancestry and arose in a monophyletic series appears untenable to most of the mycologists They support the polyphyletic origin from the various groups of algae. On the basis of multiple origin, the fungi are considered a heterogeneous aggregation, the Phycomycetes having evolved from one class of algae and Ascomycetes from another. The Basidiomycetes are thought to have been evolved from the Ascomycetes.

The supporters of the algal origin of the Phycomycetes further belong to two opposite schools of thought. One favours algal origin from a siphonaceous ancestor belonging to the green algae and the other from a Xanthophyeean ancestor such as *Vaucheria,* There are differences in metabolism and type of flagellation of the motile cells in the Phycomycetes and green algae.

Cultivated Fungi and other Fungus Foods

"By tying a Cord to the hind-leg of a Pig, and driving him before them... observing where he stops and begins to root,... they are sure to find a Trufle" 1691, RAY, *Creation,* II. (1692), 99.

When one considers the gastronomic excellence of many of the fungi, and the total disadvantages of utilising those occurring naturally, it is not surprising that early attempts should have been made to cultivate some of the more esculent kinds; but of the many species available, few have passed into general cultivation. The most important of these is the Common Mushroom *(Psalliota campestris)* which is now grown to such a degree as to provide a livelihood for large numbers of people in many different countries.

Early Attempts at Cultivation: Many sporadic attempts have been made to cultivate other species, but, with one or two exceptions, the results obtained do not appear on the whole to have warranted the labour involved, and the ancients certainly went to considerable trouble for this end. The earliest endeavours appear to have had, for their purpose, an artificial reproduction of the natural conditions under which the particular fungus grows, in the hope that, on creating the former, the latter would make its appearance.

Poplar Fungus: Thus, Dioscorides, in writing of the Poplar Fungus *(Pholiota aegeritd),* remarks: "Some people say that the bark of the white and black poplar cut into small pieces and scattered over dunged spaces will produce edible fungi at all seasons" while in the *Geoponica,* an anonymous work on rural economy, possibly published about 900 AD, it is recorded of the same species that "in order to make fungi grow one must saw off the stump of a black poplar and pour sour dough dissolved in water upon the cut pieces. Black-poplar fungi soon appear." Even if success attended these experiments, it is not surprising that the methods have rather fallen into disuse, for *Pholiota aegerita* is not considered a delicacy by modern standards.

Cultural Methods Involving Burning: The preference for charred spots exhibited by some fungi is made use of in another method recommended in the *Geoponica.* "If you would have fungi to grow from the ground, you must select a spot of light soil on a hill where reeds grow; there you must collect together twigs and other inflammable materials, and set all on fire just before rain is expected; if rain does not come you must sprinkle the spot with pure water; but the fungi thus produced are of inferior quality." Presumably the vegetable was supposed to spring incontinently from the earth with the first shower, but the fungus cultivator of old was doubtless as used to disappointments as are our agriculturists of today.

Some modern methods involving a preliminary burning, by which it is attempted to stimulate the natural production of fungi, are of interest. Thus in the Roman Campagna, a local delicacy, *Polyporus corylinus,* appears freely on charred stumps of hazel and cob-nut trees. Its growth is therefore artificially accelerated by heating the stumps and logs of the trees over burning straw, and in this charred condition the stumps are sold in the markets. By keeping them in a warm cellar, watering freely, and leaving for a time undisturbed, a plentiful crop, of excellent quality, is obtained. In the same connection, a curious custom, which formerly for a considerable time had vogue in

Germany, should be referred to. It was long ago noticed by the peasants that morels often came up abundantly where woods had accidentally been burnt, whereupon, inspired by a motive similar to that which animated the Chinaman, after his first discovery of the virtues of roast pork, they proceeded to stimulate the activities of Nature by firing the woods on purpose. Although the extent to which this was practised leaves us in no doubt as to the success of the scheme, an unsympathetic government later suppressed the practice by penal legislation.

Cultivation of the Common Mushroom: Such empirical, casual and even reprehensible methods can, however, hardly be recommended, and those who would cultivate their own fungi will be well advised to limit their activities to the Common Mushroom *(Psalliota camptstris)*, for herc they are on sure ground, and most other efforts pale into insignificance when compared with the enormous French production of this species at the present day. The Common Mushroom has been esteemed as an article of diet from very early times, but its cultivation from spawn probably did not take place before the beginning of the seventeenth century. The results of investigations appear to show that the art first arose in France, probably near Paris, the first scientific description of it being given by Tournefort in 1707.

French Industry: At the present day, Paris is still the seat of one of the most extensive and intensive mushroom-growing industries in the world, and some account of it should be given. The industry is carried on in disused quarries and mines, from which stone was formerly extracted.

There are, in the Department of the Seine, about 3,000 of these excavations, of which those no longer used for their original purpose, and almost all of which are quite close to Paris (at Montrouge, Bagneux, Vaugirard, Chatillon, Ivry, Vitry, Houilles, and Saint-Denis) are occupied by the two hundred and fifty to three hundred mushroom growers of the Department. Some of these excavations are quarries, approached by adits on the ground level, but most of them are mines, with access by a vertical shaft, usually varying between one hundred and one hundred and thirty feet in depth. The quarry entrances are often practicable for carts, when the transport question presents no difficulties.

The shafts of the mines are fitted with vertical ladders, with steps over eighteen inches apart, for the workmen; new manure is simply thrown down, while spent manure, and mushrooms, are hauled up in baskets, by means of a hand-operated winch. The interiors of these excavations consist of a number of winding corridors, varying very much in size, particularly as to height, which generally ranges from about 2ft. 6in. to 6ft. 6in.

One of the most spacious is at Mery, and has a total length of nearly two miles, with an average height of about thirteen feet. The principal entrance is at Mery, and there is another at Saint-Ouen-1'Aumone, near to the station to which manure is sent from Paris. It is quite often the case that manure has to be wheeled a distance of several hundred yards from the shaft, or entrance, and this becomes an irksome task when the roof is low, and the workman has to proceed in a bent position, or even on his knees, guided by a lamp fixed in the front of his barrow. To facilitate transport, a system of relays is used, long or shore, according to the difficulties of the course.

Ventilation: In the quarries, with the entrance on the ground level, no special means of ventilation are usually necessary; but in the mines, the breathing of men, the burning of lamps, and the emanations from the beds, would soon render these untenable, were not some definite method of ventilation in force. For this purpose, shafts of small diameter are carried up to the ground level at intervals, according to the needs of each gallery, and a small conical wooden chimney is built over each, the top of this being closed by a sort of cowl, pointing towards the north. The top of this chimney should preferably be higher than the top of the main shaft. In many cases, the ventilation shafts are those left by the stone-miners, but occasionally additional shafts have been drilled. The excavations sometimes communicate with each other, when the ventilation problem becomes simplified, but too violent draughts may result, and doors have sometimes to be installed. This varying facility or difficulty of ventilation, conjoined with the extent, the height of the galleries, and so forth, causes a considerable difference in the annual rent, which may range from one hundred and fifty to six hundred francs.

Water Supply: Some of the excavations are more favoured than others, by a regular infiltration of water, which is collected in reservoirs constructed for the purpose. Where such natural supply is not forthcoming, and where, as is usually the case, the shaft is far from buildings or wells, large barrels on wheels, holding about 132 gallons, are used for the transport, and from these the water is run down a pipe, into small wheeled buckets or tubs below.

The beds are generally supplied by an ordinary watering can, or sometimes by means of a wheeled tub, fitted with a fine "rose." The latter method is of course only suitable where the necessary roof height obtains. The whole business is thus of a somewhat laborious nature.

The Beds: As in this country, the beds are composed of stable manure, of which the kind and quality are of the first importance. The beds are smaller than is customary with us, while fresh spawn is constantly being brought in from outside, because the continual use of spawn taken from mushroom-bearing beds leads gradually to a diminution in the yield obtained. The spawn does not take the form of the compact "bricks" used in England, but of much smaller flakes of dry manure containing mycelium, and after "sowing," the surface of the bed is covered to a depth of about one inch with a layer of fine white stony soil. By keeping moist with water, mushrooms are obtained about six weeks after sowing, and the beds remain in bearing for a period of six to eight months, after which the spent material is returned to the surface for ordinary manurial purposes in agriculture. By sowing in rotation, a succession of crops is obtained. The mushrooms are not allowed to reach maturity, but are gathered in large button form. When fully expanded they are considered indigestible, and have no sale. The daily production of Paris, prior to the war, was about 25,000 kilos, worth about one franc the kilo., giving a value of 750,000 francs per month, or nine million francs per annum. Thus, not only could the needs of Paris herself be supplied, but enormous quantities were exported all over Europe.

It is to be regretted that this great industry has suffered considerably since the war. From recent information we gather that the chief causes for this would appear to be:

(1) Shortage of manure, owing to the increased use of automobiles and reduced employment of horses.

(2) Insect pests appear to be on the increase.

(3) Increased cost of labour.

(4) Mushrooms, which before the war were a common article of diet of the Parisians, have now reached such a price, 15-30 francs per kilo, that they are now quite outside the reach of the middle classes, and the supply has automatically fallen off with the decreased demand.

Cultivation in England: By comparison with the scale on which the great French industry has been developed, our own efforts, and those of most other countries in this direction, are less important, this being due partly to a lack of those cultural and distributive advantages enjoyed by Paris. The French method of raising mushrooms underground has, however, been successfully introduced into this country, where in several places, disused quarry tunnels, railway tunnels, and the like, have been pressed into service. Of these, the most extensive are at Corsham, Wilts., where hundreds of pounds of mushrooms are produced daily in underground quarries, from which were formerly extracted large quantities of Bath-stone.

The Pockeridge quarries, now used by Messrs. Agaric, Ltd., for this purpose, are entered by an inclined shaft, down the middle of which runs a tramway for drawing and lowering prepared manure, spent manure, mushrooms, sand, etc., a distance of about one hundred feet to and from the surface, for which purpose a steam engine is employed. From the bottom of the shaft to the actual beds is over three-quarters of a mile, the trucks being horse-drawn over this distance, and also on the surface. The area available for culture is about twenty-nine acres, capable, when fully developed, of yielding some 250,000 pounds of mushrooms per annum. Some thirteen acres are in cultivation, and from 3 to 4 cwt. is a quite usual day's picking.

Over forty tons of manure are used weekly in the making of fresh beds, and the total length of beds in various stages of cultivation is, roughly, five miles. A bed will remain in bearing for from one to four months, and by suitable rotation a more or less regular crop can be obtained throughout the year; the spent manure being sent weekly to the fruit-growing districts of Evesham, Stratford-on-Avon, and elsewhere.

An extensive system of water-pipes has been laid down to supply all the beds, while telephones communicate from the office on the surface to the various tables and packing stations below ground. Lighting is obtained by means of portable acetylene and other lamps. For the crushing of the Bath-stone for covering the beds, there is an installation consisting of oil engines, a stone crusher and a disintegrator.

The mushrooms are picked every morning and sent off by passenger train almost immediately, so that they generally reach their destination the same day, being sent in the usual fruit baskets containing from four to seven pounds. A large part of the crop is sent to Covent Garden, but supplies are sent also to Southampton, Liverpool, Manchester, Scotland and elsewhere, several large shipping companies taking considerable quantities. A few miles away, at Bradford-on-Avon, the same company has similar underground beds, of about half the extent of those at Corsham. The disused underground quarries at Godstone, Surrey, are employed by another company in a similar manner; as also is a disused railway tunnel at Edinburgh.

The English method, while being somewhat similar as regards cultural details, differs in the fact that the beds are prepared either under glass or in the open, although sheds, etc. are often employed. One of the largest plants, where this system is in vogue, is that of Messrs. John F. Barter, Ltd., at Alperton, near Sudbury, Middlesex, to whom we are indebted for the subjoined account.

Site for Ridges: The ground on which ridges are to be made "should have some shelter (e.g. hedgerows, buildings, etc.) from the north and east winds. On level ground the ridges should run north and south; if the ground is sloping, they should run with the slope, so that they do not interfere with the drainage. The materials used comprise horse manure, spawn of good quality, and good mould for covering the beds.

Preparation of Manure: The best manure is that from stables where plenty of straw is used, the longest litter being shaken out and dried, if possible, or thrown into a stack to dry off naturally, being then used for covering the beds. All litter less than about nine inches long should remain in the manure, which should contain about two-thirds of such litter. The manure is first made into a compact heap, in which it is allowed to remain for about four days, in order that it may heat up. It is then turned, and the outside material thrown towards the middle. This is done every other day, until it is "sweet," usually four turnings being necessary. It is then ready for making into beds.

Preparation of Beds: These consist of ridges of conical section, and are formed by a process of treading and beating the manure.

As soon as the ridge has been made, it is lightly covered with litter to prevent it from becoming sodden in wet or becoming dry in hot weather. Three days after the ridge has been built, the litter is removed, and holes nine inches apart are bored, with a one-inch iron bar, from the top of the ridge to within about nine inches of the bottom, in order to facilitate the cooling of the bed to the necessary temperature. After this operation, the litter is replaced, when, after a further period of some ten to twenty days, according to the outside temperature, the material will have become cool enough for spawning, a suitable heat being not more than 75°F. at a depth of three inches below the surface, this being ascertained by the use of a plunger thermometer. The bricks of spawn are moistened, and then broken into from eight to twelve pieces, being inserted in the sides of the ridge about nine inches apart.

Casing: After the spawning, an interval of three or four days is allowed to elapse, even up to a fortnight being not too long in mild weather, to make sure that the heat of the ridge is not rising again; after which the ridges are cased with a good loamy soil containing plenty of root and fibre, broken up fine with a fork, but not sifted. A spadeful of mould should cover about nine or ten inches square, but the thickness of the case is reduced towards the top of the ridge.

The mould is fixed in place by being lightly beaten with the spade as each spadeful is applied. The casing being finished, the litter is again put on, to a depth varying from six inches in mild weather to about fifteen when the weather is severe or very windy.

Sticks are inserted down the centre of the ridge about ten yards apart, and these, if felt occasionally, will indicate the temperature of the bed. No notice is taken of a falling temperature, however low it may become, but any rise above 75° F. is controlled by the removal of a portion of the litter. The ridge is, however, not otherwise disturbed or examined for a month from the time of casing. At the end of that time, and on a fine day, the litter is removed and well shaken up, and froth or any other matter found on the surface of the bed, is rubbed off with a brush or a handful of straw, after which the bed is recovered. This process will cause the spawn to "start," and it will show through most probably within a week or two. Crops are sometimes obtained in one month from the time of spawning. On the other hand, beds made up in November have been found to produce no mushrooms until the following April, when heavy crops were obtained. During the winter, when fallen snow might interfere with the gathering, mats or tarpaulins, which can be readily removed, are used to cover the beds.

Gathering: After the removal of a cluster of mushrooms, any uncovered mycelium is cut out with a knife and the hole is filled with fresh mould. All waste material, and all mushrooms which have withered away are removed, the paths and surroundings being kept clear of all mushroom debris. Gatherings should be made frequently, as mushrooms which are allowed to grow to a large size weaken the spawn and exhaust the bed.

The produce is separated into two grades, one consisting of "buttons" and "cups" (i.e., closed mushrooms); the other of "broilers" or "flats," of which the former are worth about one-third more than the others. Unless the stalks are cut off at the time of gathering, the mushrooms are placed in the baskets with the stems downwards, so that grit shall not enter the gills. Any dirt on the caps is removed with a piece of flannel. Care is taken to avoid touching the tender gills, on which the least pressure causes discolouration.

Growing under Glass, Cellars or Sheds

The winter temperature in a shed being much lower than that of a vinery or the usual cellar, a greater depth is necessary. The beds require a covering of two or three inches of litter in a mild temperature, and of one foot to one foot three inches in a cold one. No water need be given during the first month. After that, if the soil be getting dry, water is given in very small quantities by means of a fine "rose." The temperature of houses heated artificially should be from 55 to 60 deg. Fahr.

Cultivation of Morels: Morels are cultivated to some small extent in France on similar lines to those obtaining with the

Common Mushroom, a bed being prepared in a cellar, precisely as for the latter, under the usual conditions of darkness and dampness, and with the requisite drainage. This is sown with morels cut up into small pieces, by which it is said that production can be obtained through the spring and early summer, thus ensuring a rich harvest for the grower possessed of the necessary skill. It would appear, however, that the morel presents greater difficulties than does the Common Mushroom.

Italian Cultivated Fungi. Various fungi have been cultivated on a small scale, and the Neapolitans raise the small and delicate *Clitocybe catinus* from old coffee-grounds, which are simply kept undisturbed for a period of several months in a warm place, and watered from time to time, when the fungus makes its appearance. In the same country is found the *Pietra fungaja,* or Fungus stone, which is simply the sclerotium of a fungus, *Polypoms tuberaster,* permeating lumps of volcanic tufa, which it has the power of compacting into a hard and almost stony form. These "stones" are put in a warm place and kept well-watered, when in due course a crop is obtained. On occasions, the Fungus stone has been exported to this country and propagated in the same manner, with quite successful results. This, perhaps, can hardly be described as "cultivation."

Cultivation of Boletus edulis: Among other fungi, the cultivation of which has been attempted, is *Boletus edulis,* where the method adopted has been to select a situation resembling its natural surroundings, and to water this freely with an infusion of the fungus. Results, however, have not been very successful. Hay suggests that "the fecundity of spores depends on the presence also of the mycelium," which is rather begging the question.

Shii-take Industry of Japan: No chapter on the cultivation of fungi would be complete without some account of the Japanese "Shii-take" industry, not only because of its magnitude, but also because of the remarkable cultural methods employed. Of the large number of Japanese fungi of economic importance, most of which have previously been dealt with, the "shii-take" is that cultivated on much the largest scale, although the total crop of this is less than in the case of another species, "matsu-take," which, however, is not cultivated, but only gathered wild. The "shii-take" *(Armillaria shiitake)* is readily dried and preserved, in which process there is developed and retained a very fine aroma, this rendering it a delicate ingredient in the making of savoury soups. In the dried state it has the advantage of keeping for a long time, and large quantities are annually exported in this condition, chiefly to Hong-Kong, China and Hawaii, but also to the Straits Settlements, India, Indo-china, the Philippines, and even to Canada and the United States.

It is obvious that the latter figures are more or less nominal, the value of the total production being probably obtained by multiplying the weight by the local price per kin. Naturally, if a large proportion of this is dried and exported, its value may be considerably enhanced, with the result in question. It must further be remembered that in the case of such a crop, part of which is of natural occurrence, large quantities must be gathered and consumed without appearing in such official figures as we have quoted.

This mushroom derives its name from the "Shii" tree *(Quercus cuspidata)*, an evergreen oak of Central and Southern Japan on which it most frequently occurs; but in the artificial production, not only the "shii," but various oaks, and other trees, are employed. Among these are the "kunugi" *(Q. serratd)*, the "konara" *(Q. glandulifera)*, the "kashi" *(Q. acuta)* and the "kashiwa" while the "soro" *(Carpinus yedoensis)* is also very largely used.

In cultivation, logs of any of these trees are cut and soaked in water for some days, being afterwards pounded to soften the bark. Holes a few inches deep are then made, at short distances apart, in the outside; and "spawn," prepared by breaking up pieces of trees infected by the mycelium of the fungus, is placed therein, the logs being then left in a shaded part of the forest. The first mushrooms are produced in about two years, being usually very large and of an "umbrella" shape. According to Porter, they are "greatly superior in flavour to those of any other country. They can be grown in rotation during most of the year, and are named differently, according to the season, "haruko," "natsuko" and "akiko," being respectively spring, summer and autumn mushroom. This method of cultivation is centuries old, and large tracts of forest are thus prepared.

The profits are very large, and part of the revenue of the state forests is obtained in this manner. The districts where the industry is chiefly carried on are those of Hyuga and Bungo in Kyushu; and of Kii, Ise and Suruga in Honshu. It is obvious that such destructive methods of cultivation must be accompanied by corresponding reafforestation, and this is obtained naturally, by permitting the growth of sprouts from the stumps of the trees already felled. These sprouts reach a size suitable for "shii-take" culture in from 18 to 25 years, and by an annual rotation of felling, a new tract of forest arrives at maturity every year.

Japanese Pine Mushroom: The "matsu-take" or pine mushroom *(Armillaria edodes)* is a much-esteemed delicacy which is consumed in large quantities. It is found throughout Japan, that of Kyoto being considered the best. The following quaint account of it is contained in the official pamphlet already referred 3,079,700 kin [about 1819 tons] valued at 503,600 yen This mushroom has the highest flavour of all kinds, and people of all classes have it at dinner.

The quality, however, is very fragile, and is impossible to keep it dried for a long time, without losing its flavour, as 'Shii-take.' Therefore, we can taste its nice flavour for only one month in the Autumn which is the season of its growth. In recent years, people have begun to preserve it in tin cans, and export it to foreign countries. The most famous 'Matsu-take' producing places are the forests belonging to Osaka Major Forest Office and Hiroshima Major Forest Office, and in certain Minor Forest Offices a greater income is obtained from the proceeds of this mushroom. The gathering of this mushroom is very simple and done by women and children, and increasing demand of labour for gathering of mushrooms naturally causes the raising of the wages of poorer classes in those districts conferring on them no small benefits.

This mushroom is eaten either roasted or boiled, and is also preserved by being salted or simply dried, when, however, as we have already seen, it tends to lose flavour and become insipid. Whether the same thing occurs when it is exported in "tin cans we are unable to say.

The Chinese Cultivation of the Jew's Ear, A somewhat similar method to that of Shii-take culture is employed by the Chinese for the production of a fungus generally described as "The Jew's Ear," although, as already mentioned, it is not *Hirneola Auricula Judae,* but *H. polytricha*. This fungus, the Chinese name for which is *Mu Erh,* is an export of Manchuria, and since it is also largely cultivated in provinces as distant from Manchuria as are Shen-si and Kwang-si, it is probably fairly widely spread through China. From Shen-si, where it is grown extensively in copses of small oaks *(Quercus variabilis)* on the southern slopes of the Tsinling Mountains, it is shipped by river to Hankow for distribution.

It grows naturally on decaying oak-stumps and requires a damp atmosphere, with sufficient heat in summer, a long period of rain being particularly favourable for its development. In appearance it resembles an ear, is of a gelatinous nature, and is greatly esteemed as a delicacy, being indeed indispensable at all Chinese feasts. It is well known that the Chinese taste differs markedly from that of the Western nations, and it is therefore not surprising to read that an English traveller "did not find them very palatable, and the experiment resulted in a bad stomachache.

In the artificial production, oak copses growing on granite hills are employed, and small sapling oaks, about 6 in. in diameter, are cut down, and the branches removed, leaving the bark entire; after which the trunks are cut into poles 8 to 10 feet long. These are allowed to lie on the ground for several months, when they become infected by the mycelium of the fungus, and are finally stacked slantingly in scores or so. In the following year, the brown fungus appears on the dead poles, being collected as it grows.

Truffles: We may next consider the truffles, which act as a sort of link between the cultivated and the wild fungus foods, for they served the Greeks and Romans in the second capacity; while their production has created an enormous industry of today. They are certainly "cultivated" to some extent nowadays, but most of their interest concerns, not their cultivation, but their "hunting," which naturally presupposes that they are wild. They were, indeed, among the most famous of the fungus foods of commerce, long before attempts to cultivate them had met with any success, and they are still very largely collected in a wild state.

This industry is chiefly carried on in Southern France, over which truffles are very widely spread, some notable centres of this trade being the towns of Perigueux, Apt and Toulouse, in the Departments of the Dordogne, Vaucluse and Haute Garonne respectively, while the most celebrated French truffle is that of Perigord, *Tuber melanospermum.* These are generally found on a light and porous soil of clay-marl, resting on limestone, and at the edges and open spots in woods, while the ground must be free from undergrowth.

The trees with which they are generally associated are oaks *(Quercus coccifera Q. ilex, Q. robur* and *Q. sessiliflora),* but they are often found under beech and hazel-nut trees, occurring as curious clusters, at a depth of generally from two to eight inches, by the roots of these trees; this having given rise to the conjecture that there is some connection between the two, and that in their early stages the tubers obtain some nourishment from the roots.

Truffle-hunting: Owing to the subterranean habits of these delicacies, the ordinary methods of search are useless, but their characteristic scent enables them to be discovered by either dogs, or pigs, which have been trained for the work. In regard to this, Krombholz says: You must have a sow of five months old, a good walker with her mouth shut up by a leathern strap; recompense her for the Truffles with acorns; but as they (pigs) are not easily led, are stubborn and go astray and dig after a thousand other things, there is little to be done with them... dogs are better, of them, select a small poodle. The pig is considered to have the keener scent of the two, for the truffle, but is less easy to convince, at the psychological moment, that the find is not really for him (or rather her). Ramsbottom says that pigs, or, more generally, sows, which are more docile and also better "hunters," are employed where truffles are common, and dogs where they are more scarce, presumably because, when the crop is plentiful, it is not quite such a serious matter if the sow succeeds occasionally in maintaining her rights.

Also, a sow tires far more readily, for which reason it is always taken to the hunting-grounds in a cart. At any rate, both animals are commonly used. While the taste of the pig for truffles is a natural one, that of the dog is acquired, to which end its whole training is devoted. Thus, while still a puppy, finely-chopped truffle is mixed with its food, so that it always associates truffles with dinner. Later on, having been kept short of food for a day or so, it is taken out into the woods to endeavour to locate the spot where its master has previously hidden a plateful of chopped truffle and earth, first under some dried leaves; then beneath a light sprinkling of soil; and finally buried two or three inches deep; the reward for finding it being a small piece of meat, or of bread and cheese.

When it has become sufficiently skilled at the artificial quarry, it is used to seek out the natural one, which, having been located by the eager scratching of the dog, is dug up by means of a small fork by the *rabassier* (from *rabasso* the Provengal word for "truffle"); while the dog is similarly rewarded for each find.

Truffle Fly: In another method also used, the aid of the diminutive truffle-fly is invoked. This is *Anisotoma cinnamomea,* the larvae of which live in the truffle; while the mature flies seem to regale themselves with the scent of their erstwhile home, above which they hover in clouds. By looking horizontally along the surface of the ground, usually in the evening, these clouds of minute flies are observed by the peasants, who then seek the confirmatory evidence of the fissured ground and the truffle scent, thus finding the delicacy.

Truffle Cultivation: The French truffle industry has attained large proportions and, according to Hay, the annual yield of the Department of Vaucluse alone exceeds 60,000 Ib. The economic importance of the truffle has naturally led to attempts at cultivation, which at first seem to have had but indifferent success. These methods perhaps can scarcely be considered as a true cultivation, but rather the stimulating to a more intensive production of the natural grounds.

In a method used in Poitou, a situation on the chalk downs is selected and enclosed, being then sown with acorns. After some years' growth of the sapling oaks, truffles appear among their roots, and may be gathered for many years. It is believed by some, however, that this plan is not successful except on soil where truffles have occurred before. A second method, which has a little more claim to the title of "cultivation is also employed. Here again a situation resembling that of the natural habitat is chosen, and" the ground is cleared, the trees are thinned, and trenches are dug. Truffles are now brought from their native site, inclosed in the earth in which they grew, and are planted in the trenches. In time they propagate, and it is said that grounds have been successfully fertilized in this way." At the present day, oak plantations are planted on a large scale in Southern France, expressly for the production of truffles.

In addition to "*Tuber melanospermum,* the cultivation of which we have described above, and which does not occur in Britain, the French use also *T. aestivum,* but on a smaller scale, as it is not esteemed so highly. Ramsbottom says that this truffle probably "came into use in the courts of France and Italy some time prior to the fourteenth century. *T. melanospermum* seems to have made its appearance into Parisian cookery towards the close of the fifteenth century."

We have lately received from Bergerac in Perigord, a price-list, for 1923-24, of various delicacies ("conserves deluxe exclusivement"), of which truffles are the basis, and from this we venture to extract some evidence bearing upon the present export trade in French truffles, and their products. It had already been translated into "English," and we give it "as wrote."

"On account of the high price of life now, The fashion is spreading more and more in France of sending at Christmas an New Year, useful as well as agreeable presents especially foies-gras truffes pies, I offer this year to my foreigner customers the parcel Gift' composed of all series of my Pates de foies gras truffes.

Summer Truffle: The best of the British species is *Tuber aestivum,* our Summer Truffle, which resembles a small dark-brown or almost black apple, one to three inches in diameter, having a warty surface, at times veined with white. It is found chiefly in beechwoods from June to October.

In former days a considerable trade existed with this truffle in Sussex and Wiltshire, the dogs employed being mongrel terriers. *T. aestivum* was sold usually at two and three shillings a pound. There is still a small amount of hunting done, but the trade is said to have been ruined by the taxation of dogs. It is more likely, however, that the preference for the Perigord truffle, and its greater abundance have ousted the summer truffle from the markets.

In this connection, it is stated in *The Times* for Feb. 14, 1859, referring to poor labourers in Wiltshire, that "many of these people... live by truffling and poaching, in the absence of farmer's employment," and it is interesting to note that a few of the old truffle hunters are still to be found in Sussex. Besides the true truffles, other fungi, which somewhat resemble them, are known by the same popular name, and are also edible. Thus *Melanogaster variegatus*, the "Red Truffle," was formerly sold in the markets of Bath.

There is also the "False Truffle" or Earth Ball, *(Scleroderma vulgare)* in which a large trade is done. It is used as a substitute for the real article in various dishes, being collected in Epping Forest by gentlemen from Soho; but, as with the Red Truffle, is of very inferior flavour in comparison with the true truffles.

Food Value of Fungi: In any consideration of the fungi as foods, it is desirable to include some account of their real food value. This has unquestionably been somewhat overrated by certain writers, who, probably because the flavours of many fungi resemble those of meat, have suggested that they may rank as substitutes for flesh foods.

This is hardly correct. Langworthy lays stress on the fact that mushrooms and other edible fungi contain "a very high percentage of water—over 90 per cent, on an average. The 10 per cent, or so of nutritive material they contain, is largely carbohydrates, though a little nitrogenous material is also present. Fat is almost utterly lacking. So it is obvious that the mushroom more nearly resembles in composition such vegetables as a carrot or turnip than it does meat.

Mushrooms and some other edible fungi have flavours which to many palates suggest meat, oysters and some other animal foods. From the standpoint of flavour and palatability they are worth including in the diet if they are relished, and alone or combined with other materials they can be served in dishes which suggest meat dishes in flavour and which satisfy the palate, while the nutritive value of the meal or day's ration can be made up to the desired standard by the other dishes served at the meal with the mushrooms." The same view is expressed by Setchell, who remarks:

> "As regards nutritive value, it may be said that even the most nutritious toadstools are of little value

[as foods], but are valuable as accessories or condiments with, not in any wise as substitutes for, the meat or vegetables ordinarily consumed."

In any comparison, however, it is necessary to remember that different meats also contain from about 40 to 72 per cent, of water, the latter figure applying in the case of lean beef or mutton; while, if one neglects the water content in each case, and calculates only on the dry residue, the nitrogen contained in many fungi does not compare very unfavourably with that of meat, which again varies in different kinds. Thus lean beef (dry) would contain about 10.6 per cent, of nitrogen, and turnips (dry) about 2.1 per cent., as compared with the following figures, quoted by Schlossberger and Dopping, for various mushrooms (dried):

Nitrogen Per cent.

Cantbarellus cibarius 3-22

Lactarius deliciosus 4-68

Psalliota arvensis 7-26

As regards the nitrogenous portion of one's dietary, there is thus a fairly good case for the fungi, which are at any rate decidedly superior to either carrots or turnips. A complicating factor is, naturally, how much of this nitrogen is assimilable; but this point, in view of the scarcity of data, cannot be discussed with advantage. Again, mushrooms are not consumed in a dry state, but usually by cooking them directly after gathering. In this process, however, particularly if they are fried, a large quantity of the water is expelled, as is evidenced by the great contraction in bulk.

The question of the fat content of fungi cannot be dismissed quite as briefly as it is by Langworthy. Thus an analysis of the dried form of "shii-take," as exported, shows that it contains about 12 per cent, of albumen, 1£ per cent, of fat, nearly 70 per cent, of cellulose and other nitrogenous constituents, 4 per cent, of ash and about 12 per cent, of water. It may, therefore, be considered that fungi have some food value, and serve in some measure to make up the necessary amount of varied food required by the individual, but they should not be regarded either as a substitute for meat, or as a staple food, but chiefly as an agreeable addition to the dietary on the ground of flavour and palatability.

Cultivation of Fungi by Ants: It may possibly be a matter of surprise that fungus culture is not confined to man alone; but it is deliberately practised for food purposes by the termites and leaf-cutting ants of Ceylon, Madagascar, Brazil, and other tropical and subtropical countries. Although this method of obtaining food has been adopted by a number of different species in widely-separated regions, the general mode of cultivation followed by them appears in all cases to be substantially the same.

In their nests, the ants allocate special chambers for this work, and in these they build up spongy masses of masticated, and pounded leaf-fragments and other plant remains, which quickly become permeated with mycelial threads. This mycelium, under the methodical cultivation which enşues, produces small tufts of hyphae, which have been termed "kohl-rabi clumps," and these are cut and eaten by the ants, being apparently their chief, if not their only food. These fungus gardens are systematically tended by the workers, and not only are foreign organisms weeded out, but the fungus plants are pruned to keep them both in bearing and within reasonable proportions. If the ants are removed, the fungus ceases to produce the "kohl-rabi clumps," and throws up reproductive structures, which usually burst through the top of the nest.

That the whole proceeding is as well-ordered and intensive as our own mushroom culture, is shown by the fact that each species of ant always cultivates the same variety of fungus, which, moreover, is the only species found in the nest, and which is never found growing away from it. Most of the fungi so cultivated are agarics.

3

Respiration and Fermentation

Exercise in Pathogenic Fungi

The number of microorganisms on which studies of oxidative activities has been made increases each year, but of the fairly large number of fungi pathogenic for man only a few have been investigated. Our knowledge of cellular oxidation processes has, in large part, been developed through intensive work on a few mammalian tissues, a few bacteria, and a few yeasts. Extension of findings made with these well-studied forms to other organisms has been possible in many instances, but the complexities of mechanisms of cellular oxidation permit possible variations in pathways leading from substrate to final products. For most of the pathogenic fungi we are ignorant of their nutrition, let alone the catabolic paths of oxidation of those substrates. Investigations of the metabolism of fungi causing deep mycoses are few and particularly needed in view of the high mortality occasioned by many organisms in this group.

In the search for more active chemotherapeutic agents for the control of these organisms, work on inhibition of their respiration may prove of value. Substances presently employed in the therapy of fungus infections have seldom been subjected to an analysis of their mode of action; some of these substances may act as respiratory inhibitors.

With pathogenic organisms, other than fungi, investigations on the mechanism of therapeutic action of active substances have frequently brought to light more favourable related agents than the substance prompting the analysis. Welch (1945) in a stimulating review on the relation between chemical structure and biological action of chemotherapeutic substances, points out that the purposeful design of improved active compounds can be undertaken.

Ideally, substances employed in therapy will attack the parasite through some weak link leaving the cells of the host relatively unaffected. Frequently the weak links in the parasite occur in some metabolic process not prominently displayed by cells of the host. Fundamental investigations of all phases of the physiology of the organism are necessary for a foundation to any thoroughgoing programme aimed at the prevention, prophylaxis, and treatment of fungus diseases.

It is to be understood that the respiratory enzyme systems of the pathogens are but one site of attack in the chemotherapeutic approach. And it should be realised that inhibition of oxygen consumption may be an insignificant effect on cells of an organism that can employ hydrogen acceptors other than oxygen for the main part of its metabolism. Sevag, Richardson and Henry (1945) have emphasised this contingency as a result of their work on the mode of action of sulphonamides on the respiration of bacteria.

This discussion will be limited to two groups of pathogenic organisms for which information is available. A comparative treatment will be attempted for the pathogenic forms and related non-pathogenic organisms. Yeast-like forms, such as *Candida*, appear to have much in common in their metabolic aspects with non-pathogenic yeasts, while the dermatophytes may be compared with other filamentous fungi.

It is not believed necessary to enter into any review of the mechanisms of cellular respiration that have been uncovered in recent years. For such, reference may be made to the recent reviews by Lipmann (1943), Green and Stumpf (1944), Lardy and Elveh-Jem (1945), to the work of Green (1940), the Symposium on Respiratory Enzymes (1942) and to Goddard (1946).

Yeast-like Fungi

Candida: The study of gaseous exchange in *Candida* has been employed by nearly all students of the group as a taxonomic tool in the form of comparative qualitative observations on the anaerobic decomposition of carbohydrates by the various species. Martin, Jones, Yao and Lee (1937) and Martin and Jones (1940) presented a resume of the utility of such studies, of the difficulties involved in securing reproducible results, and of the procedures yielding reproducible results in their hands. A glance at these may be instructive.

The procedures used by these workers at the Duke University Hospital have yielded uniform results, with hundreds of different isolates of *Candida,* that conform precisely with data from pathogenicity tests and morphological examination.

Especially emphasised in their technique are: *(1)* the inoculum must be from a second or third subculture on a glucose-free medium, *(2)* the fermentation broth tubes must be sealed with vaseline, since isolates of *Candida parakrusei* regularly produce acid from sucrose and galactose if vaseline seals are omitted, but form no acid from either carbohydrate anaerobically, *(3)* the beef-extract fermentation broth must be accurately titrated to *pH* 7.2—an increase of 0.2 pH will inhibit the production of acid by *C. albicans* from sucrose and galactose and prevent the formation of both acid and gas from glucose by *C. parakrusci, (4)* results are reported after 10 days' incubation at 37° C. The findings for *C. albicans* were based on almost one hundred strains identified from a total of over five hundred isolates of *Candida* and, with the exception of one dissociated strain, all isolates of *C. albicans* were pathogenic for rabbits in 2 to 7 days.

The results obtained for the decomposition of a given carbohydrate in such large-scale, carefully performed tests possess a high degree of validity, especially when the correlations with pathogenicity and morphology are included. Of course, the observations are qualitative and less sensitive for the detection of gas production than quantitative studies. Quantitative investigations of the fermentation by *C. albicans* are desirable, but due regard must be had for the origin and culture history of the strain used, pH at which conducted, temperature and duration of the investigations, and realisation of the fact that rapid growth takes place in the procedures of Martin *et al.* since a favourable nitrogen and bios nutrition is provided.

Some data from quantitative studies by Kluyver and Custers (1940) are available for two species of *Candida* and these will be examined.

Fermentation of Sugars by Candida: An interesting group of yeasts reported to assimilate, but not ferment, disaccharides was investigated by Kluyver and Custers. Included in their study was a strain of *Candida parakrusei*, reported to assimilate both maltose and sucrose, but ferment neither, and three strains of *Candida albicans* reported not to ferment sucrose but able to assimilate it.

There is no question about the ability of any of these strains to utilise maltose or sucrose, though *C. parakrusei* is a poor user of maltose, particularly when compared with the values of *Saccharomyces cerevisiae.*

Although four different methods of detection were employed, no evolution of CO_2 by *C. parakrusei* from maltose could be detected. Quantitative measurements were made in the van Iterson-Kluyver apparatus and by the Warburg manometric technique. With a sucrose substrate, both species and all four strains showed a slight but unmistakable evolution of CO_2 anaerobically. It will be observed that with both species there is a definite production of CO_2 anaerobically in the absence of added substrate; this is in contrast to the well-confirmed findings of Stier and Stannard (1936) for *S. cerevisiae* where no gas or heat exchange has been observed anaerobically in the absence of added substrate. Additions of sucrose about doubled the rate of CO_2 production in all strains.

Although agreeing with Martin *et al.* on the lack of gas production by *C. parakrusei* from maltose, the finding of gas production by three strains of *C. albicans* from sucrose is at variance. It is to be emphasised that the amount of gas produced is small, could escape detection qualitatively, and that the experimental conditions were much different from those employed by Martin *et al.*, so there is, for all practical purposes, no conflict of statement here. The quantitative detection of €62 production from sucrose by *C. albicans* has been confirmed by Van Niiel and Cohen (1942).

Aerobic Fermentation by Candida: An aerobic fermentation of both maltose and sucrose was found by Kluyver and Custers to occur with *C. parakrusei*. This fermentation occurring in the presence of oxygen was greater than the anaerobic fermentation of sucrose. The presence of hydrolases for sucrose and maltose is thus demonstrated; these enzymes must be partially or completely inactivated anaerobically.

It was suggested by Kluyver and Custers that possibly the increased state of reduction in the interior of the cell due to the withdrawal of oxygen under completely anaerobic conditions might cause an inactivation of the respective hydrolases. This situation, wherein a substrate can be fermented only in the presence of oxygen, seems to have something in common with the observations of Stier and Stannard on the inability of *S. cerevisiae* to utilise carbohydrate reserves in the absence of oxygen.

In each case, the substrate and enzymes appear to be within

the cell but unable to form an activated complex; in the instance with *S. cerevisiae*, spatial relationships appear to be important.

Probably the occurrence of an aerobic fermentation with *C. parakrusei* is responsible for the qualitative observations of acid (and slight gas) production from sucrose and galactose referred to earlier.

Kluyver and Custers attempted to exaggerate the negative Pasteur effect by adding cysteine to suspensions of *C. parakrusei* fermenting aerobically but found no increase in CO_2 production. The cysteine was, however, readily converted to H_2S (sulphur competing with oxygen as a hydrogen acceptor) though respiration was not inhibited (the studies of Runnstrom and Brandt, 1941, on the effects of cysteine and glutathione on the Pasteur effect are of interest in this connection).

Oxidation of Substrates by Candida: It may be instructive to compare findings on the oxidation of substrates by *Candida albicans* with our information on the process as observed in other yeasts (non-pathogenic) and in other pathogenic fungi. *Candida parakrusei* was found by Kluyver and Custers to oxidise maltose readily (at about half the rate for glucose). This, then, is an organism that will not ferment maltose but which can assimilate and respire it readily.

Both *C. parakrusei* and the three strains of *C. albicans* were found to oxidise sucrose readily at slightly less than the rate for glucose. In all of these experiments the cells were obtained from growths on malt agar plates, washed, and suspended in phosphate buffer at *pH* 4.5; the cells were not reported to be starved. There is no question, from the data of Kluyver and Custers, about the increase in rate over the endogenous rate of oxygen consumption prompted by the addition of glucose, sucrose, or maltose.

Using a single strain of *C. albicans*, Van Niel and Cohen (1942) found that suspensions prepared directly from yeast-agar plates showed a high endogenous rate of oxygen consumption (rate in absence of external substrate) that declined rather rapidly with time. In contrast to the findings of Kluyver and Custers, the addition of oxidisable sugars to such suspensions brought about little increase in the rate of oxygen consumption.

With many other yeasts *(e.g.* species of *Saccharomyces, Zygosaccharomyces, Saccharomycodes,* and *Schisosaccharomyces)* the addition of oxidisable substrates to suspensions prepared directly from nutrient media is generally followed by a rapid rise in oxygen consumption to what is frequently termed the exogenous rate. Among the yeasts with these pronounced dissimilative abilities is the familiar bakers' and brewers' yeast, *Saccharomyces cerevisiae.*

Suspensions of *C. albicans* could be prepared by Van Niel and Cohen which did oxidise added substrate through the simple procedure of placing growths from nutrient media into sterile tap water or sterile iso-osmotic buffer and aerating by shaking in a shallow layer or by bubbling sterile air or oxygen through the liquid.

After several hours, during which time the cells had been respiring at their endogenous or autorespiration rate, intracellular reserve stores of oxidisable material were depleted. Addition of oxidisable substrates to such starved preparations was followed by a rapid rise in oxygen consumption to a high level. However, the quantity of oxygen theoretically necessary for complete oxidation of the added substrate was rarely consumed. Rather the oxidation process ceased with only one third of the glucose or sucrose oxidised, the remainder of the sugar being presumably converted to an assimilation product of the general formula (CH_2O).

Assimilatory Processes: Barker (1936) demonstrated that with suspensions of non-proliferating cells of the colourless alga, *Prothotheca zopfii,* only part of the disappearing non-nitrogenous substrate could be accounted for by oxidation since a large fraction of it had been converted into cellular material. Since this demonstration it has been established by several investigators for a variety of organisms that the phenomenon of assimilation can occur aerobically and anaerobically with "resting" cells.

It should be noted that Barker, Van Niel, and other authors have referred to the assimilated product as (CH_2O) but Schade and Thimann (1940) showed with *Leptomltus* that nitrogen is also involved and that the assimilation product in this organism has about the formula of alanine (incidentally this has the same state of oxidation as a carbohydrate).

In fact, the quantity expressed as (CH_2O) may not even be inside the cell, for Pickett and Clifton (1943) have shown that with suspensions of S. *cerevisiae* in a glucose solution half the material expressed as assimilated (CH_2O) from manometric studies in reality occurred free in the medium as soluble end products with an aggregate composition of (CH_2O). Even in the absence of a nitrogen source the product actually assimilated in the cells may be other than a carbohydrate, for Stier, Newton and Springe (1939) have shown starved suspensions of bakers' yeast can synthesise both fat and carbohydrate aerobically from glucose. While it is possible in this instance that the carbohydrate was transformed into fat there seems no doubt from the work of De Boer (1928) but that *Phycomyces blakesleeanus* assimilates and metabolises fat preferentially.

Inhibition of Assimilation Processes: It was found by Clifton (1937) that the addition of 2, 4-dinitrophenol (DNP) to respiring preparations of microorganisms increased the consumption of oxygen to the theoretical value for complete oxidation of the substrate present. It was believed at first that this represented an inhibition of the assimilation process, but, in at least one instance, such interpretation may not be valid. Burris and Wilson (1942) found with the nitrogen-fixing root nodule bacterium, *Rhisobia trijolii,* 205, that the addition of DNP following the levelling off of the initial high rate of O_2 consumption resulted in a rapid rise in the rate of oxygen uptake, approaching the initial rate. This return in rate persisted until the theoretical value for complete oxidation of the added substrate was reached. In this case it has been considered (Burris and Wilson, 1945) that DNP brought about a rapid oxidation of assimilated materials.

If such were true, it is hard to understand why the velocity of the catabolic reaction should decrease when the "assimilated" material from the substrate added to the resting cells was consumed, and why the process of rapid oxidation should not continue to feed on materials previously assimilated by the cells during growth unless there are decided differences in the two assimilated materials or in their utilisation. The stage at which DNP may reverse the assimilation process in *Rhisobia* (or in any other organism) has not been demonstrated, but it would seem clear that the material oxidised following the addition of DNP may not have the same "status" in the cell as previously assimilated materials.

There appears to be some parallel here with the findings of BARKER (1936) that *Prototheca zopfii* continues to oxidise, though at a reduced rate, the material immediately assimilated into the cell from a glycerol substrate. The oxidation of this assimilated, but not "fixed" substance declined in rate until the endogenous rate once again was reached.

Other substances than DNP are known that interfere with the assimilation process. It appears that a variety of substances are able to inhibit assimilation and concomitantly stimulate some catabolic process.

Runnstrom and Brandt (1941) found that glutathione inhibited assimilation from glucose by bakers' yeast, while stimulating aerobic glycolysis. The respective relations of the three processes (oxidation: assimilation aerobic glycolysis) was 4:6:0 which changed to 4:0:6 following the addition of glutathione. This negative Pasteur effect was interpreted by the authors as a diversion of glucose phosphate to fermentative decomposition away from pathways leading to glycogen synthesis. It is of interest that respiration is not stimulated in this example; it might be that the respiratory enzyme systems were working at saturation prior to the addition of glutathione and the excess "pressure" (by analogy with Mass Action) of glucose-phosphate, developed by the blocking of assimilation pathways, is shunted via aerobic glycolysis paths.

Cori (1941) had found that reducing agents, such as glutathione, markedly shift the equilibria of the reactions in favour of the formation of glucose-6-phosphate. This is away from the synthesis of glycogen, and the findings of Runnstrom and Brandt may be explainable in this manner. Along these lines Kluyver and Custers were able to inhibit the Pasteur effect in *Torulopsis dattila* with 6 mg. per cent (M/2000) cysteine (to which component glutathione owes its reducing activity) securing a pronounced aerobic fermentation of maltose which *T. dattila* oxidises and assimilates readily, but which it does not ferment anaerobically at all.

Problems of Spatial Orientation of Substrates and Enzymes in Cells: One of the most striking examples of the governing influence of spatial arrangements on metabolic processes seems to be the lack of detectable gas exchange (Stier and Stannard, 1936) or heat production (Winzler and Baumberger, 1938) in the cells of many yeasts under anaerobic conditions in the absence of external substrate. Destruction of the cellular organisation, as by grinding of the cells, is rapidly followed by a fermentation of substrate stored within the cell (Stier and Stannard). Aerobically, in the absence of added substrate, the dissimilation of stored reserves by yeasts appears to proceed only by oxidative paths. This was shown for S. *cerevisiae* by Stier and Stannard and confirmed by Spiegelman and

Nozawa (1945) for *Saccharomyces, Schizosaccharomyces,* and *Saccharomycodes*. It seems quite clear that stored reserves in most of the yeasts studied cannot be metabolised (in the absence of added substrate) either aerobically or anaerobically by fermentative paths in the intact cell; the fact that destruction of the cell organisation permits fermentation to occur would tend to support the view that stored reserves are separated spatially from the fermentation systems. As a corollary, the enzymes for endogenous respiration must be associated with reserve stores and similarly removed from the fermentative enzyme and exogenous respiration enzyme systems.

Evidence from another source supports the view of spatial separation of the endogenous and exogenous systems. Selzer and Baum-Berger (1942) showed that the endogenous respiration in *S. cerevmae* is not influenced by metallic mercury in the suspending medium for the cells, but the exogenous respiration system is markedly inhibited. They suggested that the sulphydryl groups (specifically inhibited by mercury) of the endogenous respiration systems are located within the interior of the cell, while the thiol concerned in the exogenous respiration system is located in or near the cell wall.

An interesting aspect of the relations, in the absence of substrate, of yeasts to oxygen is that while aeration reduces the endogenous oxygen consumption markedly in a few hours, suspensions can be stored anaerobically for at least 72 hours in buffer alone without decline in the endogenous rate on return to oxygen.

This was shown by Nickerson and Carroll (1943) for *Zygosaccharomyces* and confirmed by Spiegelman and Nozawa for *Saccharomyces, Schizosac-charomyces,* and *Saccharomycodes.* The question was raised by both authors regarding the energy requirements of these cells anaerobically. It has frequently been assumed that cells utilise energy to maintain their intact condition with their components in some sort of order. But here are examples of cells that apparently carry out little or no energy exchange and yet exist without impairment to a complicated internal structure that immediately responds (unlike spores) to the addition of oxygen or substrates.

A process operating with yeast cells stored aerobically is the loss (or secretion) of riboflavin or thiamin into the buffer, with consequent impairment of cellular function which can be overcome by subsequent replacement of the lost substance. Such a loss does not seem to occur anaerobically to any appreciable extent and appears to be one process distinguishing the two methods of storage. It is curious that in at least one instance the invigoration of resting (and resistant) structures is conditioned by the external supply of diffusible substances implicated, as are thiamin and riboflavin, in cellular respiration processes. This situation occurs in the stimulation of excystment in *Colpoda* (Haagen-Smit and Thimann, 1938) by members of the four carbon dicarboxylic acid series.

Three yeasts investigated by Kluyver and Custers apparently can carry out an anaerobic gas exchange in the absence of added substrate. The anaerobic conditions were carefully produced by passing a nitrogen-carbon dioxide mixture through heated copper coils; the aerobically grown cells had been washed three times and resuspended in phosphate buffer. The conditions were the same as those under which other yeasts did not produce gas. Van Niel and Cohen also found carbon dioxide (and a very slight amount of acid) produced by *C. albicans* anaerobically in the absence of substrate. It seems clear that the situation with *Saccharomyces* is not universally distributed among yeasts.

Physiological Age of Cells vs Assimilation: Assimilation of substrates by cells seems to be influenced by age of the cells and their culture history. Nickerson and Carroll (1943) pointed to the effect of age and culture history of cells on the rates of oxygen consumption by suspensions of the yeast *Zygosaccharomyces acidifaciens*. Bernstein (1944) found suspensions of young cells of *Pseudomonas saccharophila* oxidise ½ the molecule and assimilate the other half from succinic, fumaric, and malic acids. With older cells (defined by the pH to which cell growth had changed the H+ concentration of the medium) the same fraction of the substrate was assimilated but two distinct steps occurred in the oxidation; pyruvic acid accumulated as an extracellular product in the first phase, disappearing in the second phase. Van Niel and Anderson (1941), confirming and amplifying the demonstration by Winzler and Baumberger (1938) that assimilation of substrate may occur anaerobically as well as aerobically, found 30 per cent of the substrate assimilated by relatively young cells (2-5 days) of *Saccharomyces cerevisiae,* but only 7-8 per cent assimilated by old cells (28 days).

The process of producing "starved" preparations of microorganisms may have effects on the cells other than mere depletion of oxidisable reserves. Stier (1939) showed that after removal of cells from their nutrient medium, aeration of *Saccharomyces cerevisiae* in a phosphate buffer resulted in the production of suspensions exhibiting a lowered aerobic metabolism of dextrose when the suspensions were supplied with that sugar.

Giese (1942) found that the endogenous respiration of *S. cerevisiae* exhibits a decline with increase in age of the cells as well as with aeration after removal of the cells from their culture medium. Stier and Maclntyre (1942) have shown a rapid loss of riboflavin from cells of *S. cerevisiae* into the storage buffer during aeration of the yeast suspension. Van niel and Cohen found that aerobicalty grown 2-3 day old cells of *Candida albicans,* after 20 hours aeration of a suspension in tap water, are apparently deficient in cocarboxylase, as judged by the rate of oxidation of pyruvic acid by such cells. The rate of pyruvate oxidation could be doubled by the addition of thiamin to the starved cells.

Discussion: The process of liberating energy for cellular activities through the uptake of molecular oxygen by the cell in the course of oxidising its particular substrates furnishes a wealth of sites for chemotherapeutic attack on the cell if it be one whose continued existence is deemed undesirable. While the basic concepts of the mechanisms of hydrogen activation, hydrogen transport, oxygen activation, oxygen transport, and phosphorylation that have been elucidated appear in the main to hold true for cells generally, the catalytic mechanisms involved in cellular respiration are manifold, and multiple pathways of oxidation are known to exist. While even related species may carry out apparently identical oxidations, it is known that in some cases the mechanisms of oxidation may vary in important aspects, as Barron (1942) has pointed out.

The attempt has been made to compare our knowledge of the oxidative and fermentative processes carried out by pathogenic fungi with the much more extensive information available for non-pathogenic forms. From this comparative point of view we may be able to select favourable avenues for experimentation with the pathogenic fungi. To establish a basis for comparison with non-pathogenic forms it has been necessary in some instances to devote much of a section to papers of purely theoretical nature; no apology is felt necessary for this.

Blastomyces

A most interesting group of organisms for investigation of respiratory and fermentative activity would seem to be those fungi causing systemic mycoses. These serious pathogens exist in man in a single cell, yeast-like state, but in culture may be maintained either in a yeast-like or in a filamentous condition. Included among these fungi are *Blastomyces dermatitidis* and *Histoplasma capsulatum*. So far as could be learned, only one study has been made of the gaseous metabolism of a fungus causing systemic disease; indeed, little information is available on any aspect of their physiology.

In the first study of the oxidative metabolism of an organism causing systemic fungus disease, Bernheim (1942) investigated *Blastomyccs dermatitidis.* Unstarved preparations of the *yeast* form of this organism grown on Kelly's agar medium oxidised added glucose, mannose, lactate, and pyruvate in M/20 phosphate buffer at pH 6.7, or at *pH.* 7.8. The *mycelial* form of *B. dermatitidis* was reported to exhibit a lower rate of oxygen uptake and to oxidise added substrates more slowly. Sugars are not fermented by either growth-form of *B. dermatitidis.* Glucose and mannose were oxidised at the same rate and took up 2 mols of oxygen per mol, provided the high endogenous rate is considered suppressed. Of five sugars tested, arabinose, rhamnose, and, very strangely, fructose, were not oxidised.

The endogenous rate of oxygen uptake was approximately one half of the rate with added glucose.

Aliphatic aldehydes were oxidised, but aromatic aldehydes inhibited the endogenous respiration; anisaldehyde prompted a *15-25* per cent inhibition of the endogenous rate. Curiously, this substance was without effect on the oxidation of added glucose but inhibited the oxidation of pyruvate and of lactate; the effect of anisaldehyde and two other inhibitors.

As given by Bernheim, the oxidation of all substrates tested with *B. dermatitidis* is most incomplete. He did not present any analyses for these substances in the vessels or analyses on the cells after the completion of a manometric experiment.

Oxygen consumption due to added substrate had ceased with incomplete oxidation of the substrates. If one were to assume that this is a case of incomplete oxidation accompanied by assimilation, as with *Candida,* then for acetate (one atom of oxygen taken up and one molecule of CO_2 given off per molecule of acetate) a statement of the oxidation would appear to be:

$$CH_3COOH + HO_2 \rightarrow CO_2 + H_2O + (—CH_2—)$$

This would be an unusual type of assimilation reaction with the carbon residue having the state of oxidation of an unsaturated hydrocarbon. In view of the high lipid content of this organism, as revealed by Peck and Hauser, this statement might conceivably have significance.

Much more likely, however, is it that the procedure used by Bernheim of subtracting the values for endogenous oxygen consumption from the observed values in the presence of substrate gives values for oxidation that are too low. It is frequently a real problem to decide whether or not one should correct for the endogenous respiration; the discussion by Barker (1936) of the problem is valuable in this connection. In the present instance, it seems clear from the R. Q. values that the endogenous respiration (R. Q. of 0.8 indicative of fatty acid metabolism) is suppressed nearly completely on the addition of substrate since the R. Q. with glucose of 0.96 indicates oxidation of carbohydrate and hardly any (or but little) of a mixed fat-carbohydrate oxidation. Since the rate of endogenous respiration is about half of the rate with added glucose, one would expect an R.Q. of 0.87 if both the endogenous material and the glucose were being oxidised simultaneously. A recalculation based on suppression of the endogenous oxygen consumption gives values for the fraction of substrate oxidised similar to those found for *Candida* and other organisms.

The higher fatty acids are not oxidised and inhibit oxygen uptake on all substrates tested; sodium oleate and palmitate were about equally effective.

When amino acids were added to washed suspensions of *B. dermatitidis,* Bernheim reported they increased oxygen consumption over the control in the absence of substrate, but in no case did the amount of oxygen consumed indicate complete oxidation of any of the amino acids added; rather, the oxygen consumed was only sufficient to permit of an oxidative domination; but such an occurrence did not seem likely, since the amount of ammonia produced was too low to agree with an oxidative domination.

Analyses for amino-nitrogen gave equivocal results; with phenylalanine all the amino-nitrogen seemed to remain unchanged in the medium, while with glycine 20 per cent, and with alanine 12 per. cent of it disappeared. Yet all three compounds stimulated oxygen consumption. Bernheim interpreted his data to mean that the amino acids are not oxidised nor oxidatively deaminated, yet are able to stimulate the oxidation of some unknown substance in the cell. Such an interpretation is rather unsatisfying and it seems more likely that the amino acids may have been oxidised to substances of a higher oxidation state (such as alanine) without domination occurring.

The above situation has a striking parallel with that found in *Leptomitus lacteus* by Schade and Thimann (1940) with the amino acid Meucine, which was neither deaminated nor completely oxidised, yet which stimulated a considerable increase in oxygen consumption over the endogenous.

Their interpretation was that leucine was oxidised to a compound having the same oxidation state as alanine which was then directly assimilated, as shown in the following overall equation:

$$C_4H_9CHNH_2COOH + 4O_2 \rightarrow (C_3H_9NO_2) + 3\ H_2O + 3\ CO_2.$$

Dermatophytes

Introduction: With the dermatophytes the problems of handling filamentous fungi arise in making preparations for respiratory studies. By reason of their filamentous structure, these organisms are not so convenient to handle experimentally as are the unicellular organisms; a few references will be made to methods that have been used in work with these forms.

De Boer (1928), in large-scale studies on *Phycomyces blakesleeanus*, employed Pettenkofer absorption tubes and a titration procedure to measure carbon dioxide production. Oxygen consumption was determined by a modification of the Fernandes (1923) device that utilises the electrolytic hydrolysis of NaOH to supply oxygen at constant pressure, the oxygen used being calculated from the volume of hydrogen produced during the electrolysis.

Schade and Thimann studying a water mould, *Leptomitus lacteus*, utilised the Warburg technique (these methods are well described by Dixon, 1943, and by Umbreit, Burris and Stauffer, 1945) to determine oxygen consumption and carbon dioxide consumption during the respiration of "starved" mycelial preparations of this fungus on a variety of substrates. WOLF and Shoup (1943), working with four species of the water mould, *Allomyces*, found (as had Schade and Thimann) that it was necessary to employ starved suspensions of the fungi to obtain rates in excess of the control amounts when substrates were added to respiring preparations in plain buffer. Kluyver and Per-Quin (1933) cultivated moulds in flasks placed in a rotating shaking machine for the preparation of homogeneous suspensions of compact mycelia for use in respiration studies. Nickerson and Chad-wick (1946) employed a volumetric microrespirometer to study the oxygen consumption of three species of dermatophytes; the effects of alterations in environmental conditions upon oxygen uptake of these organisms were studied using vigorous, unstarved mycelia.

Oxidation of Substrates: The rates of oxygen consumption have been studied for several isolates of three species of dermatophytes by Nickerson and Chadwick. All of the strains examined were found to possess sufficient intracellular reserve stores of oxidisable materials to maintain a fairly constant rate of oxygen uptake over a period of several hours. These preparations were unstarved.

The fungus preparations studied were placed in buffer solutions without an external source of substrate. Many different carbon sources were added to respiring fungus preparations to learn which, if any, were suitable substrates for oxidation by the fungi. At hydrogen ion concentrations on the acid side of neutrality, no compound added promoted any increase in the rate of oxygen uptake over the rate in the absence of the compound. At alkaline pH, M/20 dextrose added to *T. gypseum* and M/30 glycerol added to T. *rubrum* were followed by increases of 43 and 36 per cent respectively in the rates of oxygen consumption.

Effect of Hydrogen Ion and NaCl Concentrations: With two species of dermatophytes, *E. floccosum* and *T. gypseum,* the rates of oxygen uptake vary remarkably with variation in pH (Nickerson and Chadwick, 1946). From initial levels at *pH* 2 the rate of oxygen uptake decreases to a minimum in rate in the region of pH 5.0-5.5, rising to higher levels, comparable to the initial rates, at pH 8. Pfeiffer (1929) in a review on the isoelectric point in cells and tissues pointed out there is a minimum of physiological function at the isoelectric point of plasma proteins. In view of the apparently analogous reactions of the respiratory enzyme system in these organisms to isolated protein systems with changing acidity, it was desirable to investigate other treatments to which protein systems respond characteristically. A study of the effect of increasing concentrations of sodium chloride on oxygen consumption was undertaken.

To respiring preparations of T. *gypseum* in distilled water increasing concentrations of sodium chloride were added. The rate of oxygen consumption that is highest in distilled water, falls sharply on the addition of small amounts of NaCl only to increase with increasing concentrations of the salt. This situation makes an interesting comparison with the well known behaviour of globulins in the presence of increasing concentrations of salt.

It is rather important to point out that the rates of oxygen consumption at 'any given level of NaCl or H^+ concentration were constant for the entire time the experiment was conducted at that level (always one or more hours). When a given fungus preparation was changed from one environment to another, a new rate of oxygen uptake was initiated almost immediately; the organisms responded sharply to changes in their environmental conditions.

Evidence for Duality of Cytochrome System: An alternative view of the variation in rates of oxygen consumption with changing hydrogen ion concentrations is that two enzyme systems might be involved, one active in the acid region, the other at alkaline levels. However, it is rather more difficult to account for the observations with sodium chloride on this basis; though it is possible that one system might be inhibited with increasing concentrations of salt and a second system activated.

It will be recalled that Tate (1929 a and *b*), using acetone-powder preparations, demonstrated the presence of indophenol oxidase in several species of dermatophytes, including: *T. Megnini, T. Sabouraudi, T. rubrum,* and *T. mentagrophytes.* It is of considerable interest then that in other instances a two-component system has been proposed (for organisms respiring via the cytochrome system) to account for environmentally influenced responses of the cells. These were the studies of Tang and of Borei and Lindvall, which will now be considered.

Tang (1936) found a minimum rate of oxygen uptake for the yeast, *Saccharomyces wanching*, at pH 7.2 (pH was 8.9 before adding glucose;) in a borate buffer series; he remarked about the resemblance between his curves for oxygen consumption and the characteristic behaviour of many colloidal processes with varying H^+ concentrations.

In contrast to his findings using a borate buffer series, he observed with phthalate and phosphate buffers that his organism exhibited only a slight decline in oxygen uptake towards the extreme acid or alkaline ranges, while with an acetate buffer the rate dropped very sharply with increasing acidity. Later, Tang (19360) found glucose markedly altered the pH of borate buffers and that borate affected the optical activity of glucose solutions, possibly through formation of a complex between glucose and borate, so that his measurements probably did not reflect the response of some component of the cytochrome system to pH as first believed.

Recently, Borei and Lindvall (1943) showed the endogenous respiration (rate of oxygen consumption in the absence of added substrate) of *Saccharomyces cerevisiae*, proceeding via the cytochrome (cyanide sensitive) system, to consist of two fractions. With increasing acidity (*pH* 6.5 to pH 4), the rate of one fraction remained unchanged while the second increased in rate, possibly as a result of enzyme activation. Winzler (1940) had previously found a somewhat similar situation with a strain of *S. cerevisiae* in which the endogenous respiration was stimulated about 7 fold (approaching the exogenous rate) by a phthalate buffer at pH 3.5.

As an exploratory method for determining the number and properties of the respiratory systems that may be distinguishable in cellular preparations, experiments involving alterations in cell environments have considerable value. Exploratory studies by these means, preliminary to enzyme preparation investigations, may be of material aid. Coupled with studies on specific enzyme inhibitors, the environmental-influence studies may present a more exact view of the respiration system complex as it operates within the cell.

***Q_o* of T. Rubrum vs. pH:** Against the sensitivity of the respiratory systems of *E. floccosum* and *T. gypseum* to changes in H^+ concentration, the enzyme system of T. *rubrum* controlling oxygen uptake is in marked contrast (Nickerson and Chadwick, 1946). With phthalate, citrate, and phosphate buffer series, the rate of oxygen consumption by this organism is seen to change but slightly with variation in pH.

It is the contention of Myrback and Vasseur (1943) from work on four species of lactose-fermenting yeasts that enzymes localised at the cell surface may be markedly affected by the *pH* of the cell environment, while enzymes located within the cell are not readily thus influenced. According to this interesting hypothesis, it may be that the respiratory enzyme system of *T. rubrum* is located within the cell, while those of *E. floccosum* and *T. gypseum* are near or in the cell surface, easily influenced by environmental changes.

In this vein, remembering the lower basal rate of oxygen uptake exhibited by *T. rubrum,* it is of interest to recall that *T. rubrum* is the most stubborn to eradicate of the dermatophytes causing foot infections, as Conant *et al.* (1945) point out. No assertion is made that there is necessarily any connection between the clinical behaviour of this organism and the findings on its respiration rate.

The Effect of Certain Heavy Metal Ions on Oxygen Consumption: The effect of additions of dilute concentrations of certain inorganic water-soluble salts on the respiration of the dermatophytes has been examined by Nickerson and Chadwick. The following cations were among those included in the study: silver, mercury, zinc, and cadmium. The first three of these proved to be inhibitory at concentrations as low as 1/10000 molar; cadmium had no pronounced effect on the respiratory rates of the organisms studied. A partial recovery in oxygen consumption followed the removal of M/100 $ZnCl_2$ from *E. floccosum;* an inhibition of 92-93 per cent was changed to a fairly constant rate 59-65 per cent lower than the control on washing out the zinc chloride and replacing the fungus in buffer solution.

There was no recovery of *T. rubrum* from the 100 per cent inhibition promoted by M/100 $AgNO_3$. Attempts to overcome the toxic effects of zinc salts by adding dilute concentrations of sodium thiosulphate, magnesium sulphate, or calcium chloride to zinc-poisoned preparations were unsuccessful. Reports of the success of such attempts are to be found in the literature and are frequently interpreted as indications of the mechanism of action of zinc salts.

To date, however, there seems to be no convincing scheme of the mechanism of action of zinc. The same might be said of silver, for which no specific mode of action seems to have been uncovered. From the recent work of Fildes (1940) it seems clear that the action of mercury lies in its affinity for sulphydryl (SH) groups, the dermatophytes would appear to employ active SH groups to a considerable extent in the mediation of their respiration. Heretofore, the activity of the mercuric ion against microorganisms has generally been credited to its power to precipitate proteins, but the demonstration that the mercuric ion also possesses marked affinity for sulphydryl groups makes it probable that it is by virtue of this property that the mercuric ion may interfere with metabolic processes in cells.

This interpretation brings into fold the fact that the anti-bacterial effect of Hg^{++} can be neutralised specifically by compounds containing a sulphydryl group; this reversal can be effected even after long exposure of the microorganism to the mercuric ion, indicating the probable absence of a significant non-reversible, terminal process such as a denaturative precipitation. As Dubos (1942) pointed out, the anti-sulphydryl hypothesis accounts for some facts that were incompatible with the view that mercurial germicides act by virtue of their ability to precipitate proteins, an example being the case of some organic mercurials which, although less highly ionised than inorganic salts, are often more active against bacteria. Phenyl mercuric nitrate more effectively reduces respiration than an equivalent centration of mercuric chloride.

Clinical Use of Inorganic Salts for Control of Fungus Infections: In view of the inhibitory effect of water soluble zinc salts on the respiration of the dermatophytes, Dolce and Nickerson (1947) undertook a clinical trial of zinc chloride in dilute solutions with superficial fungus infections.

No effect was noticed on a few cases of infections of the scalp caused by *Microsporum audouini* and the condition of high moisture generally associated with tinea pedis did not favour the use of zinc chloride solutions. In infections of the glabrous skin and in crural infections, there was some indication of usefulness for this salt. As the authors emphasised, the point of the work was not to add another compound to the long list of substances that have been employed in the treatment of fungus diseases but to put to trial compounds for which there was evidence on experimental grounds indicative of potential value.

The use of respiratory studies as a metabolic bio-assay method for securing information on the relative effectiveness of fungi-static chemical compounds under conditions approximating those prevailing in *situ* has been proposed by Nickerson (1946). A more complete understanding of the physiology of the organisms causing superficial and systemic fungus diseases will be of material assistance in the control of such infections.

Metabolic Products of Pathogenic Fungi

Investigations on the metabolic products of pathogenic fungi are becoming more numerous. The possibilities of clinical use of polysaccharide preparations for desensitisation of hypersensitive individuals are apparently stimulating efforts in this direction. Pigment production, in a few instances, seems to have value as a taxonomic aid for species identification.

Pigment Production

Studies on the production of pigment by members of the pathogens are frequently complicated, as with the dermatophytes, by the occurrence of pleomorphism. Pleomorphic colonies have lost their ability to produce pigments. In the genus *Trichophyton* the production of a rose-brown pigment on many media is usually characteristic of *T. mentagrophytes (gypseum)* while strains of *T. rubrum* produce a deep wine-red colour on many media. Benham (1943) mentioned observing several strains typical of *T. mentagrophytes (gypseum)* in every way except that they produced the wine-red colour on dextrose agar media.

Pigments of the Dermatophytes

Many of the dermatophytes produce pigments, especially when grown on media rich in carbohydrate. At 30° C on a glucose or maltose medium, *Trichophyton megnini (gallinae)* produces a bright red pigment which diffuses into the medium; *T. violaceum* also produces a reddish diffusible pigment. *T. mentagrophytes (gypseum)* and *T. tonsurans (crateriforme)* often develop yellowish or reddish brown pigments, especially in old cultures that are drying up. Species of *Microsporum* as *M. canis* often form yellow or brown pigments.

Sabouraud (1910) stated that the pigments of the dermatophytes are acids and precipitable by alkalis. Tate (19290) examined the pigments produced by species of *Trichophyton;* the pigments of several species *(T. Megnini, T. Sabouraudi, T. rubrum,* and T. *mentagrophytes)* were found to be red to reddish brown, easily soluble in dilute acids and acid alcohol, but only very slightly soluble in dilute alkali. On extraction the colour in acid solution was yellow changing to red or reddish brown in alkaline solution, the colour change being easily reversible, and the pigment thus as indicator of hydrogen ion concentration.

The pigments are not destroyed by boiling; they can be extracted by ether and chloroform from aqueous acid solutions. In alkaline solution, the pigment can be reduced to a pale yellow with sodium hydrosulphite and reoxidised to a red colour in contact with air; the oxidation-reduction is reversible. Tate suggested the pigments might be of an anthracene nature, resembling the yellow pigments of the lichen *Physcia.*

Thompson (1943) confirmed Tate's findings and pointed out that the pigments of *T. rubrum* and *T. mentagrophytes* are soluble in acetone; in solution, oxygen prompts a colour change of the pigment to a deep wine-red so that the two colours become indistinguishable.

On dilution of the acetone solution with water a partial precipitation occurs; the pigment remaining in solution changes to a cherry colour which deepens at pH 8 on adding alkali, becoming a blue-lavender above pH 9. On standing in alkaline solution the colour was found to turn yellow but the lavender colour might be regenerated by oxygen, though not by nitrogen, hydrogen sulphide, or carbon dioxide. This pigment seems to be a *pR* indicator with an oxidised and a reduced form. This field merits investigation of the chemical nature of these coloured compounds, and the possibility that the pigments serve for oxygen transport in these fungi deserves study.

Green Pigment of Candida Albicans

A faint greenish colour produced by colonies of some cultures of *Candida albicans* was noted by Ashford (1917). Martin, Jones, Yao and Lee (1937) and Martin and Jones (1940) described six species of *Candida,* of which it was noted that *C. stellatoidea* produced considerable amounts of a green pigment.

This pigment was investigated by Jones and Peck (1940) with the thought that it might prove an additional character by which to distinguish *C. albicans* from *C. stellatoidea* (easily separable, however, on biological differences and pathogenic action). On blood agar plates, with Sabouraud's broth as the diluent, a greenish zone was easily discernible around colonies of *Candida stellatoidea;* in these cultures the pigment resembled methemoglobin. Plates containing 25 to 50 colonies were suitable to demonstrate pigment production; too heavy seeding prevented individual colonies from reaching a large size and no pigment could be noticed in such plates with minute colonies.

Jones and Peck found that a thin greyish-green layer persisted on the surface of the masses of cells of *C. stellatoidea* that had been centrifuged from saline suspensions. The pigment was readily soluble in acetone, yielding a green solution. To obtain sufficient material for chemical investigation, fresh cells were extracted with acetone. *C. stellatoidea* produced ten times the quantity of pigment produced by *C. albicans*. The pigment proved to be pH indicator with a blue-green colour in acid solution and yellow in alkaline solutions. The substance appeared to be a weak acid, unstable in nature.

It seems fairly certain that the pigment studied by Jones and Peck was not identical with riboflavin, from the data they reported on the light absorption characteristics of an ethyl acetate solution of the crude green pigment isolated from *C. stellatoidea*. Using a spectrophotometer they observed an absorption maximum at 6550 Å. The absorption maxima for riboflavin (shown by Holliday and Stern, 1934, to be due to the alloxazine nucleus) were found by Kuhn *et al.* (1933) for nearly neutral aqueous solutions to be pronounced at 2200 and 2650 Å and of secondary magnitude at 3650 and 4450 Å.

Synthesis of Riboflavin by Species of Candida

Production of considerable amounts of riboflavin by species of *Candida* has recently been reported by Tanner, Vojnovich And Van Lanen (1945). An interesting result of their study was the influence shown for iron on the amount of riboflavin secreted into the medium.

This impression is supported by the results of Wickerham, Flick-Inger and Johnston on *Ashbya gossypii* which produced riboflavin in quantity, apparently independently of the iron concentration, but showed increasing production with the age of the culture.

The cultures were grown for 8 days with aeration in a medium containing initially 2 per cent of a commercial dextrose (Cerelose); after 4 days only 0.03 per cent dextrose remained in the medium and for the remaining 4 days the yeast cells were subjected to what was tantamount to aeration in a non-nutrient medium—a process which Stier and Macintyre (1943) showed was accompanied by considerable loss of riboflavin from the cells of *Saccharomyces cerevisiae* into the suspending medium. There is considerable evidence that the loss of growth factors into a non-nutrient buffer does not occur in the absence of oxygen.

Black Pigment of Sporotrichum

The black pigment, found exclusively in the spores, is produced by most cultures of *Sporotrichum.* From the first description of this organism by Schenck (1898) nearly every worker dealing with the group has remarked on the blackish pigment developed by these organisms in cultures.

Growth is at first white and the rate of formation of the black pigment is subject to considerable variation with strain differences, composition of the medium, oxygen supply, etc. Microscopically the pigment is seen confined to the spores; mycelia are usually colourless. The black pigment may be formed evenly over the surface of a colony or its distribution may be limited to small areas, most of the colony remaining white; in the latter case, spores in the non-pigmented areas are white. Davis (1915) isolated single spores from pigmented and non-pigmented areas and obtained cultures that were evenly black or evenly white; he did not report any sectoring occurring in ten subcultures following isolation of a spore. Both black and white types appeared to be equally pathogenic when inoculated into animals. Davis made a brief study of the black pigment and found it could not be extracted from the spores by water, dilute acids, dilute alkali, hot alcohol, ether, xylol, or chloroform.

Strong acids dispersed mycelia and spores and destroyed the pigment. No test for iron could be obtained with the pigment in the spores by staining. Cultures normally producing the pigment were unable to do so under low oxygen tension. Davis, however, did not come to any conclusion as to the nature of the black pigment.

Recent experiments by Nickerson indicate that the black pigment of *Sporotrichum schenckii* is probably a melanin. The presence of tyrosinase could be demonstrated in spore masses but not in mycelia free of spores. A substrate oxidisable by tyrosinase prepared from mushrooms *(Coprimts micaceus)* could be demonstrated in both spore masses and in mycelial masses devoid of spores.

No study was made of white-spored strains, since none was at hand. The addition under aseptic conditions of small amounts of spore masses, as nearly as possible teased free of mycelia then filtered with suction through glass wool, to a sterile solution of tyrosine (0.033 g./100 ml. water) resulted in darkening of the solution. The reaction was not rapid and darkening progressed slowly through the hallochrome stage for several weeks.

Grinding spore masses with chloroform before addition to a tyrosine solution caused the darkening to go relatively faster. Similar addition of mycelial masses practically free from spores did not cause darkening of the solution. With a preparation of tyrosinase from the mica ink cap mushroom *(Coprinus mleaceus)* there was a marked increase in oxygen uptake on adding extracts of either mycelia or spores of *Sporotrichum schenckii.* Oxygen uptake is measured in the Scholander volumetric respirometer, using the enzyme preparation and an aqueous solution extracted from fungus mycelia ground with chloroform.

Fruiting Bodies

Apparently the black pigment of *Sporotrichum schenckii* is a melanin produced by the enzymatic oxidation of some substrate by tyrosinase. Evidently the enzyme is confined to the spores, though an oxidisable substrate may be extracted from the mycelia as well as from the spores. The findings of Davis (1915) are compatible with the present finding that the black pigment is a melanin.

Melanin production is quite widespread; the pigment is found in hair, insects, fleshy fungi, cut surfaces of apples and potatoes, etc.; Wigglesworth (1938) gives an account of its production in insects. Oxygen is required for the formation of melanin since the pigment is formed by the enzymatic oxidation of a colourless precursor, aromatic in structure, termed a "chromogen." In insects the melanin-producing enzymes are generally diffused throughout the animal and it is the localisation of chromogens that determines the colour-patterns developed.

There is a considerable variety of chromogens, or substrates, upon which the melanogenic enzymes may act; tyrosine (monohydroxyphenylalanine), "dopa" (3, 4-dihydroxyphenylalanine), other polyphenols, dioxyphenylacetic acid, etc.

There is a second pigment, usually less conspicuous, produced in cultures of *Sporotrichum*. This is a yellow to yellow-brown diffusible colour. Cultures on agar come to have a yellow-brown colour with time, evidently caused by diffusion of this pigment into the agar.

Polysaccharides

At least in the case of diseases brought about by the pathogenic yeast-like fungi, the polysaccharides produced seem to occupy an important position. In the case of animals that have been immunised with whole cells of *Candida albicans*, circulating antibodies are developed which are capable of precipitating polysaccharides isolated from the cells of

C. albicans. Hypersensitivity to polysaccharides may, however, develop in some cases when the presence of such precipitins cannot be shown. Peck, Martin and Hauser (1940) pointed out that such hypersensitivity is probably of importance in certain cases of chronic infection, especially is some cases of American blastomycosis (Gilchrists's disease).

Polysaccharides of Blastomyces Dermatitidis

Peck, Martin and Hauser undertook a study of the preparation, properties, and biologic activities of some polysaccharides obtainable from *Blastomyces dermatitidis,* the causative organism of American blastomycosis. At least two different polysaccharides (one of which seemed to yield mannose after acid hydrolysis) were isolated. Cells were suspended in a saline solution, containing phenol, for 4 days, then removed by centrifugation; polysaccharides were isolated from the supernatant liquid by treatments with sodium acetate and ethanol.

The crude polysaccharide preparations obtained in yields of about 0.01 gram per gram (dry weight) of cells were faintly buff-coloured hygroscopic powders. Very dilute solutions of the purified polysaccharides fixed complement with sera prepared by immunising rabbits with ground cells of *Blastomyces dermatitidis.* Clinically, patients with American blastomycosis in the allergic stage of the disease responded to intracutaneous injection of the polysaccharide with a swelling that reached maximum size within 24 hours. With one patient intravenous injection of the polysaccharide decreased carbohydrate sensitivity and abolished the sterile abscess-formation produced in an allergic patient by injection of heat-killed whole organisms.

Immunologically Active Polysaccharide from Coccidioides immitis: Coccidioidin prepared from filtrates of old cultures of *C. immitis* is used as a diagnostic aid with coccidioidomycosis; a skin reaction is obtained on intradermal injection into sensitive persons. Hirsch and D'andrea (1927) precipitated with alcohol from filtrates of old cultures of *C. immitis* a substance yielding 20-40 per cent reducing sugars on hydrolysis; glucose was identified in the hydrolysate.

A more extensive investigation of the polysaccharide produced by this organism was undertaken by Hassid, Baker and Mccready (1943). Concentrating the filtrate from a 2-month-old culture by the use of collodion ultrafilters, through which the skin reaction substance would not pass, these authors on analysis found the polysaccharide to consist of units of galacturonic acid, glucose, and some unidentified sugar in the approximate ratio of 1:6:3 respectively. A nitrogenous compound, apparently other than protein, was found to be associated with the polysaccharide. Clinically the purified polysaccharide gave a skin reaction with sensitive individuals and a positive precipitative reaction. An acetyl derivative of the original polysaccharide, however, yielded a polysaccharide which still produced a positive precipitative reaction but no longer elicited a skin reaction.

Capsule of Torulopsis Neofornians: In the course of a study on the growth requirements of different yeasts, Aschner, Mager and Leibowitz (1945) found that *Torulopsis rotundata* and *T. neoformans* gave a steel-blue colour reaction with iodine. Cells of these yeasts had been grown in a simple inorganic salt medium with glucose, ammonium sulphate, and thiamine. The substance responsible for the colour reaction was found in the medium in which the cells had grown as well as in the large capsules with which cells of these yeasts are surrounded, but it could not be detected within the cells proper.

Several grams of the colour-responsible substance was prepared by precipitating it from cell-free media with alcohol; the substance came down as white, thread-like masses. The precipitated material was slightly soluble in the cold water, giving an opalescent solution. It was found that the material was of a carbohydrate nature and contained at least two fractions, one of which was a pentosan, the other being an amylose. Spectrophotometric analysis of the iodine complex showed it to have an extinction curve identical with an iodine complex preparation from pure potato amylose. Through the action of amylases of plant and animal origin, complete disappearance of the iodine colour reaction could be caused; a fraction remaining after this enzymatic treatment could be precipitated with alcohol, the precipitate giving a strongly positive test for pentoses.

Many carbon sources were examined for their suitability as substrates; it was found that any compound supporting growth of these yeasts was also suitable for the production of the amylose fraction. Glucose-1-phosphate, inulin, starch, and glycogen were found not to support growth. With several nitrogen sources examined it was found necessary to keep the pH at 5 (or even more acid) to permit amylose formation, though pentosan production could take place in the alkaline ranges.

Of twenty-five other yeasts examined, including uncapsulated species of *Torulopsis,* only *T. rotundata* and *T. neoformans* produced amylose. The authors mentioned are investigating the serological properties of the external polysaccharides. Benham (1935) has shown that rabbits produced antibodies following the injection of decapsulated cells of *T. neoformans* (capsule removed with acid) but did not do so following the injection of whole cells.

Miscellaneous Polysaccharides: It might be the case that the skin lesions produced by injection of spores of *Coniosporium corticole* as reported by Towey, Sweany and Huron (1932) in their study of coniosporiosis are actually prompted by a polysaccharide (or by a melanin) type substance. These authors found no appreciable skin reaction on injection of the fat soluble and protein fraction from the spores, but the black residues from the above preparations produced serious lesions similar to those produced on injection of the spores.

Proteins: There is some information available on the enzymes of the dermatophytes from the work of Tate and a little information from scattered sources on immunologically active proteins of various genera. There is a scarcity of information on structural and storage proteins of all the pathogens.

Enzymes of the Dermatophytes: Using acetone powders prepared (according to Onslow, 1923) from young cultures of representative species Tate (1929) made a general survey of the enzymes of the dermatophytes. With the method employed there is obtained from the mycelium a fine, dry powder containing most of the enzymes uninjured, which keeps for weeks without deteriorating. Cultures of *Microsporum canis (lanosum) M. Audouini, Trichophyton tonsurans, T. Schoenleiniiy* and *T. mentagrophytes (radiolatum)* were used; normal and pleomorphic strains of *T. mentagrophytes* were studied comparatively. A proteolytic enzyme active in alkaline media was present in all species; this trypsin-like enzyme hydrolysed intact proteins (casein) producing free amino acids. No proteolytic enzyme active in acid media (pepsin) was found in any strain. The amount of proteolytic enzyme present was found to vary with the species; *T. mentagrophytes* was comparatively high in its content of this enzyme. The normal strain of *T. mentagrophytes* showed a greater proteolytic activity than did the pleomorphic form; the latter strain was comparatively rich in urease and amygdalase, both poorly represented in the normal strain.

Lipids of Fungi with Special Reference to Pathogenic Fungi

The chemistry of microorganisms is attracting increased interest. Investigations dealing with the isolation, purification and characterisation of cell-constituents and metabolic products of fungi and bacteria have long been yielding data of microbiological significance.

Many of the results obtained in these studies have also been of the utmost value in facilitating the better understanding and therapy of infectious diseases. The antibiotic substance, penicillin, may be cited as an example of a metabolic product of a fungus which has achieved a very important clinical application as a therapeutic agent (Herrell, 1945). Studies carried out on the specific polysaccharides of pneumococci have resulted in the recognition of new phases of immunology, and have aided in the development of the therapy of pneumonia (Heidelberger, 1927, 1939). Investigations of the lipids of human tubercle bacilli (Anderson, 1939; Sabin, 1932) have led to the isolation of unusual types of lipid constituents which possess the property of stimulating the formation of tuberculous tissue when injected into animals. The latter studies have provided a means for relating lesions of tuberculosis to specific components of tubercle bacilli.

The lipids of pathogenic fungi represent a very important fraction of the cellular protoplasm. Investigations dealing with the chemistry of this interesting group of substances have been, however, rather limited.

The present survey is, therefore, mainly concerned with data obtained on two pathogenic, yeast-like fungi, *Blastomyces dermatitidis* (Peck and Hauser, 1938, 1940) and *Candida albicans* (Peck and Hauser, 1939, 1940) which are representative of a fairly considerable group of pathogenic organisms. Sufficient data on the lipids of non-pathogenic fungi, and pathogenic and non-pathogenic bacteria are available to permit comparison of the various types of compounds isolated.

The role of the lipid constituents of the cells of pathogenic fungi and bacteria in infectious disease mechanisms is undoubtedly significant. The lipids of pathogenic microorganisms are capable of interfering with enzyme systems as shown by Jobling and Petersen (1914) and Peck (1942). Caseation phenomena in tuberculosis, for example, may be attributed in part to such interference with a normal host mechanism. Preliminary studies of the effects of tubercle bacillus lipids on animal enzyme systems, both in *vitro* and in the living animal body by Gerstl and Tennant (1941, 1942, and 1943) and by Gerstl, Tennant and Pelzman (1945) have given further insight into this type of reaction. Inhibition appears to depend on the molecular weight, structure, and configuration of the lipid, as well as on such physicochemical factors as surface activity and adsorption. The quantitative study of the inhibition of crystalline trypsin by soaps of the fatty acids from *Blastomyces dermatitidis* and from tubercle bacilli, and by soaps of a number of other pure fatty acids by Peck (1942) brings out more clearly the effect of molecular weight of the fatty acid on the degree of inhibition. These studies suggest possibilities for further experimental work on the effects of lipids on tissue enzymes using both *in vitro* and *in vivo* systems. The results should aid in the interpretation of certain disease mechanisms.

In the case of tuberculosis, of all the substances thus far isolated from human tubercle bacilli the lipids have shown by far the most damaging effect upon the cells of the host (Sabin, 1932). The formation of lesions closely resembling those produced by the living organisms has been accomplished in rabbits and guinea pigs by the injection of the tubercle bacillus phospholipid, and in mice by means of the wax fraction (Thomas and Dessau, 1939).

The phospholipid fraction did not appear to elicit in mice the degree of response noted in rabbits; this illustrates an interesting difference in specific susceptibility. Similar studies carried out in mice with the lipids of *Blastomyces dermatitidis* by Baker (1942) showed that the phospholipid obtained from this pathogenic fungus exerted a marked effect on cellular proliferation. It nevertheless could not alone be responsible for the necrotic reactions observed on injection of the living or killed organisms.

The heterophile antigens of Forssman, lipid-protein complexes sometimes containing carbohydrates, are immunologically active substances in which the lipid fractions act as haptenes (Zinsser and Bayne-Jones, 1939). In the capacity of haptenes, lipids of pathogenic microorganisms may influence the immune and (in the case of fungi, particularly) allergic responses of the host. More data are needed in order to provide a basis for the thorough evaluation of the degree of participation of lipids of pathogenic microorganisms in the pathological reactions accompanying the diseases. The necessary investigations will require carefully purified lipid preparations from a variety of pathogenic forms, as well as information concerning the chemical structure of these lipids. Joint chemical, microbiological, immunochemical, and pathological studies are much needed.

The capacity of pathogenic microorganisms to survive in the tissues of the host depends in part on the lipid system of the invading cells. Reasoning from the findings on red cell stromata by Parpart and Dziemian (1940), it may be conceived that the cell membranes of the invading pathogenic microorganisms contain a large amount of lipid material, possibly composed of phospholipids and sterols.

The permeability of the invading cells, insofar as this is determined by the lipid components of the membrane, will be a function of the ratio of the hydrophilic colloid (phospholipid) to the hydrophobic colloid (sterol). The nature of the substances capable of penetrating the cell membrane will be determined to a considerable degree by their lipid-solubility. Penetration will also be a function of the relative amounts of acidic and basic cell components (lipid or otherwise) held in the membrane.

From the pharmacological viewpoint, the activity of certain therapeutic agents, particularly antibiotic substances, must be in part dependent on their ability to penetrate into the cells of pathogenic microorganisms in the host tissues. It becomes important to know more about the character of the cell membrane, the factors which influence its permeability to acidic or basic antibiotics of suitable lipid-solubility, and the nature of the lipids in the membrane. Studies carried out on synthetic monolayers give some insight into this problem. Physical and chemical studies of the lipid composition of pathogenic microorganisms may permit deductions as to types of antibiotic substances most likely to be specific for various pathogenic organisms.

It is greatly to the interest of a better understanding of infectious disease mechanisms to provide as much information as possible on the composition of the cells of pathogenic microorganisms. The isolation and characterisation of the components of microorganisms is in itself an arduous and complicated task. In many instances specialised technique is required. For the best evaluation of the results of these studies it is important that the greatest care be taken to avoid the occurrence of chemical changes in the lipid fractions during isolation and before physicochemical measurements are made and before pathological or immunological testing.

General Methods of Extraction and Separation of Lipid Fractions: The lipids of fungi and bacteria have been studied by gross chemical methods and by such microchemical procedures as the examination of cells stained with lipid-specific dyes. The present discussion concerns only the results obtained by the former methods.

In order to obtain the most significant results from chemical studies of the lipids of microorganisms, it is highly desirable to obtain the various fractions in a condition as similar as possible to that in which they exist in the living cells. This is also of importance whenever biological studies of the lipid fractions are to be carried out. It is, therefore, necessary to use the mildest methods in the isolation and purification of the various fractions. For this purpose, only purified neutral solvents, preferably redistilled, should be employed. In order to prevent oxidation of unsaturated lipids, air should be displaced from all solvents and containers with carbon dioxide or nitrogen. Extractions should be carried out at room temperature; concentration of extracts should be carried out under reduced pressure in an atmosphere of carbon dioxide or nitrogen. The methods recently developed by Anderson and co-workers (1, 2, 3, 4, 12, and 50) during studies of the lipids of tubercle bacilli are particularly recommended for this type of work.

Qualitatively, the composition of the lipids of microorganisms appears to be specific for the organism studied. Quantitative composition, however, depends on a number of variable factors among which are the composition of culture media, the temperature and other physical growth conditions, the methods of treatment of cells during harvesting and prior to extraction, and the extraction procedures. Failure to consider these factors, as well as the lack of sufficient material to enable satisfactory separation of fractions, has deprived much of the older quantitative data (Czapek, 1922; Zellner, 1907) of significance.

The most suitable medium for growth of microorganisms for isolation and characterisation of constituents is, of course, a synthetic medium of uniform composition containing only simple organic compounds of known constitution together with inorganic salts. Using such a medium, one can be certain that the complex constituents of the cells such as lipids, carbohydrates, and proteins have all been synthesised from very simple known compounds.

It is frequently the case, however, that virulent pathogenic fungi will not grow on such a simple medium, and more complex organic nutrients must be added to induce satisfactory growth. In such cases, especial care must be taken to prevent the medium from accompanying the cells during harvesting. Also, the interpretation of results obtained with organisms grown on the more complex media necessitates more caution. The cells should be washed free of adhering medium with water or an isotonic buffer solution during harvesting, then introduced directly into the extraction solvents.

The freshly harvested cells of the organism to be investigated are suspended in alcohol-ether and allowed to stand at room temperature for several days. The alcohol-ether extract is then removed and replaced with fresh solvent. Two or three such extractions may be followed by one or more extractions with chloroform.

The combined alcohol-ether extracts are concentrated under reduced pressure to an aqueous suspension of lipid material and the lipids are extracted into ether. The ethereal solution, after drying over anhydrous sodium sulphate or magnesium sulphate, is concentrated to small volume and treated with acetone, whereupon the phospholipids separate as an amorphous insoluble fraction. The phospholipids may be purified somewhat by repeated precipitation with acetone from ether. The acetone mother liquors on removal of solvent yield the so-called acetone-soluble lipid fraction.

The chloroform extracts on removal of solvent under reduced pressure yield a lipid fraction which in the case of fungi generally resembles the acetone-soluble lipids and may be conveniently combined with this fraction. In the case of the acid-fast bacteria (Anderson *et al.* 1937; Anderson, 1939), and of diphtheria bacteria (Chargaff, 1933), however, the chloroform extracts yield considerable amounts of wax-like fractions which differ markedly from the acetone-soluble lipids. A wax-like fraction may sometimes be separated by cooling an acetone solution of the acetone-soluble lipids, whereupon the wax fraction usually separates out as a solid which may be collected by centrifugation or filtered off and dried. It is desirable to conduct a preliminary examination of the properties of the chloroform extract before combining it with the acetone-soluble fraction.

The procedures described above represent a general method for removal of the readily extracted lipids from the cells of microorganisms. Many variations of these procedures have, of course, been employed. Insofar as possible, it is essential to use living cells for the first extractions rather than cells which have been killed, since most preliminary treatments cause more or less extensive hydrolytic and oxidative modifications of certain of the lipid fractions, particularly of the phospholipids.

After removal of lipids by means of alcohol-ether and chloroform, there still remain in the cells of bacteria and fungi considerable amounts of lipids which cannot be removed by exhaustive extraction with neutral solvents. These lipids appear to be held in part by chemical linkages to carbohydrate and/ or protein elements of the cell structure. Some mechanical occlusion also occurs in the larger or smaller cell aggregates since the cells are seldom in completely unicellular suspension (Turpeinen, 1936) during extraction. These lipid fractions have been referred to by Anderson as firmly bound lipids.

A mild method for liberating such firmly bound lipids was developed for use on human tubercle bacilli by Anderson and co-workers. After extraction with neutral solvents, the cells were treated with 25 per cent alcohol which removed considerable amounts of polysaccharides, pigments, non-protein nitrogen constituents, etc. The cells were then digested with alcohol-ether (1:1) containing hydrochloric acid at a temperature of 50° C for about five hours. The cell residues were filtered off, washed with alcohol and ether and afterwards exhaustively extracted with ether and chloroform. The extracts were worked up in the usual way.

Variations of this procedure have also been described by Peck and Hauser (1940). These methods appear to be best suited to isolation of the firmly bound lipids in relatively unchanged form for chemical study. The use of alkaline hydrolysis in liberation of bound lipids possesses the inherent drawback that the firmly bound lipids are largely saponified. From such material little chemical data of structural value can be obtained. The use of stronger acid is likewise to be recommended only where extensive hydrolysis of the lipid material does not matter.

An acid disintegration method has been worked out by Turpeinen (1935, 1936) for estimation of the total lipids present in fungi. In this method the cells are digested in 5 *N* hydrochloric acid on the steambath for twenty minutes, cooled, and extracted with ether. The ether solution is taken to dryness and the residue is dissolved in petroleum ether.

The weight of the petroleum ether-soluble lipids dried at 50°C is a measure of the total lipids. The values found in this way will be slightly low due to partial hydrolysis, but the data presented (Turpeinen, 1936) indicate that *hydrolysis* of the lipids is by no means complete. The data reported would indicate that the milder methods of extraction of lipids leave a rather significant amount of unextracted lipids in the cell residues.

Attention is, however, directed to advantages which may be gained by using micro and semi-micro procedures for isolation and characterisation in the preliminary investigation of lipids of microorganisms. These rapid and economical procedures are of value in making possible surveys of groups of organisms, and in facilitating subsequent large scale preparation of material for biological testing.

The total readily extracted lipids and the phospholipid and acetone-soluble lipid fractions were obtained in each case by the methods described above, or by minor modifications of these methods.

The properties listed for the acetone-soluble lipid and phospholipid fractions have been determined by conventional methods and should, therefore, be directly comparable. The cleavage products from hydrolysis or degradation of the lipid fractions were isolated and characterised by procedures which satisfactorily establish their identity.

Readily Extracted Lipids: The lipids extracted from fresh, living cells by means of neutral organic solvents have been designated *readily extracted lipids* to distinguish them from *bound lipids* and *total lipids.* The readily extracted lipids doubtless comprise that fraction of the total lipid material which is held loosely in the cell wall or in parts of the cell structure to which organic solvents can readily penetrate.

This fraction may be considered to include lipids which have been synthesised by the cells but which have not become bound by chemical linkage to protein or carbohydrate elements of the cell skeleton together with intermediate substances such as free fatty acids, sterols or other alcohols, and autolytic or hydrolytic products of this type involved in metabolic processes of the cells. The chemically bound lipids, on the other hand, may be considered to represent an integral part of the cell structure, possibly representing a relatively constant fraction of the cells. The limited data available do not suffice to establish this point at present. A precise measure of the total lipids is very difficult to achieve experimentally, since complete liberation of the total amount of lipid material requires procedures which involve extensive lipid degradation. The readily extracted lipids of the fungi so far studied appear to consist of glycerides, phospholipicls, free fatty acids, and sterols and are present in amounts dependent in part on the medium and culture conditions employed. The phospholipids are usually separated from the other components of the readily extracted lipids on the basis of their acetone-insolubility.

The total amounts of lipids isolated from the four batches are not significantly different; but the amount of phospholipid fraction in Batch II is very low, evidently due to hydrolytic degradation (possibly enzymatic) to acetone-soluble products. Variation in yields of lipids from several batches of *Aspergillus sydowi* treated in different ways have been reported by Strong and Peterson (1934), and by Woolley, Strong, Peterson and Prill (1935).

Properties and Composition of the Acetone-soluble Lipids: The acetone-soluble lipids constitute the major part of the lipids extracted by means of alcohol and ether from most of the fungi studied up to the present time. This lipid fraction is usually yellow-brown to red in colour, possessing an odour somewhat characteristic of cultures of the corresponding organism.

All of the acetone-soluble lipid fractions appear to contain small amounts of volatile acids. The iodine numbers show considerable variation, which corresponds roughly to the relative amounts of oleic and linoleic acids present. The variations in the relative amounts of free fatty acids and glycerides indicated by the acid and ester numbers are probably dependent in part on differences in growth conditions.

The chemical composition of the acetone-soluble lipids has usually been determined after saponification of the material with hot alcoholic potassium hydroxide.

The saturated fatty acid fractions from the acetone-soluble lipids of all of the fungi examined contained palmitic and stearic acids; in several, tetracosanoic acid was also present. The unsaturated fatty acid fractions from all of these organisms consisted essentially of oleic acid and linoleic acid in varying proportions. The ratio of oleic acid to linoleic acid was higher in the pathogenic species.

Only traces of more highly unsaturated acids were observed. Sterols were present in all of the unsaponifiable fractions and comprised roughly half of the unsaponifiable material in the case of the two pathogenic fungi. Ergosterol was identified in the case of both pathogenic and non-pathogenic organisms. The water-soluble fraction in every case contained glycerol.

It is evident that the acetone-soluble lipids of both the pathogenic and non-pathogenic fungi contain neutral glycerides, free fatty acids, and sterols. The sterols probably occur both free and as fatty acid esters. A considerable part of the unsaponifiable fraction in the case of every organism has not been characterised. The differences observed between the acetone-soluble lipids of the pathogenic fungi and those of the non-pathogenic organisms appear to be quantitative rather than qualitative.

Properties and Composition of the Phospholipids: The phospholipids of *Blastomyces dermatitidis* and *Candida albicans* have been investigated in some detail. These preparations were obtained from alcohol-ether extracts of living cells. The phospholipids are readily soluble in ether, chloroform, and benzene and are relatively insoluble in acetone and alcohol.

They are easily dispersed in water, forming slightly turbid colloidal solutions. Addition of acids or salts to the concentrated aqueous solutions causes partial coagulation. The physical properties of phospholipids differ somewhat from those of several non-pathogenic fungi.

The chemical composition of the phospholipid fraction

from *Biastomyces dermatitidis* was determined after hydrolysis with hot *5 per cent* aqueous sulphuric acid.

The presence of choline, ethanolamine, and glycerophosphoric acid in the hydrolysates of phospholipid preparations indicates that both lecithin and cephalin are present in each case. One qualitative difference is noted: the presence of small amounts of a carbohydrate in the water-soluble fractions from the phospholipid of *Blastomyces dermatitidis.*

This would indicate the presence of a carbohydrate-containing phospholipid in addition to lecithin and cephalin. No claim for homogeneity in the chemical sense can be made for the phospholipid fractions, since these fractions are amorphous, and since they are purified simply by reprecipitation from ether by means of acetone. It is, therefore, only to be expected that the phospholipid fractions represent mixtures of two or more individual phospholipids, the solubility properties of which are fairly similar. The ratio of the fatty acids to one another also indicates the presence of several phospholipids differing only in the fatty acid residues present in the molecule.

It is evident from the foregoing discussion that the phospholipid from *Blastomyces dermatitidis* is a mixture composed of lecithin, cephalin, and carbohydrate-containing types of phospholipids. The phospholipids from acid-fast bacteria are predominantly of the latter type, no cephalin or lecithin having been conclusively demonstrated as components. Like the phospholipids from the human tubercle bacillus, which produce extensive tissue reactions characteristic of tuberculosis when injected into animals (Sabin, 1932), the phospholipids of *Blastomyces dermatitidis* give rise to tissue reactions simulating those encountered in human blastomycosis (Baker, 1942).

Two isomeric structures may be written for lecithin. The isomerism depends upon the position of the linkage between the glycerol and phosphoric acid residues. Since the lecithin contains an asymmetric carbon atom, the isolation of an optically active glycerophosphoric acid fraction on hydrolysis of the lecithin will establish the presence of a lecithin. In the case of carbohydrate-containing phospholipids or other phospholipids derived from α and β glycerophosphoric acid, the biological properties may depend to a considerable degree upon the isomer studied.

No *Blastomyces dermatitidis* or *Candida albicans* which afford information on the isomers present. The lecithin fraction from *Saccharomyces cerevisiae*, however, does appear (Salisbury and Anderson, 1936) to be composed of a mixture of α and β lecithins. The evidence indicates that the α form predominates in about a 4:1 ratio. The cephalin fraction from the latter organism yielded only optically inactive glycerophosphoric acid on hydrolysis; it would thus appear to contain only α cephalin.

Brief mention is made of acetal phospholipids (Longenecker and Daubert, 1945). Acetal phospholipids have not as yet been found in lipids of fungi, but represent a possible type of component. Carbohydrate-containing phospholipids in which two isomers due to the glycerophosphoric acid moiety are possible, may also have isomeric forms as a result of the asymmetry of the carbohydrate portion. A more complete investigation of the phospholipids of various pathogenic and non-pathogenic microorganisms would appear to be profitable.

Wax-like Fractions: A well-characterised wax, ergosteryl palmitate, m.p. 107°-108°C has been isolated from a number of organisms of the *Penicillium* group (Oxford and Raistrick, 1933).

From the acetone-soluble liquids of the yeast-like pathogenic fungus, *Blastomyces dermatitidis*, there has been isolated a fraction resembling a soft wax. When 65.5 g. of crude acetone-soluble lipids of *Blastomyces dermatitidis* in 175 cc. of acetone was mixed with 75 cc. of methanol, an oily layer separated. On cooling overnight in a refrigerator, lumps of yellow wax-like solid separated from the oily insoluble layer. The yellow substance (2.79 g. recovered) was isolated by centrifuging the cold solution, decanting the two supernatant layers, and washing the residual wax-like material with cold methanol. The residue was dissolved in 50 cc. of cold acetone and reprecipitated with 50 cc. of cold methanol several times. The substance thus obtained was a solid at 10° C and a very thick, pale yellow oil at room temperature; it contained no nitrogen, phosphorus, or ash.

For determination of the chemical composition, a sample was saponified by refluxing with 4 per cent alcoholic potassium hydroxide for twelve hours under nitrogen. Unsaponifiable material, fatty acids, and water soluble material were isolated by conventional methods. The water-soluble fraction, 2.8 per cent of the wax, was characterised as nearly pure glycerol by its conversion to glycerol tribenzoate, m.p. 75°-76° C, in good yield (Einhorn and Hollandt, 1898).

The fatty acids, 89.8 per cent of the wax, were separated by the Twitchell (1921) method into solid and liquid fatty acids. The solid fatty acid fraction was a white crystalline solid representing 21.4 per cent of the total fatty acids. It was completely saturated. Recrystallised twice from acetone, it was obtained in the form of shining plates which melted at 80°-82°C, and had a neutral equivalent of 370.2. The melting point and neutral equivalent are close to the values for n-tetracosanoic acid (m.p. 83°-84°C, N. Eq., 368); thus it appears probable that the saturated acids of the wax-like fraction of *Blastomyces dermatitidis* are composed mainly of n-tetracosanoic acid.

The liquid acids represented 68.4 per cent of the wax. This fraction had an iodine number of 135. Bromination of a sample weighing 1.540 g. gave 0.7830 g. of tetrabromostearic acid. The latter was characterised by its melting point, 113°-114° C (which was not depressed by admixture with authentic tetrabromostearic acid), and by bromine analysis (calculated: Br, 53.29, found: Br, 53.01). The remainder of the bromination product appeared to be dibromostearic acid. No evidence was obtained for the presence of fatty acids more highly unsaturated than linoleic. The unsaturated fatty acids therefore consist only of oleic and linoleic acids. Oleic acid forms 30 per cent and linoleic acid 38 per cent of the wax.

The unsaponifiable fraction was shown to contain about 45 per cent of sterols by colourimetric determination using the Bloor (1916) method with ergosterol as standard. Some other higher alcohols were present, but these have not been characterised.

Bound Lipids: The bound lipids of *Candida albicans* and *Blastomyces dermatitidis* have been investigated in some detail (Peck and Hauser, 1940). After extraction of the lipid fractions, the cells of *Blastomyces dermatitidis* still contained 2.0 per cent of lipids and the cells of *Candida albicans* 4.0 per cent, as determined by solution in 6 *N* hydrochloric acid and extraction with ether (Tur-Peinen, 1935, 1936).

Fraction I represents material extracted from the cells with boiling alcohol. This is composed partly of bound lipids, partly of certain lipids soluble only in hot alcohol, and a small amount of mechanically bound lipids of the type extracted by alcohol-ether mixture. From the hot alcohol extracts, on cooling to room temperature, a white powder separated. This material was insoluble in ether, cold alcohol, water, and cold acid or alkaline aqueous solutions. It was soluble in hot alcohol and in pyridine. From the latter solvent, it could be reprecipitated by the addition of one volume of acetone. These substances were unsaturated, since they absorbed bromine.

Saponification of the substance insoluble in cold alcohol gave fatty acids and a water-soluble fraction containing carbohydrate, phosphoric acid, and nitrogenous substances. Further characterisation of these components has not been carried out, but the data available on this fraction indicate a similarity to sphingomyelin. Sphingomyelin types usually contain approximately four per cent each of nitrogen and phosphorus, the N:P ratio being 2:1. Carbohydrates, however, are not present. The cerebrosides contain a carbohydrate moiety and approximately two per cent of nitrogen, but no phosphorus.

A compound, $C_{40}H_{93}NO_5$, designated fungus cerebrin has been isolated from yeast by Reindel *et al.* (1940), and may be considered to be a bound lipid. More recently, the same type of substance was isolated from *Aspergillus sydowi* by Bohonos and Peterson (1943) after autolysis had occurred.

Analyses on the latter product and on its tetraacetate yielded the same empirical formula, $C_{46}H_{93}NO_5$, for the free compound. Degradation studies led to the conclusion that this material was identical with the fungus cerebrin previously isolated from yeast. The necessity for autolysis prior to extraction of fungus cerebrin from both organisms suggests that it exists in the cells in a tightly bound condition.

Sub-fraction from both organisms was a saturated, buff-coloured product containing about 7 per cent of nitrogen and only traces of phosphorus. It was slightly soluble in water. Saponification gave fatty acids, a polysaccharide, and water-soluble nitrogenous compounds. These fractions have not been further characterised.

Sub-fractions from both organisms were brown oils similar to the acetone-soluble lipids extracted from the fresh cells with alcohol-ether. The iodine numbers of the fractions ranged between 70 and 90. Only traces of phosphorus and nitrogen were present. Saponification gave fatty acids, glycerol, traces of carbohydrates, and unsaponifiable material containing sterols.

There is considerably more unsaponifiable material in most of the bound Hpid fractions than in the acetone-soluble fractions of the easily-extracted lipids. The unsaponifiable fractions from the bound lipids were obtained as brown oils containing crystals of sterols. In the mixed unsaponifiable fractions from *Candida albicans*, there was found 20.4 per cent of sterol, whereas in the corresponding fractions from *Blastomyces dermatitidis* there was found 8.4 per cent.

The fatty acids from the bound lipids had iodine numbers ranging between 90 and 100. There was obtained 40 per cent of saturated acids from *Blastomyces dermatitidis* and about 45 per cent from *Candida albicans.* The unsaturated fatty acid fractions were composed essentially of oleic and linoleic acids.

Miscellaneous Components of the Lipid Fractions: Pigments are present in the lipids of all pathogenic and non-pathogenic fungi, but in only a few instances have these substances been isolated or characterised.

Strains of the pathogenic yeast-like fungus, *Candida albicans,* and of the relatively non-pathogenic related organism, *Candida stellatoidea,* have been shown by Jones and Peck (1940) to produce a green pigment which is soluble in acetone, ethyl acetate, and acetic acid and practically insoluble in ether and methyl, ethyl, and amyl alcohols. This pigment exhibits an absorption band at about 6550Å in ethyl acetate solution. *Candida stellatoidea* produces about ten times as much of the pigment as *Candida albicans.* The pigment is an indicator, being blue-green in acid solution and yellow in alkaline solution. It is unstable in the presence of oxidising or reducing reagents.

Of known types of pigments which have been shown to be present in lipid fractions obtained from fungi, carotinoids and quinones may be mentioned. It is also possible that antibiotic substances may be present in the lipid fractions obtained from pathogenic and non-pathogenic fungi. Recent work has demonstrated the widespread production of antibiotic substances by fungi.

The solubility properties of a number of these substances are such that they could be carried along with the lipid fraction, and indeed might be present in the cells in close association with lipids. The classic example in the case of bacteria is pyocyanase, the crude lipid of *Psendomonas aeruginosa,* which contains much antibiotic material.

Discussion: In this survey, attention has been focused mainly on the lipids of two pathogenic, yeast-like fungi, *Blastomyces dermatitidis* and *Candida albicans,* which appear to be representative of a fairly considerable group of pathogenic microorganisms. The lipids of these two fungi have been shown to be essentially the same as far as qualitative composition is concerned. Neutral glycerides, sterol esters, free fatty acids and sterols, lecithin and cephalin types of phospholipids, and small amounts of carbohydrate-containing phospholipids represent the major constituents of the readily extracted lipids. The bound lipids, on the other hand, appear to contain considerable amounts of more complex substances, including fractions resembling the sphingomyelins and cerebrosides. The differences observed between the lipids of the two organisms are mainly quantitative.

The readily extracted lipids of the non-pathogenic fungi show qualitative similarity to the corresponding lipid fractions from the two pathogenic forms. The presence of carbohydrate-containing phospholipids in the pathogenic group and the absence of this type of phospholipid in all but one of the non-pathogenic fungi constitutes, perhaps, the most significant variation in the readily extracted lipid fractions. The phospholipids of a number of pathogenic bacteria, particularly the acid-fast bacteria, contain polysaccharides (1, 2, 3, 13). These polysaccharides have been shown in many cases to be immunologically specific, and may be intimately concerned with phases of the disease processes.

Insufficient data are available to permit comparisons between bound lipid fractions of pathogenic and non-pathogenic fungi, but it appears reasonable to suppose that considerable similarity exists among these forms. It is believed possible that the bound lipids, also, may be concerned with the disease processes, since these fractions must be liberated in the tissues when lysis of the invading organisms occurs in the animal body.

The capacity of host tissue cells for removal of such breakdown products can doubtless be influenced markedly by various properties of the lipid constituents themselves. The enzyme-inhibiting power of certain lipid types has already been mentioned. During lysis of killed pathogenic organisms in host tissues, there may also be liberated enzymes capable of modifying otherwise harmless lipids (as well as other substances). For example, there might thus be set free a lecithinase which would convert lecithin to the very cytolytic lysolecithin and thereby interfere with host cells endeavouring to remove the dead pathogenic organisms.

The lipids of a number of pathogenic bacteria contain considerable amounts of a new series of branched-chain fatty acids of fairly high molecular weight. These unusual fatty acids, when injected into normal animals, stimulate tissue reactions (formation of monocytes, epithelioid cells, and giant cells) similar to those produced in the animals by the corresponding pathogenic organisms. Acids of this type have not so far been observed among the cleavage products of the lipids of any of the fungi studied.

It is possible, however, that small amounts of these substances may be present. Much work remains to be done on the fractionation and characterisation of the cleavage products of lipids of pathogenic fungi, and likewise on the pathology of these substances. It is not impossible that in such fungi there may be present complex lipid fractions (inactive after hydrolysis or other chemical modification) which in the natural state are capable of eliciting pathologic responses by reason of their chemical structure, specific linkages, or particular configuration.

There appear to be significant differences of a qualitative nature between the lipids of bacteria and fungi. The differences in the composition of the lipids of the various fungi studied are of a less striking nature. There is a great need for further data on the qualitative and quantitative nature of the lipids of both pathogenic and non-pathogenic organisms.

Nutrition and Metabolism of Pathogenic Fungi

It is a general premise that the more known about a situation or an object, the more possible it becomes to cope with the situation or object. Such may be said for the fungi causing disease in man. It is a matter of considerable importance and interest to have experimental data on the nutrition and metabolism of these organisms. Such information will permit comparison with the numerous non-pathogenic forms that have been, in many cases, very thoroughly studied. The comparisons may lead us to some hypotheses, admitting of experimental test, concerning the metabolic reactions of the pathogen that might distinguish it from related non-pathogenic fungi. Some insight may thus be gained into the necessity, cause, or "reason" for the parasitic life of the pathogen and further progress towards the specific prevention or treatment of the disease caused by the pathogen.

Among workers concerned with the pathogenic fungi one aim has been the isolation of an organism for purposes of identification and establishing proof of its pathogenicity. Any medium repeatedly permitting good growth of an organism was satisfactory for these purposes.

Such work will of course continue to occupy many workers in the field, but attention from an increasing number is being directed towards obtaining chemically defined media for each organism in order to elucidate the nutritional requirements (inorganic as well as organic) of each pathogen. Knowledge of the requirements of a pathogen from a host may permit attention to be focused on means of preventing the pathogen from obtaining its requirements. Growth in chemically defined media permits one to experiment in search of inhibitory compounds without complications from the buffering, adsorbing, and protective effects of proteins, colloids, and unknown substances generally. The leads thus gained may be checked by *in vitro* experiments in the presence of natural products, thereby eliminating from *in vivo* testing all but the more promising substances.

From the work of Knight (1936) and Lwoff (1938) there has developed a general concept that the growth factor requirements of microorganisms are very much the same, and the demand by any individual for an exogenous supply of a given factor reflects a loss of synthetic capacity on its part.

As stated by Tatum (1944), "One of the main contributions of this concept is that the losses in synthetic apacities are frequently correlated with an adaptation or modification to more and more complex environments as far as vitamin supplies are concerned." An essential metabolite for which an organism loses synthetic ability must then be obtained from some exogenous source, and an animal (or plant) could be such a source. It is evident that any further modification that is inheritable (powers of synthesis once lost are apparently rarely regained) in the metabolic nature of the microorganism increasing its invasive ability will have survival value for the pathogen. The work of Lwoff referred to above was with the nutrition of protozoa and that of Knight with bacteria.

It has been considered that the fungi also fit in with this concept. So far as the plant pathogens are concerned, it seems clear that most of these have absolute deficiencies for one or more growth factors (Robbins and Kavanagh, 1942). But it is not clear that there is any correlation between the organic growth factor requirements of the fungi pathogenic for man and their tendency towards a pathogenic existence.

Some of even the most virulent pathogenic fungi may be grown on very simple chemically defined media. The absence of organic growth factor requirements at least makes permissible an independent non-exacting existence in nature as a saprophyte for such forms as *Coccidioides.*

Indeed, all evidence indicates that the fungi causing systemic infections have a saprophytic existence in nature. Such a situation may require a shift of emphasis away from some of the implications of the growth factor-pathogenicity work that has been so valuable with bacteria and protozoa and previously assumed to hold for the fungi pathogenic for man; corresponding increases in emphasis may be placed on other metabolic characteristics of the pathogenic fungi. For instance, the real requirement by many species of relatively high nutritional levels of certain heavy metals might bear closer scrutiny.

Growth on Natural Products and Complex Media

Since the site of attack of the dermatophytes is the hair, nails, and epidermis of animals, it was natural that interest was early developed in the ability of these fungi to grow *in vitro* on such products. Growing cultures of species of *Trichophyton* were found by Roberts (1894) to attack, to a varying degree, hair of human and animal origin; some species destroyed the cuticle and cortex simultaneously, others worked on the cortex first and but little on the cuticle. Macfayden (1894) was able to grow *Trichophyton tonsurans* on purified keratin obtained from quills; the keratin was made more soluble in strong alkali by the action of the fungus, but Macfayden did not demonstrate a definite enzyme acting on keratin. Nannizzi (1926) obtained luxuriant growths of dermatophytes on hair and on feathers and suggested that the organisms are able to hydrolyse keratin. But Tate (19296), employing acetone powder preparations of enzymes from several dermatophytes, was unable to observe any hydrolysis of purified keratin with such preparations.

Keratin

There are a variety of keratins from animal sources, but available evidence indicates a considerable degree of similarity among them. The few immunological studies that have been carried out indicate little species specificity among keratins; for example, the lens protein from the eye of a given animal will react with this source of keratin from widely diverse species of animals. Keratins are characterised by their chemical stability; Block and Vickery (1931) have emphasised this in their definition, "A keratin is a protein which is resistant to digestion by pepsin and trypsin, which is insoluble in dilute acids and alkalies, in water and in organic solvents, and which, on acid hydrolysis, yields such quantities of histidine, lysine, and arginine that the molecular ratios of these amino acids are respectively approximately 1:4:12." Many keratins, human hair notably, are also remarkable for their high content of the sulphur containing amino acid, cystine; human hair may contain up to

14 per cent of this compound. The resistance of keratins to attack by pepsin and trypsin has called forth theories to explain their stability; Ssadi-Kow (1926) has postulated different types of linkages between the constituent amino acids from what may occur in other classes of proteins.

No enzyme has yet been found that actually hydrolyses a keratin, but Linderstrom-lang and Duspiva have shown that an enzyme in the intestine of the clothes moth readily digests a reduction product of keratin. Sulphydryl compounds secreted into the intestine of the clothes moth bring about the reduction of the keratin. It may well be that the action of the dermatophytes on keratin is not through primary enzymatic digestion but secondarily by enzymatic attack on a reduction product, as shown for the clothes moth.

In the preparation of keratin for experimental purposes, it appears from the work of Cohen (1944) that grinding to a powder is contraindicated since the process was shown to split the disulphide linkages, other linkages, and to destroy tryptophane. Goddard and Michaelis (1934, 1935) were able to reduce keratin by splitting the disulphide bonds with the formation of sulphydryl groups to yield a product they called kerateine. The product of reduction of wool by thioglycolic acid was found by Routh and Lewis (1938) to be digested by both trypsin and pepsin; and the alkali solubility of reduced wool is much greater than that of untreated wool (Patterson *et al.* 1941). Jones and Mecham (1943) have described a combination method of reduction and dispersion (by urea, acetamide, etc.) for keratins which permits the preparation of a product with a minimum of hydrolysis at neutral pH and at as low as 40° C.; they believe the polypeptide chains to be relatively unaltered after this treatment. In light of this recent work it is interesting to recall the early findings of Macfayden cited previously on the increase in solubility of keratin in alkali following the action of *T. tonsurans.* He may have been witnessing an example of increased dispersion following reduction.

Extractions and Hydrolysates of Hair and Skin

J. W. Williams (1934a) found *Microsporum Audouini* made an appreciable growth in 7 days in distilled water to which hair from a child had been added; with hair from an adult, there was no appreciable growth before 30 days or more. It was pointed out that this is of interest since the pathogenicity of *M. Audouini* is manifest only before adolescence. Williams subjected hair from pre-adolescents to continuous ether extraction for 24 hours. Hair with the ether-soluble fraction removed was then added to distilled water and tested as a culture medium with several pathogenic fungi.

Trichophyton (Achorion) schocnleinii was found to grow upon the ether-extracted hair but would not grow upon untreated hair. Many of the fungi examined grew as rapidly on the extracted-hair medium as on the conventional Sabouraud medium; evidently the hair of children is an excellent culture medium for these forms.

Blastomyces dermatitidis (Endomyces capsulatus) grew only on untreated hair; possibly some essential growth factor was removed in the ether extraction process. In comparison studies on ether-extracted hair from adults, Williams (1935) found again that *Blastomyces dermatitidis*, as well as *Monosporium apiospermum (Indiella americana)*, showed only scant growth on the extracted hair after 18 days' incubation, though each of these organisms grew well on plain hair. In the other direction, *T. Schoenleinii* was conceivably inhibited by some substance, removable in the ether extraction process, since it grew well on extracted hair within 3 days, but showed only scant growth in 10 days on untreated hair— only slightly better growth than reported earlier on untreated hair from children.

Employing acid hydrolysates of skin or of hair in place of peptone in culture media, Williams (1936) obtained good growth with several pathogenic fungi. The growth was of a subsurface type, however, not on the surface as usually obtained with solid media.

Similar encouragement of a subsurface type of growth was obtained when peptone was replaced by a combination of alkaline-hydrolysate of hair (cysteine-cystine thus removed) and cysteine-cystine, in a dextrose agar medium. Robbins and Ma (1945) also observed that growth of *T. mentagrophytes* with cystine as the sole source of nitrogen was poor and markedly subsurface in character.

Cellulosic Materials

It has been stated, but not with the support of experimental data, that species of dermatophytes may grow on cotton; Bonar and Dreyer (1932) could find no evidence that representatives of this group could attack cellulosic materials. The latter authors also examined the ability of new sound wood, including oak, Douglas fir, and pine, to support the growth of species of dermatophytes.

Blocks of wood, after being moistened with distilled water, were sterilised in flasks and inoculated; after 6 months' incubation at a favourable temperature and relative humidity there was no detectable development of any species. With old weathered wood similarly handled, *T. mentagrophytes* showed appreciable development in 3 out of 12 cultures; viable fungi, however, could be recovered even from the flasks without appreciable growth. Floor boards from showers accumulate a mixture of organic materials in a short time; such material, after sterilisation, supported a luxuriant growth of species of dermatophytes.

Nitrogen Nutrition

Although considerable has been written since the time of Sabouraud on the ingredients for culture media for the pathogenic fungi (with regard to dermatophytes especially), little information has been available until recently on the specific nitrogen requirements of any of the pathogens. Peptone has been the source of nitrogen commonly employed in culture media for the pathogenic fungi; peptone is a complex mixture of widely variable composition (witness the many papers on the cultural differences obtained using different peptones). Peptones (not especially purified) contain assorted inorganic compounds in varying amounts, inconstant amounts of amino acids and protein split products, and varying amounts of substances with growth nutrilite functions. Since it is desirable to know, in terms of definite chemical substances, what will satisfy the nitrogen requirements of an organism, it is clear that more definitely characterised substances must be used.

Nutrition of Dermatophytes

One of the first attempts to learn of the nitrogen nutritional requirements of a pathogenic fungus was the careful investigation of Mosher, Saunders, Kingery and R. J. Williams (1936) on *Trichophyton mentagrophytes (interdigitale)*. These authors found that the fungus would not grow in a medium with only inorganic nitrogen; the organism could not grow in a synthetic medium unless it were supplied with amino acids. There was no evidence for the indispensability of any single amino acid, although leucine seemed to fall nearly in the indispensable category, while asparagine and threonine promoted growth considerably.

A varied assortment of amino acids was found to be superior to any single amino acid or group of three or four. There was evidence from colour tests on the media deficient in tryptophane and tyrosine wherein the fungus had grown that these amino acids (and therefore possibly others) had been synthesised by the fungus. Though growth with a few amino acids was slower and less extensive than on a casein digest, the fungus was evidently able to modify and synthesise the requisite units for its protein synthesis.

The nitrogen nutrition of *T. mentagrophytes* was again investigated by Robbins and Ma (1945) in an extensive series of experiments. The findings of Mosher *et al.* were mainly confirmed. Inorganic nitrogen as NH_4NO3 was found to be almost completely unavailable to this organism and, while no single amino acid was found to be indispensable, no single amino acid promoted growth as well as did a suitable mixture of amino acids. The fungus could transform a single amino acid such as asparagine into all of the various amino acids necessary for the synthesis of its proteins.

Apparently the value of protein hydrolysates lies in the fact that the fungus can assemble the amino acids into its requisite proteins faster than it can transform any single amino acid into such protein. The near essentiality observed by Mosher *et al.* for leucine was not apparent in the experiments of Robbins and Ma. Leucine did promote growth in the early stages (through 7 days' incubation) better than any other amino acid but was not outstanding after that.

Nutrition of Variant Strains

A pleomorphic strain that developed from the isolate of *T. mentagrophytes* used by Robbins and Ma was found by them to differ physiologically as well as morphologically from the parent or "normal" strain. The pleomorphic strain was observed to have a higher rate of growth than the parent when supplied with asparagine or casein hydrolysate and was able to utilise ammonium nitrate as the sole source of nitrogen.

Growth with NH_4NO_3 was not so rapid though as with casein hydrolysate. Robbins and Ma speculated in an interesting fashion on the apparent similarities between the sudden appearance of a rapidly growing variant in a culture and of a rapidly growing cell in a tissue. They did not remark, however, of the importance to be attached to what is apparently a gain in synthetic abilities by a fungus through mutation. It would seem that the ability to assemble cell constituents from NH_4NO_3 represents a distinct gain in synthetic mechanisms and it is most unusual in light of recent genetic studies on fungi for mutations to be other than losses of synthetic ability.

The mutants that can appear suddenly in cultures of dermatophytes appear to be permanent in nature; the mutant designated as the "pleomorphic" form *sensu* Sabouraud is a white, fluffy, non-sporulating growth which appears very commonly in old cultures of most species (it is probably this form which Robbins and Ma used). In an extensive study of variation in dermatophytes, Cousins (1936) isolated mutants from 4 species: *E. floccosum, T. mentagrophytes, M. canis,* and *M. gypseum.* Among 4 mutants obtained from *Microsporum canis* one was the pleomorphic form and notable for its rapid growth, while a second one was slow growing and glabrous. No fewer than 20 stable mutants were obtained from *M. gypseum;* the pathogenicity of 3 of these was compared with that of the parent strain by inoculations into guinea pigs—the variants were fully as pathogenic as the parent.

Cousins obtained indicative evidence that the production of these variants possibly results from irregularities in nuclear behaviour, possibly genetic rearrangements; four identical mutants were isolated from 4 separate normal strains of a species and some variants were isolated repeatedly from their respective normal strains. In addition to the possible gains in synthetic ability in the variants already mentioned Cousins found variants notable for their pigment production.

It may be that these forms would repay a thorough investigation; there would not be the complications attendant with some of the few other instances where gains in synthetic ability have apparently arisen by mutation (utilisation of ammonia by bacteria, Knight, 1936; growth factor synthesis by yeasts, Leonian and Lilly, 1943; and phage resistance in *E. coli,* Luria and Delbruck, 1943). For in cultures of single cell organisms the possibility of selection of one exceptional individual in a population (within the normal distribution limits of the population) is most difficult to rule out (discussed further in Nickerson, 1943).

Metabolism of Proteins and Peptones by Dermatophytes: Gluisms the possibility of selection of one exceptional individual in a population (within the normal distribution limits of the population) is most difficult to rule out.

Glucose was found by Goddard (1934) to exert a "sparing" effect on the hydrolysis of calcium caseinate by cultures of *T. mentagrophytes (intcrdigitde)*. In his study of the metabolism of non-pleomorphic strains, Goddard investigated some of the changes taking place in peptone and in peptone-glucose media during growth of the fungus. The presence of glucose had no effect on the increase in alkalinity of the cultures, nor on the increase in ammonia. Evidently the increasing alkalinity in cultures of dermatophytes can be generally attributed to the production of ammonia from amino acids by what is probably an oxidative domination process.

Nitrogen Nutrition of Coccidioides Immitis: Of several nitrogen sources tested by Baker and Smith (1942), all were suitable as the sole source of nitrogen for *C. immitis* when added to a basal medium containing salts, and 1.5 per cent glucose with the exception of nitrite (added as KNO_2).

Growth was only fair with nitrate (added as KNO_3) but heavy growth occurred with NH_4Cl, urea, acetamide, asparagine, glycine, alanine, glutamic acid, tyrosine, cystine, and peptone. These findings (which held for 15 strains) were in good agreement with those of Stewart and Meyer (1938) with a possible exception in regard to the utilisation of nitrate, which the latter authors classed as supporting only poor growth, but this may be only a relative matter. As with the carbon-nutrition of this organism, requirements with respect to the nitrogen source are singularly non-exacting for a serious pathogen and at least make it permissible that C. *immitis* may exist as a saprophyte in nature. *C. immitis* is no more exacting in its requirements for growth than the common contaminant *Aspergillus niger*. Traditional concepts of the exacting nature of serious pathogens based on bacteria and protozoa find little support from pathogenic fungi.

Carbon Nutrition: One of the earliest studies of the metabolism of pathogenic fungi was that carried out by Verujsky (1887) in Duclaux's famous laboratory. *Trichophyton*

(Achorion) schoenleinii was found to be unable to utilise sugar from any medium but assimilated nitrogenous substances readily. On the other hand, *T. tonsurans* did not invert sucrose but was found to utilise glucose.

On a malt medium the ratio for *T. tonsurans* of weight of fungus formed to weight of glucose consumed was 1:2, while the addition of glycerin increased the ratio to 2:3 (in neither case was nitrogen utilisation measured, nor was glycerin disappearance estimated).

In these experiments, when *T. tonsurans* was grown on impoverished media such as Raulin's medium, oxalic acid was detected as a product; Verujsky regarded this as a definitely abnormal product and no subsequent investigation has found this compound as a metabolic product of this species or of any other dermatophyte. Bodin (1899) found *Microsporum canis (equinum)* did not utilise sucrose but assimilated glucose, dextrin, and maltose.

Trichophyton mentagrophytes (Achorion Quinckeanum) was found by Bodin (1902) to utilise glucose more readily than lactose or maltose. Tate (1929a) reported none of the dermatophytes he studied could utilise sodium acetate, formate, or lactate as a carbon source. Mallinckrodt-Haupt (1927, 1933) found species of the dermatophytes could utilise neutral fats as the sole source of carbon if they were of animal origin, but showed little or no growth with vegetable oils.

Tate (1929) studied the enzymes of several of the dermatophytes, using acetone-powder preparations; of the carbohydrases examined, invertase, lactase, zymase, and inulase were absent in all forms. Maltase and diastase were present to a varying degree in all species examined. The general carbohydrase activity appeared to vary inversely with the proteolytic activity, the species with the least trypsin having the greatest carbohydrase activity.

Using freshly isolated non-pleomorphic strains of *Trichophyton mentagrophytes (inter digitale)* and *Microsporum canis (lanosum)* Goddard (1934) measured the increase in dry weight of the growth obtained on adding various sugars to a peptone-distilled water medium. *T. mentagrophytes* was found to assimilate glucose, man-nose, and galactose, but not lactose. *M. canis* had very similar assimilative abilities but did not utilise either lactose or galactose. These findings are in general agreement with those of Tate on the enzymes of dermatophytes, although Tate did not find sucrase in the organisms he studied.

Carbon Nutrition of Coccidioides immitis: Some conflicts appear in the literature concerning the utilisation of certain carbon compounds by *C. immitis* for growth, but all authors are in agreement that the metabolism is purely oxidative, there being no report of the fermentative decomposition of carbohydrates. In an effort to clarify the situation, Baker and Smith (1942) investigated the availability of over 40 carbon compounds for several strains of *C. immitis.*

A variety of compounds will serve as a suitable carbon source when added to a basal medium containing salts and NH_4Cl. No special growth factors were found to be necessary. Tubes of media were inoculated with suspensions of chlamydospores and hyphal fragments for each of 15 strains; little strain variation was observed except for slight growths in some of the sucrose and lactose tubes. While the sodium salts of formic, butyric, oxalic, tartaric, gluconic, and citric acids were not utilised, these compounds were not toxic since spores in the inoculation germinated and developed briefly (at the expense of reserve stores within the spore) just as did the controls in the basal medium alone.

These results were in agreement with the earlier findings by Hirsch and Benson (1927) that lactate forms a suitable substrate and with Stewart and Meyer (1938) who found acetate, but not formate (and ammo acids, but not urea), to be suitable as a carbon source. *C. immitis* is thus seen to be non-exacting in regard to its carbon source and to possess no special growth factor requirements.

Mineral Nutrition: In a general study of the nutritional requirements of *Trichophyton mentagrophytes (interdigitate),*

Mosher *et al.* (1936) examined its mineral requirements, including the "micronutrient" or "trace" elements. Although the nitrogen requirements of this fungus could not be satisfied with ammonium nitrate, it was still found to require the ammonium ion for growth. Other ions in the "macronutrient" class were potassium, magnesium, and calcium. The essential micronutrients were: zinc, iron, copper, manganese, phosphate, and sulphate. Of the ions tested, four seemed to be non-essential for *T. mentagrophytes;* these were thallium (Richards, 1932, has shown the essentiality of thallium for a strain of *Saccharomyces cerevisiae),* iodide, chloride, and borate.

Medium Growth

The difficulties surrounding investigations of the micronutrient requirements of higher plants have their counterpart in comparable investigations with fungi. Impurities in even the highest quality chemicals and dissolution of ions from glass culture vessels are possible sources of the ion under investigation and must be carefully watched.

The extensive work of Steinberg since 1918 on the mineral nutrition of *Aspergillus niger* has shown the essentiality of iron, zinc, copper, manganese, molybdenum, and gallium for a strain of this mould. There is evidence that molybdenum has an important role in *A. niger* in the reduction of nitrates and nitrites—of critical importance when these compounds were the sole source of nitrogen. His researches have also brought to light the extreme care that must be exercised during such work since the minute amounts of the heavy metal required can easily be supplied as an impurity unless guarded against. Many purification techniques designed to rid nutrient solutions of traces of heavy metals are in use today; some, including the 8-hydroxy quinoline method of Waring and Werkman (1943), act by complex formation with heavy metals; others, including Steinberg's (1919) use of $CaCO_3$, work by adsorption onto finely particulate surfaces. The latter method is simple and satisfactory. Excess calcium carbonate (15 g./litre) is added to the nutrient solution, autoclaved at 15 Ib. for 15 minutes and filtered while hot; the nutrient solution is then distributed as desired and sterilised, care being taken not to reintroduce heavy metals.

In an excellent review of the comparative aspects of heavy metal nutrition of fungi, Foster (1939) included a discussion of the relationships between certain of these metals (notably iron, copper, and manganese) and the pigments elaborated by fungi.

Coccidioides Immitis: Roessler *et al.* (1946) grew the fungus submerged in liquid cultures by continuous mechanical agitation; the formation of free spores was thereby encouraged. Growth was abundant in a chemically defined medium with ammonium acetate as the source of nitrogen, glucose, $MgSO_4$, K_2HPO_4, KH_2PO_4, and $ZnSO_4$. No accessory growth factors were required. Magnesium (used as O.OQ$_8$M $MgSO_4$) and zinc (0.00003M $ZnSO_4$) were the only cations found to be required for growth.

Growth Factor Requirements: A growing literature on this topic in the nutrition of pathogenic fungi permits some comparisons among the various groups and with related non-pathogenic forms. A recent excellent discussion of the comparative aspects of the growth factor requirements for microorganisms generally is that of Knight (1945). In the genus *Candida*, the one pathogenic species, *C. albicans*, is not distinguished notably from the non-pathogenic species. There is as wide variation among different isolates of *C. albicans* as between species of *Candida*. All species are indicated to require biotin for growth, and several have their growth stimulated on the addition of thiamin. *C. albicans* has been shown by Van Njiel and Cohen (1942) to utilise the thiamin-containing co-enzyme cocarboxylase, as evidenced by the accumulation of pyruvic acid with suspensions that had become thiamin deficient through aeration in a non-nutrient medium.

Thiamin: With a few interesting exceptions, growth factor requirements among the filamentous pathogenic fungi are limited mainly to a need for thiamin. Some of the dermatophytes, for example strains of *Trichophyton mentagrophytes,* have been shown by several authors to have no growth factor requirements, while *Trichophyton Schoenleinii (T. discoides)* has been shown by Robbins and Ma (1943) to require thiamin and to be unable to synthesise thiamin when supplied with both the pyrimidine and thiazol moieties of thiamin. Oyama (1937) has described a few instances wherein growth of an organism is not stimulated by thiamin but is stimulated by an extract of rice polishings (a source rich in thiamin).

Riboflavin: There are no reports of any fungi requiring riboflavin for growth, though there is increasing indication that the level of riboflavin nutrition in the cell may vary during the growth cycle in certain organisms and actually be a limiting factor under certain conditions. Such a situation has been shown by Doudoroff (1938) with strains of the luminescent bacterium, *Photobacterium phosphorescens;* concentrations of riboflavin sufficient to support growth of some strains were insufficient to permit luminescence in the same strains. Mosher *et al.* (1936) studied a strain of *T. mentagrophytes* for which riboflavin was stimulating. The concentration level of riboflavin has also been implicated by Nickerson and Thimann (1943) in the sexual reproduction of a conjugating yeast, *Zygosaccharomyces acidifaciens,* which has no requirement for added riboflavin during growth but which exhibits a marked increase in the relative number of cells conjugating in cultures to which riboflavin has been added. (The riboflavin stimulation was intensified on the addition of glutaric acid.)

Pyridoxin: Pyridoxin has been implicated in the nutrition of a few of the lower fungi, chiefly through the work of Robbins and Ma (1942) and of Fries (19430, fc). *Trichophyton schoenleinii (T. discoides)* remains the only pathogenic fungus for which pyridoxin has been reported as a growth factor (Robbins, Mackinnon and Ma, 1942), though pyridoxin is stimulatory for *Pityrospomm ovale* when oleic acid is present to permit growth (Benham, 1941).

Biotin: Biotin has not been shown to be a growth factor for any of the filamentous pathogenic fungi, though it is required by many yeasts and bacteria. Schopfer and Blumer (1942; schopfer, 1943) showed that, under certain conditions, biotin can serve as a growth factor for *T. schoenleinii (T. album).* In a medium with glucose, asparagine, MgSO4, and KH_2PO_4, only thiamin is required as a growth factor. Though biotin is stimulatory in the early stages of development, growth in controls without biotin was found to overtake cultures supplied with biotin. When $(NH_4)_2SO_4$, rather than asparagine, was used as the nitrogen source, biotin was required even up to advanced stages of development; with ammonium citrate as the nitrogen source, biotin promoted better growth than in controls without biotin over a 24 day period. Apparently the ability of the organism to synthesise biotin varies with the age of the culture and with the nitrogen source.

Fatty Acids: The report by Benham (1941) of the requirement of oleic acid by *Pityrosporum ovale* for growth in subcultures was the first demonstration of such a need for a fungus. Fleming (1909) had shown that *Corynebacterium acnes* (the acne bacillus), ordinarily grown only with difficulty, grew well on nutrient agar with 1 to 5 per cent of oleic acid added.

Benham discusses her work elsewhere in this volume, but a few figures will be used here for comparative purposes. In a medium with NH_4NO_3, salts, and glucose, it was found that 100 mg. oleic acid per litre produced a slight effect with growth increasing up to 10 grams per litre, though growth still was not heavy. Addition of asparagine permitted good growth in 1-2 weeks. In contrast to the concentration levels of oleic acid required by *P. ovale,* it has been found for *Clostridium tetani* by Feeney, Mueller and Miller (1943) that its oleic acid requirement was satisfied optimally at a concentration of only 0.75 mg. oleic acid per litre.

Against these requirements of a fatty acid for growth it has been found by Rothman, Smiljanic and Shapiro (1945) that the hair fat of adults contains free saturated aliphatic fatty acids which inhibit the growth of *Microsporum audouini* in concentrations of 2 to 10 mg. per litre. Preliminary analyses indicated that the active fraction comprised fatty acids containing 7 to 11 carbon atoms. The fat extracted from children's hair had only 1/5 the fungistatic action of adults' hair fat. It is tempting, of course, to see in this development an explanation of the relative incidence of *Microsporum* infections of the scalp in adults *vs.* children. Rothman (1945) pointed out that the non-hairy skin of adults remains susceptible to infection by *M. audouini.*

Physicochemical Factors in Growth: In addition to the demands an organism may have for growth in regard to carbon and nitrogen sources, minerals, and growth factors, it can be more or less exacting with respect to the *pH* range within which nutrients can be assimilated and its enzymes act. In similar fashion, temperature limits and redox potential limits operate to influence the activity of any protoplasmic system. A consideration of some of these factors and related phenomena for the pathogenic fungi reveals some interesting relationships. The effects of temperature on morphogenesis in some of the fungi that are yeast-like in the body offers a fertile field for further experimentation.

Hydrogen Ion Concentration: None of the pathogenic fungi is particularly exacting in its pH requirements for *in vitro* cultivation. Although great extremes of acidity are not tolerated, each of the pathogens is able to grow over a considerable range of hydrogen ion concentrations.

The dermatophytes are possibly the least exacting of the pathogens in this regard. It is possible to culture *Trichophyton mentagrophytes* easily in a range from pH 3 to 10.5. In fact, Leise and James (1945) have made use of the ability of the dermatophytes to grow well under alkaline conditions in devising a medium buffered to pH 10.5 for the isolation of dermatophytes from sources such as shoes, floors, shower baths, etc., where the presence of saprophytes in large numbers has in the past complicated the problem of isolating the slower growing pathogens. There is apparently no sharp pH optimum for *T. mentagrophytes,* especially in a good nutrient medium many workers grow the dermatophytes at an initial pH of 5.5 where it does well. This pH is the same generally found for healthy skin (Blank, 1939). It has been shown that pH 5.5 is the region of lowest rate of respiration for at least two of the dermatophytes.

Mosher *et al.* (1936) remarked that the production of bases during growth of T. *mentagrophytes* in Sabouraud's medium changed the pH from 5.5 to 8.0, which is a considerable rise in view of the buffering capabilities of peptones and amino acids. Robbins and Ma (1945) found only a slight increase in alkalinity in 10 days' growth of T. *mentagrophytes* on a peptone medium; increases in alkalinity become more pronounced in older cultures—after the buffering effect has been overcome. The pH indicator action of the pigments produced by many of the dermatophytes would lead one to believe that the pH of the medium is frequently on the alkaline side, since these pigments are reddish under alkaline conditions. The good growth of most of the dermatophytes under conditions as acid as *pH* 3.0 is curious in light of Tate's (1929&) failure to find any pepsin-like proteolytic enzyme, though all species investigated possessed trypsin-like enzymes active in alkaline conditions.

As part of a comparative study on *Histoplasma capsulatum* and morphologically related form-species, Howell (1941) investigated the influence of hydrogen ion concentration of the medium on growth and spore formation. Comparing *H. capsulatum* and *Sepedonium chrysospermum* (Bull.) Link—a parasite on other fungi, this strain having been isolated from a *Boletus* sp.

The optimum pH for growth for each is about 6.5. At this pH *H. capsulatum* produced the most aerial mycelia and achieved the greatest radial growth; aleuriospore production was also maximum. The same relationships held generally for *S. chrysospermum,* but this species maintained good growth to pH 5.0 although aerial mycelium and aleuriospore production were poor at the more acid range. The high H-ion concentrations attained by cultures of each species from growth at an initial pH of 5.0 indicates the formation of considerable amounts of acid in each case. The metabolic products of neither species appear to have been investigated, however.

Temperature: The pathogenic fungi may conveniently be divided into two groups on the basis of the temperature range of growth *in vivo*. The organisms causing superficial skin infections inhabit an environment which when dry reflects atmospheric temperatures to some extent, so that even *in vivo* the dermatophytes are subjected to a considerable range of temperature. Organisms causing deep mycoses are maintained at more constant temperatures.

In vitro the dermatophytes grow well in a range from 25°C to 38°C showing little variation from temperature *per se,* but the slow growth rates necessitate guarding against the drying out of cultures on agar or a retarding of growth by dehydration will occur (though there is a tendency for spore formation to be less abundant above 35° C as Verujsky, 1887, early observed). With some of the fungi causing deep mycoses, the temperature of incubation for *in vitro* growth is quite critical. Characteristic growth forms that develop at 25°C may be entirely lacking at 37°C at which temperature the growth *in vitro* may more closely approximate the appearance of the fungus *in vivo,* but be less differentiated and of less value for diagnostic purposes.

Blastomyces dermatitidis may develop in culture into one of two quite different forms, depending on the temperature of incubation. At 37°C the yeast-like cells found in lesions will develop on cultures on blood agar, but at 25°C a cottony filamentous growth develops on Sabouraud's agar. Many workers had contended that once the mycelial type of growth had been allowed to develop at room temperature the yeast-like growth could not be recovered, but Conant (1939), confirming Ciferri and Redaelli (1935), demonstrated that heavy inoculations from mycelial cultures on blood-agar slants incubated at 37°C developed typical yeast-like growths.

CONANT mentioned that on inoculating yeast-like growths on Sabouraud's agar and incubating at room temperature the growth at first is moist and glabrous, but soon becomes dry and cottony with aerial projections giving the colony a prickly appearance which eventually is replaced by an all over cottony appearance.

Numerous conidia were produced in both the prickly and filamentous stages of growth. A fatal infection in mice could be produced by inoculating growths obtained at either room temperature or at 37°C.

The effect of temperature on the development in culture of *Histoplasma capsulatum* shows a striking parallel with the case for *Blastomyces*. Conant (1941) has shown that the saprophytic filamentous form of *H. capsulatum* developed on Sabouraud's agar at room temperature can be converted, without intervention of animal passages, to the yeast-like form found in tissues by cultivation on sealed blood-agar slants at 37° C. Conant made a strong point in this work that classification of an organism should be based on the phase of the life-cycle exhibiting a spore form; *H. capsulatum* develops characteristic tuberculate chlamydospores on all media at 25°C, but develops only a yeast-like stage at 37°C.

Heat Resistance of Dermatophytes: In connection with an extensive study on the public health aspects of dermatophyte infections, Bonar and Dreyer (1932), examined the effects of different temperatures on four species growing on various natural products.

Trichophyton mentagrophytes is the most resistant to heat of the species examined; spores of this organism are also more resistant than were spores of the other species studied. Although an exposure to 75°C for 10 minutes kill all the spores, approximately 25 per cent of the spores of *T. mentagrophytes* germinate after 10 minutes at 70°C.

With cultures growing on hair, *T. mentagrophytes* proved again the most resistant of the four species, but no subcultures could be obtained after a 10 minute exposure of cultures to 55° C., with only 60 per cent of the subcultures showing growth after being held the same length of time at 50°C. Weidman (1927) also found.

T. mentagrophytes (interdigitale) the most heat resistant of the dermatophytes he studied. The skin scales in which mycelia of *Epidermophyton floccosum* were embedded afforded some protection against the lethal effects of elevated temperatures. Bonar and Dreyer emphasised that infectious fungus material embedded in hairs or in skin scales is enclosed in structures highly impermeable to aqueous solution. They also demonstrated markedly increased resistance by mycelia in skin scales to chemical disinfectants as compared with the resistance of spores.

Moisture Requirements of Dermatophytes: In an important study, Davidson and Gregory (1934) demonstrated that the highly differentiated spore forms commonly found in cultures growing on media also developed from infected hairs and skin scales detached from the host and maintained by *in situ* culture. Hairs naturally infected with *Microsporum Audouini, M. canls (felineum),* or *Trichophyton mentagrophytes (gypseum)* and skin scales containing *Trichophyton (Achoriori) Schoenleinii* were placed without added nutrient in van Tieghem cells maintained at different relative humidities.

While parasitising an animal body, these fungi produce only hyphae or thalospores but under suitable conditions the *in situ* cultures produced the highly differentiated structures characteristic of the species when grown on nutrient media. As the relative humidity decreased from a completely water saturated atmosphere to 97 per cent R. H. there was an increase in the time required for the fungi contained in the infected hairs or skin scales to "germinate" or show signs of growth. A relative humidity as high as 95.8 per cent is insufficient to permit the growth of cultures of *M. Audouini in situ,* while the radial growth of *M. canis* decreases markedly below 97 per cent R. H. The lowest R. H. at which the growth of

T. mentagrophytes and *T. schoenleinii* is observed under these conditions is 97 per cent and 93.3 per cent R. H. respectively.

The values presented place these fungi at the upper end of the scale of minimum relative humidity for growth in comparison with the requirements of other fungi. Bonner (1945) listed the range of minimum relative humidity for 58 fungi at from 70 to 98 per cent with a fairly even distribution of species over that range.

It is possible that a slight reduction in the relative humidity from saturation is sufficient to stop the growth and development of several dermatophytes. The beneficial results on the incidence and severity of foot infections prompted by the routine wearing of open-toed sandals as found in a large-scale test with nearly 2500 soldiers was attributed by Nickerson, Irving and Mehmert (1945) largely to the free access of air and consequent removal of saturated water vapour from the vicinity of the feet.

Radiations: It appears to be a matter of individual preference in each laboratory whether cultures of pathogenic fungi are grown in the light or the absence thereof; no incontrovertible evidence has accumulated to make a real brief for either procedure. The use of x-rays in therapy of fungus diseases and the use of ultraviolet in detection of fungus infections of the hair and scalp are well known.

Few intensive studies of the effects of radiations on these organisms have been conducted, but one excellent study by Hollaender and Emmons (1939) and Emmons and Hollaender (1939) deserves attention. They found monochromatic ultraviolet light between 2537 and 2650 Å to be fungicidal for spores of *Trichophyton mentagrophytes*.

This same range was also found to increase the rate of appearance of mutations in the spores that survived. The maximum rate of mutation appearance among surviving spores was reached when 50 to 99 per cent of the irradiated spores did not survive.

These mutations differed from the parent strain in rate of growth, colony form, pigment production, and other characteristics. The mutations persisted through many transfers and could be called permanent alterations.

Since sexual reproduction is not known in the dermatophytes, it was impossible for the authors to present positive proof of the genetical nature of these mutants, but by comparison, from experiments on *Neurospora* and other fungi, it seemed that the origin of the mutations in *Trichophyton* was definitely genetical. Some of the mutant strains were observed to produce much larger amounts of pigments than the controls.

Comparative aspects of the problem of induction of mutations in fungi with ultraviolet irradiation are presented by Hollaender, Raper and Coghill (19450) on *Aspergillus niger* and by Hollaender *et al.* (1945) on *Neurospora crassa.*

Action of Sulphonamides and Antibiotic Agents on the Pathogenic Fungi

The discovery by Domagk in 1935 of the chemotherapeutic activity of prontosil (sulphamido chrysoidine) in bacterial infections marked the beginning of great developments in medical bacteriology and chemotherapy. Study of the action of similar compounds—now generally known as the sulphonamides—upon bacteria was soon broadened to include a consideration of the effects of the sulpha drugs upon fungi.

Mycological work with the sulphonamides began with the experiments of Fourneau, Trefouel, Nitti and Bovet (1936), which demonstrated that the active ingredient of prontosil is the substance known as sulphanilamide. Working primarily with *Aspergillus niger* and to a lesser extent with *A. janselmi, A fumigatus* and *Lichtheimia (Mucor) italica,* Fourneau *et al.* found that the addition of 0.01 *per cent* sulphanilamide to Raulin's solution resulted in a 24-hour delay in the beginning of growth by the fungi. One part of sulphanilamide per thousand checked growth of *A. niger* for 8-10 days. This organism was able to grow even in media containing 1 *per cent* sulphanilamide, but only after a delay of two months. It was shown by these experiments that the action of sulphanilamide was fungistatic in character. No evidence of adaptation of the fungi to the drug was noted.

It is perhaps important to realise that these experiments with fungi provided the first proof that sulphanilamide is effective *in vitro* as well as *in vivo,* and antedate all experiments with bacteria on the *in vitro* activity of sulphonamides (Long and Bliss, 1939).

In a footnote to the paper of Fourneau *et al.* is to be found the following quotation: "Nous poursuivons actuellement des essais sur Tactivite du 1162F (sulphanilamide) dans les mycoses experimentales."

This study by the workers of the Pasteur Institute is important in that it first clearly established that the active principle of Domagk's prontosil is sulphanilamide, that sulphanilamide is effective *in vitro,* that sulphanilamide has fungistatic properties, and it suggested the possibility of using this compound in the treatment of mycotic diseases. The pioneer work of Fourneau *et al.,* although mentioned by Long and Bliss, has unfortunately been overlooked by subsequent investigators and therefore has not been given the prominence which it so justly deserves.

Mechanism of Action: Any discussion of the manner by which sulphonamides are effective against bacteria must consider the similarity in chemical structure between the sulphonamides and p-aminobenzoic acid.

Fildes (1940) proposed the hypothesis that anti-bacterial substances, such as the sulphonamides, act by interfering with an "essential metabolite," *i.e.* any substance which takes an essential part in the metabolic reactions necessary for bacterial growth. Woods and Fildes (1940) and Woods (1940) found evidence indicating that p-aminobenzoic acid is essential for the growth of streptococci, and demonstrated that this compound very effectively counteracts the inhibition of streptococci by sulphanilamide. They therefore suggested that the inactivation of bacterial cells by sulphanilamide involves a competition for some enzyme between two chemically similar enzyme substrates, the "essential metabolite" (p-aminobenzoic acid) and the inhibitor (sulphanilamide).

Isolation of p-aminobenzoic acid from natural sources was soon reported by Rubbo and Gillespie (1940) and Blanchard (1941) from yeast. Rubbo and Gillespie (1941) demonstrated, through failure of growth in its absence, the essential nature of p-aminobenzoic acid. Selbie (1940) first demonstrated the anti-sulphonamide action of p-aminobenzoic acid *in vivo*.

As it pertains to bacteria, the "p-aminobenzoic acid theory" of the action of sulphonamides has been rather generally accepted. In a recent review, Welch (1945) applies the "lock and key" simile to the mechanism of sulphonamide action. According to this hypothesis, the metabolite is compared to a key, which operates perfectly the "lock" of some enzymatic synthesis.

The analog or inhibitor is likened to an imperfect key, which enters the lock but is unable to operate the mechanism, and further, prevents the entrance of the proper key. It was shown by Lockwood and Lynch (1940) that peptone, a common ingredient of bacteriological and mycological culture media, gives rise to products (not chemically characterised, but other than p-aminobenzoic acid) which likewise interfere with the action of sulphanilamide.

For a further consideration of the theories which attempt to explain the anti-bacterial action of the sulphonamides, reference is made to the review of Henry (1943).

In the case of fungi, Dimond (1941) has studied the relationship between p-aminobenzoic acid, sulphanilamide, and the growth of *Trichophyton purpurenm.* The addition of 1 part per thousand of sulphanilamide to an agar medium upon which *T. purpureum* is inoculated permitted spore germination to occur, but mycelial development was completely inhibited. When, however, p-aminobenzoic acid was added (1 part in 500,000), the inhibition due to sulphanilamide was entirely overcome, and normal growth of the fungus resulted.

Similarly, Landy and Dicken (1942) have reported that inhibition of the growth of yeast, brought about by sulphanilamide, sulphapyridine, sulphaguanidine or sulphathiazole in concentrations of 10-25 mg. per cent, can be completely neutralised by p-aminobenzoic acid. When the weight of yeast cells (a function of the number of cells) per unit volume of medium containing a sulphonamide is plotted logarithmically against the concentration of p-aminobenzoic acid, sigmoid curves are obtained.

Tatum and Beadle (1942) found that the growth of *Neurospora crassa* is inhibited by sulphonamides, and that this inhibition is overcome by p-aminobenzoic acid. This species is normally able to synthesise p-aminobenzoic acid, but as a result of X-ray treatment an "aminobenzoicless" mutant was produced, which was incapable of growth upon media lacking this substance. The respiration of both the normal strain of *N. crassa* and the "aminobenzoicless" mutant has recently been studied by Tatum and Giese (1946), who found that the respiratory rate is unaffected by sulphanilamide.

Further experiments with p-aminobenzoic acid in relation to the growth of other pathogenic fungi have not been performed. The behaviour of other pathogenic fungi requiring p-aminobenzoic acid may be no different from that reported for yeast and for *T. purpureum.*

In general, therefore, the inhibition of fungus growth by the sulphonamides may, in all probability, be regarded as follows: (1) the fungus, in the course of its normal metabolism, requires p-aminobenzoic acid, (2) p-aminobenzoic acid is very similar in chemical structure to the sulphonamides, (3) the sulphonamide is absorbed and saturates an enzyme system of the fungus, and (4) some of the syntheses required for growth of the fungus cannot then be performed.

Lwoff, Nitti, Trefouel and Hamon (1941) have shown that the anti-fungal activity of a sulphonamide is greatly influenced by the acidity of the medium. Using *Aspergillus niger* grown in Czapek's solution, the concentration of sulphanilamide which would produce a delay in growth of the fungus for 24 hours more than control cultures was determined at various pH levels. At pH 7.1, sulphanilamide 1:200,000 produced such a delay in the growth of *A. niger;* at pH 5.8, only 1:20,000 sulphanilamide was required; at *pH* 4.6, 1:10,000; and at pH 3.7, 1:5,000. Thus the activity of sulphanilamide against *A. niger* decreased as the medium became more acid.

In Vitro Experiments: Studies of the action of sulphonamides on pathogenic fungi, *in vitro,* although admittedly inadequate in several respects, nevertheless constitute a rational approach to the problem of successful chemotherapy of mycotic infections. It has already been noted that Fourneau, Trefouel, Nitti and Bovet (1936) performed the first experiments concerning the *in vitro* effects of sulphonamides on fungi. The addition of sulphanilamide to a medium in which *Aspergillus niger* was grown resulted in a marked inhibition of growth. 0.01 per cent sulphanilamide inhibited growth completely for one day, 0.1 per cent sulphanilamide delayed growth for 8-10 days. Essentially similar results were noted with *A. janselmi, A. fumigatus,* and *Lichtheimia (Mucor) italica.* The action of sulphanilamide was therefore primarily one of fungistasis.

The first observations concerning the effect of a sulphonamide upon a pathogenic fungus were made by Dimond who reported that 0.1 per cent sulphanilamide checked growth of *Trichophyton purpureum,* and that this inhibitory action could be overcome by antagonism with p-aminobenzoic acid.

The effects of sulphanilamide, sulphathiazole and sulphadiazine upon two strains of *Actinomyces bovis (hominis)* were studied by Cutting and Gebhardt (1941). The organisms were grown on Krainsky's glucose agar and on a thioglycollate medium under both aerobic and anaerobic conditions. Sulphonamides were added to the media in quantities equivalent to 10, 50 and 100 mg. per cent. Slight inhibition was obtained with 10 mg. per cent of sulphanilamide, while 50 or 100 mg. per cent checked growth more or less completely. In similar concentrations, both sulphathiazole and sulphadiazine were more effective than sulphanilamide.

Simultaneously, but independently of other workers, Lewis and Hopper (1941) made a thorough study of the action of six sulphonamides upon *Trichophyton gypseum* and *Candida albicans.* The sulphonamides available to them were sulphanilamide, sodium sulphapyridine, sulphathiazole, sodium sulphathiazole, sulphadiazine and sodium sulphadiazine.

None of the compounds proved effective in causing any decrease in the growth of *Candida albicans.* Each of the compounds exhibited some growth-inhibitory action upon *Trichophyton gypseum,* but sulphanilamide was the most effective compound. With 1 *per cent* sulphanilamide, three weeks were required for minute colonies to appear.

These results were confirmed and extended by the work of Dimond and Thompson (1942) upon *Trichophyton gypseum* and *T. purpureum.* Experiments were performed in which spores were suspended in 0.1 per cent solutions of sulphanilamide, sulphapyridine, sulphathiazole, sulphaguanidine or sodium sulphadiazine for various periods of time, after which they were washed and inoculated upon an agar medium. Such cultures grew normally after the removal of the drug, and no signs of injury as a result of this treatment were apparent. The action of the sulphonamides was thus again shown to be a fungistatic rather than a fungicidal one.

Further experiments, with agar media containing sulphonamides in concentrations ranging from 1:500 to 1:10,000, demonstrated that sulphanilamide was far more effective than sulphapyridine, sulphathiazole, sulphaguanidine or sodium sulphadiazine in inhibiting growth of *T. gypseum* and *T. purpureum.*

Measurements of the growth rate of *T. purpuremn* on honey-peptone agar containing varying amounts of sulphanilamide showed that the fungistatic action of sulphanilamide was to be interpreted as an extension of the lag phase of the growth curve. In other words, sulphanilamide will delay the beginning of growth of fungi for a time, but has absolutely no influence upon the rate of growth, once growth has begun.

It was further shown by Dimond and Thompson that the fungistatic action of sulphanilamide was much more pronounced on a peptone-free medium, since peptone gives rise to anti-sulphonamide factors.

Noojin and Callaway (1943) have determined the action of seven sulphonamides upon *Blastomyces dermatitidis* grown on Sabouraud's agar. Each drug was tested in concentrations ranging from 50 to 250 mg. per cent.

Thus, sulphathiazole, sodium sulphathiazole and sulphaguanidine were completely without fungistatic activity against *B. dermatitidis* in all levels tested, sulphapyridine and sodium sulphapyridine were slightly fungistatic in high concentrations, and sulphadiazine and Sulphanilamide were the most effective agents against *B. dermatitidis*.

Since 10 mg. per cent represents the maximal sulphonamide blood level which can be maintained for considerable periods and tolerated by the patient, clinical use of sulphonamides in the treatment of blastomycosis is necessarily limited to local applications to cutaneous lesions.

Noojin and Callaway (1944) subsequently studied the action of sulphonamides upon the growth of *Sporotrichum schenckii*. The sulphonamides used were the same as in their previous work with *Blastomyces dermatitidis*, but with the addition of two new derivatives, sulphapyrazine and sodium sulphapyrazine. Sabouraud's agar was used, containing the drugs in concentrations of 50-1,000 mg. per cent.

Senturia and Wolf (1945) studied the action of sulphonamides applied to cultures of fungi isolated from cases of otomycosis. The growth of *Aspergillus fumigatus, A. niger, A. glaucus, A. sydowi,* and *Mucor corymbijer* upon plates of Sabouraud's agar was markedly inhibited by Sulphanilamide, when the quantity of the drug approximated 20-30 mg. per culture. Similar amounts of sulphathiazole, sulphadiazine, sulphaguanidine or sulphamerazine were ineffective.

Candida albicans was unaffected by any of these sulphonamides, confirming the finding of Lewis and Hopper (1941). It is interesting to compare these results with those of Fourneau *et* al. *(l.c.). Aspergillus niger* and *A. fumigatus* were employed as test organisms in each case, with identical results upon treatment with sulphanilamide.

Keeney, Ajello and Lankford (1944) have investigated the action of sodium salts of sulphathiazole, sulphadiazine and sulphamerazine upon many of the common pathogenic fungi and *Actinomyces bovis,* taking into account the action of sulphonamide antagonists.

The sodium salts of the sulphonamides were selected for study because of their greater solubility and theoretically greater promise of successful systemic use. The fungi studied were *Trichophyton mentagrophytes, T. rubrum, Epidermophyton floccosum, Microsporum audouini, M. jelineum, Candida albicans, Cryptococcus neoformans, Blastomyces dermatitidis, Sporotrichum schenckii, Coccidioides immitis, Phialophora pedrosoi, Histoplasma capsulatum* and *Actinomyces bovis.* All organisms were grown in dextrose broth except *A. bovis,* for which a thioglycollate medium was used.

The quantity of each of the sulphonamides necessary to neutralise their respective antagonists present in the media was determined by adding sulphonamides in different quantities to the media, inoculating with *Staphylococcus aureus,* and determining, for each medium and each sulphonamide, the lowest sulphonamide concentration which would just permit apparent growth. These concentrations varied from 100 to 800 mg. per cent.

Tubes of dextrose broth containing 50-1,000 mg. per cent of the various sulphonamides, in excess of the amounts required to neutralise the sulphonamide antagonists, were then inoculated with fungi, incubated, and examined after three weeks

Little or no significant fungistatic activity was obtained except with *Microsporum audouini, C. neoformans, P. pedrosoi* and *H. capsulatum.* Inhibition of growth of *M. audouini* and *C. neoformans* largely disappeared, however, when the organisms were grown on agar media containing whole blood or serum. The clinical trial of sodium sulphamerazine in the treatment of chromoblastomycosis and of sodium sulphathiazole in the treatment of histoplasmosis was recommended as a result of these experiments.

It was further shown by Keeney, Ajello and Lankford that these sulphonamides were fungicidal when a sufficiently high concentration was employed. With *Actinomyces bovis* in thioglycollate media, concentrations of 500 or 1,000 mg. per cent of sodium sulphathiazole, sodium sulphadiazine and sodium sulphamerazine produced both fungistatic and fungicidal effects.

Wolf (1945) studied the action of sulphanilamide, sulphathiazole, sulphadiazine and sulphaguanidine upon the common pathogenic fungi. These were *Candida albicans, C. krusei, C. tropicalis, C. parakrusei, C. pseudotropicalis, Cryptococcus neoformans, Epidermophyton floecosum, Trichophyton mentagrophytes, T. rubrum, Monosporium apiospermum, Hormodendrum pedrosoi, H. compactum, Phialophora verrucosa, Microsporum canis, M. gypseum* and *Sporotrichum schenckii.*

Sulphanilamide was very markedly fungistatic to all of the dermatophytes *(Trichophyton mentagrophytes, T. rubrum, Epidermophyton floccosum, Microsporum canis* M. *gypseum)* and to *Sporotrichum schenckii.* Sulphanilamide had no pronounced activity against any of the remaining fungi tested, and sulphathiazole, sulphadiazine and sulphaguanidine were not effective against any of the pathogens.

The basis for the great anti-fungal activity *in vitro* of sulphanilamide, noted by Fourneau *et al.* (1936), Lewis and Hopper (1941), Dimond (1941), Dimond and Thompson (1942), Noojin and Calla-way (1943, 1944), Senturia and Wolf (1945), and Wolf (1945) has not been adequately explained, but it seems evident that sulphanilamide has a greater anti-mycotic activity, against many of the pathogenic fungi, than does any other sulphonamide.

One possible theoretical criticism may be made of studies which attempt to compare the action of the sulphonamides *in vitro* with that of their sodium salts. Owing to the high pH of solutions of the sodium derivatives of the sulphonamides (8.2 in the media used by Lewis and Hopper, 1941) it may well be that a portion of the fungistatic activity of these compounds is a direct effect of pH rather than a specific chemical action upon the fungus by the compound concerned. This objection was recognised both by Lewis and Hopper (1941) and Noojin and Callaway (1944). It is probable that studies which now consider optimal media and optimal temperatures for the growth of pathogenic fungi, and the question of sulphonamide antagonists present in the media, must also consider optimal pH, when drugs of differing pH are to be compared *in vitro.*

No data are yet available concerning the action of Marf anil (homo-sulphanilamide) upon fungi. The study of the action of this drug, as well as that of other newer sulphonamide derivatives, upon fungi must await future laboratory experimentation.

Animal Experimental Studies: From the time of discovery of a new drug until it is placed into general use, there are usually several important steps which intervene including: *(1)* "test tube" experiments to determine activity *in vitro, (2)* more stringent *in vitro* studies, in which blood or serum is added to the system, to determine the degree of interference under these conditions, *(3)* animal experimental studies, to test activity of the compound *in vivo* and to determine its toxicity to body tissues, and *(4)* clinical trials, in the treatment of selected human cases.

It has been mentioned that Fourneau *et al.* visualised the possibility of using the sulphonamides in the therapy of experimental fungus diseases in animals. In view of the numerous clinical trials which have been made of sulphonamides in cases of mycotic infections, it is surprising that more attention has not been devoted to animal experimental studies.

Dosa (1941) investigated the effect of a sulphonamide upon experimental torulosis, blastomycosis and actinomycosis in animals. Rats were chosen as the experimental animal, and the sulphonamide used was the proprietary preparation "Neosanamid" (p-acetylaminobenzol-sulphon) iminobenzol-sulphonamid), given orally.

Upon inoculation, the animals were divided into four groups. Group I included animals to which the drug was administered immediately upon inoculation, and thereafter in decreasing dosages for 12 days. The animals in group II were treated similarly, but the dosage was only one half of that for group I. The drug was administered to group III animals only upon the appearance of disease symptoms. The control animals, group IV, were not treated.

Mortality in the control group was 100 per cent, 10-12 days being the duration of life in animals with torulosis or blastomycosis, and 21-25 days in the case of animals inoculated with *Actinomyces*. Postmortem pathologic examination disclosed numerous lesions and the characteristic organisms within the tissues. No instance of sickness due to administration of the sulphonamide was noted, and the fact that the symptomatic animals recovered in large measure upon treatment clearly indicates possible success in the use of sulphonamide chemo therapy in the systemic mycoses.

Jones and Klinck (1945) have recently reported the results of studies upon experimental torulosis in mice, in which sulphadiazine was found to be ineffective.

Clinical Results: A detailed discussion of the treatment of mycotic infections by use of sulphonamides is beyond the scope of this work, and may be found in medical and clinical texts. A few general remarks concerning the relation between the results of laboratory research on the pathogenic fungi and their practical application may not, however, be inappropriate. Likewise, it is here our intention to refer briefly to the more important clinical papers, and to state an opinion as to the probable value of sulphonamides in each of the various mycotic diseases.

The basis of sulphonamide treatment of fungus diseases has been largely empirical. In many instances, clinical results have antedated *in vitro* studies, animal experimental studies have been largely non-existent, and in a few cases, reports of successful clinical cures are not in agreement with the results of laboratory *in vitro* studies.

For some diseases, the number of cases upon which reports of cures are based appears to be too small to justify any definite conclusion. Reported failures, in some instances, may be due to the fact that sulphonamide therapy was instituted only *in extremis,* when more accepted means of therapy had failed.

Of far greater importance, however, is the fact that successful therapy of a fungus disease implies far more than merely killing the causative organism. The pathologic changes which occur in the tissues are largely manifestations of reactions of the tissues to the fungus, and generally the question of physical damage due to the mere presence of a foreign organism is of relatively minor importance. Hence the emphasis on autogenous vaccines and other measures whose primary purpose is one of desensitisation.

Further, it is to be recognised that successful therapy usually is not attained through any single agent, no matter what its properties, except in conjunction with surgery, desensitisation procedures, measures for control of secondary infection, and others when indicated.

Spectacular results have been obtained by means of

sulphonamides in the treatment of actinomycosis. Abdominal, thoracic and cervico-facial types of this disease have been rapidly and completely cured.

Sulphanilamide or sulphapyridine has been the drug most often employed. Following the pioneer work of Walker (1938), a great number of case reports have appeared. These include those of Miller and Fell (1939), Hall (1939), Darling and Eckhoff (1940), Ogilvie (1940), Christopher and Karabin (1940), Wilkinson (1941), Dobson, Holman and Cutting (1941), Mitchell (1942), Mccloy (1943), Lyons, Owen and Ayers (1943), Ladd and Bill (1943), Hollenbeck and Turnoff (1943), Billington (1944), and Benbow, Smith and Crimson (1944). The results obtained upon sulphonamide therapy of actinomycosis, when employed in conjunction with surgical measures, have been sufficiently good to justify the use of sulphonamides as a standard method of treatment of this infection.

The successful treatment of a case of maduramycosis (Madura foot), in which the causative organism was an actinomycete, by the oral administration of sulphanilamide was reported by Dixon (1941).

Reeves, Butt and Hammack (1941) and Marshall and Teed (1942) reported recovery, following sulphonamide therapy, in cases of torulosis (cryptococcosis, European blastomycosis) of the meninges and central nervous system, in which the prognosis is always extremely poor.

According to Navarro-Martin (1940), sporotrichosis also responds to sulphonamide therapy.

Schroeder (1940) and Albert (1943) report the cure of blastomycosis as a result of sulphonamide therapy. Other instances of successful treatment of blastomycosis are to be found in some of the more obscure South American journals. In the case of South American blastomycosis caused by *Blastomyces brasiliensis*, Iriarte (1945) reports successful therapy in two cases, in one of which sulphapyridine was used, and sulphathiazole in the other instance.

Pfalzgraf (1941) employed a prontosil solution in the treatment of epidermophytosis. The results of more recent large scale tests, however, would appear to show that sulphonamides may do a great deal of harm in "athletes' foot" and allied conditions, and are definitely contraindicated.

Van bree (1941) reported marked improvement in a case of generalised moniliasis, an observation which is difficult to understand in the light of *in vitro* studies. Wessler and Browne (1945) obtained no beneficial results in a case of *Candida albicans* infection of the blood stream upon sulphonamide treatment.

Various bronchopulmonary mycoses, in which *Blastomyces, Aspergillus, Penicillium, Mucor, Sporotrichum* or *Actinomyces* was involved, respond to sulphonamide therapy in the experience of De Almeida and Lacaz (1942).

Moore and Jorstad (1943) report the failure of sulphonamides in cases of histoplasmosis, while Alonso and Freijo (1944) obtained recovery following sulphadiazine therapy.

Goldstein and Mcdonald (1944) obtained no beneficial results with sulphonamides in a case of primary pulmonary coccidioidomycosis.

Otomycosis or external otitis, an infection of the external ear canal caused by various fungi including species of *Aspergillus* and *Mucor,* while by no means a serious infection in comparison with other mycoses, has been a problem of some importance to military and naval personnel in tropical and subtropical areas. Senturia (1944) of the Army Air Forces reported success in treatment of this disorder by local applications of sulphanilamide, sulphathiazole and zinc peroxide. This information was made available to naval medical officers through an anonymous article in the Burned News Letter (1944). More recently Simon (1945) of the Navy has experienced good results with a combination of systemic sulphathiazole and local sulphanilamide therapy.

In conclusion, it would appear that laboratory studies of

the pathogenic fungi have played an important part in the development of methods of therapy, which in some fungus diseases are quite successful, but in other cases show the need for further fundamental laboratory and clinical experiments.

Antibiotic Agents

The finding that certain metabolic products of microorganisms are inhibitory to the growth of pathogenic fungi is by no means a new one. The phenomenon of antagonism between fungi and other organisms has been known for many years, and the pathogenic fungi form no exception in this regard.

As examples of early work on microorganisms which have an antagonistic action upon the pathogenic fungi, we may consider the work of Weidman (1927) and Chambers and Weidman (1928).

Weidman (1927) states, "Several years ago I noted that those portions of a certain ringworm fungus colony which were adjacent to a bacterial colony did not devélop as luxuriantly as the more remote parts; the result was a lopsided fungus colony. A bacterium was present which restrained the growth of the fungus, but did not destroy it. ... I tested out a number of yeast species obtained from normal feet, and finally secured one which inhibited the growth of *Epidermophyton cruris* when mixed with it and planted on culture mediums."

Chambers and Weidman (1928) isolated a hemolytic dull green organism identified as *Bacillus subtilis* from the skin of the normal toes of a large number of individuals. When inoculated upon suitable media together with the spores of pathogenic fungi, it was found that this organism completely inhibited, for a period of 2-3 months, the growth of several dermatophytes. *Sporotrichum schenckii* was similarly inhibited.

The antibiotic agents are substances formed as products of the metabolism of microorganisms, differing greatly from one another in origin, and in chemical and biological properties.

Some, like penicillin, clavacin and gliotoxin, are produced by fungi; others, such as streptothricin and streptomycin, are produced by actinomycetes; still others, including tyrothricin and pyocyanine, are formed by bacteria. In the case of certain antibiotic agents, the conditions required for their production and even the chemical composition of the antibiotic principle are known, so that they can be exactly characterised as definite chemical compounds.

In other instances, the preparations in use are merely concentrates of high activity, admixed in unknown amounts with extraneous substances, from which isolation of the active compound in a pure state has not yet been accomplished.

The necessity of providing a "yardstick" of the activity of such preparations has resulted in the establishment of "Oxford units," "streptothricin units," and the like, by which the activity of a particular antibiotic agent is defined in terms of its effect upon a particular test organism, such as *Staphylococcus aureus*. However necessary the system of unitage may be, it suffers a serious disadvantage in its inability to permit direct comparison of the activity of two different antibiotic agents upon any organism other than the one upon which the standards are based, a fact which must be constantly borne in mind by the reader. This is a consequence of the specific selective action of antibiotic agents.

According to Reilly, Schatz and Waksman (1945) the effects of antibiotic substances upon fungi may include modifications in the morphology of the organism, changes in its physiology, pigment production or rate of growth, or complete destruction of the fungus.

Finally, the development of many of the antibiotic agents has been so recent that very little clinical information, dealing with the treatment of mycotic diseases, is yet available.

Penicillin: The best known of the antibiotic agents, penicillin is produced by *Penicillium notatum*, *P. chrysogenum*, and certain other fungi. The discovery of penicillin by Fleming, and the successes which have attended its use in bacterial infections, originally by Florey and later by many other workers, are so well known as not to require comment here.

Shortly after its anti-bacterial properties became generally known, experiments testing the action of penicillin against fungi were performed.

Abraham *et al.* (1941), using a preparation with an activity of 40-50 Oxford units per milligram, noted that complete inhibition of *Actinomyces bovis* occurred at a dilution of 1:1,000,000. Partial inhibition of this organism was obtained with penicillin 1:2,000,000 but none at 1:4,000,000. *A. bovis* is, therefore, very susceptible to penicillin. This observation was confirmed by Fisher (1943) in experiments using crude penicillin, before the purified material became available generally and in large quantities.

Hobby, Meyer and Chaffee (1942) reported that *Cryptococcus hominis* was penicillin sensitive, but that the growth of *Candida (Monilia) albicans, C. krusei* and *C. Candida* was not affected. The result of these workers with respect to *C. hominis* has not been verified by the subsequent results of others.

Foster and Woodruff (1943) found that *Candida albicans, C. stellatoidea, Trichophyton inter digitale, T. gypseum, T. schoenleinii, Epidermophyton floccosum, Microsporum felineum* and *Blastomyces dermatitidis* were unaffected by penicillin, even in dosages as high as 30 Oxford units per cc.

The most extensive work, however, concerning the action of penicillin upon pathogenic fungi is that of Keeney, Ajello and Lank-Ford (1944). The organisms studied were *Trichophyton mentagrophytes, T. rubrum, Epidermophyton floccosum, Microsporum audouini, M. felineum, Candida albicans, Blastomyces dermatitidis, Sporotrichum schenckii, Cryptococcus neoformans, Coccidioides immitis, Phialophora pedrosoi* and *Histoplasma capsulatum.* The organisms were grown in dextrose broth containing 0.1, 0.5, 1, and 10 Oxford units of penicillin per cc. Because penicillin is unstable under some conditions, fresh penicillin was added every 48 hours, so that the original concentrations were maintained for the duration of the 14 day period of observation. No inhibition of growth of any of these fungi was noted.

Keeney *et al.* (1944) also studied the action of penicillin upon *Actinomyces bovis,* grown in thioglycollate broth containing 0.005-10 Oxford units per cc. This organism was inhibited and apparently killed by a concentration of 0.01 Oxford unit per cc., thus confirming the results of previous workers as to the susceptibility of *A. bovis* to penicillin.

The clinical use of penicillin in cases of actinomycosis, especially when employed as an adjunct to surgical intervention, has met with some success (Keeney *et al.*, 1944; Shulman, 1944) but it is perhaps too early as yet to form a definite opinion as to its value. The failure of attempted penicillin therapy in the treatment of coccidioidomycosis has been reported by Michael, Mclaughlin and Cenac (1944).

Streptothricin and Streptomycin: It is perhaps desirable to consider together the results obtained with streptothricin and streptomycin, two of the antibiotic agents derived from actinomycetes. Streptothricin, produced by *Actinomyces lavendulae,* and streptomycin, from *A. griseus,* were isolated and studied by Waksman and his co-workers.

The action of streptothricin upon a considerable number of yeasts, saprophytic fungi and pathogenic fungi was studied by Foster and Woodruff (1943). The "streptothricin unit" employs *Bacillus subtilis* as the test organism.

Thus, streptothricin has considerable activity against pathogenic fungi. *Blastomyces dermatitidis* and *Trichophyton schoenleinii* were inhibited by streptothricin in concentrations comparing favourably with that required for the inhibition of *Escherichia coli.* The remaining pathogenic fungi tested were relatively insensitive to streptothricin, some being unaffected by 100 times the concentration required to inhibit *E. coli.*

Streptomycin, on the other hand, is almost completely

inactive against fungi. Robinson, Smith and Graessle (1944) compared the action of streptothricin and streptomycin upon a number of pathogenic fungi. The "streptothricin unit" was defined as the quantity, in 1 cc. of nutrient broth or agar, which will just inhibit growth of *E. coli.* The streptomycin preparation used by Robinson *et al.* (1944) had a potency of 27-100 units per milligram.

Although streptomycin has little effect upon pathogenic fungi generally, it would appear from the work of Schatz and Waksman (1944) that certain actinomycetes are highly sensitive to this compound. *Actinomyces (Streptomyces) albus,* isolated from a case of otomycosis, was completely inhibited for three days by 0:4-1.25 units of streptomycin per cc. *Actinomyces (Nocardia) asteroides* and *A. (N.)gypsoides* however, were much more resistant to streptomycin. No data are yet at hand concerning the action of streptomycin on *Actinomyces boms.*

Reilly, Schatz and Waksman (1945) have compared the activity of several antibiotic agents upon *Candida albicans Trichophyton mentagrophytes,* and *Cryptococcus neoformans.* Streptomycin was less toxic to these fungi than gliotoxin, actinomycin, clavacin, fumigacin, streptothricin, or chaetomin. Streptothricin, however, is far more active against fungi, and shows promise clinically.

Gliotoxin: The antagonism between *Trichoderma* and soil microorganisms has long been known to microbiologists. The potent antibiotic agent produced by *Gliocladium fimbriatum,* strains of *Trichoderma* and certain other fungi was isolated by Weindling (1941), and is now known as gliotoxin.

Weindling reported that the fungicidal activity of this substance to *Rhisoctonia solani* was intermediate between that of copper sulphate and that of mercuric chloride. Thus from the very beginning gliotoxin was known as an anti-fungal compound, differing in this respect from other antibiotic agents effective against bacteria, whose anti-fungal properties were regarded as incidental or of definitely secondary importance.

Methods for the large-scale production of gliotoxin, information as to its physical and chemical properties, and the first data as to its action upon pathogenic fungi were provided by Johnson, Bruce and Butcher (1943). Tests showed that 15.6 mg. of gliotoxin per cc. inhibited the growth of *Epidermophyton* in broth culture, and 3.9 mg. per cc. inhibited the growth of this organism in an agar medium. *Blastomyces dermatitidis* failed to grow in a medium containing gliotoxin 1:8,000.

Herrick (1945) has determined the fungistatic and fungicidal effects of gliotoxin upon *Trichophyton gypseum.* Growth of this organism on Sabouraud's agar was completely inhibited by 0.001 per cent gliotoxin, but no inhibition occurred at 0.0001 per cent. Spores suspended in gliotoxin solutions, removed by centrifugation, washed and transferred to agar media were unharmed after 24 hours exposure to 0.001 per cent gliotoxin. Two hours exposure to a 0.01 per cent solution, however, was lethal to more than 99 per cent of the spores of *T. gypseum.*

Reilly, Schatz and Waksman (1945) have studied the fungistatic activity of gliotoxin, in comparison with other antibiotic agents from fungi and actinomycetes, upon *Candida albicans, Cryptococcus neoformans,* and two strains of *Trichophyton mentagrophytes.*

Clavacin: Clavacin is an antibiotic agent derived from *Aspergillus clavatus* by Waksman, Horning and Spencer (1943). It has been isolated in a pure state, and its structure determined as that of an unsaturated ketone. The empirical formula of clavacin is $C_7H_6O_4$.

Waksman, Horning and Spencer (1943) noted that clavacin has rather limited fungistatic properties. The fungi with which clavacin was tested by them are not named, but it is noted that they "appeared to overcome the effect in a short time, either by inactivating the clavacin or by adjusting themselves to its action." Katzman *et al.* (1944) reported that clavacin had considerable activity against *Rhizopus nigricans, Saccharomyces cerevisiae, Candida albicans* and *Sporotrichum schenckii.*

The fungistatic and fungicidal properties of clavacin upon *Candida albicans, Oidium asteroides* and *Trichophyton gypseum* have been recently studied by Herrick (1945). The clavacin was produced from *A. clavatus* according to the method of Waksman, Horning and Spencer; in addition, the product of commercial drug houses was used. *T. gypseum* proved to be more susceptible to clavacin than either *Candida* or *Oidium,* a 0.001 *per cent* solution causing slight inhibition of growth, and 0.02 or 0.04 per cent concentration checking growth of *T. gypseum* completely. These concentrations had only a slight inhibitory effect on *C. albicans* and *O. asteroides* 0.1 per cent clavacin did not kill *C. albicans* or *O. asteroides* even after 24 hours, but was lethal to *T. gypseum* in 2-5 hours. 0.5 and per cent solutions were fungicidal for all organisms.

Reilly, Schatz and Waksman (1945) noted that clavacin was an effective fungistatic agent against *Candida albicans, Cryptococcus neoformans* and *Trichophyton mentagrophytes,* inferior only to gliotoxin and actinomycin, but they see little promise for the clinical use of this compound because of its high toxicity to animals.

Geiger and Conn (1945) determined that the antibiotic activity of clavacin was due to the unsaturated group in the molecule. Synthetic unsaturated ketones such as acrylophenone were found to resemble clavacin in fungistatic properties. These workers presented evidence supporting the hypothesis that the antibiotic activity of clavacin is due to its reaction with sulphydryl groups of either enzyme systems of the microorganisms or some of their essential metabolites.

Pyocyanine and Hemipyocyanine: *Bacillus pyocyaneus,* more properly known as *Pseudomonas aeruginosa,* is an organism from which a number of substances having antibiotic activity have been isolated. Among these are pyocyanine and hemipyocyanine (alphahydroxy phenazine), which have been known for many years.

Stokes, Peck and Woodward (1942) have studied the

growth-inhibitory action of pyocyanine and hemipyocyanine upon yeasts and pathogenic fungi. The hemipyocyanine used by them was prepared by synthetic methods, while the pyocyanine was isolated as the per-chlorate from cultures of *Ps. aeruginosa.* The results of this study upon *Candida albicans, Trichophyton (Achorion) schoenleinii, Microsporum gypseum* and *Trichophyton gypseum,* grown on Sabouraud's agar.

Thus, pyocyanine has relatively little action upon the growth of pathogenic fungi. Hemipyocyanine, however, in dilutions as high as 1:20,000 completely inhibited the growth of all the fungi tested, and *T. schoenleinii* was inhibited by hemipyocyanine 1:60,000. Essentially similar results were obtained with a number of yeast species.

Hemipyocyanine has a fungistatic potency superior to most of the commonly used disinfectants, including mercuric chloride, phenol, cresol, salicylic acid and mercurochrome, but inferior to crystal or gentian violet. It cannot be used intravenously, according to Stokes, Peck and Woodward (1942) but possibly could be employed in topical applications in the treatment of superficial fungus infections.

Tyrothricin: Tyrothricin, an antibiotic substance isolated from the spore-forming soil organism *Bacillus brevis,* was discovered by Dubos in 1939. Chemical studies soon disclosed that tyrothricin was not a single substance, but a mixture of two materials known as tyrocidine and gramicidin, both of which are polypeptides.

It has appeared undesirable to separate these two fractions for clinical use because of their complementary spectra of activity. Tyrothricin produces hemolysis and hence cannot be employed parenterally. Stokes, Peck and Woodward (1942) have studied the action of tyrothricin upon the growth of *Trichophyton (Achorion) schoenleinii, Microsporum gypseum, Trichophyton gypseum* and *Candida albicans* on Sabouraud's agar.

Tyrothricin in dilutions of 1:5,000-1:20,000 completely inhibited the growth of all of the pathogenic fungi tested.

The fungistatic activity of tyrothricin is thus approximately equal to that of commonly used disinfectants, and Stokes, Peck and Woodward (1942) suggest that it may be of value in the treatment of superficial fungus infections.

Miscellaneous Antibiotics

Actinomycin is an antibiotic agent produced by *Actinomyces antibioticus.* Its activity against *Candida albicans, Cryptococcus neoformans* and *Trichophyton mentagrophytes* was found by Reilly, Schatz and Waksman (1945) to be very high, inferior only to gliotoxin. Because actinomycin is highly toxic to animals, however, it probably cannot be used clinically.

Fumigacin, from *Aspergillus fumigatus,* was found by Waksman, Horning and Spencer (1943) to possess a limited activity against fungi. Reilly, Schatz and Waksman (1945) found that the activity of fumigacin against *Candida albicans, Cryptococcus neoformans,* and *Trichophyton mentagrophytes* was generally less than that of gliotoxin, actinomycin or clavacin, and do not regard fumigacin as having activity sufficient for clinical use.

Chaetomin, from *Chaetomium cochlioides,* was found by Reilly *et al.* (1945) to be almost completely inactive against the pathogens *Candida albicans, Cryptococcus neoformans* and *Trichophyton mentagrophytes.*

The activity of penicillic acid, an antibiotic substance produced by *Penicillium puberulum* and *P. cyclopium*, upon a number of saprophytic fungi has been determined by Geiger and Conn (1945) as being much less than that of clavacin.

Subtilin, an antibiotic agent from *Bacillus subtilis,* was found by Salle and Jann (1945) to inhibit the growth of *Actinomyces asteroides, A. pelletieri,* and *Nocardia mexicana* when present in Sabour-aud's agar in the proportion 1:1000. *Candida albicans, Cryptococcus neoformans, Sporotrichum schenckii* and *Trichophyton gypseum*, on the other hand, were not affected by this quantity of subtilin.

Thus, during the very brief period in which attention has been devoted to the action of antibiotic agents upon the pathogenic fungi, substances have been found which are effective *in vitro* against a number of pathogens.

Actinomyces bovis is inhibited by penicillin, *Candida albicans* by gliotoxin, clavacin, actinomycin and hemipyocyanine, *Cryptococcus neoformans* by gliotoxin, clavacin and actinomycin, *Blastomyces dermatitidis* by gliotoxin and streptothricin, *Sporotrichum schenckii* by clavacin, *Trichophyton (Achorion) schoenleinii* by streptothricin, tyrothricin and hemipyocyanine, *Trichophyton mentagrophytes (gypseum)* by gliotoxin, clavacin, actinomycin, tyrothricin and hemipyocyanine, and *Microsporum gypseum* by tyrothricin and hemipyocyanine. The development of new antibiotic agents, and the possibilities of practical use of findings such as these must be left to the future.

4

Various Contents

Parasitic Activity

Influence of Soil Organic Content: Diseases reported to be encouraged by the application of organic material to the soil.

The species of *Rosellinia* are well-known saprophytes on leaf mould and other organic detritus lying in and on the soil; such organic material provides these fungi with a substrate not only for subsistence but also for spread, and at the same time with a food-base from which their attack on such plants as they may chance to encounter can most expeditiously develop. Nevertheless, the parasitism of these fungi is not confined to this type of environment; Waterston (1941) has described comparatively their parasitic activity in the two contrasting situations in which they cause patches of disease in tropical plantation crops, stating: "The areas affected with *Rosellinia* vary between two environmental extremes. The first consists of regions of unrestricted spread, where conditions of high rainfall and lack of sunshine due to intense shade encourage humus formation. Under these conditions, the fungus attack is in the nature of an advancing sheet. The fungus in this phase may not only derive its food saprophytically from decaying leaf mould but appears to behave as a dangerous though incidental parasite to the stems and roots of trees and shrubs encountered during the course of its advance.

These conditions exist in some of the wetter areas as in Dominica where root diseases caused by *Rosellinia* spp. may act as limiting factors to lime production. In cases like these, the conditions which best suit the host plant, best suit the fungus also, with the result recorded by Fawcett (1915), No Well (1916) and Stell, (1929) that the finest trees are most liable to attack and are most frequently killed.

"The second region which may be recognised is one of slow and restricted spread, found where rainfall is lighter, with little accumulation of humus or surface litter and where shade trees may be absent (South, 1913, Stockdale, 1908, and Briton-jones, 1934). Under these conditions, the spread of the fungus is confined to root contacts or from fragments of wood or root stumps of susceptible hosts which have been aptly described by Gadd (1936) as 'food-bases'."

The higher incidence of infection by some cereal smuts in soils well supplied with organic matter seems to be due to a direct effect of the organic material on spore germination. Thus Rabien (1927) found that spore germination of the *Tilletia* spp. and incidence of bunt in wheat were both favoured by a high humus content of the soil.

Noble (1924) reported that spore germination in *Urocystis tritici* was stimulated by the roots of non-susceptible plants (peas, beans, and rye), and by traces of certain volatile substances, such as benzaldehyde, salicylaldehyde, butyric acid, and acetone. Platz *et al.* (1927) found that crushed plant tissue encouraged the germination of spores of *Ustilago zeae* in closed chambers, and that this stimulating effect was due to the carbon dioxide given off; an atmospheric concentration of 15 per cent carbon dioxide was found optimum for spore germination.

The action of organic matter in encouraging the parasitic

activity of *Verticillum alboatrum* seems likely to be an indirect one, operating through the physiology of the host plant; Roberts (1943) has found that the percentage of established infections in tomato increases with the amount of nitrogen supplied to the plant by the soil.

Much attention has been paid of recent years to the possibilities of "biological control" of soilborne diseases by application of fresh organic material to the soil. It is assumed that increased activity of the saprophytes developing on the fresh organic material must occur at the expense of, and to the detriment of, the plant parasites.

The assimilation, respiration and excretion of the saprophytes must compete with those of the parasites; metabolic products which are toxic to other microorganisms may be produced by some saprophytes, and, in addition, the development of a high population of microorganisms may encourage the parasitism of certain microorganisms upon others. Historical reviews of the subject of microbiological antagonism and biological control have been published by Fawcett (1931), Garrett (19346, 1939), Waksman (1937,1941), Porter and Carter (1938), Weindling (1938), and Garrard and Lochhead (1938); discussion here will be confined to the three most important of the diseases.

Control of potato scab by green manuring with grass cuttings was first reported by Millard (1921). Sanford (1926) suggested that such control was brought about through the antagonism to *A. scabies* of saprophytic species of bacteria developing on the decomposing green manure (rye grass). Further experiments by Millard and Taylor (1927) showed that the development of scab on potatoes grown in sterilised soil inoculated with *A. scabies* could be reduced by simultaneous inoculation of the soil with *A. praecox,* an obligate saprophyte of more vigorous vegetative growth than *A. scabies*. By increasing the proportion of *A. praecox* to *A. scabies* in the inoculum, the degree of scabbing on the test potatoes was reduced from 100 per cent to nil.

The sterilised soil evidently provided sufficient food material for a good development of *A. praecox,* and only a rather small increase of control was given by the admixture of grass cuttings, sterilised along with the soil.

Millard and Taylor concluded that, whilst their tests were concerned only with the biological control of *A. scabies* by *A. praecox,* soil bacteria might exercise a similar controlling effect, as suggested by Sanford. Experiments by Goss (1937) convincingly demonstrated the controlling effect of the general soil microflora upon the development of scab, though he failed to secure any result with an inoculation of *A. praecox* alone.

Field control of the take-all disease of wheat in Kansas by the application of such organic materials as chicken and horse manure, green alfalfa, boiled oats and barley kernels, and potato flour was very briefly reported by Fellows (1929).

Garrett (1936, 1937, 1938 and 1940) has attempted to analyse this controlling effect of organic materials upon the take-all disease; as a result of his study of the spread of *Ophiobolus graminis* along wheat roots under different soil conditions, he suggested that addition of organic material to the soil would depress the parasitic activity of *O. graminis* insofar as it increased the concentration of carbon dioxide in the microatmosphere around the roots. He obtained striking reductions in the survival period of *O. graminis* in infected wheat straw by incorporating fresh organic material, low in nitrogen content, with the soil. Working on the same disease, Stumbo *et al.* (1942) and Clark (1942) have recently emphasised the beneficial manurial effect of such organic supplements in increasing the resistance to attack of the wheat plant.

The efficacy of organic manure for the control of root rot in irrigated cotton under continuous cultivation in Arizona has been established by King *et al.* (1934a). These authors supported their hypothesis of microbiological antagonism as the mechanism of control by an extensive study of the soil microflora in the treated and control plots by the Rossi-Cholodny slide technique (Eaton and King, 1934;, and by soil respiration determinations. They were able to demonstrate in this way that whereas the development of saprophytic organisms was most profuse in the slides buried in the manured plots, the mycelium of *Phymatotrichum omnivorum* was most abundant on the slides in the unmanured plots.

Further light has been thrown upon this controlling effect of organic amendments by the laboratory experiments of Mitchell *et al.* (1941). Quart jars of clay soil, untreated and with additions of superphosphate, 3 per cent farmyard manure, and 3 per cent chopped sorghum, respectively, were incubated at "optimum" moisture content for periods of 0, 5, 10, 15, 20 and 30 days, and then inoculated with agar blocks taken from a culture of *P. omnivorum,* or with infected roots.

Mycelium of *P. omnivorum* quickly developed over the soil surfaces in jars filled with untreated soil or with soil + superphosphate; the mycelium showed little apparent disintegration during the 16-day observation period. In the soils receiving fresh organic material, however, mycelial growth either failed to occur, or else was followed by disintegration. In the soil -f- sorghum, growth of *P. omnivorum* was inhibited for the full 30 days after addition of the organic material to the soil; in the soil + farmyard manure, growth of *P. omnivorum* from the inoculum block occurred in those series that had been incubated for 20 days or more following the original addition ci the organic material.

Ability of *P. omnivorum* to grow out from the inoculum seemed to be inversely correlated with bacterial numbers in the different soil series. As a further indication of the susceptibility of *P. omnivorum* to the competition of other microorganisms, Mitchell *et al.* reported that the fungus failed to grow upon non-sterile cotton roots in culture flasks, even when the roots were quite fresh; growth was also inhibited if sterile cotton roots were inoculated with a pinch of field soil. The addition of 3 per cent organic material to the soil also resulted in the rapid disappearance of a high proportion of *P. omnivorum* sclerotia; Mitchell *et al.* demonstrated that whilst this effect was due in part to attack and decomposition of the sclerotia by other soil microorganisms, it was chiefly brought about by spontaneous germination of the sclerotia, under the stimulus of the organic amendments. In further experiments along the same lines reported by Clark (1942), the antibiotic effect of the organic materials upon the viability of the sclerotia was more important than the direct stimulus given by such materials to sclerotial germination.

According to Rea (1939), experiments on the use of organic manures for the control of cotton root rot under dry-land farming conditions in the Blackland region of Texas have not so far given encouraging results; the success of King and his collaborators in Arizona may be attributed in part to the fact that their experiments were carried out on irrigated cotton, and in part to the special methods of manure placement and timing that King (1937) has recommended.

Concentration of Plant Nutrients: Generous application of artificial fertilizers is undoubtedly an effective way of reducing the proportion of crop lost through the agency of certain root diseases. According to Garrett (1942), the proportional loss occasioned by the take-all disease of wheat is greatly aggravated by poverty of the soil in one or more of the three major plant nutrients.

The common association between serious outbreaks of this disease and phosphate deficiency of the soil is well known from accounts by investigators in Australia, where the soils are apt to be very deficient in this nutrient. A generous supply of plant nutrients probably increases plant resistance chiefly by promoting a more rapid production of new crown roots to replace those destroyed by the disease, though an additional direct effect upon the resistance to infection of individual crown roots cannot be precluded.

Since the appearance of this review by Garrett, Stumbo *et al.* (1942) have demonstrated that development of the take-all disease in potted wheat plants grown in a naturally infected Kansas soil might be frustrated by adequate applications of nitrogenous and phosphatic artificial fertilizers. They attributed the control of take-all obtained by Fellows (1929) with certain organic materials largely to the plant nutrients, especially nitrates, set free in the soil by decomposition of these materials; such control was experimentally demonstrated by Clark (1942), whose results revealed a correlation between nitrate nitrogen set free in the soil through decomposition of the organic material, and the degree of disease control obtained.

Especially striking was Clark's demonstration of the effect of volume of soil per plant upon resistance of wheat to take-all; plants grown in 1 kg. of an infected Kansas soil suffered severely from the disease, whereas others grown in 9 kg. of the same soil escaped serious injury. A paradoxical situation, which serves further to emphasise the paramount effect of plant nutrients upon the development of take-all in potted plants, has been reported from Australia by Angell (1943).

A second crop of wheat grown in drums of soil inoculated with *O. graminis* in the previous season was perfectly healthy, whereas that grown in the non-inoculated drums (which served as controls in the previous season) was badly affected by take-all. This reversal of results in the second season was attributed on the one hand to depletion of nutrients under the heavy crop grown in the non-inoculated drums in the first season, and, on the other, to their conservation under the diseased crop grown in the inoculated drums.

Conservation of plant nutrients under patches of severely diseased plants in an annual field crop may be partly responsible for the failure of a disease to reappear in severe form on the same areas in the following season's crop. This phenomenon of disappearance of a disease from old areas and appearance in new ones under continuous cropping is well established, and has been described in connection with root rot of cotton by Taubeniiaus and Killough (1923), Mcnamara *et al.* (1931) and by many others, and in connection with take-all of cereals by Glynne (1935) and by Fellows and Ficke (1939). Discussing this problem, Garrett (1938a) favoured the explanation put forward by Taubenhaus and Killough, that the chances of survival of a root-infecting fungus from one season to the next were reduced by the early death of infected host plants; a disease was therefore less likely to reappear on sites occupied by severely infected and early killed plants in the previous season than on sites occupied by apparently healthy plants infected relatively late in the season.

Such a reduction of over-wintering inoculum unclear sites of severe disease patches must be held partly responsible for failure of the patches to reappear on the same sites in the following season's crop; at the same time, conservation of plant nutrients under the scanty crop in such patches must tend to increase plant resistance to the take-all disease, and also, under some circumstances, to cotton root rot.

Nitrogen: Nitrogenous fertilizers were found by Jordan *et al.* (1939) to reduce losses due to root rot *(Phymatotrichum omnivoruwi)* in almost-continuous cotton sequences in Texas, whereas phosphatic fertilizers increased mortality. These opposing effects of nitrogen and phosphate upon incidence of root rot were most pronounced on the Wilson very fine sandy loam soil, and were less striking on the heavier soils of the Wilson series; they were smallest and least consistent on the Houston clays.

From their analyses of root bark of plants taken from the same field experiments, Adams *et al.* (1939) obtained a highly significant negative correlation between nitrogen content and root rot mortality, and a highly significant positive correlation between phosphoric acid content and mortality. An inverse correlation, also highly significant, was found between phosphoric acid and nitrogen contents of root bark samples. These results have been confirmed by Blank (1941), who found nitrogenous fertilizers to reduce root rot mortality; phosphatic fertilizers had no significant effect upon incidence of root rot on Houston clay soils of Washington County (Texas), but on the less calcareous soils of Burleson County they significantly increased both incidence of disease and crop yield.

Reduction in losses from pea root rot, due to *Aphanomyces euteiches*, following application of nitrogenous fertilizers, has been reported by Haenseler (1931) and by Walker and Musbach (1939) from the USA, and by Geach (1936) from Australia. This effect of nitrogenous fertilizers was attributed by Geach to direct action upon *A. euteiches*, which he found to be especially sensitive to the nitrogen content of the medium when growing on agar plates.

Thus on cornmeal agar with 2 *per cent* peptone, formation of oogonia and antheridia by the fungus was suppressed, and the mycelium alone was no longer culturable after 33 days at laboratory temperature. On corn-meal agar without peptone, however, oospores were formed, and these remained viable during the whole test period (60 days).

Nevertheless, further evidence seemed to be needed to bridge the rather wide gap between the observed behaviour of *A. euteiches* in pure culture on the agar plate, and the reaction of the disease to the application of nitrogenous fertilizers to the soil. A more likely explanation has been suggested by Smith and Walker (1941), as a result of their experiments on the effect of nutrients upon development of the disease in continuous-flow nutrient sand culture.

The amount of disease varied inversely as the total salt concentration of the solution, irrespective of its composition; contrary to results of fertilizer trials in the field, the nitrogen-containing salts were no more effective in disease control than were those containing potash and phosphate. Smith and Walker therefore suggested that the greater effectiveness of nitrogenous compounds in disease control in field trials was due to the fact, established by White and Ross (1939), that they increased the salt concentration of the soil solution much more than did corresponding quantities of the potassium and phosphorus salts used in commercial fertilizers. These results of Smith and Walker suggest, therefore, that no directly toxic effect of nitrogenous compounds, as such, upon *A. euteiches* is involved, as believed by Geach, Smith and Walker further demonstrated that, once infection had occurred, its development could not be inhibited by changing the concentration of the basal nutrient solution from a low to a high level.

In experiments on agar, *A. euteiches* was found to show little reduction in linear growth rate even at 5 X concentration of the basal solution, but growth of the mycelium on agar can scarcely be considered a satisfactory criterion of the behaviour of the zoospores in a solution without agar. Smith and Walker have also incidentally pointed out that GEACH'S observations on the poor growth of *A. euteiches* on pea-decoction agar with the addition of various nitrogenous compounds are most simply explained by reference to the high concentrations which he employed.

Since the publication of these experiments by Smith and Walker, Schroeder and Walker (1942) reported that the development of pea wilt due to *Fusarium oxysporum* f. *pisi* race 1 in nutrient sand culture was similarly checked by increase in concentration of the nutrient solution, when peas were grown at the favourable temperature of 21°C. On the other hand, at a temperature of 27°C, which is unfavourable to normal development of peas, disease development was greatest at the highest concentration of the nutrient solution.

Another disease that can be much reduced by application of nitrogen is root rot of sugar-beet due to *Sclerotium rolfsii*. In field experiments carried out in California, Leach and Davey (1942) found that, on an average of all trials, 50 Ibs. of nitrogen per acre reduced infection by some 28 per cent, 100 Ibs. by 54 *per cent*, and 200 Ibs. by 65 per cent. Ammonium sulphate, anhydrous ammonia, calcium nitrate and cyanamide proved equally effective at equivalent nitrogen levels. Leach and Davey found no evidence for a directly toxic effect of the nitrogenous compounds upon the parasite, since calcium nitrate gave as effective a control as did equivalent-nitrogen amounts of anhydrous ammonia; the former was non-toxic to the mycelium and sclerotia of *S. rolfsii*. The effect of nitrogenous fertilizers in reducing the disease was nearly as pronounced in fields of high fertility as in fields of low fertility, which suggested that increased resistance of the sugar-beet was associated with a luxury consumption of nitrogen rather than with the correction of a deficiency, in the usual sense of this term.

Deficiency of nitrogen, on the other hand, has been claimed by Clayton (1923) and Fisher (1935) to inhibit infection of tomato by the vascular wilt fungus, *Fusarium bulbigenmn* var. *lycopersici*. Cook (1937), however, concluded from his experiments that nitrogen deficiency did not inhibit infection by the fungus, but did check its development in the plant. The position has been further clarified by Roberts (1943), working with another vascular parasite of tomato, *Verticillium alboatrum*. Roberts found that the number of infected plants increased with the amount of nitrogen supplied to the soil; in some experiments, infection in some of the nitrogen-deficient plants was confined to the root system or even to a single root, suggesting that nitrogen deficiency may act as much, or more, by hindering or inhibiting development of the fungus in the vascular tissues as by preventing initial infection of the root from the soil. It is pertinent to note here that a bacterial vascular parasite, *Phytomonas stewarti*, was found by Spencer and McNew (1938) to cause most severe wilting in maize plants that had received excess nitrogen; under conditions of nitrogen deficiency, little or no wilting occurred.

The damping-off of pine seedlings was found by Hartley (1921) to increase directly with the nitrogen content of the soil. Wright (1941) has correlated damping-off (due chiefly to *Rhizoctonia, solani* and *Pythium ultimum)* amongst seedlings of deciduous trees with nitrate nitrogen content of the soil; he reported that the stand of seedlings was increased almost threefold by the application of dextrose to the soil, whereby the nitrate nitrogen content was temporarily reduced.

Phosphate: Browning root rot of cereals due to *Pythium arrhenomanes* and other *Pythium* spp. in Canada was reported by Vanterpool (1930-40) to be associated with a low phosphate and high nitrate status of the soil; Vanterpool has stressed the analogy between behaviour of this disease and that of sugar-cane root rot, studied by Carpenter (1934), which is also caused, apparently, by *Pythium arrhenomanes* (Drechsler, 1936). The cereal disease can be controlled by application of a soluble phosphate, which does not seem to reduce the proportion of roots infected by the fungus, but, by greatly increasing the total number of roots produced, increases the number of functional roots.

The disease was stated by Vanterpool to be worse in crops after fallow than in those sown on stubble land, in contrast to experience with other cereal root rots. This was attributed by Vanterpool to increase in the available nitrogen content of the soil after following, as a result of nitrification. A possibly analogous case has been reported by Grandfield *et al.* (1935), who found in Kansas that stands of alfalfa seedlings were progressively decreased through damping-off by *Pythium* spp. with increase in the preceding period of fallow from two to five years. Thus only 33 *per cent* of the original stand of alfalfa seedlings survived till the following spring on the five years fallow plot, as compared with a survival of 70 per cent on a plot under rotation of sorghum, corn, wheat, oats and alfalfa.

Wilt disease (due to *Fusarium vasinfectum)* of pigeon pea, on the other hand, was reported by McRae and Shaw (1933) to be aggravated by applications of superphosphate on the manurial plots at Pusa, although the crop gave a positive response in yield to this fertilizer *(cf.* the results of Jordan *et al.* (1939) and Blank (1941) with *Phymatotrichum omnivorum*).

Potash: The effect of potash in reducing incidence of cotton wilt, due to *Fusarium vasinfectum,* has been studied by Young (1938), in a nine years' series of field experiments carried out at 15 different centres in Arkansas. Field observations at an earlier date had indicated a correlation between incidence of "rust", correctly attributed by Atkinson (1892) to potash hunger, and that of the *Fusarium* wilt.

Young found that fertilizers containing potash gave control

of rust and at the same time reduced wilt in every year and at every location, except two. Increasing applications of potash gave, up to a point, increased control of the wilt.

Whilst nitrate alone had little or no effect on the incidence of wilt, the application of acid phosphate and nitrate together appeared to aggravate both rust and wilt; where acid phosphate was used alone, the effect was still more pronounced.

Young's results have been confirmed in Alabama by Dick and Tis-Dale (1938) in a factorial experiment involving all possible combinations of N, P and K at three different levels; joint applications of nitrogen and potash were found to reduce incidence of wilt, whereas phosphate increased it, and substantially reduced the beneficial effect of applying potash. Smith (1940), announcing the results of two seasons' cooperative manurial trials in 13 localities of 9 States of the USA, confirmed this beneficial effect of potash in reducing cotton wilt. In Uzbekistan, Zaprometoff (1929) found the incidence of wilt to be very slight on soils with a high content of common salt, which suggests that salt might be able to replace potash for control of wilt and rust in cotton, just as it can be used to replace potash for the manuring of sugar-beet.

A reduction in the percentage of tubers infected by the potato wart organism, *Synchytrimn endobioticum,* has been claimed by Leszczenko and Szymanski (1938) to follow application of potash salts to artificially inoculated field plots in Poland. In experiments on the relation between development of clubroot in crucifers and the concentration of plant nutrients, Pryor (1940) found the percentage of plants forming clubs to be markedly reduced by a deficiency of potash.

Minor Elements: Secondary infection by fungi and other microorganisms may complicate and aggravate the symptoms of minor element deficiencies in plants. Thus, Gerretsen (1937) has claimed that bacterial infection of the roots is concerned in the production of the "grey leaf" symptoms characteristic of manganese deficiency in oats. In Victoria, Millikan (1938) found that applications of zinc sulphate at rates of 15 and 30 Ibs. per acre considerably improved the growth of wheat crops suffering from a combined attack of eelworm *(Heterodera schachtii)* and of a number of root-rotting fungi, including *Helminthosporium sativum, Curvularia ramosa, Fusarium culmorum,* and *Rhizoctonia solani.*

The treatment did not apparently give any appreciable reduction in the number of roots infected by the parasites, but, by greatly increasing the rooting capacity of the plant, minimised the damage due to such infection. Millikan (1942) has since demonstrated that soil microorganisms may compete with a wheat crop for available supplies of minor elements. A deficiency of available zinc was made good by steam sterilisation of the soil; in steamed soils, wheat no longer responded to application of this minor element. The responsiveness to zinc could be restored, however, by reinoculation of the soil with unsterilised soil two months before sowing wheat. From the results of such experiments, Millikan concluded that the poor growth of wheat on certain Victorian soils was due more to competition by saprophytic soil microorganisms for limited supplies of zinc and other minor elements than to root infection by parasites.

Influence of Soil Moisture Content

Diseases known to be favoured by high soil moisture content.

Diseases

- *Pythium arrhenomanes* Root rot of sugarcane
- *P. arrhenomanes*
- *Sclerospora graminicola*
- *Phytophthora cactorum*
- *P. cambivora*
- *P. cinnamomi*
- *P. parasitica*

- *Fusarium annuum,*
- *F. cubense*
- *F. lycopersici*
- *F. orthoceras* var. *pisi*
- *Fusarium* sp.
- *Armillaria mellca*
- *Calonectria graminicola*
- *Hclminthosporium sativum*
- *Sclerotinia sclerotiorum*
- *Sphaerostilbe repens Root rot of corn*
- *Seedling blight of Italian*
- *Crown rot of rhubarb*
- *Ink disease of chestnut*
- *Pineapple wilt*
- *Chilli pepper wilt*
- *Banana wilt*
- *(Panama disease)*
- *Tomato wilt*
- *Pea wilt*
- *Celery yellows*
- *Mushroom root rot*
- *Snow mould of cereals*
- *Foot rot of cereals*
- *Collar rot of lettuce*
- *Violet root rot of tea*

The first eleven diseases are due to phycomycetous fungi and other organisms which infect the host by means of free-swimming zoospores; when soil moisture content falls below a certain value, infection is usually completely inhibited. Thus Monteith (1924) found in controlled soil moisture experiments that infection of cabbage seedlings by *Plasmodiophora brassicae* occurred at moisture contents of *60 per cent* saturation and above, but not at those of 45 per cent saturation or below; the severity of infection increased with rise in moisture content. Similarly, Smith and Walker (1941) found practically no infection by *Aphanomyces euteiches* at a soil moisture content of 45 per cent saturation, whereas *72 per cent* infection occurred at a moisture content of 75 per cent saturation. Hickman (1940) has recently demonstrated that the occurrence of red core root disease of strawberries, due to *Phytophthora fragariae,* depends upon high soil moisture content.

The next five diseases on the above list are vascular wilts due to species of *Fusarium;* to these may be added *Fusarium vasinfectum,* which was found by Tharp and Young (1939) to produce maximum disease at a moisture content of 80-90 per cent saturation.

Pontis (1940) reported percentage of wilt in chillies due to the same fungus to be increased by frequency of irrigation in a field experiment; satisfactory control of this disease was secured by planting "on the ridge".

Evidence is accumulating to show that infection by the vascular wilt fungi, including *Verticillium alboatrum,* is favoured by the "soft" type of growth promoted by high soil moisture content and heavy nitrogenous manuring (organic or inorganic); the opposite conditions seem to reduce the incidence and severity of this disease not so much by reducing the original number of root infections as by restricting the active development of the vascular parasite within the plant.

This has been demonstrated by Clayton (1923), Fisher (1935) and Cook (1937) in experiments the effect of nitrogenous manures upon the susceptibility of tomato to infection by *Fusarium bulbigenum* var. *lycopersici,* whilst very similar results have recently been obtained by Roberts (1943), working with the same host but with another vascular parasite, *Verticillium alboatrum.*

In some situations, high soil moisture content may increase the prevalence of a soilborne disease in another way, *i.e.* by flood dispersal. This factor is said by Wardlaw (1935) greatly to have influenced the distribution of Panama disease of bananas. Very high soil moisture content may also, by reducing soil aeration, injure the roots of the host plant and thus afford ingress to such parasites as *Sphaerostilbe repens,* the cause of violet root rot in various tropical and subtropical crops.

Factors predisposing sugar-cane to *Pythium arrhenomanes* root rot under the conditions of high water table and impeded drainage in the heavy clay soils of the Louisiana sugar belt have been investigated by Rands and Dopp (1938).

Hydrogen sulphide and salicylic aldehyde were selected as examples of substances accumulating in soils under semi-anaerobic conditions, and the effect of these two substances upon the predisposition of sugar-cane to *Pythium* root rot in nutrient sand culture was tested.

Hydrogen sulphide was not found significantly to intensify the effects of the disease. On the other hand, salicylic aldehyde in concentrations of 20 to 40 p.p.m., which had little if any influence on cane growth in the absence of the fungus and showed no effect on the growth of the fungus in culture, apparently so predisposed the roots to infection in the presence of the fungus as to result in a reduction in weight of the plants from 2 to 7 times as great as that caused by the fungus alone.

Next for consideration comes a group of diseases that do not require a high soil moisture content for development, but are yet more severe in wet seasons, and may be virtually absent in unusually dry ones; root rot *(Phymatotrichum omnivoruwi)* of cotton and take-all *(Ophiobolus graminis)* of cereals fall within this class.

Ezekiel (1938) reported that whereas the presence or absence of cotton root rot in different sections of Texas seemed to be determined by the coincidence of favourable neutral to slightly alkaline soils with mean normal temperatures above 60° F., its prevalence from year to year was apparently limited chiefly by the rainfall from the beginning of April to the end of July.

Multiple correlation analyses of root rot losses in 10 selected counties of Texas, in which the soils were approximately equally favourable for development of this disease, showed that an increase of one inch of rainfall in April or May was associated with an increase of about *15 per cent* in root rot, one inch in June with an increase of about *20 per cent,* and one inch in July with an increase of some *30 per cent* in the disease.

Variation in August rainfall did not seem to affect the incidence of root rot. The prevalence of the take-all disease of wheat in Saskatchewan over the 5-year period 1923-27 was correlated by Russell (1930) with aggregate rainfall for June, July and the first half of August, whilst Garrett (1934a), from a historical survey of the 34-year period 1900-33 in South Australia, was able to correlate abundance of the same disease with aggregate rainfall during the spring months of August, September and October. Commenting upon the distribution of take-all in the Murray Mallee district of the same State, Griffiths (1933, 1940) declared that little damage was done by the disease in localities with an average rainfall of 10-12 inches, but that as the rainfall increased, so did the disease become more prevalent.

We find the cereal smuts conspicuous as a group. As soil moisture content decreases, aeration improves, and this may account for the increased activity of the *Actinomyces* spp. (Sanford, 1926) and of the cereal smut fungi (Rabien, 1927). Ling (1941) has demonstrated, however, that soil moisture content affects the development of the stripe smut fungus *(Urocytis occulta)* within the rye plant, in addition to any effect that it may have upon the incidence of initial infection. A sufficiently low soil moisture content must also favour infection of young cereal.

Dickson *et al.* (1923) demonstrated that the greater development of seedling blight of wheat and corn due to *Gibberella saubinetii* in drier soils was due to a retarded development of the normal host resistance; Muskett (1937) has recently shown a similar relationship to exist between *Helminthosporium avenae* and the oat plant.

Texture: Relatively few soilborne diseases appear to be favoured by soils of heavy texture.

The eye-spot disease of cereals is not a disease of the roots, but affects the stem just above soil level; the inoculum, however, may be partly soilborne.

Recent work by Sprague (1937) indicates that heavy soils favour the disease insofar as they promote a higher relative humidity of the atmosphere around the stem bases of the plants just above soil level, owing to impeded drainage. The snow mould diseases of cereals are also favoured by high soil moisture content; drainage of water from the melting snow is slower on the heavier soils. The relation between the occurrence of violet root rot and high soil moisture content, due to impeded drainage in heavy soils, has been noted by numerous observers.

The majority of soilborne diseases seem to be favoured by light-textured soils.

Diseases favoured by light soils:

- *Actinomyces scabies*
- *Botryodiplodia theobromac*
- *Femes annosus*
- *F. lucidus*
- *F. noxius*
- *Fusariunt cubense*
- *R lini*
- *F. hyperoxysporum*
- *F. batatis*
- *F. orthoceras* var. *pisi*
- *F. vasin fee turn*
- *Fusarium* sp.
- *Ophiobolus graminis*
- *Sclerotinia sclerotiorum*
- *S. trifoliorum*

- *Spongospora subterranca*
- *Tilletia tritici*
- *T. levis Potato scab*
- *Internal root rot of tea*
- *Red rot of pines*
- *Root rot of coconut palms*
- *Brown root rot of tea and rubber*
- *Panama disease of bananas*
- *Flax wilt*
- *Stem rot of sweet potatoes*
- *Pea wilt*
- *Cotton wilt*
- *Collar rot of peas*
- *Take-all of wheat*
- *Collar rot of lettuce*
- *Clover rot*
- *Powdery scab of potatoes*

The root-infecting fungi, and indeed fungi in generally are, strong aerobes, and their activity must therefore be favoured by soils of light texture, insofar as these are characterised by a good natural diffusion of the soil gases. Potato scab and take-all of cereals appear to be diseases particularly affected by soil aeration. Garrett (1936) demonstrated a close correlation between those soil conditions favouring the occurrence of take-all in the field and those favouring most rapid growth of *Ophiobolus graminis* along the roots of the wheat plant, *viz.* light texture, loose condition and alkaline reaction; the paramount importance of soil aeration was subsequently proved by direct experiment (Garrett, 1937).

Reinking (1935) was able to correlate numbers of *Fusarium oxysporum cubense* in the soil with incidence of Panama disease of bananas on soils of different type in Central America; numbers of the organism and severity of the disease both increased with increasing percentage of sand and decreasing percentage of clay in the soil.

Reinking comments upon his results as follows: "Apparently in heavily infected soil root infections are so numerous that the plant is unable to ward off attack, as has been shown to be possible by Wardlaw (1930), and severe disease is produced. When the organism is not present in the soil in large numbers the infections are less, and undoubtedly the plant is capable of warding off attack to a certain degree resulting in less severe disease.

The effect of the different soil types upon the possible production or non-production of resistance within the plant to attack by the parasite apparently is not of such great importance." Reinking did not discover any connection between pH value of the soil and prevalence of the disease; Wardlaw (1941), on the other hand, recently voiced the following general conclusions concerning the distribution of Panama disease with soil type:

> "Spread by contact takes place more rapidly in some soils than in others; the more open the texture, and the more acid the soil, the more rapid is the spread of disease. Alternatively, clay soils and slightly alkaline soils appear to possess more 'resistance' to the disease. Recent years have also shown that clay soils, for many years considered 'resistant', have eventually shown an even more rapid spread a phenomenon for which no adequate explanation has so far been forthcoming.... The collective experience, practical and scientific, in Central America is that hydrogen ion concentration is the master factor determining the seventy of Panama disease, followed by texture."

It is dangerous to conclude too readily that because the activity of a root-infecting fungus is favoured by good soil aeration, the disease that it causes will be more severe on light-textured soils, or, vice versa, that greater prevalence of the disease on such soils is to be attributed to enhanced activity of the parasite. It is necessary to remember that light-textured soils are usually naturally poor in plant nutrients and often acid in reaction, whereas heavy-textured soils tend to be better supplied with plant nutrients, and are more often neutral or alkaline in reaction. This point is brought out by consideration of the distribution of cotton root rot on different types of soil in Texas and Arizona.

The mode of spread of *Phymatotrichum umnirorum* closely resembles that of *Ophiobolus graininis.* Yet whereas take-all is favoured by soils of light texture, and by any other condition making for improved soil aeration, cotton root rot, according to Fraps (1936), is more prevalent in Texas on heavy-textured alkaline soils than on light-textured soils of neutral or acid reaction.

The association of this disease, however, may not be with heavy texture, but rather with the alkaline reaction of these soils (Taubenhaus *et al.*, 1937), or with their high water-holding capacity (Ezekiel, 1938).

The importance of the last-named factor is suggested by Streets (1937), in commenting upon the distribution of cotton root rot in Arizona:

> "In Arizona root rot is noticeably more prevalent in the finer textured soils along streams and river bottoms and in alluvial valleys than on the coarser soils of the mesas. Under irrigation, however, it thrives in the coarser soils when their moisture content is maintained by frequent applications of water. In fact, the superior aeration in the mesa soils favours the growth of the organism."

Cotton wilt, due to *Fusarium vasinectum,* is greatly aggravated by potash deficiency (Young, 1938), which is notoriously more common on light than on heavy soils. Internal root rot of tea, formerly attributed to *Botryodiplodia theobromae,* was shown by Gadd (1928, *1929b)* in Ceylon to be associated with lack of starch reserves in the root, due to excessive plucking and pruning; he further demonstrated that this condition actually *preceded* infection by *B. theobromae.*

These instances show, therefore, the need for caution in interpreting field observations on the association between soil texture, or any other soil character, and prevalence of a soilborne disease.

Reaction: Diseases are favoured by acidic soils and others favoured by alkaline soils.

Here, again, the need for caution in interpreting such observations is apparent; field observations may be misleading owing to the common association between soil acidity, lightness of texture, and poverty in plant nutrients.

Conclusions are more trustworthy for diseases of great economic importance; field observations on such diseases have usually been supplemented by experiments, Such experiments are more easily performed with diseases of annual or field crops than with diseases of plantation crops; for this reason, evidence available for correlating incidence of disease with soil conditions is both more abundant and more reliable for field crops than for plantation crops.

Prominent amongst the diseases favoured by acid soils are the vascular wilts caused by various species of *Fusariuin,* to which, according to Wardlaw (1941), may be added banana wilt, better known as Panama disease.

The better development of tomato wilt in acid soils, experimentally demonstrated by Sherwood (1923) and Scott (1926), has been confirmed by Phillips (1940) and Harrison (1940). It is perhaps remarkable that so few experiments seem to have been undertaken to analyse the action of so clear-cut a factor assoil reaction upon the incidence of disease.

The preference of club-root of crucifers for acid soils, and of potato scab, take-all of cereals, root rot of cotton and black root rot of tobacco for alkaline soils is widely known, and has been repeatedly confirmed by experiment.

Diseases Favoured by Acidic Soils

Fungus	Disease
Armillaria mellea	Mushroom root rot
Botryodiplodia	Internal root rot of tea
theobromae	Red rot of pines
Fomes annosus	Wilt of china asters
Fusarium	
conglutinans	
var. *Callistephi* and	Tobacco wilt Tomato wilt
var. *Majus*	Cotton wilt
Fusarium oxysporum	Clubroot of crucifers potatoes
var. *nicotianae*	Powdery scab of Potatoe
F. Lycopersici	potatoe wart
F. Vasinfectum	Covered smut of barley
	Charcoal root rot
Plasmodiophora	Root rot of sugar-beet
brassicae	
Spongospora	
subterranea	
Synchytrium	
endobioticum	
Ustilago hordei	
Ustulina sonata	
Aphanomyces levis	
Phoma betae	
Pythium dcbaryanum	

Diseases Favoured by Alkaline Soils

Fungus	Disease
Actinomyces poolensis	Pox of sweet
A. Scabies	Potatoes
Calonectria	Potato scab
graminicola	Snow mould of cereals
	Pea wilt
Fusarium orthoceras	
var. *pisi*	
Monilochaetes infuscans	Scurf of sweet potatoes
Ophiobolus graminis	Take-all of wheat
Phymatotriehum omnivorum	Cotton root rot
Thielavia basicola	Black root rot of tobacco
Urocystis tritici	Flag smut of wheat
Verticillium alboatrum	Tomato, etc., wilt

The greater prevalence of the take-all disease on alkaline soils was correlated by Garrett (1936) with a more rapid growth of *Ophiobolus graminis* along the roots in such soils; he suggested that rate of growth of the fungus along the roots in soils of heavy texture and in acid soils was checked by the accumulation of respiratory carbon dioxide, and that alkaline soils increased fungal growth rate by acting as acceptors of carbon dioxide. By forced aeration, Garrett (1937) secured as rapid a growth of this fungus along the roots in acid (pH 5.2) as in alkaline (pH 8.0) soils.

Evidence in support of Garrett's hypothesis has been published by Blair (1943), working with *Rhizoctonia solani,* the mycelium of which grows freely through the soil.

Blair found that whereas the growth of this fungus along Rossi-Cholodny slides was normally most rapid in soils of neutral reaction, yet when 1 per cent of ground wheat straw or dried grass had been added to the soil growth was most rapid at a reaction of pH 8.

In this instance, the respiratory carbon dioxide was provided not by the root and its rhizosphere, but by the microorganisms decomposing the fresh organic material. In another experiment, Blair was able to eliminate the depressing effect of ground wheat straw upon growth of *R. solani* through the soil by keeping the soil container, open at top and bottom, in a closed vessel containing dishes filled with saturated solutions of strong alkali, which must rapidly have absorbed the carbon dioxide given off by the respiring microorganisms decomposing the wheat straw.

An interesting sequel has followed the report of Fokster and Vasey (1929) that flag smut of wheat was encouraged by application of lime in field plot experiments in Victoria. Millikan (1939) was able to show in sand culture experiments that a deficiency of calcium markedly reduced the development of the disease; if the degree of deficiency was sufficiently great, the disease was completely inhibited. On the other hand, the addition of excess calcium to the nutrient solution significantly increased the incidence of infection. From the results of chemical analyses made on the plants at the end of their growth period, Millikan postulated an optimum concentration of calcium in the plant for the development of this disease.

These pot experiments, which were carried out in three consecutive years, were supplemented by field experiments at 4 different centres in Victoria during 1933-38.

A variety of manurial and mineral treatments were tested in these experiments; wide variation in the mineral composition of the plants was induced by the treatments, as revealed by chemical analyses carried out 6 and 9 weeks after germination (considered by Millikan to be the critical period for the development of flag smut, in view of the fact that sori consistently appeared in these field experiments at some 11 weeks after germination of the wheat). As in the pot culture experiments, the only correlation between incidence of disease and mineral composition that could be established was that with lime content of the tissues.

Millikan again concluded that the evidence indicated an optimum content of calcium in the plant for the establishment of infection, and that any treatment which shifted the calcium content of the plant further away from this optimal value would result in a decrease of the disease. This optimum concentration of calcium was not the same in all the experiments, but appeared to change with variation in other soil factors. In none of Millikan's field experiments did he repeat the results of Forster. and Vasey (1929) by securing increased development of flag smut after liming; a simple explanation of this apparent anomaly, however, is suggested.

The pH value of the soil in the untreated plots at Werribee, where Forster and Vasey made their observations, was 5.5; frequent applications of lime had changed the pH value to a maximum of 8.3 on those plots showing the highest percentage of flag smut.

The majority of Millikan's experiments, on the other hand,

were carried out at Rutherglen, with soil reaction of pH 7.7, and at Walpeup, with soil reaction of pH 8.6. Millikan remarks: "The most severe flag smut damage occurs in the Mallee district of Victoria, in which Walpeup is situated, where the soil contains a high percentage of limestone".

It is worth noting at this point an experiment carried out by Ling (1941) on the development of stripe smut *(Urocystis occulta)* in rye plants grown in nutrient sand cultures. Ling employed a three-salt nutrient solution comprising potassium dihydrogen phosphate, calcium nitrate, and magnesium sulphate; the incidence of infection was lowest in the series receiving maximum potassium phosphate, and highest in that receiving equimolecular proportions of calcium nitrate and magnesium sulphate, but minimum phosphate.

The two diseases of wheat cited above, take-all and flag smut, which may both be encouraged by liming the soil, afford an interesting contrast; in the take-all disease, with activity of the fungus partly external to the host, soil reaction seems directly to affect the parasite, whereas in the flag smut disease, with development of the fungus almost wholly internal, soil reaction seems to operate upon the fungus through the physiology of its host plant.

Parasitic Activity of the Root-infecting Fungi

A distinction was drawn between primitive parasites of the *soil inhabitant* type, such as the species of *Rosellinia,* in which parasitic activity was merely incidental to a saprophytic existence, and the more specialised parasites of the *soil invader* type. Variation in growth habit and parasitic activity of the latter group may now be considered. The more specialised parasites spread underground in three ways:

(1) Through the soil and over the underground parts of the host plant,

(2) Only over the underground parts of the host plant,

(3) Inside the vascular system of the host plant.

Fungi belonging to the first two classes invade the internal tissues of the host roots, of course, but external spread never lags appreciably behind internal spread, and is often considerably in advance of it.

The fungi of class (3) do not travel either independently through the soil, or over the outside of the host roots; their spread is limited to the vascular cylinder of the host. Spread of fungi in classes (2) and (3) from one host plant to another is, therefore, limited to points of root contact between host plants; this constitutes no hindrance to spread amongst crops where the ground is occupied by a pure stand of a susceptible host plant, except where the crop is still young, and root contacts have not yet been established.

In natural vegetation, however, plants susceptible to a particular root parasite are often isolated from one another by the root barriers of immune plants, and so the spread of the parasite is restricted. At one time it was thought that a fungus able to spread independently through the soil by free mycelial or rhizomorphic growth possessed a considerable advantage over a fungus which could spread from host to host only by root contact; this distinction is now realised to have been overdrawn.

Spread External to the Host: A feature of particular interest in the root-infecting fungi, though by no means limited to this group, is the rhizomorph an organ whereby the fungus is enabled to spread from one host (or substrate) to another, some distance away.

The evolution of the rhizomorph through aggregation of individual hyphae into strands, followed by differentiation of tissues into an external protective skin or rind, and internal conducting tissue, seems intrinsically probable; the rhizomorph serves to protect the mycelium of the fungus from the hazards of the environment, and especially from that of desiccation, and to facilitate its spread from one substrate (the food-base) to another, some distance away.

The following fungi may be cited as illustrating different, but not necessarily consecutive, hypothetical stages in evolution of the rhizomorph:

(i) Spread by individual hyphae *Rhisoctonia solani* (causing black scurf and stem canker of potatoes, etc.).

(ii) Spread by individual hyphae occasionally aggregating into strands. *Ophiobolus graminis* (causing the take-all disease of cereals).

(iii) Some aggregation of hyphae, with copious secretion of mucilage binding a covering of soil closely around host root *Pomes noxius* (causing brown root disease of rubber, tea, etc.).

(iv) Differentiated strands composed of large central hypha surrounded by several layers of fine hyphae. *Phymatotrie humomnivorum* (causing Texas root rot of cotton).

(v) Aggregation into definite rhizomorphs, but no tissue differentiation. *Fomes lignosus* (white root disease of rubber, etc.).

(vi) Aggregation into definite rhizomorphs with toughened outer skin. *Ganoderma pseudoferreum* (causing red root disease of rubber) and *Poria hypolateritia* (causing red root disease of tea).

(vii) Aggregation into definite rhizomorphs with well differentiated outer rind. *Armillaria mellea* (causing a root disease of tea, cacao, etc.) and *Sphaerostilbe repens* (causing violet root rot of rubber, tea, citrus, etc.).

The mucilage-secreting power of *Fonics noxius* was discovered by Sharples (1922), who considered that the cementing of soil around the mycelium by this means provided a valuable safeguard for the fungus against desiccation Sharples (1936) has laid great stress upon the function of the rhizomorph in protecting the fungus from drying-out whilst external to the host tissues. The outer hyphal layers of the differentiated strands of *Phymatotrichum omnivorum*, the development of which has been described in detail by Rogers and Watkins (1938), no doubt help to protect the large central hypha from desiccation, though the strands do not seem to be particularly resistant to drying-out; even the sclerotia of *P. omnivorum,* unlike those of most other species, possess comparatively little resistance to desiccation.

In *Armillaria mellea,* rhizomorphs are developed not only external to the host but also internally as well, between the bark and the wood; they were distinguished as *Rhizomorpha subterranea* and *R. sub cor tic alls,* respectively, by Persoon. In *Sphaerostilbe repens,* rhizomorphs are formed only between the bark and the wood of the host root, though they show a degree of differentiation comparable to that in *Armillaria mellea.*

The capacity of a root-infecting fungus for making a free and independent spread through the soil from one host plant to another has frequently been a subject for controversy. Thus Fetch (1921, 1923) was convinced that both *Fomes lignosus* and *Poria hypolateritia,* behaved in this way in the rubber and tea plantations of Ceylon. As a result of field observations on rubber estates in Malaya, Napper (1932-34) disputed Fetch's assertion that *F. lignosus* had more than a very limited capacity for free growth through the soil, and insisted that growth was almost entirely confined to solid surfaces; the fungus could travel over the surface of stones and dead wood buried in the soil, but could only infect and obtain nourishment from living roots.

In the USA, Taubenhaus and Killough (1923) declared that *Phymatotrichum omnivorum,* the agent of cotton root rot, spread through the soil only along the roots of its host plants, and from one plant to another by root contact. This view was disputed by Peltier *et al.* (1926), who asserted that *P. omnivorum* spread through the soil in the manner of a fairy ring fungus; their view, apparently, was based upon the observation that "the point of attack is nearly always on the tap root within the first foot of soil, whereas the lateral or side roots are usually diseased only at the point of attachment with the tap root. If the spread of *Ozoniurn* occurred through the contact of a diseased root with a healthy one, then we should expect to find that the rot began at the ends of the laterals and worked up to the main tap root.

Later, Taubenhaus and Ezekiel (1930a) reported experiments designed to elucidate this question. Cubes of soil were taken from next to recently wilted plants in the active zones of root rot spots in cotton fields, and used to inoculate healthy cotton plants. Inoculations were made by removing soil from beside the healthy plant, watering heavily, and filling the hole with the soil inoculum from which all larger roots were first removed.

Controls consisted of similar healthy plants which were inoculated with freshly infected cotton roots. Not one of the 2367 plants inoculated with the presumably infective soil cot tracted the disease, whilst 180 of the 220 control plants, inoculate with infected roots, became infected.

In another experiment, soil was again secured from the active zones of root rot spots in fields on the different types of soil; part of the soil of each type was passed through a X-inch sieve, and part was left unsifted, *i.e.* with all the origins roots. Root rot appeared only in cotton planted in the unsifted soil containing the roots.

Finally, Taubenhaus and Ezekiel obtained further evidence by excavating the soil from about the roots of 117 cotton plants in early spring. Against selected plants, an isolation trench was dug, and the soil immediately around the root system was then carefully dissected away with stiff needles and with ice-picks. Each soil particle, as it was removed, was carefully examined under the hand lens for hyphae of *P. omnivorum*, and all roots and rootlets were similarly searched; doubtful material was studied under the microscope.

No free spread of the mycelium away from the roots was discovered, though the fungus was found travelling along the finest rootlets, and spreading from plant to plant by complete or próximate root contact. Taubenhaus and Ezekiel pointed out that the finer infected rootlets might decay and disappear, leaving strands of the fungus apparently spreading freely through the soil; only close to the centres of well advanced spots were occasional isolated strands of the fungus found away from roots, and most of these were following insect or earthworm channels through the soil.

The mycelium spread more readily along the wall of the glass container than through the soil itself, suggesting that *P. omnivorum,* like *Fomes lignosus,* kept to the continuous surface of a solid body where such was available.

Another well-known root parasite which has been shown to spread only by root contact is *Ophiobolus graminis,* cause of the take-all disease in cereals. Padwick (1935) in Canada demonstrated through glasshouse experiments that whereas no appreciable spread of the fungus from buried inoculum occurred in fallow soil, a spread of as much as 10 inches in 52 days took place under growing wheat and species of *Agropyron.* A very similar experiment was carried out in the field by Adam and Colquiioun (1936) in South Australia; they compared the spread of *Ophiobolus* through a clay-loam soil and through sand, in which wheat seeds were planted at distances apart of 2, 4, 8 and 12 inches. At harvest, no detectable spread of the fungus had occurred where the plants were spaced 12 inches apart; a maximum spread of 20 inches from the original inoculum was recorded under the plants spaced 2 inches apart. Fellows and Ficke (1939) in Kansas, have confirmed this conclusion as to spread of *O. graminis.* Garrett (1936) was unable to demonstrate mycelial growth from inoculum of this fungus placed on the surface of unsterilised soil in covered glass dishes. Nevertheless, it has since been found that such growth will occur under special conditions, *e.g.* when freshly-infected straw is buried in moist sand against a glass surface; the mycelium of the fungus then grows out over the glass, and may even form perithecia. Here, again, the fungus spreads over a continuous surface, but less readily, or not at all, over and through a discontinuous medium.

Root excavations carried out under tropical crops have led to a similar conclusion. Thus, concerning the spread of *Armillaria mellea* in tea and coffee plantations in East Africa, Wallace (1935a) writes:

> "Hundreds of infected coffee and tea roots have been examined in Tanganyika and without exception infection has taken place from diseased roots lying against or in fairly close proximity to the dead bush. It is doubtful whether rhizomorphs have an effective range of more than a few feet, but this requires confirmation."

A similar observation has been made by Bade (1927) concerning the same parasite on cacao in the Gold Coast; he states:

> "In each of the hundreds of cases which have been carefully examined infection has been found to be due to actual contact of the host's roots with other diseased roots or rotting wood. It must be added, however, that Dade found no rhizomorphs of *A. mellea* in the soil of these cacao plantations on the Gold Coast; he considered that the heavy texture of the soil may have inhibited their development".

Concerning the spread of *Fomes lignosus* on rubber estates in Malaya, Sharples (1936) declares:

> "Hundreds of diseased trees have been carefully opened up, even to depths of four feet, and in every case, without a single exception, the diseased material, from which infection has spread, has been found in contact with the diseased roots at some point. There has been such a large number of serious outbreaks which have been dealt with entirely successfully by simply keeping in mind that digging must be continued until the actual source of infective material has been discovered and extracted".

How, then, are we to reconcile such observations and

experiments on the subterranean spread of root disease with earlier observations such as those of Fetch (1921)? The answer is simple enough; *it is essential to distinguish between spread of the fungus and spread of the disease.*

Taking *Armillaria mellea* as an example, we may recall Ellis' (1929) report that the rhizomorphs of this fungus had been found to extend for as much as 22 yards from an infected pit-prop in a mine working. Yet Wallace considered that the effective range of the rhizomorphs probably did not exceed a few feet. Further experimental evidence upon this point is obviously required.

The Food-base: The importance of the food-base behind the host-invading mycelium is coming more clearly to be recognised. Thus Bliss (1941) obtained infection with rhizomorphs of *A. mellea* only when these were attached to a suitable substrate; detached rhizomorphs were ineffective. From the first, the importance of organic connection between the infecting hyphae and the food-base behind them has been generally admitted.

Thus, concerning the spread of *Fames lignosus* through the soil, Fetch (1921) observes:

> "It is always attached to its base, *i.e.* the stump on which it originated, and it must derive its food from that source until it meets with other dead wood, or a living plant which it can attack. In all probability it will die if separated from its base, unless it soon meets with fresh material from which it can derive nourishment."

And some years later, concerning the difficulty of replacing the "natural" food-base with pure-culture inoculum of a root-infecting fungus, Fetch (1928) writes:

> "It is not possible to imitate these conditions from pure culture. If the fungus is grown in pure culture on agar, the amount of mycelium which can be used for inoculation is small, and the agar substratum usually soon disappears. If, on the other hand, the fungus is

grown in pure culture on sterilised wood blocks, a similar difficulty is encountered. The volume of a block of wood which can be sterilised with certainty is small; about 1 cubic inch is usual. Moreover, although a vigorous growth of mycelium may occur, it will be found that this is chiefly external, and unless the cultures are allowed to develop for a long period, there will be little penetration of the wood block".

Tunstall (1930) reported that he had successfully inoculated tea bushes with *Fonics noxius* by tying pieces of dead wood permeated by a pure culture of the fungus onto the roots, but not with any other form of inoculum. De Jong (1933) has performed some illuminating inoculation experiments with *Tomes lignosus* on rubber trees.

When pure cultures of the fungus on media other than wood, or the tissue of a fructification, were used as inocula, epiphytic growth of the fungus occurred on the surface of the root, but no decay, and the rhizomorphs eventually disappeared without having achieved infection. Successful infection was obtained only through the use of naturally or artificially infected pieces of wood for inoculation, and then only if the inoculum block was sufficiently large. Inoculation with pieces of wood carrying the fungus as an epiphyte failed to achieve infection more than occasionally, and even then the infection subsequently failed to progress.

De Jong noted that an infection might frequently be established and then fail to progress further; the decayed patches became surrounded by callus and eventually healed over completely. It was noticeable that the process of recovery usually started at about the time when the wood used for inoculation had itself disappeared through decay. Sometimes, however, the decay continued to progress after the exhaustion of the original inoculum, presumably owing to the weakening of the root by the infection. De Jong's results are of particular significance, inasmuch as they underline the essential fact that infection is a "struggle" between the offensive forces of the fungus and the defensive forces of the root; the issue of the "struggle" is likely to be decided by the extent of the reserves available on either side.

Similar results have been reported by other investigators. Thus Bliss (1941) reported failure to infect citrus roots with pure cultures of *Armillaria mellea* on nutrient agar or bran; he achieved success only with naturally or artificially infected wood. Cooley (1942) was unable to infect apple roots with cultures of *Xylaria mali* on agar, or on small chips of infected wood, but secured success by the use of 2-inch lengths of infected twig.

Similar difficulties were encountered by investigators working on *Phymatotrichum omnivorum*. Thus Taubenhaus *et al.* (1929), reporting on the use of freshly-infected roots as inoculum in experiments with *P. omnivorum*, remark:

> "Though the disease has been known for more than 30 years and investigated extensively, definite progress with it has been hindered by the nearly complete failure of experimental inoculations. Thus it was not until 1923 that the first experimental proof was furnished that *Phymatotrichum omnivorum* is the cause of root rot. King (1923) inoculated cotton plants in the field by inserting cotton roots infected naturally or from artificial cultures, in a trench dug next to the roots of the plants, and then filling the trench with earth. Taubenhaus and Killough (1923) succeeded in inoculating cotton plants in sterilised soil with pure cultures on mulberry stems or cotton roots."

It seems probable, therefore, that infection of the host root by mycelium or rhizomorph will be favoured by proximity to the food-base, by efficient translocation of food material along the mycelial strands or rhizomorphs, and by ample reserves in the food-base; it will be hindered by both passive and active defences of the host, *i.e.* by thick cork barriers on the one hand, and by active wound reactions on the other. As Gadd (1936a) has pointed out in a general discussion of this subject, fungi such as the species of *Rosellinia,* which can use leaf-mould and other decaying organic detritus on the surface of the soil as a food-base, possess an advantage in their resulting ability to make a rapid surface spread in favourable *(i.e.* moist and shady) situations; surface spread in such a situation is necessarily more rapid than underground spread.

Experiments are certainly needed to determine the effective parasitic range of rhizomorphs from a food-base; the translocation of food material along the rhizomorphs of *Armillaria mellea* must be efficient enough to account for a spread of such distance over an inhospitable surface as that recorded by Ellis (1922), cited above.

Conduction must be remarkably efficient even along the undifferentiated rhizomorphs of *Pomes lignosus,* for Napper (1938) states that they may be epiphytic on the rubber root for as great a distance as 15 feet ahead of actual penetration by the fungus; infection, of course, takes place immediately in advance of preceding infection, in a continuous wave.

The rhizomorph cycles in the three important root-infecting fungi found on rubber in Malaya are of much comparative interest; whereas the rhizomorphs of *F. lignosus* commonly extend for a distance of 5-15 feet epiphytically along the root in advance of penetration, those of *F. noxius* and *Ganodcnna pseudoferreum* are usually found only some 18 inches in advance of the infected zone (Napper, 1938). In consequence of this difference, the proportion of total infected trees to dead trees will be higher with F. *lignosus* than with the other two parasites.

Concerning this process, Napper (1934) writes:

> "In nature the process is more or less continuous. Infection progresses in a steady wave, the advancing column of rhizomorphs being the sum of the contributions from numerous successive cycles at small phase intervals".

"The production of rhizomorphs from infected tissue

is probably a reaction to starvation conditions, the stimulus to production being brought into play when the fungus has completely gutted its host of all available food material and, although continuing to inhabit the empty skeleton, is forced to draw upon its own reserves for the necessities of life. Rhizomorphs are never produced from newly-infected tissue where there is ample food material available for the nourishment of the fungus".

In view of the nutritional requirements for root infection, it is not surprising that no investigator appears to have reported successful inoculation of any tree root with a spore suspension. Leach (1939) failed to achieve infection either of tree roots or of the cut surfaces of stumps by inoculation with basidiospore suspensions of *Armillaria mellea*.

An apparently similar limitation was found by Garrett (1939) to prevent infection of wheat roots by the ascospores of *Ophiobolus graminis* under natural soil conditions. In numerous experiments with natural soils, with steamed (*i.e.* partially sterilised), soils, and with sand, Garrett was unable to secure any ascospore infection of the roots, either in seedlings or in older plants.

The ascospores germinated well on agar, and the agar cultures thus obtained were able to infect wheat seedlings in a normal manner. Satisfactory infection by ascospores was finally obtained by using sterile wheat seedlings growing in sterilised soil or sand, to which was added a suspension of ascospores in sterile water. By reference to the work of Brown (1922) on leaf infection by the spores of *Botrytis cinerca,* Garrett concluded that the ascospores of *O. raminis* required an external food supply before they could establish root infection. In sterile sand, such food material was assumed to be provided by the root excretions of the wheat seedlings, which would be wholly available to the germinating ascospores; in unsterilised sand, on the other hand, such root excretions would be quickly assimilated by other microorganisms around the roots.

Spread Internal to the Host: A different method of subterranean spread is found amongst the vascular parasites, such as *Verticittium alboatrum* and the wilt-producing species of *Fusarium.*

After passing through the cortex opposite the point of invasion, the fungus enters the water-conducting elements of the xylem (hence the name trichomycoses applied to diseases caused by these fungi), along which it spreads in either direction. It does not leave the vascular cylinder until death of its host plant is imminent.

No means of active spread through the soil except from plant to plant by root contact has been demonstrated for any fungus belonging to this group; quarantine measures based on this assumption appear to have been completely successful in limiting the active spread of Panama disease *(Fusarium oxysporum cubense)* of bananas in Jamaica (Cousins and Sutherland, 1930). From these observations, it follows that *a root does* not become infectious to other roots in contact with it as soon as it has become infected, because the fungus does not emerge from the vascular cylinder until the disease reaches its penultimate phase in the plant.

In an experiment to test this assumption, Roberts (1943) found that approximately twice as many tomato plants became infected in a given time around *Verticillium mizcted* plants that had died as a result of bark-ringing as became infected around similar *yerticillium-inkcted* plants that had not been ringed and still remained, alive. Some data obtained by McKay (1926) in field departments on *Ferticillium* wilt of potatoes may be interpreted by the same hypothesis. McKay reported that the periodical rouging of wilted plants throughout the growing season did not decrease the incidence of infection amongst immediately adjacent plants in the same row, but actually increased it, from 19 to 24.4 per cent (the percentage infection was estimated at the end of the season by determining the proportion of infected tubers). Single-plant rouging of infected individuals would not, of course, be expected to check the spread of the disease, but the slight though probably significant increase in spread was perhaps rather surprising. Mckay himself suggests the following explanation:

> "It seems reasonable to suppose that where diseased plants are rouged the roots of adjoining plants would grow more freely into the area formerly occupied by the diseased plant, and since in rouging diseased plants many *Verticillium* roots would be broken off and left in the ground the chances of spread might be increased rather than decreased. On the basis of Roberts' experiment, Mckay's explanation may perhaps be supplemented by suggesting that the violent removal of wilted plants from the ground, in the act of rouging, hastened the death of those portions of infected roots left behind in the soil, with the result that the fungus emerged from the vascular cylinder of such roots earlier than it would have done had the plants been left undisturbed".

Influence of Soil Temperature

Soil conditions may influence the occurrence of soilborne disease in a variety of ways. Most important is their direct influence upon the parasitic activity of the root-infecting fungi, and their indirect influence through the medium of host resistance. Of less importance is their effect upon survival of the fungi during periods between susceptible crops. If the soil is favourable to the development of disease, the amount of infectious material left behind in the soil for infection of the next susceptible crop is much increased.

It thus happens that soil conditions often influence the survival of a root-infecting fungus more strongly in an indirect way, *viz.* by increasing or by decreasing the amount of inoculum left in the soil, than in a direct way by their effect on the survival of a given quantity of inoculum. To take a specific example, Taubenhaus *et al.* (1937) showed that both incidence of cotton root rot in the current season, and the amount of

Phymatotrichum omnivorum surviving in the soil to the next cotton season, were greater in alkaline than in neutral or acid soils; a greater bulk of the fungus survived the winter in such soils because the amount of inoculum left behind in the soil after cotton harvest was greater.

It is quite possible, therefore, that survival of unit quantity of inoculum was greater in acid or neutral soils than in alkaline soils. Indeed, in experiments upon the take-all disease of wheat, Garrett (1938-1940) and Fellows (1941) showed that survival of *Ophiobolus graminis* was greatest under those conditions of soil temperature, aeration and moisture content least favourable (*i.e.* within the experimental ranges employed by them) to parasitic activity of the fungus.

Another way in which soil conditions may affect the incidence of disease is through their influence upon dispersal or passive spread of the parasite; fungi may be spread, for instance, through the agency of floods, or by irrigation.

The influence of soil conditions upon the incidence of soilborne disease may now be examined and discussed under the following heads:

(i) temperature
(ii) moisture content
(iii) texture
(iv) reaction
(v) organic content
(vi) concentration of plant nutrients.

For this purpose, free use will be made of a previous compilation by Garrett (1938a). In analysing these observations and experimental results, the following generalisation may be kept in mind: soil conditions will exert the greatest direct influence upon those root-infecting fungi that spread chiefly by mycelium external to the host, and the greatest indirect influence, through the physiology of the host upon those fungi that spread only inside the host.

Temperature: Soil temperature exercises an important effect upon the occurrence of soilborne diseases, often determining not only their relative prevalence at different times of the year, and from one year to another, but also their geographical distribution. Our knowledge of the operation of this factor is due very largely to the work of L. R. Jones and his collaborators at the University of Wisconsin, a monographic account of which has been given by Jones *et al.*

The Wisconsin school developed the well known soil temperature tanks for the study of soilborne diseases under controlled temperature conditions; they did not confine themselves to a study of the temperature factor alone, but concurrently investigated the effect of variation in soil moisture content and other component factors of the soil environment. The accurate control of soil temperature that they obtained was later followed by control of air temperature as well, though this refinement has not added much to the important results previously obtained by the use of the soil temperature tanks alone. Experiments carried out in the temperature tanks on development of a particular soilborne disease were supplemented by a study of the growth of the parasite at different temperatures in pure culture; last but not least, extensive records of soil temperatures were collected in the field, for correlation with prevalence of soilborne diseases at different times of the year, and from one year to another.

The importance of such supplementary field observations is emphasised by Jones *et al.*, who remark:

> "The environment under which such plants are grown experimentally is never duplicated under natural conditions of growth. For example, the experimental soil temperatures here discussed were, as a rule, held approximately constant or else at two regularly alternating points as contrasted with the irregular fluctuations in nature. The avoidance of hasty and unsound judgments has been, therefore, best insured in our experience through coupling field studies with

those of the greenhouse. Close observations as to disease development under natural conditions in the field correlated with field temperature records greatly increases one's confidence in the interpretation of such experimental data."

Such a conjunction of extensive field data with the results of controlled temperature experiments in the Wisconsin tanks may be considered to have established certain important soilborne diseases as high temperature diseases, and others as low temperature diseases; diseases favoured by high temperature are cabbage yellows *(Fusarium conglutinans)*, tomato wilt *(F. bulbigcnum* var. *lycopersici)*, flax wilt *(F. lini)*, and seedling blight of wheat *(Gibberella saubinctii)*, whilst diseases favoured by low temperature are onion smut *(Urocystis cepulae)*, tobacco root rot *(Thielaviopsis basicola)*, stem canker of potatoes *(Rhizoctonia solani)*, and seedling blight of corn *(Gibberella saubinetii)*.

The optimum temperature for development of cabbage yellows, a vascular wilt, in the Wisconsin tanks, was found by Tisdale (1923) to be approximately 26°C, whilst the optimum temperature for growth of *Fusarium conglutinans* on agar was the same. The temperature optima for development of two other vascular wilts were only a degree or two. The optimum for tomato wilt was found by Edgerton and SpEland (1920) to be 29°C, and by Clayton (1923a) the optimum temperature for growth of *Fusarium bulbmfium* var. *lycopersici* on agar was determined by Clayton as Jo C. The optimum for flax wilt was found by Jones and Tisdjle (1922) to be 24°-28°C, with an optimum of 26°-28°C for grown of *Fusarium lini* on agar.

A and interesting contrast between the behaviour of the tomato wilt due to *Fusarium bulbigenum* var. *lycopersici* and that of another vascular wilt of tomato due to *Verticillium alboatrum* was demonstrated by Bewley (1922) in England. Bewley found the optimum temperature for development of *Verticillium* wilt to be 21°-23°C; at 25°C and above, development of the disease was inhibited. He inoculated his plants by hypocotyl stab and not through the soil; infection occurred at 25°C, but made comparatively little progress in the vascular cylinder, and seemed to have no adverse effect upon growth of the tomato plant. Bewley also found that wilted plants could be "cured" by transferring them to a glasshouse in which the average temperature was not less than 25°C; if such plants were later moved back to the lower temperature house, the persistence of the "cure" appeared to vary directly as the duration of the high temperature treatment.

The beneficial effect of high temperature in assisting recovery was found to be augmented by shading the plants. At the same time, Bewley gave data concerning the occurrence of these two vascular wilts of tomato in England, which supported both his own conclusions concerning *Verticillium* wilt and those of the American workers concerning *Fusarium* wilt. *Verticillmtn* wilt had been found to occur more frequently in England than *Fusarium* wilt, which was comparatively rare. *Verticillium* wilt normally appeared about the middle of April and increased in intensity up to the second or third week in May. The attacks usually died down during the second half of June, were scarce or absent altogether in July and August, and reappeared at the end of September, when the plants died prematurely. *Fusarium* wilt, on the other hand, occurred at the hottest part of the season, usually in July and August.

The optimum temperature for development of onion smut, a low temperature disease, was found by Walker and Jones (1921) to be 19°-22°C; the optimum temperature for germination of chlamydospores and hyphal fragments was shown by Walker and Wellman (1926) to lie between 15° and 20°C, a range which is lower than the optimum temperature for vegetative growth of the majority of fungi (25°C).

Walker and Jones further discovered that the onion seedling was susceptible to infection by *U. cepulae* only via the young parts of the cotyledonary leaf, which later became covered by the sheathing bases of the outer leaves. The period during which infection of the young plant could occur from the soil was limited to about three weeks; higher soil temperatures were therefore thought perhaps to diminish the chances of infection by reducing the length of this susceptible period.

A similar effect of high soil temperatures in diminishing infection by hastening the emergence of sprouts was thought by Richards (1921, 1923a) to play some part in reducing incidence of stem canker of potatoes due to *Rhlzoctonm solani* at higher soil temperatures.

Thus Frederiksen *et al.* (1938) recently demonstrated in Denmark that deep planting (12 cm.) of tubers aggravated the incidence of stem canker, whereas shallow planting reduced it to a minimum.

Richards determined the optimum temperature for development of stem canker as 18°C; above 24°C, little damage was caused by this parasite. The optimum temperature for growth of *R. solani* in pure culture is approximately 25 °C, but stem canker damage was found to be worst at 18°C not only in potato but also in cotton, pea and bean (Richards, 1923).

In view of the fact that development of some of these hosts was better at temperatures above 20°C, and that of others was better below this temperature, Richards concluded that the temperature optimum of 18°C for stem cankering was connected in some way with the physiology of the parasite rather than with that of the host, nor was it to be wholly explained by the "disease escaping" result of more *rapid* growth (a high temperature may promote more rapid emergence of a plant from the soil, but the same plant may eventually make its greatest growth at a lower temperature). Richards, therefore, suggested that, although 25°C might be the optimum temperature for (most rapid) growth of *R. solani* on agar, a lower temperature might promote the best production of enzymes or toxins essential for development of parasitism. Some strains of this fungus are known, however, to have a much higher optimum temperature for disease production; Leclerg (1934) found the optimum temperature for development of root rot in sugar-beet to be 25°-33°C, and Vasudeva and Ashraf (1939) demonstrated that the root rot of cotton due to *R. solani* in India was most severe at 35°C.

Tobacco root rot, another low temperature disease, was found by Johnson and Hartman (1919) to develop best at 17°-23°C; at 26°C the damage was considerably reduced, and at 29°C the amount of infection was negligible. The optimum temperature for growth of *Thielaviopsis basicola* in pure culture, however, was found to be 28°-30°C. Conant (1927) studied the reaction of varieties of tobacco, susceptible and resistant to *T. basicola,* to infection at temperatures of 20°, 25° and 30° C. He observed a close correlation between cork formation and resistance to infection, both in the resistant variety, which was resistant at all three temperatures, and in the susceptible variety, which was resistant at 30°C. Jones *et al.* (1926), commenting upon Conant's (then unpublished) results, remark:

> "The difference between susceptible and resistant plants may not be primarily in relative potential phellogen activity but in the fact that the resistant plant has some inherent quality or characteristic which so slows down the fungus invasion as to permit time for such cork formation to occur. In any case, however, the 'corking-out' of the fungus is the final and evident expression of the fungus-resisting reaction."

This reservation made by Jones *et al.* in accepting Conant's conclusions in full has been confirmed by Jewett (1938), who was unable to substantiate Conant's claim that resistance to *T. basicola* could be correlated with cork formation; she concluded that resistance was not associated with any Such anatomical factor.

A comparison was made above between two vascular wilt parasites of the tomato, *Fusarium bulbigenum* var. *lycopersici* and *Verticillium alboatrum,* with widely different temperature optima for disease production in the same host. Dickson (1923), in a now classical study of seedling blight in wheat and corn, demonstrated that a single fungus, *Gibberella saubinetii*, might have widely different temperature optima for disease production in different hosts, *viz.* wheat and corn.

Whereas seedling blight developed to the greatest extent in corn at low soil temperatures (8°-16°C), the disease affected wheat seedlings most severely at relatively high soil temperatures (16°-28°C). The optimum temperature for development of seedling blight in each cereal was therefore that furthest removed from the optimum temperature for development of the underground parts of the host plant. Dickson, and Dickson *et al.* (1923), further elucidated the reasons for this difference in temperature optimum for disease development, as follows.

The cell membranes of the seedling sheath of both wheat and corn are at first composed chiefly of pectic substances, and are easily penetrated by *G. saubinetii;* under conditions optimum for plant development, the membranes pass quickly through the susceptible pectin stage to the mature condition, in which celluloses and even lignin or suberin predominate. These less easily hydrolysable cell wall constituents were found to resist penetration by the parasite.

Under adverse conditions of temperature, *i.e.* in corn sown at low temperatures and in wheat sown at high temperatures, this natural development of resistance is retarded; other adverse conditions, such as too low a moisture content of the soil, or too low a light intensity, were found to produce the same result. Dickson and Holbert (1926) demonstrated a similar sequence of events in their study of resistance to seedling blight in susceptible and resistant lines of corn at a range of temperatures.

Results obtained in the investigation of two other soilborne diseases of wheat show a curious discrepancy, upon which JONES *et al.* comment. The optimum temperature for development of the disease caused by *Helminthosporium sativum* in wheat seedlings was found by Mckinney (1923) to be about 28°C; the optimum temperature for development of the take-all disease, due to *Ophiobolus graminis*, was found by Mckinney and Davis (1925) to lie between 12° and 16°C. The optimum temperature for growth of both fungi on agar was 25°C.

Mckinney and Davis pointed out that whereas the high temperature optimum for the disease caused by *H. sativum* was readily explicable as being most favourable for activity of the fungus and least favourable for development of the wheat plant, this expectation was contradicted by the behaviour of the take-all disease. This puzzling anomaly was brilliantly resolved by Henry's (1932) experiments on development of the take-all disease at different soil temperatures.

In unsterilised soil, Henry obtained a disease-temperature curve similar to that of Mckinney and Davis for *O. graminis*, with a peak of 18°C; in sterilised soil, however, the curve more nearly resembled that of Mckinney for *H. sativum*, inasmuch as the temperature optimum had shifted upwards. Henry therefore suggested that the fundamental disease-temperature curve for *O. graminis* had been obtained in sterilised soil, but that in unsterilised soil an uncontrolled factor, *viz.* microbiological antagonism to the parasite, had increased with rise of temperature and, by curtailing the activity of *O. graminis*, had masked the effect of rising temperature upon the host-parasite relationship.

Henry's findings were confirmed by Garrett (1934a), who found the optimum temperature for development of the disease to be approximately 24°C. In a review of published and unpublished work on the relation between cereal foot rot diseases and soil temperature, Garrett (1934) applied Henry's conclusions to other data, and showed that the masking effect of microbiological antagonism, which increased with rise in temperature, appeared to be more pronounced where cultures of the parasites on cooked cereal grains had been used as inoculum than where spore suspensions had been employed. In the ultimate analysis, therefore, the difference between the results of Mckinney (1923) with *L. sativum* and those of

Mckinney and Davis (1925) with *O. graminis* might be attributed to the fact that *H. sativum* produces conidia profusely in culture (a spore suspension was used by Mckinney as inoculum), whereas O. *graminis* is usually sterile (a mycelial culture on oat and barley kernels was therefore employed by Mckinney and Davis as inoculum).

From a review of all available experimental evidence, Garrett (1942) has concluded that the take-all disease is favoured by high soil temperatures, with an optimum around 25 °C. A similar interaction of microbiological antagonism with temperature has been suggested by Schroeder and Walker (1942) to explain their observation that whereas wilt disease of pea due to *Fusarium oxysporum* f. *pisi* race 1 developed in an unsterilised virgin soil most rapidly at 18°C, in sterilised sand with nutrient solution it developed most rapidly at 27°-30°C. As a postscript to the discussion of this particular problem, a quotation from Jones *et al.* must now be credited with prognostic significance: "A second difficulty constantly requiring consideration in such work is that of eliminating the occurrence of variables other than the one under consideration which was, of course, soil temperature in most of the cases here discussed."

Temperature optima may be widely different not only for the development of different parasites in the same host and for the development of the same parasite in different hosts, but even for different phases of infection by one parasite in a single host. For example, Porter and Melhus (1932) found that whereas pre-emergence rot of watermelon seedlings due to *Fusarium bulbigenum* var. *nivcum* was more severe at a low temperature (16°-18°C) than at a high temperature (25°-28°C), the wilting of the young seedlings was more serious at the higher temperature. An almost identical observation has been made by Greeves and Muskett (1936), who found that whereas pre-emergence killing of swede seedlings by *Pythium debaryanum* was greatest at the lowest temperature (6°C) in their experiment, the damping-off of the emerged seedlings was greatest at the highest temperature (23°C). Both pairs of investigators attributed the high incidence of pre-emergence killing at low temperatures to delayed emergence of the seedlings. It follows, therefore, that temperature may determine not only the intensity but also the type of infection within the host plant. Thus Schroeder and Walker (1942) have pointed out that two vascular wilt fungi, *Fusarium oxysporum* f. *pisi* race 1 and *F. niveurn,* cause extensive cortical decay of the roots when the host plant is grown under adverse temperature conditions. Thus *F. oxysporum* f. *pisi* produces root necrosis in peas grown at too high a temperature (27°-30°C), whereas *F. niveum* behaves similarly in water melons grown at too low a temperature (16°-18°C).

Lastly may be considered the effect of soil temperature upon the occurrence of a soilborne disease in the field; this may be expressed in at least three different ways, (i) by variation in incidence of the disease according to the time of year, whereby control can sometimes be secured in practice through alteration in date of planting, (ii) by fluctuation in prevalence of the disease from one year to another, (iii) by limitation of the geographical distribution of the disease. No better examples of the last effect can be given than those quoted by Jones *et al.*: "It is strikingly exemplified in the case of onion smut. Wherever this occurs in the United States, it is an introduced parasite. The fungus spores are regularly transported with diseased bulbs and especially with onion sets from smutted fields, as well as in other ways with such soil. As a result, since first detected in Massachusetts about fifty years ago, it has insidiously spread and established itself across the northern States, Its complete absence in the Gulf States, where it has been regularly introduced with onion sets finds a surprisingly simple explanation when one knows how temperature, during the germination period, functions as a limiting or 'conditioning' factor. The cabbage *Fusarium* (cabbage yellows disease) is quite regularly serious in the second tier of northern States from New Jersey to Iowa and occurs in the southern parts of the most northern States, *e.g.* New York, Wisconsin. It disappears, however, as one proceeds northwards in these northernmost States primarily, we believe, because the summer temperatures are not regularly quite warm enough".

5

Impact on Human-beings and Plants

Damage Caused by Fungi

When a survey is made of the extensive feeding-grounds utilised by the fungi, encompassing as they do every corner of the earth, and every berm of organic material upon it, one is immediately truck by the general similarity between their wants and our own.

Man himself and his animals are of course organic, while the enormous use made by him of other organic materials hardly needs elaboration: his plants md trees, his food, his clothing, and his furniture, are but a few of the appurtenances which fall within this category. When, therefore, the fungi use any of these articles as a means of sustenance, they naturally encroach in what man regards as rightly his—an encroachment which he, equally naturally, resents and combats.

In the warfare which ensues, the casualties are not wholly with the attackers. On the contrary, the losses sustained by ourselves, not only of material and money, but even of human life, are sufficiently grave to afford ground for furious thought.

The operations of the fungus forces, which are in this way arrayed against us, are dominated by their method of attack—in other words, by the manner in which they feed. Thus, from our point of view, they may be divided into three armies, comprising respectively:

(1) Those fungi which attack and cause disease in man and animals.

(2) The fungi which carry out these same processes on plants.

(3) The fungi which confine their attentions to the inanimate organic substances which man employs, viz. timber, foods, clothing, etc.

Fungus Diseases of Man and other Animals

Of these three classes, the most insidious foes—and therefore, in one sense, the most important—are the fungi which, by direct infection or by the contamination of staple foods, jeopardise the health, or even the life, of man himself; or the health and well-being of the domestic animals with which he has surrounded himself.

Fortunately, this class is not large, and their evil effects have been greatly reduced by modern prophylaxis. Yet their activities, even now, are by no means negligible; while the harm which they have wrought in the past has often had very serious consequences. So far as man himself is concerned, the disease which has been most epidemic is undoubtedly ergotism, a malady brought about by the continued use of cereal foods, notably rye flour, which have become contaminated by the sclerotia of Ergot *(Claviceps purpurea)*. This disease is a form of chronic poisoning, which is frequently accompanied by mortification of the fingers and toes. In earlier times, when people were wholly dependent on the produce of local harvests, which were frequently scanty and of poor quality, and when the milling of flour was not carried out with the care which is now exercised, this disease was extremely common. During the Middle Ages, many of the malignant plagues, which ravaged Europe after seasons of rain and scarcity, may be attributed to the presence of ergots in the cereals consumed by the inhabitants.

Those recorded by chroniclers of the 6th and 8th centuries were possibly due, in several instances, to this cause; while less doubt exists concerning the epidemics which prevailed in the 10th century in France and the 11th in Spain.

In 1596, Hesse and the adjoining regions experienced a terrible pestilence, which was attributed by the Medical Faculty of Marburg to ergotism. The same disease appeared in France in 1630; in Voigtlandia (Saxony) in the years 1648, 1649, and 1675; and again in various parts of France, as Aquitaine and Sologne, in 1650, 1670, and 1674. Freiburg and its surroundings were visited by the disease in 1702, other parts of Switzerland in 1715-16, Saxony and Lusatia in 1716; and many districts of Germany in 1717, 1722, 1736, and 1741-42.

The last epidemic in Europe occasioned by ergot is apparently that which, after the rainy season of 1816, descended on Lorraine and Burgundy, where it proved fatal to many of the poorer classes. Today, although cases from time to time occur, ergotism in man is comparatively rare—thanks to the more careful milling of grain, and to the transport facilities which tend to make famine, and the use of damaged grain to which it leads, a thing of the past.

Less preventible, but also less dangerous, are those skin diseases of man, the *dermycoses,* which arise from infection by fungi. The "ringworm" of children—that human fairy-ring—and the "barber's itch" of adults, are both due to minute fungi, *Trichophyton tonsurans* and other species. That form of eczema known as *Favus* is brought about by another, *Achorion Schonleinii;* while thrush, the white pellicle which develops in the mouths of infants, is merely the crust-like mycelium of

Oidium albicans. All these diseases are almost universal, and are spread by the use of insanitary appliances, e.g. dirty towels, feeding bottles, etc. The leprosy mentioned in Leviticus, as causing baldness of the head and beard, may possibly have been brought about by a parasitic fungus.

The fungus diseases of the tropics are much more severe than those common to temperate regions, for the causal organism is frequently more acute and more deeply seated. Of this type is the dreaded "Madura foot/ a disease especially common in Southern India, but occurring also in other tropical regions, and due to various different species of *Madurella, Indiella,* etc. This organism attacks the feet, in which it causes deformity and fungus tumours. The details are singularly revolting to the non-medical mind and may well be passed over. It is, however, of interest to note that the disease known as "Trench feet," suffered by our troops during the late war, was of fungus origin and therefore allied, in a measure, to "Madura foot." These are but a few of the fungus diseases of man which might be cited, but they will serve to show that many of these plants affect mankind a little more intimately than is generally supposed.

The fungus diseases of animals only tend to become obvious to man when they appear on animals of economic importance to him. It is for this reason that, except in the case of those infecting domestic animals, they are in general little known. Many of these fungi are the same as, or similar to, those which attack man. For example, Favus *(Achorion schonleinii)* occurs also in rabbits, cats, fowls, and other birds and mammals. Ringworm *(Irichophyton tonsurans,* etc.) is common to man, oxen, horses, dogs and rabbits. Thrush *(Oidium albicans]* attacks young herbivora and birds as well as children; while ergotism is of course produced in any animals which are fed on grain containing ergots.

A disease of cattle, sheep, and pigs, which has at times had very serious consequences in this and other countries, is "Lumpy Jaw" or "Hard Tongue," due to *Actixomyces bovis.*

This fungus grows on many grasses, and, when eaten with these foods, attacks the animal, forming ulcerating tumours, which seriously interfere with the functions of the organs affected. The chief seat of the trouble is the mouth, where are developed the symptoms suggested by the name of the disease, which runs a chronic course and is very often fatal. In 1907, it was ascertained that on various farms in Norfolk this disease affected from two per cent, to as much as thirty per cent, of the cattle kept. Lumpy Jaw occasionally appears in man, such cases usually arising from chewing grain, or blades of grass—a habit often practised by country people.

A number of gastric diseases of herbivorous animals are due to the eating of fodder, the component plants of which are infected with various parasitic fungi. The symptoms vary, but they are usually akin to those of gastroenteritis. Various bronchial troubles in animals may also be traced to the action of fungi, arising from the inhalation of spores. Of this type is aspergillosis, caused by species of *Aspergillus*. This disease, which is especially common in birds, is accompanied by such pneumonia-like symptoms as purulent inflammation of the lungs, etc. Fishes and other amphibians are very liable to attack from various of the water-moulds *(Saprolegnieae),* of which the effects on goldfish, etc., are quite familiar to breeders of such creatures. From an economic standpoint, the most destructive member is *Saprolegnia jerax,* which has been responsible for a number of serious epidemics among adult salmon in Scottish and English rivers.

Insects, of all animals, are the favourite hosts of parasitic fungi. In one sense this is very fortunate, for in this way are destroyed millions of insects noxious to mankind. The "cholera" of the common housefly, due to *Empusa muscae,* is a case in point. An insect disease which has, however, led to serious financial losses is the "muscardine" disease of silkworms, caused by *Botrytis bassiana.* First discovered in 1835, this affection has on several occasions been so widespread as to threaten the existence of the French silk industry.

An enormous number of other animal diseases of various kinds caused by fungi could be detailed, if space permitted; but enough have been cited to demonstrate that many of the fungi concern mankind more closely than is often realised, and these certainly cannot be dismissed, as many would the "toadstools" with a shrug of the shoulders or a kick of the boot.

Fungus Diseases of Plants

Turning to the parasitic fungi, which are either the primary or secondary cause of disease in plants, we find that they are so multitudinous in number and so far-reaching in effect that it is difficult to decide how much or how little to discuss them. It is scarcely an exaggeration to say that there is no plant of any description which is not a potential victim; while it is in no way wide of the mark to assert that no cultivated crop, no forest, no similar collection of plants in bulk, escapes their ravages. This is undoubtedly by far the most serious aspect of fungus depredation, for the direct attacks of fungi on mankind and the domestic animals are somewhat isolated and sporadic, while the march of modern civilization has naturally diminished enormously the effects of ergotism and other food-transmitted diseases.

On the other hand, the fungus diseases of plants, particularly of the various crops on which mankind is so dependent, have, owing to a variety of reasons, considerably increased. This aspect of the matter, therefore, deserves a more detailed consideration, which may first be given from the historical standpoint.

There is no reasonable doubt that fungus diseases attacked plants long before the advent of plant cultivation—an art which was certainly acquired at a very early stage in man's most primitive evolution, and which, indeed, was one of the first things which differentiated him from his simian forerunners. It is hardly to be expected, however, that these plant diseases should impress themselves upon him at that early stage. There are few records of early husbandry and agriculture, and the extent to which these were affected by parasitic fungi must remain a matter for conjecture. As, however, these arts came more and more into being, improved cultivated varieties were developed from the more useful of the wild plants—a process which incidentally has given us the bulk of the vegetable necessities and luxuries which we enjoy today—and collaterally with this process, the respective fungi which were originally harboured by the wild plants adapted themselves, and continued to affect these cultivated forms.

But, as the latter, by means of intensive cultivation, were gradually relieved of the need of battling for their existence, a battle which their wild confreres are facing daily, and in which the power of resisting disease plays no small part, they have naturally become more vulnerable, and the attacks of the parasitic fungi have, therefore, gained in intensity. In addition, the cultivated plants have become liable to infection by new fungi, whose attacks they had formerly successfully withstood, the damage wrought being thus correspondingly augmented. The history of the evolution of plant diseases is, therefore, in a measure coincident with that of the development of agriculture in its numerous forms.

Early References to Plant Diseases: The first written records of plant diseases which have come down to us occur in early Hebraic writings, where they are treated as if they were then of almost everyday occurrence. The first reference to a disease of crops is found in Genesis, and, if the usual Biblical chronology be accepted, dates back to about the year 1715 BC Here the disease is ascribed to the evil effects of the east wind. In various other books of the Bible, "blasting" and "mildew" of the crops, which were certainly caused by the attacks of parasitic fungi, are referred to in very familiar fashion. As with all scourges, then strange and inexplicable, these pests were considered to be expressions of the wrath and disfavour of the Deity, and this role they filled for many centuries.

Classical References to Plant Diseases: The next large group of citations relating to crop diseases is to be found in classical writings. The early Greeks and Romans had unpleasant practical experience of the evils wrought by these pests, particularly by the red rust of wheat, but were certainly quite unaware as to the cause.

If Pliny is correct, the losses caused by the wheat rust induced Numa Pompilius, the second king of Rome, some six centuries before the coming of the Christian era, to appoint a special day of intercession to, and propitiation of the Corn God, to avert the possible ravages. Whether this is correct or not, we have more definite evidence as to crop diseases from Aristotle, who was born in 384 BC, and a little later from the writings of his famous pupil, Theophrastus the latter, who flourished about 322 BC, appears to have made a special study of them, for in his celebrated history of plants, which appeared a little later than 307 BC, he remarks:

> "As to diseases—they say that wild trees are not liable to diseases which destroy them. Cultivated kinds, however, are subject to various diseases, some of which are, one may say, common to all or to most, while others are special to particular kinds. Generally speaking, cereals are more liable to rust than pulses, and among these barley is more liable to it than wheat; while of barleys, some are more liable than others, and most of all, it may be said, the kind called 'Achillean' Moreover, the position and character of the land make no small difference in this respect; for lands which are exposed to wind and elevated are not liable to rust, or less so, while those that lie low and arc not exposed to wind are more so. And rust occurs chiefly at full moon".

It may be noted that the remarks on the effect of the situation of the land are as true today as when they were written. Here, then, are the first speculations as to the origin of the crop diseases, and although the ancients made much conjecture concerning this, little real progress was made in the elucidation of the matter for many centuries, indeed until comparatively recent times. Some account should, however, be given as to classical belief on these points. In ancient Greece, half a century before the Christian era, so destructive were the diseases of the staple crops of the country that libations were offered regularly to Apollo or to some other god, to gain relief. Pliny, that prodigious compiler of miscellaneous data upon natural history, who lived from 23 to 79 AD, thought that rust, which he describes as "the greatest pest of the crops" was due to frost, although others considered that it was caused by the of Pliny, Ovid and others, is the modern *Puccinia graminis*. It should be noted that Pliny does not agree with Theophrastus and considers wheat to be the cereal most severely affected.

Mediaeval References to Plant Diseases: From the time of the final passing of the Roman Empire, in 476 AD, until the classical revival of the fourteenth and fifteenth centuries, there is an almost complete absence of records of plant diseases. There is, however, one bright light which suddenly shines out in these dark ages, in the writings of Ibn-al-Awam, an Arabian country gentleman, who lived at Seville in Spain during the tenth century. He wrote on agriculture and appears to have been quite conversant with the works of Theophrastus and Pliny, but was also a critical and independent observer of plant diseases, who writes with accuracy and gives much consideration to their control.

In the fourteenth century began in Europe a great recrudescence of interest in Greek and Roman literature, which continued during the next century, and this brought to the attention of the naturalists and agriculturists of that day, the data supplied by the two classical authors in question on plant diseases. It is, however, to be regretted, that the few writers on agriculture in this era of darkness contented themselves with merely repeating in blind acceptance the statements of the Greek and Roman philosophers, no attempts being made to confirm or refute them by actual experiment.

Even so much later as the year 1600, Coler in his Oeconomia includes information on plant diseases which is merely a rehash of the views of ancient writers; and a few years later Lauremberg expressed the opinion that the so-called secret evils, among which he included rust, mildew and carbuncle, were due to the evil influence of certain stars, such as Orion, the Pleiades and others.

The seventeenth century certainly showed the arising of a new spirit, and, among other contributions to scientific thought, numerous references to plant diseases are to be found in the agricultural literature of this period, although the views of most writers who dealt with their supposed causes were still dominated by the classical legend. It was the agriculturists and gardeners, rather than the botanists, who led the way in this revival, and, in addition to descriptions of the symptoms of various diseases, there are given empirical, but often original, methods by which these might be controlled. As an example, we may refer to Parkinson's advocacy of the use of vinegar against canker on fruit trees.

This century is also noteworthy for the promulgation at Rouen, in 1660, of the first legal enactment aimed at the prevention of a plant disease. This directed the grubbing-up and destruction of all barberry bushes, as they were held, as we now know rightly, to have some mysterious influence on the distribution of wheat rust. There is little doubt that this deduction was in the first case due to the practical observation of farmers, and there were probably scientific men in that day who were amused at such superstitious beliefs. Towards the end of this century, Heinrich Hesse stated the three chief causes of the blighting of trees to be as follows:

(1) Superfluous sap with inflammation of the same.

(2) Transplanting a tree so as to give it an orientation differing from its original one, (He says that such trees may be preserved from attack by coating the south side of the trunk with a mixture of cow manure, oat chaff, glue and ashes.)

(3) Using a breadknife in grafting a tree.

With further reference to grafting, he believed that cankers on trees were the result of performing this operation at a time when the moon lies in the sign of the crab or scorpion.

The Study of Plant Diseases in the Eighteenth Century showed further development. The causes were still generally considered to be various occult influences, such as winds, low temperatures, unfavourable soils and the like, but there were seen the beginnings of the theory that these were autogenetic, i.e., that they arose from an unhealthy condition of the affected individuals rather than from any outside influence. Some attempts at classification, based on the supposed causes, were made, and further efforts at control are to be noted.

Thus, in 1755, the state legislature of Massachusetts passed an Act compelling the inhabitants to extirpate all Barberry bushes. In 1790, Forsyth recommended the following mixture to cure disease, defects, and injuries of plants.

> "Take one bushel fresh cow dung, one half bushel lime rubbish from old buildings, one half bushel wood ashes, one-sixteenth bushel pit or river sand. The last three are to be sifted fine before they are mixed. Then work them well together with a spade, and afterward with a wooden beater until the stuff is very smooth, like fine plaster used for the ceilings of rooms."

Soap-suds or urine was employed so as to give the material the consistency of plaster or paint, and after application it was to be covered with a sifting of fine powder, made of "dry powder of wood ashes, mixed with the sixth part of the same quantity of the ashes of burnt bones."

Modern Developments in the Study of Plant Diseases: During the first quarter of the nineteenth century, the classification of plant diseases was still based on their supposed causes, while the autogenetic theory of their origins received general support. As early as 1807, the efficacy of copper salts as a preventive of smut was discussed by Provost; while in 1821, Robertson first recommended sulphur as "the only specific that can be named for the treatment of mildew in peaches. It should be mixed with soap-suds and then applied by dashing it violently against the trees by means of a rose syringe." A variety of this remedy is still widely used against the attacks of the powdery mildews, e.g. of the grape.

Towards the middle of the century the autogenetic theory of plant diseases was shown to be entirely fallacious by the work of Fries, Berkeley, Tulasne, de Bary and others, and it became recognised that the causal agents of disease were certain classes of fungi. This demonstration of the true relation of fungi to plant diseases gradually enabled the methods of prevention to be established on a sound basis, and heralded the dawn of the modern era of disease control. The years between 1878 and 1882 were noteworthy for severe epidemics of plant diseases in Europe, especially in the case of the downy mildew of the vine, caused by *Plasmopara viticola,* which appears to have invaded Europe from America in the first of these years. The European vines were found to be more liable to attack than the American varieties.

These epidemics led to the inauguration of an extensive campaign to discover the most effective remedies, and many experiments were made. We may associate with this new era the name of Millardet, who was responsible for the introduction of the now-celebrated Bordeaux Mixture.

The original discovery was the result of a happy chance, and arose from the practice, obtaining in some vineyards, of sprinkling a few rows of vines, adjoining the road, with a mixture of milk of lime and blue vitriol (copper sulphate), in order to give them a poisonous appearance, and thus avert the inroads of the hungry wayfarer.

It was observed by Prillieux and Millardet, in 1882, that the vines which had been thus treated were less attacked by the mildew than were the others, which fact they naturally ascribed to the treatment.

As a result of further experiments with varying mixtures,

the efficacy of the lime-bluestone wash was demonstrated by Millardet, who published, in 1885, the following recipe:

Bordeaux Mixture

Water—130 litres (34 gallons)

Bluestone—8 kilos (176 Ib.)

Lime—15 kilos (33 Ib.)

This was to be shaken on the plants with a broom. Later tests have shown that much weaker solutions are quite effective. The last forty years have seen steady progress in means of inhibition and control, particularly in the development of methods of spraying much more effective in distribution than was Millardet's somewhat primitive broom, in the devising of special washes for particular diseases, and in the production of disease-resisting varieties.

Effects of Plant Diseases on Mankind: It might not at first be supposed that these depredations of parasitic fungi would be on sufficiently large a scale to imperil human life. Yet such has too often been the result, and a vivid example is to be found no further away than Ireland. In that country, in 1846, the potato crop was devastated by a severe epidemic of the Blight or Downy Mildew *(Phytophthora infestans)*, the results of which can have no more graphic description than that of an eyewitness, one Father Mathew, who wrote:

> "On July 27 I passed from Cork to Dublin, and the doomed plant bloomed in all the luxuriance of an abundant harvest. Returning on August 3rd, I beheld with sorrow mere wastes of putrefying vegetation."

In those seven days the stage had been set for one of the most appalling famines which that country has ever known—a famine which cost the inhabitants tens of thousands of lives, and brought misery and privation to millions more.

Epidemics of such devastating proportions are of course infrequent, but, nevertheless, every season in every country witnesses the exaction by plant diseases of a toll on the food crops of the people which, in the aggregate, is exceedingly heavy. Some slight impression of the extent of this spoliation is gained from the fact that in the United States alone, in the year 1919—a season which was by no means exceptional—plant diseases were responsible for the loss of 190,000,000 bushels of wheat, of 78,000,000 bushels of oats, of 200,000,000 bushels of maize, of 86,000,000 bushels of potatoes, of 58,000,000 bushels of sweet potatoes, and of 18,000,000 bushels of apples; quantities which represent from two per cent, to as much as fifty per cent, of the individual harvests.

When such losses, even in normal years, are possible, it will be readily understood that after really adverse seasons—especially when they are experienced in several countries simultaneously—a condition of scarcity may, and frequently does, arise, so as to place many of the common necessities of life almost beyond the reach of those people whose purchasing powers are limited, and may in extreme cases even lead to that state of semi-starvation which paves the way for so many further ills.

Monetary Losses Occasioned by Plant Diseases

Turning to the financial aspect of the matter, and extending our field so as to include other than food crops, we are brought face to face with a situation which is staggering in its immensity. Thus, considering a few particular crops in one country, the United States, the ravages, measured by a monetary standard, perpetrated in different years by parasitic fungi, are given below, the authority responsible for each estimate being also cited:

- The California vine disease in 1892 $10,000,000 Pierce.
- Wheat rust in the United States, 1898 $67,000,000 Galloway.
- Wheat rust in Illinois, 1885 $1,875,000 Burrill.
- Violet leaf spot in the United States, 1900 $200,000 Dorsett.

- Peach leaf curl in the United States, 1900 $2,335,000 Pierce.
- Potato late blight in New York, 1904 $10,000,000 Stewart.
- Oat smut in the United States, annual $6,500,000 Orton.
- Wheat loose smut, United States, annual $3,000,000 Orton.
- Wheat bunt in the United States, annual $11,000,000 Orton.
- Potato blight in the United States, annual $36,000,000 Orton.

Taking cereal crops alone, the Prussian Statistics Bureau estimated the losses in Prussia, during the year 1891, from grain rust attacking wheat, oats and rye, at £20,628,147. In Australia in the same year, the loss to the wheat harvest caused by rust was estimated at £2,500,000. In New South Wales, in 1916, stem rust *(Puccinia graminis)* alone accounted for damage exceeding two millions sterling. Other crops in various parts of the world exhibit similar losses. Data concerning the ravages perpetrated in this manner in our own country are not available, but would probably be of the same order of magnitude.

Over the crops of the whole world, we envisage damage which can only be reckoned in hundreds of millions sterling. In 1912, no less an authority than the late George Massee stated that "the annual loss throughout the world due to injury caused by parasitic fungi exceeds £ 150,000,000," and then, as if fearing to underestimate, he adds "*Probably double this amount would be nearer the truth*" Such stupendous totals represent, of course, the activities of very many different species of fungi and are, therefore, a little difficult to visualise; but the losses, as a whole, undoubtedly constitute a drain upon the world's vegetable resources which is very disquieting.

Nor is this feeling materially relieved by the additional information that plant diseases "are increasing by importation from other countries and other states; that diseases formerly insignificant are, in many cases, becoming serious; and that long intensive culture of one crop in a locality permits new diseases to develop." In this factor alone we can find evidence of the destructive capabilities of fungi which demands the most earnest attention of every citizen,

Control of Plant Diseases: A question which naturally arises is, whether many of these losses are preventible; and the answer must be emphatically in the affirmative. Unfortunately, the spirit of apathy, which so long has dogged the footsteps of the fungi, is still abroad, and it is this spirit which, to a large extent, is nullifying the efforts of scientific men. Agriculturists are a notoriously conservative race and do not take over kindly to new methods. There is, however, now no doubt as to the efficacy of many methods of disease control, some of which may be briefly indicated.

It is, of course, of the utmost importance to investigate the life history of the parasite (so as to be able to know when to catch the fungus "napping"), for by such knowledge one is enabled to institute preventive, rather than curative, measures, and it should be hardly necessary to state that the former methods are the only useful ones, the value of an infected individual being generally too small to permit of the labour and expense involved in the latter.

On the other hand, many diseases can largely be prevented by spraying the plants, at suitable times, with the wash found most efficacious for the particular disease; while the cost of so spraying the whole of the crop may be only a small fraction of the loss involved if the plants later on become seriously blighted. This practice is, therefore, very generally adopted with certain crops, e.g. potatoes, apples, grapes, gooseberries, and fruit of various kinds.

The question as to the compounds best suited for the individual crops, their strengths, and the times of application, are matters which are too complex to describe here, but are fully dealt with in special treatises. As a typical example of the increased yields obtainable by treatment, we may cite, however, those obtained with apples by the use of ordinary 5-5-50 Bordeaux mixture of the following formula:

Ordinary Bordeaux Mixture 5-5-50

Copper Sulphate (bluestone) ... 5 pounds.

Quicklime ...... 5 pounds.

Water .... 50 gallons.

(This strength is suitable for most plants, whose foliage is not specially sensitive to copper salts.)

Other methods of control involve the destruction of the fungus on the seed (smuts of oats and wheat, or the black rot of cabbage); the grubbing-up of complementary hosts (wheat rust); the avoidance of the use of infected hay or manure (black scab of potato) or of methods which will distribute the disease (e.g., unsuitable location, ground being already invaded by infecting organism); the arranging of a proper rotation of crops, the use of disease-resisting strains, and so on.

By these means, many diseases, which have in the past committed extensive ravages, have been very considerably diminished in extent, and there is little doubt that, in the future, practically all diseases of cultivated crops due to fungi will, if the proper methods be employed, be very largely subjugated.

Depredations of Saprophytic Fungi

The activities of saprophytic fungi in causing damage of various kinds is the aspect most obvious to the layman, but even these are considerably greater, and cover an even wider range, than he would probably imagine, for between such widely-separated items as the expenditure of the thousands of pounds required to renew the woodwork of some public building attacked by dry rot, and the merely nominal loss involved in the discarding of a mouldy crust of bread, occur an enormous number of deteriorations, the variety of which is too great to tabulate, and the value impossible to compute. It is thus idle to attempt any full description of the machinations of the thousands of fungi which in this way are working adversely to man's interests, but a consideration of some of them will serve to show that their effects are more serious than would generally be considered possible.

One or two of the main classes of saprophytic fungi are, therefore, dealt with briefly below:

> The Wood Rots are responsible for enormous damage to stored timber, the wood of buildings, etc. The "leprosy of the house" referred to in some detail in Leviticus, ch. xiv., is probably one of these, although there is not sufficient evidence to identify it with any species that we now know. The extent of the damage that may be done is well illustrated in the case of the old Louis XIII Wing of the Palace of Versailles. Here the oak beams of the roof were recently found to be attacked by the polypore, *Polyporus cryptarum.* The conditions for the development of this fungus, which generally occurs in cellars and mines, were singularly favourable, for the beams in question had been embedded in plaster, thus excluding light and air. The attacked timbers, especially the ends fixed in the masonry, were reduced almost to the consistency of lint, and were easily crushed by slight pressure, but did not crumble.

There are two main classes of fungi which affect timber, differing in their manner of attack:

> In the first class, spores coming in contact with the outside of the timber, germinate and give rise to minute thread-like hyphae, which grow into and spread through the wood, always keeping at some distance from the exterior. The wood may thus show no external sign of attack, except when broken, or when a fruiting body emerges.

In the second class, the attacks of which are more severe, the hyphae grow not only into the wood, but also develop rapidly on the outside, forming matted masses, skins or long strings (*rhizomorphs*from all of which are developed other hyphae which penetrate in their turn into the wood, By means of the rhizomorphs, these fungi are enabled to climb over brick, stone or metal, in their apparent search for more distant woodwork.

Their development, for which water is necessary, sometimes proceeds at an extraordinarily rapid rate. Some, e.g., *Coniophora cerebella* cannot flourish except in really damp wood, and can thus readily be destroyed, even when they have gained a footing, by drying the wood and ensuring a good ventilation for the future. Others, although requiring moisture for the germination of their spores, can, when established, produce the necessary water themselves. This is the case with *Merulius lacrymans*, the fungus responsible for "dry-rot".

This requires air for respiration, during which process it takes up oxygen and gives out water, which may thus be produced to the extent of half as much by weight as the original wood. The presence of these drops of water on the hyphae have been responsible for its name of *lacrymans*—"weeping." This faculty, in conjunction with its frequency of occurrence, and its habit of growth, renders *Merulius lacrymans* probably the most dangerous of any timber-rot, for it can attack the driest wood, and cannot be eradicated by drying or ventilation, unless heat is applied. It affects not only soft woods but also hard woods, including teak, oak and mahogany, which in the space of a few years, or under exceptional circumstances in a much shorter time, may be practically destroyed.

The woody tissue permeated by mycelium is left as a spongy mass of brownish material, which absorbs water so as largely to retain its original dimensions while wet.

The drying-up of the material produces, however, the well-known "dry-rot" effect, showing a multiplicity ot cracks often disposed more or less at right angles to each other. The power of "locomotion" of this fungus is really remarkable, for it has been known to travel for yards along thin tubes containing bell-wire; it can climb up a wall from one floor of a building to the next, and it can even penetrate brickwork through the mortar, involving in a common disintegration the walls no less than the woodwork.

Merulius lacrymans appears to be almost always found in buildings, and very rarely in woods, for which reason it is known to the Germans as "haus-schwamm," or house-fungus.

Among other fungi of the same type are *Polyporus vaporarius* and *Coniophora cerebella*. The first is a dangerous enemy to the woodwork of dwellings, having effects very similar to those of *Merulius*. It, however, is frequently found in woods, both on the roots and stems, gaining a footing by means of a wound. It attacks chiefly soft coniferous woods. The second has already been mentioned as more amenable to treatment than *Merulius*. As the latter, it produces water in respiration, but to a lesser degree.

Since the most effective method of prevention of dry-rot differs with the species responsible for it, it is necessary to identify the individual accurately, and in the case of the three species in question, this is best accomplished by a comparison of their fruiting bodies or sporophores, which are characterised as follows:

(a) That of *Merulius* is rusty, with irregular furrows and ridges.

(b) That of *Polyporus* is white, with regular and definite pores.

(c) That of *Coniophora* is brown, with the surface embossed all over with small pimple-like lumps.

As it is obviously not always a satisfactory procedure to wait for the appearance of fruiting bodies, the strands are also used for this purpose. As counter measures, the timber used should first be thoroughly seasoned so as to render it as dry as possible, and an effective circulation of air should also be arranged within the structure, so as to prevent the conditions from becoming suitable for the development of these fungi. A counsel of perfection, where dry-rot is feared, is to soak the wood, before use, in a solution of creosote or a derivative, or other solution of a fungicidal nature.

Affected wood should be removed and replaced by treated wood, and the adjoining portions must also be treated by coating with creosote or other suitable compound, of which a large number are employed. In addition to coating, drilling and injection is also practised. The adjoining brickwork, etc. should also be sterilised.

This may be accomplished by spraying with a creosote derivative heated to 140-160° F., or with formalin diluted with water. The brickwork should be allowed to dry, and sprayed again as before. It should be noted that some wood-destroying fungi only attack particular timbers, while others are quite immune, and another method is to replace the wood infested by one of these "specialists" by another wood, not subject to attack by the particular fungus. The question of the method of infection, often brought about by the use of timber already affected before felling, must perforce be passed over.

In view of the dangerous consequences entailed in the case of the attack of mine timbers, railway sleepers, etc. the importance of the subject can hardly be exaggerated, while the question of the annual cost of replacing the rotten timber is a most serious one. Thus in the United States alone, about the year 1909, some 110,000,000 ties were being used by the railroads annually; while to maintain 4,000,000,000 posts in use, an annual replacement of 500,000,000 was necessary. In contrast with US practice, in which the ties are not treated, they are in this country creosoted under pressure, and data as to the comparative "tie mortality" would be of interest. Other factors, besides fungus attacks, play a part in the deterioration of timber, but fungi are responsible for no small proportion of the damage. In view of the enormous annual charge thus involved, the question of the prevention of these fungus depredations by impregnation has become of the first importance, and, as with the ravages of the parasitic fungi, increasing attention is being devoted to the subject.

The Meat-Moulds: These have frequently been responsible for considerable damage to imported meat kept in cold storage, this being particularly the case in the year 1918, when, owing to war conditions, the normal length of time during which the meat was in transit was often considerably increased. These moulds are quite common on vegetable debris and animal excreta in the southern hemisphere, where they are found around abattoirs.

Airborne spores are deposited on the carcasses before and during storage, and under favourable conditions develop into the moulds in question. The spores retain their vitality for extended periods, e.g. for as long as two years at a temperature of - 6 °C, in the case of several species.

Much investigation on the nature of these fungi has been carried out for the Food Investigation Board of the Department of Scientific and Industrial Research. The fungus most commonly found on cold-store meat is *Sporotrichum carnis* which occurs in the form of a large number of small white somewhat-woolly patches, which are entirely superficial, and apparently require a temperature not sensibly lower than about freezing-point for effective development.

The same temperature is required for the profuse greyish-white growths, known in the meat-trade as "whiskers," due to species of *Thamnidium* and *Mucor,* which grow very little below the temperature of o° C. These patches do not penetrate into the flesh at all, do no harm to it, and can readily be removed by means of a cloth.

A very common mould found on chilled meat gives rise to "Black Spot," caused by the ramifícation in the tissues of the dark hyphae of *Cladosporium herbarum* which possesses the somewhat remarkable power of growth at a temperature of-6° C. The germination of its spores can also take place at this low temperature, but subsequent growth is more rapid if germination has taken place at the ordinary temperature.

Several other species also occur on cold-store meat, of which fungi *Penicillium expansum* and *Torula botryoides* will also grow slightly at-6° C, but readily at freezing-point, and possibly to an intermediate degree between these temperatures. When profuse growths of any of these fungi are found on meat, it usually indicates that the temperature has been raised to 0°C, or slightly higher, at some time or other during storage—for example, by a temporary breakdown of the refrigerating system. All the growths are merely superficial, the maximum recorded penetration, in the case of "Black Spot," being 4 mm. Further, no poisonous properties are imparted to the meat, which, unless the fungi are associated, as sometimes, with putrefactive bacteria, is quite fit for human consumption. Such meat, however, has frequently been condemned.

In this connection, it should be remembered that at temperatures a few degrees above freezing-point, bacterial growth may proceed so vigorously that the development of moulds is prevented. Such conditions, however, can very seldom occur. The occurrence of the moulds can be prevented by a proper control of humidity and temperature, and by avoiding too long a period of storage, for "Black Spot" may occur with prolonged storage at a temperature several degrees below zero. It is noteworthy that similar moulds are found on meat coming from widely-separated countries, this being evidence of the very cosmopolitan nature of the fungi, which is referred to later.

The Fruit Moulds: The effects of these are familiar to everyone, although the extent of the losses that they occasion is probably not always so well-known. In the case of "Bitter rot," or "ripe rot," of apples, due to *Glomerella rufomaculans* it was estimated by the president of the National Apple Shippers Association that a loss of $10,000,000 was caused to the US apple crop in the year 1900, while there is frequently in certain localities a loss of from 50 to 75 per cent, of the crop.

The most prevalent apple rot, however, is that which attacks stored apples late in the year, caused by a species of *Penicillium* and known as "soft rot" or "blue mould." This requires a bruise or fracture of the skin before it can gain a foothold, uninjured fruit being never affected. The means of prevention are thus obvious.

The result of infection is a light tan-coloured rotten area, which is soft and watery, and which spreads rapidly over the whole fruit; in the case of apples in store or in transit, which are in contact with one another, throughout the whole mass of fruit. Later appear delicate tufts of hyphae, breaking through cracks in the skin over the diseased tissue, and finally growing over the whole mass. These hyphae are at first white, and later bluish green, owing to the development of large numbers of the coloured spores.

A most destructive fungus on European apples and other fruits is *Monilia fructigena*, responsible for "Brown rot," which rapidly causes an entire decay of the whole fruit, which becomes brown, soft and wrinkled, and shows the usual tufts of hyphae. The fruit finally dries up to a hard mass, which may remain on the tree or fall to the ground, and in which the spores winter. The "Brown rot" of plums, cherries, and peaches is caused by the same fungus, which does enormous damage to these crops in the United States however, it appears to affect apples, pears and quinces in a decidedly less degree. This variability of attack of the same species in different countries is by no means uncommon.

The estimated damage inflicted by "Brown rot" on the Ohio peach crop alone in one year was $250,000; while in one year, in Pennsylvania, twenty car-loads of fruit were ruined. It is frequently the case that fruit which appears quite sound on picking becomes seriously damaged in a day or two during transport to market.

This disease is often to be seen on cherries on sale in the shops. As protective measures, spraying at intervals with lime-sulphur wash is of great service, although the injury of fruit by insects will permit the entry of "Brown rot," even when spraying is carried out quite effectively. In addition, all affected fruit must be removed and destroyed, so as to prevent further infection in the following spring. Warm damp weather appears to provide the best, conditions for development of this disease, the ravages of which have been so severe in the United States that many peach and plum cultivators have had to give up growing these fruits.

Among other fruits liable to be seriously damaged by fungus attack during transport is the tomato, which is affected by the dreaded potato blight or downy mildew *(Phytophthora infestans)*, and where, as frequently, no indication of unsoundness may have been seen on packing. The whole plant is often destroyed by this fungus, which has been responsible for much damage. It should be noted that two other species of *Phytophthora* also attack the tomato. Another disease of the tomato, anthracnose, caused by *Colletotrichum phomoides*, occurs chiefly on the ripe fruit, as sunken, discoloured and wrinkled areas, with black specks.

Other Saprophytic Fungi: Probably the most obvious, and, therefore, best known forms of fungus depredation, are the various kinds of "mouldiness" which attack foods of every description when left too long exposed to the air. The blue moulds of bread are familiar to all, as also are those of cheese. In the case of the latter is found one of the few cases where such fungi are not only tolerated, but even encouraged, for, as is well known, special cheeses are deliberately inoculated with special species of *Penicillium*, the consumption of which appears to lead to no ill-effects.

The ordinary species attack any cheese, when their inroads are not regarded with the same favour. In certain cases of the consumption of mouldy provisions, serious effects have ensued; whether due to fungus poisoning, or to the products of decomposition of the matrix, it is impossible to say. In the case of bread, the development of *Penicillium* has been known to take place within a few hours, the mould being in active growth almost before the bread was cold. Among other saprophytes which are the bane of the housewife are those moulds which attack jams, books, linen, silk, and, in fact, any organic material when stored under damp conditions.

It is unnecessary to discuss these in detail; enough having already been given to show that the depredations, of which fungi are the prime or secondary cause, are of an extensive and far-reaching character. The need for dealing more thoroughly with these various depredations in their multitudinous forms—a need even now evident to many—must become more and more pressing as time goes on, with its resultant steady increase of population.

It is admitted generally by those who are in a position to judge, that the world's supply of foods, and of the other necessities of life, cannot be extended, nor even augmented, at a rate commensurate with that of man's increasing needs. The conservation of existing supplies must, therefore, be pressed to its limits, and consequently, there must inevitably come a time when fungus inroads upon staple products can no longer be tolerated.

In view of these facts—even if these plants themselves have no attractions for the layman—the damage which they do to him and his possessions should be sufficient inducement for him to learn more of their life habits. For only in this way can he hope to apply intelligently the various preventive and remedial measures, the devising of which is absorbing the whole energies of numerous investigators in various parts of the world, and which constitute the sole means of reducing the losses which we have indicated.

Bibliography

Aberdeen, J. E. C.: *Introduction to the Mushrooms: Toadstools and Larger Fungi of Queensland,* Queensland Naturalists Club, Brisbane, 1979.

Ahmadjian, V. and Paracer, *Symbiosis: An Introduction to Biological Associations,* University Press, New England, 1986.

Ainsworth, G. C.: *The Fungi: An Advanced Treatise,* Academic Press, New York, 1973.

———————— : *Ainsworth and Bisby's Dictionary of the Fungi,* Commonwealth Mycol, Kew, 1971.

———————— : *Introduction to the History of Mycology,* Cambridge University Press, Cambridge, 1976.

Ajello, L. Georg L.K. Kaplan W. and Kaufman L.: *Laboratory Manual for Medical Mycology,* US Publ., New York 1963.

Amaranthus, M. P. and Luoma D. L.: *Diversity of Ectomycorrhizalfungi in Forest Ecosystems,* International Symposium on Microbiology—Santos, Brazil, 1997.

Ames, L. M.: *A Monograph of the Chaetomiaceae,* MacMillan, New York, 1963.

Ammirati, Joseph F. James A.: *Poisonous Mushrooms of the Northern United States and Canada,* University of Minnesota Press, Minneapolis, 1985.

Arnold, Katya and Swope Sam: *Katya's Book of Mushrooms,* Henry Holt and Co., New York, 1997.

Arora, David and Bowers J.: *Mushrooms of the World Coloring Book,* Dover Publications, New York, 1984.

Arora, David: *All that the Rain Promises,* Ten Speed Press, Berkeley, 1991.

——————— : *Mushrooms Demystified,* Ten Speed Press, Berkeley, 1986.

Arx, J. A.: *The Genera of Fungi Sporulating in Pure Culture,* J. Cramer, Vaduz, 1981.

Barnes, R. S. K.: *Diversity of Living Organisms,* Blackwell Publishers, New Jersey, 2002.

Barnett, J. A. and R. J. Pankhurst: *A New Key to the Yeasts,* North Holland Publ. Co., Amsterdam, 1974.

Barnett, H. L. and Hunter B. B.: *Illustrated Genera of Imperfect Fungi,* MacMillan, New York, 1987.

Barron, G. L.: *The Genera of Hyphomycetes from Soil,* Williams and Wilkins, Baltimore, 1968.

Barron, George: *Mushrooms of Northeast North America,* Lone Pine Publishing, Renton, 1999.

Bell, A.: *Dung Fungi: An Illustrated Guide to Coprophilous Fungi in New Zealand,* Victoria University Press, New Zealand, 1983.

Beneke, E. S.: *Medical Mycology Laboratory Manual,* Burgess Publ. Co., Minneapolis, 1958.

Benjamin, D. R.: *Mushrooms: Poisons and Panaceas,* WH Freeman and Company, New York, 1995.

Bessette, Alan and Sundberg, Walter J.: *Mushrooms: A Quick Reference Guide to Mushrooms of North America,* MacMillan Publishing Co., New York, 1987.

——————— : *Mushrooms of North America in Color: A Field Guide Companion to Seldom-Illustrated Fungi,* Syracuse University Press, Syracuse, 1995.

Bessette, Alan E. Roody, William C.: *North American Boletes,* Syracuse University Press, Syracuse, 2000.

—————— : *North American Boletes: A Color Guide to the Fleshy Pored Mushrooms*, Syracuse University Press, Syracuse, 2000.

Bigelow, Howard E.: *North American Species of Clitocybe*, J. Cramer, Vaduz, 1985.

Bissett, J.: *Mariannaea* and *Paecilomyces*, Canada Agric., Ottawa, 1979.

Boerema, G. H. and Dorenbosch M. M. J.: *The Phoma* and *Ascochyta Species Described by Wollenweber and Hochapfel in Their Study on Fruit-Rotting*, C.B.S. Stud. Mycol, 1973.

Booth, C.: *Methods in Microbiology*, Academic Press, New York, 1971.

Bougher, N. and Syme, K.: *Fungi of Southern Australia*, University of Western Australia Press, Perth, 1998.

Breitenbach, J. and Kranzlin F.: *Entolomataceae, Pluteaceae, Amanitaceae, Agaricaceae, Coprinaceae, Strophariaceae*, Verlag Mykologia, Luzern, 1995.

—————— : *Strobilomycetaceae, Boletaceae, Paxillaceae, Gomphidiaceae, Hygrophoraceae, Tricholomataceae, Polyporaceae (lamellate)*, Verlag Mykologia, Luzern, 1991.

—————— : *Fungi of Switzerland: Non-gilled Fungi*, Verlag Mykologia, Lucerne, 1986.

Brodie, Harold J.: *The Bird's Nest Fungi*, University of Toronto Press, Toronto, 1975.

Brusca, R. C. and Brusca G. J.: *Invertebrates*, Sinauer Associates, Sunderland, 1990.

Buchanan, E. E. and Gibbons N. E.: *Bergey's Manual of Determinative Bacteriology*, Williams and Wilkins, Baltimore, 1974.

Burgess, L. W., Liddell C. M. and Summerell B. A.: *Laboratory Manual for Fusarium Research*, University of Sydney, Sydney, 1988.

Cameron, Eleanor: *Stowaway to the Mushroom Planet*, Little Brown and Company, Boston, 1956.

Carmichael, J. W. and Kendrick W. B.: *Genera of Hyphomycetes*, Alberta Press, Edmonton, 1980.

Chamuris, George P.: *The Non-Stipitate Steriod Fungi in the Northeastern United States and Adjacent*, Berlin, J., Canada, 1988.

Cole, M. Fuhrer B. and Holland A.: *A Field Guide to the Common Genera of Gilled Fungi of Australia*, Inkata Press, Melbourne, 1984.

Cooke, R. C.: *Fungi, Man and His Environment*, Longman, London, 1977.

Corner, E. J. H.: *A Monograph of Cantharelloid Fungi*, Oxford University Press, London, 1996.

—————— : *A Monograph of Clavaria and Allied Genera*, Oxford University Press, London, 1950.

—————— : *Supplement to a Monograph of Clavaria and Allied Genera*, Oxford University Press, London. 1970.

Currah, R. S.: *An Annotated Key to the Genera of the Onygenales*, Systema, Ascomycetum, 1988.

Dashorst, G. R. M. and Jessop J. P.: *Plants of the Adelaide Hills*, Kangaroo Press, Adelaide, 1990.

Delafosse, Claude and Heliadore: *Le Champignon*, Academic Press, France, 1995.

Dennis, R. W. G.: *British Ascomycetes*, J. Cramer, Vaduz, 1981.

Desjardin, Dennis E.: *The Agaricales (Gilled Fungi)*, Mad River Press, California, 1987.

—————— : *Tricholomataceae I: Marasmioid Fungi*, Mad River Press, California, 1987.

Dix, N. J. and Webster, J.: *Fungal Ecology*, Chapman & Hall, London, 1995.

Domsch, K. H. W.: *Compendium of Soil Fungi*, Academic Press, London, 1980.

Dring, D. M.: *Techniques for Microscopic Preparation, Methods in Microbiology*, Academic Press, New York, 1971.

Eberhart, J. L. and Luoma, D. L.: *Truncocolumella Citrina: Pseudotsuga Menziesii*, Victoria B.C., Canadian, 1996.

Ellis, M. B.: *More Dematiaceous Hyphomycetes*, Commonwealth Mycol, Kew, 1976.

Findlay, W. P. K.: *Fungi: Folklore, Fiction and Fact*, Richmond Publishing Company, London, 1982.

Fischer, David W. and Alan E. Bessette: *Edible Wild Mushrooms of North America*, University of Texas Press, Austin, 1992.

Frederick, K. Sparrow and Alfred S.: *A Taxonomic Review with Keys: Ascomycetes and Fungi Imperfecti*, Academic Press, New York, 1973.

Fuhrer, B.: *A Field Companion to Australian Fungi*, Bloomings Books, Melbourne, 2001.

Fuhrer, B. and Robinson R.: *Rainforest Fungi of Tasmania and south-east Australia*, CSIRO, Melbourne, 1992.

Fuller, M. S.: *Lower Fungi in the Laboratory*, Georgia Press, Athens, 1978.

Gilbertson, R. L. and Ryvarden L.: *North American Polypores*, Oslo, Fungiflora, 1987.

————————: *Fungi That Decay Ponderosa Pine*, University of Arizona Press, Tuscon, 1974.

Gilman, J. C.: *A Manual of Soil Fungi*, Iowa State Coll. Press, Ames, 1957.

Ginsburg, Mirra: *Mushrooms in the Rain*, MacMillan Publishing Co., New York, 1974.

Glick, Phyllis G.: *The Mushroom Trail Guide*, Holt, Rinehart and Winston, 1979.

Gregory, P. H.: *The Microbiology of the Atmosphere*, Leonard Hill, Aylesbury, 1973.

Griffiths, K.: *A Field Guide to the Larger Fungi of the Darling Scarp and South West of Western Australia*, Privately Published, Perth, 1985.

Guba, E. F.: *Monograph of Monochaetia and Pestalotia*, Harvard University Press, Cambridge, 1961.

Halling, Roy E.: *The Genus Collybia (Agaricales) in the Northeastern United States and Adjacent*, Cramer, Canada, 1983.

Hawksworth, D. L. Kirk P. M.: *Ainsworth & Bisby's Dictionary of the Fungi*, CAB International, New York, 1995.

Heller, Ruth: *Plants that Never Bloom*, Grosset & Dunlap, New York, 1984.

Hennebert, G. L.: *Echinobotryum*, *Wardomyces* and *Mammaria*, Trans. Brit. Mycol, 1968.

Hesler, L. R. and Alexander H. Smith: *North American Species of Crepidotus*, Hafner Publishing Company, New York, 1965.

______________ : *North American Species of Hygrophorus*, University of Tennessee Press, Knoxville, 1963.

______________ : *North American Species of Lactarius*, University of Michigan Press, Ann Arbor, 1979.

______________ : *North American Species of Gymnopilus*, Hafner Publishing Company, New York, 1969.

Hoog, G. S. de and J. Guarro: *Atlas of Clinical Fungi*, C.B.S., Baarn, Netherlands, 1995.

______________ : *The Black Yeasts and Allied Hyphomycetes*, C.B.S. Stud. Mycol, 1977.

Hoog, G. S. de, M. TH. Smith and E. Guého: *A revision of Geotrichum*, C.B.S. Stud., Mycol,1986.

Hudler, George: *Magical Mushrooms, Mischievous Molds*, Princeton University Press, Princeton, 1998.

Huffman, Donald M.: *Mushrooms and Other Fungi of the Midcontinental United States*, Iowa State University Press, Iowa, 1989.

Ingold, C. T.: *Dispersal in Fungi,* Oxford University Press, London, 1953.

Inui, T. Takeda Y. and Iizuka H.: *Taxonomical Studies on Genus Rhizopus,* MacMillan, New York, 1965.

Jenkins, David T.: *Amanita of North America,* Mad River Press, Eureka, 1986.

Johnson, Emily: *North American Mushrooms Photo Postcards,* Dover Publications, New York, 1988.

Johnson, Sylvia: *Mushrooms,* Lerner Publishing Company, Minneapolis, 1982.

Jong, S. C. and Davis E. E.: *Contibution to the Knowledge of Stachybotrys and Memnoniella in Culture,* Mycotaxon Press, Sydney, 1976.

Jordan, Michael: *The Encyclopedia of Fungi of Britain and Europe,* David & Charles, Devon, 1995.

Jordan, Peter and Wheeler, Steven: *The Ultimate Mushroom Book,* Smithmark Publishers, New York, 1995.

Katsaros, Peter: *Familiar Mushrooms, North America,* The Audubon Society Pocket Guides, New York, 1990.

——————: *Illustrated Guide to Common Slime Molds,* Mad River Press, Eureka, 1984.

Kaveler, Lucy: *The Wonders of Fungi,* John Day Co., New York, 1964.

Kaye, T. Liston A.: *Conservation and Management of Native Plants and Fungi,* Native Plant Society of Oregon, Corvallis, 1997.

Kendrick, B.: *The Fifth Kingdom,* Mycologue Publications, Sidney, 2001.

Kendrick, Bryce: *A Young Person's Guide to the Fungi,* Mycologue, Canada, 1986.

Kibby, Geoffrey: *An Illustrated Guide to Mushrooms and Other Fungi of North America,* Dragons World Ltd., London, 1993.

Kirk, P. M. Cannon, P. F.: *Ainsworth and Bisby's Dictionary of the Fungi*, CABI Publishing, Eureka, 2001.

Krahn, Fernando: *Sebastian and the Mushroom*, Delacorte Press, New York,1976.

Kulik, M. M.: *A Compilation of Descriptions of New Penicillium Species*, U.S. Dept., Sydney, 1968.

Kuyper, Thomas W.: *A Revision of the Genus Inocybe in Europe*, Rijksherbarium, Leiden, 1986.

———————— : *Subgenus Insperma and the Smooth—Spored Species of Subgenus Inocybe*, Rijksherbarium, Leiden, 1986.

Laessoe, Thomas and Lincoff, Gary: *Eyewitness Handbooks/ Mushrooms*, D.K. Publishing, New York, 1998.

Laessoe, Thomas: *The Mushroom Book*, D.K. Publishing, New York, 1996.

Largent, David L. and Harry D.: *How to Identify Mushrooms to Genus II: Field Identificaton of Genera*, Mad River Press, Eureka, 1987.

Largent, David L.: *Entolomatoid Fungi of the Western United States and Alaska*, Mad River Press, Eureka, 1994.

———————— : *How to Identify Mushrooms to Genus?: Macroscopic Features*, Mad River Press, Eureka, 1986.

———————— : *The Agaricales (Gilled Fungi) of California: Hygrophoraceae*, Mad River Press, Eureka, 1985.

Larsen, M. J.: *Phellinus (Hymenochaetaceae): A Survey of World Taxa*, Fungiflora, Oslo, 1990.

Larson, D. W.: *Laboratory Response of Net CO_2 Exchange to Moisture, Temperature and Light*, J. Bot, Canadian, 1987.

Lilly, V. G.: *The Chemical Environment for Fungal Growth*, Academic Press, New York, 1965.

Lincoff, Gary H.: *The Audobon Society Field Guide to North American Mushrooms*, Alfred A. Knopf, New York, 1981.

———————— : *Simon and Schuster's Guide to Mushrooms*, Simon and Schuster, New York, 1981.

Lingelbach, Jenipher: *Hands-on Nature,* Vermont Institute of Natural Science, Woshington, 1985.

Lodder, J.: *The Yeasts: A Taxonomic Study,* North Holland Publ. Co., Amsterdam, 1970.

Luoma, D. L. and Trappe J. M.: *Mycorrhizal Fungi in Ecosystems,* Annual Meeting & Symposium, Bend, 1992.

Luoma, D. L. Eberhart J. L.: *Biodiversityof Ectomycorrhizal Types from Southwest Oregon,* Native Plant Society of Oregon, Corvallis, 1997.

————————— : *Communitystructure and Dynamics of Ectomycorrhizal Fungi in Managed Forest Stands,* USDA Forest Service Gen.Tech., Portland, 1996.

Macdonald, R. and Westerman, J.: *A Field Guide to Fungi of South-eastern Australia,* Nelson, Melbourne, 1979.

Margulis, L. and Schwartz K. V.: *Five Kingdoms: An Illustrated Guide to the Phyla of Life on Earth,* W. H. Freeman, New York, 1998.

Marr, Currie D. and Daniel E. Stuntz: *Ramaria of Western Washington (Bibliotheca Mycologica, Band 38),* J. Cramer, Vaduz, 1973.

Martin, G. W.: *Revision of the North Central Tremellales,* State University of Iowa, Iowa, 1952.

Matsushima, T.: *Lcones Fungorum a Matsushima Lectorum,* Kobe, Japan, 1975.

Matthes-Sears, U.: *In Situ Diurnal Field Measurements of Two Sites on a Coast-inland Gradient,* J. Bot, Canadian, 1986.

McCann, I. R.: *Australian Fungi Illustrated,* Macdown Productions, Melbourne, 2003.

McGinnis, M. R.: *Laboratory Handbook of Medical Mycology,* Academic Press, New York, 1980.

McKenny, Margaret and Daniel E.: *The New Savory Wild Mushroom,* University of Washington Press, Seattle, 1987.

——————— : *The New Savory Wild Mushroom*, University of Washington Press, Washington, 1987.

McKnight, Kent H. and McKnight, Vera B.: *Peterson Field Guides, Mushrooms*, Houghton—Mifflin, Boston, 1987.

Methven, Andrew S.: *The Agaricales (Gilled Fungi) of California*, Mad River Press, Eureka, 1997.

——————— : *The Genus Clavariadelphus in North America*, J. Cramer, Berlin, 1990.

Metzler, Susan and Metzler, Van: *Texas Mushrooms: A Field Guide*, University of Texas Press, Austin, 1992.

Miller, Orson K.: *Mushrooms of North America*, E. P. Dutton, New York, 1980.

Molina R. J.: *Developingan Inventory and Monitoring Protocol for Commercially Harvested Forestmushrooms*, Hillsboro, New Jersey, 1994.

Monsell, Mary Elise: *Crackle Creek*, Atheneum, New York, 1990.

Morquer, R. G. Viala, J.: *Contributiona l'étude Morphogénique du Genre Gliocladium*, Mycol, France, 1963.

Morton, F. J. and Smith G.: *The Genera Scopulariopsis, Microascus* and *Doratomyces*, Mycol, Mycol, 1963.

Moser, Meinhard: *Keys to Agarics and Boleti (Polyporales, Boletales, Agaricales, Russulales)*, Roger Phillips, London, 1983.

Nag, Raj T. R.: *Coelomycetous Anamorphs with Appendage-bearing Conidia*, Mycologue Publications, Canada, 1993.

O'Dell, T. E.: *Diversity and Conservation of Forest Fungi*, USDA Forest Service Gen. Tech., Portland, 1996.

Orr, F. G.: *The Genus Gymnoascus Baranetzky*, Mycopathol, Mycol, 1963.

Orr, Robert T. and Dorthy B. Orr: *Mushrooms of Western North America*, University of California Press, Bereley, 1979.

Orton, P. D.: *British Fungus Flora, Agarics and Boleti, Pluteaceae, Pluteus & Volvariella*, Royal Botanic Garden, Edinburgh, 1986.

Pegler, D. N. Læssøe T. and Spooner B. M.: *British Puffballs, Earthstars and Stinkhorns*, Royal Botanic Gardens, Kew, 1995.

——————: *British Truffles: A Revision of British Hypogeous Fungi*, Royal Botanic Gardens, Kew, 1993.

—————— : *The Genus Lentinus: A World Monograph*, HMSO, London, 1983.

Perry, Phyllis J.: *Let's Learn About Mushrooms*, Harvey House, New York, 1974.

Petersen, Ronald H.: *The Genus Clavulinopsis in North America*, Hafner Publishing Co., New York, 1968.

—————— : *Ramaria subgenus Echinoramaria*, J. Cramer, Vaduz, 1981.

Phillips, Roger: *Mushrooms of North America*, Brown, & Co., Boston, 1991.

Pitt, J.: *The Genus Penicillium*, Academic Press, New York, 1980.

Ramirez, C.: *Manual and Atlas of the Penicillia*, Elsevier Biomedical Press, New York, 1982.

Raper, K. B. and Fennell D. I.: *The Genus Aspergillus*, Williams and Wilkins Co., Baltimore, 1965.

Rebell, G. and Taplin D.: *Dermatophytes: Their Recognition and Identification*, University Miami Press, Coral Gables, 1970.

Rice, Miriam: *Mushrooms for Color*, Mad River Press, Eureka, 1980.

Rifai, M. A.: *A Revision of the Genus Trichoderma*, Mycol, 1969.

Robinson, R.: *Fungi of the South-West Forests*, Department of Conservation and Land Management, Perth, 2003.

Romanova, Natalia: *Once There Was a Tree*, Dial Books, New York, 1985.

Samson, R. A. Hoekstra E. S.: *Interoduction to Food-borne Fungi*, C.B.S., Baarn, Netherlands, 1995.

____________ : *Revision of the Subsection Fasciulata of Penicillium and Some Allied Species*, C.B.S. Stud, Mycol, 1976.

____________ : *Paecilomyces* and *Some Allied Hyphomycetes*, C.B.S. Stud, Mycol, 1974.

Schaechter, Elio: *In the Company of Mushrooms*, Harvard University Press, Boston, 1997.

Schipper, M. A. A. and Stalpers J. A.: *A Revision of the Genus Rhizopus*, C.B.S. Stud., Mycol, 1984.

____________ : *Zygospore Ornamentation in the Genera Mucor and Zygorhynchus*, Persoonia, New York, 1975.

Selsam, Millicent: *Mushrooms*, William Morrow and Co., London, 1986.

Seth, H. K.: *A Monograph of the Genus Chaetomium*, Beih. Nova Hedw, 1970.

Shanks, Kris M.: *Gilled Fungi*, Mad River Press, Eureka, 1997.

Shepherd, C. J.: *Mushrooms and Toadstools of Australia*, Inkata Press, Sydney, 1988.

Shoemaker, R. A. and Kokko E. G.: *Trichurus Spiralis*, Canada Agric., Ottawa, 1977.

Short, P. S.: *History of Systematic Botany in Australasia*, Australian Systematic Botany Society, Melbourne, 1990.

Shuttleworth, Floyd S.: *A Golden Guide, Non-Flowering Plants*, Golden Press, 1967.

Singer, Rolf: *The Agaricales in Modern Taxonomy*, Koeltz Scientific Books, Koenigstein, 1986.

Sissón, Edith H.: *Nature With Children of All Ages*, Prentice Hall Press, New York, 1982.

Smith, A. H. and Zeller S. M.: *A Preliminary Account of the North American Species of Rhizopogon*, Botanical Gardens, Bronx, New York, 1966.

Smith, A. H. and Thiers H. D.: *The Boletes of Michigan*, University of Michigan Press, Ann Arbor, 1971.

Smith, Alexander H.: *The North American Species of Pholiota*, Hafner Publishing Company, New York, 1968.

———————— : *North American Species of Mycena*, University of Michigan Press, Ann Arbor, 1947.

————————: *A Field Guide to Western Mushrooms*, University of Michigan Press, Ann Arbor, 1975.

Snell, Walter H. and Esther A. Dick: *A Glossary of Mycology*, Harvard University Press, Cambridge, 1971.

———————— : *The Boleti of Northeastern North America*, J. Cramer, Lehre, 1970.

Stamets, Paul: *Psilocybin Mushrooms of the World*, Ten Speed Press, Berkeley, 1996.

States, Jack S.: *Mushrooms and Truffles of the Southwest*, University of Arizona, Tucson, 1990.

Stevenson, Steven L. and Stempen, Henry: *Myxomycetes, A Handbook of Slime Molds*, Timber Press, Portland, 1994.

St.Germain, G. and R. Summerbell: *Identifying Filamentous Fungi*, Star Publ. Co., Belmont, California, 1996.

Stolk, A. C. and Samson R. A.: *The Ascomycete Genus Eupenicillium and Related Penicillium Anamorphs*, C.B.S. Stud. Mycol, 1983.

Storer, T. I., Usinger R. L.: *General Zoology*, McGraw-Hill Book Company, New York, 1979.

Tesar, Jenny: *Fungi, Our Living World Series*, Blackbirch Press, Woodbridge, 1994.

Thiers, Harry D.: *Gomphidiaceae*, Mad River Press, Eureka, 1985.

Tokumasu, S.: *Notes on Japanese Oidiodendron*, Trans. Mycol, Japan, 1973.

Tylutki, Edmund E.: *Mushrooms of Idaho and the Pacific Northwest, Discomycetes*, University of Idaho Press, Moscow, 1979.

————————: *Mushrooms of Idaho and the Pacific Northwest: Non-Gilled Hymenomycetes*, University of Idaho Press, Moscow, 1987.

Ulloa, Miguel and Richard T. Hanlin: *Illustrated Dictionary of Mycology*, APS Press, Paul, 2000.

Van Oorschot, C. A. N.: *Taxonomy of the Dactylaria Complex, V. A Review of Arthrobotrys and Related Genera*, C.B.S. Stud., Mycol, 1985.

Waksman, S. A.: *The Actinomycetes*, Williams and Wilkins Co., Baltimore, 1961.

Wang, C. J. K.: *Identification Manual for Fungi from Utility Poles in the Eastern United States*, Type Culture Collection, Rockville, 1990.

Watling, Roy and Evelyn Turnbull: *British Fungus Flora: Agarics and Boleti*, Royal Botanic Garden, Edinburgh, 1998.

Watling, Roy, Norma M. Gregory and Orton P. D.: *British Fungus Flora, Agarics and Boleti*, Royal Botanic Garden, Edinburgh, 1993.

Watling, Roy: *Fungi*, Smithsonian Institution Press, Washington, 2003.

——————: *How to Identify Mushrooms to Genus?: Cultural and Developmental Features*, Mad River Press, Eureka, 1987.

Weber, Nancy Smith: *A Morel Hunter's Companion*, Two Penninsula Press, Lansing, 1988.

Webster, Vera R.: *Plant Experiments*, Childrens Press, Chicago, 1982.

Young, A. M.: *Common Australian Fungi*, New South Wales University Press, Sydney, 2000.

Zambettakis, C.: *Recherches Anatomiqueset Biologiques Sur Sphaeropsidales-Phaeodidymae Des Fungi Imperfecti*, Arch. Mus. Nation, Nat, 1955.

——————: *Recherches Sur la Systématique Des Sphaeropsidales-Phaeodidymae*, Mycol, France, 1954.

Zycha, H. and Siepmann R.: *Mucorales, Eine Beschreibung Aller Gattungen and Arten Dieser Pilzgruppe*, PHBS Press, Lehre, 1969.

Index

D

E

F

G

H

I

L

M

N

O

P

Q

R

S

T

U

V

W

X

Y

Z

ENCYCLOPAEDIA
OF
FUNGI

ENCYCLOPAEDIA OF FUNGI

Vol. 3

Edited by

A.M. Bagulia

ANMOL PUBLICATIONS PVT. LTD.
NEW DELHI - 110 002 (INDIA)

ANMOL PUBLICATIONS PVT. LTD.
H.O.: 4374/4B, Ansari Road, Darya Ganj,
New Delhi-110 002 (India)
Ph.: 23278000, 23261597
B.O.: No. 1015, Ist Main Road, BSK IIIrd Stage
IIIrd Phase, IIIrd Block,
Bangalore - 560 085 (India)
Visit us at: www.anmolpublications.com

Encyclopaedia of Fungi

First Published, 2008

ISBN 978-81-261-3526-4 (Set)

PRINTED IN INDIA

Printed at Mehra Offset Press, Delhi.

Contents

Preface

Fungi are a kingdom of eukaryotic organisms. The fungi are heterotrophic organisms characterised by a chintinous cell wall, and in majority of species, filamentous growth as multicellular hyphae forming a mycelium; while some fungal species grow as single cells. Sexual and asexual reproduction is via pores, often produced on specialised structures or in fruiting bodies. For example, yeasts, moulds, and mushrooms are fungi.

Fungi are found all over the world, and grow in a wide range of habitats, including desert. Most fungi grow in terrestrial environments, but several species occur only in aquatic habitats. Fungi along with bacteria are the primary decomposers of organic matter in almost all terrestrial ecosystems, worldwide. As per estimates there are 1.5 million fungal species of which around 70,000 have been marked. Most fungi grow as thread-like filaments called hyphae, which form mycelium, while others grow as single cells. Human use of fungi for preparation or preservation and other purpose is extensive: yeasts are required for fermentation of beer and bread, some other fungal species are used in the production of soy sauce and tempeh.

Mushroom farming and gathering is a large industry in many countries. Many fungi are producers of antibiotics, including β-lactam antibiotics, such as a pencillin and cephalosporin.

Some of the most well-known types of fungi are edible or poisonous mushrooms. Many species are commercially raised but others are grown in forests. *Agaricus bisporus*, sold as button mushrooms, when small and Protobello mushroom when grownup, are the most commonly eaten species, used in salads, soups, and many other dishes.

Many fungi have important symbiotic relationships with organisms from most, if not all kingdoms. The mycorrhiza symbiosis is most ancient and its existence dates back to 400 million years ago.

This eminent work covers all perfect and realistic features of fungi in an interesting manner. Expectedly, this work would create new waves in the discipline and would prove to be useful for all—students, scholars and general readers.

Editor

1

Diseases Caused by Fungi

Different Fungal Diseases

The Fungi (singular fungus) are a kingdom of eukaryotic organisms. They are heterotrophic and digest their food externally, absorbing nutrient molecules into their cells. Yeasts, moulds, and mushrooms are examples of fungi. The branch of biology involving the study of fungi is known as mycology.

Fungi often have important symbiotic relationships with other organisms. Mycorrhizal symbiosis between plants and fungi is particularly important; over 90 per cent of all plant species engage in some kind of mycorrhizal relationship with fungi and are dependent upon this relationship for survival. Fungi are also used extensively by humans: yeasts are responsible for fermentation of beer and bread, and mushroom farming and gathering is a large industry in many countries.

Fungi and bacteria are the primary decomposers of organic matter in almost all terrestrial ecosystems worldwide. There are an estimated 1.5 million species of fungi with around 70,000 of them having been described.

Phylogeny and Classification

Fungi were originally classified as plants, however they have since been separated as they do not fix their own carbon through chlorophyll based photosynthesis. Traditionally, they are considered heterotrophs, organisms that rely solely on carbon fixed by other organisms for metabolism. Cutting edge research at the Albert Einstein College of Medicine suggests that at least some fungi have the capability to extract energy from ionising radiation.

They apparently accomplish this by using the pigment melanin in lieu of chlorophyll. Ekaterina Dadachova has been studying certain fungi that thrive near Chernobyl in the midst of ecological catastrophe. Her experiments demonstrated the surprising result that their growth is enhanced rather than undermined by radiation. Plans are underway to extend the investigation to other species of fungi.

Fungi are now thought to be more closely related to animals than to plants, and are placed with animals in the monophyletic group of opisthokonts. For much of the Paleozoic Era, the fungi appear to have been aquatic. The first land fungi probably appeared in the Silurian, right after the first land plants appeared even though their fossils are fragmentary. For some time after the Permian-Triassic extinction event, the fungi went into a fungal spike because they were the dominant life forms—nearly 100 per cent of the available record. Fungi absorb their food while animals ingest it; also unlike animals, the cells of fungi have cell walls. For these reasons, these organisms are placed in their own kingdom, Fungi, or Eumycota.

The Fungi are a monophyletic group, meaning all varieties of fungi come from a common ancestor. The monophyly of the fungi has been confirmed through repeated tests of molecular phylogenetics; shared ancestral traits include chitinous cell walls and heterotrophy by absorption, along with other shared characteristics.

The taxonomy of the Fungi is in a state of rapid flux at

present, especially due to recent papers based on DNA comparisons, which often overturn the assumptions of the older systems of classification.

There is no unique generally accepted system at the higher taxonomic levels and there are constant name changes at every level, from species upwards. Fungal species can also have multiple scientific names depending on its life cycle.

Morphology

Though fungi are part of the opisthokont clade, all phyla except for the chytrids have lost their posterior flagella. Fungi are unusual among the eukaryotes in having a cell wall of chitin. All fungi are made up of many thin thread-like structures called hyphae. These hyphae can be one of two types: septate, or coenocytic. Septate hyphae have "walls" between their cells, called septa, though these septa have holes that allow cytoplasm, organelles, and sometimes nuclei to pass through. Coenocytic hyphae have no such marked divisions between cells. Coenocytic hyphae are essentially multinucleate supercells. Parasitic fungi have special structures on their hyphae called haustoria, which penetrate directly into a host organism's cells, allowing nutrients to be taken by the fungus. All of a fungus's hyphae form a structure called the mycelium. In mushroom forming fungi, the mycelium is normally underground. In moulds, the mycelium forms directly on the food source. The only fungi that do not form hyphae or mycelia are yeasts, which are unicellular.

Unlike animals and vascular plants, fungi do not spend the majority of their life cycle in a diploid condition. When a spore begins to grow into a mycelium, the organism is haploid. The haploid mycelium may or may not produce haploid spores asexually. When one haploid organism encounters another, through growth of the mycelium, since fungi are not motile, the two may merge, in a process called plasmogamy. The fungi then enter a heterokaryotic, or multinucleate stage. Usually, one nucleus from one parent fungus will pair off with one nucleus from the other parent. Some fungi spend most of their life cycle in this stage. At a given time, the paired off nuclei will merge, in a process called karyogamy, producing a diploid nucleus. This will normally happen in a separate reproductive structure; in basidiomycetes and ascomycetes, the mushroom. The diploid nucleus will then undergo meiosis to produce haploid nuclei, which are then released as spores to start the cycle once again.

Reproduction

Fungi may reproduce sexually or asexually. In asexual reproduction, the offspring are genetically identical to the "parent" organism (they are clones).

During sexual reproduction, a mixing of genetic material occurs so that the offspring exhibit traits of both parents. Many species can use both strategies at different times, while others are apparently strictly sexual or strictly asexual. Sexual reproduction has not been observed in some fungi of the Glomeromycota and Ascomycota. These are commonly referred to as Fungi imperfecti or Deuteromycota.

Yeasts and other unicellular fungi can reproduce simply by budding, or "pinching off" a new cell. Many multicellular species produce a variety of different asexual spores that are easily dispersed and resistant to harsh environmental conditions. When the conditions are right, these spores will germinate and colonise new habitats.

Sexual reproduction in fungi is somewhat different from that of animals or plants, and each fungal division reproduces using different strategies. Fungi that are known to reproduce sexually all have a haploid stage and a diploid stage in their life cycles. Ascomycetes and basidiomycetes also go through a dikaryotic stage, in which the nuclei inherited by the two parents do not fuse right away, but remain separate in the hyphal cells.

In zygomycetes, the haploid hyphae of two compatible

individuals fuse, forming a zygote, which becomes a resistant zygospore. When this zygospore germinates, it quickly undergoes meiosis, generating new haploid hyphae and asexual sporangiospores.

These sporangiospores may then be distributed and germinate into new genetically-identical individuals, each producing their own haploid hyphae. When the hyphae of two compatible individuals come into contact with one another, they will fuse and generate new zygospores, thus completing the cycle.

In ascomycetes, when compatible haploid hyphae fuse with one another, their nuclei do not immediately fuse. The dikaryotic hyphae form structures called asci (*sing.* ascus), in which karyogamy (nuclear fusion) occurs. These asci are embedded in an ascocarp, or fruiting body, of the fungus. Karyogamy in the asci is followed immediately by meiosis and the production of ascospores.

The ascospores are disseminated and germinate to form new haploid mycelium. Asexual conidia may be produced by the haploid mycelium. Many ascomycetes appear to have lost the ability to reproduce sexually and reproduce only via conidia.

Sexual reproduction in basidiomycetes is similar to that of ascomycetes. Sexually compatible haploid hyphae fuse to produce a dikaryotic mycelium. This leads to the production of a basidiocarp. The most commonly-known basidiocarps are mushrooms, but they may also take many other forms. Club-like structures known as basidia generate haploid basidiospores following karyogamy and meiosis. These basidiospores then germinate to produce new haploid mycelia.

Ecological Role

Although often inconspicuous, fungi occur in every environment on Earth and play very important roles in most ecosystems. Along with bacteria, fungi are the major decomposers in most terrestrial (and some aquatic) ecosystems, and therefore play a critical role in biogeochemical cycles and in many food webs.

Many fungi are important as partners in symbiotic relationships with other organisms, as mutualists, parasites, or commensalists, as well as in symbiotic relationships that do not fall neatly into any of these categories.

One of the most important of these relationships are various types of mycorrhiza, which is a kind of mutualistic relationship between fungi and plants, in which the plant's roots are closely associated with fungal hyphae and other structures.

The plant donates to the fungus sugars and other carbohydrates that it manufactures from photosynthesis, while the fungus donates water and mineral nutrients that the hyphal network is able to find much more efficiently than the plant roots alone can, particularly phosphorus.

The fungi also protect against diseases and pathogens and provide other benefits to the plant. Recently, plants have been found to use mycorrhizas to deliver carbohydrates and other nutrients to other plants in the same community and in some cases can make plant species that would normally exclude each other able to coexist in the same plant community. Such mycorrhizal communities are called "common mycorrhizal networks".

Over 90 per cent of the plant species on Earth are dependent on mycorrhizae of one type or another in order to survive, and it is hypothesised that the presence of terrestrial fungi may have been necessary in order for the first plants to colonise land. Research in 2005 showed that mycorrhizal fungi facilitate significant nitrogen transfer to their plant hosts.

Lichens are formed by a symbiotic relationship between algae or cyanobacteria (referred to in lichens as "photobionts") and fungi (mostly ascomycetes of various kinds and a few basidiomycetes), in which individual photobiont cells are embedded in a complex of fungal tissue. As in mycorrhizas, the photobiont provides sugars and other carbohydrates while the fungus provides minerals and water.

The functions of both symbiotic organisms are so closely intertwined that they function almost as a single organism.

Certain insects also engage in mutualistic relationships with various types of fungi. Several groups of ants cultivate various fungi in the Agaricales as their primary food source, while ambrosia beetles cultivate various kinds of fungi in the bark of trees that they infest. Termites on the African Savanna are also known to cultivate fungi.

Some fungi are parasites on plants, animals (including humans), and even other fungi. Pathogenic fungi are responsible for numerous diseases, such as athlete's foot and ringworm in humans and Dutch elm disease in plants. Some fungi are predators of nematodes, which they capture using an array of devices such as constricting rings or adhesive nets.

Human Uses of Fungi

Fungi have a long history of use by humans. Many types of mushrooms and other fungi are cultivated or picked from the wild eaten directly, while yeast cultures are used in many foodstuffs. The study of the historical uses and sociological impact of fungi is known as ethnomycology.

Cultured Foods

A type of single-celled yeast fungus called *Saccharomyces cerevisiae* yeast is used in baking bread. Yeast is also used to create alcoholic beverages through fermentation. Mycelial fungus is used to make Shoyu (soy sauce) and tempeh.

Fungi are also used to produce industrial chemicals like lactic acid, antibiotics and even to make stonewashed jeans. Some types of fungi are ingested for their psychedelic properties, both recreationally and religiously.

Stilton cheese veined with Penicillium roqueforti.

Some of the most well-known types of fungi are the edible and poisonous mushrooms. Many species are commercially raised, but others must be harvested from the wild. *Agaricus bisporus,* sold as button mushrooms when small or Portobello mushrooms when larger, are the most commonly eaten species, used in salads, soups, and many other dishes.

Other commercially-grown mushrooms that have gained in popularity in the West and are often available fresh in grocery stores include straw mushrooms (*Volvariella volvacea*), oyster mushrooms (*Pleurotus ostreatus*), shiitakes (*Lentinula edodes*), and enokitake (*Flammulina* spp.). There are many more mushroom species that are harvested from the wild for personal consumption or commercial sale. Milk mushrooms, morels, chanterelles, truffles, black trumpets, and porcini mushrooms (also known as king boletes) all command a high price on the market. They are often used in gourmet dishes.

It is also a common practice to permit the growth of specific species of mould in certain types of cheeses that give them their unique flavour. This mould is non-toxic and is safe for human consumption. This accounts for the blue colour in cheeses such as Stilton or Roquefort which is created using *Penicillium roqueforti* spores.

Hundreds of mushroom species are toxic to humans, causing anything from upset stomachs to hallucinations to death. Some of the most deadly belong to the genus *Amanita,* including the destroying angel *(A. virosa)* and the death cap *(A. phalloides),* the most common cause of mushroom poisoning. Stomach cramps, vomiting, and diarrhoea usually occur within 6-24 hours after ingestion of these mushrooms, followed by a brief period of remission (usually 1-2 days). Patients often fail to present themselves for treatment at this time, assuming that they have recovered.

However, within 2-4 weeks liver and kidney failure leads to death if untreated. There is no antidote for the toxins in these mushrooms, but kidney dialysis and administration of corticosteroids may help. In severe cases, a liver transplant may be necessary (Kaminstein 2002). It is difficult to identify a "safe" mushroom without proper training and knowledge, thus it is often advised to assume that a mushroom in the wild is poisonous and leave it alone.

Fly agaric mushrooms (*A. muscaria*) are also responsible for a large number of poisonings, but these cases rarely result in death. The most common symptoms are nausea and vomiting, drowsiness, and hallucinations. In fact, this species is used ritually and recreationally for its hallucinogenic properties. Historically Fly agaric was used by Celtic Druids in Northern Europe and the Koryak people of northeastern Siberia for religious or shamanic purposes.

Fungi in Biological Control of Pests

Many fungi compete with other organisms, or directly infect them. Some of these fungi are considered beneficial because they can restrict, and sometimes eliminate, the populations of noxious organisms like pest insects, mites, weeds, nematodes and other fungi, such as those that kill plants. There is much interest on the manipulation of these beneficial fungi for the biological control of pests.

Some of these fungi can be used as bio-pesticides, like the ones that kill insects (entomopathogenic fungi). Specific examples of fungi that have been developed as bio-insecticides are *Beauveria bassiana, Metarhizium anisopliae,* Hirsutella, *Paecilomyces fumosoroseus,* and *Verticillium lecanii.*

Chytridiomycota

Chytridiomycota is a division of the Fungi kingdom and contains only one class, Chytridiomycetes. The name refers to the chytridium (from the Greek chytridion, meaning "little pot"): the structure containing unreleased spores. In older classifications, chytrid orders (except the recently established Neo-callimasticales and Spizellomycetales) were placed in the Class Phycomycetes under the Subdivision Myxomycophyta of the Kingdom Fungi. Also, in an older and more restricted sense (not used here), the term "chytrids" referred just to those fungi in the order Chytridiales.

The chytrids are the most primitive of the fungi and are mostly saprobic (degrading chitin and keratin). Many chytrids are aquatic (mostly found in fresh water). There are approximately 1,000 chytrid species, in 127 genera, distributed among 5 orders.

Both zoospores and gametes of the chytrids are mobile by their flagella, one whiplash per individual. The thalli are coenocytic and usually form no true mycelium (having rhizoids instead). Some species are unicellular.

Some chytrid species are known to kill amphibians in large numbers. The actual process, however, is unknown. The infection is referred to as chytridiomycosis. Decline in frog populations led to the discovery of chytridomycosis in 1998 in Australia and Panama. Chytrids may also infect plant species; in particular, maize-attacking and alfalfa-attacking species have been described. *Synchytrium endobioticum* is an important potato pathogen.

Batrachochytrium Dendrobatidis

Batrachochytrium dendrobatidis is a chytrid fungus that causes the amphibian disease chytridiomycosis.

Systematics: It is the only known species of the genus Batrachochytrium. The initial classification of the pathogen as a Chytrid was based on zoospore ultrastructure. DNA analysis of the ssu-rDNA has corroborated the view, with the closest match to *Chytridium confervae*.

Morphology: The fungus in the epidermis has a thallus bearing a network of rhizoids and smooth-walled, spherical to subspherical, inoperculate sporangia. Each sporangium produces a single discharge tube. Zoospores have an elongate-ovoidal body and a single, posterior flagellum and possess a core area of ribosomes often with membrane-bound spheres of ribosomes within the main ribosomal mass. A small spur has been observed, located at the posterior of the cell body, adjacent to the flagellum, but this may be an artifact in the formalin-fixed specimens.

The core area of ribosomes is surrounded by a single cisterna of endoplasmic reticulum, two to three mitochondria, and an extensive microbody-lipid globule complex. The microbody closely appose and almost surround four to six lipid globules (three anterior and one to three laterally), some of which appear bound by a cisterna. Some zoospores appear to contain more lipid globules (this may have been a result of a plane-of-sectioning effect, because the globules were often lobed in the zoospores examined). A non-functioning centriole lay adjacent to the kinetosome.

Nine interconnected props attach the kinetosome to the plasmalemma, and a terminal plate is present in the transitional zone. An inner ring-like structure attached to the tubules of the flagellar doublets within the transitional zone has been observed in transverse section. No roots associated with the kinetosome have been observed.

In many zoospores, the nucleus lay partially within the aggregation of ribosomes and was invariably situated laterally. Small vacuoles and a Golgi body with stacked cisternae occurred within the cytoplasm outside the ribosomal area. Mitochondria, which often contain a small number of ribosomes, are densely staining with discoidal cristae.

Life cycle: *Batrachochytrium dendrobatidis* has two life stages—a sessile, reproductive zoosporangium and a motile, uniflagellated zoospore released from the zoosporangium.

Neocallimastigales

Neocallimastigales is a chytrid order in the fungi, found within the stomach of ruminants. Neo-callimastigales is typified by the genus Neo-callimastix.

Reproduction

These fungi reproduce in the stomach of ruminants through the use of zoospores, and have been known to utilise horizontal gene transfer in their development of xylanase (from bacteria) and other glucanases.

Metabolism

Neocallimastigales uses hydrogenosomes to oxidise NADH to NAD+, evolving H_2 as a product. All evidence indicates it as an evolved mitochondrion with mitochondrial DNA.

The Greek termination, "-mastix", referring to "whips", i.e. the many flagella on these fungi, is changed to "-mastig-" when combined with additional terminations in Latinised names. The family name Neo-callimastigaceae was originally incorrectly published as "Neo-callimasticaceae" by the publishing authors which led to the coinage of the misspelled, hence incorrect "Neo-callimasticales", an easily forgiven error considering that other "-ix" endings such as *Salix* goes to Salicaceae. Correction of these names is mandated by the International Code of Botanical Nomenclature, Art. 60. The corrected spelling is used by Index Fungorum.

Synchytrium Endobioticum

Synchytrium endobioticum is a chytrid fungus that causes the potato wart disease or *black Scab*. It also infects some other plants of the Solanum genus, though potato is the only cultivated host.

Systematics: Traditionally, *Synchytrium endobioticum* has been placed to the subgenus *Mesochytrium*, but it has been suggested that on the basis of the mode of germination it should be transferred to the subgenus *Microsynchytrium*.

At least 18 pathotypes of the fungus exist, most of them with quite limited ranges in Central Europe. The most widely distributed is the pathotype 1.

Morphology

Like other Chytridiales, *Synchytrium endobioticum* develops no mycelium. The fungus produces a thick walled structure known as a winter sporangium. It is 25-75 μm in diameter and contains 200-300 spores. Sporangia are clustered into thin-walled soruses. The motile life stage, zoospore is about 0.5 μm in diameter and has one flagellum.

Life cycle

In spring, at higher temperature and moisture, overwintering sporangia germinate to release mobile zoospores which infect suitable host epidermal cells. In infected cells, the summer sporangia develop, which quickly release new populations of zoospores. The infection cycle may be repeated as long as infection conditions are suitable. The infected plant cells swell, divide and surround the dividing zoospores resulting in the wart. Under certain stress conditions some zoospore pairs fuse, resulting in a zygote. The zygote bearing host cells divide, forming eventually the walls of a new winter sporangium. In autumn, the warts rot and disintegrate, releasing new thick-walled resting spores of the fungus into the soil. The diploid resting spores (pro-soruses) undergo a dormancy period and before germination a meiotic division and several mitotic divisions, becoming a sorus.

Ecology

Synchytrium endobioticum is an obligate parasite that infects several plants of the genus *Solanum*. The most favourable conditions for its development are warm temperatures (but not over 20°C) with enough humidity. Winter sporangia can remain viable for up to 20-30 years. It can survive at depths of 50 cm in the soil. Three different fungi have been observed to parasitise the resting sporangia.

S. endobioticum originates from the Andean region of South-America, with now almost worldwide distribution in areas where potatoes are cultivated (absent in most of tropical Africa, Middle East, most of Canada, Japan and Australia).

Zygomycota

Zygomycota, or zygote fungi, are a phylum of fungi. The name of the phylum comes from zygosporangia, resistant spherical spores formed during sexual reproduction. Approximately 600 species of zygomycetes are known. They are mostly terrestrial in habitat, living in soil or on decaying plant or animal material. Zygomycete hyphae may be coenocytic, forming septa only where gametes are formed or to wall off dead hyphae.

Reproduction: A common example of a zygomycete is black bread mould (Rhizopus stolonifer), a member of the Mucorales. It spreads over the surface of bread and other food sources, sending hyphae inward to absorb nutrients.

In its asexual phase it develops bulbous black sporangia at the tips of upright hyphae, each containing hundreds of haploid spores. If the mycelia of complementary mating types are present, the fungus reproduces sexually and produces zygosporangia. Zygosporangia are typically thick-walled, highly resilient to environmental hardships, and are metabolically inert. When conditions improve, however, they germinate to produce a sporangium or vegetative hyphae.

Some zygomycetes disperse their spores in a more precise manner than simply allowing them to drift aimlessly on air currents. *Pilobolus,* a fungus which grows on animal dung, bends its sporangiophores towards light with the help of a light sensitive pigment and then "fires" them with an explosive squirt of high-pressure cytoplasm. Sporangia can be launched as far as 2m, placing them far away from the dung and hopefully on vegetation which will be eaten by an herbivore, eventually to be deposited with dung elsewhere.

Different mechanisms for forcible spore discharge have evolved among members of the zygomycete order Entomophthorales.

Phylogeny: The Zygomycota are generally placed near the base of the fungal phylogenetic tree, having diverged from other fungi after chytrids.

However, molecular phylogenetic investigation reveals that zygomycetes form a paraphyletic group of a number of different evolutionary lineages. The order Glomales was removed in 2001 and elevated to Division Glomeromycota due their lack of zygospore formation, their mycorrhizal habit, and evidence from DNA sequence analysis.

Glomeromycota

Glomeromycota is a phylum within the kingdom Fungi. It includes those species that form arbuscular mycorrhizae with plants. Originally placed in the order Glomales of phylum Zygomycota, these fungi were elevated to their own phylum in 2001 by Walker and SchuBler.

Glomeromycota are characterised by the formation of dichotomously-branching arbuscules in the roots of angiosperms, in an obligate symbiotic relationship known as a mycorrhiza (*pl.* mycorrhizae or mycorrhizas). Arbuscules serve as the main sites of nutrient exchange between the plant host and the fungus. Some species may also form balloon-like structures known as vesicles within host roots. The structures are thought to be for storage of fungal metabolites.

All species of the Glomeromycota are known to reproduce asexually by the formation of blastospores. Blastospore development and morphology and DNA sequence data are used to separate groups within the Phylum.

Ascomycota

Ascomycota is a Division/Phylum of Fungi, and subkingdom Dikarya, whose members are commonly known as the Sac Fungi. Characteristically, when reproducing sexually, they produce non-motile spores in a distinctive type of microscopic cell called an "ascus" (from Greek: askos, meaning "sac" or "wineskin").

These spores are called ascospores. However, some members of the Ascomycota do not reproduce sexually and do not form asci or ascospores. These members are assigned to Ascomycota based upon morphological and/or physiological similarities to ascus-bearing taxa, and in particular by phylogenetic comparisons of DNA sequences.

This monophyletic grouping is an extremely significant and successful group of organisms. Familiar examples of sac fungi include morels, truffles, Baker's yeast, Dead Man's Fingers, cup fungi, and the majority of lichens (loosely termed "ascolichens") such as *Cladonia*. Many plant-pathogenic fungi belong to the Ascomycota.

Commonly seen examples include apple scab, ergot, black knot, and the powdery mildews. Species of ascomycetes are also popular in the laboratory. *Sordaria fimicola, Neurospora crassa* and several species of yeasts are used in many genetics and cell biology experiments. *Penicillium* species on cheeses and in the antibiotic industry are examples of asexual taxa, otherwise known as anamorphs, that belong in the Ascomycota. Prior to definitive phylogenetic research, moulds such as *Penicillium* were sometimes classified in an artificial phylum, called the Deuteromycota.

Ascomycetes versus Ascomycota

In the past, before the recognition of the fungal kingdom, the sac fungi were considered to be a *Class*, not a *Phylum*. The original collective term for them was "Ascomycetes", a label first coined in the 1800s for a rankless non-lichenised taxon based upon the presence of asci. "Ascomycetes" was soon used to include lichenised taxa, and became the standard term, at the class level, for all ascus-bearing species, just as the term "Basidiomycetes" became used for their basidium-bearing counterparts.

Elevation of the taxonomic rank of the Ascomycetes resulted in the names Ascomycetae, Ascomycotina, and finally Ascomycota. The names Ascomycota, Ascomycetes, etc. are based upon the term "ascus". Together, the Ascomycota and the Basidiomycota form the subkingdom Dikarya. The more familiar term, Ascomycetes, is still loosely used, e.g. at fungal forays it is often said of a fungus, such as *Peziza*, "It is an ascomycete, not a basidiomycete" in reference to their sexual reproductive mode. The terms are further abbreviated to "ascos" and "basidos" which are not officially sanctioned technical names.

Modern Classification of Ascomycota

There are 3 subphyla:

- The subphylum Pezizomycotina is the largest subphylum and contains all the Ascomycota that produce ascocarps (fruiting bodies), except for one genus, *Neolecta,* in the Taphrinomycotina. Therefore, it includes virtually all macroscopic "ascos" such as truffles, ergot, ascolichens, cup fungi (discomycetes), pyrenomycetes, lorchels, and caterpillar fungus, as well as many microscopic fungi, e.g. powdery mildews, ring worm fungi, chalkbrood fungus, Laboulbeniales, and most black moulds around sinks and tubs. The older named taxon Euascomycetes is roughly equivalent.
- The subphylum Saccharomycotina comprises most of the "true" yeasts, such as Baker's yeast and Candida which are in general single-celled, or short chains of cells, and reproduce vegetatively by budding rather than by the production of hyphae. As a result, most

were classified in a vaguely defined taxon with the older name Hemiascomycetes.

- The subphylum Taphrinomycotina includes a disparate group of Ascomycota and were only recognised as a distinctive group after the advent of molecular (DNA) analyses. The group is basal to the other subphyla and hence is considered to be more primitive. Consequently the taxon was originally named Archiascomycetes alternatively spelled Archaeascomycetes. It includes both hyphal fungi (Neolecta, Taphrina), and fission yeasts Schizosaccharomyces and the peculiar mammalian lung parasite, *Pneumocystis* that was originally believed to be a protozoan.

Commonly Used but Obsolete Morphologically Defined Class Names

Several obsolete class names, based upon morphology, are still used in informal or introductory discussions. Among those based upon the sexual fruitbodies (teleomorphs) are: the Discomycetes which included all species forming apothecia; the Pyrenomycetes which included all sac fungi that formed perithecia or even pseudothecia, or any structure approaching these morphological structures; and the Plectomycetes which included those Ascomycota that formed cleistothecia. Hemiascomycetes included the yeasts and yeast-like fungi that are now split between Saccharomycotina and Taphrinomycotina, while the Euascomycetes covered the rest of the Ascomycota, now in the Pezizomycotina and Neolecta in the Taphrinomycotina.

Some Ascomycota never reproduce sexually, or are not known to produce asci. These are sometimes called "Mitosporic Ascomycota" because of the production of conidia otherwise known as mitospores, and other asexual structures, all collectively called anamorphic taxa. In some classifications these would have been placed in a separate artificial phylum, the Deuteromycota (also known as Fungi Imperfecti). Molecular analyses can now be used to place these genera and species among ascus-bearing taxa (if they are Ascomycota), or amongst other phyla such as the Basidiomycota (if related to them).

No mitosporic taxa have been found that form a phylum distinct from the other major phyla of Fungi. However, it is extremely difficult in the absence of molecular analysis to link most anamorphic (asexual morphs) fungi to their teleomorphs (sexual counterparts). There are over 250 years of names available (since Linnaeus' Species Plantarum, 1753) for both asexual and sexual components of the same fungi.

For instance, the sexual form of the kerosene fungus is known as *Amorphotheca resinae* while the asexual stage is called *Hormoconis resinae*. Most anamorphic fungi are Ascomycota, and therefore the obsolete classification of the Deuteromycota is largely that of Ascomycota anamorphs.

The Deuteromycota were classified as Coelomycetes if they produced their conidia in small seed-like, fly-speck sized, flask-shaped conidiomata, or structures resembling or approximating them in structure. The Hyphomycetes were those species where the conidiophores (i.e. the hyphae which carry conidia-forming cells on the end) are free or loosely organised. They are mostly isolated but sometimes also appear as bundles of cells aligned in parallel (described as *synnematal*) or as cushion-shaped masses (described as *sporodochial*).

Physical Make-up

The adjective which describes these fungi is "ascomycetous". The majority of ascomycetous fungi grow as a thallus, called a mycelium, consisting of many hyphae which are microscopic multi-branched filaments. If the hyphae of some typical mycelia were laid end to end, they could reach a length of several kilometres. Ascomycota typically produce great numbers of asci at any one time, and these may be contained in a multicellular, often readily visible structure called an "ascocarp" (also called an "ascoma", the fruiting body of ascomycetes). Many exceptions to the structure described above occur, for example in one extreme these fungi are single-celled yeasts, and there is no mycelium, no fruitbody, and the entire cell is converted into an ascus in such ascomycetous yeasts such as Baker's Yeast (*Saccharomyces cerevisiae*).

In the case of lichenised species, the thallus of the fungus defines the shape of the symbiotic colony. Other Ascomycota are dimorphic, which can mean that they can appear either in single- or multicellular form. Other species are pleomorphic, exhibiting multiple asexual forms (i.e. anamorphs as well as a sexual form (a teleomorph).

The ascoma come in multiple forms: cup-shaped, club-shaped, potato-like, spongy, seed-like, oozing and pimple-like, coral-like, nit-like, golf-ball-shaped, perforated tennis ball-like, cushion-shaped, plated and feathered in miniature (Laboulbeniales), microscopic classic Greek shield-shaped, stalked or sessile, solitary or clustered, etc. They can be fleshy or carbonaceous (like charcoal), leathery, rubbery, gelatinous, slimy, powdery, or cob-web-like, etc. They come in multiple colours such as red, orange, yellow, and rarely green or blue, although brown or black are more common.

Except for lichens, the mycelium (if produced) is usually inconspicuous because it is subterranean or embedded in the substrate, and only the ascoma is seen in season. But spectacular, bizarre, or otherwise noteworthy exceptions occur.

Many ascomatous fungi have melanised hyphal walls (referred to as dematiaceous walls), and therefore are black or brown. Black spots on bathroom caulking are often colonies of Ascomycota, e.g. *Cladosporium*. Many moulds that grow on spoiled foods are Ascomycota, and therefore the pellicles or skins that develop on jams, juices, and other foods in containers at home are in fact the thalli of Ascomycota (occasionally Mucoromycotina, and almost never Basidiomycota). Sooty moulds that develop on plants, especially in the tropics are the thalli of many species.

Sometimes it is the mass of asci or ascus-like cells, or conidia or yeast cells that are the conspicuous elements. Pneumocystis species fill lung cavities causing a form of pneumonia (visible in x-rays).

Ascosphaera cysts (asci) fill honey bee larvae and pupae making them appear mummified and chalk-like, hence the name "chalkbrood". Free living yeasts form yeast colonies. Excessive Candida yeast growth in the mouth or vagina is called "thrush" or candidiasis.

The cell walls of these fungi are almost always formed of Chitin and β-Glucans; individual cells are formed from divisions of the hyphae called "septa". These give stability to the hyphae and prevent a great loss of cytoplasm in the event that the cell membrane should be locally damaged.

Mostly the cell divisions are centrally perforated, so they have a small opening in the middle, through which cytoplasm and also nuclei can move more or less freely throughout the system of hyphae. Often hyphae have only one nucleus per cell, and are therefore described as *uninucleate,* but some ascomycetous fungi can also be multinucleate at times.

Metabolism

Like most fungi the Ascomycota principally digest living or dead biomass. To achieve this, they secrete into their surroundings powerful digestive enzymes which break down organic substances into small molecules, which are then absorbed through the cell wall. Many species live on dead plant material such as fallen leaves, twigs, or logs. Others attack plants, animals, or other fungi as parasites and derive their metabolic energy, as well as all the nutrients they need, from the cell tissue of their hosts. Especially in this group extreme specialisation appears; for instance certain species of Laboulbeniales attack only one particular leg of one particular insect species. The Ascomycota also often take up symbiotic relationships – for instance some combine with green algae or cyanobacteria, from which they obtain photosynthetic nutrients, to form lichens; others form symbioses with tree roots as mycorrhizal fungi. There are also carnivorous fungi, which have developed hyphal traps in which they can catch small protists such as amoebae, as well as roundworms (*Nematoda*), rotifers, tardigrades, and small arthropods such as springtails (*Collembola*).

Through their long evolutionary history the Ascomycota have developed the capability to break down almost every organic substance. Unlike most organisms they are able to use their own enzymes to digest plant cellulose and the lignin contained in wood.

Collagen, an abundant structural protein in animals, and keratin (which hair is made of), can also serve as food sources. Exotic examples are given by the ascomycete *Aureobasidium pullulans,* which metabolises wall paint, and the kerosene fungus *Amorphotheca resinae,* which (to the misfortune of the airline industry) feeds on aircraft fuel, and in tropical regions sometimes blocks fuel pipes. Others resist osmotic stress to grow on salted fish, and a few live in water.

Distribution and Living Environment

The Ascomycota are present in all land ecosystems worldwide – they even occur in Antarctica – and their spores and hyphal fragments are distributed through the atmosphere and fresh water environments, as well as ocean beaches and tidal zones. The distribution of individual species is very variable: some are found on all continents, while for example the white truffle *Tuber magnatum,* which is much sought after for culinary purposes, only appears in isolated locations in Italy and France. Plant parasitic species are often restricted by their host distributions. *Cyttaria* is only found on *Nothofagus* (Southern Beech) in the Southern Hemisphere.

Reproduction

Asexual Reproduction: Asexual reproduction is the dominant form of propagation in the Ascomycota, and is responsible for the rapid expansion of these fungi into areas which were previously not colonised. It occurs through reproductive structures, the "conidiospores" (or "conidia"), which are genetically identical to the parent and mostly have just one nucleus. They are also called "mitospores" due to the way they are generated through the cellular process of mitosis. They are generally formed on the ends of specialised hyphae, the "conidiophores". Depending on the species they may be dispersed by wind or water, or also by animals.

Asexual Spores

In order to further classify the Ascomycota in the asexual stages, it is important to consider the spores, which can be distinguished by colour, form and the way they are separated into cells. The most frequent types are the single-celled spores which are designated *amerospores*. If the spore is divided into two by a cross-wall (septum), it is a *didmyospore*. When there are two or more cross-walls the classification depends on the shape. If the septa are *transversal*, like the rungs of a ladder, it is a *phragmospore* whilst if they form a net-like structure it is a *dictyospore*. In *staurospores* ray-like "arms" radiate from a central body; in others (*helicospores*) the entire spore is wound up in a spiral like a spring. Finally very long worm-like spores, of which the ratio length: diameter is more than 15:1, are called *scolecospores*.

Conidiogenesis and Dehiscence

One distinguishes:

- Acervular conidiomata, or *acervuli*, which develop in the host and can thus be:
  1. *Subcuticular*, lying under the outer layer of the plant (the cuticle),
  2. *Intraepidermal*, inside the outer cell layer (the epidermis),
  3. *Subepidermal*, under the epidermis, or
  4. Deeper inside the host.

Mostly they develop a flat layer of relatively short conidiophores which then produce masses of spores. The increasing pressure finally leads to the splitting of the epidermis and cuticle and so allows the conidia to escape.

- Pycnidial conidiomata or pycnidia, which unlike the acervuli form in the fungal tissue itself, and which are generally shaped like a bulging vase. The spores are released through a small opening at the apex, the ostiole.

Two further important characteristics of the anamorphs of the Ascomycota are the *conidiogenesis*, the fashion in which the spores are formed, and their *dehiscence*, i.e. how they separate from the parent structures.

The former corresponds to Embryology in animals and plants and can be divided into two fundamental forms of development: *blastic* conidiogenesis, where the spore is already evident before it separates from the conidiogenic hypha which is giving rise to it, and *thallic* conidiogenesis, where first a cross-wall appears and then the thus created cell develops into a spore.

These two basic types can be further classified as follows:

- *Blastic-acropetal* (repeated budding at the tip of the conidiogenic hypha, so that a chain of spores is formed with the youngest at the tip),
- *Blastic-synchronous* (simultaneous spore formation from a central cell, sometimes with secondary acropetal chains forming from the initial spores),
- *Blastic-sympodial* (repeated sideways spore formation from behind the leading spore, so that the oldest spore is at the main tip),
- *Blastic-annellidic* (each spore separates and leaves a ring-shaped scar which is concentrically inside the scar left by the previous spore),
- *Blastic-phialidic* (the spores arise and are ejected from the open ends of special conidiogenic cells called

phialides which remain constant in length; an example is the anamorph of *Penicillium*),

- *Basauxic* (where a chain of conidia, in successively younger stages of development, is emitted from the mother cell),
- *Blastic-retrogressive* (spores separate off by formation of crosswalls near the tip of the conidiogenic hypha, which thus becomes progressively shorter),
- *Thallic-arthric* (double cell walls split the conidiogenic hypha into cells which develop into short, cylindrical spores called arthroconidia; sometimes every second cell dies off, leaving the arthroconidia free),
- *Thallic-solitary* (a large bulging cell separates from the conidiogenic hypha, forms internal walls, and develops to a *phragmospore*).

Essentially dehiscence can happen in two different ways. In the schizolytic variant a double dividing wall with a central lamella (layer) forms between the cells; the central layer dissolves to release the spores. In the case of rhexolytic dehiscence on the other hand the cell wall which joins the spores on the outside simply degenerates and sets free the conidia.

Heterocaryosis and Parasexuality

A significant number of Ascomycota species either have no sexual stage or none is known. In spite of this, there are two ways in which they can conserve their genetic diversity: Heterocaryosis and Parasexuality.

The former happens simply through the merging of two hyphae belonging to different individuals, a process known as anastomosis. As a result, there are more cell nuclei than normal in the mycelium and they come from genetically different parent organisms.

Parasexuality, on the other hand, refers to a phenomenon where two cell nuclei merge without any sexual process and the chromosome count is doubled. This involves a complex form of the type of cell division called mitosis, where there is *crossing over* or *recombination*, i.e. an exchange of genetic material between corresponding pairs of chromosomes. In sexual reproduction, in contrast, crossing over occurs only during meiosis. Finally the chromosome count will be restored to normal by haploidisation, whereby the nucleus splits into two parts each having a single set of chromosomes, with each daughter genetically different from the original parents.

Sexual Reproduction

Sexual reproduction in the Ascomycota is marked by a characteristic structure, the *ascus*, which distinguishes these fungi from all others. An ascus is a tube-shaped vessel, a *meiosporangium*, which contains the sexual spores produced by meiosis. The latter are called *ascospores* in contrast to the asexual *conidiospores*.

Apart from exceptions such as Baker's Yeast (Saccharomyces cerevisiae), almost all fungi of the Ascomycota are haploid, so their nuclei only contain one set of chromosomes, which makes them especially susceptible to mutations. During sexual reproduction there is a diploid phase (with two sets of chromosomes), which as a rule is very short. Then meiosis occurs, generally very soon, so that the haploid state is re-established.

Formation of Sexual Spores

The sexual part of the life cycle commences when two suitable hyphae meet each other. These come from the same web of hyphae which can also generate asexual spores. The first deciding factor as to whether conjugation—that is, sexual merging—will occur, is whether the hyphae belong to the same organism, or whether they come from different individual fungi. Whilst many species are thoroughly capable of self-propagation, i.e. they are homothallic, others need non-identical partners and so are heterothallic. Besides this, the two hyphae in question must also belong to the same mating type. Mating types are a peculiarity of the fungi and correspond roughly to the sexes in plants and animals; however one species may have more than two mating types.

In the case of compatibility, gametangia form on the hyphae; these are the generative cells for the gametes, in which numerous nuclei gather. A very fine hypha, called the trichogyne, which grows out of one gametangium, now termed the ascogonium, makes a passage to a gametangium of the other individual, which is then the antheridium. Nuclei then pass from the antheridium (playing a 'male' role) to the ascogonium (playing a 'female' role).

Unlike the process in animals and plants, after the union of the cytoplasms of the two gametangia (plasmogamy), the merging of the nuclei (karyogamy) does not usually occur immediately. Instead, the nuclei which have migrated in from the antheridium pair up with the nuclei of the ascogonium, but remain separate next to their partners.

With this the dikaryophase of the life cycle begins; during this time the pairs of nuclei repeatedly synchronously divide, so that a great number are produced. In all probability the dikaryophase is an evolutionary adaptation which serves to exploit the potential of sexual reproduction to the full in circumstances where it is a rare event for different individuals to meet each other.

After the genetic raw material has been increased by repeated division, recombination will take place independently in each pair during meiosis, so that the greatest possible quantity of genetically different spores will arise. In the red algae (Rhodophyta), a similar solution to the corresponding problem evolved independently.

Next millions of new dinucleate hyphae, into each of which two nuclei migrate, emerge from the fertilized ascogonium. They are also called ascogenous or fertile. They are fed by ordinary uni- or mononucleate hyphae (with only one nucleus), which are also called sterile. The tissue of sterile and fertile hyphae now grows in many cases into a macroscopically visible fruiting body, the ascocarp, which may contain millions of fertile hyphae.

In the actual fruiting layer, the hymenium, the asci now appear. At one end of an ascogenous hypha, there develops a U-shaped hook, which points back opposite to the general growth direction. The two nuclei contained in the terminal cell then divide in such a way that the threads of their mitotic spindles run parallel, and thus two pairs of genetically different daughter nuclei arise, with one daughter of each pair near the point of the hook, and the other in the base part of the hypha. Then two parallel cross-walls appear, dividing the hypha into three sections: that at the point of the hook with one nucleus, that at the base of the original hypha with one nucleus, and the middle U-shaped part with two nuclei.

If the positioning in the fruiting layer is right, the karyogamic fusion of the nuclei finally takes place in the U-shaped cell, creating the diploid zygote. It lengthens to form an elongated tube-shaped or cylinder-shaped capsule, the actual ascus. Then meiosis occurs, giving rise to four haploid nuclei. This is almost always followed by a further mitotic division, so that the ascus ultimately has eight daughter nuclei.

These become enclosed, together with some of the cell plasma, each by their own membranes, and generally with a hard cell wall. Thus the dissemination cells (the ascospores) develop, lying initially like peas in a pod inside the ascus. Later, when an appropriate opportunity presents itself, they are liberated.

Not having flagella, ascospores are disseminated in various other ways: some are spread by wind and with others the ripe ascus breaks open on contact with water to set free the spores. Certain species have evolved regular 'spore cannons' which can eject them up to 30 cm. away. When the spores reach a suitable substrate, they germinate, form new hyphae, and so restart their life cycle, which has come full circle.

The form of the ascus is important for classification and is divided into four basic types: unitunicate-operculate, unitunicate-inoperculate, bitunicate, or prototunicate.

Ecology

The Ascomycota fulfil a central role in most land-based ecosystems. They are important decomposers which break down such organic materials as dead leaves, twigs, fallen trees, etc. and help the detritivores (animals which live off this decomposing material) to obtain their nutrients. By processing substances like cellulose or lignin, which are otherwise difficult to exploit, they take on an important place in the natural nitrogen cycle and the carbon cycle.

Inversely the fruiting bodies of the Ascomycota provide food for a very diverse set of animals from insects and slugs and snails (*Gastropoda*) to rodents and larger mammals such as deer and wild boars.

Fungi of the Ascomycota are also known for their numerous symbiotic relationships with other organisms.

Lichens

Probably since early in their evolutionary history the Ascomycota have "domesticated" green algae (Chlorophyta), as well as occasionally other types of algae and cyanobacteria. Together they form the mutualistic associations known as lichens, which can survive in the least hospitable regions of the earth, including the Arctic, the Antarctic, deserts and mountaintops, and can withstand temperature extremes from -40°C to +80°C. While the photoautotrophic algal partner creates metabolic energy through photosynthesis, the fungus offers a stable supportive framework and protects from radiation and drying out. Around 42 per cent of the Ascomycota (numerically about 18,000 species) form lichens, and almost all the fungal partners of lichens belong to the Ascomycota—the proportion of Basidiomycota is probably only two to three per cent.

Mycorrhizal Fungi and Endophytes

Members of the Ascomycota make two particularly important types of relationship with plants: as *mycorrhizal* fungi and as *endophytes*. The former make symbiotic associations with the root systems of the plants, which for some trees, especially conifers, can be of vital importance, enabling the uptake of mineral salts from the soil.

The fungal partner is in a much better position to absorb minerals due to its finely divided mycelium, whilst the plant provides it with metabolic energy in the form of photosynthetic products. Cases are even known where mycorrhizal fungi can transport nutrients from one plant to another, stabilising the recipient. It is likely that mycorrhizal associations enabled the conquest of the land by plants–in any case the earliest known fossils of land plants have mycorrhizae.

Endophytes on the other hand live inside plants, especially in the stem and leaves, but generally do not damage their hosts. The exact nature of the relationship between endophytic fungus and host is not yet well understood, but it seems that this form of colonisation can bestow a higher resistance against insects, roundworms (nematodes), and bacteria; also it can enable or augment the production of poisonous alkaloids, chemicals which can affect the health of plant-eating mammals.

Symbiotic Relationships with Animals

A series of Ascomycota species from the genus *Xylaria* are found in the nests of leafcutter ants and other fungus-growing ants of the tribe *Attini* and in the fungal gardens of termites (*Isoptera*). Since they do not generate fruiting bodies until the insects have left the nests, it is suspected that, as confirmed in several cases of Basidiomycota species, they may be cultivated.

On the other hand bark beetles (*Scolytidae*) are certainly important symbiotic partners. The female beetles transport the spores to new hosts in characteristic tucks in their skin, the *mycetangia*.

There they eat tunnels in the wood, which lead into large chambers in which they lay their eggs. At this time the spores are released and give rise to hyphae which unlike the beetles can digest the wood. The beetle larvae feed on the fungus and after they have metamorphosed into the adult state they again carry spores with them to renew the cycle of infection. A well-known example of this is Dutch elm disease, caused by fungus *Ophiostoma ulmi*, being carried by the European elm bark beetle *Scolytus multistriatus*.

Importance for Humans

Ascomycetes make many contributions to the good of humanity, and also have many ill effects.

Harmful Interactions

One of their most harmful roles is as the agent of many plant diseases. For instance:

- Dutch Elm Disease, caused by the closely related species *Ophiostoma ulmi* and *Ophiostoma novo-ulmi*, has led to the death of many elms in Europe and North America.
- The originally Asian *Cryphonectria parasitica* is responsible for attacking Sweet Chestnuts (*Castanea sativa*), and virtually eliminated the once-widespread American Chestnut (*Castanea dentata*).
- A disease of Maize (*Zea mays*), which is especially prevalent in North America, is brought about by *Cochliobolus heterostrophus*.
- *Taphrina deformans* causes leaf curl of peach.

Claviceps Purpurea on Rye (Secale Cereale)

- *Uncinula necator* is responsible for the disease Powdery mildew, which attacks grapevines.
- Species of *Monilia* cause brown rot of stone fruit such as peaches (*Prunus persica*) and sour cherries (*Prunus ceranus*).

- Members of the Ascomycota such as *Stachybotrys chartarum* are responsible for fading of woollen textiles, which is a great problem especially in the tropics.
- Blue-green, red and brown moulds attack and spoil foodstuffs - for instance *Penicillium italicum* rots oranges.
- Cereals infected with *Fusarium graminearum* contain mycotoxins like deoxynivalenol (DON), which can lead to skin and mucous membrane lesions when eaten by pigs.
- Ergot (*Claviceps purpurea*) is a direct menace to humans when it attacks wheat or rye and produces highly poisonous and carcinogenic alkaloids, causing ergotism if consumed. Symptoms include hallucinations, stomach cramp, and a burning sensation in the limbs ("Saint Anthony's Fire").
- *Aspergillus flavus*, which grows on peanuts and other hosts, generates aflatoxin, which damages the liver and is highly carcinogenic.
- *Candida albicans*, a yeast which attacks the mucous membranes, can cause an infection of the mouth or vagina called thrush or candidiasis, and is also blamed for "yeast allergies".
- Fungi like *Epidermophyton* cause skin infections but are not very dangerous for people with healthy immune systems. However if the immune system is damaged they can be life-threatening; for instance, *Pneumocystis jiroveci* is responsible for severe lung infections which occur in AIDS patients.

Positive Effects

On the other hand, ascus fungi have brought some important benefits to humanity.

- The most famous case may be that of the mould *Penicillium chrysogenum* (formerly *Penicillium notatum*), which, probably to attack competing bacteria, produces an antibiotic which, under the name of Penicillin, triggered a revolution in the treatment of bacterial infectious diseases in the 20th century.

- The medical importance of *Tolypocladium niveum* as an immunosuppressor can hardly be exaggerated. It excretes Ciclosporin, which, as well as being given during organ transplants to prevent rejection, is also prescribed for auto-immune diseases such as multiple sclerosis, although there is some doubt over the long-term side-effects of the treatment.
- Some ascomycete fungi can be altered relatively easily through genetic engineering procedures. They can then produce useful proteins such as insulin, human growth hormone, or TPa, which is employed to dissolve blood clots.
- The red bread mould *Neurospora crassa* is an important model organism in biology, of which the genome has now been fully sequenced.
- Baker's Yeast (*Saccharomyces cerevisiae*) is used to make bread, beer and wine, during which process sugars such as glucose or sucrose are fermented to make alcohol and carbon dioxide. In the case of bread-making, the alcohol disappears and the carbon dioxide serves to make the dough rise.
- Enzymes of *penicillium camemberti* play a role in the manufacture of the cheeses Camembert and Brie, while those of *Penicillium roqueforti* do the same for Gorgonzola, Roquefort and Stilton.
- In Asia *Aspergillus oryzae* is added to a pulp of soaked soya beans to make soy sauce.
- Finally, some members of the Ascomycota are eaten with relish; morels (*Morchella*) and truffles (*Tuber*) are some of the most sought-after fungus delicacies.

Basidiomycota

Basidiomycota is one of two large phyla, that together with the Ascomycota, comprise the subkingdom Dikarya, which were in general what were called the "Higher Fungi" within the Kingdom Fungı. More specifically the Basidiomycota include mushrooms, puffballs, stinkhorns, bracket fungi, other, polypores, jelly fungi, boletes, chanterelles, earth stars, smuts, bunts, rusts, mirror yeasts, and the human pathogenic yeast, *Cryptococcus*.

Basically, Basidiomycota are filamentous fungi composed of hyphae (except for those forming yeasts), and reproducing sexually via the formation of specialised club-shaped end cells called basidia that normally bear external spores (usually four), which are specialised meiospores called basidiospores. However, some Basidiomycota reproduce asexually, and may or may not also reproduce sexually. Asexually reproducing Basidiomycota can be recognised as members of this phylum by gross similarity to others, by the formation of a distinctive anatomical feature (the clamp connection, cell wall components, and definitively by phylogenetic molecular analysis of DNA sequence data.

Classification

The most recent classification adopted by a coalition of 67 mycologists recognises 3 subphyla (Pucciniomycotina, Ustilaginomycotina, Agaricomycotina) and 2 other class level taxa (Wallemiomycetes, Entorrhizomycetes) outside of these, among the Basidiomycota. As now classified, the subphyla join and also cut across various obsolete taxonomic groups previously commonly used to describe various Basidiomycota.

The Basidiomycota had traditionally been divided into 2 obsolete classes, the Homobasidiomycetes (including true mushrooms); and the Heterobasidiomycetes (the Jelly, Rust and Smut fungi). Previously the entire Basidiomycota were called Basidiomycetes, an invalid class level name coined in 1959, as a counterpart to the Ascomycetes, when neither of these taxa were recognised as phyla. The terms basidiomycetes and ascomycetes are frequently used loosely to refer to Basidiomycota and Ascomycota. They are often abbreviated to "basidios" and "ascos" as mycological slang.

The Agaricomycotina includes what had previously been called the Hymenomycetes (an obsolete morphological based class of Basidiomycota that formed hymenial layers on their fruitbodies), the Gasteromycetes (another obsolete class that included species mostly lacking hymenia and mostly forming spores in enclosed fruitbodies), as well as most of the jelly fungi.

The Ustilaginomycotina are most (but not all) of the former smut fungi and along with the Exobasidiales.

The Pucciniomycotina includes the rust fungi, the insect parasitic/symbiotic genus *Septobasidium*, a former group of smut fungi (in the Microbotryomycetes, which includes mirror yeasts), and a mixture of odd, infrequently seen or seldom recognised fungi, often parasitic on plants.

Two classes, Wallemiomycetes and Entorrhizomycetes cannot at present be placed in a subphylum.

Typical Life-cycle

Unlike higher animals and plants which have readily recognisable male and female counterparts, Basidiomycota (except for the Rust (Pucciniales)) tend to have mutually indistinguishable, compatible haploids which are usually mycelia being composed of filamentous hyphae.

Typically haploid Basidiomycota mycelia fuse via plasmogamy and then the compatible nuclei migrate into each others mycelia and pair up with the resident nuclei. Karyogamy is delayed, so that the compatible nuclei remain in pairs, called a dikaryon. The hyphae are then said to be dikaryotic. Conversely, the haploid mycelia are called monokaryons. Oftentimes, the dikaryotic mycelium is more vigorous than the individual monokaryotic mycelia, and usually the dikaryotic mycelium then takes over the substrate in which they are growing. The dikaryons can be long-lived, lasting years, decades, or centuries. The monokaryons are neither male nor female. They have either a bipolar (unifactorial) or a tetrapolar (bifactorial) mating system.

This results in the fact that following meiosis, the resulting haploid basidiospores and resultant monokaryons, have nuclei that are compatible with 50 per cent (if bipolar) or 25 per cent (if tetrapolar) of their sister basidiospores (and their resultant monokaryons) because the mating genes must differ for them to be compatible. However, there are many variations of these genes in the population, and therefore, over 90 per cent of monokaryons are compatible with each other. It is as if there were multiple sexes.

The maintenance of the dikaryotic status in dikaryons in many Basidiomycota is facilitated by the formation of clamp connections that physically appear to help coordinate and re-establish pairs of compatible nuclei following synchronous mitotic nuclear divisions.

Variations are frequent and multiple. In a typical Basidiomycota lifecycle the long lasting dikaryons periodically (seasonally or occasionally) produce basidia, the specialised usually club-shaped end cells, in which a pair of compatible nuclei fuse (karyogamy) to form a diploid cell. Meiosis follows shortly with the production of 4 haploid nuclei that migrate into 4 external, usually apical basidiospores.

Variations occur, however. Typically the basidiospores are ballistic, hence they are sometimes also called ballistospores. In most species, the basidiospores disperse and each can start a new haploid mycelium, continuing the lifecycle. Basidia are microscopic but they are often produced on or in multi-celled large fructifications called basidiocarps or basidiomes, or fruitbodies), variously called mushrooms, puffballs, etc. Ballistic basidiospores are formed on sterigmata which are tapered spine-like projections on basidia, and are typically curved, like the horns of a bull. In some Basidiomycota the spores are not ballistic, and the sterigmata may be straight, reduced to stubbs, or absent. The basidiospores of these non-ballistosporic basidia may either bud off, or be released via dissolution or disintegration of the basidia.

Summary: Meiosis takes place in a diploid basidium. Each one of the four haploid nuclei migrates into its own basidiospore. The basidiospores are ballistically discharged and start new haploid mycelia called monokaryons. There are no males or females, rather there are compatible thalli with multiple compatibility factors.

Plasmogamy between compatible individuals leads to delayed karyogamy leading to establishment of a dikaryon. The dikaryon is long lasting but ultimately gives rise to either fruitbodies with basidia or directly to basidia without fruitbodies. The paired dikaryon in the basidium fuse (i.e. karyogamy takes place). The diploid basidium begins the cycle again.

Variations in Life-cycles

Many variations occur. Some are self compatible and spontaneously form dikaryons without a separate compatible thallus being involved. These fungi are said to be homothallic, versus the normal heterothallic species with mating types. Others are secondarily homothallic, in that two compatible nuclei following meiosis migrate into each basidiospore, which is then dispersed as a pre-existing dikaryon. Oftentimes such species form only two spores per basidium, but that too varies. Following meiosis, mitotic divisions can occur in the basidium. Multiple numbers of basidiospores can result, including odd numbers via degeneration of nuclei, or pairing up of nuclei, or lack of migration of nuclei.

For example, the chanterelle genus *Craterellus* often has 6-spored basidia, while some corticioid *Sistotrema* species can have 2-, 4-, 6-, or 8-spored basidia, and the cultivated button mushroom, *Agaricus bisporus* can have 1-, 2-, 3- or 4-spored basidia under some circumstances. Occasionally monokaryons of some taxa can form morphologically fully formed basidiomes and anatomically correct basidia and ballistic basidiospores in the absence of dikaryon formation, diploid nuclei, and meiosis. A rare few number of taxa have extended diploid life-cycles, but can be common species.

Examples exist in the mushroom genera *Armillaria* and *Xerula*, both in the Physalacriaceae. Occasionally basidiospores are not formed and parts of the "basidia" act as the dispersal agents, e.g. the peculiar mycoparasitic jelly fungus, *Tetragoniomyces* or the entire "basidium" acts as a "spore", e.g. in some false puffballs (*Scleroderma*). In the human pathogenic genus *Filobasidiella* 4 nuclei following meiosis remain in the basidium but continually divide mitotically, each nucleus migrating into synchronously forming non-ballistic basidiospores that are then pushed upwards by another set forming below them, resulting in 4 parallel chains of dry "basidiospores".

Other large variations occur, some as standard life-cycles (that themselves have variations within variations) within specific orders.

Typical Rust Life-cycle: Rusts (Pucciniales, previously known as Uredinales) at their greatest complexity produce five different types of spores on two different hosts in two unrelated host families. Such rusts are heteroecious (requiring 2 hosts) and macrocyclic (producing all 5 spores types). Wheat stem rust is an example. By convention the stages and spore states are numbered by Roman numerals.

Typically, basidiospores infect host one, the mycelium forms pycnidia, called spermagonia, which are miniature, flask-shaped, hollow, submicroscopic bodies embedded in host tissue (such as a leaf). This stage, numbered "0", produces single-celled, minute spores that ooze out in a sweet liquid and that act as non-motile spermatia, and also protruding receptive hyphae. Insects and probably other vectors such as rain carry the spermatia from spermagonia to spermagonia, cross inoculating the mating types. Neither thallus is male or female. Once crossed, the dikaryons are established and a second spore stage is formed, numbered "I" and called aecia, which form dikaryotic aeciospores in dry chains in inverted cup-shaped bodies embedded in host tissue. These aeciospores then infect the second host genus and cannot infect the host on which they are formed (in macrocyclic rusts).

On the second host a repeating spore stage is formed, numbered "II", the uredospores in dry pustules called uredinia. Urediospores are dikaryotic and can infect the same host that produced them. They repeatedly infect this host over the growing season. At the end of the season, a fourth spore type, the teliospore, is formed. It is thicker-walled and serves to overwinter or to survive other harsh conditions.

It does not continue the infection process, rather it remains dormant for a period and then germinates to form basidia, sometimes called a promycelium. In the Pucciniales, the basidia are cylindrical and become 3-septate after meiosis, with each of the 4 cells bearing one basidiospore each. The basidospores disperse and start the infection process on host 1 again. Autoecious rusts complete their life-cycles on one host instead of two, and microcyclic rusts cut out one or more stages.

Typical 'Smut' Life-cycle: The most characteristic part of the life-cycle of smuts is the thick-walled, often darkly pigmented, ornate, teliospore that serves to survive harsh conditions such as overwintering and also serves to help disperse the fungus as dry diaspores. The teliospores are initially dikaryotic but become diploid via karyogamy.

Meiosis takes place at the time of germination. A promycelim is formed that consists to a short hypha (equated to a basidium). In some smuts such as *Ustilago maydis* the nuclei migrate into the promycelium that becomes septate, and haploid yeast-like conidia/basidiospores sometimes called sporidia, bud off laterally from each cell. In various smuts, the yeast phase may proliferate, or they may fuse, or they may infect plant tissue and become hyphal. In other smuts, such as *Tilletia caries*, the elongated haploid basidiospores form apically, often in compatible pairs that fuse centrally resulting in "H"-shaped diaspores which are by then dikaryotic.

Dikaryotic conidia may then form. Eventually the host is infected by infectious hyphae. Teliospores form in host tissue. Many variations on these general themes occur.

Dimorphic Basidiomycota: Smuts with both a yeast phase and an infectious hyphal state are examples of dimorphic Basidiomycota. In plant parasitic taxa, the saprophytic phase is normally the yeast while the infectious stage is hyphal. However, there are examples of animal and human parasites where the species are dimorphic but it is the yeast-like state that is infectious.

The genus *Filobasidiella* forms basidia on hyphae but the main infectious stage is more commonly known by the anamorphic yeast name *Cryptococcus*, e.g. *Cryptococcus neoformans* and *Cryptococcus gattii*.

Anamorphs and Anamorphic Basidiomycota: The dimorphic Basidiomycota with yeast stages and the pleiomorphic rusts are examples of fungi with anamorphs, which are the asexual stages. Some Basidiomycota are only known as anamorphs.

Many are yeasts, collectively called basidiomycetous yeasts to differentiate them from ascomycetous yeasts in the Ascomycota. Aside from yeast anamorphs, and uredinia, aecia and pycnidia, some Basidiomycota form other distinctive anamorphs as parts of their life-cycles. Examples are *Collybia tuberosa* with its apple-seed-shaped and coloured sclerotium, *Dendrocollybia racemosa* with its sclerotium and its *Tilachlidiopsis racemosa* conidia, *Armillaria* with their rhizomorphs, *Hohenbuehelia* with their *Nematoctonus* nematode infectious state, and the coffee leaf parasite, *Mycena citricolour* and its *Decapitatus flavidus* propagules called gemmae.

Deuteromycota

The Deuteromycota (Greek for "second fungi") were once considered a formal phylum of Fungi. The term is now used only informally, to denote species of fungi in the phyla Ascomycota and Basidiomycota in which sexual reproduction is unknown. All fungi formerly classified in the Deuteromycota are now considered to be asexually-reproducing members of the fungal phyla Ascomycota and Basidiomycota.

Other names for the Deuteromycota have included Deuteromycetes and *Fungi imperfecti*. Currently these fungi are most often called anamorphic fungi, or mitosporic fungi, but these are informal terms without taxonomic rank. Most moulds were once considered members of the Deuteromycota.

Problem of the Deuteromycota

Although the taxon Deuteromycota is no longer formally accepted, many of the fungi it included have yet to find a place in modern fungal classification. This is because most fungi are classified based on characteristics of the fruiting bodies and spores produced during sexual reproduction, but members of the Deuteromycota have only been observed to produce asexual spores.

For this reason, mycologists are unique among those who study extant organisms in using a dual system of nomenclature. Dual naming is permitted by Article 59 of the International Code of Botanical Nomenclature (which governs the naming of plants and fungi). Under this system, a name for an asexually-reproducing fungus is considered a 'form taxon.' For example, the ubiquitous and industrially-important mould *Aspergillus niger* has no known sexual form.

Thus *Aspergillus niger* is considered a form taxon. In contrast, the sexual state of its close relative *Aspergillus nidulans* was found to be *Emericella nidulans*. When a teleomorph is known, its name takes priority over the name of the anamorph, hence the latter fungus is properly called *Emericella nidulans*.

Placing the Deuteromycota

The phylogenetic classification of asexually reproducing fungi can be resolved using the methods of molecular systematics. The resulting phylogenetic trees yield hypotheses of relationships among asexually-reproducing fungi and their sexually-reproducing counterparts. In this way, many asexually-reproducing fungi have found their place in the tree of life. However, this approach requires a significant investment of time and money, and can't be applied to species that are not available as cultures or fresh specimens. Furthermore, under the current system of fungal nomenclature, teleomorph names cannot be applied to fungi that lack sexual structures. Classifying and naming asexually-reproducing fungi is the subject of ongoing debate in the mycological community.

Historical Classification of the Deuteromycota

These groups are no longer formally accepted because they do not adhere to the scientific principle of monophyly. The taxon names are sometimes used informally. In particular, the term 'hyphomycetes' is often used to refer to moulds, and the term 'coelomycetes' is used to refer to many asexually-reproducing plant pathogens that form discrete fruiting bodies. Other systems of classification are reviewed by Kendrick (1981).

- Class Hyphomycetes (lacking fruiting bodies).
  - Order Moniliales (producing spores on simple conidiophores).
  - Order Stilbeliales (producing spores on synemmata).
  - Order Tuberculariales (producing spores in sporodochia).
- Class Coelomycetes Spores (produced in fruiting bodies).
  - Order Melanconiales (producing spores in acervuli).
  - Order Sphaeropsidales (producing spores in pycnidia).
- Class Agonomycetales (lacking spores).

Achorion

Achorion is an obsolete genus of dermatophyte fungus. The species previously belonging to genus *Achorion* have been moved to other genera. Some notable species include:

- *A. quinckeanum*: now named *Trichophyton quinckeanum*, the species which causes favus in mice.
- *A. schoenleinii*: now named *Trichophyton schoenleinii*, or "Schonlein's tricophyton", the species which causes favus in humans.
- Other pathogenic fungi occasionally found in humans:
  - — *A. arloingi*, now called *Aleurisma arloingi*
  - — *A. gypseum*, now called *Microsporum gypseum*

Acremonium

Acremonium is a genus of fungus. *Acremonium* species are usually slow growing and are initially compact and moist. *Acremonium* hyphae are fine and hyaline and produce mostly simple phialides. Their conidia are usually one-celled (i.e. ameroconidia), hyaline or pigmented, globose to cylindrical, and mostly aggregated in slimy heads at the apex of each phialide.

Clinical Significance: The genus Acremonium currently contains approximately 100 species, of which most are saprophytic, being isolated from dead plant material and soil. many species of acremonia are recognised as opportunistic pathogens of man and animals, causing mycetoma, onychomycosis, and hyalohyphomycosis. Clinical manifestations of hyalohyphomycosis caused by Acremonium include arthritis, osteomyelitis, peritonitis, endocarditis, pneumonia, cerebritis and subcutaneous infection.

Important Diseases

The history of root disease investigation falls into a series of well-marked though overlapping phases, and may be considered to have started with the second half of the nineteenth century. During these fifty years, the economic importance of root diseases was increasingly appreciated by growers and professional botanists alike, but many and varied were the hypotheses put forward in explanation of such diseases; as Thaxter (1891) remarked of potato scab:

> "It is a well-known fact that the theories and explanations which have been advanced to account for certain diseased conditions of the surface of potato tubers generally known as scab are nearly as numerous as the experimenters who have studied the disease."

Especial weight was given not only by farmers but also by men of Science to the non-living factors of the environment as a probable cause of root disease; either soil or season might be blamed, though insect pests.

Parasitic origin of disease is a commonplace to us today, but to botanists of the nineteenth century, it was a novel and even a startling idea; It is not surprising, therefore, that farmers were at first unwilling to abandon their former hypotheses of soil, climate or insect pests as responsible for root disease in their crops, in favour of hypotheses postulating a microscopic parasite, which they were often unable even to see.

Rules of proof were formulated by Kocjt (1882), through the application of which the connection between a bacterial parasite and the disease which it caused in man or animal might experimentally be established.

These rules are: (1) constant association of the organism with the disease, (2) isolation of the organism in pure culture, (3) reproduction of the disease by inoculation with the pure culture of the organism, (4) reisolation of the organism from the inoculated diseased host, and identification of it with the original inoculant.

Plant pathologists have generally accepted Koch's postulates as binding on them, too, though with obligate parasites such as *Plasmodiuphora brassicae,* which cannot be cultured on artificial media, the procedure cannot be followed in full.

Nevertheless, strict compliance with these rules of proof was by no means universal amongst the early investigators of root diseases. The first rule was always followed, but the second and third rules were not, naturally-infected host tissues often being substituted for pure cultures in inoculation experiments; as remarked above, this procedure was unavoidable where an obligate parasite was concerned. The fourth rule was most commonly neglected of all. Fortune favoured many of the earlier investigators, inasmuch as they discovered the primary parasite responsible for a root disease in spite of incomplete compliance with Koch's postulates.

This combination of good fortune with intuition, as a supplement to scientific method, has been rewarded by the fame always attaching to discovery, however made!

One of the first soilborne diseases to be correctly attributed to the activity of a parasitic fungus was stem canker and tuber scurf of potatoes, which Kuhn (1858) in Germany showed to be due to infection by *Rhizoctonia solani.*

Not long afterwards, Reinke and Berthold (1879) showed that a wilt of potatoes in the same country was due to infection by another fungus, *V'erticillium alboatrum.* Just before Reinke and Berthold's discovery, Hartig (1873), the well-known German forest pathologist, had shown *Armillaria mellea* to cause a serious root disease in forest trees.

These early discoveries by German investigators reflect the well-known pre-eminence of Germany at that time in the sciences of botany and plant pathology. Then, however, the famous Russian mycologist, Woronin (1878), published his discovery of *Plasmodiophora brassicae* as the cause of clubroot in crucifers, a widespread disease which had been only too well-known all over Europe for many years previously. Then Prillieux and Delacroix (1890) correctly proclaimed that *Ophiobolus graminis* was at least one of the parasites causing foot rot or "pietin" of wheat in France; not long afterwards, Mcalpine (1904) declared that the two diseases familiar to Australian farmers as "take-all" and "whiteheads" were in reality different phases of the same disease, due to infection by *Ophiobolus graminis,* the cause of the "pietin" in France.

In the USA, Pammel (1890) was the first investigator to assign the cause of Texas root rot of cotton to infection by *Ozonium auricomum;* the validity of Pammei's discovery was not affected by Shear's (1907) transfer of the cotton root rot fungus to a new species, 0. *omnivorum.*

With Duggar's (1916) discovery of a sporing stage of this fungus, the name was changed once more to *Phymatotrichum omnivorum,* by which the cotton root rot fungus is still known. At about the same time, another American investigator, Thaxter (1891, 1892), demonstrated that *Oospora scabies* was the cause of potato scab; this organism is now classified not with the Fungi, but with the Actinomycetes, and its name has, therefore, been changed to *Actinomyces scabies.*

The cause of another soilborne disease well-known at that time in the USA, black root rot of tobacco, was discovered not there but in Italy by Peglion (1897), who determined it (albeit not quite correctly) as *Thielavia basicola;* the tobacco root rot fungus is now known as *Thielairiopsis basicola.*

Somewhat later, investigators turned their attention to the root diseases of tropical crops; Ridley (1904) determined *Fames semitostus* (now corrected to *F. lignosus)* as the cause of white root disease of rubber in Malaya, whilst some twelve years later Belgrave (1916) ascribed red root disease of the same plant to another hymenomycete (now known as *Ganoderma pseudojerreum).* Credit for discovering the parasite responsible for the widespread Panama disease of bananas should go to the well-known American plant pathologist, Erwin F. Smith (1910), who obtained pure cultures of a fungus which he named *Fusarium cub ens e* (now reclassified as *F. oxysporum cub ense)* from tissues of wilted banana plants sent from Cuba.

He induced symptoms of disease by inoculation of this fungus into glasshouse-grown bananas at Washington, DC Most of the credit for establishing *F. oxysporum cubense* as the cause of Panama disease has rightly gone, however, to Brandes (1919), who in 1915, in ignorance of the report by E. F. Smith, set up extensive inoculation experiments in Porto Rico to determine whether this fungus was the responsible pathogen; in Brandes' experiments, it's rules of proof were rigidly followed in every particular.

It is no matter for surprise that the parasitic origin of most soilborne diseases, at first slow to gain recognition, should later have come to be enthusiastically accepted by botanists and plant pathologists, to the exclusion for some years of other considerations. The very name of mycologist, by which some plant pathologists are still officially known, is eloquent of this excessive preoccupation with the parasite.

Control measures were directed particularly towards elimination or inactivation of the parasite, whether in the soil or in the tissues of the host. This is not to say that plant pathologists during this period entirely neglected the possibility of indirectly controlling soilborne diseases, through the influence of soil conditions. Halsted (1900), for instance, reported that the clubroot disease of crucifers was favoured by acid soils, whereas potato scab was encouraged by alkaline soils.

At length, however, investigators began to realise that the soil environment exercised a profound effect upon the development of most soilborne diseases, and that the suitability or not of the environment for development of disease in the plant might be a factor as important as the presence or absence of the parasite.

The credit for bringing about this change of view must go especially to the Wisconsin school of investigators, led by L. R. Jones; a summary of this work is given by Jones *et al.* (1926) in their outstanding monograph on the relation of soil temperature to plant disease.

Particular attention was paid by the Wisconsin workers to the effect of soil temperature upon soilborne diseases, and a technique for accurate control of soil temperature under experimental conditions was evolved. The American investigators took full advantage of the opportunity which climatic variations of continental extent in the USA afforded them, in correlating incidence of disease with soil and air temperatures in different regions.

Microbiological Equilibrium of the Soil: The importance of the non-living factors of the soil and aerial environment in their effect upon the development of soilborne disease became well established by work at Wisconsin and elsewhere. In the late twenties of the present century, another factor of the soil environment, of profound importance, at last received recognition, when it came to be realised that development of a soilborne disease might be affected not only by the behaviour of the particular parasite concerned, but also by that of other microorganisms as well.

Commenting upon the control of potato scab by green manuring with grass cuttings, reported by Mi Llard (1921), Sanfq-rd was the first to suggest at such control might be thought about through the antagonism to *Antinomyces scabies* of saprophytic species of bacteria. Millard and Taylor (1927) later showed that development of scab on potatoes grown in sterilised soil inoculated with *A. scabies* could be reduced by simultaneous inoculation of the soil with *A. praecox*, an obligate saprophyte of a more vigorous habit of growth.

In nature, the host-parasite relationship must be affected by the biotic as well as by the non-biotic factors of the environment. In infected roots, the fungus causing the disease is rapidly followed into the host tissues by other microorganisms, with which it must thereafter be associated. The pure-culture method of experimentation,, as postulated by KOCH, is essential or solution of the original problem, *i.e.* determination of the primary parasite responsible for production of disease in the host. Once this question has been settled, however, experimentation should no longer be restricted to work with pure cultures of the parasite introduced into sterilised soil; some experiments, at least, should be carried out in natural unsterilised soil, under conditions permitting full play of the microbiological factor.

A timely warning concerning the false conclusions which might be derived from such methods of pure culture experimentation was voiced by Fawcett (1931) in his Presidential Address to the American Phytopathological Society; Fawcett suggested that experiments with cultures of single microorganisms should be supplemented by others in which known mixtures of microorganisms were employed.

Attention was also drawn to the importance of the microbiological factor in the soil environment of the root-infecting fungi by the experiments made with cereal root rot fungi by certain Canadian investigators, to whom particular credit is due. For the purpose of making pot and field experiments with one of these fungi, it was customary to culture the organism upon some organic medium, such as cooked wheat kernels, a mixture of oats and pearl barley, oat hulls, or a mixture of soil or sand with cornmeal, etc. The infected material was buried in the experimental soil, often in rather large quantities.

Such inoculum did not prove very satisfactory, however, owing to its rapid deterioration in virulence; after a few months, inoculated soils often proved completely harmless to cereal seedlings planted therein.

Since this deterioration was more rapid in field soil than in sterilised or partially sterilised soil in pots, Sanford and Broadfoqt (1931) suggested that it was due to the operation of microbiological antagonism upon the pathogenic inoculum.

Broadfoot (1933a) showed; that infection of wheat seedlings by "*Ophiobolus graminis*, initially much less in unsterilised than in sterilised soil, fell off markedly with the progress of recolonisation of the sterilised soil by the returning microflora.

Moritz (1932) observed that the severity of the take-all (O. *graminis)* disease of wheat on three German soils could be inversely correlated with the "protective" action of the general soil micro-flora, which was estimated by comparing infection in the unsterilised soil with that obtained in the sterilised soil. The interaction of microbiological antagonism with temperature was first demonstrated for wheat seedling infection by *O. graminis.* Henry's findings were confirmed by Garrett (1934a), and extended to other cereal foot rot fungi (Garrett, 1934b). Rising soil temperature was found by these authors to increase microbiological antagonism to the parasitic fungus; the fundamental effect of high soil temperature in accelerating infection could be seen under sterile conditions, but it was masked in the presence of other soil microorganisms by their inactivation of the pathogen inoculum, which increased with rising temperature.

From these and later experiments, it appeared that total exclusion of the microbiological factor from all experiments on soilborne diseases was undesirable, as being too far removed from natural conditions in the field; still further removed from natural conditions, however, were those experiments in which this factor had been unduly emphasised, *i.e.* by the use as inoculum of fungal cultures on rich media containing unassimilated carbohydrates and other nutrients, mixed in high proportion with the soil. In sterilised soil, with or without supplementary carbohydrates or other organic nutrients, a much simplified microflora in the presence of relatively high nutrient concentrations is obtained.

These conditions approach those of experiments on plates of nutrient agar, in which two selected fungi or other microorganisms are grown in juxtaposition.

In such plate cultures, striking demonstrations of antagonism have been reported by many investigators, for an account of which the reviews of Waksman (1937, 1941), Porter and Carter (1938), Weindling (1938) and Daeth (1939) should be consulted. Not merely antagonism, but even actual parasitism of one microorganism upon another has been demonstrated in such agar plate experiments.

A noteworthy example is the parasitism of *Trichodenna lignorum* upon *Rhizoctonia solani* and other fungi; the mechanism of this parasitism has been brilliantly elucidated by the researches of Weindling and his collaborators (Weindling, 1932, 1934, 1937, 1941; Weindling and Emerson, 1938; Weindling and Fawcett, 1936).

In so far as conditions in sterilised soil, *vis.* a much simplified microflora and abnormally high nutrient concentration, repeat those of the agar plate, similar results have been obtained; witness (1931) demonstration that growth of *Helminthosporium sativum* on sterilised soil might be completely suppressed by the addition of a mere pinch of unsterilised soil, and a similar demonstration by Mitchell *et al.* (1941) with *Phymatotrichum omnivorum* on a substrate of sterilised cotton roots.

In natural soil, on the other hand, the microflora is far more complex, and the concentration of free nutrients is very much lower. The special conditions favouring development of antagonism, or parasitism of one microorganism upon another, on the agar plate and in sterilised soil are, therefore, absent in unsterilised or natural soil.

This may be concisely expressed by saying that *natural soil is biologically buffered.* For this reason, "biological control" has often failed to fulfil in field trials its extravagant promise on the agar plate or in pots of sterilised soil.

The most important effect of the microbiological factor upon the root-infecting fungi is probably to be found not so much in restriction of their parasitic activity as in curtailment of their saprophytic existence. From the ease with which most root-infecting fungi could be cultured for indefinite periods as saprophytes on artificial media, it was often formerly supposed that they might persist indefinitely in the soil, existing arid even spreading upon the organic matter of the soil.

Such a conclusion ignored, of course, the very effective

control of most soilborne diseases by the practice of crop rotation. It was undermined by Waksman's (1917) discovery that there is a cosmopolitan microflora of soil-inhabiting fungi, composed of regular saprophytes; amongst this flora of *soil inhabitants,* other fungi were present as exotics, or *soil invaders.*

Waksman's conception was applied by Reinking and Manns (1933, 1934) and Reinking (1934) with conspicuous success to their conclusions concerning the distribution of species of *Fusarium* in the soils of Central America.

Certain *Fusarium* spp. were present in all, or nearly all, the soils examined, and were designated as *soil inhabitants;* other *Fusarium* spp. were present in the soil only in close association with tissues of their host plants, and eventually disappeared, thus meriting Waksman's term of *soil invaders.*

This distinction made by Reinking and Manns between saprophytic and parasitic species of *Fusarium* was later extended by Garrett (1938a) to differentiate between two contrasting types of behaviour amongst root-infecting fungi in general.

Root Disease Control: Some consideration must now be given to the application of these discoveries to the problems of root disease control in crops. Certain measures of root disease control are applicable to root diseases of all crops in all countries, and some of these, indeed, to almost all diseases of crops.

Such, for example, are the production of disease-resistant or disease-immune varieties of the crop plant, and the practice of crop sanitation, *e.g.* prevention of disease introduction through infected planting material, etc. Little further space will be devoted to such measures, important as they are, in the following pages, because they belong to the province of plant pathology in general, rather than to that of root disease control in particular. Special control measures in use against root diseases are best considered in relation to the type of crop, *viz.* field crops, plantation crops, and glasshouse crops.

The root diseases of field crops are controlled by crop rotation, by application of the principles of plant sanitation to the disposal of infected residues, and by checking the development and spread of root disease under the growing crop. Crop rotation is the oldest and most effective method of root disease control in field crops; its effectiveness is least against those diseases caused by fungi forming long-lived resting spores or sclerotia at the conclusion of periods of parasitic activity. A noteworthy study of the relation between length of crop rotation essential for root disease control and survival of the causal organism, *Phymatotrichum omnivorum,* in the soil was carried out by Rogers (1937), working on cotton root rot in Texas.

Recent evidence, presented by Bose (1938) in India and by Hildebrand and West (1941) and Richardson (1942) in Canada, suggests that certain crops in a rotation may depress the incidence of soilborne disease in a following crop more than others, even though all such preceding crops are alike in being non-susceptible to attack by the particular parasite or parasites concerned. A notable advance in crop sanitation was made by King *et al.* (1934a) and King (1937) when they demonstrated that losses due to cotton root rot *(P. omnivorum)* in continuous cotton under irrigation in Arizona could be greatly reduced by incorporation of organic material in the soil some months before cotton was sown.

This reduction in disease was tentatively attributed by King *et al.* to biological control of *P. omnivorum* by saprophytic soil microorganisms developing on the added organic material; Mitchell *et al.* (1941) and Clark (1942) have obtained fresh evidence in support of this hypothesis, and have also shown that sclerotia of *P. omnivorum* may be induced to germinate by the presence of decomposing fresh organic material in the soil. It seems likely, too, that the plant nutrients liberated during decomposition of the green manure may enhance the resistance of the cotton plant to this disease (Jordan *et al.,* 1939).

Variation in the incidence of soilborne diseases with soil conditions is now well known, and has been exploited in a variety of ways. In the field, some control of soil temperature, at least during the early stages of crop growth, can be secured by variation in planting time. Thus Richards (1921, 1923a) working in Utah showed that the incidence of stem canker *(Rhizoctonia solani)* of potatoes, a disease favoured by low soil temperatures, could be greatly reduced by delaying the date of potato planting.

Direct control of soil moisture is impossible in the field except under irrigation; in Egypt, Jones and Seif-El-Nasr (1940) demonstrated a striking control of certain cereal smut diseases through very shallow sowing in very wet soil. Soil aeration can be directly controlled, to a considerable extent, by cultivation; in Australia, the preparation of a very firm seed-bed for wheat was found many years ago by pioneer farmers to be an effective means of reducing the ravages of the take-all *(Ophiobolus graminis)* disease.

The effect of soil acidity in controlling scab *(Actinomyces scabies)* of potatoes, and of soil alkalinity in controlling clubroot *(Plasmodiophora brassicae)* of crucifers is well known. The control of scab by green manuring (Millard, 1921) is surprisingly effective in smallholdings, where very large amounts of fresh, chopped-up organic matter, such as grass cuttings, can be thoroughly incorporated with the soil.

The concentration of plant nutrients in the soil may have a paramount effect upon the occurrence of certain soilborne diseases, such as browning root rot *(Pythium* spp.) of cereals in Canada, which can be effectively controlled by application of phosphate (Vanterpool, 1940).

In tropical plantation crops, special care was formerly advised in eradicating all sources of root disease inoculum. Conscientious adoption of this advice entailed great, and expensive, labour in the excavation of buried roots before planting up. A good account of the earlier methods of root disease control in rubber plantations is given by Fetch (1921). Principles of root disease control in rubber plantations were later revolutionised through the brilliant investigations of Napped ,(1932-40) at the Rubber Research Institute of Malaya;

a well-documented history of rubber root disease investigations has been compiled by Sharples (1936). By Napper's method, a minimum of clearing is carried out on a virgin jungle site in preparation for rubber, but a collar and proximal root inspection of all the young rubber trees is made at intervals of a few months by gangs of labourers trained to recognise the mycelia of the root disease fungi.

When the mycelium of one of the root parasites is discovered, at such a routine inspection, on the proximal part of a tap or lateral root, the epiphytic mycelium is peeled off, and the root lightly rubbed with copper sulphate solution; the infection is traced to its source, a piece of infected jungle timber, and both infected root sections of the young rubber tree and the source of infection are removed and destroyed. The method depends on its success upon the fact that the mycelia of these root parasites of the rubber tree are epiphytic for some distance ahead of penetration; that of *Ponies lignosus*, the most important root parasite of young rubber on jungle sites, is epiphytic for as much as 5-15 feet ahead of penetration. A similar method can be used for eradication of root disease in young rubber on the site of former plantations, but if root disease has been severe in the old stand, a preliminary excavation of infected roots is advised before planting.

Roots of jungle or plantation trees infected by root disease before felling are not the only source of danger to the young plantation crop, however; the act of felling may so lower the natural resistance of the roots as to render them susceptible to rapid infection by a parasite to which they were formerly somewhat resistant.

This fact was first demonstrated by Leach (1937, 1939), working with *Armillaria mellea* on tea in Nyasaland; this fungus attacks few trees in the standing jungle, but develops freely as a parasite in the roots of many species once the tops have been felled. Leach further demonstrated that if the trees were killed by ring-barking some time before felling, their roots were invaded not by *A. incllea*, but by harmless saprophytes, such as *Rhisoctonia bataticola*. Leach's discoveries will undoubtedly exercise a profound effect upon the development of methods for root disease control in tropical plantation crops.

Nappek (1939, 1940) has since shown that a substantial proportion of the infections by *F. lignosus* in young rubber plantations must be due to infected roots of old jungle or plantation trees which are healthy at the time of felling, but which rapidly become infected by *F. lignosus* after felling, consequent upon the sharp lowering of host resistance. Napper found that killing of the trees or stumps by injection with sodium arsenite before or immediately after felling appeared to eliminate such post-felling infection of the roots by *F. lignosus;* injection by sodium arsenite in Malaya thus fulfilled the same function as ring-barking in Nyasaland. The practical importance of these discoveries is too obvious to require further emphasis.

The greater degree of control over soil and aerial environment possible in glasshouse practice should permit the development of more direct methods of root disease control than is possible in the field. A good example of such is furnished by Bewley's (1922) method of controlling the wilt of tomatoes due to *Verticillium alboatrum* by raising glasshouse temperature to a level exceeding 25°C.

Control of root diseases under intensive glasshouse cropping is chiefly effected, however, by partial sterilisation of the soil through heat or chemicals. Such partial sterilisation, which has become a routine in good commercial glasshouse practice, greatly benefits crop growth, apart from its valuable effect in controlling root diseases and pests.

The benefits of partial sterilisation were discovered independently by Girard (1894) in France and by Oberlin (1894) in Alsace, using carbon disulphide; Darbishire and Russell (1907) later demonstrated that heat treatment of soil produced a similarly beneficial effect upon crop growth in "sick" glasshouse soils. At present, heat is more commonly employed for soil sterilisation in glasshouses than are chemicals, though a cheapening of antiseptics in present use, or the discovery of cheaper or still more effective ones, may reverse this situation. It seems safe to predict, however, that treatment of soil by fungicidal chemicals is unlikely to find any general application outside its present limited sphere of usefulness, *vis.* in glasshouses, in nursery beds, and in intensive or market garden cropping. Some development in the "placement" of soil fungicides can be expected, in parallel with the development now proceeding in "placement" of artificial fertilizers.

By the placement method, one of the chief difficulties involved in fungicidal treatment of soil, *viz.* the great bulk of soil to be treated, is evaded. A good example of this procedure is afforded by Preston's (1928, 1931) use of mercuric chloride for watering in young cabbage plants at transplanting, in soils infected by the clubroot parasite *(Plasmodiophora brassicae).*

A further development along such lines may be the use of soil fungicides in conjunction with mechanical planters and transplanters, as in the well-known formalin drip treatment for control of onion smut originally devised by Selby (1902), and greatly improved by Anderson and Osmun (1923).

Parasitic Specialisation in Root-infecting Fungi

When a plant dies, its roots are quickly invaded by various soil microorganisms, prominent amongst which are species of soil fungi. The competition between members of the soil microflora for colonisation of this freshly opened substrate is intense, but the results are by no means determined by chance; the organisms which become dominant colonists are those best fitted to decompose that particular substrate under those particular environmental conditions. With the progress of colonisation and decomposition of the substrate, the habitat is changed, so that other organisms become dominant in their turn, and so there is a tendency for a succession of microorganisms to develop. The decomposition of dead root tissues in the soil is thus merely a particular instance of a general phenomenon well known to plant ecologists.

This succession of microorganisms developing in dead root or other plant tissues buried in the soil has been studied, either deliberately or incidentally, by a number of workers. The plant tissues have usually been surface sterilised, washed, and then plated out on a nutrient agar. The number of fungus species forming colonies on the agar plate may constitute only a small fraction of the total number engaged in decomposition of the plant tissues; the colony-forming species must be those that are best suited by the nutrient agar substrate.

The particular species forming colonies must be determined not only by the nature of the nutrient agar substrate, but also by other conditions of incubation, such as that of temperature; Walker (1941), moreover, has demonstrated that the choice of surface sterilising agent may strongly influence the results. Walker buried short lengths of fresh wheat straw in the soil, and took samples for surface sterilisation and plating-out at regular intervals.

The dominant fungi producing colonies on the agar plates were *Fusarium culmorum* and *Penicillium* spp.; the ratio between the numbers of *F. culmorum* colonies and those of *Penicillium* colonies developing from any sample of straws varied within very wide limits according to the type of surface sterilising agent employed and the time of treatment.

F. culmorum, a fungus of a vigorous and rapid habit of growth, showed low resistance to the action of the more severe sterilising agents, such as mercuric chloride and silver nitrate, but developed better after surface sterilisation of the straws with calcium hypochlorite, a mild sterilising agent, and best of all after a mere washing in sterile water. *Penicillium* spp. were apparently crowded out by the more vigorous growth of *F. culmorum* after these mild treatments of the straws; on the other hand, they were very tolerant of the more severe surface sterilising agents, mercuric chloride and silver nitrate, and after the longer period of treatment were often the only organisms developing on the plates.

Further to pursue the concepts of plant ecology, we may recall the development in certain plants of an ability to colonise certain virgin habitats unfavourable to general plant growth. The most frequently quoted examples of such habitats are sand dunes, shingle banks, and bare rock faces.

By the evolution of a specialised growth habit, such plants are considered to have *escaped* from the intense competition characteristic of habitats more favourable to the development of vegetation. Amongst the soil fungi, this role of escapers has been taken by the root-infecting fungi, which have developed ability to open up and colonise a habitat not available to saprophytic fungi.

To continue the analogy with above-ground plant ecology, we may further recall that in due course the primary colonisers of a sand dune, shingle bank or bare rock face so change the habitat that it permits growth of a wider range of species, with the result that the primary colonisers are gradually ousted and finally disappear completely. In the same way, once a root-infecting fungus has opened up a new habitat, the plant root, it is rapidly followed by other soil microorganisms. The root-infecting fungus obtains, however, an initial advantage in colonisation which it may retain for longer than was formerly supposed, from results of attempts to isolate it on the culture plate. It is well known that attempts to isolate a root-infecting fungus from diseased tissue, by surface sterilising selected portions of tissue and plating-out on the surface, of nutrient agar, often give disappointing results. This is due to the fact that the nutrient agar plate is a selective medium, not for the primary plant parasite, but for the saprophytes which are its cohabitants in the infected tissues.

Investigators studying the survival of parasitic fungi therefore, now generally employ the ideal selective medium, *i.e.* a host plant of the particular fungus concerned. Examples of this technique are furnished by the work of Taubenhaus and Ezekiel (1930&) with *Phyma-totrichwn omnivorum,* of Padwick (1935), Garrett (1938k), and Fellows (1941) with *Ophiobolus graminis,* of Thung (1938) with *Phytophthora parasitica* var. *nicotianae,* of Gibbs (1938) with *Phoma lingam,* of Nagel (1938) with *Cercospora bcticola,* and of Matzule-vitch (1936) with *Urocystis cepulae.*

In their studies of the root rots of strawberry and tobacco in Ontario, Hildebrand (1934), Koch (1935), and Hildebrand and Koch (1936) have endeavoured to overcome this shortcoming of the isolation plate technique by supplementing it with direct microscopical study of the succession of microorganisms developing in the diseased roots.

Presumably, it should be possible to obtain a complete picture of the succession of fungi in such diseased plant tissues by employing a sufficient number of selective isolation media (in the widest sense of this term); as the number of specific isolation techniques approached the number of fungi involved, so the investigator's knowledge of the succession should approach totality.

Soil-inhabiting and Soil-invading Fungi: Serious study of the fungus flora of the soil may be said to date from Waksman's (1917) postulate that there is a definite fungus flora of the soil. From his examination of the fungi found in 25 different soils collected from widely separated regions of the North American continent, and from the Hawaiian Islands, and from the results of European investigators, Waksman concluded that the basic fungus flora of the soil was essentially cosmopolitan, though the relative importance of different groups might vary with latitude and soil type. Thus soils of the northern latitudes were characterised by abundance of *Mucorales* and *Penicillia,* whilst *Aspergilli* were predominant in the warmer soils.

Again, the genus *Trichoderma* was found especially in acid soils *(e.g.* forest soils) and in waterlogged soils. As a result of this study, Waksman introduced the fundamental conception of a basic cosmopolitan fungus flora of *soil inhabitants,* amongst which were to be found exotic fungi or *soil invaders.* This conception of Waksman's was applied by Reinking and Manns (1933, 1934) and Reinking (1934) to their findings on the distribution of *Fusarium* spp. in the soils of Central America. As the result of a very large series of isolations from soils, these authors concluded that the tropical American soil *Fusaria* could be divided into two groups, *soil inhabitants* and *soil invaders*. The *soil inhabitants* were those species of *Fusarium* found in all, or nearly all, of the soils examined, and were hence regarded as common soil saprophytes. Certain other species, on the other hand, were found to have only a local distribution, and this distribution was further found to coincide with the presence, or past presence, of the appropriate host plant. With the death of the host plant, the numbers of such a species gradually dwindled, until it disappeared from the soil, and was hence designated a *soil invader*.

This conception of *soil inhabitants* and *soil invaders* was in turn employed by Garrett (1938) to distinguish between two contrasting types of behaviour amongst the root-infecting fungi, as follows:

> "The *soil inhabitants* are considered to be primitive or unspecialised parasites with host range; these fungi are distributed throughout the soil, and their parasitism appears to be incidental to their saprophytic existence as members of the general soil microflora. The *soil invaders*, to which class the majority of the root-infecting fungi seem to belong, are more highly specialised parasites; the presence of such fungi in the soil is generally closely associated with that of their host plants. In the continued absence of a host plant, such fungi die out in the soil, owing to their inability to compete with the soil for an existence on non-living organic matter. This close as between the *soil invaders* and their host plants".

Any consideration of the qualifications of a particular fungus for its role of a *soil inhabitant* involves recognition of the fact that two factors, of a spatial and a temporal nature, respectively, are concerned. By the spatial factor is implied the number of kinds of plant tissue, or other substrate, which the fungus can colonise; the species of plant, the part of the plant, its age and other causes of variation in tissue composition must all affect not only the rate and course of decomposition in the soil (Waksman, 1931) but also the species of decomposing microorganisms.

By the temporal factor is meant the average period for which the fungus can actively maintain itself on a substrate, during the passage of that substrate from the virgin condition to that of relatively inert soil humus. To take an extreme instance, a root-infecting fungus which was a truly omnivorous plant parasite could certainly be regarded as a *soil inhabitant,* even if its powers of survival as a saprophyte were strictly limited, because at almost all times and in almost all places it could find a substrate.

With a fungus of more restricted host range, however, the ability to function as an active saprophyte for a longer period in the decomposing plant tissues becomes correspondingly more important. The development of long-lived resting spores and sclerotia by root-infecting fungi may be regarded as an example of one way by which the disadvantage of nutritional specialisation has incidentally been overcome.

Mycorrhizal Fungi: If symbiosis be regarded as the evolutionary end-point of the parasitic habit, then the mycorrhizal fungi must be regarded as the most highly specialised of all the root-infecting fungi. This group of fungi has stimulated continuous interest, and not a few controversies, amongst mycologists ever since the discovery of the mycorrhizal habit in plants.

The work of the earlier investigators has been discussed by Rayner (1927), in her monograph on mycotrophy in plants. More recently, Surges (1936) has published a concise but comprehensive, and commendably objective, review of the whole subject. Two groups of mycorrhizal fungi are usually distinguished; the *ectophytes,* which develop mainly externally, forming a fungus mantle around the host roots, and the *endophytes,* which develop inside the host tissues. The differences between ectophytic and endophytic mycorrhizas have been lucidly summarised by Rayner (1939), in the following paragraph:

> "It is still convenient to classify mycorrhizas into two groups, the extreme types in which show marked structural differences, correlated with the distribution and character of fungus infection. The *ectotrophic* mycorrhizas of most trees and shrubs are readily recognisable as distinct from ordinary rootlets; a sheath of mycelium is formed about the tip and younger part of the emerging rootlet and the hyphae composing it extend inward, forming an intercellular network between the cortical cells known as the 'Hartig net'. In *endotrophic* mycorrhizas, a more or less extensive distribution of intercellular and intracellular mycelium within the root is associated with a variable but usually scanty development of hyphae on the surface. In general, endotrophic mycorrhizas resemble ordinary roots in external appearance, although slight modifications such as irregularities of diameter or differences in opacity of the tissues may betray their character to the experienced observer."

Rayner goes on to point out that the distinction between these two types of mycorrhiza is an artificial one and can be condoned only on grounds of convenience; intermediate types are now known. The association of the ectophytes with their host root may be conditioned by root excretions from the host; Melin (1925) demonstrated the excretion of phosphatides by the roots of various forest trees, and showed that these stimulated the development of the mycorrhizal fungi.

Evidence for an obligate mycotrophic habit is strongest in the case of certain species of forest trees forming ectophytic mycorrhiza. Foresters have frequently experienced difficulty in establishing exotic coniferous trees; lack of success has usually been associated with absence of typical mycorrhiza formation by the trees. Inoculation of nursery beds with soil taken from under a vigorous stand of the same species of tree has frequently changed failure into success, and the young trees have grown away well in the inoculated soil.

Rayner (1938) has summarised information collected by the Imperial Forestry Institute of Oxford on the use of soil or humus inoculum as an aid to establishment of exotic species of forest trees; the empirical use of such inoculum appears to be widespread, and to give satisfactory results in the majority of trials.

Hatch (1937) has recently published substantial experimental evidence in support of Stahi's (1900) hypothesis that ectophytic mycorrhizal fungi facilitate absorption of mineral salts by their hosts, especially in infertile soils. He demonstrated a negative correlation between production of mycorrhizae by pine seedlings and the availability of mineral salts in a number of forest soils. In his experiments, mycorrhizae were most abundantly produced under shortage of available nitrogen, phosphorus, potassium or calcium, or under lack of balance between these nutrients.

Hatch experimentally demonstrated striking increases in growth and nutrient absorption by seedlings of *Pinus strobus* in a prairie soil after inoculation with pure cultures of appropriate mycorrhizal fungi. Hatch's observations have been confirmed and extended by Routien and Dawson (1943), using an artificial soil compounded of pure quartz sand and a purified colloidal clay.

This artificial soil possessed base exchange properties, and its content of available nutrient ions could be experimentally varied within wide limits. Working with this artificial soil, Routien and Dawson found that substantial growth and salt absorption by *Pinus echinata* took place even in the absence of mycorrhizae at the higher levels of base saturation of the soil. But if absorption of nutrients was impeded by replacing the nutrient ions with adsorbed hydrogen ion, then formation of mycorrhizae substantially increased growth and salt uptake by the pine seedlings. Hatch has stressed the increase in total absorbing surface resulting from conversion of short roots into mycorrhizae; Routien and Dawson observed that development of such short roots into mycorrhizae doubled or even quadrupled production of respiratory carbon dioxide. They, therefore, concluded that mycorrhizae increased the salt-absorbing capacity of the roots chiefly by adding to the supply of exchangeable hydrogen ion, derived in part at least from respiratory carbonic acid.

Infection by the appropriate endophyte was formerly thought to be essential for germination of the seed in many species of the *Ericaceae* and *Orchidaccae;* evidence for and against this view has been summarised by Burges (1936), who states: "Rayner (1929), as the result of her pure culture experiments, concludes that, unless the seedlings of *Calluna vulgaris* were infected by the endophyte, arrest of growth and the inevitable symptoms of malnutrition followed.

This result was in discordance with those of Christoph (1921) and Knudson (1929), who considered that infection was not essential. In a later paper Knudson (1933) records the result of his reinvestigation of the problem in the light of Rayner's criticisms of his previous technique, and has shown that *Calluna* seedlings can develop normally without the endophyte. Freisleben (1934) has reached a similar conclusion in respect to the genus *Vactinium.*" In a later paragraph concerned with mycotrophy in seedling orchids, Burges comments: "It used to be thought that orchid seedlings were unable to germinate and grow oh their own account.

Bernard (1904) showed that if the endophytes were added to the cultures growth frequently followed. Knudson (1927) has extended this work and has shown that the important feature is not the penetration of the orchid by the endophyte, but the bringing of organic material into solution. For example, on a medium containing starch, the carbohydrate is unavailable to the orchid. If the fungus present causes hydrolysis of the starch, some of the resulting sugars, once in solution, are utilised by the orchid, which is then able to establish itself and later may become independent of this auxiliary food supply. If the organic material were supplied in solution, *e.g.* as fructose, no fungal agent was required. He further showed that other fungi such as *Phytophthora* could replace the endophyte and bring the organic material of the medium into solution.

The fungi also caused a change in the hydrogen-ion concentration of the medium. This on its own did not lead to germination, yet it was shown to be a contributory factor. Other workers have also shown the need for correct adjustment of the hydrogen-ion concentration of the medium.

Various nutritional benefits have been suggested as accruing to the host plant of an endophyte. It is supposed that nutrients are absorbed from the soil by the external hyphae of the endophyte, and translocated thence into the internal hyphae, which are eventually digested by the host cells. Burges, however, agrees with the criticism that the connections between external and internal hyphae are insufficient for the host plant to benefit much in this way.

Ternetz (1904) demonstrated nitrogen fixation by certain species of *Phoma* acting as endophytes in members of the *Ericaceae;* the amounts of nitrogen so fixed were, however, too low to be of much nutritional significance to the host plants, and endophytes with a greater nitrogen-fixing capacity have yet to be discovered. Butler (1939) has made the interesting suggestion that endophytes of the vesicular-arbuscular type may benefit their host by the manufacture of accessory growth factors, which the host is unable to make for itself.

The relationship between the endophyte and its host seems to be one of controlled parasitism. Under certain conditions, the resistance of the host may be broken down, and the controlled parasitism become uncontrolled, as observed by Reed and Fremont (1935) and Samuel (1926). Reed and Fremont found that the roots of Californian citrus trees receiving no fertilizers during the preceding seven years had little power to resist invasion or to digest the intracellular mycelium, in contrast to the roots of trees which had been manured annually with cover crops and stable manure. Samuel observed that the roots of oats growing in a South Australian soil deficient in available manganese were heavily infected by an endophytic fungus, whereas the roots of oats growing in normal soils were free from infection.

Samuel's observation recalls an analogous one by Brenchley and Thornton (1925), who found that root-nodule bacteria became destructively parasitic in the root tissues of broad bean plants deprived of boron. Again, a fungus forming a mycorrhiza with one host may act as a destructive parasite in others.

An example of such behaviour is furnished by a dangerous parasite of forest trees and other plants, *Armillaria mellea,* which was demonstrated by Kusano (1911) to form a mycorrhiza with the Japanese orchid *Gastrodia elata.* Kusano observed, however, that under certain conditions the symbiosis might be upset, and that infection of the orchid tuber by *A. mellea* was then accompanied by a necrotic reaction of the invaded cells of the host.

From the evidence adduced above, Burges has concluded that the chief benefit which mycorrhizal plants derive from this association is that due to the decomposition of soil organic matter by the mycorrhizal fungi. Plants with ectophytic mycorrhizal fungi are, of course, exceptionally well situated for the reception of such benefit. As Burges has pointed out, however, it is not necessarily the mycorrhizal fungi which are always of greatest service to the higher plant in this way; common soil-inhabiting fungi may conceivably be of greater importance to a mycorrhizal plant than its own particular fungal associate.

The enhanced activity of certain common soil-inhabiting fungi, and especially of *Mucor* spp. and *Penicitkum* spp., in the "rhizosphere" has been thought by Jahn (1934) to constitute a "peritrophic mycorrhiza." This interpretation of mycotrophy by Surges has been criticised by Hatch (1937), in a masterly review of the subject to which he has himself made numerous important experimental contributions. Hatch has studied chiefly the ectophytic mycorrhizae of *Finns,* and has interpreted his data by the "mineral salt hypothesis' of Stahl (1900).

Elsewhere, Burges (1939) has commented upon the paucity of information concerning the saprophytic life of the mycorrhizal fungi. This would be the more surprising were it not for the fact that a similar obscurity surrounds the saprophytic life of a closely analogous but better known group of microorganisms, the root nodule bacteria of leguminous plants. The mode of survival of mycorrhizal fungi outside their host plants therefore remains an unsolved problem, or series of problems, concerning which Burges (1939) has commented: "This class of soil fungi is the most difficult of all to study. Many of its members appear to be almost obligate parasites in their relations to their hosts, yet their ubiquitousness leads one to believe that they must have some existence in the soil apart from their host roots. In isolation of soil fungi, it is rare to find any member of this group. Many are Basidiomycetes and these are conspicuous by their absence from lists of soil fungi. The typical ectotrophic mycorrhizas are produced by Hymenomycetes and many species have been identified. Butler (1939), in his review of work on the vesicular-arbuscular or phycomycetous type of endophyte, has observed that this endophyte is more common and more easily to be found in the roots of perennial or plantation crops than in the roots of annual field crops, *viz.:* "The regularity of its occurrence in some perennial plants and it's incidental presence in some annual field crops is probably merely the result of the greater chance which an organism growing in close association with living roots has of persisting indefinitely and passing from the older to the more newly developed roots in the former class. In annuals, much of the fungus perishes with the roots, and survival is limited to the thick-walled resting mycelium and vesicles found in the soil when the new crop is sown".

2

Diseases Affecting Human-beings

Fungal diseases are called mycoses and those affecting humans can be divided into four groups based on the level of penetration into the body tissues:

Superficial mycoses are caused by fungi that grow only on the surface of the skin or hair.

Cutaneous mycoses or dermatomycoses include such infections as athlete's foot and ringworm, in which growth occurs only in the superficial layers of skin, nails, or hair.

Subcutaneous mycoses penetrate below the skin to involve the subcutaneous, connective, and bone tissue.

Systemic or deep mycoses are able to infect internal organs and become widely disseminated throughout the body. This type is often fatal. The most common type of subcutaneous mycosis seen worldwide is sporotrichosis, which occurs most often in gardeners and farmers who come in direct contact with soil. This is a chronic infection caused by the fungus *Sporothrix schenckii,* occurring in three forms. The 'cutaneous lymphatic form' is characterised by a single pustule or nodule that forms at the site of invasion. This is followed by lymphatic spread and the development of numerous subcutaneous lesions. This 'disseminated form' is marked by multiple, painless cutaneous or subcutaneous nodules, which can form into ulcers or abscesses involving the muscles, joints, bones, eyes, gastrointestinal system, mucous membranes, and nervous system. The 'pulmonary form' results from the inhalation of spores, but produces much the same forms of the disease.

Other forms of subcutaneous mycoses occur mostly in the tropics and subtropics and are caused by several fungal species. These conditions are called chromomycosis (producing wartlike nodules that can ulcerate) and maduromycosis (or mycetoma—a chronic slowly progressive destructive infection involving several layers of skin, producing abscessing granulomas). Treatment is difficult and often requires surgical removal of the offending tissues.

Dermatomycoses is a superficial fungal infection that penetrates only the epidermis, hair, or nails. About thirty different species of the genera *Epidermophyton, Microsporum,* and *Trichophyton* (collectively known as dermatophytes) cause infections commonly known as athlete's foot, jock itch, and ringworm. The term tinea, along with the area of body involved, is also used when referring to these infections.

Tinea is Latin for worm or grub because the infections were originally thought to be caused by worm-like parasites. For example:

- Tinea capitis is an infection of the scalp common in school children;
- Tinea corporis is an infection on the body such as ringworm;
- Tinea pedis occurs on the feet especially the classic athlete's foot;
- Tinea cruris is an infection of the genital area commonly called 'jock itch;'

- Tinea barbae affects the area of the beard;
- Tinea unguium or onychomycosis is an infection of the toenail or fingernail;
- Tinea manvum affects the hands.

Different species afflict different anatomical sites in different ways. Dermatophyte fungi typically causes scaly lesions that tend to be circular (ringworm). Dermatophyte fungi are moulds, but there are yeasts that also cause skin infections. *Trichophyton rubrum* and *T. mentagrophytes* are the dermatophytes that cause athlete's foot.

Malassezia furfur causes a type of dandruff, as well as an infection known as *Pityriasis versicolour* (formerly *Pityrosporum ovale* and *P. orbiculare*). *M. furfur* is actually a complex of seven species that are normally harmless, but, for some reason, turn pathogenic at times. Infection produces a flaking skin rash and either a bleaching of the skin or a deposit of more pigment to the area. Lesions are generally found on the trunk, and sport a golden yellow fluorescent colour when an ultraviolet light is shone on them.

Systemic mycoses occur in two basic forms—respiratory and disseminated tissue. If left untreated, the disseminated form is usually fatal. Surgical removal of large pulmonary lesions have been useful in some cases. Other systemic mycoses include:

Coccidioidomycosis (valley fever) is caused by *Coccidioides immitis,* a fungus found in desert areas and prevalent in the southwestern US and in some areas of Central and South America. These spores are carried by desert winds and infect the pulmonary spaces. As many as 80 per cent of the population of the central valleys of California test positive for this fungus.

Histoplasmosis is caused by the fungus *Histoplasma capsulatum,* which grows in soil fertilized by bat droppings. It is most prevalent in the Ohio River and Mississippi River valleys, where positive skin tests reveal that at least 80 per cent of the population have been infected. Mild lung infections are usually the result of the disease and often go unnoticed, but, on healing, small calcified nodules that are often mistaken for tuberculosis lesions on chest X-rays are left on the lungs.

Blastomycosis is caused by the fungus *Blastomycoses dermatitidis*. It grows in soil as a mould producing conidia that infect human by the airborne route. Along with common mild respiratory infections, skin lesions can also appear. The fungus grows as a budding yeast in human tissues.

Cryptococcosis is caused by the yeast *Cryptococcus neoformans*, normally found in soil and habitats of pigeons. It exists only in the yeast form. Scientists speculate that it is millions of years old.

Most infections are of a mild respiratory nature, but, in compromised persons, the infection may spread throughout the body affecting mostly the central nervous system. *Cryptococcus neoformans gattii* is usually found in tropical regions, but has now made an appearance on Vancouver Island in Canada, with several deaths attributed to it.

The fungus, *Coccidioides immitis* or cocci, for short, is an infectious, but not contagious, disease contracted by inhaling spores produced in soil. Anything that disturbs the soil will send these spores into the air—construction, earthquakes, off-roading, dust storms, archaeological digs, etc. So far, it is found only in the Western Hemisphere, particularly in the American southwest.

The strong winds of southern California between Thanksgiving and Christmas produce numerous cases of these deadly cocci illnesses. The fungi grow so rapidly that death can result very quickly. Once inhaled, the spores often migrate to one area and lay dormant in a pocket of tissue, looking like a hanging beehive. The only known drug proven effective has been amphotericin B, but it has horrendous side effects, killing healthy human cells as well as the invasive agent. This leaves the kidneys and bone marrow especially sensitive, and usually results in kidney damage and anaemia.

Mucormycosis refers to the fairly rare diseases produced by a variety of common fungi of the order Mucorales. These, too, are seen only in the severely immunocompromised patient. The fungi penetrate the respiratory or intestinal mucosa and can enter through breaks in the skin as well. Localised lesions may develop, followed by a spreading to the blood, which carry it to all organs. Death often results.

Fungi most commonly associated with specific immunocompromised patients:

- *Candida species, Aspergillus* species, *Phycomyces* species (Leucopenia or bone marrow failure);
- *Candida, Cryptococcus, Coccidioides, Histoplasma* (Cellular immunity or tissue transplants);
- *Zygomyces, Rhizopus, Mucor, Absidia* (Diabetes);
- *Zygomyces* (steroid therapy);
- *Candida, Cryptococcus, Histoplasma* (malignancies as in leukaemia and lymphoma, and Hodgkin's disease);
- *Candida, Cryptococcus, Histoplasma* (AIDS);
- Clinically significant fungi and the areas of disease they cause:
- *Malassezia furfur* and *Exophiala werneckii* (superficial skin)
- *Piedraia hortae* and *Trichosporon beigelii* (hair)
- *Microsporum* species, (skin and hair)
- *Epidermophyton* species (skin and nails)
- *Trichophyton* species (skin, hair, and nails)
- *Sporothrix schenckii, Cladosporium* species, *Phialophora* species, and *Fonsecaea* species (subcutaneous/lymphatic tissues - chromoblastomycosis)
- *Histoplasma capsulatum, Coccidioides immitis, Fusarium* species, *Penicillium* species (systemic respiratory)
- *Blastomyces dermatitidis* (subcutaneous/respiratory)
- *Cryptococcus neoformans* (respiratory/CNS)

- *Aspergillus* species, *Mucor* species, *Candida* species, and *Rhizopus* species (opportunistic involving various body sites)
- Allergic bronchopulmonary aspergillosis
- Allergic bronchopulmonary aspergillosis (ABPA) is a hypersensitivity response to the fungus *Aspergillus fumigatus,* the spores of which are ubiquitous in soil and are commonly found in the sputum of healthy individuals. *A. fumigatus* is responsible for a spectrum of lung diseases commonly grouped under the heading of aspergilloses.

Epidemiology

Estimating the prevalence of ABPA has been made difficult by lack of uniform diagnostic criteria and standardised tests. It usually occurs as a complication of other chronic lung disease, in particular asthma or cystic fibrosis.

It is estimated that in the US ABPA may be present in between 0.5 and 2 per cent of all asthma patients.

Immunology and Pathophysiology

There is both a type I (atopic) and type III hypersensitivity response. Precipitating antibodies incite a type I acute hypersensitivity reaction with release of immunoglobulin E (IgE) and immunoglobulin G (IgG), resulting in mast cell degranulation with bronchoconstriction and increased capillary permeability.

Immune complexes and inflammatory cells are then deposited within the bronchial mucosa leading to tissue necrosis and eosinophilic infiltrate, a type III reaction. The subsequent damage to the bronchial wall causes (proximal) bronchiectasis.

Repeated acute episodes left untreated can result in progressive pulmonary fibrosis that is often seen in the upper zones and can give rise to a similar radiological appearance to that produced by tuberculosis.

The main features are therefore: bronchospasm increased mucus production and plugging of distal airways, leading to their collapse bronchiectasis

Clinical Picture

Symptoms

- Wheeze
- Cough
- Fever
- Malaise
- Recurrent pneumonia (eosinophilic, therefore not responding to antibiotics)
- Expectoration of mucoid plugs (containing the fungal hyphae)
- Unexplained worsening of asthma or cystic fibrosis.

Investigations

A full blood count usually reveals eosinophilia more than 10 per cent and there is a raised serum IgE more than 1000ng/ml.

Chest radiography shows various transient abnormalities:

- Consolidation or collapse
- Thickened bronchial wall markings
- Peripheral shadows
- Signs of central bronchiectasis

Aspergillus specific tests:

- Precipitating antibodies to aspergillus species in >90 per cent of cases
- Aspergillus-specific IgE Rast test
- Skin-prick test is almost always positive to Aspergillus fumigates

Management

The aim of treatment is to suppress the immune reaction to the fungus and to control bronchospasm. The immune reaction is suppressed using oral corticosteroids:

- A high dose of prednisolone or prednisone (30 to 45 mg per day) in acute attacks
- A lower maintenance dose (5-10 mg per day)

Mucus plugs may be removed by bronchoscopic aspiration. It is almost impossible to eradicate the fungus but sometimes itraconazole (an anti-fungal) is used in combination with steroid therapy. Regular monitoring of the condition includes chest x-rays, pulmonary function tests, and serum IgE. The antibody levels usually fall as the disease is controlled, but they may rise again as an early sign of flare-ups.

Alternative Treatments

Topical oils: Users report instant relief from itching after topical application of tea tree oil or crocodile oil, allowing lesions to heal due to the cessation of scratching the itch.

Vinegar

There is no scientifically collected evidence showing that any type of vinegar, including that which is made from apple cider, produces direct fungicidal effects. The confusion exists because it is a well known food preservative. However, vinegar may have some utility in reducing the pH of the epidermis, where the fungus of tinea pedis dwells. Thus it could create an inhospitable environment for fungus, while functioning as a pharmaceutic adjuvant for some other (acid stable) anti-fungal agent.

A treatment entails soaking the affected areas of the foot in a shallow container of vinegar. Typically the soaking period is about one hour (at room temperature), in order to saturate the epidermis and penetrate beneath the toe nails.

Urine Therapy

Proponents of urine therapy claim that urine is very effective at killing athlete's foot. Urea, the "active ingredient" in urine, is already used in many drugs and treatments made by pharmaceutical companies to treat athlete's foot. This controversial treatment method recommends urinating on the infected area once a day in the shower. According to supporters, urine therapy not only kills existing fungi, it prevents new fungi from growing in the infected area.

One biochemist states that urea is only used to soften the outer layers of skin so that anti-fungal drugs can reach fungi below the surface, and that the urea must be concentrated and applied for a long period of time in order to be effective. According to another article about high-concentration urea cream, the compound is used to "dissolve proteins and [as a] denaturant.

The ability of urea to macerate [tissue] has been attributed to a 'proteolytic effect', but others attribute the maceration to the hydrating properties of urea." This use requires a high concentration of urea, up to 40 per cent, and extended exposure. Urea itself without the presence of an additional anti-fungal drug is not referred to in scholarly literature as having anti-fungal properties. Thus, it is unlikely that urinating on one's feet in the shower will significantly improve a case of athlete's foot. This theory was recently covered on the Straight Dope for March 30th, 2007.

Basidiobolus Ranarum

Basidiobolus ranarum is a microscopic fungus in the order Entomophthorales. It produces large, spherical, asexual spores that are forcibly discharged. Those spores can germinate directly to produce hyphae. When they land in an unfavourable location, they can alternately germinate to produce a long, slender capilliconidiophore that bears a single, falcate capilliconidium. At the distal tip of the capilliconidium is a sticky mucous drop that is presumed to aid in dispersal. *B. ranarum* can also produce thick-walled resting spores that are presumably the sites of meiosis.

B. ranarum can be isolated from decaying leaf litter and the excrement of frogs and terrestrial, insect-eating reptiles. It has been considered a commensal of frogs, although there is no evidence that it multiplies or persists in a frog's gut. It can be a human pathogen, causing a disease called basidiobolomycosis (formerly entomophthoromycosis).

Basidiobolomycosis

Basidiobolomycosis is a rare disease caused by the fungus Basidiobolus ranarum, member of the class Zygomycetes, order Entomophthorales, found worldwide. Usually basidiobolomycosis is a subcutaneous infection but it has been associated with gastrointestinal disease.

Clinical Features

Basidiobolomycosis is usually a superficial infection of skin, but may very rarely cause lesions of the bowel or liver, mimicking bowel cancer, or Crohn's disease. In patients with deep involvement, the eosinophil count may be raised, falsely suggesting a parasitic infection.

Diagnosis

Diagnosis is by laboratory culture of the organism, usually from pieces of tissue taken from the patient. It grows easily on most media, but risks being discarded as irrelevant or being reported as a contaminant because laboratory staff are unfamiliar with it. Diagnosis is often difficult because it is a rare disease and therefore often not recognised. The lesions often look like tumours rather than infection, so often no sample is sent for microbiology, however, the histopathology is characteristic: the Splendore-Hoeppli phenomenon describes the presence of fungal hyphae (which may exist only as ghosts on the slide) surrounded by eosinophilic material.

Treatment

Treatment for skin lesions is traditionally with potassium iodide, but itraconazole has also been used successfully.

Blastomycosis

Blastomycosis is a fungal infection caused by the organism *Blastomyces dermatitidis*. Endemic to portions of North America, blastomycosis causes clinical symptoms similar to histoplasmosis.

Blastomycosis was first described by Thomas Casper Gilchrist in 1894 and sometimes goes by the eponym *Gilchrist's disease*. It is also sometimes referred to as Chicago Disease.

Epidemiology

In the United States, blastomycosis is endemic in the Mississippi river and Ohio river basins and around the Great Lakes. The annual incidence is less than 1 case per 100,000 people in Mississippi, Louisiana, Kentucky, and Arkansas. The cases are greater in northern states such as Wisconsin, where cases occur roughly 1 in 100 people.

In Canada, most cases of blastomycosis occur in northwestern Ontario, particularly around the Kenora area. The moist, acidic soil in the surrounding woodland harbours the fungus.

Blastomycosis is distributed internationally; cases are sometimes reported from Africa.

Pathology

Infection occurs by inhalation of the fungus from its natural soil habitat. Once inhaled in the lungs, they multiply and may disseminate through the blood and lymphatics to other organs, including the skin, bone, genitourinary tract, and brain. The incubation period is 30 to 100 days, although infection can be asymptomatic.

Clinical Features

Blastomycosis can present in one of the following ways:

- A flulike illness with fever, chills, myalgia, headache, and a non-productive cough which resolves within days.
- An acute illness resembling bacterial pneumonia, with symptoms of high fever, chills, a productive cough, and pleuritic chest pain.
- A chronic illness that mimics tuberculosis or lung cancer, with symptoms of low-grade fever, a productive cough, night sweats, and weight loss.
- A fast, progressive, and severe disease that manifests as ARDS, with fever, shortness of breath, tachypnea, hypoxemia, and diffuse pulmonary infiltrates.
- Skin lesions, usually asymptomatic, appear as ulcerated lesions with small pustules at the margins
- Bone lytic lesions can cause bone or joint pain.
- Prostatitis may be asymptomatic or may cause pain on urinating.
- Laryngeal involvement causes hoarseness.

Diagnosis

Once suspected, the diagnosis of blastomycosis can usually be confirmed by demonstration of the characteristic broad based budding organisms in sputum or tissues by KOH prep, cytology, or histology. Tissue biopsy of skin or other organs may be required in order to diagnose extra-pulmonary disease. Commercially available urine testing appears to be quite sensitive in suggesting the diagnosis in cases where the organism is not readily detected. While culture of the organism remains the definitive diagnostic standard, its slow growing nature can lead to delays in treatment of up to several weeks.

However, sometimes blood and sputum cultures may not detect blastomycosis; lung biopsy is another option, and results will be shown promptly.

Treatment

Itraconazole given orally is the treatment of choice for most forms of the disease. Cure rates are high, and the treatment over a period of months is usually well tolerated. Amphotericin B is considerably more toxic, and is usually reserved for critically ill patients and those with central nervous system disease.

Prognosis

Mortality rate in treated cases:

0-2 per cent in treated cases among immunocompetent patients

29 per cent in immunocompromised patients

40 per cent in the subgroup of patients with AIDS

68 per cent in patients presenting as acute respiratory distress syndrome (ARDS)

Butternut Canker

Sirococcus clavigignenti juglandacearum (Nair et al., 1979) is a fungus that causes the Butternut canker disease. It is known to parasitise members of the family Juglandaceae.

Distribution

Butternut canker was first reported in Wisconsin in 1967. It has since spread to other states. It was reported that 84 per cent of all butternut trees in Michigan as well as 58 per cent of all trees from Wisconsin have been affected; later surveys revealed that 91 per cent of all living trees were diseased or cankered. In Virginia and North Carolina, the butternut population has been reduced from 7.5 million to 2.5 million. It has since been reported in Canada.

Symptoms

Lenial and broad cankers on the main stem, branches, young twigs, and exposed roots. Most cankers are covered with bark cracks and shredded bark above cankers in late stages.

Life Cycle

Pycnidiospores are released during rainy periods. When the spores make contact with wounds or broken branches, they germinate and penetrate deep into the tree to produce cankers. Infection hyphae typically penetrates through the parenchyma phloem intracellularly but they can also penetrate intercellularly through uni- and multiceliate xylem ray cells and paranchyma cells. Later, the fungus will produce mycelial mats of stroma and mycelial pegs.

Stroma mats will produce uni or multilocular pycnidia. Inside the pycnidia are branched and unbranched conidiophores with two-celled pycniospores, which later are ejected from the pycnidial ostiole.

Additionally, the stroma will produce a peg of interwoven mycelium. These pegs put pressure on the outer peridium of the host bark, which exposes the pycnidia below. These pegs also produce pycnidia that are smaller than the pycnidia in the stroma. While different in size, the spores produced are identical.

Resistance

Many species of tree show varying degrees of resistance, such as the heartnut, butternut, and the Japanese, Black, and Persian walnuts. The heartnut and Japanese walnut trees show the highest degree of resistance, as both have a peridium layers between 35-45 cells think which act as a barrier for pathogen penetration. Additionally, both trees produce phenolics immediately upon attack, later producing gums and tyloses to surround the pathogen.

Control

Breeding for resistance is important for fighting butternut canker. While standard practice has been that infected trees should be removed to prevent further spread, there is a growing opinion that the time for this is past. The disease has now been found in virtually all parts of the butternut range.

Additionally, it is suggested that "removing diseased trees" is a guarantee that infected, but not dying trees, i.e. those that are specifically "partially resistant" to the fungus, will be killed; eliminating any chance of increased resistance in progeny. Instances are known of long-term survival of pure butternuts infected by the canker.

Recent reports have shown that the fungus can be internally seed-borne, so seeds should be subjected to intense quarantine protocols; most especially if destined for plantings where the disease is not already established.

Candidiasis

Manifestation: In immunocompetent people, candidiasis can usually only be found in exposed and moist parts of the body, such as:

- The oral cavity (oral thrush).
- The vagina and/or vulva (vaginal candidiasis or thrush).
- Folds of skin in the diaper area (diaper rash).
- The nipples while breastfeeding.
- The penis or foreskin in men.

Candidiasis is the second most common cause of vaginal irritation, or vaginitis, and can also occur on the male genitals. In immunocompromised patients, the *Candida* infection can involve the esophagus and can become systemic, causing a much more serious condition: fungemia.

Children, mostly between the ages of 3 and 9 years, can be affected by chronic mouth yeast infections, normally seen around the mouth as white patches. However, this is not a common condition.

Causes

Yeast organisms are always present in all people, but are usually prevented from "overgrowth" (uncontrolled multiplication resulting in symptoms) by other naturally occurring microorganisms.

In a study of 1009 women in New Zealand, the fungus, *Candida albicans*, was isolated from the vaginas of 19 per cent of apparently healthy women. Carriers experienced few or no symptoms. However, external (use of irritants, such as some detergents or douches) or internal (hormonal or other physiological changes) disturbances can perturb the normal flora, constituting lactic acid bacteria, such as lactobacilli, and an overgrowth of yeast can result in noticeable symptoms. Pregnancy, the use of oral contraceptives, engaging in vaginal sex immediately and without cleansing after anal sex, and using lubricants containing glycerine have been found to be causally related to yeast infections.

Diabetes mellitus and the use of antibiotics are also linked to an increased incidence of yeast infections. Candidiasis can be sexually transmitted between partners. Diet has been found to be the cause in some animals. Hormone Replacement Therapy and infertility treatments may also be predisposing factors.

Symptoms

Symptoms include severe itching, burning, and soreness, irritation of the vagina and/or vulva, and a whitish or whitish-grey discharge, often with a curdlike appearance.

Many women mistake the symptoms of the more common bacterial vaginosis for a yeast infection. In a 2002 study published in the *Journal of Obstetrics and Gynaecology*, only 33 per cent of women who were self treating for a yeast infection actually had a yeast infection. Instead they had either bacterial vaginosis or a mixed-type infection.

In men, symptoms include red patchy sores near the head of the penis or on the foreskin. The sores may feel irritated and itchy, and sometimes they will burn as well.

Diagnosis

Medical professionals use two primary methods to diagnose yeast infections: microscopic examination, and culturing.

For the microscope method, a scraping or swab of the affected area is placed on a microscope slide. A single drop of 10 per cent potassium hydroxide (KOH) solution is then also placed on the slide. The KOH dissolves the skin cells but leaves the *Candida* untouched, so that when the slide is viewed under a microscope, the hyphae and pseudo spores of *Candida* are visible. Their presence in large numbers strongly suggests a yeast infection.

For the culturing method, a sterile swab is rubbed on the infected skin surface. The swab is then rubbed across a culture medium. The medium is incubated for several days, during which time colonies of yeast and/or bacteria develop. The characteristics of the colonies provide a presumptive diagnosis of the organism causing symptoms.

Treatment

It is important to consider that *Candida* species are frequently part of the human body's normal oral and intestinal flora. Candidiasis is occasionally misdiagnosed by medical personnel as bacterial in nature, and treated with antibiotics against bacteria. This can lead to eliminating the yeast's natural competitors for resources, and increase the severity of the condition.

In clinical settings, candidiasis if commonly treated with anti-mycotics—the anti-fungal drugs commonly used to treat candidiasis are topical clotrimazole, topical nystatin, fluconazole, and topical ketoconazole.

In severe infections (generally in hospitalised patients), amphotericin B, caspofungin, or voriconazole may be used. Local treatment may include vaginal suppositories or medicated douches. Gentian violet can be used for breastfeeding thrush, but pediatrician William Sears recommends using it sparingly, since in large quantities it can cause mouth and throat ulcerations in nursing babies, and has been linked to mouth cancer in humans and to cancer in the digestive tract of other animals.

Some home remedies for candidiasis include the consumption or direct application of yogurt (which contains lactobacillus), probiotics, acidophilus tablets or salves, Pau d'arco tea, and even lightly crushed cloves of garlic, which yield allicin, an anti-fungal agent.

Boric acid has also been used to treat yeast infections (by inserting gelcaps filled with boric acid powder into the vagina at bedtime for three to four consecutive nights). Eating a diet consisting primarily of green, fresh, raw vegetables also may give relief. Other alternative treatments consist in consuming a fermented beverage called Kefir.

While home remedies may offer relief in minor cases of infection (although a peer-reviewed study in Australia found yogurt ineffective as treatment for *Candida albicans*), seeking medical attention may be necessary, especially if the extent of the infection cannot be judged accurately by the patient. For instance, oral thrush is visible only at the upper digestive tract, but it may be that the lower digestive tract is likewise colonised by *Candida* species.

Treating Candidiasis solely with medication may not give desired results, and other underlying causes require consideration. As an example, oral candidiasis is often linked to the use of inhaled corticosteroids in asthma medication. Patients on long-term inhaled corticosteroids should rinse their mouths after each dose of steroids to counteract this effect. Oral candidiasis can also be the sign of a more serious condition, such as HIV infection, or other immunodeficiency diseases. Following the health tips at vulvovaginal health can help prevent vaginal candidiasis.

Babies with diaper rash should have their diaper areas kept clean, dry, and exposed to air as much as possible. Sugars assist the overgrowth of yeast, possibly explaining the increased prevalence of yeast infections in patients with diabetes mellitus, as noted above. As many *Candida spp.* reside in the digestive tract, dietary changes may be effective for preventing or during a *Candida* infection.

Due to its requirement for readily fermentable carbon sources, such as mono- or dimeric sugars (e.g., sucrose, glucose, lactose) and starch, avoiding foods that contain these nutrients in high abundance may help to prevent excessive *Candida* growth. Breast milk is a suitable growth substrate for yeasts, and both nursing mother and baby need to be treated (even if both are not symptomatic) to prevent thrush from being passed between mother and child.

History and Taxonomic Classification

B. Lagenbeck in 1839 in Germany was the first to demonstrate that a yeast-like fungus existed in the human oral infection "thrush." He also found that a fungus was able to cause thrush.

The genera *Candida*, species *albicans* was described by botanist Christine Marie Berkhout. She described the fungus in her doctoral thesis, at the University of Utrecht in 1923. Over the years the classification of the genera and species has evolved. Obsolete names for this genus include *Mycotorula* and *Torulopsis*. The species has also been known in the past as *Monilia albicans* and *Oidium albicans*. The current classification is *nomen conservandum*, which means the name is authorised for use by the International Botanical Congress (IBC).

The full current taxonomic classification is available at *Candida albicans*.

The genus *Candida* includes about 150 different species. However, only a few of those are known to cause human infections. *C. albicans* is the most significant pathogenic (=disease-causing) species. Other *Candida* species causing diseases in humans include *C. tropicalis, C. glabrata, C. krusei, C. parapsilosis, C. dubliniensis*, and *C. lusitaniae*.

Alternative Views

Alternative medicine proponents frequently diagnose people with "systemic candidiasis" using methods not deemed valid by mainstream western medicine. Belief in widespread "systemic candidiasis" originated from a book published by Dr. William Crook, which hypothesised that a variety of common symptoms such as fatigue, PMS, sexual dysfunction, asthma, psoriasis, digestive and urinary problems, multiple sclerosis, and muscle pain, could be caused by subclinical infections of *Candida albicans.*

Dr. Crook suggested a variety of remedies to treat these symptoms, ranging from dietary modification to colonic irrigation. Mainstream Western medicine has ignored these methods, since they have not been proved using scientific methods. However, nutritionists have also proposed that these symptoms may be due to intestinal wall damage (known as leaky gut syndrome) caused by *Candida* overgrowth or other effects.

Chytridiomycosis

Chytridiomycosis is a fatal infectious disease that affects amphibians, caused by the chytrid—*Batrachochytrium dendrobatidis*. Chytridiomycosis has been linked to dramatic population declines or even extinctions in western North America, Central America, South America, and eastern Australia. There is no effective measure for control of the disease in wild populations. The disease is contributing to a worldwide decline in frog populations, a worldwide decline of species that apparently has affected 30 per cent of the amphibian species of the world.

Although some declines have been credited to the fungus *B. dendrobatidis*, there are species that resist the infection and some reports have found that some populations can survive with a low level of persistence of the disease. In addition, most of the laboratory studies are not ecologically-relevant, which makes it difficult to take their conclusions to the field. The first reports of infection of *Batrachochytrium* are from African clawed frogs of the genus *Xenopus*.

Because *Xenopus* has been widely transported around the world, it is one potential vector for transmission of *B. dendrobatidis*. Other studies, however, suggest that *B. dendrobatidis* has been present in North and Central America for decades.

Laboratory studies suggest that the fungus performs poorly at high temperatures, and that exposure of infected frogs to high temperatures will kill the fungus. This may explain why chytridiomycosis occurs primarily in cool climates.

Coccidioidomycosis

Coccidioidomycosis (also known as Valley fever, California valley fever, and (incorrectly) coccidiomycosis) is a fungal disease caused by *Coccidioides immitis* or *C. posadasii*. It is endemic in certain parts of Arizona, California, Nevada, New Mexico, Texas, Utah and northwestern Mexico.

C. immitis resides in the soil in certain parts of the southwestern United States, northern Mexico, and a few other areas in the Western Hemisphere.

Infection is caused by inhalation of airborne, fungal particles known as arthroconidia, which are a form of spore. The disease is not transmitted from person to person. *C. immitis* is a dimorphic saprophytic organism that grows as a mycelium in the soil and produces a sherule form in the host organism.

Symptomatic infection (40 per cent of cases) usually presents as an influenza-like illness with fever, cough, headaches, rash, and myalgia (muscle pain).

Some patients fail to recover and develop chronic pulmonary infection or widespread disseminated infection (affecting meninges, soft tissues, joints, and bone). Severe pulmonary disease may develop in HIV-infected persons.

Biological Warfare

C. immitis was investigated by the United States during the 1950s and 1960s as a potential biological weapon. The Cash strain received the military symbol OC, and original hopes were for its use as an incapacitant. As medical epidemiology later made clear, OC would have lethal effects on several segments of the population, so it was later considered a lethal agent. It was never standardised, and beyond a few field trials, it was never weaponised. Most military work on OC was on vaccines by the mid-1960s.

Diagnostic Test

The fungal infection can be demonstrated by microscopic detection of diagnostic cells in body fluids, exudates, sputum and biopsy-tissue. With specific nucleotide primers *C.immitis* DNA can be amplified by PCR. It can also be detected in culture by morphological identification or by using molecular probes that hybridise with *C.immitis* RNA. An indirect demonstration of fungal infection can be achieved also by serologic analysis detecting fungal antigen or host antibody produced against the fungus.

Cryptococcosis

Cryptococcosis is an infection acquired by inhalation of soil contaminated with the encapsulated yeast (fungus) Cryptococcus neoformans.

Etiology and Incidence

Cryptococcosis is a defining opportunistic infection for AIDS, although patients with Hodgkin's or other lymphomas or sarcoidosis or those receiving long-term corticosteroid therapy are also at increased risk.

Distribution is worldwide. The prevalence of cryptococcosis has been increasing over the past 20 years for many reasons, including the increase in indicence of AIDS and the expanded use of immunosuppressive drugs.

In humans, *C. neoformans* causes three types of infections:

- Wound or cutaneous cryptococcosis
- Pulmonary cryptococcosis, and
- Cryptococcal meningitis.

Cryptococcal meningitis (infection of the brain) is believed to result from dissemination of the fungus from either an observed or unappreciated pulmonary infection. *Cryptococcus gattii* causes infections in immunocompetent people (those having a functioning immune system), but *C. neoformans v. grubii*, and *v. neoformans* usually only cause clinically evident infections in persons who have some form of defect in their immune systems (immunocompromised persons). People who have defects in their cell-mediated immunity, for example, people with AIDS, are especially susceptible to disseminated cryptococcosis. Cryptococcosis is often fatal, especially if untreated.

Diagnosis

Symptoms include chest pain, dry cough, swelling of abdomen, headache, blurred vision and confusion. Although symptoms may not be able to be detected. Detection of cryptococcal antigen (capsular material) by culture of CSF, sputum and urine provides definitive diagnosis. Blood cultures may be positive in heavy infections.

Treatment

The standard regimen of treatment in non-AIDS patients intravenous Amphotericin B combined with oral flucytosine.

AIDS patients often have a reduced response to Amphotericin B and flucytosine, therefore after initial treatment as above, oral fluconazole can be used.

Dermatophyte

A dermatophyte is a large parasitic fungus upon the skin. The term embraces the imperfect fungi of the genera *Epidermophyton, Microsporum* and *Trichophyton*.

Dermatophytes (name based on the Greek for 'skin plants') are a common label for a group of three types of fungus that commonly causes skin disease in animals and humans. These anamorphic (asexual or imperfect) genera are: Microsporum, Epidermophyton and Trichophyton. There are about 40 species in these three genera. Species capable of reproducing sexually belong in the teleomorphic genus, Arthroderma, of the Ascomycota.

Dermatophytes cause infections of the skin, hair and nails due to their ability to obtain nutrients from keratinised material. The organisms colonise the keratin tissues and inflammation is caused by host response to metabolic by-products. They are usually restricted to the non-living cornified layer of the epidermis because of their inability to penetrate viable tissue of an immunocompetent host.

Invasion does elicit a host response ranging from mild to severe. Acid proteinases, elastase, keratinases, and other proteinases reportedly act as virulence factors. The development of cell-mediated immunity correlated with delayed hypersensitivity and an inflammatory response is associated with clinical cure, whereas the lack of or a defective cell-mediated immunity predisposes the host to chronic or recurrent dermatophyte infection.

Some of these infections are known as ringworm or tinea. Toe- and fingernail infection are referred to as onychomycosis. Dermatophytes usually do not invade living tissues, but colonise the outer layer of the skin. Occasionally the organisms do invade subcutaneous tissues, resulting in kerion development.

Identification

Microscopic morphology of the micro and macroconidia is the most reliable identification character, but a good slide preparation is needed, and also needed is the stimulation of sporulation in some strains. Culture characteristics such as surface texture, topography and pigmentation are variable so they are the least reliable criteria for identification.

Clinical information such as the appearance of the lesion, site, geographic location, travel history, animal contacts and race is also important, especially in identifying rare non-sporulating species like Trichophyton concentricum, Microsporum audouinii and Trichophyton schoenleinii.

Transmission

Dermatophytes are transmitted by direct contact with infected host (human or animal) or by direct or indirect contact with infected exfoliated skin or hair in clothing, combs, hair brushes, theatre seats, caps, furniture, bed linens, towels, hotel rugs, and locker room floors.

Depending on the species the organism may be viable in the environment for up to 15 months. There is an increased susceptibility to infection when there is a pre-existing injury to the skin such as scars, burns, excessive temperature and humidity. Adaptation to growth on humans by most geophilic species resulted in diminished loss of sporulation, sexuality, and other soil-associated characteristics.

Classification

Dermatophytes are classified as anthropophilic (humans), zoophilic (animals) or geophilic (soil) according to their normal habitat.

Anthropophilic dermatophytes are restricted to human hosts and produce a mild, chronic inflammation.

Zoophilic organisms are found primarily in animals and cause marked inflammatory reactions in humans who have contact with infected cats, dogs, cattle, horses, birds, or other animals. This is followed by a rapid termination of the infection.

Geophilic species are usually recovered from the soil but occasionally infect humans and animals.

They cause a marked inflammatory reaction, which limits the spread of the infection and may lead to a spontaneous cure but may also leave scars.

Classification

About 58 per cent of the dermatophyte species isolated are Trichophyton rubrum:

- 27 per cent are T. mentagrophytes.
- 7 per cent are T. verrucosum.
- 3 per cent are T. tonsurans.

Infrequently isolated (less than 1 per cent) are Epidermophyton floccosum, Microsporum audouinii, M. canis, M. equinum, M. nanum, M. versicolour, Trichophyton equinum, T. kanei, T. raubitschekii, and T. violaceum.

Fungal Diseases of Plants with Major Impact on Humans

There are many examples of plant diseases that have made a major impact on society and have even changed human history. More than 70 per cent of all major crop diseases are caused by fungi Late blight of potato was responsible for the loss of 25 per cent of the population of Ireland; during the 1840s, more than 1 million people died from starvation or famine-related diseases, and more than 1.5 million emigrated from Ireland A more recent epidemic that resulted in large-scale famine was caused by a fungus responsible for brown spot of rice; 2 million people died of starvation during the great Bengal famine of 1942.

A related fungus, which attacks corn and causes southern leaf blight, resulted in a widespread epidemic in the US in 1970; ca. 15 per cent of the total corn crop was lost, with yields in some states reduced 50 per cent Late blight of potato are attacked by many pathogens and pests, and the total amount of agricultural chemicals applied to the potato crop is greater than for any other food crop.

Late blight of potato has been the most important disease of potato since the 1840s, when it cause the destruction of the potato crop in Ireland.

The pathogen responsible for late blight of potato is the

fungus *Phytophthora infestans*, a member of the water mould class, oomycetes. The fungus invades host tissue, resulting in the rapid death of infected parts; it can also infect the tubers themselves.

Various Diseases

Some of the most important cereal diseases are caused by these fungi, which have produced serious epidemics throughout history Coffee, apple, and pine trees, and the economies depending on these crops, have been devastated by rust fungi; e.g. coffee rust destroyed the coffee plantations in Sri Lanka (then Ceylon) in the 1870s and 1880s Epidemics caused by rust fungi have been noted for millennia.

These epidemics were recognised in ancient Greece and described in the writings of Aristotle and Theophrastus. The Romans held a religious ceremony/festival, the *Robigalia*, to appease the gods Robigo and Robigus, whom they believed responsible for the rust epidemics In ancient times, wheat was the most important crop in these Mediterranean countries, and wheat rusts were major threats.

In North America today, the most important rust pathogen is the fungus *Puccinia graminis* subsp. *tritici*, which is responsible for stem rust of wheat.

Wheat plants infected by the fungus are severely weakened but not destroyed, and the grain yield is significantly reduced; it is estimated that worldwide more than 1 million metric tons of wheat are lost annually to stem rust.

Cedar-apple rust causes yellow spots on the leaves and fruits of apple and crab apple (and related species), and produces galls on red cedar (juniper) trees. Corn smut; The smut fungi, like the rusts, are Basidiomycotina plant pathogens that cause significant losses of grain crops. Incredible numbers of spores form, usually within galls; these masses of dark-brown or black spores, resembling soot or smut, give the common name to these pathogens. Corn smut is a widespread disease that is more common on sweet corn than on other varieties.

Corn smut is caused by *Ustilago maydis*, which forms galls on any above ground part but is most conspicuous when the galls form on the ears Recently, the young galls have become popular in the US as a gourmet delicacy, smoky maize mushroom, which sells for $50 per pound; the fungus has been cultivated for centuries in Mexico as a food item.

Dutch Elm Disease

During the first half of the 20th century, the chestnut blight fungus caused the death of 80 per cent of the native American chestnut trees throughout eastern forests from Maine to Georgia The Dutch elm disease fungus, *Ophiostoma ulmi*, has led to the destruction of American elm trees and has altered urban landscapes by killing ornamental elms across the country.

The disease appeared in Europe around 1919, and the fungus was described in Holland in 1921; it was first found in the US in Ohio in 1930 It quickly spread across the continent, and it has been estimated that more than 77 million elms have died.

Although the fungus is responsible for the disease, elm bark beetles are the vectors that carry the fungal spores from infected elms to healthy trees. Various measures have been employed to control spread of the disease; e.g., removal of infected trees, biological control (using another fungus) to interrupt the reproductive cycle of the bark beetle, systemic fungicides.

A promising new approach is a preventive treatment that is being called "elm vaccine". A glycoprotein produced by the fungus is injected into the tree, and the tree produces mansonones, which are phytoalexins. Phytoalexins are secondary products produced by plants in response to bacterial and fungal pathogens; they are known to kill or inhibit the growth of pathogenic organisms.

Powdery Mildew (Erysiphe graminis)

Powdery mildews are common on many kinds of plants. Although they may look the same, each kind of plant is infected by a different species of powdery mildew fungus. Powdery mildew fungi on grasses will not infect lilacs, phlox, roses or other garden plants.

Symptoms: Powdery mildew may appear quite suddenly, usually in shaded areas, and most commonly on Kentucky bluegrass. The grass blades look as if they were dusted with flour or lime. The white to gray powder is a combination of the mycelium and spores of the powdery mildew fungus. The mycelium grows over the surface of the leaf, absorbing nutrients from the plant.

Later, the leaf may turn yellowish and begin to dry up and die, but the leaves often support the presence of the powdery mildew fungus for some time without significant injury. Powdery mildew is most common in turf from July to September, and occasionally in the spring. It is most common during overcast periods of cool, moist weather.

Life Cycle: Powdery mildew fungi overwinter on infected grass plants and in survival structures on dead grass. Spores can infect leaves in less than two hours, and new spores are produced in abundance in about a week. Air currents carry the spores to new grass plants. Disease development can be so rapid that powdery mildew may seem to appear very suddenly.

Cultural Management: Powdery mildew is usually a disease of shaded turfgrass. Similar grasses growing nearby in full sun usually will remain free of powdery mildew. To reduce shade and increase air circulation, prune shrubs and tree branches. If grass is thin or weakened in shady areas, reseed with shade-tolerant grass species and select cultivars resistant to powdery mildew.

If grass is growing very poorly due to shade, consider its replacement with shade-loving ground covers such as hosta, pachysandra or lily-of-the-valley. In moist, shaded areas of lawns, powdery mildew may be prevented or reduced by careful lawn care. Avoid excess nitrogen fertilizer, raise the mowing height and water deeply but infrequently. Avoid frequent, light sprinkling.

Chemical Management: Fungicides are not recommended for powdery mildew control on lawns.

Red Thread (Laetisaria fuciformis) and Pink Patch (Limonomyces roseipellis)

Symptoms: Red thread and pink patch are caused by two different fungi, but they often appear together and under the same environmental conditions. Lawns infected by these fungi may have a pink-to-reddish cast when viewed from a distance. If moist grass is viewed more closely, pinkish gelatinous growth of the fungi and tiny cotton-candy tufts of spores may be seen. In drier conditions, tiny red threads of the red thread fungus may be observed extending from the leaf tips.

The grass itself may show irregular patches of dead and dying leaves, giving the lawn a ragged appearance. These diseases can be found on many turfgrass species but are seen most often on perennial regresses and fine leaf fescues. The diseases are restricted to the leaves and are not usually very destructive. Both fungi occur mostly in the spring and fall in cool (65-70°F), moist weather.

Disease Cycle: The gelatinous masses of both fungi serve as survival structures in the thatch when dry. The fungi can also survive as mycelium in infected leaves and thatch. Some spores or pieces of fungus may be airborne, but the fungi are commonly spread by water, feet and mowing. Infections are most common on slow-growing grass stressed by compaction, drought, low temperatures or inadequate fertility.

Cultural Management: Although we have no control over periods of prolonged cool, moist weather, it is possible to prevent or reduce red thread and pink patch through good lawn care practices.

Soil pH should be maintained at 6.5-7.0 or as appropriate for the turf species. Watering should be infrequent and deep. Avoid frequent sprinklings late in the day that extend the time that the grass blades stay wet.

Prune trees and shrubs to increase light penetration and air circulation. Soil fertility must be balanced. These diseases are most common in nitrogen-deficient turf, but be cautious about applying excess nitrogen fertilizer, especially in spring. Other more serious diseases, such as leafspots and melting out, can be enhanced by excess nitrogen.

If red thread or pink patch diseases are found in slow growing grass, it may be helpful to temporarily collect clippings during mowing to reduce the amount of fungus returned to the lawn. Aeration of compacted soil will improve turf growth and reduce these diseases.

Chemical Management: Fungicides are not recommended for red thread or pink patch on lawns except in extreme cases.

Rust (Puccinia spp.)

Symptoms: There are many different rust fungi that can infect lawn grasses, but they all have in common the production of reddish, yellowish or orange spores that give "rusts" their name. If only a small amount of infection has occurred, the rust spores will only be seen by close observation of the tiny, powdery pustules on the leaf blades.

Rust fungus on grass blades

In severe cases, there may be enough spores present to leave a reddish dust on mowers, pants and shoes. The turfgrass itself may not look particularly diseased until after the infection is well developed. In fact, because spore production does not occur until one to two weeks after infection, rust-infected turfgrass leaf blades are usually mowed away before rust spores can be produced. In rare severe infections, the lawn may look thin and be weakened and more susceptible to other stresses such as drought or winterkill. Rust is most common on Kentucky bluegrass and perennial ryegrass, and occurs mostly in late summer and early fall.

Disease Cycle: Most rust fungi have complex life cycles that involve two unrelated host plants and many spore stages. The disease cycle on lawns is usually limited to the rust-coloured spore stage which rarely survives New England winters. As a result, most rust infections are initiated by spores that are blown up from warmer southern areas.

Rust fungi cannot grow without a living host plant. Infection only occurs when water is on the leaf surface and can occur at most normal summer temperatures. There is usually not enough time for spores to be produced on lawns that are mowed regularly, so epidemics are not usually severe.

Cultural Management: The most important way to avoid rust infection on lawns is to keep the turfgrass growing vigorously so that it will be mowed before spores can be produced. Balanced fertility based on *soil tests,* aeration to relieve compaction, *thatch removal* and *adequate water supply* are among the important factors that will optimise turfgrass growth. Watering practices that keep foliage dry as much as possible will help reduce infections by rust and other foliar fungi.

Chemical Management: Fungicides are not recommended for rust diseases on lawns except in extreme cases.

Slime Molds

Symptoms: A superficial fungus-like growth may appear suddenly on leaf blades of turfgrass in warm, wet weather.

Slime moulds are initially slimy but mature to a powdery stage when spores are produced. They vary in colour from black to powdery gray, pink, yellow or orange. They may occur in large quantities in rainy weather, enough to be observed even at a distance.

Disease Cycle: Slime moulds are saprophytes, obtaining their nutrients from dead organic matter in thatch and soil. They are unable to infect turf or other plants as parasites and do not cause any injury beyond that caused by coating the leaf blades. Slime moulds are initiated from spores and develop into a slimy film. As they mature, the slime is converted into a mass of powdery spores which may be airborne to new sites. Slime moulds are commonly observed on tree leaf litter, rotting wood, turfgrass, wood chips and other organic mulches used in landscaped areas.

Cultural Management: Because slime moulds do not infect turfgrass and simply cover the surface of the leaves, they can be removed by raking or hosing the superficial substance away.

Chemical Management: Fungicides are not recommended for slime moulds because they do not cause turf diseases and can be easily removed mechanically.

Leaf Smuts (Stripe smut (Ustilago striiformis) and Flag smut (Urocystis agropyri))

Symptoms: Leaf smuts are named for the powdery black spore masses produced by the smut fungi that infect Kentucky bluegrass and, occasionally, perennial rye-grass. In the spring and fall, at temperatures of 50-65 °F, plants may look stunted and yellow, turning greyish just before the streaks of black spores break through the leaf surface. Later, leaves appear twisted and frayed. During hot summer weather, smut symptoms often disappear, but the plants are still infected. They are very susceptible to various stresses, such as heat and drought, and many will die. Smut symptoms may return in the fall, and the fungus will continue to live inside the plant until the plant dies. Smut is usually not reported until lawns are three to four years old, when yellowing and thinning is first noticed and black streaks of spores are discovered.

Disease Cycle: The black dusty smut spores survive in soil, thatch, and on seed. They are easily transported via air, soil, water, feet, seed and equipment. Turfgrass plants are infected by smut only at growing points, and they will remain infected until they die.

Cultural Management: Only smut-free seed and sod should be used to establish new lawns. A few infected plants are easy to overlook, and smut is usually not discovered until infections are more extensive in older lawns. In established lawns, smut will be worse with excess thatch, soil pH below 6.0, frequent watering and excess nitrogen fertilizers. These same factors encourage other diseases as well and should be avoided. After infected plants die out during summer stress, reseed in the fall with a blend of resistant cultivars.

Chemical Management: Only penetrant systemic fungicides are useful against smut fungi. They are expensive and usually must be applied more than once. Fungicide applications are not economical or practical for most lawn situations compared to reseeding damaged areas.

3

Diseases Affecting Plants

Banana Diseases

Black Sigatoka: Black sigatoka is a leaf spot disease of banana plants caused by fungus Mycosphaerella fijiensis Morelet. Plants with leaves damaged by the disease may have up to 50 per cent lower yield of fruit.

In commercial export plantations, Black Sigatoka is controlled by frequent applications of fungicides. Removal of affected leaves, good drainage, and sufficient spacing also help to fight the disease.

Although fungicides improved over the years, the pathogen developed resistance. Therefore higher frequency of application is required, increasing the impact on the environment and health of the banana workers.

Small farmers growing bananas for local market cannot afford expensive measures to fight the disease. However, some cultivars of banana are resistant to the disease. Research is done to improve productivity and fruit properties of these cultivars.

Mycosphaerella fijiensis

Mycosphaerella fijiensis is an ascomycete fungus that is an important banana pathogen. It causes the Black Sigatoka (black leaf streak disease). The anamorph of the fungus is traditionally called Paracercospora fijiensis or Pseudocercospora fijiensis.

M. fijiensis reproduces both sexually and asexually, and both conidia and ascospores are important in its dispersal. The conidia are mainly water-born to short distances, while ascospores are carried by wind to more remote places (the distances being limited by their susceptibility to ultraviolet light). Over sixty distinct strains with different pathogenetic potentials have been isolated. In order to better understand the mechanisms of its variability, the "Genetic diversity of *Mycosphaerella fijiensis* Project" has been initiated.

Panama Disease

Panama disease, also known as Fusarium wilt, is a banana plant disease caused by the fungus Fusarium oxysporum. The fungus attacks the roots of the banana plant. The disease is resistant to fungicide and, so, cannot be controlled chemically.

Gros Michel or 'Big Mike' was an early export cultivar of banana. This cultivar was wiped out by Panama disease in the 1950s. The disease first appeared in Suriname, then made its way to the Caribbean, and by the 1920s, to Honduras, the world's largest producer of bananas at the time. Although there are many banana cultivars, Gros Michel was especially suitable for export to non-tropical nations. A search for a substitute located the Vietnamese Cavendish cultivar which is resistant to the disease. However, more care is required for shipping the Cavendish banana and its quality compared to Gros Michel is debated.

Recently, a new strain called 'tropical race four Panama disease' has begun to attack Cavendish banana plants in south Asia. Given the high volume of modern international trade, banana producers expect this strain to spread through Africa and into South America and the Caribbean. Plant breeders and geneticists are trying to develop new cultivars that are resistant to this new strain of Panama disease.

Unfortunately, such efforts are progressing slowly because the banana cultivars selected for human consumption are seedless and reproduce asexually which decreases genetic variation and makes breeding difficult.

Alternaria

Alternaria is a genus of ascomycete fungi. Alternaria species are known as major plant pathogens. They are also common allergens in humans, growing indoors and causing hay fever or hypersensitivity reactions that sometimes lead to asthma. They readily cause opportunistic infections in immunocompromised people such as AIDS patients.

There are 44 definite, well-known species, but there may be hundreds more. They are ubiquitous in the environment and are a natural part of fungal flora most everywhere. They are normal agents of decay and decomposition. The spores are airborne and found in the soil and water, as well as indoors and on objects. The club-shaped spores are single or form long chains. They can grow thick colonies which are usually black or gray.

At least 20 per cent of agricultural spoilage is caused by Alternaria species. Many human health disorders can be caused by these fungi, which grow on skin and mucous membranes, including on the eyeballs and within the respiratory tract. Allergies are common, but serious infections are rare, except in people with compromised immune systems.

However, species of this fungal genus are often prolific producers of a variety of toxic compounds. The effects most of these compounds have on animal and plant health are not well known. The terms alternariosis and alternariatoxicosis are used for disorders in humans and animals caused by a fungus in this genus.

Not all Alternaria species are pests and pathogens; some have shown promise as biocontrol agents against invasive plant species.

Selected species:

- *Alternaria Alternata:* causes upper respiratory infections in AIDS patients, asthma in people with sensitivity, and also infests plants.
- *Alternaria arborescens:* causes stem canker of *tomato*.
- *Alternaria arbusti:* causes leaf lesions on *Asian pear*.
- *Alternaria blumeae-* causes lesions on *Blumea aurita*.
- *Alternaria brassicae:* infests many vegetables and roses.
- *Alternaria brassicicola:* grows on *cole crops*.
- *Alternaria carotiincultae:* causes leaf blight on *carrot*.
- *Alternaria conjuncta:* grows on *parsnip*.
- *Alternaria dauci:* grows on carrot.
- *Alternaria euphorbiicola:* infests cole crops.
- *Alternaria gaisen:* causes ringspot disease of pear.
- *Alternaria infectoria:* infests *wheat*.
- *Alternaria japonica:* infests cole crops.
- *Alternaria molesta:* may cause skin lesions on *porpoises*.
- *Alternaria panax:* causes *ginseng* blight.
- *Alternaria petroselini:* causes *parsley* leaf blight.
- *Alternaria selini:* causes parsley crown decay.
- *Alternaria smyrnii:* infests *alexanders* and parsleys.

Alternaria-alternata

Alternaria-alternata has been recorded infecting over 380 host species. It is opportunistic pathogen on numerous hosts causing leaf spots, rots and blights on many plant parts. It also causes upper respiratory tract infections in AIDS patients and asthma in people with sensitivity.

Banana and Planting Diseases

Anthracnose	*Colletotrichum musae*
Armillaria corn rot	*Armillaria mellea, Armillaria tabescens*
Black cross	*Phyllachora musicola*
Black leaf streak (black sigatoka)	*Mycosphaerella fijiensis, Paracercospora fijiensis* [anamorph]
Black root rot	*Rosellinia bunodes*
Brown blotch	*Pestalotiopsis leprogena*
Brown spot	*Cercospora hayi*
Ceratocystis fruit rot	*Ceratocystis paradoxa, Chalara paradoxa* [anamorph]
Cigar-end	*Verticillium theobromae, Trachysphaera fructigena*
Cladosporium speckle	*Cladosporium musae*
Corm dry rot	*Junghuhnia vincta*
Cordana leaf spot	*Cordana johnstonii, Cordana musae*
Crown rot	*Fusarium pallidoroseum, Colletotrichum musae, Verticillium theobromae, Fusarium* spp. *Acremonium* spp.
Cylindrocladium root rot	*Cylindrocladium* spp.
Damping-off	*Deightoniella torulosa*
Deightoniella fruit speckle, leaf spot and tip rot	*Deightoniella torulosa*
Diamond spot	*Cercospora hayi Fusarium* spp.
Dwarf Cavendish tip rot	*Nattrassia mangiferae* = *Hendersonula toruloidea*
Eyespot	*Dreschslera gigantea*
Fruit freckle (freckle)	*Guignardia musae, Phyllosticta musarum* [anamorph]
Fruit rot	*Botryosphaeria ribis*
Fungal root-rot	*Fusarium solani, Nectria haematococca* [teleomorph] *Fusarium oxysporum, Rhizoctonia* spp.
Fungal scald	*Colletotrichum musae*

Fusarium wilt (Panama disease)	*Fusarium oxysporum* f. sp. *cubense*
Leaf rust	*Uredo musae Uromyces musae*
Leaf speckle	*Acrodontium simplex*
Leaf spot	*Curvularia eragrostidis*
Leaf spot	*Drechslera musaesapientum*
Leaf spot	*Leptosphaeria musarum*
Leaf spot	*Pestalotiopsis disseminata*
Main stalk rot	*Ceratocystis paradoxa*
Malayan leaf spot	*Haplobasidion musae*
Marasmiellus rot	*Marasmiellus inoderma* = *Marasmius semiustus*
Moko	*Pseudomonas solanacearum* (race 2)
Peduncle rot	*Lasiodiplodia theobromae Fusarium pallidoroseum Fusariumoxysporum Verticillium theobromae*
Pestalotiopsis leaf spot	*Pestalotiopsis palmarum*
Phaeoseptoria leaf spot	*Phaeoseptoria musae*
Pitting	*Pyricularia grisea*
Pseudostem heart rot	*Fusarium moniliforme, Gibberella fujikuroi* [teleomorph]
Root & rhizome rot	*Cylindrocarpon musae*
Sclerotinia fruit rot	*Sclerotinia sclerotiorum*
Sheath rot	*Nectria foliicola Mycosphaerella musicola, Pseudocercospora musae* [anamorph]
Sooty mould	*Limacinula tenuis*
Speckle	*Mycosphaerella musae*
Squirter (black end disease)	*Nigrospora sphaerica*
Stem-end rot	*Colletotrichum musae*
Trachysphaera finger rot	*Trachysphaera frutigena*
Tropical speckle	*Ramichloridium musae* = *Veronaea musae* = *Periconiella musae*
Verticillium tip rot	*Verticillium theobromae*

Canola Diseases

Albugo Candida: White rust, Albugo candida is species from family Albuginaceae. However, all the other species in this family are called "white rust".

Synonyms:

- *Albugo candidans*
- *Aecidium candidum* Pers., (1797)
- *Albugo cruciferarum* (DC.) Gray, (1821)
- *Cystopus candidans*
- *Cystopus candidus* (Pers.) Lev., (1847)
- *Uredo candida* (Pers.) Fr., (1832)
- *Uredo candida* Rabenh.
- *Uredo cruciferarum* DC.

Hosts:

- Shepherd's Purse *(Capsella bursa-pastoris).*
- Brassicaceae species.

Mustard Diseases

Fungal Diseases	
Alternaria black spot	*Alternaria brassicae, A. brassicicola, A. raphani*
Anthracnose	*Colletotrichum gloeosporioides Glomerella cingulata* [teleomorph], *C. higginsianum*
Black leg (leaf, root and stem rot)	*Leptosphaeria maculans, Phoma lingam* [anamorph]
Black root	*Aphanomyces raphani*
Cercospora leaf spot	*Cercospora brassicicola*
Clubroot	*Plasmodiophora brassicae*
Damping-off	*Fusarium spp. Rhizoctonia solani, Thanatephorus cucumeris [teleomorph]*

Downy mildew	*Peronospora parasitica*
Head rot	*Rhizoctonia solani*
Leaf spot	*Myrothecium roridum Phyllosticta brassicae = P. brassicina*
Powdery mildew	*Erysiphe polygoni*
Sclerotinia stem rot	*Sclerotinia sclerotiorum*
Southern blight	*Sclerotium rolfsii, Athelia rolfsii* [teleomorph]
White rust	Albugo candida (*Peronospora* sp. commonly present in staghead phase)
White leaf spot	*Pseudocercosporella capsellae*
Wirestem	*Rhizoctonia solani*
Yellows	*Fusarium oxysporum*

Fruit Spots of Tomato

There are many fungi that cause leaf and fruit spots on tomatoes. The ones described here are similar in that they do not have obvious fungal growth on either the leaf or the fruit spots.

Early Blight

Early blight is caused by *Alternaria solani*. The fungus causes disease in tomato, potato, eggplant, horse nettle, and black nightshade. First described in 1882, it is now found on all continents. It is very destructive in temperate humid climates. Although the disease is called "early" blight, it can occur on the plant at all stages of development. Early blight can cause a decrease in fruit quantity and quality.

Symptoms: All above ground parts of the plant can have symptoms of this disease. The fungus can kill seedlings before or after they break the soil surface. It can also cause collar rot in young plants, in which a dark, slightly sunken spot develops on one side of the stem. This spot will either grow completely around the stem or weaken the stem, causing the plant to fall over and die.

Leaf spots are circular, up to 1/2" in.diameter, and dark to light brown. They have a distinct pattern of concentric rings, like a target. Spots may occur singly or in large numbers on the leaf. The leaf may turn yellow, then brown, and fall off. Older leaves are usually affected before the disease works up the plant. In a severe infection, the entire plant may be defoliated. Defoliation may cause the fruit to become sun scalded.

The fruit is usually affected at the stem end. One or more firm, depressed rot spots appear on either the green or ripe fruit. The spots usually have distinct concentric rings; they may appear leathery and may be covered with a velvety mass of black spores.

Disease Identification: Target spots on the leaves are very characteristic.

Prevention: Use disease-free seed and transplants. Resistant varieties are available. Avoid planting near potatoes or other tomatoes. Avoid injury to the plant, especially during transplanting. Long crop rotations are important. High soil fertility reduces the severity of this disease, as does maintaining plant vigour. Limit leaf wetness by watering early in the day and by watering at the base of the plant. Control weeds, especially horse nettle and black nightshade. Eradicate volunteer tomato plants. Remove or destroy debris immediately after harvest, or at least plough under deeply.

Septoria Leaf Spot

Septoria leaf spot is caused by *Septoria lycopersici*. It is one of the most destructive diseases of tomato foliage. The fungus can cause disease on tomato, potato, eggplant, petunia, nightshade weeds, horse nettle, and ground cherry. Defoliation of the plant leads to a decreased yield.

Symptoms: Symptoms of this disease can be found on the leaves and other green parts of the tomato plant. Small, round, yellowish spots appear on the leaves, usually after the first fruit sets. Symptoms generally start on the lower leaves first. The spots later become brown or greyish and may have a yellowish or dark border.

On susceptible tomato varieties, leaf spots may be up to 1/8" in diameter; but they are smaller on resistant varieties. Dark specks or dots appear in the centre of the spots on susceptible varieties. These are usually fewer or absent on resistant varieties. Heavily infected leaves fall off, normally starting at the bottom of the plant and progressing upward. Fruit infection is rare.

Disease Identification: Leaf spots are small and may have dark specks in them; no spots on fruit.

Similar Diseases: Similar diseases exhibit somewhat different symptoms. Early blight leaf spots are generally larger (1/2"), have a target appearance, and may have a light-coloured ring around each spot. Early blight causes large target-like spots near the stem of green or ripe fruit. Leaf spots caused by bacterial speck and bacterial spot are dark brown to black. They may have a yellow ring around them. These diseases cause dark specks or spots on the fruit.

Prevention: Use disease-free seeds and transplants. Resistant varieties are available and should be used whenever possible. A crop rotation of 1 to 3 years, controls weed hosts and , therefore, decreases the level of disease. Planting early may help reduce disease, since the pathogen prefers warm temperatures.

Avoid working in plantings when leaves are wet. Stake plants to keep foliage dry and off the ground. Avoid overhead irrigation. Immediately after harvest, remove all debris from the field and bury it. As an alternative, use a rotary 'brush-hog' mower and plough under debris. These techniques prevent the fungus from moving to other hosts to overwinter on them. See current recommendations for chemical control measures.

Anthracnose Ripe Rot

Anthracnose ripe rot is caused by *Colletotrichum coccoides, C. dematium,* and *C. gloeosporioides*. This major tomato disease was first noted in 1879. It is now found in North America, Europe, and Asia.

There are many plants affected by these fungi, including tomato, pepper, eggplant, many melons and beans, grape, apple, and many nightshade weeds. The fungi may invade the roots of cabbage, chrysanthemum, lettuce, wheat, and other plants. This disease causes a rot of ripe fruit, which may cause huge losses in the field and during storage. Anthracnose ripe rot is favoured by moist weather.

This disease affects ripe or overripe fruit. Fruit spots are round, slightly sunken, and up to 1/2" in diameter. Under the skin, the spot may be lighter, dry, and granular. The centre of the spot may become tan and flecked with small black specks and, in wet conditions, may produce a mass of slimy salmon-coloured spores.

Leaf spots are small, circular, and brown. Yellow rings often surround them. The root system may be damaged so severely that the plant can be easily pulled.

Disease Identification: This disease affects ripe fruit only, causing almost perfectly round sunken spots with dark centres.

Similar Diseases: Early blight has different symptoms. Early blight fruit spots are dark and have a target appearance. They are usually near the stem. They can be found on green fruit.

Prevention: Use disease-free seed. Resistant varieties are available and should be used whenever possible. Rotate with a non-host for at least three years if anthracnose is a problem. Avoid overhead irrigation. Control weeds, since the pathogen has a broad host range that includes many common weeds. Avoid root injury. Harvest fruit as soon as possible after ripening.

Wheat Diseases: Leaf Rust (Brown Rust)

Puccinia Recondita

Symptoms: The postules are circular or slightly elliptical, smaller than those of stem rust, usually do not coalesce, and contain masses of orange to orange-brown urediospores. Infection sites primarily are found on the upper surfaces of leaves and leaf sheaths (1), and occasionally on the neck and awns.

Development: Primary infections usually are light and develop from wind-borne urediospores that may have travelled long distances. The disease can develop rapidly when free moisture is available and temperatures are near 20°C. Successive generations of urediospores can be produced every 10-14 days if conditions are favourable. As plants mature or when environmental conditions are not favourable, masses of black teliospores may become evident.

Hosts/Distribution: Leaf rust can affect wheat, triticale and many other related grasses. The disease is found wherever temperate cereals are grown. The alternate hosts are *Thalictrum*, *Isopryum*, *Anemonella*, and *Anchusa* spp.

Importance: Severe early infections can cause significant yield losses, mainly by reducing the number of kernels per spike, test weights, and kernel quality.

Stem Rust (Black Rust)

Puccinia graminis f.sp. tritici

Symptoms: Pustules (containing masses of urediospores) are dark reddish brown, and may occur on both sides of the leaves, on the stems, and on the spikes. With light infections the pustules are usually separate and scattered, but with heavy infections they may coalesce. Prior to pustule formation, "flecks" may appear. Before the spore masses break through the epidermis, the infection sites feel rough to the touch; as the spore masses break through, the surface tissues take on a ragged and torn appearance.

Development: Primary infections are usually light and develop from wind-borne urediospores that may have travelled long distances. The disease can develop rapidly when free moisture (rain or dew) and moderate temperatures prevail. If temperatures average about 20°C or more, the first generation of urediospores will be produced in 10-15 days. As plants mature, masses of black teliospores may be produced.

Hosts/Distribution: Stem rust can affect wheat, barley, tritlcale, and many other related grasses; it is found wherever temperate cereals are grown. The alternate hosts are *Berberis* and *Mahonia* spp.

Importance: If infection occurs during the early crop stages, the effects can be severe: reductions in tillering and losses in grain weight and quality. Under favourable conditions, complete crop loss can occur.

Stripe Rust (Yellow Rust)

Puccinia Striiformis

Symptoms: The pustules of stripe rust, which, contain yellow to orange-yellow urediospores, usually form narrow stripes on the leaves. Pustules also can be found on leaf sheaths, necks, and glumes.

Development: Primary infections are caused by wind-borne urediospores that may have travelled long distances. The disease may develop rapidly when free moisture (rain or dew) occurs and temperatures range between 10-20°C. At temperatures above 25°C, the production of urediospores is reduced or ceases and black teliospores are often produced.

Host/Distribution: Stripe rust can attack wheat, barley, triticale, and many other related grasses. The disease is found in all highland and/or temperate areas where cereals are grown. No alternate host is known.

Importance: Severe infections can cause yield losses, mainly by reducing the number of kernels per spike, test weights, and kernel quality.

Common and Dwarf Bunt (Stinking Smut)

Tilletia caries, T. foetida, T. controversa

Symptoms: The main symptoms caused by these three species are fungal structures called "bunt balls," which resemble kernels but are completely filled with black teliospores. The bunt balls of common bunt, caused by *T. caries* and *T. foetida*, are about the same size and shape as the kernels they replace; those of dwarf bunt, caused by T. controversa, are more nearly spherical. When bunt balls are crushed, they give off a fetid or fishy odour. Infected spikes tend to be bluish green in colour (or darker), and the glumes tend to spread apart slightly; the bunt balls often become visible after the soft dough stage. A slight reduction in plant height is typical of common bunt, while a pronounced reduction in height is typical of dwarf bunt.

Development: Spores lying dormant in the soil or on seed germinate and infect emerging seedings. Infection is favoured by cool temperatures during germination. The disease develops systemically, with visible symptoms appearing after heading.

Hosts/Distribution: Wheat and (less commonly) triticale are affected by these diseases, as are several other related grasses. Common and dwarf bunts can occur worldwide. Both are limited to temperate climates; dwarf bunt occurs in areas having prolonged snow cover.

Importance: Considerable yield losses can occur when susceptible cultivars are grown or chemical seed treatments are not used.

Karnal Bunt (Partial Bunt)

Tilletia indica (syn. Neovossia Indica)

Symptoms: Karnal bunt is not easily detected prior to harvest, since it is usual for only a few kernels per spike to be affected by the disease. Following harvest, diseased kernels can be easily detected by visual inspection: a mass of black teliospores replaces a portion of the endosperm, and the pericarp may be intact or ruptured. Diseased kernels give off a fetid or fishy odour when crushed.

Development: Karnal bunt is a seed- or soilborne, floral infecting disease inoculum (teliospores) on or near the soil surface germinates, producing sporidia, which are carried by wind to the floral structures.

These sporidia in turn germinate and penetrate the glumes, rachis, or the ovary itself. The fungus enters the newly formed kernel and develops in the intercellular space between the endosperm and seed coat. The degree of disease establishment and development depends on environmental conditions from spike emergence through grain filling.

Hosts/Distribution: Karnal bunt can affect wheat, triticale, rye, and several other related grasses, but not barley. The disease is endemic in the Asian Subcontinent and now in Mexico.

Importance: Karnal bunt is a relatively minor disease. Actual losses in yield are minimal, but the disease is on the quarantine lists of many countries, and therefore of importance in world grain trade.

Loose Smut

Ustilago Tritici

Symptoms: The entire inflorescence, except the rachis, is replaced by masses of smut spores. These black teliospores often are blown away by the wind, leaving only the bare rachis and remnants of other floral structures.

Development: Wind blown teliospores that land on the flowers of wheat plants can germinate and infect the developing embryo of the kernel. The mycelium of the loose smut fungus remains dormant in the embryonic tissues of the kernel until the kernel begins to germinate. The mycelium then develops along with the growing point of the plant, and at flowering time replaces the floral parts of the spike with masses of black spores. Infection and disease development are favoured by cool, humid conditions, which prolong the flowering period of the host plant.

Hosts/Distribution: The disease can occur wherever wheat is grown.

Importance: Yield losses depend on the number of spikes affected by the disease; incidence is usually less than one per cent and rarely exceeds thirty per cent of the spikes in any given location.

Flag Smut

Urocystis Agropyri

Symptoms: Masses of black teliospores are produced in narrow strips just beneath the epidermis of leaves, leaf sheaths and occasionally the culms. Diseased plants often are stunted, tiller profusely and the spikes may not emerge. A severe infection usually induces the leaves to roll, producing an onion-type leaf appearance. The epidermis of older diseased plants tends to shed, releasing the teliospores.

Development: Germinating kernels or very young seedlings are infected by germinating spores on the seed or in the soil. The disease continues to develop systemically, and the black subepidermal strips of teliospores become visible near heading. Infection is favoured by low soil moisture and cool soil temperatures.

Hosts/Distribution: Bread wheats are the primary hosts of flag smut fungi, and the isolates attacking bread wheat tend to be so exclusively. There are few reports of flag smut on durum wheats and triticales. The disease is found in most winter wheat areas and in cool, fall-sown spring wheat areas.

Importance: Flag smut generally is not an economically important disease, but where present, yield losses can range from trace amounts to moderate levels (when susceptible cultivars are grown).

Powdery Mildew

Erysiphe graminis f. sp. tritici

Symptoms: On all hosts, the first visible symptoms of this disease are white to pale gray, fuzzy or powdery colonies of mycelia, and conidia on the upper surfaces of leaves and leaf sheaths (especially on lower leaves), and sometimes on the spikes. Older fungal tissue is yellowish gray. This superficial fungal material can be rubbed off easily with the fingers. Host tissue beneath the fungal material becomes chlorotic or necrotic and, with severe infections, the leaves may die. Eventually, black spherical fruiting structures (cleistothecia) may develop in the mycelia, and can be seen without magnification.

Development: The development of powdery mildew is favoured by cool (15-22°C), cloudy, and humid (75-100 per cent relative humidity) conditions.

Hosts/Distribution: The fungus has a high degree of host specificity. Isolates infecting wheat do so exclusively; the same appears to be true for isolates infecting barley, oats, and rye. Further specialisation exists in the form of races. Powdery mildew occurs worldwide in cool, humid, and semi-arid areas where cereals are grown.

Importance: Powdery mildew can cause major yield losses if infection occurs early in the crop cycle and conditions remain favourable for development so that high infection levels are reached before heading.

Septoria Diseases

Septoria Tritici Blotch

- *Mycosphaerella graminicola* (*Septoria tritici*)
- Septoria Nodorum Blotch
- *Leptosphaeria nodorum* (*S. nodorum*)
- Septoria Avenae Blotch
- *Leptosphaeria avenaria* f.sp. *triticea* (*S. avenae* f.sp. *triticea*)

Symptoms: Initial infection sites tend to be irregular in shape, oval to elongated chlorotic spots or lesions. As these sites expand, the centres of the lesions become pale, straw coloured, and slightly necrotic, often with numerous small black dots (pycnidia).

The lesions of septoria tritici blotch tend to be linear and restricted laterally, while those of septoria nodorum blotch and septoria avenae blotch are more lens shaped. All above ground plant parts can be affected.

Light infection produces only scattered lesions, but heavy infection can kill leaves, spikes *(S. nodorum)*, or even the entire plant. Identification of species in the field can be difficult, and microscopic examination is often necessary.

Development: Initial infections tend to be on the lower leaves, progressing to the upper leaves and spikes if environmental conditions remain favourable. Cool temperatures (10-15°C) and prolonged wet, cloudy weather favours the development of these diseases.

Hosts/Distribution: These are primarily diseases of wheat, but other cereals are somewhat susceptible. The diseases are limited to temperate wheat-growing areas where cool and moist conditions prevail.

Importance: Major losses can occur, through seed shrivelling and lower test weights, if these diseases reach severe levels prior to harvest.

Helminthosporium: Leaf Blotch (Spot Blotch)

Cochliobolus sativus (*Helminthosporium sativum*) Syn. *Bipolaris sorokiniana, Drechslera sorokiniana*

Symptoms: Lesions caused by this disease are elongated to oval in shape and are generally a dark brown colour. As lesions mature, the centres often turn a light brown to tan colour, surrounded by an irregular dark brown ring (21 on leaf; 22 on spike).

Development: Primary infections tend to be on the lower leaves, beginning as chlorotic flecks or spots. These infection sites enlarge, turn dark brown, and often coalesce. When the disease is severe, affected leaves or leaf sheaths may die prematurely.

Hosts/Distribution: Spot blotch affects wheat, triticale, barley, and most grasses. It is found worldwide, but is especially prevalent in more humid and higher rainfall areas.

Importance: If infection occurs early in the crop cycle and conditions remain favourable for development, complete defoliation is possible; major reductions in yield and severely shrivelled kernels will then results.

Tan Spot (Yellow Leaf Spot or Blotch)

Pyrenophora trichostoma (Helminthhosporium tritici-repentis) Syn. *P. tritici-repentis, Drechslera tritici-repentis*

Symptoms: At first, lesions appear as tan to brown flecks, which expand into large, irregular, oval- or lens-shaped tan blotches with a yellow or chlorotic margin. As these spots coalesce, large blotches are formed. The development of a dark brown to black spot in the centre of the lesion is characteristic of the disease. As the disease progresses, entire leaves, spikes, and even whole plants may be killed.

Development: Initial infections come from diseased crop debris in the soil, or from diseased grass hosts. Usually the lower leaves are infected first, and the disease progresses to the upper leaves and leaf sheaths if conditions are favourable. This disease develops over a wide range of temperatures and is favoured by long periods (18 hours or more) of dew or rain.

Hosts/Distribution: Tan spot can affect wheat and several related grasses; triticale, barley, and rye are less frequently affected. The disease is found in the major temperate wheat-growing areas.

Importance: When severe, tan spot can cause premature death of the leaves, thereby reducing yield by lowering test weights and producing a high degree of kenel shrivelling. This disease has become more important with the increased adoption of minimum and/or no-till agronomic practices, which leave crop debris in place.

Alternaria: Leaf Blight

Alternaria Triticina

Symptoms: Small, chlorotic, oval- or elliptical-shaped lesions appear and, as they enlarge, these lesions become irregular in shape. The chlorotic borders of the lesions may become diffuse and turn light to dark brown in colour.

Lesions are difficult to distinguish from those caused by Helminthosporium spp. Infection usually starts on the lower leaves, but symptoms can be found on all plant parts.

Development: The fungus survives as conidia on seed or as mycelia within seed. Sporulation on lower leaves provides inoculum that can be dispersed by wind, leading to secondary spread of the disease.

Seed-borne inoculum often results in spike infections late in the crop cycle. High humidity or irrigation, as well as warmer temperatures (20 to 25°C) favour infection and disease development.

Hosts/Distribution: Bread wheat and durum wheat, as well as several related grasses, are the primary hosts. The disease is common in the eastern and central areas of the Asian Subcontinent.

Importance: Alternaria leaf blight can be very severe if environmental conditions are favourable for disease development; major losses can result when susceptible cultivars are grown.

Fusarium: Leaf Blotch (Snow Mould)

Symptoms: The blotching caused by this organism becomes evident on leaves at about late-joint to early-boot growth stage.

Young lesions occur as oval to elliptical, greyish green mottled areas, usually located where the leaf bends. The lesions enlarge rapidly, developing into large, "eyespot" blotches with bleached or light grey centres; the leaves tend to split or shred, beginning at the centres of the lesions. The fungus also can cause seedling blight, foot rot, head scab and, in winter cereals, pink snow mould.

Development: Spores are produced on crop debris left on or near the soil surface. These spores are transmitted to leaves by the wind or by splashing rain. Disease development is favoured by cool, moist weather.

Hosts/Distribution: Generally, the disease affects durum wheat and triticale more than bread wheat or rye; oats and barley appear to be immune. Reports indicate that the disease is restricted to East Africa, the highland areas of Mexico, the Andean region of South America, and parts of southern China.

Importance: Severe disease development can cause complete defoliation, resulting in poor grain development, shrivelling, and low test weights.

Scab (Head Blight)

Fusarium spp.

Symptoms: Infected florets (especially the outer glumes) become slightly darkened and oily in appearance. Conidiospores are produced in sporodochia, which gives the spike a bright pinkish colour. Infected kernels may be permeated with mycelia and the surface of the florets totally covered by white, matted mycelia.

Development: Several species of Fusarium can attack the spikes of small grain cereals; the ovaries are infected at anthesis, and infection is favoured by warm and humid weather during and after heading. Temperatures between 10 and 28°C are required for infection. Once primary infection has occurred, the disease can spread from floret to floret by mycelial growth through the spike structure.

Hosts/Distribution: All small grain cereals may be affected by this disease. Fusarium spp. are present in nearly all soils and crop residues.

Importance: Severe levels of infection can cause yield losses of more than 50 per cent and significant reductions in grain quality. Kernels from diseased spikes are often shrivelled. Harvested grain containing more than 5 per cent infected kernels can contain enough toxin to be harmful to humans and animals.

Ergot

Claviceps Purpurea

Symptoms: At flowering, infected florets produce a yellowish, sticky, sweet exudate (containing conidia) that is visible on the glumes. As the spike matures, kernels of infected florets are replaced by brown to purplish black fungal structures (sclerotia or "ergot bodies"). These ergot bodies can reach 20 mm in length.

Development: The primary infection originates from ascospores in fruiting bodies produced by sclerotia from the previous year's crop. Ascospores infect the florets, which then produce the sticky exudate containing conidia. Insects are attracted to the sweet exudate, and carry conidia to healthy florets in the same spike or to adjacent spikes.

Rainy or humid weather favours the production of exudate and spores. An ergot body develops in each infected floret; these fungal structures can survive in the soil from one season to the next, and under dry conditions they can remain viable for many years. Sclerotia require cold temperatures before they can germinate.

Hosts/Distribution: Ergot is found in all small grain cereal crops, especially if sterility occurs for some reason (e.g., frost). Sterile florets tend to open and thus become more liable to infection. The disease is more prevalent in cool, humid climates.

Importance: Yield losses tend to be small, but losses due to discounted grain quality can be significant and occur worldwide.

Black Point

Alternaria, Helminthosporium, and Fusarium spp.

Symptoms: The pericarps of maturing wheat kernels turn dark brown to black, with the discolouration usually restructed to the germ-end of the kernel (33; 34 is healthy seed). If caused by Alternaria spp., the dark colour affects only the pericarp; if caused by *Helminthosporium* or *Fusarium* spp., the germ may be invaded and injured or killed. There are other fungi that can cause black point, but the three noted here are the most common.

Development: Usually, kernels are infected by these fungi during the dough stage. If humid weather prevails for a few days to a week just prior to harvest, the incidence of infection will increase and black point will develop in many cultivars.

Hosts/Distribution: Wheat is the principal host; triticale and several related grasses also can be affected. Distribution is worldwide, wherever small grain cereals are grown.

Importance: Losses are due mainly to discounted prices paid for discoloured grain; if *Fusarium* or *Helminthosporium* spp. are involved, the viability of the seed also may be reduced.

Downy Mildew

Sclerophthora macrospora (Sclerospora macrospora)

Symptoms: Diseased plants tiller profusely; they have short, erect, irregular, or crooked yellowish-green culms; and the leaves are thick, erect, and usually in whorls. Tillers die prematurely or never head. If formed, the heads may be branched, and some of the floral tissues grow into leaflike structures.

Development: The disease in cereals is generally associated with waterlogged or excessively irrigated fields. Development is enhanced if temperatures range between 10 and 25°C. Infection may be initiated by inoculum in the soil, or from diseased weed hosts, and water must be present for infection to occur. Symptoms are most obvious during the tillering/ stem elongation growth stages of the host plant.

Hosts/Distribution: Small, localised epidermics can occur when conditions are favourable. There have been no reports of widespread and destructive epidemics.

Importance: Small, localised epidemics can occur when conditions are favourable. There have been no reports of widespread and destructive epidemics.

Take-All

Gaeumannomyces graminis f. sp. tritici Syn. Ophiobolus graminis

Symptoms: This fungus causes rotting of the roots and lower stems. Basal stem and leaf sheath tissues, as well as roots, may turn a shiny black colour. When examined with a hand lens (10x), dark fungal hyphae may often be found on the subcrown internode beneath the old leaf sheaths. Coarse, black runner hyphae or conspicuous on roots. Severe disease development is indicated by stunted plants with whitened stems and spikes. When infection occurs early in the crop cycle, the number of tillers is often reduced and spikes are often sterile.

Development: The fungus persists on crop debris in the soil. Initial infections come from contact with hyphae or ascospores in the soil. Infection can occur throughout the crop cycle, but is favoured by cool (12-18°C) soil temperatures and alkaline or nutrient deficient soils.

Nitrate also appears to enhance disease development. Infections of the roots occurring in the fall and early spring generally progress to the crown and lower stem tissues; infections occurring later in the crop cycle cause less damage since they usually are confined to the roots.

Hosts/Distribution: The take-all fungus displays a degree of specialisation for wheat, triticale and several related grasses. The disease appears to be restricted to temperate wheat-growing areas.

Importance: Take-all is widespread in monocropped areas and has been known to cause considerable yield losses in winter wheat and fall-sown spring wheat areas, especially where liming or minimum tillage is practised.

Pseudocercosporella herpotrichoides Syn. Cercosporella herpotrichoides

Symptoms: The most obvious symptoms of this disease are the eye-shaped, elliptical lesions produced on the internodes of the lower stem. The lesions are bordered by dark brown to greenish brown rings, have straw-coloured centres, and frequently develop on the leaf sheath at soil level.

These lesions may coalesce and lose their distinct "eye-spot" appearance. When disease development is severe, the stem or culm may break near the ground or through the lesion where the stem is weakened. Symptoms do not appear on the roots.

Development: Primary infection occurs from conidia or mycelia produced on crop debris on or near the soil surface; contact with the developing coleoptile or basal areas of young culms is required. The fungus is limited to the basal areas of the plant. Disease development is favoured by cool, damp weather, and by high humidity at the soil level.

Hosts/Distribution: Wheat, triticale, rye, oats, and other related grasses can be affected by the disease, with wheat being the most susceptible; winter wheat and fall-sown spring wheat are more frequently damaged. Eyespot occurs in cool, moist climates where fall-sown cereals predominate.

Importance: Eyespot may kill individual tillers or even whole plants. More commonly, yield losses result from reduced kernel size and number, and from lodging.

Sharp Eyespot and Rhizoctonia Root Rot

Rhizoctonia Solani

Symptoms: The primary symptoms of sharp eyespot are the lesions that develop on basal leaf sheaths; these lesions are similar to those caused by *Pseudocercosporella herpotrichoides* (eyespot). Sharp eyespot lesions are more superficial and more sharply outlined than those typical of eyespot. The margins are dark brown with pale, straw-coloured centres. The mycelia often present in the centres of lesions are easily removed by rubbing. Roots can also be affected, usually becoming brown in colour and reduced in number. The disease can cause stunting and a reduction in the number of tillers.

Development: Infection is highly dependent upon environmental conditions. Disease development is favoured by dry, sandy soils, cool temperatures, and high humidity. The fungus persists in soil and in crop debris, and invades root and crown tissues.

Hosts/Distribution: *Rhizoctonia solani* has perhaps the widest host range of any pathogen, attacking most cultivated crops; virtually all members of the Graminae family are susceptible. It is a ubiquitous fungus, present almost everywhere in soil and crop debris.

Importance: The disease is usually more severe in fields that are continuously sown to cereals, especially winter wheat. However, no major or widespread epidemics have been reported.

Common Root Rot, Foot Rot, and Crown Rot

Helminthosporium, Fusarium, and Pythium spp.

Symptoms: These fungi produce a darkening or browning of the root, crown, and basal culm tissues. Individual plants or groups of plants may lodge. White spikes often are visible just prior to normal physiological maturity. Infection early in the crop development can cause pre-or post-emergence "dampong off" of seedlings. Since each fungus can attack a different plant part at a different growth stage, positive field identification of the causal agent is difficult.

Development: The roots or crown tissues are infected by conidia or mycelia present on crop debris. Infection by and development of common root rot (*Helminthosporium* spp.) is favoured by warm, dry soils when the plants are under stress. The root, foot, and crown rots caused by *Fusarium* spp. Are generally favoured by cooler, moist soils, and wet soils favour *Pythium* spp.

Hosts/Distribution: These diseases affect all major small grain cereals grown in the temperate regions of the world.

Importance: Major epidemics have not been reported. However, localised losses can result from thinned stands, a decrease in the number of tillers, and from reductions in head size and tests weights.

Sclerotium Wilt (Southern Blight)

Corticum Rolfsii (Sclerotium rolfsii)

Symptoms: If infection occurs early in the crop cycle, pre- or post-emergence "damping off" of seedlings can result. Diseased tissues will frequently have white, fluffy fungal mycelia on the surface, which often permeate the soil surrounding the plant. Subsequent disease development results in rotted culms, crowns and roots, and the eventual death of the plant; this leads to the appearance of "white heads" or spikes in the green crop. Sclerotia are commonly found on the crown tissues, culms, or near the soil surface. Young sclerotia are whitish and turn brown to dark brown with age.

Development: Sclerotium rolfsii can attack the plant at any stage of development. Fungal mycelia on crop debris or sclerotia serve as primary inoculum. Infection and disease development are favoured by warm (20°C +) temperatures, excessive moisture, and acid soils.

Hosts/Distribution: Most cereals and grasses, plus many broad-leaf plant species are susceptible, and the fungus is widespread in tropical and subtropical environments.

Importance: The disease is seldom a problem in traditional wheat-growing areas.

Black Moulds (Sooty Moulds)

Alternaria, Cladosporium, Stemphylium, Epicoccum and other Species

Symptoms: The typical symptom is the blackened appearance of mature or dead spikes, caused by a superficial accumulation of mycelia and sporulating fungal tissue.

Development: When wet or humid weather occurs at or near crop maturation, or when plants are heavily infested with aphids, or when plants die prematurely, they may be invaded by one or more of these fungi. Technically, black moulds do not constitute a disease, since the fungi are saprophytic and invade only dead or dying plant tissue.

Hosts/Distribution: Black moulds affect any kind of dead or dying plant tissue; distribution is worldwide.

Importance: Black moulds generally are not economically important. Under humid or rainy conditions, the fungi can invade mature kernels, causing discolouration, black point, or smudge.

Symptoms of Disease

It is highly important that all persons engaged in general farming, truck gardening, fruit raising, floriculture or forestry should be able to recognise the presence of disease in plants. Every plant in its particular way shows when it is suffering from disease. Crop producers should not only know how plants live and grow but should be familiar with those outward signs or symptoms which exist when plants are sick. The study of symptoms is the first step in making a diagnosis or the determination of the nature or identity of a trouble. A correct determination can be made in many cases by the observation of symptoms alone, but in others it is necessary to study environmental factors or to dissect diseased plants or portions of plants in the search for a pathogene.

In the nature of the resulting symptoms, there are no hard and fast lines that can be drawn between parasitic and non-parasitic diseases or between various "insect" troubles and those due to bacteria or fungi. The first thing of importance is to detect the presence of disease and then to endeavour to find out its cause, nature and probable outcome and the treatment or preventive practices which should be employed.

The great diversity of symptoms, the numerous non-parasitic disturbances and the multitude of pathogenes so complicate diagnosis that the grower must frequently seek the services of the trained plant doctor.

Summary of Symptoms: The following outline of symptoms will be of service to the practical grower and also to the plant-disease specialist:

1. Discolouration or change of colour from the normal:
   - A. *Pallor:*
     - a. Pale green.
     - b. *Yellow:*
       Etiolation.
       Chlorosis.
     - c. Silvering.
     - d. White—albinism.
   - B. *Coloured spots or areas:*
     - a. General difference between parasitic and non-parasitic troubles; exceptions.
     - b. *Colours:* Whitish or gray; red or purple; brown; black; variegated and concentrically zonate.
   - c. Autumnal colourations or spring colourations.

2. *Shot hole:* Perforations of leaves:
   - A. *Physiological:* Toxic substances; frost; drought; etc.
   - B. Fungous or bacterial origin
3. *Wilting:*
   - A. *Juvenile condition:* "Damping-off" of seedlings.
   - B. *Adult plants:*

     Physiological wilting with recovery.

     Wilting Without Recovery: "Wilt" diseases; thrombosis of woody plants.
4. *Necrosis:* Death of parts:
   - A. *Organs:* Leaves, stems, flowers, fruits—blight.
   - B. *Localised Areas of Tissue:* Internal necrosis, e.g., bitter pit of apple; internal brown spot, net necrosis and blackheart of potato.
5. Dwarfing or atrophy—reduction in size:
   - A. *Entire Individual:* Non-parasitic; parasitic.
   - B. *Parts or Organs:* Leaves, fruits, etc.
6. *Hypertrophy or Increase in Size of Parts:* Roots, stems, leaves, flowers, fruits:
   - A. Increase in size of cells.
   - B. Abnormal multiplication of cells or increase in number—Hyperplasia.
7. Transformation of organs or replacement of organs by new structures:
   - A. *Replacement:* By sclerotium, as in ergot of grasses.
   - B. *Transformations of Floral Organs:* Petals to foliage leaves, etc.; "phyllody."
8. *Mummification:* Transformation of fruits into shrivelled structures, called mummies, which process is initiated by rotting, followed by resistance to decay. Mummies contain either dormant mycelium or overwintering fruits of fungi.

9. *Alteration in Habit and Symmetry:*
   A. Change of position from prostrate to erect or ascending.
   B. Change from rosette to cauline type.
   C. Change of simple leaves to lobed leaves.
   D. Reduction or increase in branching.
   E. Change in form of inflorescence.
   F. Change in symmetry of flowers.
10. *Destruction of Organs:* e.g., flowers or flower parts; seeds; seeds and pits in "bladder plums"; fruit (caryopsis) in cereals; inflorescence, as in loose smuts of cereals.
11. Dropping of leaves, blossoms, fruits or twigs.
12. Production of excrescences and malformations:
   A. Abnormal trichome growth—Erineum.
   B. Intumescences on leaves and stem blisters or swellings involving groups of cells but without the presence of a parasite.
   C. Pustules, warts, tubercles, galls or tumours. Parasite included. Minute to large and fleshy or woody; crown gall, etc.
   D. *Cankers:* Localised lesions generally resulting in corrosion of tissue with final production of an open wound, generally on woody structures:
      a. Parasitic.
      b. Non-parasitic.
   E. Witches' brooms.
   F. *Hairy Root:* Simple; broom-root type; woolly-knot type; aerial form.
   G. *Production of Rosettes:* Woody plants; herbaceous plants.
   H. Development of dormant or rudimentary structures or of new organs.

I. Prolification.

J. Rolling, curling or crinkling of leaves.

K. Fasciation and spiralism.

L. *Roughening of Surfaces:* Local or extended.

M. Deforming of fruits.

13. *Exudations:* Guttation and bleeding contrasted with formation of exudates:

A. *Bacterial oxudates:* e.g., fire blight.

B. *Slime flux:* Deciduous trees; exudate semi-fluid, does not set into solid masses.

C. *Gummosis:* Product of decomposition of tissues; sets into solid masses, clear or amber coloured. Common in citrus and stone fruits. Parasitic and non-parasitic.

D. *Resinosis:* Abnormal production of resin or pitch from coniferous trees.

E. *Latexosis*: Exudation of milky juice or latex.

14. Rotting:

A. Dry rot and soft rot, the "gangrene" of plant tissue. Fleshy structures more generally show soft rot; woody structures, dry rot:

a. *Root rots:* Fleshy or woody roots, e.g., alfalfa, cotton; root crops—beets, carrots, etc.; woody roots—apple, cherry, etc., and shade or forest trees.

b. *Stem rots:* Herbaceous stems—carnation, aster, potato; modified stems—tubers, rhizomes, bulbs or corms (storage organs); woody stems—dry rots, as sap rot or heart rot.

c. *Bud rots:* e.g., carnation and coconut bud rots, cabbage black rot.

d. *Fruit rot:* Fleshy fruits of many kinds.

(1) *Non-parasitic:* Blossom-end rot of tomato, blossom-end rot of water-melon.

(2) *Parasitic:* Many on various hosts. Mostly fungi. Why few bacteria?

1. *Discolouration or Change of Colour from the Normal:* This deviation from the normal may be noted on various plant structures as illustrated by the pink colouration of the roots in pink root of the onion or by the red-spotted fruits characteristic of peach yellows, but the discolourations are very frequent and striking on parts which are normally green, such as herbaceous stems and foliage leaves.

   Plants, like people, frequently look pale when they are sick. Nutritional disturbances preventing the constant production of the green pigment chlorophyll cause normally green structures to become pale green or yellowish green or to exhibit pallor.

   The normal shade of green for any particular plant may be recognised, and the practised eye soon learns to tell when something is wrong. The green pigment may disappear entirely and its place be taken by a yellow pigment. When this yellowing is brought about by lack of light or prolonged exposure to darkness, this condition is spoken of as *etiolation,* and the yellowed organs may be said to be *etiolated.* It may be noted in this connection that the etiolation or blanching of certain garden plants, such as asparagus, endives, artichokes, celery, etc., is practised to secure desired flavour or tenderness and that in certain cases self-blanching varieties have been secured by selection.

   Somewhat similar effects may result from the operation of other factors, such as low temperatures, lack of iron, excess of lime, excess of alkali, presence of a virus disease like peach yellows or mosaic disease or from the disturbances caused by bacterial or fungous parasites. In these latter cases, it is the custom to speak of the condition as *chlorosis,* and the affected structures are said to be *chlorotic*. We speak, then, of iron or lime chlorosis, alkali chlorosis, infectious chlorosis, etc.

Chlorosis as a disease symptom should not be confused with "panaschiering," in which the leaves are part green and part yellow or spotted with white and green or white bordered. Plants showing these peculiarities have been selected and propagated for their ornamental value. Silvering of the foliage, or the assuming of a dull, metallic lustre, is not uncommon in plums, apples and other woody plants and is a striking symptom of the infectious silver-leaf disease or may result from non-parasitic disturbances. Among plants, as among animals, there are the *albinos*, or those entirely devoid of pigment. Such constitutional albinos must soon succumb, since they have no power to produce carbohydrate food because of their lack of a chlorophyll apparatus.

Discolourations, instead of being general or diffuse, may show as more or less definite or circumscribed discoloured spots or areas. In such cases, the spotting may result from localised disturbances from parasites or chemical agents or from deep-seated nutritional disorders in which leaf spotting is but a phase.

There is no certain character which will differentiate leaf spots of parasitic or non-parasitic origin, but many fungous infections, especially on netted-veined leaves, produce circular or sub-circular areas or lesions which are soon limited in size. A spot comes wherever a spore of a pathogene lodges and is able to establish its mycelium; hence parasitic lesions are generally scattered irregularly over the surface of a leaf and show no definite arrangement with reference to the veins, either large or small.

Apple foliage injured by arsenical spray containing free arsenic will sometimes show a brown spotting very similar to that caused by the black-rot fungus *(Physalospora malorum).*

With numerous spotting, the lesions may coalesce to form more extended discoloured areas. Some parasites that invade the leaf spread unchecked and so produce extended dead areas or may involve the entire leaf as in the late blight of the potato *(Phytophthora infestans)* or in leaf invasions by the bacteria of fire blight *(Bacillus amylovorus).* Leaf spots due to certain kinds of non-parasitic disturbances are frequently located between the main veins of pinnately veined or palmately veined leaves.

We can even note that these spots start at the blind ending of the veins in the smallest areolae of the leaf or at points the most distant from the water-conducting channels. Drought injury or poisoning from smelter fumes may show this type of spotting, or, under other conditions, terminal or rusts *(Albugo spp.);* to the groups of aerial conidiophores, as in the downy mildews of the grape, cucurbits, lettuce and also in various diseases caused by hyaline-spored imperfect fungi, as illustrated by species of Ramularia and Cercosporella; and finally to changes in the host tissue which lift the epidermis and admit air into the cells or intercellular spaces and so cause a reflection of the light. The centres of old leaf spots which were at first brown finally become a dull gray in various cases.

Yellow or orange spots may be due either to tissue changes or to the colour of spore groups or masses. There may be a localised chlorosis, as in the leaf spot of the violet *(Alternaria violce),* also in various downy mildews *(e.g., Peronospora trifoliorum* on alfalfa) and in many rusts *(e.g., Tranzschelia punctata* on peach), in which the sporulation is confined to the under surface. The spore fruits of rusts, especially the secial or cluster-cup stages, and the paler forms of the so-called "red-rust" stage exhibit varied shades of yellow or orange.

Red or purple colouration in spots may also be due to tissue changes or to the masses of spores developed by the pathogene. In the first type, there is a decomposition of chlorophyll and the formation of red pigment *(anthocyaniri),* which is dissolved in the cell sap.

Chlorophyll disappears entirely or is masked by the presence of the red pigment. This red colouration is a host peculiarity and is frequently the first response of injured cells. Incipient infections may show the red colouration first, with change to brown later, while the occurrence of red or purple border may characterise many more advanced lesions. The colour variations are due in part to different degrees of acidity or alkalinity of the cell sap.

Some of the rusts producing the more highly coloured types of spore fruits would probably be classed as red (both aecial and red-rust stages), but in such cases more or less raised or powdery pustules of spores are evident. Brown is the characteristic colour of dead tissue and is the final colour of the great majority of leaf spots, although some finally become greyish with age.

Many rust lesions will show brown powdery pustules, representing either darker red-rust stages or paler black-rust sori. Superficial fungi with dark mycelium (sooty moulds) or various dark-spored, imperfect fungi may also cause the brown colouration due to the accumulation of mycelium or conidiophores or both *(Cladosporium fulvum* on tomato; *Venturia incequalis* on apple).

The black colouration is well illustrated in the so-called tar spots *(Rhytisma spp.)* of the maple, willow and oak, the black spot *(Phyllachora spp.)* of grasses and other hosts, the black spot *(Diplocarpon rosce)* of roses and especially in some covered and naked forms of black-rust sori *(Puo cinia spp.* on cereals). In all of these examples the colouration is due to fungous tissue or to spores, but in a few cases the black spotting may be due to changes in the host tissue. This is well illustrated in the pear, in which injured leaf tissue frequently assumes a very dark or almost black colour *(Fabrcea maculata)* or in the blackspot *(Bacterium pruni)* of plum fruits.

Leaf spots may show concentric bands of different shades of brown, as in the so-called target-board spots characteristic of early blight *(Alternaria solani)* of the potato or in the frog-eye spots *(Sphceropsis malorum)* of apple leaves. Concentric zones of different colours are a noticeable feature in the leaf spot *(Mycosphcerella fragarice)* of the strawberry, in which well-matured lesions are white or grayish in the centre, surrounded by a brown zone bordered by purple and red, shading out into the surrounding healthy tissue.

Black and gray or brown concentric zones alternate in the anthracnose *(Colletotrichum lagenarium)* of some squashes. Many leaf spots of hosts in which the red pigment is not commonly developed show a halo of yellow surrounding the brown, dead tissue. Some leaf spots show entirely different colours on the two surfaces, as is well illustrated in fruiting lesions of grape downy mildew which are yellow above and white below or in alfalfa downy mildew with yellow upper surfaces and grayish or faintly purple under surfaces.

The autumnal colourations of foliage may be briefly mentioned at this point. Chlorophyll is decomposed by the sunlight, and sometimes yellow and red colours of many and varied shades are formed, before the final assumption of the somber brown characteristic of dead-leaf tissue. The low temperatures which prevail check the normal life processes and prevent the reformation of chlorophyll. In a very similar way, the low temperatures of early spring are responsible for yellow or red colouration of young leaves.

Shot Hole or Perforation of Leaves: The formation of localised lesions on the leaves is frequently followed by the falling out of the dead or diseased tissue, leaving circular or slightly irregular perforations, which have suggested the descriptive term "shot hole." Certain troubles have been called shot-hole diseases because of the peculiar effect upon the foliage. The occurrence of shot hole is not a diagnostic feature of certain species of fungi but is more properly to be viewed as a host peculiarity. Some varieties are prone to shot hole whenever localised areas of leaf tissue are killed, whether this death of tissue results from the presence of parasitic bacteria or fungi or from the operation of non-parasitic factors, such as toxic chemical agents, drought or frost.

Among our stone fruits, the tendency to the shot-hole symptom is very marked, as in bacterial black spot *(Bacterium pruni),* the leaf spots or blights *(Coccomyces spp.)* of cherries and plums and the California blight *(Coryneum beijerinckii)* of apricots, peaches and cherries. Localised action of strong blue-stone solution, free arsenic or other toxic chemicals on peach leaves will cause more or less perforation.

The behaviour of the leaves of pome fruits —apple, pear and quince, in which perforation is rare—is in marked contrast to the behaviour of the stone fruits. Sometimes extreme cases of shot hole may resemble insect work so closely as to be confusing (Cercospora and Phyllosticta spots of the Virginia creeper). Sometimes the leaves fall before the perforations are complete, while in other cases severely affected leaves will remain hanging until the shot holing has run its course.

Wilting: Two types of wilting may be recognised: the sudden wilting of seedlings, or "damping-off"; and the wilting of growing or adult plants. The damping-off of seedlings is most frequent and severe in crops grown under glass, also in garden crops or in seedling trees in the nursery, but is less injurious in most field crops.

In these cases, the young plant droops over and lies prostrate on the ground or "drop^ dead/' as it were, for a parasite, generally a fungus, has entered the stem at about the ground level and has struck a vital blow, interrupting essential physiological functions.

The rigid stem becomes flaccid, and the flow of water from root to seed leaves is interrupted. Recovery is impossible, for the tissues have been killed at about the ground level by the parasite which has entered the young stem from the soil. Various soil-inhabiting fungi may cause damping-off, the most prominent being *Rhizoctonia spp., Pythium debaryanum, Thielavia basicola, Sclerotinia spp.* and *Sclerotium spp.* Damping-off by killing numerous seedlings in sugar-beet fields is frequently the explanation for poor stands. A damping-off disease of cotton seedlings is prevalent throughout the South and is known to the planters as "sore shin." The rotting of herbaceous cuttings at the ground line in greenhouse benches is generally due to some of the damping-off fungi.

The normal or physiological wilting of growing plants, followed by recovery, should not be confused with the wilting of disease. During periods of bright sunshine in the heat of the day, succulent shoots droop, and leaves may become limp or roll up, due to excess of water evaporation over root absorption, but when lowered temperatures and darkness check evaporation-the flaccid parts again become turgid, and growth continues. In certain cases, the leaves of plants wilt rather suddenly, sometimes with slight premonitory symptoms, and recovery does not follow.

Notable cases are the so-called "wilt diseases" of herbaceous plants or the thromboses of woody plants, in which the water-conducting vessels of the hosts are plugged up with either bacteria or fungous mycelium, so that the transpiration current is completely interrupted. The wilt *(Bacillus tracheiphilus)* of cucurbits, the wilt *(B. solanacearum)* of Solanacese or speçies of the nightshade family, the Fusarium wilts of various hosts and the thrombosis *(Verticillium)* of the maple are typical illustrations.

Necrosis, or Death of Parts: In some diseases, the death of special parts or organs of the plant, as leaves, stems or twigs, buds, flowers or fruits, is the first symptom of disease that is noted, although in many cases wilting is an antecedent phenomenon. The affected structures generally assume the characteristic dark or brown colourations of dead tissue, and the accompanying disease is frequently characterised as a *blight*. This is well illustrated in the fire blight of the apple and pear, in which leaf blight, body blight, twig blight, blossom blight and fruit blight are but phases of a single disease. Necrosis may be localised in certain organs, or it may be more general and result in the death of the entire plant (*e.g.*, severe attacks of late blight, *Phytophthora infestans* of potato).

Localised areas or groups of cells in certain organs may die, with the result that small or extensive masses of brown, frequently collapsed cells may be found in otherwise normal structures. Internal necroses are well illustrated in the bitter pit or Baldwin spot of the apples and in the internal brown spot, net necrosis and blackheart of the potato, all of which are non-parasitic troubles.

Dwarfing or Atrophy: Either the entire individual or special organs may be reduced in size as a result of unfavourable factors, either environmental or parasitic. Early attacks of parasites which are general upon the foliage, like mildews, rusts and other fungi; diseases which are systemic, like some of the seed-borne smuts *(Tilletia tritici)*; or the curly dwarf of the potato may cause the affected plants to remain stunted, or a localised disease like crown gall *(Pseudomonas tumefaciens)* of the apple may cause a permanent retardation of growth. Leaves may be atrophied or reduced in size as the result of localised parasites (some rusts), in witches' brooms; in trees suffering from mushroom root rot *(Armillaria mellea)*, crown or collar rot or from lowered vitality due to winter injury; also from various other nutritive disturbances.

Reduction in size sometimes suggests the common name of a trouble, as in the little-leaf disease of the apple. Tubers may be reduced in size in the Rhizoctonia disease and in the various degeneration or virus diseases of the potato, and there may be similar results in any root crop. The small size of the fruits is such a characteristic feature that it has suggested the name "little peach/" which is applied to a serious contagious disease of that fruit tree, while severely rusted cereals produce shrivelled and shrunken grain. It is necessary only to compare normal and diseased plants to note numerous but less striking examples of reduction in size.

Increase in Size or Hypertrophy: Practically all plant organs may be stimulated by the action of parasites so that they are increased in size. The result may be shown either, first, by the increase in the size of the component cells or, second, by increased cell division resulting in the abnormal multiplication of cells (hyperplasia).

In some cases, increase both in size and in number of cells may occur. (See also item 12 *B* and C.) Hypertrophied roots are well-illustrated in various mycorrhizas, in the club root, or finger-and-toe disease *(Plasmodiophora brassicce)* of the cabbage and other crucifers and in the root knot or eelworm disease *(Caconema (Heterodera) radicicola)* of numerous hosts; hypertrophied stems as in white rust *(Albugo Candida)* of crucifers, stem invasions of the Vaccinium rust *(Calyptospora columnaris)* and in some Exobasidium diseases of certain species of the heath family; hypertrophied leaves in the well-known leaf curl *(Taphrina deformans)* of the peach, in the rose bloom *(Exobasidium vaccinii)* of the cranberry, in which the pink enlarged leaves are grouped to resemble a flower, various rusts and in many other fungi; hypertrophied flowers or flower parts, as in the white rust *(Albugo Candida)* of radish, infections caused by various Exoascales or leaf-curl fungi; and in certain species of Exobasidiales; and hypertrophied fruits, as illustrated by plum pockets or bladder plums *(Taphrina pruni)* and the leaf-curl fungus *(T. farlowii)* of the chokecherry.

Transformation of Organs or the Replacement of Organs by New Structures: In the ergot of rye and other grasses *(Claviceps spp.)*, certain of the ovaries are destroyed by the action of the parasite, and, in the place of the seeds, horny, elongated, dark-purple, spur-like bodies, the sclerotia or "ergots," appear.

These are resting structures composed of a dense aggregate of fungous tissue and serve the purpose of carrying the parasite over the winter period. In normal flowers the sepals, petals, stamens and pistil occupy definite positions, but under abnormal conditions one kind of floral organ may be transformed into another kind or into ordinary leaves. Stamens may become leafy in the green-ear or downy-mildew disease *(Sclerospora graminicola)* of *Pennisetum typhoideum* and other grasses; the whole staminate head, or tassel, or the ear becomes a leafy structure in certain phases of the head smut *(Sorosporium reilianum)* of corn; petals may become like sepals, stamens like carpels and carpels leaf-like in the white rust *(Albugo Candida)* of crucifers; all the floral parts are transformed into foliage leaves in the Japanese plum affected with rust *(Cceoma makinoi)*. The term "phyllody" is applied to the change of floral organs into leafy structures. While phyllody is frequently caused by the presence of a parasite, other disturbances may produce similar effects.

Mummification: The transformation of fruits into shrivelled structures called "mummies" is a phase of numerous diseases affecting our commercial fruits and may also occur in the fruits of wild plants. The fruit generally undergoes an initial change which would be characterised as rotting, but during this process the tissues become filled with the mycelium of the parasite, and the dry, shrivelled structure which results consists of the remains of host cells mingled with this mycelium alone or also with spore fruits which have been organised.

Mummies remain hanging on the tree or fall to the ground but resist further decay until a later time or after they have produced a crop of spores. Spores may develop at once or not until the next spring or in some cases not until the mummies have passed through two winters. In cases where mummies produce spores in the fall, this process is continued into the following spring, so that mummies may be looked upon as devices to provide for the overwintering of those parasites by which they are produced.

The formation of mummies is a very characteristic feature of the brown rot *(Sclerotinia spp.)* of stone or pome fruits, like apricots, peaches, plums, cherries and apples or pears. In plums or peaches, the brown, shrivelled mummies, frequently cemented into clusters which remain hanging on the trees, are a familiar sight in regions where the brown rot of these hosts is prevalent. Other typical illustrations may be found in the bitter rot *(Glomerella cingulata)* and black rot *(Physalospora malorum)* of the apple and in the black rot *(Guignardia bidwellii)* of the grape.

Alteration in Habit and Symmetry: Some plants which under normal conditions are more or less prostrate or creeping become ascending or even erect when attacked by a fungous parasite and, as it were, signal their distress. This is notably true in purslane (pusley), a common garden weed *(Portulacaoleracea)*, when attacked by white rust *(Albugo portulacece)*, and in certain species of spurges *(Euphorbia spp.)* harbouring the clustercup stages of rust. The same symptom is shown in some tree diseases in which normally horizontal limbs or branches become grouped into more or less erect clusters (see Witches' Brooms under Art. 12). In these cases, a dorsiventral symmetry is changed to a more or less evident radial symmetry. Closely related to the above is the change from the rosette to the cauline type in *Launea asplenifolia*, the unbranched stem and radical leaves being changed to a much branched axis with cauline leaves, as the result of a rust *(Puccinia butleri)*.

The vegetative organs may show various alterations; *leaves* may be changed from simple to irregularly lobed, as in a South American barberry attacked by a rust (Ecidium). *Sempervivum hirtum* with obovate leaves normally twice as long as broad may be changed to produce leaves seven times as long as broad when infected with a rust-like fungus *(Endophyllum sempervivi)*. Leaves may be variously twisted or deformed by various parasitic attacks (numerous rust infections); stems may be twisted or deformed, and internodes elongated or shortened; and branching may be reduced (various rusts) or increased, as in the degeneration disease of the potato known as witches' broom (see also Hairy Root and Witches' Brooms).

The reproductive structures may be transformed by the action of disease: In the club types of wheat, the head is changed to the elongated or "vulgare" type by the presence of bunt or stinking smut *(Tilletia tritici);* in a species of Acacia, the inflorescence is changed from a head to a spike by a rust *(Ecidium esculentum);* flowers may be changed from regular to irregular (actinomorphic to zygomorphic) or *vice versa,* cyclic types to strobilate types, dioecious to perfect, and various other modifications of the flowers or flower parts may result.

Destruction of Organs: The complete destruction of organs may result either from non-parasitic causes or from the inroads of a pathogene. The rudiments of seeds—the ovules—may dry up without the production of seed, owing to low-temperature injuries, lowered vitality or failure of fertilization, as in the empty or partially filled cells of an apple, or seeds may be destroyed by the action of a parasite, as in the smut of sheep sorrel *(Oxalis spp.),* in which the spores of the parasite *(Ustilago oxalidis)* are forcibly expelled from the seed capsule just as *if* true seeds were present.

Entire flowers may be destroyed, or only flower parts, as in the anther smut of pinks caused by *U. violacea.* In the case of "bladder plums "due to *Taphrina prunij* the pit fails to develop and the rudiment of the seed is destroyed, producing the empty inflated "fools" characteristic of this disease.

The so-called "seeds" or fruits of cereals (the caryopses) are frequently destroyed by the operation of a parasite. This is notably true in the covered or kernel smuts—kernel smut *(U. levis)* of oats, bunt or stinking smuts *(Tilletia tritici* and *T. levis)* of wheat and kernel smut *(Sphacelotheca sorghi)* of sorghum. In these kernel smuts, the young mycelium of the fungus enters the young developing ovary and sporulates there, using up the food that would normally be stored in the seed. The final result is the "smut ball" or spore mass enclosed by a remnant of the wall of the kernel. All or part of the grains in a head may be destroyed, and all or only part of the heads on an infected plant may be smutted. In other cases, the destruction may be more complete and the entire inflorescence may be involved. This is well illustrated in the loose smuts of cereals—loose smut *(Ustilago avence)* of oats, loose smut *(U. tritici)* of wheat and loose smut *(U. nuda)* of barley. In a typical case, as in the loose smut of wheat, all parts of the head except the central axis, or rachis, are destroyed and replaced by a black powdery mass which has been largely dissipated by harvest time, leaving the rachis naked or with only a few remnants of the spore mass.

Dropping of Leaves, Blossoms, Fruits or Twigs: This is, of course, to be considered as a symptom of disease only when it occurs prematurely or in excessive amount. It may first be illustrated by the normal behaviour of leaves, which are all shed in deciduous species at the end of the growing season, or by the loss of a part of the needle leaves each season by coniferous evergreens.

Under such conditions, the leaf falls because of certain changes which take place in the *abscission layers,* or cleavage zone, at the base of the petiole, causing a separation. Non-parasitic influences or parasites acting upon the growing plant may bring about the same changes in the abscission layers as the cold of autumn, thus producing premature leaf fall. Blossoms, twigs, fruit spurs or fruits may be separated from their points of attachment in a similar manner.

Owing to a sudden change of external conditions from a moist greenhouse to a dry room, from moderate to intense light or from cool conditions to warm, such house plants as fuchsias, foliage begonias, azaleas, rubber plants and many others will drop their leaves. The dropping of the leaves is such a characteristic symptom of certain diseases as to suggest the common name, as in the leaf casts *(Hypoderma* or *Lophodermium spp.)* of the pine and larch. The shedding of lower leaves which have become spotted and chlorotic is a very characteristic behaviour of alfalfa plants affected with leaf spot *(Pseudopeziza medicaginis)* or yellow leaf blotch *(Pyrenopeziza medicaginis)*. In severe attacks, some cherries affected by the yellows or leaf spot *(Coccomyces hiemalis)* may be nearly defoliated by midsummer. The shedding of blossoms may be illustrated in the blossom drop of the tomato, which is frequently very noticeable during the first part of the growing season when cool, cloudy weather prevails; and in the so-called shelling of grape blossoms or of partially developed berries. It is probable that the shelling or dropping of the reproductive structures is the result of insufficient food to supply both vegetative and reproductive activities. In hyacinths given too much water and subjected to high temperatures at the beginning of the growing season, the stalk bearing the entire inflorescence may separate at the base.

Dropping of fruits may be illustrated by shedding of bolls in cotton or the June drop of stone or pome fruits, the former following extremes of weather, either dry or wet, while the latter may be induced by various factors affecting the nutrition of the developing fruits. Parasites localised on the fruits or fruit pedicels may cause dropping of young fruits, as in scab *(Venturia inaequalis)* of apples or California blight *(Coryneum beijerinckii)* of the cherry. The casting of fruit spurs may occur in pears or apples, or there may be an abnormal abscission of twigs and small branches in such trees as willows and poplars due to excess of water or to temporary drought followed by an excessive supply of moisture.

Production of Excrescences and Malformations: A great variety of abnormal formations may be grouped under this heading. Attention may be directed first to a condition known as *erineum,* which is the name applied by early mycologists to an abnormal development of hairs or trichomes from the surface of leaves giving felt-like patches. While these formations were at first supposed to be of fungous origin, they are now known to be due to the effects of certain parasitic mites and are designated as *erinose.* Common instances may be found in the erinose of the vine, in which young trichome felts suggest the white spots of fruiting downy mildew, and in the erinose of the mountain maple with its striking patches of red or scarlet hairs.

Knot-like or pustule-like distensions of tissue, occurring most abundantly on leaves but also on stems and fruits, due to the abnormal elongation of groups of cells, with or without increased cell division, are called *intumescences.* No parasite is concerned in the formation of these structures, which result from environmental factors.

In a typical leaf intumescence, groups of palisade parenchyma cells are abnormally elongated, so that a raised, blister-like pustule is formed. In other cases, they may rupture the epidermis of leaf or twig and make a rough, hairy papilla, which may later become brown and collapsed, due to the death of the cells. Intumescences are not uncommon on the unripe pods of peas.

They are in the form of pale, irregular warts, which by their appearance almost suggest a fungous infection, especially if they have been exposed to very moist conditions. Woolly streaks which appear in the core walls of apples under certain conditions are formed by thick bunches of elongated cells or cell rows and are quite similar in origin to external intumescences.

If the overdevelopment of cells is somewhat general, rather than distinctly localised, and extensive swellings of organs result, the condition is spoken of as "oedema" or dropsy. This is not uncommon when there has been pronounced stimulation of growth or a lessened consumption of water with unabated absorption. The condition on apple branches and roots in which the elongated cells separate and dry to make a fine, brown powder—the so-called "tan disease".

Localised enlargements on various organs in the form of small pustules or warts, larger tubercles, tumours or masses of cells making a morbid outgrowth of either fleshy or woody character, in which host tissues and parasites mingle, are generally designated as *galls.* These abnormal growths may be due either to plant parasites, mostly fungi or bacteria, or to various insects.

The study of galls and their causal agents constitutes a special field of science known as *cecidology.* A few illustrations will be cited to show something of the variety of gall formation by plant parasites: Very small reddish galls, *1^5* inch in diameter or slightly less, are produced on stems, leaves and other parts of the cranberry in the gall disease *(Synchytrium vaccinii)*; the large, rounded or irregular tumours of corn smut *(Ustilago zece)* up to 6 inches or more in diameter represent about the extreme in size of galls on herbaceous structures. The so-called "cedar apples" of our common red cedar, evident as brown, more or less spherical enlargements, *per cent* to 2 inches in diameter and quite fruit-like in appearance, are the result of a rust *(Gymnosporangium juniperivirginiance)*.

The more or less elongated, rough, black outgrowths—the black knots—on the twigs and branches of plums and cherries illustrate a different type of fungous gall *(Plowrightia morbosa)*. Bacterial galls *(Pseudomonas tumefaciens)* are well illustrated in the woody or fleshy outgrowths from the crown of fruit trees and numerous other herbaceous or woody plants, hence the name *crown gall*. While the graft union is a very common place for such galls, they may appear on any part of the root system and also upon aerial parts, as, for example, on the blackberry, where they involve long stretches on the canes and are sometimes referred to as cane galls.

The galls on the leaves, twigs, branches or trunk of the olive, very small on the leaves but varying in size from that of a garden pea to others 2 to 3 centimetres in diameter on the branches or trunk, are also of bacterial origin *(Pseudomonas savastanoi)*. The club root or finger-and-toe disease of cabbage, turnips and other species of the mustard family, characterised by enlarged roots or tumour-like outgrowths from the main root, is the evident feature in one of the important chytrid diseases *(Plasmodiophora brassicce)*.

Localised lesions on woody or more rarely herbaceous stems which generally result in the corrosion and sloughing away of tissue with the final production of an open wound, exposing or penetrating the wood, are designated as *cankers*. Young cankers are at first evident as localised discoloured areas upon the bark, but with the death of the tissue involved there are shrinking, cracking and shredding and the lesion comes to be more pronounced.

A fungous canker may advance slowly or rapidly and may reach its full size in the course of a single growing season (infection in fall of previous year), as in the black spot or Pacific Coast canker *(Neofabrcea malicorticis)* of the apple, or it may be perennial, spreading in an ever widening zone from year to year till the infected branch or trunk is girdled and killed as in the Endothia canker *(Endothia parasiticd)* of the chestnut. In a slow-growing, perennial canker, the host will make efforts to heal the wound or to head off the progress of the parasite by the development of callus around the advancing border of the lesion. In time, the building up of these ridges of callus will be evident as more or less regular or broken concentric ridges which mark either years or periods of growth, as is well illustrated in the European canker *(Nectria galligena)* of the apple or in the Strumella disease *(Strumella corynoidea)* of oaks and chestnuts.

The majority of canker-producing fungi develop mainly in the bark or with a very limited penetration of the wood, although in other cases (the last mentioned) the parasite may eat into the wood and cut the trunk in two or so weaken it that it is unable to withstand the force of strong winds. Cankers of bacterial origin are well illustrated in the canker stage of the fire blight *(Bacillus amylovorus)* of the apple, pear and other hosts. The bacteria in the majority of cankers die during the first season, but they survive the winter in a limited number, which are then called "holdover cankers." While the majority of cankers are of fungous or bacterial origin, they may be caused by non-parasitic influences, as is well illustrated in the "winter-sun-scald" cankers which appear on the southwest side of tree trunks, on the exposed faces of larger limbs or in the crotches of the larger limbs.

Various woody plants may produce closely grouped clusters of fine, slender branches, generally arranged more or less parallel to each other and frequently originating from an enlarged axis. These broom-like clusters are called "witches' brooms/" which is a literal translation of the German name "Hexenbesen," originally applied to these structures by the superstitious peasants of central Europe, who attributed their origin to the machinations of witches. In the majority of cases, the branches that normally would be horizontal, or only slightly ascending, become erect or nearly so.

Various leaf-curl fungi which become perennial in the host tissue are responsible for witches' brooms, as is well illustrated by the leaf curl and witches' broom *(Taphrina cerasi)* of the cherry; rust of various species may give rise to very characteristic brooms on various conifers *(e.g., Gymnosporangium libocedri* on the incense cedar); *Melampsorella elatina* on the silver fir; while the scaly or dwarf mistletoes *(Razoumofskya spp.)* may be the inciting agents in pine, larch and fir trees. The majority of witches' brooms are due to plant parasites, but in some cases insects may play a part, as in the branch knot of the hackberry *(Celtis occidentalis)*, caused by a mite (Phytoptus) followed by a fungus.

A development somewhat similar to the witches' brooms of aerial structures may occur on the root systems of woody plants, giving rise to an abnormal number of fine fibrous roots, frequently making compact clusters, the so-called "hairy root." These roots may originate from the main root or the crown, from some lateral root, or they may come from the surface of a crown gall, giving what is known as the woolly-knot form of hairy root. The primordia of adventitious roots may sometimes be organised in closely aggregated groups on trunk or branches but, failing to develop into true roots, may produce rough, papillate, gall-like formations—the aerial hairy root of apple, pear or quince.

Just as some plants which normally have the rosette habit give up this type of growth as a result of disease, others which normally never produce rosettes adopt the rosette habit when they are diseased. In the formation of these rosettes, axes which normally are much elongated remain short and congested due to failure of the internodes to elongate, and as a result the leaves are closely grouped.

This rosette symptom is so characteristic in certain diseases that it has given rise to the common name of the trouble, as in the apple rosette of the Pacific Northwest, the pecan rosette of the southern United States and the contagious rosette disease of peaches and plums of the southeastern states. Somewhat similar modifications of growth may accompany diseases of herbaceous plants.

Disease may cause the development of dormant or rudimentary structures, or entirely new organs may appear. Dormant buds may be started into growth, producing new shoots, as may be noted on ringing or girdling diseases of trees, in which the new development is just below the zone of attack, as in silver leaf *(Stereum purpureum)* of fruit trees. Stamens that normally are rudimentary in the pistillate flowers of *Lychnis dioica* when attacked by anther smut *(Ustilago violacea)* grow to full size but produce smut spores instead of pollen; or buffalo grass *(Buchloe dactyloides)* may form ovaries in its staminate flowers when parasitised by smut *(Tilletia buchlceana)*. The formation of new organs or structures, either similar to or entirely unlike any normal parts of the host, may occur on leaves, stems or flowers. Striking illustrations are found in the production of slender, branched, leaf-like outgrowths from the fronds of a fern (Pteris) infected with one of the leaf-curl fungi *(Taphrinalaurencia)*; or in the pallid, leafless branched structures which appear on shoots and leaves of a Japanese conifer (Thujopsis), when infected with a rust *(Cceoma deformans)*. New flower parts may appear, as in *Viola silvestris*, which produces extra petals when affected with rust.

In certain cases, an organ may continue to produce new growths after it has reached the form or stage which would normally end its development. This symptom—*proliferation* or *prolification*—is well illustrated in the conifers, especially the larch, in which the cone axis may grow out into an elongated, needle-bearing branch; in roses, in which a new blossom may sometimes come from the centre of an old one; in certain composites, in which peripheral florets may grow into elongated axes bearing small but perfect flower heads; in the so-called "sprouting pears" (also other pome fruits), in which one pear may be formed directly above another or a leafy twig may continue to grow from the calyx end; or in the second growth or sprouting of potato tubers while still in the soil. In such cases, there is a disturbed nutrition giving an oversupply of plastic food substances.

Somewhat similar disturbances may result from the attacks of fungi, as in the "green-ear disease" or downy mildew *(Sclerospora graminicold)* of certain grasses in which the head is transformed wholly or in part into a loose green head, due to the growth of florets into small, leafy shoots; or in corn affected with head smut *(Sorosporium reilianum)*, in which flowers or an infected inflorescence sometimes grows out into either leafy structures or small leafy shoots with the elimination of spore production.

Rolling, crinkling or curling of leaves may result from the operation of either parasitic or non-parasitic factors. Rolling of leaves is the first signal of thirst in many plants and may serve as a temporary check to loss by evaporation.

This symptom is so characteristic in certain troubles that it has suggested the common name, as in the leaf curl *(Taphrina (Exoascus) deformans)* of the peach; or the leaf roll and curly dwarf of the potato, belonging to the degeneration or virus diseases. In other cases, diseased leaves, instead of showing the normal smooth surface, may be variously crinkled or puckered, as in numerous mosaic diseases or in apple leaves showing frost blistering.

Cylindrical organs may become broad, flattened and more or less band-like, as though a number of branches had grown together. Such banded *orfasciated* structures may be more or less twisted or thrown into coils or spirals. The entire inflorescence of some plants may undergo general fasciation, as has occurred in the cockscomb *(Celosia cristata)*, in which the abnormality has been propagated for ornamental purposes.

Cylindrical stems may show little or even no evident flattening but may grow in such a manner as to produce an irregular scraggly growth with a flat or open spiral form, or the spiral form may be noted in the *spiralism* of certain woody plants *(e.g.,* Yellow Newtown apple or the Gravenstein, where it has come to be almost a variety characteristic).

The abnormal roughening of surfaces, which under normal conditions should remain smooth, is another symptom of disease. This roughening may be very slight, as in the russeting of the skin of apples or pears, which may be due to frost injury, to use of a spray like Bordeaux or to the attacks of a parasite as in the net russeting of apples from powdery mildew *(Podosphcera leucotricha).*

Russeting may be general or diffuse, or it may be localised in the form of small patches, rings or bands. In some varieties, russeting in part or throughout has come to be a normal variety character. If the roughening of the surface is more pronounced, the name "scurf" or "scab" is frequently applied to the condition, as may be illustrated by the rough-bark disease of the apple and pear (non-parasitic), the rough-bark canker *(Phyllosticta prunicola),* the scab of citrus fruits, the scab of apples and pears, the freckle or scab of peaches and the common scab of potatoes, each an infectious or parasitic disease due to a specific pathogene.

Fruits may be variously deformed, or root crops may be cracked, corroded or irregular in shape. Deformed fruits may result from such parasitic attacks as the apple scab fungus *(Venturia incequalis)*, or sucking insects like the rosy aphis may produce irregular, gnarly apples (Stigmonose), while freezing injury when the fruits are young may cause other deformities. Pears may develop an excessive stoniness, or an increased number of grit cells may be formed (Lithiasis), either scattered or in localised groups. Potato tubers that are cracked, corroded or irregular in form are common in plants affected with Rhizoctonia and also in the case of some of the degeneration diseases.

Exudations: The forcing out of water or cell sap upon free surfaces is a normal physiological process in many leaves (guttation), and *bleeding* from the cut ends of stem or branches may be expected in many cases, but in certain diseases the abnormal oozing out of watery or slimy products which may or may not set into hard masses is a very striking symptom. The character of the exudate varies with the nature of the causal factors and the peculiarities of the affected plants.

Bacterial Exudates: This symptom is well illustrated in the fire blight of apple, pear and other hosts, in which the ooze or exudate consists of myriads of the minute fire-blight bacteria mingled with decomposition products from the affected tissue. The bacterial ooze may burst out through a break in the bark of an active canker and run down the side of the trunk or branch. At first, or under conditions of abundant moisture, it is a sticky, fluid mass, but with evaporation of moisture it may set into firm masses which remain until washed away by rains.

On succulent structures like 1-year-old twigs, leaf stalks, flower pedicels or fruits, the ooze may solidify into minute, pearl-like droplets, which are clear or slightly tinged with yellow. In various other parenchyma types of bacterial diseases, exudate will be formed, even in bacterial leaf spots spreading as a thin film or crust over the surface of a lesion.

Slime Flux: This name is applied to the fluid or semi-fluid outflow, from the bark or wood of various deciduous trees, which does not set into solid masses. Wounds may bleed, and the exudate, which is really a good nutrient solution, may support various bacteria or fungi which delay or prevent the healing process, while in other cases decomposition products of the diseased tissue may be formed and finally poured out upon free surfaces, sometimes with the development of pockets due to the breaking down of affected cells. Parasitic organisms may be associated with slime flux, but in other cases they are to be viewed as accompaniments of tissue derangement.

Gummosis: This term is applied to the formation of clear or amber-coloured exudates which set into solid masses upon the surface of affected parts. Internal gum pockets may be formed which become filled with the decomposition products of the diseased tissue, or the decomposition products may be forced out upon some free surface where they accumulate. With continued disintegration of tissue, the gum masses may increase in size, since they are not soluble in water.

Gummosis is a peculiarity of certain plants and is therefore, no reliable indicator of any specific parasitic or non-parasitic disturbance. It is an accompaniment of many diseased conditions in stone fruits, especially in cherries and peaches, and is also common in citrus stock.

In the above mentioned plants, gummosis is likely to follow a localised death of cells from any cause whatever. In the cherry, for example, it may result from growth in heavy, poorly aerated soils, from too deep setting, from winter injury, from excessive pruning, from the presence of pathogenic bacteria—*e.g.*, bacterial gummosis *(Pseudomonas cerasus)*—from a definite fungous disease—*e.g.*, California blight *(Coryneum eijerinckii)*—or from the injurious effects of an insect pest—*e.g.*, San Jose scale. It should then be emphasised that gummosis, in itself, is not a disease but a symptom of disease which may be brought about in a great variety of ways.

Resinosis: The production of resin by special secretory canals is a normal feature in pines, firs, spruces and other coniferous trees, and the limited occurrence of pitch as a surface exudate is frequent, but the excessive outflow of resin may indicate the presence of disease. Resinosis is an attendant symptom in certain coniferous rusts *(Peridermium spp.)* and may also result from the attacks of wood-destroying fungi, the abnormal production of resin sometimes giving the first signal of the presence of an internal parasite.

Latexosis: The production of latex, or a milky fluid, in special lacticiferous vessels is a characteristic of certain species of plants, and this milky fluid will ooze out from cut surfaces. An abnormal outflow of latex may accompany certain diseases—*e.g.*, the older stages of rubber canker.

Rotting: Succulent or woody stems and roots, fleshy leaves, flower buds or fruits may be affected with either dry or soft rot—the "gangrene" of plant tissue.

The character of the rot may depend on the structures involved, the causal factors or complications and external conditions. In the majority of cases, soft or pulpy fruits undergo a soft, or wet, rot, because of the abundance of water in their tissues, while firmer structures are affected with dry rot unless the initial changes are followed by the attacks of putrefactive bacteria.

Root Rots: Either fleshy or woody roots may be attacked by bacteria or fungi which may cause a rapid or a slow disintegration of the invaded tissues.

This effect is well illustrated in the destructive Texas root rot *(Ozonium omnivorum)* and the root rots of alfalfa due to various fungi *(Fusarium spp. Rhizoctonia spp.).* Root crops like beets, carrots, turnips, parsnips, sweet potato, etc., may develop soft rots due to bacteria *(Bacillus carotovorus* and others), or either dry or soft rots may be caused by fungi *(e.g., Phoma betce* on beet, Rhizopus rot of sweet potato).

Woody roots of our forest and shade trees, or of bush fruits or ornamental shrubs, may be rotted by fungi. The shoe-string fungus or honey agaric *(Armillaria melled)* as the cause of mushroom root rot is a parasite of wide distribution; the Ozonium of cotton attacks many other hosts, both herbaceous and woody; and various wood-destroying fungi may work in the roots and trunk of affected trees.

Leaf or Stem Rots: Succulent or *herbaceous* leaves or stems may rot from the inroads of either bacteria or fungi, as in the late blight *(Phytophthora infestans)* of the potato, the slimy soft rot of lettuce (bacterial), the stem rot or wilt' *(Sclerotinia sclerotiorum)* of various garden and field crops, the stem rot *(Fusarium sp.)* of the cultivated aster, the blackleg *(Bacillus atrosepticus)* of potato and numerous other illustrations.

Modified stems, or tubers, rhizomes, bulbs and corms, are storehouses of reserve food and as such furnish an excellent pabulum for bacteria or fungi, and since their tissues are in a dormant or languid condition they frequently fall an easy prey to weak parasites.

The potato with its soft and dry rots due to various bacteria and fungi will present typical illustrations: dry rots due to the late-blight fungus *(Phytophthora infestans)* and various species of Fusarium; soft rots due to bacteria following late-blight attacks, to the tuber phase of blackleg or to the inroads of a soil fungus *(Pythium debaryanurn),* which produces the disease known as watery soft rot, melters or leak.

Woody stems of living hosts—standing dead timber or posts, poles and lumber—may undergo decay or disintegration by the action of wood-destroying fungi which grow through or between the wood cells. In living hosts, these fungi are generally wound parasites, entering through some opening which has exposed the wood.

Once on the inside, they may continue to grow, sapping the life of the tree or destroying its mechanical support. Certain fungi grow primarily in the heartwood *(heart rots),* while others grow primarily in the sapwood and bark *(sap rots).* Wood that is invaded by fungi may be discoloured or made brittle, soft or punky, the exact effects varying with host and fungus. The appearance of fruiting bodies on the bark of trunk or branches or at the crown of a tree in the form of brackets ("punks or conchs" of lumbermen), flat or prostrate, smooth or toothed structures or toadstool-like forms is a signal that the branch or trunk is already pervaded by the vegetative body of the fungus, which is now emerging from its hiding place in order that it may send its offspring in the form of spores to conquer new worlds. Many wood-destroying fungi are pure saprophytes and in such a role cause the disintegration of posts, poles and structural timber whenever it comes in contact with the ground or is sufficiently moist to support fungous growth.

Bud Rots: Fleshy buds sometimes offer a congenial home for rot-producing organisms.

This may be seen in the bud rot of the carnation, a disease in which the unopened buds may have the enclosed petals and other flower parts rotted by the combined action of a fungus *(Sporotrichum anthophilum)* and a mite. The spores of the pathogene are apparently carried by the young mites which migrate to unopened buds and soon find the soft, rotted petals a suitable substratum in which to complete their development.

A second illustration is afforded by the bud rot of the coconut, produced by a fungus *(Phytophthora faberi)* with secondary invasions by bacteria. It is claimed that this disease is spread by vultures which are attracted by the odour of putrefaction. The cabbage, which is really a specialised bud, may be entirely destroyed by the bacteria of black rot *(Pseudomonas campestris)* and secondary invaders which work down the leaf veins and get into the centre of the head, during a putrid, foul-smelling mass.

Fruit Rots: The rotting of fruit may result in a few cases from the operation of non-parasitic factors, but the great majority of cases of fruit rot are due to either bacteria or fungi. The best example of a non-parasitic fruit rot is illustrated by the blossom-end rot of the tomato, which generally begins on immature fruits and produces a dry, black rot on the blossom end. The rotted tissue is frequently invaded by fungi, which were at first thought to bear a causal relation to the disease; but it is now known that these are entirely secondary factors and that the decomposition of the tissue is the result of a deranged nutrition. A somewhat similar blossom-end rot of the watermelon has been observed in the South.

The great majority of fruit rots of parasitic origin are due to fungi, rather than to bacteria, since the chemical reaction of most fruit juices seems to be unfavourable to the growth of bacteria but favourable to the growth of many different fungi. The soft rot *(Bacillus melonis)* of the muskmelon is due to bacteria which gain an entrance through some wound and cause a complete decay with an offensive odour. Bacteria of fire blight will grow in young fruits of susceptible hosts but are never responsible for the rotting of mature fruit or fruit in storage. None of the transportation or storage rots of small fruits, stone fruits or pome fruits is caused by bacteria, but numerous fungi belonging to widely separated groups are very destructive.

The fruit-rotting fungi may gain an entrance through the unbroken skin in numerous cases—*e.g.*, bitter rot of apple *(Glomerella cingulata)*— but many fungi can enter only through a wound or bruise. The rotting may begin when the fruit is still immature and hanging on the tree and continue during transportation to market or during storage.

Many fruit-rotting fungi are weak parasites and attack only ripe fruit, just at the time of maturity or after it has been harvested. Oft fruits like strawberries suffer rapid decay from fungi, while fruits like apples with a protective epidermis may be preserved for a long time. If it were possible to prevent the inroads of fungi, the drying of fruits from loss of moisture would be the factor limiting the keeping of fresh fruit in many cases, although it must be remembered that the cells of fruits are still alive and that chemical breakdown will follow in due time even though moisture is retained and fungi are excluded *(e.g.*, internal breakdown of the apple).

Rots of fruits may be soft or dry, varying with the nature of the fruit, the identity of the causal organism and the environmental factors which are operative. Strawberries attacked by *Rhizopus nigricans* undergo a very soft watery rot which is called "leak/" while fruits affected with Botrytis remain firmer and finally dry to shrivelled mummies.

Apples rotted by blue mould (Penicillum) are soft and decay completely, while attacks of brown rot (Sclerotinia) may finally produce slightly shrunken, coal-black mummies which resist final decay.

Fruits which are not fleshy may be attacked by fungi and undergo decay or dry rot, as is well shown in the dry rot *(Diplodia zew)* of corn, which is very destructive in the corn belt of the United States.

The most important fungi causing rots of our perishable fruits are species of the following genera: Pythiacystis, Sclerotinia, Penicillium, Alternaria and Phomopsis on citrus fruits; Glomerella, Physalospora, Neofabnea, Sclerotinia, Penicillium, Botrytis and Alternaria on apples and pears; Sclerotinia on stone fruits; Botrytis, Rhizopus and Penicillium on strawberries and bush fruits; and Guignardia, Plasmopara and Glceosporium on grapes overgrowths, cankers, tubercles, tumours or galls are formed.

These may be illustrated by the crown gall on both woody and herbaceous hosts, the olive knot, the tuberculosis of the oleander, the galls of several species of pines, citrus canker and other less perfectly known canker diseases (poplar, oak).

In these cases it is probable that some chemical substance produced by the bacteria or by the irritated cells excites the cells to cell division, which proceeds with great rapidity. In many of the overgrowths the cells divide so rapidly that they remain small, and little differentiation takes place, with reduction of vascular elements. In other cases the same gall or tumour may exhibit hyperplasia and also enlargement or hypertrophy of some of the cells.

Dissemination of Bacterial Diseases: Bacterial diseases may be introduced into a new environment or spread from diseased to healthy plants in a great variety of ways.

Diseased plants may harbour the disease and serve as centres of infection when transported to distant points or when allowed to remain in fields or orchards.

Some of the more important agents of transmission may be mentioned:

(1) *Seed* used in the broadest sense to include true seeds, fruits, bulbs, tubers or other propagating stock. The black rot of cabbage, Stewart's wilt disease of sweet corn, black chaff of wheat, yellow disease of hyacinths, and blackleg of potato are recognised seed-borne bacterial diseases.

(2) *Insects and other animal life,* including birds, molluscs and worms.

Here may be mentioned the insect transmission of fire blight of apple, pear and other hosts (by various sucking and chewing insects) and the wilt of cucurbits (by striped cucumber beetles, *Diabrotica spp.)',* the transmission of the olive tubercle by snails; and the work of root-knot nematodes or round worms, which open the way for the entrance of the bacterial wilt or brown rot of tobacco, tomato and other susceptible hosts.

(3) *Contaminated fertilizer.* The compost heap may harbour pathogenes which are later introduced into the field. Scab and rots of potato, black rot of cabbage and other troubles may be disseminated in this way.

Geographic Distribution of Systemic Fungus Diseases

Data on the geographic distribution of the systemic mycoses are very meagre when compared with the extensive knowledge that has accumulated concerning the distribution of diseases caused by other types of infectious agents. Systemic fungus infections are transmitted from man to man only under exceptional circumstances, and the lack of contagiousness is one of the more important reasons why so few epidemiologic studies of fungus diseases have been made.

However, during recent years it has been recognised that some of the mycoses are sufficiently widespread to present real and important public health problems. Intensive studies of coccidioidomycosis have shown that infection with *Coccidioides immitis* has a much wider geographic distribution than was thought formerly and that the incidence of the infection in endemic areas is extremely high (Smith, 1940).

Evidence available at the present time points to the possibility that histoplasmosis may be relatively common in a benign form in the Mississippi Valley area of the United States (Palmer, 1945).

Except for actinomycosis few, if any, of the fungus diseases are reportable to various health departments, and sources of information such as the master tables of causes of death published by the Bureau of the Census are of no aid because none of the reference either to a specific fungus disease or to mycoses as a group.

Although published case reports are of value in attempting to determine areas in which fungus infections may be found, they are of little value in estimating the prevalence of any one disease in such areas.

Undoubtedly, many cases are not reported because the disease either is unrecognised or, if recognised, the physician or laboratory shrinks from the task of untangling the numerous systems of terminology and methods of classification that have been advocated by various workers in the field. For this reason, there is a tendency for reported cases to be clustered geographically around certain areas in which there are clinics or laboratories with special interests in the problem of fungus infection.

The lack of standard methods of identification and classification introduces additional confusing elements into the problem of mapping the distribution of specific fungous diseases. In some of the more important mycoses, the etiologic agents are described under a variety of different names, depending upon the opinion of the author or collaborating mycologist. On the other hand, certain infections caused by specifically different and unrelated fungi are grouped together under a common clinical term. For example, the name blastomycosis is used commonly to designate infections caused by *Blastomyces dermatitidis*, *Cryptococcus neoformans* and *Candida (Monilia) albicans* in spite of the fact that the lesions in each infection are quite different and the fungi are separated easily by cultural methods. The use of the term "European blastomycosis" to designate infections caused by *Cryptococcus* serves as a method of identifying the infection, but the term is misleading in that it implies that the disease has a predominantly European distribution.

With the above qualifications, an attempt has been made to estimate the areas in which specific fungus diseases have been reported.

For convenience in discussion, the systemic mycoses are discussed under three general headings: (1) infections, presumably endogenous in origin, in which the distribution is worldwide; (2) infections of exogenous origin in which the geographic distribution is worldwide but spotty; and (3) infections of exogenous origin which are characterised by being reported only from rather sharply limited geographic areas.

Endogenous Infections: Included in this group are three fungus diseases in which the evidence indicates that the infection is acquired from some focus in the body rather than from an extraneous source in nature. Such infections would be expected to have a worldwide distribution.

Actinomycosis: Actinomycosis is the most ubiquitous of all systemic mycoses, and cases have been reported from every part of the world.

The anaerobic fungus *A. bovis* has been shown to be a common inhabitant around the teeth, gums, and tonsils in a high percentage of individuals who show no evidence of serious infection (Slack, 1942). The organism, thus, is in a position to initiate severe infection under a variety of circumstances, such as invasion of the mucosa to produce cervicofacial actinomycosis or through aspiration or swallowing, to result in pulmonary or abdominal infections.

Of all the systemic mycoses, actinomycosis is probably the easiest infection to diagnose clinically because it often presents very characteristic clinical features which are well known to most physicians. Furthermore, the diagnosis is established readily by hematoxylin and eosin-stained sections obtained at biopsy or necropsy. In such sections the sulphur granules are large and brilliantly stained and are not apt to be confused with any other infection.

It is felt that both of the factors mentioned above play a contributory role in making actinomycosis a disease which is known and reported from all parts of the world.

Moniliasis: Infection with the yeast-like fungus *Candida (Manilla) albicans* occurs also in all parts of the world, but the organism produces such a variety of clinical syndromes that it is impractical to attempt an allocation of any type of infection to any particular geographic area.

Bronchopulmonary moniliasis, first thought to be a tropical disease, now is known to occur in many other climates (Conant, Martin, Smith, Baker, and Callaway, 1944). Predisposition to infection is dependent more upon personal factors than climatic conditions. Like *Actinomyces,* the fungus is harboured by apparently normal individuals; it has been isolated from the skin, oral cavities, and faeces. The fungus can produce thrush, vaginitis, and various types of infection in the skin, nails, lungs, and meninges, and is often a secondary invader in other conditions. The organism may be present in the sputa of tuberculous patients, in the faeces of individuals with sprue, and in the lesions of perleche.

Although *C. albicans* is the primary etiologic agent in many clinical syndromes, it often is difficult to evaluate the specific role of the fungus and one is left with the impression that *Candida (Monilia)* is capable of causing almost any type of infection in almost any part of the world.

Cryptococcosis: It is not known whether or not diseases caused by *Torulopsis (Cryptococcus) neoformans* should be included among infections of endogenous origin.

Strains of the fungus, pathogenic for animals, have been isolated from certain fermenting fruit juices, but there is no direct evidence that man has contracted an infection from such a source (Sanfelice, 1895).

Species of *Torulopsis* have been isolated from normal skin and faeces. Insufficient work has been done on these strains isolated from normal individuals to establish their relationship to the disease. Some of the species of *Torulopsis* are pigmented; others are known to be non-pathogenic, but it is not unlikely that a few of them may acquire pathogenicity under certain conditions and initiate infection.

Cases have been reported from the United States, mostly from the eastern and southern parts, from Germany, Italy, France, Brazil, Argentina, and Paraguay.

In the southwest Pacific region, cases have been reported from Japan, the Philippines, Dutch East Indies, and Australia. There is considerable clinical evidence that the portal of entry may be the lungs, and we have seen cases with pulmonary lesions by x-ray in which symptoms were minimal. It is possible that many pulmonary infections exist unrecognised because of the lack of symptoms. Unfortunately, there are no simple methods of testing for sub-clinical infections analogous to those which have proven so successful in studying the epidemiology of coccidioidomycosis. The spotty distribution in various parts of the world can be correlated with a known interest in the disease and does not necessarily indicate that the infection is limited to these areas.

Exogenous Infections with Widespread Geographic Distribution: Although the exact sources in nature of the infections included in this group have not been established, certain features of case histories and the location of lesions suggest an exogenous origin.

Sporotrichosis: This infection has been reported from practically all parts of the world, but the recorded cases do not indicate a coverage as complete as that of actinomycosis. Large numbers of cases have been reported from some areas, reflecting again the interest of certain clinics in the particular mycosis. Thus, many cases have been reported from France, South Africa, and the upper Mississippi Valley in the United States. As an illustration of the intensity of infection in certain areas or under peculiar conditions, it is of interest to cite a recent publication where 650 cases were reported in men working in a mine shaft in the Transvaal area in South Africa.

Like actinomycosis, Sporotrichosis has certain characteristics by which a clinical diagnosis can be made quite easily. The response to treatment is so satisfactory that it is probable that only rare or unusual cases now are considered worth reporting.

Chromoblastomycosis

Like Sporotrichosis, chromoblastomycosis has been reported from widely scattered parts of the world, but unlike infections due to *Sporotrichum,* the number of cases reported from any one area is small. Instances of chromoblastomycosis have been found in various parts of the United States, Cuba, Puerto Rico, Santo Domingo, Guatemala, Costa Rica, the Canal Zone, Venezuela, Brazil, Argentina, Paraguay, and Uruguay. The only case reported from Europe was found in Russia (Tschernjawski, 1929). In Africa, there have been instances of the disease reported in Algeria and southern Africa. In the Pacific area, cases have been recorded in Java, Sumatra, and Japan.

That the infection is exogenous in origin has been suspected from clinical histories and the location of the primary lesion which is usually on an extremity.

Furthermore, a fungus morphologically indistinguishable and antigenically related to *Phialophora verrucosa* has been isolated from wood pulp (Martin, 1938).

Several different fungi can cause the clinical picture of chromoblastomycosis, and at one time it was thought that the etiologic agent of the disease in North America was *Phialophora verrucosa* and that the South American form of the infection was caused by *Hormodendrum pedrosoi.* As more cases have been studied, it has been learned that there is no relationship between the fungus and the area in which the infection is acquired. A third fungus, *Hormodendrum compactum,* has been isolated from only one patient whose infection was acquired in Puerto Rico.

Histoplasmosis: Like chromoblastomycosis, the reports of infection with *Histoplasma capsulatum* have come from many parts of the world, such as the United States, Central America, South America, England, South Africa, the Philippines, and Java. In North America, by far the greater number of cases have been found in the North Central States with a scattering of cases in the East and South. All of the above data have been based on case reports of severe infections, most of which terminated fatally.

A recent survey made of student nurses in the North Central States revealed a high percentage of individuals with calcified lesions in the lungs and hilar lymph nodes who showed a negative tuberculin but a positive skin test to an extract of *Histoplasma capsulatum* (Palmer, 1945). Further work will have to be done before concluding that such positive skin tests represent actual proof of past infection with the fungus. However, if such proof can be obtained, additional surveys should show the true geographic distribution of infection and may reveal that primary invasion by the fungus is much more common than now realised.

Exogenous Infections with Limited Geographic Distribution: Coccidioidomycosis — Except for three cases reported from Italy and South America, Coccidioidomycosis is limited essentially to the southwestern part of the United States. The infection definitely is acquired from an exogenous source, and the fungus has been isolated from the soil in the endemic area. It has been shown recently that certain rodents are naturally infected with the fungus and may represent a reservoir in nature from which the soil is contaminated (Emmons, 1943).

The discovery that *Coccidioides immitis* causes such an extremely high incidence of infection in the population of endemic areas was a startling observation, and was the first demonstration of a mycosis which was sufficiently important to present a public health problem (Smith, 1940). Skin tests with coccidioidin have shown that the endemic areas extend as far east as western Texas (Marcuse and Karotkin, 1944). Although surveys have not been conducted in all parts of the United States, tests in sample areas in the northern and eastern parts of the country have shown positive reactions very rarely. As far as is known, surveys of this nature have not been made in other parts of the world.

North American Blastomycosis: This infection is limited essentially to the United States. Except for several cases in Canada and one in England, there have been no authenticated cases reported outside the continental United States. The authors recording the English case commented on the fact that their patient was engaged in the occupation of opening packing boxes which had been shipped to England from the United States (Bowling and Elsworthy, 1925).

In the United States, there is a rather peculiar distribution, most cases having been reported from the Central and South Atlantic States. Few cases have been reported from the New England and Middle Atlantic States. There have been several reported in the Pacific States, but we are unaware of any originating in the Mountain States.

The high incidence of the infection in males and the clinical histories suggest that the infection is acquired from some source in nature, but proof of this assumption is lacking.

South American Blastomycosis: As far as is known at present, infection with *Blastomyces brasiliensis (Pctracoccidioides* *brasiliensis)* is limited to South America where it has been reported from Venezuela, Brazil, Argentina, Peru, and Paraguay. The prevalence of the infection in certain of these areas is high as judged by the large numbers of cases which have been reported by individual investigators studying the disease.

As in North American blastomycosis, the majority of patients are males and the infection is found most frequently in manual labourers whose occupation requires direct contact of the skin with vegetative material.

Miscellaneous Fungus Infections: Nocardiosis (actinomycosis): The aerobic species of the family *Actinomycetaceae* now are called *Nocardia* — The disease produced by these aerobic organisms is indistinguishable from actinomycosis caused by *A. bovls,* but the organisms occur in nature throughout the world and must, therefore, be included among the mycoses of exogenous origin. Pathogenic *Nocardia* have been recovered in a few instances from the soil.

Geotrichosis: Several species of the genus *Geotrichum* occur normally in the mouth and intestinal tract of man. They produce a disease of the oral mucous membranes which simulates thrush; they also produce bronchitis, pulmonary infections and possibly, under rare conditions, infections of the intestinal mucosa. Geotrichosis is mild, as a rule, but may be confused with North American blastomycosis.

Aspergillosis, Mucormycosis$_t$ and Penicilliosis: These moulds are present in abundance in nature in all parts of the world. They produce disease frequently in the ears, nose, and sinuses, and in the lungs and other organs occasionally. In France, squab feeders and fur cleaners not infrequently develop aspergillosis of the lungs from the inhalation of large numbers of these spores.

In the wheat-growing regions of the United States, *Aspergillus* occurs in abundance about the wheat threshers and may be inhaled with the dust into the lungs, giving rise to a peculiar type of miliary calcification (Schneider, 1930).

Maduromycosis: Maduromycosis is an example of an exogenous infection of worldwide distribution. Twenty-two species of fungi belonging to nine genera have been isolated from the lesions of maduromycosis. The aerobic *Actinomyces,* or *Nocardia,* and *Monosporium apiospermum* are encountered most frequently. The disease occurs most often in tropical and subtropical zones when the feet come into direct contact with the soil. Instances have been reported from India, Africa, Europe, South America, Mexico, Canada, and the United States where it has been recognised in Massachusetts, Indiana, Iowa, Minnesota, Maryland, North Carolina, Georgia, Arizona, California, and Texas.

Rhinosporidiosis: Rhinosporidiosis is an exogenous infection caused by *Rhinosporidium seeberi.* It occurs with great frequency in India and Ceylon; sporadically in Argentina, Paraguay, Brazil, South Africa, the Philippines, and the Malay States; and occasionally in Persia, Italy, England, Scotland, and the United States. Thirteen cases have been found in the United States. Most of the patients give a history of swimming or diving in stagnant water. The lesions are pedunculated or sessile polyps which occur on the mucous membranes of the nasopharynx or the conjunctivae of the eye. Death may occur from laryngeal obstruction or secondary infection, but the disease does not spread through the body.

Coniosporiosis: Coniosporiosis is a relatively rare disease. It was found in northern Michigan in lumberjacks who were engaged in peeling the bark from maple logs (Towey, Sweany, and Huron, 1932). Clouds of spores of *Coniosporium corticale* were inhaled by the workers, and severe pulmonary symptoms developed with broncho-pneumonic patches in the lungs. The patients recovered slowly after a period of weeks or months; on re-exposure to the spores, they immediately relapsed. These patients gave positive skin tests to an extract of the spores, and experimental work in guinea pigs suggests that the disease is primarily an allergic manifestation and not a true progressive infection.

Diseases Due to the Sphere Fungi and Allies Pyrenomycetes

The fungi belonging to this group have no distinctive feature of either mycelium or conidial fruits. The latter generally constitute a very prominent stage in the life history of these pathogenes, while the ascigerous stage, which is frequently formed independent of the host or as a saprophyte on dead parts, is in many cases much less conspicuous. Various types of conidial fruits may be developed and when these alone are known the fungi forming them are referred temporarily to the proper groups of the *Fungi Imperfecti.*

Ascigerous Fruits: The distinctive feature of the sphere fungi and allies is the ascigerous fruit, which is typically an ostiolate *perithecium.* This is a globular or sac-like structure enclosing the asci, which generally originate in a group from the basal portion of the cavity. Paraphyses or sterile filaments may be present or lacking. The perithecia may be simply sac-like cavities immersed in an aggregate of fungous tissue, a stroma, or they may be provided with definite walls. In the latter case, the perithecia may be single and embedded in the substratum, seated upon it, or they may be grouped in stromata that are either embedded in the substratum or are superficial. The sphere fungi and allies as here considered constitute a large part of the group of *Pyrenomycetes,* while the cup fungi and allies are the forms included under the *Discomycetes.*

The asci are typically eight-spored and the ascospores may be hyaline or dark, show great diversity of form and vary from continuous to many-septate. As in the cup fungi, the ascospores are forcibly expelled at maturity, but only a single ascus explodes at once. In a typical case an ascus elongates until the tip protrudes through the ostiole, and when the hydrostatic pressure within the ascus becomes sufficient the explosion occurs and the spores are shot forcibly upward into the air. Following the collapse of the old ascus wall another ascus may elongate to take its place and soon another spore load is shot into the air. Spore expulsion continues during favourable conditions of temperature and moisture until the asci have all exploded. Perithecia are thus repeating spore guns which are sending ascospores into the air to be carried away by air currents. In a few cases, the asci do not explode but the ascospores are set free by the gelatinisation of the asci.

Hypocbeales

The perithecia are soft leathery, fleshy or membranous and are *never black* but show various other colours, such as buff, yellow, brown, red or purple. The mycelium is also light or bright coloured but never black. The perithecia are globular or flask-shaped, free on the substratum, embedded in a superficial mycelial weft or subiculum or seated on or sunken in a stroma. In general, the ostiolate perithecia have a thick wall of more or less pseudoparenchymatous character, but in Claviceps the walls of the perithecia are not differentiated from the stroma.

Hypocreaceae: A single family with the characteristics of the order.

Nectria: Perithecia soft membranous or fleshy, yellow, red or brown, single or grouped on the substratum, on a subiculum or in or on a fleshy stroma. Sporea two-celled, ellipsoid, blunt or pointed, generally hyaline, but rarely reddish. Conidial stages prominent and various.

Calonedria: Perithecia red or yellow, free or seated in a white cottony radiate subiculum. Spores elongated, hyaline, four-celled.

Gibberella: Perithecia transparent blue or violet, free or seated on an effused or fleshy stroma, and scattered or grouped. Spores ellipsoid or fusoid, four-celled, hyaline. Conidial stage a Fusarium.

Claviceps: Develops in the ovaries of grasses or sedges and forms horny sclerotia which replace the seed. Overwintered sclerotia produce a number of cylindrical stalks each bearing a globular head (stroma or receptacle) in the periphery of which the asci are borne in perithecial cavities without differentiated walls. Asci eight-spored, ascospores continuous, fascicled, needle-shaped and hyaline: A conidial or Sphacelia stage is formed during the early period of development of the sclerotia.

Ustilaginoidea: Conidia-bearing sclerotia replace grain of host; conidia continuous, globose, greenish, echinulate; perithecia in a stroma as in Claviceps.

Epichlce: Stromata smooth, fleshy membranous, surrounding the culms and leaf sheaths of various grasses, white at first, becoming reddish brown. Perithecia immersed, spores filamentous, breaking up within the asci.

Cordyceps: Develops in the bodies of adult or larval stages of certain insects, a few in subterranean fungi, transforming them into mummies or sclerotia, from which stalked stromata are developed. Perithecia embedded in the enlarged terminal portions. Ascospores filamentous, breaking up within the asci.

Dothidiales

Mycelium forming a sclerotia-like stroma, generally within the substratum, but becoming free by the rupture of overlying host tissue. *Stromata black* superficially but pale or light coloured in the interior (stromata are sometimes superficial from the first or remain permanently immersed). The perithecia are simply ascus-containing locules in the stromata and are without differentiated walls.

Dothidiaceae: Characters as indicated for the order.

Phyllachora: Stromata in the mesophyll of leaves, remaining covered by the epidermis. Ascospores continuous, hyaline.

Plowrightia: For the most part in branches of trees or shrubs and forming a compact superficial stroma on which the perithecia are borne. Ascospores ovate, two-celled and hyaline.

Dothidella: Similar to Phyllachora, but ascospores hyaline, unequally two-celled.

Speleriales

The Sphaeriales, or the sphere fungi, are characterised by the formation of typically ostiolate, mostly globular perithecia with well-developed walls, which constitute a distinctive character as contrasted with the Hypocreales and the Dothidiales.

The perithecia are carbonous, soft leathery or membranous, brittle, tough or delicate, single or grouped and may be immersed, erumpent or superficial on the substratum or in connection with either a subiculum or a well-developed stroma. The ostiole of the perithecium is sometimes only a circular opening in the free end but more frequently it is at the tip of a short papilla, an elongated beak or a long, slender neck, many times the length of the body of the perithecium proper. The various forms of ostioles may be found in both free and stromatic species. The spores are either hyaline or dark and great diversity is shown in form and septation, varying from globular to acicular and from continuous to septate forms, either linear aggregates or muriform masses. The light or dark (mostly the latter) mycelium is either in or on the substratum and may be filamentous or compacted into stroma-like aggregates.

The Sphaeriales consist of numerous families which differ from each other by only slight characters. The number of genera and species is very large, and while the number of parasitic species is small as contrasted with the saprophytic forms, the group furnishes a considerable number of very important plant pathogenes. The saprophytic forms may occur on various substrata, but they are especially abundant on decaying herbaceous stems, woody twigs or leaves.

Of the parasitic species, the great majority have retained the saprophytic habit for the perithecial stage, while a conidial stage develops on the living parts of the host. The perithecial stages may reach maturity only after the old dead host tissues (stems, twigs, cankers, fruits or dead leaves) have been subjected to winter temperatures, and for this reason the ascospores have sometimes been designated as the "winter spores," while the conidia have been called the "summer spores."

This condition does not always prevail, but in some forms ascospores are matured whenever conditions for growth are afforded, while conidia may be developed even during the winter period when the temperatures are moderate.

The most important genera of the various families furnishing plant pathogenes will be given brief characterisation in the following key, based principally upon spore characters:

I. Spores one-celled, hyaline:

A. Perithecia clothed with hairs: *Trichosphceria.*

AA. Perithecia smooth:

B. Stroma lacking:

Perithecia long-beaked and single: *Ceratostometta.*

Perithecia long-beaked and in a valsoid cluster: *Calosphceria.*

Perithecia papillate or apapillatc

Paraphyses present: *Physalospora.*

Paraphyses lacking: Ascospora

BB. Stroma valsoid:

Spores slender, curved: *Valita.*

Spores elongated, cylindrical: *Cryptosporella.*

Spores fusoid: *Botryosphosria.*

BBB. Stroma not valsoid, perithecia rostrate: *Glomerella.*

II. Spores one-celled, dark:

A. Perithecia superficial or sunken in the substratum, separate:

Apapillate *RoseUinia.*

Spiny rostrate: *Acanthorhynchus.*

A A. Perithecia in a stroma:

Stroma disk-like: *Numuttaria.*

Stroma pulvinate: *Hypoxylon.*

Stroma club-shaped or branched: *Xylaria.*

III. Spores two-celled, hyaline:

A. Stroma dark, diatrypoid or valsoid: *Diaporthe.*

A A. Stroma orange to brown, valsoi: d*Endolhia.*

A A A. Stroma lacking:

B. Perithecia smooth, immersed or protruding:

C. Asci thickened at the end and provided with a terminal canal.

CC. Asci not thickened or without a terminal canal. Spore cells very unequal, small lower cell cut off shortly before maturity: *Guignardia.*

(Mycosphcerella Spore cells equal or nearly so*):* *Didymellina.*

BB. Perithecia setose at apex, immersed and later protruding or super-ficial: *Veniuria*

IV. Spores two-celled, smoky or dark:

A. Perithecia many paraphysate, carbonous, superficial: *Neopeckia*

Black Knot

Plowrightia morbosa (Schw.) Sacc: This characteristic disease of plums and cherries is readily recognised by the black, elongated galls which appear on twigs and branches. It has been called plum wart, but black knot is the name most generally used.

History and Geographic Distribution: Black knot was noted in Massachusetts in destructive form over a hundred years ago (1811), and since that time has been treated in popular and scientific literature.

During the earlier years of its consideration two different

ideas prevailed as to the origin of the knots: (1) that they were due to a diseased condition of the sap; (2) that they were the result of insect stings.

The first intimation that the knots were of fungous origin was in 1843, when it was shown that the curculio merely lays its eggs in young juicy tissues that are already diseased. The ascigerous stage of the pathogene had previously been described by Schweinitz (1821), but he appeared to regard the combined action of a gall fly and the fungus as the inciting cause of the hypertrophies.

The idea of the insect relation was gradually abandoned as more attention was given to the fungus. The first complete account of the structure of the fungus was published by Peck (1872), and a few years later (1876) Farlow presented a fuller account of the life history of the fungus which left little doubt as to its causal relation to the knots.

Later studies, by Crozicr, Scribner (1890), Humphrey (1891, 1893), Halsted (1891), Beach (1892) and Lodeman (1894) added somewhat to our knowledge of the disease and considered its prevention or control. During recent years but little detailed study has been devoted to the disease.

Black knot was especially abundant in the New England states when attention was first directed to it, and it is believed to be of American origin. It probably first affected wild species and spread from these to the cultivated varieties of plums and cherries.

Whether the disease originated on the Atlantic seaboard arid gradually spread westward is uncertain. There are some indications that it was present in western localities long before attention was directed to it. By 1879, there were reports that the disease had appeared for the first time in the vicinity of Cincinnati. While it spread westward it never became so common or so severe in the Middle West as on the Atlantic seaboard. The disease is rare in the regions west of the Rocky Mountains, except on wild species. It appears very abundantly on the native chokecherry *(Prunus demissa)* in Washington and adjacent territory, but does not attack cultivated species even when they are growing in close proximity to thickets of the wild host. During an experience of 17 years in Washington not a single authentic case of the occurrence of the disease on cultivated species has been brought to our attention. The disease is not known to occur outside of America.

Symptoms and Effects: The black knot is first in evidence as a slight swelling of a twig or branch, either adjacent to an old knot or separate from it. While this swelling may sometimes be observed in the fall, it becomes conspicuous in the spring after growth starts and the bark soon ruptures and a straw-coloured or light yellowish-brown, granular growth fills the crevices. As the season progresses the overgrowth becomes more pronounced and somewhat darker.

In the late spring or early summer the smooth surface of these excrescences shows a pale-greenish tinge at places and soon the entire surface becomes an olive green and appears covered with a velvet-like pile. This velvety surface soon disappears, the knot becomes darker and by late fall it has become perfectly black.

The knots vary in location, shape and size. They may appear on young twigs or on older branches up to about 2 inches in diameter, and may originate at any of the following places: (1) crotches of limbs; (2) at the union of the growth of consecutive seasons; (3) on fruit spurs; (4) at the tips of twigs; and (5) in the axils of leaves or about an axillary bud. Some of the knots may be short, but they generally extend for some distance along a branch, and are more or less fusiform, although they may sometimes be nearly uniform in diameter for the entire length.

Adjacent knots may sometimes fuse to form much more extended excrescences. Their diameter depends largely on the size of the structure on which they form, so they may be only a fraction of an inch or 1 or 2 inches in diameter. It is typical for the knot to be confined to one side of the twig or branch, yet at times the branch is completely encircled. The affected structures may be straight or curved or even thrown into more or less of an irregular spiral. If knots are formed at the forking of branches, they generally extend along both branches and also down on the main axis.

In a tree in which the disease has been allowed to run its course undisturbed for a number of years the following should be found if the tree is examined in June: (1) young excrescences developed during the spring growth; (2) old knots that matured during the early spring; and (3) old knots developed during previous seasons, which are likely to be more or less eaten by insects and are frequently infested with saprophytic fungi. Some of the new knots may be extensions of old ones rather than representing new infections. The character of the injury may be expressed as follows:

Small twigs are frequently killed outright, for in them the disease soon cuts off all communication between the parts above and those below the affected portion. Large limbs do not succumb so rapidly. The disease gradually extends from year to year and it may be a long time before the death of the parts beyond takes place, for such, with possibly rare exceptions, is the result which eventually follows the appearance of the trouble (Lodeman, 1894).

Black knot has been an important factor in the production of plums and cherries, especially east of the Alleghanies and from North Carolina to southern Ontario and Maine.

In some parts of New England, particularly in Maine and along the seacoast, the raising of cherries has been almost abandoned in consequence of the ravages of the black knot. An idea may be formed of the small crop of plums now raised in New England from the fact that $2.50 was given in Boston for a peck of Damsons for preserving (Farlow, 1876).

The economic importance of the disease at a later date in New York may be judged by the fact that the Western

New York Horticultural Society in 1892 passed resolutions including the "demand that the Legislature of the state shall, without delay, enact such a law as shall, in its enforcement and execution, thoroughly and effectually exterminate that infectious and incurable disease known as the black knot" (Beach, 1892). It is also stated that the growing of cherries was nearly abandoned in North Carolina about 1906, because of the ravages of black knot.

Etiology: This disease is caused by an ascomycetous fungus which produces at least one *conidial stage* when the young knots appear velvety and olive green, and an *ascigerous stage* in the surface of the black matured knots. The pathogene was first named *Sphceria morbosa* by Schweinitz in 1821, and some other generic names were proposed later, but none came into general use until Saccardo assigned the fungus to the Dothidiales and called it *Plowrightia morbosa* (Schw.) Sacc., by which name it has been known until recent years. Some differences of opinion have arisen as to its real relationship, as witnessed by the recent assignment to a newly recognised order, the Pseudosphseriales (Theissen and Sydow, 1918), in which it appears as *Dibotryon morbosum* (Schw.) T. & S. More recently it has been assigned to the Melogrammatacese and the genus Botryospheria (Sorauer, fourth edition, 2: 328-330, 1921). The causal relation is based largely on the presence of mycelium in the young knots and on the perennial nature of the infections.

Aside from the early statement of Farlow (1876) that "we have made direct experiments to show that the spores of the knot on the chokecherry will germinate and produce the knot in healthy plum trees," successful inoculations have not been reported. With our present information there are several blanks in the life history of the pathogene, the part played by the different spore forms, and the exact time and manner of infection being based on observational data.

Knots may be *primary,* that is, the direct result of infection from spores, or *secondary,* due to the invasion of new portions by the mycelium that is already present in the tissues.

Mycelium comes to be present in the cambium, and when growth starts, the cells are stimulated to an abnormal activity and both wood and bark are affected. The mature gall is a complex of disarranged cells, the wood and bark undergoing about proportional increase in thickness. According to Stewart (1914), the cambium retains its normal position between xylem and phloem, except opposite the broad medullary rays, where it may be broken up and pushed out into the cortex, and then give rise to isolated xylem elements or groups of xylem cells. Many of these misplaced xylem elements are scalariform tracheids, while the normal wood produces only pitted elements. The mycelium grows outward in bundles or strands following the medullary rays (Farlow, 1876) and finally forms a stromatic layer, or pseudoparenchyma, of closely aggregated fungous cells which occupies the surface.

At this time, the knots are solid but rather fleshy or pulpy. The stromatic layer soon becomes covered with a dense coating of brownish, erect, septate conidiophores (May and June) which give the velvety appearance noted under Symptoms. The conidiophores are to 60, simple or sometimes branched and produce terminal or lateral, brown, continuous, ovate conidia, about 6µ long.

The conidiophores are generally somewhat flexuous, and may become geniculate due to the extension of the terminal cell beyond the point of origin of the conidia. The conidia are very easily separated from the points of attachment and a copious production follows until about midsummer. From their position and powdery character they are admirably suited to wind dissemination.

Even before conidial production has ceased, the primordia of perithecia have been forming beneath the conidiophores, their position being indicated by minute hemispherical protuberances, visible with a hand lens. As summer passes, the conidiophores dry up, the surface papillse become more prominent, the knots become harder and more brittle and assume the characteristic black colour of the mature galls.

During this time, internal organisation of the perithecia is proceeding, but maturity is not reached until nearly spring. The asci grow slowly during the winter months, the actual time of maturity varying with the temperatures.

Farlow found mature ascospores in abundance in February, but observations as to the exact time when they are discharged seem to be lacking. Each perithecium seems to be a single, ostiolate cavity within a stroma-like aggregate of fungous tissue which makes up the body without any well-defined wall. Each locule or perithecial cavity is filled with asci and sterile, filiform, non-septate filaments with slightly enlarged tips.

The asci contain eight, hyaline, one-septate spores, the lower cell being uniformly shorter and narrower than the terminal cell. The ascospores are obliquely uniseriate or irregular in arrangement. Under suitable conditions of moisture and temperature the mature ascospores are able to germinate in 1 to 5 days. One or more hyphae may grow out from the large cell and none or only one from the small cell.

Farlow (1876) described three types of pycnidia which he found mingled with perithecia: (1) one with yellowish septate spores, since named *Hendersonula morbosa* by Saccardo; (2) a Phoma-like form with hyaline spores liberated in tendrils; and (3) a Phomopsis-like type, which he called spermogonia. Humphrey (1891) later described a pycnid with single-celled brown spores which he thought belonged to the black-knot fungus but he was unable to reproduce the disease by inoculations made with these spores from pure cultures. Some or all of these pycnidial forms are either saprophytic on the old knots or possibly parasitic on the black-knot stromata.

It is the prevailing opinion that infections, whether originating from conidiospores or from ascospores, take place soon after the dissemination of these reproductive bodies, since they are capable of immediate germination. Lodeman (1894) showed that in certain cases the external cork layer may be interrupted by a fissure leading into the cortex in the axil of a branch, and expresses the opinion that infection takes place through such fissures or at some other points where the external cork layer is poorly developed.

However infection takes place, the formation of a knot apparently does not begin until the mycelium of the pathogene reaches the cambium. In most cases it seems that cambial activity has subsided by the time the mycelium reaches the cambium, and consequently only a few of the knots become evident in the fall following the time of infection. These, however, grow rapidly in the spring and produce the conidial stage and later the perithecia which are mature and ready for ascospore dissemination in the spring of the next season.

Host Relations: The black-knot pathogene attacks various species of plums and cherries. It is especially in evidence on many wild species, but is sometimes noted on one species when an adjacent species, perhaps, with interlocking limbs remains free, and wild and cultivated species may be in proximity without both becoming infected.

Stewart (1914) records that "the chokecherry and wild plum (P. *americand)* are infected in the vicinity of Madison, Wis.," and further states that the wild black cherry sometimes growing in thickets of badly infested chokecherries remains free from knots. Reference may be made again to the conditions in the Pacific Northwest, where the native chokecherry (P. *demissa)* is very generally affected, while cultivated plums and cherries remain free from the disease.

The probable explanation for this behaviour is the existence of biological species or strains that have become so adapted to certain hosts that they are unable to infect others. Farlow (1876) called attention to the fact that no morphological difference could be detected between the fungus on the cherry and the one on the plum.

The opinion has been expressed that there is no variety of cultivated plum that is not subject to the disease. Farlow (1876) believed that the disease spread from either the bird cherry

(P. *pennsylvanica)* or the chokecherry (P. *virginiana)* to the cultivated plums, since these species are common in the vicinity of Boston, while the wild plum is very rare there or does not occur at all.

The Trifloras are said to be affected less than any other group of plums, and the Institias rank next in immunity, although the Damson is said to be very susceptible. On the other hand, the Domesticas are susceptible, except possibly the Middleburg and Palatine, which are relatively free from black knot (Hesler and Whetzel, 1917).

In general, cherries are reported to suffer less than plums, although their behaviour seems to vary with the environment. In Ontario, cherries are said to suffer more than plums. Sour cherries, especially the English Morello, are very susceptible to black knot, while the Mazzard comes next (Farlow, 1876). Sweet cherries are so resistant that they are seldom infected, although a few cases have been recorded.

Control: The practices which have been recommended for the elimination or control of the disease are: (1) the cutting out of groups or thickets of wild cherries or plums adjacent to cultivated orchards; (2) surgical treatment of affected cultivated trees; (3) spraying to prevent new infections.

The need of removing the wild hosts will vary with the environment, since in some localities they are not a menace to the cultivated species. Whenever it is known that the disease is passing from the wild hosts to the cultivated species, other attempts at control will be partially nullified if infected wild hosts are allowed to remain nearby.

In the treatment of cultivated trees several features should be kept in mind: (1) that the fungus is perennial in affected branches and extends beyond the external evidences of the knots; (2) that conidiospores are matured when the young knots show the olive-green velvety coating; (3) the ascospores are matured early in the spring on old knots; (4) that spores are said to ripen on the brush pile.

The objective in surgical treatment is, first, to remove all of the perennial mycelium so that secondary knots cannot be formed from its extension, and, second, to remove and destroy the knots previous to the liberation of spores in order to prevent new infections.

If pruning is started in time, the sacrifice of the affected parts will not seriously thin out the branches. The cuts should be made several inches below the external evidence of the knots, and the prunings should be collected and burned. The knots should be removed when they show yellowish brown in the spring, that is, before the conidia have matured.

A second inspection should be made in the fall and any knots that have been overlooked should be cut out, but special emphasis is to be placed on thorough work in the spring. If the trees are gone over carefully twice previous to the formation of the conidia, few knots should escape. In regions in which outside sources of infection can be largely eliminated, pruning alone should serve to hold the disease in check or to eradicate it. The neglected orchard is a menace, but in many states the painstaking orchardist is protected by legislation for the suppression of the black knot, and inspectors should have no hesitation in enforcing such laws as do exist.

In environments in which surgical treatment is not effective the grower is left the alternative of spraying or of giving up the growth of susceptible trees. The latter has happened in a number of cases, but it has been shown that spraying will very greatly reduce infections. In no case should main reliance be placed on spraying, since at its best it must be considered supplemental to the surgical treatment.

The recommendations for spraying are based largely on the work of Lodeman (1894), who suggested the following applications of Bordeaux: (1) during late March or early April; (2) when buds are beginning to swell; (3) during the latter part of May, or when the new knots begin to show the velvety coating; (4) two or three subsequent applications at suitable intervals.

While this treatment did not give perfect control, the number of new knots in test plots was reduced from 2000 to 165. It should be realised that these recommendations rest on very meagre experimental results, and that carefully controlled tests are needed. It would seem that more certain knowledge as to the time when infections occur, coupled with the use of stickers as employed in modern spraying, should increase the effectiveness of the applications and perhaps make possible a reduction in their number.

Apple Scab

The disease of the apple known as scab manifests itself upon either leaves, blossoms or fruits and sometimes appears on the young or 1-year-old twigs. It causes a spotting, discolouration or distortion of both fruit and foliage, but it is never directly responsible for a rotting of the fruit. It is the fruit attacks which cause the serious and frequently enormous losses.

The disease is also referred to as apple scab, scurf, black spot, black-spot fungus, Tasmanian black spot, black-spot scab and rust. The last is probably derived from the German name, "Rostflecken," and should never be used, since there are true rusts of the apple. "Black spot" is the name most used in England, Australia and South Africa, while "scab" is the common name most frequently employed in America. It is apparently the production of scab-like spots on the fruit, which suggested the common name. The disease should not be confused with the scab diseases of other hosts, which are produced by entirely different organisms. Mention may be made of the common scab of potatoes due to bacteria, powdery scab of potatoes caused by a chytrid and citrus scab and peach scab, two entirely distinct fungous diseases.

History: Scab has been known to botanists since the first part of the nineteenth century, and the causal organism was first described and named by Fries in Sweden in 1819. It was first reported in Germany by Wallroth in 1833. Schweinitz gives the first authentic record of the disease in America, recording its presence in New York and Pennsylvania in 1834, while Curtis mentions it as common in North Carolina as early as 1867. The earliest report from England seems to have been that of Berkeley in 1855. Experiments on the control of the disease in America were started under the direction of Galloway of the US Department of Agriculture in 1886 and 1887, and by 1891 Bordeaux mixture was reported to be the most efficient fungicide.

Since that time the disease has been treated by numerous writers and a voluminous literature has accumulated. In 1914, Morris listed 505 different publications dealing with some phase of scab. Much of the published material added but little to our knowledge of the disease, the bulk of the work being empirical tests of various fungicides, or merely spraying demonstrations.

Several of the more important researches which contributed to the advancement of our knowledge were the work of Aderhold in Germany (1894 and later) and of Clinton in Illinois (1898). Both workers demonstrated the presence of an overwintering stage of the scab fungus on the fallen leaves, and made other additional contributions to the life history of the causal organism. Bordeaux continued to be the standard fungicide for scab for many years, but in 1908 Cordley first reported the successful use of lime sulphur, and this soon came into general use in Oregon arid Washington, while its value in the East was established by the work of Wallace in New York (1909-1911) and by Scott for the U. S. Department of Agriculture (1909-1911). Since that time lime sulphur has come into general use for the scab disease in the United States, except in regions where it causes serious injury, and the use of Bordeaux has been very largely discontinued.

Geographic Distribution: Apple scab is prevalent in severe form in all the European countries where the apple is grown and has been sufficiently severe in Australia, New Zealand and Soutii Africa to demand control measures. It is probably of foreign origin and occurred on the wild parents of the apple, accompanying that host ever since it has been under cultivation. It undoubtedly came to America from some European importations. At the present time scab is present in all the states from the Atlantic to the Pacific as well as in British Columbia and the eastern provinces of Canada.

The importance of the disease may be evidenced by the fact that nearly all of the Agricultural Experiment Stations in the northern half of the United States have issued one or more bulletins dealing with the character of the disease and its method of prevention. It is undoubtedly the most serious apple disease in the northern portion of the United States and southern Canada, while other diseases like bitter rot and blotch replace it somewhat in importance in the southern range.

In the earlier days of Nebraska orcharding, the scab disease of apples was absent and horticulturists at first believed that the mecca for apple growers had been found, but the disease soon appeared and developed with its customary virulence. None of the important apple sections east of the Rockies are now free from the disease, the region of severest infections including eastern Nebraska, Iowa, northern Illinois, Indiana, Ohio, southern Minnesota, Wisconsin, Michigan and Ontario, Pennsylvania, New York and the apple sections of the New England states as well as Nova Scotia.

The disease manifests unusual severity in the humid coast regions of the Pacific Northwest, including Oregon, Washington and British Columbia but is somewhat more variable in its occurrence in the Inland Empire country of Washington and Idaho arid has almost disappeared from this section during recent years. It is severe in the famous Bitter Root Valley of Montana at an elevation of 3000 to 4000 feet. Scab also finds favourable conditions for its development in the mountain sections of the southern states, especially Virginia and Arkansas. There are but few important apple-growing regions in the United States in which scab is either absent entirely or present in such slight amounts as to render control measures unnecessary. The most important of these favoured regions are the famous Yakima and Wenatchee valleys of central Washington. In both of these regions, the growers have not yet found it necessary to spray for scab. Cultural practices and climatic conditions in these irrigated valleys have been unfavourable for the spread of the scab disease, and it seems probable that the trouble there will never become an important factor in apple production.

Symptoms: Scab upon the leaves may show definite, more or less circular spots which are brownish or gray in the early part of the season, but later, at least under favourable conditions, may become even olive green or nearly black.

These spots when young frequently show the radiating dendritic ramifications of the scab fungus, easily visible with a hand lens. Later in the season the scab spots on leaves not affected sufficiently to fall from the tree may show considerable thickening of the leaf tissue, the brown, velvety surface becomes smoother and dull and occasionally dead areas drop out leaving irregular perforations.

Under certain conditions the fungus spreads diffusely over the surfaces of the leaves, causing a rather uniform brown colouration. When the scab spots are abundant or when the diffuse form is severe there may be more or less reduction in the size of the leaves, with curling and distortion. In the severe diffuse type much leaf tissue is killed and the affected leaves or portions of them may appear as if scorched or burned. In some cases, an entirely different symptomology may be noted, the affected trees often showing much yellow foliage.

The leaves may be gradually shed and a partial or almost complete defoliation may result. Some of the affected leaves will be of a uniform yellow colour, while others may be more or less variegated with yellow, brown and green. The scab fungus may be found on the isolated green spots, the affected spots showing a conservation of the chlorophyll. The diseased areas may appear on either surface of the leaves, but the earliest infections are most frequently upon the lower surfaces.

The scab lesions on the fruits first show as small, more or less raised, brown or black, somewhat circular spots with no break in the skin. As they increase in size the cuticle ruptures and the exposed surface shows a brown or almost black velvet-like appearance surrounded by an irregular whitish band, which represents the frayed and torn cuticle.

In this condition, which is generally reached by early July, the surface of the fruit lesions has much the same appearance as the leaf spots. As the fruit develops, the spots increase slowly in size and the central and older part may become bare and brown and show a more corky character, while the dark peripheral zone and the band of cuticle still persist.

Still later in the season the cuticular band may disappear entirely, and the whole scab lesion becomes slightly raised, and shows a rough corky surface, not at all resembling the early or midseason appearance. Scab spots may be few in number or they may be sufficiently numerous to coalesce and form extended lesions. Severe infections are likely to cause more or less cracking of the fruit.

The above is the normal or typical development of the scab spots. In late-autumn infections they are generally darker, being dense and black, and do not break through the cuticle as soon as the early infections. These autumnal lesions seldom reach the stage of development when the centre becomes bare and brown, and they are frequently much smaller than the spring infections. For this reason, the spots of this type are sometimes designated as "pin-point scab".

Apples which are apparently free from scab when they go into storage sometimes develop infections of the pin-point type. These infections may remain as covered black spots or in some cases break through the cuticle and expand somewhat. *Storage scab* may result from incipient infections that were too small to be seen at picking time, or there may be new infections during storage from conidia originating from some of the older scab spots. Scab lesions may occur on the blossoms during favourable years.

These appear generally on the pedicels, calyx or the young fruits about the time the petals fall. There may be single small lesions or the diffuse type of infections may predominate.

Scab spots may appear on the twigs, being confined to the bark of 1-year-old shoots. Twig infections were formerly considered rare for the apple in America, but they have been observed in Maine, New York, New Jersey, Nova Scotia, Washington and Montana. Young lesions on the twigs are very similar in appearance to those on the fruits, showing the central spore surface bordered by the uplifted epidermis. Later in the season, the spores may disappear and the bark show a more scaly character, due to the peeling of the bark in flakes. This type of injury is more common on pear twigs than on apple and has been called "grind or scurf" by German writers. The twig lesions may be few and scattered and thus may be easily overlooked, or they may be so numerous as to coalesce and produce more extended affected areas.

Effects: Scab offers an illustration of a disease which may cause a serious loss without endangering the life of the tree. This is due to a lowering of both quality and quantity of the commercial product, the fruit. The injury may be current, that is, affecting the immediate production, or it may be in part delayed till the season following a serious epiphytotic.

As a result of the foliage attacks there may be more or less defoliation. Many leaves showing severe scab of the diffuse type or numerous distinct lesions fall from the trees in early summer. Many of the affected leaves, however, persist through the growing season, but they are only partially effective in carrying out their normal work of food manufacture, since their producing power is lessened in proportion to the amount of tissue (surface area) killed by the scab fungus. Severe diffuse forms of scab may kill extended areas of leaf tissue and in epidemic years many leaves appear as if scorched or burned. Severe foliage infections must, therefore, have a pronounced devitalising effect by lessening the available carbohydrate food, and influence to some extent "the formation of fruit buds for the following year and hinder the normal wood growth which is the basis for future crops".

It is by direct attacks on blossoms or fruit that scab takes its largest toll. Severe infections on blossoms, fruit pedicels or young fruits cause a complete blighting or dropping of the fruits while still very small.

Since this phase of scab is more obscure than the others, and thus escapes their notice, orchardists frequently attribute this failure to set fruit to adverse climatic conditions, such as frosts or cold rains during blossoming time. Authentic cases are on record in which the apple crop was a complete failure due to the destruction of blossoms and young fruit by scab, but such a wholesale destruction is not very common even in unsprayed orchards. In ordinary scab years, however, the standard spraying practices increase the set of fruit, showing that the disease is nearly always causing some blighting of blossoms and young fruit.

Fruits that escape the early severe attacks may persist on the trees through the growing season, though exhibiting one to numerous lesions. According to the degree of infection, the affected fruits may be reduced in size or variously deformed, but even though normal size may be attained the presence of the scab lesions is a disfiguring feature which cannot be overlooked. In the fall, there may be a premature dropping of scabby fruit, but it is difficult to say just how much of the dropping is due to scab.

Fruits which show large lesions, or numerous ones which coalesce, frequently develop rather deep cracks or fissures extending into the pulp. These are especially pronounced in such susceptible varieties as the Early Harvest and McIntosh. All infected fruits must be sorted out and placed in a lower grade and sold for less than the clean fruit.

The effect of scab on the keeping qualities of the fruit must not be overlooked. Scabby fruit shrivels in storage faster than normal fruit, since the breaks in the skin caused by the fungous lesions increase the water loss. The losses from rotting are increased by scab, since the lesions offer easy avenues for the entrance of various rot-producing fungi. Cases are on record in which the development of scab in storage has caused heavy loss, since it made necessary a resorting and packing. The careful culling out of infected fruits at packing time is an added burden and expense to the grower.

Losses from Scab: The aggregate loss from scab each year is enormous and even on the most conservative estimates runs into millions of dollars.

The loss to be charged to scab is due to; first, the reduction in the total yield of the country; second, the lowering of quality and consequently a reduction in the financial returns to the orchardists; and, third, to the increased cost of production due to the necessity for carrying out control measures and the increased labour of sorting and packing.

Spraying is not a perfect control of the disease and besides there are many small home orchards in which spraying is not practised. It has been estimated that there is an average annual loss of more than $40,000,000 as a result of failure to spray apples. Of course, this should not all be charged to scab, since other troubles like blotch, bitter rot, rust, etc. are also controllable by spraying.

A number of specific instances will serve to emphasise the loss from scab and show the necessity for practising the known control measures. It is estimated that the average annual loss from scab in New York State is not less than $3,000,000; that the damage done in Illinois amounted to $6,000,000 in a single year; that there was a loss of 15 per cent of the crop in Montana in 1911 and 1912; while $48.50 per acre is given as the average annual loss in Australia. A loss of $47 per acre has been estimated as a result of failure to spray for scab in New York State, while Minnesota records place the loss for that state at $51.40 per acre.

Etiology: Scab is due to an ascomycetous fungus, *Venturia* *incequalis* (Cke.) Wint., which lives as a parasite on the leaves, twigs and fruits of the apple (the imperfect or Fusicladium stage), and completes its life cycle as a saprophyte (the perfect, perithecial or Venturia stage) upon the dead and fallen leaves of the same host.

All of the publications on scab in America dealt entirely with the parasitic or Fusicladium stage *(Fusicladium dendriticum)* until the work of Clinton in 1901, when attention was directed to the saprophytic stage on the overwintering leaves, and the connection between the two definitely established.

The connection had long been suspected (as early as 1887), but definite proof of the relationship was first published in the exhaustive researches of Aderhold in 1894 and later. The earlier work of the German botanists was at first overlooked and the conclusions of Clinton were arrived at without a knowledge of their discoveries. The work of Clinton and Aderhold has been confirmed and extended by other observers in this country and in Europe.

Some of the most important facts which have supported the relationship of the two forms are as follows:

1. The constant association of the perithecia with leaf lesions known to represent the parasitic stage during the previous season.
2. The development of the Venturia mycelium within the fallen leaves directly from the stromata of the Fusicladium stage.
3. The typical Fusicladium stage resulting from artificial inoculations made with ascospores.
4. Production of typical Fusicladium mycelium and characteristic summer spores of the scab from cultures started from ascospores on artificial media.
5. Production of perithecia of Venturia in artificial cultures originating from the Fusicladium stage.
6. The relation of the first spring infections of scab to the period of expulsion of the ascospores.

The mycelium on the leaves is at first nearly hyaline, subcuticular and develops radially with a characteristic dendritic branching which suggests the specific name applied to the conidial stage.

This mycelium later develops one or more layers of closely aggregated cells, forming a stroma-like growth which breaks the cuticle and permits the production of numerous erect conidiophores exposed to the surface. The older mycelium and conidiophores are brownish and the numerous spores and conidiophores give the scab spots a powdery or velvety appearance and an olivaceous or almost black colour.

On the fruit, the strornatic cushions of mycelium frequently completely destroy the epidermal cells and extend deeper into the pulp tissue. The conidiophores are short, erect, straight when young, but more or less flexuous with age, olivaceous and continuous or one- or two-septate.

The conidia are lanceolate or ovate, with the base generally somewhat truncate, continuous or more rarely one-septate and are produced in indefinite numbers from each conidiophore. The conidia are "resistant to detachment from the conidiophores when dry but quickly become detached in water." For this reason, their chief mode of dissemination is meteoric water.

While the conidial or Fusicladium stage is produced on all parts of the host, the ascigerous or perithecial stage is formed only on the fallen overwintering leaves. The mycelium develops as a saprophyte on the dead fallen leaves, and penetrates into the mesophyll from the superficial stromatic layers.

The perithecia appear mostly on the under surface as minute black dots or pimples barely visible to the naked eye, but are really spherical or subspherical bodies 90 to 1CKV in diameter embedded in the leaf tissue and protruding on the surface as minute dome-shaped pimples. The mature perithecium is ostiolate and the short beak is generally armed with several short bristles; the wall is dark and composed of several layers of cells; and each produces an indefinite number of oblong spatulate asci or spore sacs. The asci are eight-spored, the ascospores two-celled, olive brown, with the upper cell wider than the lower cell. Paraphyses are lacking.

The perithecia on the fallen overwintered leaves begin their development in the fall and early winter, and mature ascospores are ready to be expelled by the time the petals show pink (often earlier). The spores are forcibly expelled following the warm spring rains. Each ascus elongates, protrudes its tip through the ostiole, bursts with the forcible expulsion of its spore content and then the process is repeated by other asci until all the asci are discharged.

The period of ascospore expulsion varies in different localities. On the Pacific Coast, the first expulsion occurs in late February or early in March and discharge continues for about 3 months before all the perithecia are exhausted. This means that ascospores are expelled before the foliage has even started. In the eastern United States (New York) ascospore expulsion does not begin so early and according to available data does not continue for much longer than 1 month. Recent studies of ascospore discharge in Delaware (Adams, 1925) and in Virginia (Schneiderhan and Fromme, 1924) have shown conditions more comparable to those of the Pacific Coast.

Spore discharge begins very soon after the beginning of rainfall. In Virginia, in 1922, 16 discharge periods were recorded, beginning on Apr. 18 and ending June 12, while in 1923 there was a total of 13 discharges from Apr. 28 to July 30. The studies in Delaware emphasise the great variation in the time of initiation of spore discharges and the length of the period during which discharge continues. For example, in one locality discharge began about May 19 and continued for little more than a month, while in a test made in another county during the same season the discharge began on Mar. 28 and continued into the month of June.

Studies in Wisconsin have shown, in general, a shorter

duration of spore discharge more comparable to the New York experiences: At Madison beginning Apr. 23 and ending June 12, while at Sturgeon Bay the first ascospores were discharged on May 20 and the last on June 30 (Keitt, 1921). The time of ascospore discharge has a bearing on the success of spraying. Recent studies have emphasised that it is difficult to time scab sprays correctly.

According to Adams (1925), although conditions favouring spore discharge are beyond one's control, they can be predicted to a certain extent. "However, it is not possible or practical to attempt spraying on the prediction of weather conditions that may favour all periods of possible spore discharge. In other localities (Young and May, 1927) very satisfactory results have been obtained by timing the sprays in accordance with the reports of the local weather service.

Ascospore expulsion is controlled by moisture and temperature conditions, and consequently we may expect variations in different regions. While it is a common thing to find abundant discharge of spores about the time the blossoms show pink, activity may begin earlier or it may even be delayed until after the blossoms have opened. In regions or seasons of early spore expulsion, scab is likely to be severe, while a delayed expulsion may lessen the amount of the disease that develops.

The expelled ascospores are carried upward by wind or air currents and reach the young leaves or blossom buds and produce the first infections after a period of incubation. These are the primary infections and they soon give rise to conidia which may be scattered to other host parts and give rise to secondary infections.

The primary lesions from the ascospores furnish conidia which may cause very abundant secondary infections on fruit and leaves. Primary or secondary lesions may furnish conidia throughout the growing season and new infections may result whenever moisture and temperature conditions are favourable, but it should be understood that young leaves and fruits are much more readily infected than older and more mature structures.

Late infections may develop during some seasons and they are likely in those regions of abundant rain or fog and moderate temperatures. Apples that were free from scab at the time of picking but were taken from scabby orchards may develop infections after they go into storage. It seems probable that infection occurred in such cases in the orchard, and that the incubation period had not expired at the time of picking.

The life history of the scab fungus may be summarised as follows:

> Scab infected leaves fall to the ground in the autumn, and the overwintering structures, the *perithecia,* develop within the leaves from a saprophytic mycelium which originates from the superficial stromatic cushions. The perithecia mature their ascospores in the early spring, and these are forcibly expelled following the warm spring rains and produce the first or primary lesions which soon produce conidia; these conidia are scattered by wind or rain and produce secondary infections, and these secondary lesions are, in turn, the source of other conidia, from which still other infections may arise.

The part which twig infections play in carrying the fungus through the winter period is rather uncertain. There is some evidence that in the milder climates twig lesions may resume spore production in the spring, but all evidence points to the ascospores as the important source of the primary spring infections in most scab-infested regions.

Predisposing Factors: The abundance and severity of scab in any locality are dependent on the proper combination of climatic factors, especially temperature and moisture, which influence the development and expulsion of ascospores, the primary infections by ascospores and secondary infections by conidia. The temperatures of January, February and March are believed by some to be more important to ascospore development than the amount of precipitation in the period immediately preceding the maturity of the ascospores. Germination of ascospores or conidia may take place from 0.5 to 32°C, with infection resulting at temperatures ranging from 6 to 26°C. Continuous wetting of surfaces for a certain length of time is necessary for infection, the period of wetting required being longer for the low than for the higher temperatures.

During the periods of low temperatures, it seems that the lengths of time that susceptible surfaces are held moist will be more likely to be those required for infection than when the temperatures are relatively high. In general, it can be said that cool, moist weather is favourable, while higher temperatures are unfavourable even though moist conditions prevail. Any time after primary or secondary sources of infection are available, a cool rain period followed by conditions which favour the retention of moisture is very favourable to infection. Showers followed by winds are unfavourable, since the moisture evaporates rapidly, while showers in the evening followed by a calm night are very conducive to infection. The relatively high spring temperatures and bright sunshine of some of our important apple districts, like the Yakima and Wenatchee valleys of Washington, are probably the principal factors which have excluded scab.

It should also be noted that poorly pruned trees with dense branching and heavy foliage generally suffer more than well-cared-for orchards with open growth which permits a good circulation of air and rapid evaporation of moisture. Trees in low pockets are more subject to scab than those on hill tops where the air drainage and evaporation are good.

Host Relations and Varietal Resistance: The conidial stage of *Venturia incequalis* occurs on many *Pyrus* species of the *Mains* group, but the scab of hawthorn (Cratsegus) and Sorbus represent distinct species which cannot infect apple varieties. The scab of pears *(Venturia pyrina)* agrees, in general, in its life history with the scab of apples, but here also cross-infections are impossible and the fungus represents a distinct species. The much more frequent occurrence of twig lesions in pear scab is one of the striking differences.

No varieties of apples are immune to scab, but great variation in susceptibility is shown. It is found that certain varieties may be resistant 1 year and heavily scabbed another year, under conditions favourable for scab. It is also to be noted that varieties noted as resistant in one locality may prove very susceptible in another region.

Various writers have listed varieties as resistant, moderately resistant, moderately susceptible or susceptible, but it may be noted that the Baldwin, which is generally classed as resistant, has shown as high as 98 per cent scab. No attempt will be made to explain this and many other similar variations. The general reputation of a variety can, however, be taken into consideration in making selections for commercial orchards. In the light of recent studies on other parasites and work of Wiesmann (1931) it seems probable that physiological strains may explain the variable reports on resistance. Five strains of apple scab and four of pear scab are reported differing in cultural and morphological characters and possibly in pathogenicity.

Control: While the use of fungicides constitutes the principal means of protecting orchards from scab, there are certain other practices that should receive attention.

Planting Practices: Since scab sprays are applied with reference to time of blossoming, it is important that commercial plantings should consist of solid blocks of certain varieties rather than of mixed plantings. Even in the small home orchard the mixed planting of varieties makes timely spraying doubly difficult. A recognition of the predisposing influence of pockets or lowlands with poor air drainage would suggest the selection of orchard sites to avoid this factor.

Sanitary Measures: Observations in many sections agree in reporting scab at its worst in neglected orchards. Failure to keep the trees properly pruned makes a dense growth of branches and heavy foliage, which excludes sunshine and retards evaporation, so that spore germination, and consequently infection, are favoured. If there were no other reasons for pruning, the effect of failure to do so upon scab alone would be sufficient argument for attention to this orchard practice.

Our knowledge of the part which hibernation of the scab fungus on fallen leaves plays in the life history of the disease suggests that their destruction would lessen the primary infections. Clean cultivation and close mowing of cover crops or weeds, either to bury the leaves or to give less favourable conditions for the development of perithecia and ascospore dissemination, are to be looked upon as measures to supplement the use of fungicides.

Use of Fungicides: The carrier for the toxic chemical may be either water (spraying) or air (dusting). Spraying generally gives better control and is more generally practised. The time of application of the fungicide may be based on the stage of growth of the trees or upon the maturing of the ascospores in connection with reports by a local weather service if such spray warnings are issued.

Conditions are so variable in different regions that the advice of local authorities should be followed as to the spray programme to be adopted and the fungicide to be used (also the insecticide).

The following are the more important sprays which have been used:

1. Bordeaux, at strengths varying from 4-4-50 to 2-10-50.
2. Liquid lime sulphur, 1 gallon to 40 to 50 gallons of water.
3. Dry lime sulphur, 3-50 or 4-50.
4. Dry-mix lime sulphur.
5. Calcium sulphide, 8-50 to 12-50 (Hurt and Schneiderhan, 1929).

The dust mixtures which have given most general satisfaction are:

1. Finely powdered or dusting sulphur 85 to 90 per cent plus 15 or 10 per cent fluffy.
2. Dusting sulphur 40 per cent, lead arsenate 20 per cent and 40 per cent hydrated lime, the latter used as a filler.
3. Hydrated lime 86 per cent, dehydrated copper sulphate 10 per cent and calcium arsenate 4 per cent (Sanders' dust).

The period up to 1908-1911 when lime sulphur first came into use has been designated the Bordeaux and Paris-green period in apple spraying, since this combination of fungicide and insecticide was generally used. Repeated experiments continuing up to the present have shown the superior fungicidal value of Bordeaux over lime sulphur for scab control, but because of the less injury to foliage and fruit, lime sulphur has come to be generally used especially in America, except in regions where lime-sulphur injury is extreme or where certain other diseases must be controlled for which lime sulphur is not so effective.

In earlier tests dry lime sulphur gave poorer control than liquid lime sulphur, but in more recent trials it has given protection equal to the liquid lime sulphur (Britton *et al.*, 1921; Doran and Osmun, 1924; Keitt and Jones, 1926; Ballou and Lewis, 1927).

Dry-mix lime sulphur and calcium sulphide have yielded results equal to any other sulphur-containing sprays and with less injury (Hurt and Schneiderhan, 1929). Floatation sulphur has given adequate control. According to Button (1930), dry-mix lime sulphur, wettable sulphur, colloidal sulphur, activated sulphur and sulphocide have proved inferior to liquid lime sulphur. Proprietary remedies such as Adheso, B.T.S., Spra-sulphur, Fungi-Bordo and Soluble Sulphur are not to be recommended, either because of injury to fruit and foliage or because of greater cost.

Earlier tests of the dusting method gave very poor results, but since the employment of finely divided dusts and improved dusters, this method of control has given results in some sections that have justified its use. It is now generally conceded that dusting is somewhat inferior to spraying (Folsom, 1926; Giddings *et al.*, 1927; Moore, 1930; Hamilton, 1931 Spraying of fallen leaves to prevent the formation of ascospores has been tried with promising results. Curtis (1924) obtained as much as 40 per cent reduction from three dead-leaf sprays. The use of calcium arsenite for this purpose has also been suggested.

Does it Pay to Spray for Scab: Spraying must be looked upon as an insurance, and growers must decide from results whether it is a paying proposition. There are at least three types of localities, so far as scab control is concerned:

1. Those in which scab is unknown or relatively rare. In such localities spraying would never be profitable.
2. Regions in which severe scab years are infrequent. There are places where scab may be severe one season out of four or five. In such cases the actual money value of yearly spraying is doubtful.
3. Regions in which scab-free years are uncertain and relatively infrequent. In such localities the successful orchardist must spray each year. Careful cost and production records have shown net profits, due to spraying, of $65 to $400 per acre.

Diseases Due to Chytrids: Chytridiales

The species belonging to this order are mostly obligate parasites of very primitive character which live on other microscopic organisms, such as protozoa, rotifers, algae, water moulds and some other fungi, pollen grains that fall into the water and a smaller number which live in the cells of seed plants. Many of the seed-plant hosts are aquatic forms, but a few are terrestrial plants of economic importance and suffer serious disturbances from the attacks of these minute forms. The name of pond-scum parasites is sometimes applied to the order, because of the large number of species which attack freshwater algae.

General Characters: The fungous body consists of a single cell or a single cell with a few root-like or hypha-like outgrowths. In a few genera a plasmatic body of delicate fibrils penetrates from one cell of the host to another, to form a plasmacomplex, or a branched mycelium may be formed.

The plant body generally organises the reproductive stage or passes into a resting form after a short existence in the vegetative condition. The fungous body may be transformed directly into a zoosporangium, or zoosporangia may be formed from specialised portions of the plant body.

The zoosporangia are either thin walled and capable of organising zoospores or swarm spores at once, or they are thick-walled resting sporangia, which develop swarm spores only after a period of dormancy. The swarm spores are mostly uniciliate, occasionally biciliate and generally of definite form, although those of a few species have an amoeboid character.

In most species asexual reproduction by swarm spores predominates. In some forms two separate cells unite, the contents of one passing into the other, to form a zygote or zoosporangium, or in another case resting sporangia are formed by the conjugation of two swarm spores.

The Chytridiales are probably primitive forms which occupy a place close to the base of the fungous family tree, but the opposing view sees them as degenerates in which the power of sexual reproduction has generally been lost.

The order includes six families and many genera, of which the following are the most important as furnishing parasites of crop plants:

- *Plasmodiophoracese:* Plasmodiophora and Spongospora.
- *Olpidiacese:* Olpidium and Olpidiaster.
- *Synchytriacese:* Synchytrium.
- *Cladochytriacese:* Physoderma and Urophlyctis.

Plasmodiophora: The vegetative body within the host is a naked protoplast, at first uninucleate, but increases in size and becomes multinucleate within the host cells. These form free spherical, smooth-walled uninucleate resting spores, which when set free germinate to form amoeboid zoospores by which new infections result.

Spongospora: The vegetative body is a multinucleate naked, plasmodium-like body. The resting spores are grouped in sponge-like balls, consisting of many uninucleate cells. Each cell of the spore ball germinates to form a single, uninucleate amoeba and these fuse to form plasmodia-like bodies previous to penetration of a new host.

Olpidium: The vegetative body is a naked protoplast which later surrounds itself with a thin cell wall and becomes a zoosporangium.

It then forms a long tubular neck extending from the host cell to the outside through which the uniciliate swarm spores are set free. Resting sporangia with thicker walls may be formed, but these also germinate to form swarm spores.

Olpidiaster (Asterocystis): Very similar to Olpidium, but the zoosporangium not forming a neck. Resting sporangia showing a star-like folding of the wall, hence the old name Asterocystis.

Synchytrium: The vegetative body of the fungus is a large cell occupying an epidermal cell of its host which it nearly fills. The content-is generally coloured yellowish or yellowish red by oil drops. This vegetative cell, which is early surrounded by a cell wall, may be transformed into a resting sporangium or it may divide to form a group of sporangia, a sporangial sorus. The sporangia germinate to form uniciliate swarm spores. The invaded host cell is frequently enlarged and surrounding cells also hypertrophied to form galls, which are generally bright coloured.

Physoderma: Fungous body consisting of fine fibrils and enlarged portions ("Sammelzellen") occupying the host cells and frequently spreading, from cell to cell by means of the fibrillar hyphae. Coloured resting zoosporangia formed from the "Sammelzellen," and set free by the disintegration of the host cells. Zoosporangia germinate to form uniciliate swarm spores.

Urophlyctis: Fungous body of fine branched hyphae provided with enlargements; confined to a single hypertrophied host cell or spreading to many cells. Resting cells formed by proliferation from segmented turbinate cells and globular or ellipsoid-flattened, and towards one side provided with a circle of hyaline appendages (haustoria) which disappear with maturity. Parasitic on the subterranean or aerial organs of higher plants, and frequently forming conspicuous malformations or galls.

Club Root of Cabbage and Other Crucifers Plasmodiophora Brassicce

The cabbage and other species of the mustard family are frequently attacked by a disease which first produces swellings or distortions of the root, followed frequently by decline in vigour or by death of the affected plants.

The characteristic effect upon the root system has suggested such names as "club foot", "clump foot," "club root," "clubbing," "finger-and-toe disease," "anbury," "Kohlhernie" (German), "maladie digitoire" and "gros-pied" (French).

History and Geographic Distribution: The origin of the disease is uncertain, but it was the cause of concern in Scotland nearly 100 years before it attracted special attention in Russia, where it was so widespread and destructive in the region of St. Petersburg that the Russian Gardeners' Association offered a prize for its study in 1872. As a result Woroni began the study of the disease in 1873 and by 1876 had completed studies which were published in the fall in 1878. Previous work by Caspary had failed to establish the presence of a parasite as the cause of the disease, the true nature of the disease as due to a simple slime mould being first determined by Woronin.

Previous to the work of Woronin (1878) finger-and-toe was frequently discussed in garden literature between 1800-1860 in various English journals. As late as 1853-1854, Buckman claimed that the disease was due to a reversion to the original wild forms, while other writers held that soil and climatic conditions were responsible.

The disease has become world wide and attention has been given to it in England and various European countries, while it has been frequently reported in American disease literature. Recent important contributions to our knowledge of the disease have been made by Lutman (1913), Cunningham (1914), Chupp (1917), Kunkel (1918), Bremer (1924), Jones (1928), Cook and Schwartz (1930) and Wellman (1930).

The disease has given most concern because of its attacks on cabbage and turnips. In 1892, club root was common in New England and the Middle Atlantic states where it was considered one of the worst enemies of the market gardener. It spread southward and westward, reaching Virginia and the Carolinas on the south and Wisconsin, Illinois and Iowa on the west. It appeared in the truck section of the Puget Sound country in sufficient abundance to call forth a special bulletin in 1910. It is now known in 36 states and is reported as important in 21 (Wellman, 1930).

Symptoms and Effects: The disease may affect seedlings, which after 3 to 5 weeks show "flagging" or the leaves assume a pale-green or yellowish colour and the roots will show swellings ten to twelve times the diameter of normal ones. Seedlings that are infected early are usually killed before the season is half over, while later attacks are less serious.

The distorted roots fail to absorb nourishment from the soil and are often unable to transfer the plant food and water collected by the healthy roots to the cabbage head or other storage place. The plant starves, little or nothing being stored either in a head, as in the case of cabbage and such plants, or in a succulent taproot as in the case of turnips and radishes.

This functional failure of the root stunts the plant. The outer and older leaves become yellow and sickly, soon drop and, in the case of cabbage and cauliflower, the head is always small. At first they wilt only in the hottest weather or during midday and recover in the cool of the day, appearing perfectly normal the next morning, but as the disease advances, the outer leaves wither and fall one after the other until in severe attacks the whole plant dies, a result noticeable in July or August, but most severe in September (Cunningham).

The swollen distorted roots begin to decay in the soil towards the latter part of the growing season due to invasion by bacteria and soil-inhabiting fungi, and thus the injury is increased.

The injury to cabbage and similar crops caused by this disease can only be appreciated by those who have seen it destroy whole crops. Cabbage, cauliflower and radish suffer most here, but in Europe, where stock turnips or rutabagas are largely raised, these crops suffer severely. Certain Scottish lands were long ago abandoned for raising turnips because of its prevalence, and many American gardeners no longer raise cabbages for the same reason (Cunningham, 1914).

The type of hypertrophy of the root system varies in the different susceptible hosts. Six general types are recognised by Cunningham as follows:

1. *Complete Clubbing of Main and Lateral Roots:* Brassica oleracea.
2. *Clubs on Main Roots, Laterals Free:* Sisymbrium altissimum.
3. *Clubs on Lateral Roots, Main Root Free:* S. officinale and Erysimwn cheiranthoides.
4. *Clubs on Main and Lateral Roots with Club-free Rootlets above the Diseased Portions:* Lepidium sativum.
5. *Clubs as Tumours of the Root:* Raphanus sativus.
6. *Dark, Decomposing Spots on the Root:* R. sativus.

The hypertrophied roots in group one are very frequently somewhat elongated or fusiform, and thus the malformed root system is of such a nature as to be well characterised by the descriptive name of finger-and-toe disease. In turnips or rutabagas the swellings of the roots are very frequently globular and grouped mainly on the laterals, with the fleshy taproot more nearly normal, but in other cases turnips may be so badly clubbed as to present only a group of branched hypertrophied roots.

Attention should be directed to the occurrence of galls or hypertrophies on the root system of Crucifers and many other species, due to the eel worm or nematode *(Caconema radicicola).* These root knots may easily be mistaken for the effects of club root, as superficial appearances are frequently quite similar, but it is generally quite easy to determine the presence of the nematodes in the cortex of the hypertrophied tissues by microscopic examination.

Etiology: Club root was first shown by Woronin in 1878 to be caused by a species of slime mould to which he gave the name of *Plasmodiophora brassicce* more recently grouped with the Chytrids. The occasional occurrence of bacteria in the tissue of clubs has led to the theory that the disease was caused by bacteria working in symbiotic relationship with the plasmodia of the slime mould, but cultural tests from clubs have failed to substantiate this theory.

The bacteria are to be considered only as secondary invaders, which follow rather than accompany the causal organism. Although pure cultures of Plasmodiophora cannot be maintained, successful inoculations have repeatedly been made by the use of soil contaminated with the spores, or of soil filtrates containing a suspension of the spores and more recently by bringing together cabbage seedlings grown under aseptic conditions and spores obtained directly from the interior of young clubs. This behaviour, with the constant occurrence of plasmodia in the cells of affected roots, is sufficient proof of the active pathogenicity of *P. brassicce.*

According to recent investigations by Chupp (1917), no bacteria can be obtained by cultural methods from roots showing young swellings, but medium-sized swellings yielded a few colonies, and larger galls, especially those with broken epidermis, yielded numerous colonies.

It seems probable that, while the bacteria are not the primary invaders, they follow the slime mould, and by their activities bring about the disintegration or rotting of the diseased roots. In so doing they contribute to the injury and cause the spores which have developed within the roots to be set free into the surrounding soil.

The spores which are set free by the decay and disintegration of the diseased roots may germinate at once or after a period of rest. According to Chupp, 1 to 5 per cent will germinate in muck soil filtrate when taken from fresh roots, more will germinate if they have been exposed to a freezing temperature for 2 weeks or more, while drying also increases the germination if not too severe.

The optimum temperature for germination is 27 to 30°C, but germination and infection will take place at 16 to 21°C if susceptible seedlings are present. The spores are spherical or slightly hexagonal due to crowding, and somewhat variable in size. In the process of germination, the spore swells to be one-third larger, bulges on one side, the cell wall ruptures and the content escapes as a naked, uninucleate mass of protoplasm, provided with a single long cilium and a slowly contracting vacuole.

The details of form and behaviour of these myxamoabae or swarm spores as given by different observers are somewhat at variance. According to Woronin, they are long, spindle-shaped structures, of variable form, with the single cilium at the narrow end, and endowed with the power of locomotion by the lashing of the cilium and by the protrusion of pseudopodia in characteristic amoeboid fashion. As described by Chupp, they are 1.7 to 3.5μ in diameter, more or less pyriform, with a thick flagellum from the anterior or pointed end, never amoeboid, but with definite form, and move entirely by the action of the flagellum. These two different descriptions are explained by the ability of the zoospores to change form (Wellman, 1930).

The manner in which these myxamoabae or swarm spores bring about infection has been one of the disputed points in the etiology of the disease. Various observers have found amoebae or plasmodia containing two to six nuclei in the young infected tissue, but the exact behaviour between the uninucleate, free, swarm-spore stage and the several-nucleate plasmodium within the host cells seems to be somewhat obscure. Chupp claims to have found uninucleate amoebae within the root hairs of young roots, but Kunkel (1918) seems to be of the opinion that these were not the amoebae of Plasmodiophora.

More recently Cook and Schwartz (1930) report that the swarm spores first penetrate root hairs and form a small plasmodium with up to 30 nuclei, each of which organises a zoosporangium. These then germinate, forming 4 to 6 zoospores (gametes) much smaller than typical swarm spores. These gametes then migrate to the epidermal and cortical cells of the root and there fuse in pairs, and the resulting zygotes give rise to plasmodia of typical clubs. No evidence of fusion of zoospores was found by Wellman (1930). Whether the infecting amoebae are uninucleate or become several-nucleate before penetration into the host cells is relatively unimportant. There is no evidence of the fusion of myxamoebae previous to penetration, as in Spongospora, and nuclear division must begin very soon after the amoebae have entered the host cell if it does not begin earlier.

The idea had generally prevailed that infection could take place only through young tissue of roots. Kunkel (1918) has presented rather conclusive evidence that "old tissues are readily penetrated by the parasite and that root hairs are by no means necessary to infection." He was able to secure infections whether inoculations were made near to the root tips or at points far back from the regions producing the root hairs. Further, he was able to produce typical clubs by making inoculations on rather old cabbage stems.

When infection takes place through young roots, it is possible that the amoebae may penetrate through root hairs and other epidermal cells of the primary cortex and pass on into the deeper tissues, but it seems probable that this method of penetration is not common, but rather that infection takes place through portions of roots from which the primary cortex has already been lost.

Various workers have generally agreed that there are two methods by which the plasmodia may be spread throughout the affected root: first, by the division of cells in which two or more amoebae are present; and, second, by the migration of plasmodia from cell to cell.

First importance has generally been given to the former, with migration as secondary, but recently Kunkel has shown that, in the distribution of the parasite in the tissues, migration is more important than cell division. After the amoeba are within the cells they begin to increase in size and the nuclei divide, making multinucleate plasmodia which produce new plasmodia by division or separation.

In this way, a single cell may contain a considerable number of plasmodia. The amoebae penetrate into deeper-lying cells passing from one cell to another, at the same time migrating somewhat along the longitudinal axes of the root, and finally reach the cambium. They also penetrate into cells on the inner face of the cambium, but migrate more rapidly in the cambium and thus spread the infection lengthwise from the original centre of infection.

The plasmodia then migrate from the cambium into adjacent tissues of cortex or medullary rays. Carefully prepared sections will show plasmodia passing through the separating cell walls of adjacent cells. As a result of the presence of the parasite certain cells are greatly increased in size, and cell division is stimulated, the stimulus extending beyond the cells actually occupied by the parasite, and in this way a swollen or distorted root is produced.

A single, spindle-shaped club is to be considered as a morphological unit resulting from a single infection that occurred at some point near its middle, rather than as a result of multiple infections.

Finally, some cells remain permanently infected, that is, the plasmodia do not migrate from them. These may be single, in longitudinal chains or in groups called "Krankheitsherde." Affected cells may be crowded full of plasmodia, which are highly granular from the presence of reserve food material in the form of oil globules. Whether these plasmodia finally merge with each other does not seem to be quite clear, but by the time the parasite is ready to start spore formation the individual plasmodia are packed so closely together as to appear like larger fusion plasmodia. In the formation of spores, the plasmodia become filled with large and small vacuoles, the cytoplasm collects around the nuclei and the whole mass is thus cut up into uninucleate masses which round off and surround themselves with protecting walls. The mature spores are always uninucleate, filled with minute oil drops and contain one or more small vacuoles.

A single diseased root thus produces millions of spores which are set free into the soil by its decay, and these are capable of infecting succeeding crops on the same ground or the disease may be spread to new fields. There seems to be little evidence that the spores are wind borne, although this might seem theoretically possible. The migration of the motile swarm spores in the soil is also believed to be a minor factor in the spread of the disease. It may be spread by soil carried on farm implements, on the feet of animals, by earthworms (Gleisberg, 1922), through the use of contaminated fertilizer, by drainage water flowing from contaminated soils to healthy fields or by the use of infected seedlings.

Host Relations: The statement is generally made that all species and varieties belonging to the mustard family (Cruciferab) are susceptible to club root. Although some reports have been published of the occurrence of the disease on other than cruciferous hosts, these appear to have been due to confusing root swellings caused by nematodes (root knot) with the somewhat similar malformations caused by club root. Extensive tests of susceptibility were carried out by Cunningham (1914), using over 100 species belonging to 28 genera. In these tests extending through three seasons, the different species showed a variation from 100 per cent susceptible to 100 per cent immune.

The 11 species that remained free from club root are rather rare and very seldom cultivated. Practically all of our cultivated species showed a susceptibility to the disease: varieties of cabbage, cauliflower and brussels sprouts *(Brassica oleracea),* turnip *(B. campestris),* rutabaga *(B. rapa),* rape *(B. napus),* various mustards *(B. spp.),* radish *(Raphanus sativus),* pepper grass *(Lepidium spp.),* alyssum *(Alyssum spp.)* and various other less frequently cultivated species. More recently Naoumoff (1926) has reported the results of infections on 180 species in 49 genera, including the first report on species of the subfamily Thelypodieae.

Only a few of the more important cultivated species have been tested for comparative resistance of varieties. Of 13 varieties of cabbage tested by Cunningham in 1911, the four showing the most resistance were: Hollander, 26.5 per cent; Stone Mason, 14.4 per cent; Large Late Flat Dutch, 9.9 per cent; and Early Jersey Wakefield, 9.4 per cent free from clubs when grown on heavily contaminated soil. The most susceptible varieties, like Mammoth Red Rock and Perfection Savoy, gave 100 per cent clubbed. Thirteen varieties of radish tested in 1912 and 1913 showed Early Long Scarlet the most susceptible and Early Giant Stuttgart and Early Scarlet Turnip the most resistant. Per cent of clubbing varied from 1.8 per cent in Early Scarlet Turnip to 92.3 per cent in Early Long Scarlet.

Turnips and rutabagas were also tested, with the result that Sweet Gorman, White Swede, Early White Milan, Early Snowball and Purple Top Aberdeen can be counted as relatively resistant, while Southern Curled, Early Purple Top Strap-leaved and Improved Purple Top Strap-leaved may be classed as relatively susceptible. Turnips as a class are somewhat more susceptible than the rutabagas, although some varieties of turnips are more resistant than certain varieties of rutabagas.

Predisposing Factors: Club root is especially favoured by acid soils, the number of zoospores liberated reaching the highest numbers under such conditions. Spore germination and infection are not exclusively dependent upon the H-ion concentration, but no infection will ordinarily result above pH 7.2 to 7.4 (Chupp, 1928). A survey of 116 fields (Well-man, 1930) showing club root gave a range of pH 5 to 7.8. The excessive use of acid fertilizers or highly nitrogenous manures and the withdrawal of lime by the action of smoke gases in industrial centres may be expected to favour the development of club root (Bremer, 1924).

Spores germinate at a temperature range of 6 to 27°C with the maximum germination at 25°C. Club-root development occurs from 12 to 27°C, with the optimum from 18 to 25°C. "The optimum temperature for host-root development, 20°C, is distinctly lower than the optimum temperature for spore germination and disease development, 25°C" (Wellman, 1930).

The moisture factor seems to be more important than temperature, in its relation to infection. Experimental tests have shown that the disease does not develop where the moisture content of the soil is down to 45 to 50 per cent of its water-holding capacity but will occur when the moisture content is higher, reaching heavier infection as saturation is approached. Low-lying, poorly drained soil might then be expected to favour club-root, and well-drained soils to inhibit it. It has, however, been shown that infection of the host results in 18 hours' exposure to favourable moisture relations; consequently heavy, prolonged rains may offer conditions for infection even in the best-drained soils.

Preventive or Control Measures: The various practices bearing on prevention or control may be briefly enumerated:

1. *Sanitary practices:* There are designed to prevent the contamination of new areas. Diseased roots if fed to live stock should be thoroughly boiled before feeding, since the spores of the organism will survive passage through the digestive tract of animals, and might be carried to the fields with contaminated manure. When plants are grown in seed beds some method of sterilisation should be practised, especially if club root is known to be prevalent in the environment. Corrosive sublimate, 1 ounce to 10 gallons of water, applied five times to badly infested seed beds has given excellent protection (Chupp). Later trials by others have been varied in methods of application, but in general fair control has resulted (Clayton, 1926; Preston, 1928; Blunck, 1928). Good results from the use of uspulun have been reported (Bremer, 1923; Clayton, 1926; Preston, 1928) but in general, less satisfactory than with mercuric chloride. Control in seed was obtained for three seasons by watering with a 10 per cent solution of washing soda (Osterwalder, 1929). Consideration should also be given to the fact that soil from contaminated fields may be carried on cultivators and the feet of horses, and such transport guarded against as much as possible.
2. *Crop Rotation:* Cabbages generally grow best on certain types of bottom land or muck soil, and since this type of land is restricted in area in many localities, there is a tendency to grow cabbages on the same land for a period of years. Under such conditions club root is likely to increase in severity until the entire soil becomes heavily contaminated. Since it has been shown that the club-root organism can live in a soil for 3 or more (6) years, a comparatively long rotation should be adopted in handling contaminated soils. Four or five,

and preferably six years should intervene between cabbage crops and no cruciferous crop of any kind should be grown in the interim. Attention should also be given to the elimination of all weeds belonging to the mustard family.

3. *Use of Fertilizers:* Consideration should be given to kinds of fertilizers not to use as well as to those which may be used with profit. The immediate application of barnyard manure is favourable to the disease, hence this fertilizer should be applied during the season preceding a susceptible crop, but not to a cabbage or turnip crop itself. The same rule would apply if acid phosphate is to be used as a fertilizer. Since most soils in which club root becomes severe are distinctly acid, the use of lirrie has long been practised with marked success, and should certainly be adopted for contaminated soils on which rotations of sufficient duration are not practical. Lime in various forms has been used as raw-ground limestone, caustic lime, air-slacked lime and hydrated lime, the last giving the best results for field use, 1500 to 2000 pounds per acre for commercial control in badly contaminated soils (Wellman, 1930). Best results will be obtained by treatment several months previous to use of the ground. Cunningham reports an increase in yield of cabbage from 672 pounds per acre where no lime was applied to 23,082 pounds per acre when treated with lime. Calcium cyanamide, which is also of value as a fertilizer, has been recommended as a substitute for lime (Kindshoven, 1928).
4. *Drainage:* Since an excess of moisture in the soil is favourable to the development of club root, low-lying or waterlogged soils, whether acid or non-acid, should be thoroughly drained and the physical condition improved as much as possible, but drainage alone cannot be counted on to control the disease.

Powdery Scab of Potatoes

Powdery scab is a disease that attacks roots, stems, stolons and tubers, producing small hyperplastic galls on the first three structures and small, circular, scab-like lesions or, more rarely, cankers on the tubers. Only parts that are underground are infected. Secondary rots may enter the tubers through the scab lesions.

Various common names have been applied to the disease, such as corky end, corky scab, powdery scab, Spongospora scab and potato canker, but powdery scab has been most generally used, especially in America. The name is descriptive of the character of the lesions and serves to distinguish the trouble from the more widespread and common scab *(Actinomyces scabies).*

In Germany, it has been called "Kar-toffelraude," "Knollenbrand" and "Schorf" or "Grind" at various times; in Sweden it is known as "Skoro"; and in Ecuador as "Cara," which is equivalent to the English word "scab."

History: The disease was first described in 1841 as occurring in Germany. Berkeley discovered the disease in England in 1846 in connection with his studies of late blight, and Brunchorst studied it in Norway in 1886, but did not then recognise it as identical with the trouble known to be in other countries. While the disease was spread more or less throughout Great Britain and on the Continent, it did not attract much attention until many years later. It is worthy of note that Lagerheim found the powdery scab in 1891 in Ecuador, where it seemed to be well known to the natives.

In 1913, the disease was found in the United States on potatoes imported from both the Netherlands and Belgium, which would indicate that the disease was well established in those countries. Early in the spring of 1913 it was reported from a number of provinces of Canada by Giissow, and from Maine in 1913 and 1914 by Melhus. Since then, there have been sporadic occurrences of the disease in a number of the other northern states, from Maine to Washington and Oregon, and in one locality in Florida, but it has not spread in these localities to any great extent. The discovery of the powdery-scab organism in 1915 on potatoes from Peru lends additional support to the belief that the disease is endemic in South America. More recently the disease has appeared in Australia, New Zealand, Tasmania and Kenya Colony.

Following the discovery of the disease in Canada and northern Maine, a Federal quarantine was established to protect the other sections that were still free from the trouble.

This quarantine prohibited importations of potatoes except under a strict system of certification. A few years' study sufficed to show, however, that climatic barriers were more effective than the most rigid quarantines in preventing the spread of the disease throughout the United States, and the quarantine was lifted.

Symptoms and Effects: The disease appears first on very young tubers and is then evident as small, slightly raised pimples or swellings with a slight discolouration of the surface. The invaded tissue, when cut into, appears purplish. The swellings may occur in groups or patches, or they may be very well scattered over the surface of the tubers. With the progress of the disease, the swellings enlarge, and become somewhat more raised, resembling nodules.

Finally, the nodules or swellings break down, leaving a cavity filled with a mass of brownish powder surrounded by the lighter-coloured, frayed-out periderm or skin. These scab spots are usually circular or oval in shape, but generally less than the maximum size. They may be few in number or so numerous as to coalesce in groups and thus obscure their normal form. The circular form, the ragged margins of the periderm and the central powdery mass are characteristic features which seem to distinguish the disease from the common scab, in which the lesions are larger, more irregular in shape, sometimes raised, sometimes depressed, but always lacking the brown powder. The characteristic appearance of the lesions may be modified by friction in handling so that tubers examined in storage may have the top and the powder rubbed away, leaving only slightly depressed, empty lesions. Some lesions may not have matured sufficiently to rupture the epidermis and in many such cases the brown powder may be recognised by breaking away the external covering. It has recently been claimed by Shapovalov (1923) that under certain conditions the lesions of powdery scab become aborted and never advance beyond the simple condition found on young tubers, and that these aborted infections have been described as "skin spot" due to *Oospora pustulans.* This opinion concerning skin spot does not seem to have found acceptance by English workers.

In severe attacks, especially in moist soil, a distinctly warty appearance may develop, quite different from the ordinary type of powdery scab. These warts are several times larger than the scab pustules already described, and are usually smooth and roundish, although somewhat irregular, varying in colour from a light to a dark brown and more frequently occurring at the terminal or "seed" end of the tuber (Morse, 1914).

It is stated that this "warty condition of the tuber is not so evident after the tuber has been removed from the soil for some time, as the warts become flattened and discoloured, so that raised, more or less chocolate-coloured scars remain in their stead" (Home).

In the most severe form of powdery scab, the *cankerous stage,* there is actual destruction of the flesh of the tuber, leaving hollowed-out, eroded areas. These cankers appear to come beneath areas originally occupied by closely aggregated groups of sori of the typical form. The canker stage is generally due to growth in a wet, poorly drained soil, and has been produced experimentally. It is also favoured by an excess of lime. The canker stage has not been found in the United States, but it has been reported from Canada and is not uncommon in the British Islands.

The powdery scab affects other underground parts of the potato plant, producing white galls on roots, stolons and stems, These resemble bacterial nodules and vary in size from minute tubercles the size of a pinhead to others as large as garden peas. The extent to which they may be present may be illustrated by a single plant which showed 149 on the roots, 19 on nine stolons and 8 on three stems. The root galls do not appear to exercise any markedly injurious effect upon the growth of the plants, and may be present in certain cases when the tubers are entirely free from infection.

The injury from powdery scab varies with the type of the attack. In mild cases, it consists only of surface scabbing, which disfigures the crop and depreciates its market value, with but little reduction of real value as table stock.

The seed value of affected stock is greatly lowered, unless it is to be grown in areas climatically unsuited to the development of the disease. In such regions, very severely infected seed has given excellent results. In more severe cases of the common type as well as the canker stage, there may be heavy losses beginning in the field and extending into storage. It is reported that portions of some fields in Maine showed over 90 per cent of the hills affected, while there were several cases in which the infection ranged from 50 to 75 per cent.

The scab sori leave open wounds which permit more rapid loss of water, and serve as the avenues through which wound parasites may enter and cause rot or decay, while in certain cases the scab organism resumes activity and destroys cells adjoining the original lesion. Each of the factors may operate singly or in combination to produce effects which may be designated as powdery-scab dry rot. As a result of desiccation the tissues adjoining sori may become discoloured, shrivelled and shrunken. The renewed activity of the parasite destroys cells in the immediate vicinity of the sorus, producing a hard, dry spot. This is apparently due to a behaviour of the parasite similar to that which occurs in the development of the canker type, but of a milder form. The entrance of wound parasites causes the most destructive type of powdery-scab dry rot. The bottoms of the scab pits are protected by but little or no wound cork, so that penetration of fungi is relatively easy. In Maine, a species of Phoma, described as new *(P. tuberosa)*, was the most common of the wound parasites. The lesions caused by this parasite are sunken, dark, often hard and bony, vary from 2 millimetres to 5 centimetres in diameter, may penetrate to a depth of 2 to 4 centimetres and, when removed, leave a clean smooth cavity which has suggested the name of "button rot." Other rot-producing organisms, especially bacteria and various Fusarium species, may enter the Phoma lesions and complete the destruction with some confusion of symptoms.

Etiology: Powdery scab is caused by *Spongospora subterranea* (Wallr.) John., one of Plasmodiophoraceae. While the organism has been recognised as a parasite since the work of Wallroth in 1842, it was some years before its true character was understood.

It was first named *Erysibe subterranea*, and even as late as 1850 it was considered one of the smut fungi by Berkeley, as indicated by the name which he gave it, *Tuburcinia scabies*. In 1877, Fischer von Waldheim transferred the organism to Sorosporium, and it was not until the work of Brunchorst in Norway in 1886 that it was placed in the Myxomycetes. More recently the concepts of the relationship of the Plasmodiophoracese to the Myxomycetes have changed, so that the powdery-scab pathogene is now generally grouped with the chytrids.

The powdery content of the scab sori consists of numerous minute brown bodies known as spore balls, which are in reality collections of spores held together in more or less sponge-like, spherical or ovoid masses, varying in size. Under favourable conditions these spores of the ball (all or part of them) may germinate. The germination has been observed by Kunkel on artificial media and it seems probable that the spores will behave in a similar way in the soil or on the surface of young developing tubers. Each cell of the spore ball may produce a single, small, hyaline, uninucleate amoeba which generally escapes through an opening in the spore wall, leaving the spore ball intact, but in some cases the entire spore ball disintegrates, setting free as many amoebae as there were cells in the spore balls.

These amoebae are actively motile and soon crawl away from the spore ball from which they are produced. Under dry conditions, these amoebae round up and become surrounded by a thick, rough wall, or become encysted, in which state they are resistant to conditions which would be fatal to the naked amoebse. With the return of favourable conditions, these cysts or resting cells may germinate and set free the amoebae again. This behaviour will explain how the parasite may persist in the soil from year to year.

The exact manner of infection has been a disputed point, and until the work of Kunkel it was claimed by Osborn and others that single amoebse entered the host cells and later coalesced to form plasmodia. According to Kunkel, numerous amoebae external to the host coalesce to form a plasmodium which becomes the infecting body, which then passes down through and between the epidermal cells.

Usually a considerable number of cells are killed at the point where the plasmodium enters. Once beneath the epidermis, it spreads out in all directions and forms a rather flat, disk-shaped mass which separates the epidermis from the tissue beneath. In this way it comes to occupy a space between the uplifted epidermis and sound tissue beneath. Soon, however, a number of projections of pseudopodia begin to extend downward, push in between the cells of the sound tissue and seem to crowd them apart.

According to Wild (1929), the lenticels constitute the principal infection courts, rather than the unbroken skin of the tuber, with some penetration through wounds.

Small strands of protoplasm, the "infecting pseudopodia are pushed through the softened walls of the host cells, and in some manner become separated from the remainder of the plasmodium. Shortly after the cells become infected they enlarge and elongate radially to five to ten times normal size, forming giant cells which are responsible for the raised condition of young lesions. Ultimately the giant cells are cut up into smaller cells, which are all infected.

Finally each nucleus of the intracellular, multinucleate plasmodium organises a spore, and these become grouped to form the characteristic spore balls. There will then be left in the sorus a group of spore balls, mingled with fragments of old cell walls of host cells or other remains of host cells. It may be noted from this account that the first effect of primary invasion by the plasmodium is the stimulation of the young cells of the tuber to increased cell division, and that later these cells are killed by the plasmodia which become intracellular. In these primary infections, the action of the parasite seems to be limited to a small group of cells.

The plasmodium dry rot which always starts around old sori is due to secondary invasions by plasmodia formed from the germination of spore balls in the base of the sorus. In such cases, the old cells of the tuber with which the plasmodium comes in contact are not stimulated to a new growth, but are quickly killed. The plasmodium enters a cell, consuming its cytoplasm and nucleus and then passes on to the next, leaving behind it a mass of broken-down cell walls, starch grains and other debris. In this way the dry rot advances by the direct killing effect of the plasmodia, rather than by the production of toxic substances which diffuse into the cells in advance of the line of attack.

The source of an infection is likely to be due to (1) the planting of affected tubers; or (2) to planting clean tubers in a previously contaminated soil. The spore balls carried by infected seed tubers germinate in the soil under the same conditions as the seed, and plasmodia are produced which may cause the infections in the manner outlined above. Spore balls generally remain dormant during the winter, but in the spring many will germinate, and under favourable conditions the plasmodia are probably able to lead a saprophytic life if a susceptible host is not available, or under conditions of stress single amoeba) may become encysted as resting spores and so prolong the life of the parasite. It is uncertain how long the powdery-scab organism is able to live in the soil, but English authorities believe that they have evidence that a new crop may be infected after a lapse of 3 to 5 years.

Predisposing Factors: Careful study of the powdery-scab disease has shown that infection takes place only under favourable climatic conditions. The essentials for infection seem to be rainfall periods during the young stages of tuber formation, followed by cool, damp, cloudy weather.

If in addition the soil is poorly drained, the chances of infection are increased. The proper temperature, coupled with the right moisture conditions, must be provided for infections to take place. It seems that the favourable relations are found only in American sections near the Canadian boundary and farther to the north, with the possible exception of one locality in Florida where winter potatoes are grown.

It is worthy of note that heavily infected seed planted at fifteen different places on the Atlantic Coast from Massachusetts to Florida, and at six different points in Washington from Everett southward, yielded an absolutely clean crop. These and other experiences point to a very effective climatic barrier to the spread of the disease. An ideal soil for infection is said to be one with large pore spaces, a high humus content, a high methyl-pentosan content and large water-holding capacity. The incidence of the disease is not affected by a pH range of 5.9 to 7.6 (Wild, 1929).

Host Relations: For many years the potato was supposed to be the only host of *Spongospora subterranea*, but the discovery of galls on the roots, stolons and stems of the potato led Melhus to test the susceptibility of other species of Solanaces. Of 16 species planted in contaminated soil, seven developed infections, and it is significant that the tomato was one of the susceptible varieties. In no case were mature spore balls produced in any of the root galls, but this was attributed to the shortness of the season. *Solanum nigrum*, a common weed, remained free from infection.

There seem to be marked differences in the susceptibility of different potato varieties, but it is uncertain whether any are immune. In planting tests in Maine in 1915, four named varieties and seven seedlings remained free from infection, and some varieties showed very slight infections, while others were severely affected. Since the control plantings of the variety, Green Mountain, showed very wide fluctuations in the per cent of infection, it is believed that the variation in varietal response was not due entirely to resistance, but rather to the fact that the tubers of certain varieties escaped infection.

Control Practices: The following control practices may be emphasised:

1. Select potatoes free from the disease for seed purposes in regions climatically favourable for the disease.
2. Avoid contaminated land. In fields known to be infested, a long rotation (3 to 5 years or more) should be followed as a means of starving out the parasite.
3. If contaminated land must be used, heavy applications of sulphur up to 900 pounds per acre will very materially reduce the disease. Attention to drainage of the land may also be beneficial. Lime should not be used, as it increases the severity of the disease.
4. Infected seed or tubers suspected of being contaminated should be disinfected. No treatment of infected seed has given perfect control, but best results have been obtained by using hot formaldehyde, 2 pints to 30 gallons of water at 46 to 50°C for 5 minutes, or mercuric chloride, 4 ounces to 15 gallons water at 44 to 45°C for 5 minutes. These treatments have given better results than the standard long, cold treatments with either formaldehyde or mercuric chloride.

5. Attention should be given to all possible sanitary practices to guard against the introduction or spread of the disease. It should be borne in mind that the spore balls may be carried by contaminated bags or other articles which have come in contact with infected tubers or with contaminated soil. Farm implements or contaminated manure may harbour the organism. Infected tubers or parings may be boiled and fed to hogs.

Diseases Due to Downy Mildews and Allies Oomycetes

The parasitic fungi which form a non-septate mycelium and reproduce by the development of oospores, unless sexual spores are omitted, and by either swarm spores or conidia (the Oomycetes) may be placed in the three following orders:

1. The chytrids (Chytridiales) with a plant body consisting mostly of a single unbranched or only slightly branched cell. While this order includes many species which are parasitic upon algae, it also furnishes a number of very important pathogenes of our crop plants (see special treatment of Diseases Due to Chytrids, Chap. XVII).
2. The water moulds (Saprolegniales) with well-developed mycelium, principally saprophytic in habit. Many of the species may be found in the fresh waters of streams or lakes living upon dead insects, fish or plant remains, while others are common in the soil. A few forms are parasitic on freshwater algae, fish or other aquatic animals.

   Asexual reproduction is by biciliate swarm spores produced in specialised sporangia which are mostly persistent. Sexual reproduction is by antheridia and oogonia containing one to several naked oospheres or egg cells. The order comprises two families:

   The Leptornitacece including six genera, all species saprophytic. One species, *Leptomitus lacteus,* is noteworthy because of its development in drain pipes and sewer water.

The Saprolegniacece including fourteen genera, of which only one, Aphanomyces, is of importance as furnishing destructive plant parasites. Plectospira is of minor importance.

3. The pythiaceous fungi, white rusts and downy mildews *(Peronosporales)*, which are principally parasitic in habit, and for the most part obligate parasities. The chief distinguishing features of this order are:
   (a) a well-developed non-septate mycelium which is intercellular, with the exception of forms of Pythiacese, which are also intracellular;
   (b) asexual reproduction by *conidia (zoosporangia)*, which in the more primitive forms produce *swarm spores* and hence are sporangia, while in the highest types an infection hypha is the first product of germination; and
   (c) sexual reproduction, when this is not suppressed, by the union of unequal and dissimilar gametes to form *oospores*. A large female cell or *odgonium* produces either a uninucleate or multinucleate gamete, while the male cell or *antheridium* is smaller and gives rise to several male gametes or sperm nuclei. The three following families are recognised: *Pythiacece, Albuginacece* and *Peronosporacece.*

Pythiace

This family represents transitional forms between the water moulds (Saprolegniales) and the downy mildews (Peronosporacese). They have specialised in the development of zoosporangia. In the more primitive forms definitely differentiated conidiophores are lacking, the swarm sporangia being formed from ordinary hyphae, to which they remain attached. Aerial conidiophores, when formed, may be delicate or robust, and either simple or branched. Typical oospores may be formed, or in certain species they may be either rare or entirely suppressed. The important genera may be briefly characterised.

Pythium: Swarm sporangia on branches similar to the mycelium, spherical, oval, obpyriform, filamentous, etc. Sporangia germinate by a rupture or by a beak through which the protoplasmic contents are extruded into a thin-walled vesicle, after which they are differentiated into laterally biciliate (uniciliate according to some authorities) swarm Some species produce termibrancned mycelium.

Phytophthora: Sporangiophores simple or branched, generally emerging through the stomata of the host, but sometimes through epidermal cells. Sporangia at first terminal, but becoming lateral by the further growth of the branch, and thus produced in succession.

The germination may be pythiaceous, without the formation of a vesicle, or by the direct formation of an infection hypha. The genus was formerly included with the Peronosporacese, but certain species can be scarcely distinguished from Pythium. Phytophthora species differ from the downy mildews in producing an intracellular growth as well as intercellular, in not being obligate parasites and by the successive development of the sporangia. Oospores lacking or, when formed, of the typical oomycetous type, or in certain species "amphigenous," that is, with an antheridium in the form of a basal collar. (Phythiacystis, Kawakamia, Belpharospora and Pythiomorpha formerly described as distinct genera are now merged with Phytophthora.)

Trachysphsera: Sexual stage similar to Phytophthora with amphigenous antheridia; one or more globose echinulate conidia (sporangia) on sterigmata borne on an enlarged vesicle at the tip of the sporangiophore. Germination by germ tube; no swarm spores known.

Albuginaceje

General Characters: This family has received the common name of the white rusts because of the abundant production of whitish fruiting pustules which burst through the host epidermis in much the same way as the coloured pustules (reddish or black) or sori of the true rusts. The special features which characterise this group are:

1. The production of conidiophores in associated groups, or sori, beneath the host epidermis, which later break through and expose the powdery or dusty mass of spores (conidia—in reality, sporangia).
2. The unlimited production of conidia (sporangia) in chains.
3. The germination of both conidia and oospores by the production of zoospores or swarm spores.

Development of Mycelium and Asexual Reproductive Structures: First infection results in the development of a copious growth of internal, intercellular, non-septate mycelium, the hypha showing considerable irregularity in diameter.

The mycelium soon organises the characteristic groups of conidiophores which develop beneath the epidermis, raising it, to make whitish pustules or extended blister-like areas due to the merging of adjacent sori. As soon as the covering epidermis ruptures, the conidia or sporangia are set free and may germinate at once if favourable conditions are offered. These conidia are the so-called "summer spores" and serve for the rapid dissemination of the fungus during the growing season.

The conidiophores are short, basally branched, club-shaped structures and give rise to simple chains of spores. The number of spores produced is indefinite, and they are formed in basipetal succession, that is, the conidiophore forms a cross-wall or septum, cutting off that portion which is to become a spore. This gradually takes on the characteristic form, the conidiophore increases in length, a second spore is cut off and the process continues, resulting in the simple chains of multinucleate conidia, generally separated one from another by short neck-like projections. As conidial production continues, the older, terminal portions of the chains break, setting free the individual conidia.

The conidium does not normally germinate by the production of a hypha or germ tube, but becomes a *zoosporangium*. The protoplasmic contents divide into six or more portions, which emerge through an opening as naked protoplasts or swarm spores, each provided with two unequal lateral cilia or organs of locomotion. They are able to swim about actively, but soon come to rest, absorb their cilia, surround themselves with a cell wall and germinate by the formation of a germ tube. If this germination occurs on a susceptible host, the infection thread will penetrate to form an internal intercellular mycelium.

Formation of Oospores: The same mycelium which has produced son will a little later in the season organise the antheridia and oogonia and produce oospores which remain within the host tissue. These are the so-called "winter spores," and serve the important function of carrying the fungus over the winter period, as they will not germinate until the following spring.

The oogonia and antheridia are formed on the internal intercellular mycelium, each appearing on separate but adjacent hyphae. The oogonium is a large globular multinucleate cell, the antheridium a smaller, more or less globular cell. One or more antheridia come to occupy a position close to an oogonium. There are two different types of egg organisation within the oogonium. In certain species *(A. Candida)* the protoplast becomes differentiated into a peripheral or external zone, the *periplasm,* which contains many nuclei, and a central mass, the *egg cell,* or ooplasm, which contains a single nucleus. In other species of white rusts the central ooplasm remains multinucleate *(A. bliti* and *A. portulacce).*

The antheridium, which is a multinucleate cell, produces a short, tube-like outgrowth, the *fertilizing tube,* which penetrates the periplasm and comes in contact with the egg. The antheridial or male nuclei are discharged through this tube into the egg cell. In the uninucleate egg the female nucleus fuses with a single male nucleus, and in the multinucleate egg female and male nuclei fuse in pairs, this nuclear union constituting the process of fertilization.

Following fertilization the egg is gradually transformed into a thick-walled oospore. The periplasm is absorbed, the oospore wall darkens and thickens and develops characteristic external ridges, reticulations or knobs, while the interior of the oospore becomes filled with an abundance of reserve food in the form of oily or fatty globules. The fully developed oospore lies within the old empty oogonial cell.

The oospores are set free only by the weathering and decay of the host parts in which they were formed. They are resting spores and will -not germinate until they have been subjected to winter temperatures. Under favourable spring conditions they germinate by the production of swarm spores. The oospore wall splits and allows the internal, transparent, sac-like membrane containing the swarm spores to protrude.

This sac finally ruptures and the biciliate swarm spores escape. These swarm spores behave in the same manner as those formed from the conidia and will produce new infections under favourable conditions. Albugo is the only genus.

Peronosporacae

This group has received its common name of the downy mildews from the characteristic production in the typical species of aerial conidiophores in loose white tufts or downy aggregates on the surface of the substratum. The special features which characterise the group are:

1. The production of simple or branched conidiophores which generally emerge in small groups from the stomatal openings.
2. The limited production of conidia (sporangia), a single spore being developed from each branch of the compound sporophores, the maturing being simultaneous, rather than successive as in Phytophthora.
3. The germination of conidia either by the formation of

swarm spores or by direct development of infection threads.

4. Oospores germinate by direct production of a germ tube or infection thread.

The downy mildews generally produce only an internal, intercellular, non-septate mycelium, very similar to that of the white rusts. The aerial conidiophores constitute the most distinguishing feature, and clearly separate the downy mildews from the closely related white rusts.

Following an infection there is the development of the characteristic internal mycelium which frequently becomes aggregated in the substomatal chambers. One or several aerial conidiophores grow out through the stomata, and become branched sporophores bearing more or less pear-shaped conidia (sporangia) singly, but never in chains as in the white rusts. The conidia germinate by the formation of an infection thread or germ tube in Peronospora and Bremia species, but in the other genera swarm spores are produced *(Bremia lactucce* may also form swarm spores).

These are either set free through a germ pore as fully matured swarm spores, or escape surrounded by a delicate vesicle within which they complete their development. Oospore formation is essentially similar to that described for the white rusts. A deviation may be noted in Sclerospora, in which the oospore completely fills the oogonium and becomes closely adherent with the oogonal wall. Oospores which have been observed to germinate give rise to a germ tube instead of forming swarm spores (except in Sclerospora), or produce a promycelium which forms a single zoosporangium.

The following are the most important genera:

Sclerospora: Mycelium intercellular, with vesicular haustoria; conidial stage inconspicuous in most species; conidiophores stocky and solitary or in groups of two or three; oospores generally conspicuous, permanently united with the walls of the oogonia. Oospore germinating by swarm spores.

Plasmopara: Sporangiophores slender, tree-like in form, solitary or fasciculate from the stomata of the host, monopodially branched, more or less at right angles, the ultimate branches obtuse. Germination by swarm spores, or the entire protoplasmic mass may escape and then send out a germ tube.

Peronoplasmopara: Sporangiophores as in Plasmopara, but pseudo-monopodially branched, the branches forming more or less acute angles, with the ultimate portions acute. Germination of sporangia by swarm spores.

Bremia: Sporangiophores dichotomously branched, the branches terminated by disks or swellings which give rise to short radiately arranged, conidium-bearing branches. Germination of conidia direct or sometimes by swarm spores.

Peronospora: Conidiophores dichotomously branched, at acute angles, the ultimate branches acute. Germination of conidia direct by lateral germ tubes.

4

Morphology and Anatomy of Fungi

Fungal Morphology and Anatomy

Agaric meaning 'tree-fungus': from Latin agaricum. An agaric is a type of fungal fruiting body characterised by the presence of a pileus that is clearly differentiated from the stipe, with lamellae (gills) on the underside of the pileus. "Agaric" can also refer to a basidiomycete species characterised by an agaric-type fruiting body.

Most species of agarics are classified in the Agaricales, however, this type of fruiting body is thought to have evolved several times independently, hence the Russulales, Boletales, Hymenochaetales, and several other groups of basidiomycetes also contain agaric species.

Older systems of classification place all agarics in the Agaricales, and the some (mostly older) sources still use "agarics" as a common name for the Agaricales. Contemporary sources now tend to use the term *euagarics* when referring only to members of the Agaricales. "Agaric" is also sometimes used as a common name for members of the genus *Agaricus*, as well as for members of other genera, for example, *Amanita muscaria* is sometimes called "fly agaric".

Amadou

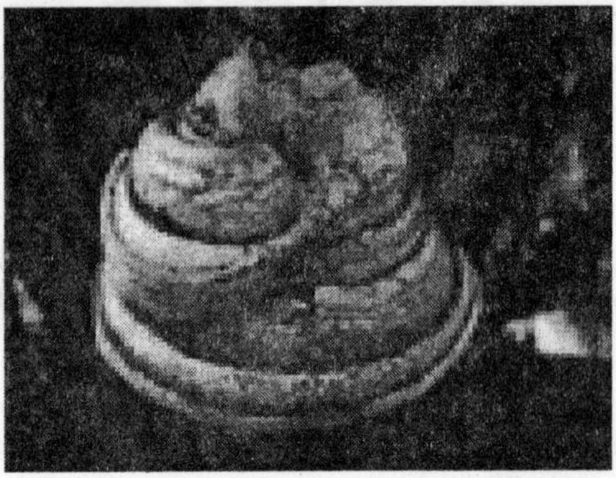

Fomes fomentarius

Amadou is a spongy, flammable substance prepared from bracket fungi. The species generally used is *Fomes fomentarius* (formerly *Ungulina fomentaria* or *Polyporus fomentarius*) which in English is also called horse's hoof fungus or tinder fungus. It is used as tinder (especially after being pounded flat, and boiled or soaked in a solution of nitre). It is also used in fly fishing for drying out fly lures. It is sometimes also used to form a felt-like fabric used in the making of hats and other items. It has great water-absorbing abilities.

Amadou was a precious resource to ancient people, allowing them to start a fire by catching sparks from flint struck against iron pyrites.

Annulus (Mycology)

An annulus is the ring like structure sometimes found on the stipe of a mushroom. The annulus represents the remaining part of the partial veil, after it has ruptured to expose the gills or other spore-producing surface. An annulus may be thick and membranous, or it may be cobweb-like. The latter type of annulus is referred to as a cortina. An annulus may be persistent and be a noticeable feature of a mature mushroom, or it may disappear soon after the emergence of the mushroom, perhaps leaving a few remnants on the stipe as an "annular zone".

An annulus can be any of various shapes, which can be important in identification of the mushroom:

- *Pendant:* drooping or skirt-like
- *Flaring:* spreading outward from the stipe
- *Sheathing:* opening upwards around the stipe

Antheridium

An antheridium (plural: antheridia) is a haploid structure or organ producing and containing male gametes (called *antherozoids* or sperm). It is present in the gametophyte phase of lower plants like mosses and ferns, and also in the primitive vascular psilotophytes. Many algae and some fungi, for example ascomycetes and water moulds, also have antheridia during their reproductive stages.

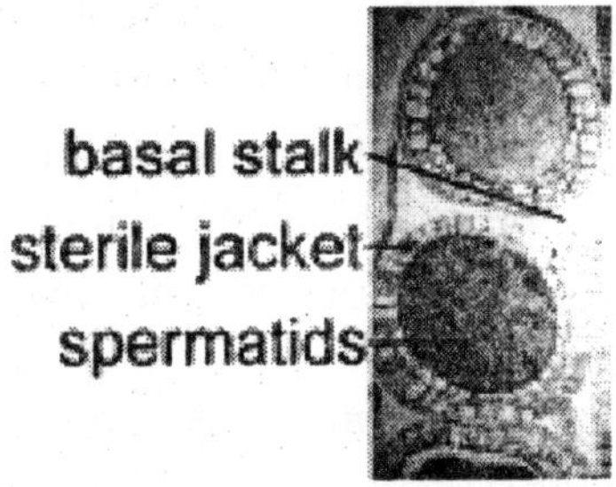

Diagram of antheridium anatomy

An antheridium typically consists of sterile cells and spermatogenous tissue. The sterile cells may form a central support structure or surround the spermatogenous tissue as a protective jacket. The spermatogenous cells give rise to spermatids via mitotic cell division. In bryophytes, the antheridium is borne on an *antheridiophore*, a stalk-like structure that carries the antheridium at its top.

In many gymnosperms and all angiosperms, the male gametophytes have been reduced to pollen grains and their antheridia have been reduced to a single generative cell within the pollen grain. During pollination, this generative cell divides and gives rise to two sperm nuclei. The female counterpart to the antheridium is the archegonium.

Arthroconidium

Arthroconidia are a type of fungal spores typically produced by segmentation of pre-existing fungal hyphae.

These spores are asexual and are generally not as durable and environmentally persistent as, for instance, bacterial endospores or chlamydospores.

Some medically significant pathogens, such as *Coccidioides immitis*, and *Coccidioides posadasii*, both causative agents of Coccidioidomycosis (also known as Valley fever), are transmitted through airborne arthroconidia.

Arum Type

Arum-type refers to the morphology of fungal hyphae living in, or around plant root cells.

- Forms in arbuscular or tree-like fashion, branching off dichotomously at predetermined junctions.

This is a type of mycorrhizal infection whereby the fungus in question invaginates the cell membrane of a plant cell, and branches in arbuscular manner. Arum-type growth of hyphae is used in endomycorrhizal symbiosis with a plant.

- Often, but not always, accompanied by intercellular hyphal growth.

Ascocarp

An ascocarp, or ascoma (plural: ascomata), is the fruiting body (sporocarp) of an ascomycete fungus. It consists of very tightly interwoven hyphae and may contain millions of asci, each of which contains typically eight ascospores. Ascocarps are most commonly bowl-shaped, but may take on a number of other designs.

Classification of Ascocarps

The ascocarp is classified according to its placement (in ways which are not fundamental to the basic taxonomy). It is termed epigeous if it grows above ground, as with the morels, whilst underground ascocarps, such as truffles are hypogeous.

The form of the hymenium is divided into the following types (which are important for classification). Apothecia can be relatively large and fleshy, whereas the others are microscopic — about the size of flecks of ground pepper:

- *Apothecium:* Here the ascocarp is open above like a cup. The fertile layer is free, so that many spores can be dispersed simultaneously. The morel, *Morchella*, an edible mushroom favoured by gourmets, is a mass of apothecia fused together in a single large structure or cap. The genera *Helvella* and *Gyromitra* are similar.
- *Cleistothecium:* In this case the ascocarp is round with the hymenium enclosed, so the spores do not automatically get released, and fungi with cleistothecia have had to develop new strategies to disseminate their spores. The truffles, for instance, have solved this problem by attracting animals such as wild boars which break open the tasty ascocarps and can spread the spores inside over a wide area. Cleistothecia are found mostly in fungi which have little room available for their ascocarps, for instance those which live under the bark of trees or under the ground like truffles. Also the dermatophyte *Arthroderma* forms cleistothecia.
- *Perithecium:* This has the shape of a skittle or a ball. Its distinguishing feature is that on top it has a small pore, the ostiole, through which the spores are released one by one when ripe (in contrast to apothecia where they

are released together). Perithecia are found for example on *Xylaria* (Dead Man's Fingers, Candle Snuff) and *Nectria*.

- *Pseudothecium:* This is similar to a perithecium, but the asci are not regularly organised into a hymenium and they are bitunicate, having a double wall which expands when it takes up water and shoots the enclosed spores out suddenly to disperse them. Example species are Apple scab (*Venturia inaequalis*) and the horse chestnut disease *Guignardia aesculi*.

Ascospore

An ascospore is a spore contained in an ascus or that was produced inside an ascus. This kind of spore is specific to fungi classified as ascomycetes (Ascomycota). Typically, a single ascus will contain eight ascospores. The eight spores are produced by a combination of a meiosis division followed by a mitotic division. The meiosis division turns the original diploid zygote nucleus into four haploid ones. That is, the single original cell from which the whole process begins contains two complete sets of chromosomes. In preparation for meiosis, all the DNA of both sets is duplicated, to make a total of four sets. The nucleus that contains the four sets divides in two stages, separating into four new nuclei—each of which has one complete set of chromosomes. Following this process, each of the four new nuclei duplicates its DNA and undergoes a division by mitosis. As a result, the ascus will contain four pairs of spores.

Ascus

An ascus (plural asci) is the sexual spore-bearing cell produced in ascomycete fungi. On average, asci normally contain 8 ascospores, produced by a meiotic cell division followed, in most species, by a mitotic cell division. However, asci in some genera or species can number 1 (e.g. *Monosporascus cannonballus*), 2, 4, or multiples of four. In a few cases, the ascospores can bud off conidia that may fill the asci (e.g. *Tympanis*) with hundreds of conidia, or the ascospores may fragment, e.g. some *Cordyceps*, also filling the asci with smaller cells. Ascospores are non-motile, usually single-celled, but not infrequently may be septate, and in some cases septate in multiple planes. Mitotic divisions within the developing spores populate each resulting cell in septate ascospores with nuclei.

In many cases the asci are formed in a regular layer, the *hymenium*, in a fruiting body which is visible to the naked eye, here called an *ascocarp* or *ascoma*. In other cases, such as single-celled yeasts, no such structures are found. In rare cases, asci of some genera can regularly develop inside older discharged asci one after another, e.g. *Dipodascus*.

Asci normally release their spores by bursting at the tip, but they may also digest themselves passively releasing the ascospores either in a liquid or as a dry powder. Typically, actively discharging asci have a specially differentiated tip, either a pore or an operculum.

In some hymenium forming genera, when one ascus bursts, it can trigger the bursting of many other asci in the ascocarp resulting in a massive discharge visible as a cloud of spores—the phenomenon called "puffing". This is an example of positive feedback. A faint hissing sound can also be heard for species of Peziza and other cup fungi.

Asci, notably those of Neurospora crassa, have been used in laboratories for studying the process of meiosis, because the four cells produced by meiosis line up in regular order. By modifying a gene coding for spore colour, the biologist can study crossing over and other phenomena.

Asci of most Pezizomycotina develop after the formation of croziers at their base. The croziers help maintain a brief dikaryon. The compatible nuclei of the dikaryon merge forming a diploid nucleus that then under goes meiosis and ultimately internal ascospore formation. Members of the Taphrinomycotina and Saccharomycotina do not form croziers.

Ascus Classification

The form of the ascus, the capsule which contains the sexual spores, is important for classification of the Ascomycota. There are four basic types of ascus.

- A unitunicate-operculate ascus has a "lid", the Operculum, which breaks open when the spores ripen and in this way sets them free. Unitunicate-operculate asci only occur in those ascocarps which have apothecia, for instance the morels. 'Unitunicate' means 'single-walled'.
- Instead of an operculum, a unitunicate-inoperculate ascus has an elastic ring that functions like a pressure valve. On ripening it briefly expands and so lets the spores shoot out. This type appears both in apothecia and in perithecia; an example is the illustrated *Hypomyces chrysospermus*.
- A bitunicate ascus is enclosed in a double wall. This consists of a thin brittle outer shell and a thick elastic inner wall. When the spores are ripe the shell splits open so that the inner wall can take up water. As a consequence this begins to extend with its spores until it protrudes above the rest of the ascocarp so that the spores can escape into free air without being obstructed by the bulk of the fruiting body. Bitunicate asci occur only in pseudothecia and are found only in the classes *Dothideomycetes* and *Chaetothyriomycetes* (which were formerly united in the old class *Loculoascomycetes*). Examples: *Venturia inaequalis* (Apple Scab) and *Guignardia aesculi* (Brown Leaf Mould of Horse Chestnut).
- Prototunicate asci are mostly spherical in shape and they have no active dispersal mechanism at all. The ripe ascus wall simply dissolves so that the spores can escape, or it is broken open by other influences such as animals. Asci of this type can be found both in

perithecia and in cleistothecia, for instance with Dutch elm disease (*Ophiostoma*). This is something of a catch-all term for cases which do not fit into the other three ascus types, and they probably belong to several independent groups which evolved separately from unitunicate asci.

Basidiocarp

A basidiocarp, basidiome or basidioma (plural: basidiomata), is the sporocarp of a basidiomycete, the multicellular structure on which the spore-producing hymenium is borne. Basidiocarps are characteristic of the hymenomycetes; rusts and smuts do not produce such structures. As with other sporocarps, epigeous (above-ground) basidiocarps that are visible to the naked eye (especially those with a more or less agaricoid morphology) are commonly referred to as mushrooms, while hypogeous (underground) basidiocarps are usually called false truffles.

Structure

All basidiocarps serve as the structure on which the hymenium is produced. Basidia are found on the surface of the hymenium, and the basidia ultimately produce spores. In its simplest form, a basidiocarp consists of an undifferentiated fruiting structure with a hymenium on the surface; such a structure is characteristic of many simple jelly and club fungi. In more complex basidiocarps, there is differentiation into a stipe, a pileus, and/or various types of hymenophores.

Types

Basidiocarps are classified into various types of growth forms based on the degree of differentiation into a stipe, pileus, and hymenophore, as well as the type of hymenophore, if present.

Growth forms include:

- *Jelly Fungus:* Fruiting body is an undifferentiated mass of jelly-like tissue.

- *Club Fungus and Coral Fungus:* Erect fruiting body without a distinct stipe and pileus, either unbranched (club fungus) or profusely branched (coral fungus).
- *Polypore:* Fruiting body is hard, woody, and perennial, and often grows shelf-like on the side of a tree or log. Polypores have a pileus, and usually (but not always) tubes and no stipe.
- *Cantharelloid Fungus:* Fruiting body with shallow fold-like lamellae running over most of the lower surface of the fruiting body and not much differentiation between the stipe and pileus.
- *Tooth Fungus or "Hydnoid Fungus":* Fruiting body with tooth-like hymenophores.
- *Gasteromycete or "Gastroid Fungus":* Fruiting body has a ball-like shape and in which the hymenophore has become entirely enclosed on the inside of the fruiting body.
- *False Truffle:* Like a gasteromycete, however, but with a hypogeous (underground) fruiting body.
- *Secotioid Fungus:* Like a gasteromycete, but with stipe. Though to be an evolutionarily intermediate stage between a gasteromycete and an agaric.
- *Agaric or "Agaricoid Fungus":* Fruiting body with a pileus, lamellae, and (usually) a stipe.
- *Bolete:* Fruiting body with a pileus, a stipe, and tubes.

Basic divisions of Hymenomycotina were formerly based entirely upon the growth form of the mushroom. Molecular phylogenetic investigation (as well as supporting evidence from micromorphology and chemotaxonomy) has since demonstrated that similar types of basidiomycete growth form are often examples of convergent evolution and do not always reflect a close relationship between different groups of fungi. For example, agarics have arisen independently in the Agaricales, the Boletales, the Russulales, and other groups, while secotioid fungi and false truffles have arisen independently many times just within the Agaricales.

Basidiospore

A basidiospore is a reproductive spore produced by Basidiomycete fungi. Basidiospores typically each contain one haploid nucleus that is the product of meiosis, and they are produced by specialised fungal cells called basidia. Most basidiospores are forcibly discharged, and are thus considered ballistospores.

When basidiospores encounter a favourable substrate, they may germinate, typically by forming hyphae. These hyphae grow outward from the original spore, forming an expanding circle of mycelium. The circular shape of a fungal colony explains the formation of fairy rings, and also the circular lesions of skin-infecting fungi that cause ringworm. Some basidiospores germinate repetitively by forming small spores instead of hyphae.

Basidium

A basidium (pl., basidia) is a microscopic, spore-producing structure found on the hymenophore of fruiting bodies of basidiomycete fungi. The presence of basidia is one of the main characteristic features of the Basidiomycota. A basidium usually bears four sexual spores called basidiospores; occasionally the number may be two or even eight. In a typical basidium, each basidiospore is borne at the tip of a narrow prong or horn called a sterigma (pl. sterigmata), and is forcibly discharged upon maturity. The word *basidium* literally means *little pedestal*, from the way in which the basidium supports the spores. However, some biologists suggest that the structure more closely resembles a club. An immature basidium is known as a basidiole.

Basidium Structure

Most basidiomycetes have single-celled basidia (holobasidia), but in some groups basidia can be multicellular (a phragmobasidia). For instance, rust fungi in the order

Uredinales have four-celled phragmobasidia that are transversely septate; some jelly fungi in the order Tremellales have four-celled phragmobasidia that are cruciately septate. Sometimes the basidium (metabasidium) develops from a probasidium, which is a specialised cell which is not elongated like a typical hypha.

Mechanism of Basidiospore Discharge

In most basidiomycetes, the basidiospores are ballistospores—they are forcibly discharged. The propulsive force is derived from a sudden change in the centre of gravity of the discharged spore.

Important factors in forcible discharge include Buller's drop, a droplet of fluid that can be observed to accumulate at the proximal tip (hilar appendage) of each basidiospore; the offset attachment of the spore to the subtending sterigma, and the presence of hygroscopic regions on the basidiospore surface.

Upon maturity of a basidiospore, sugars present in the cell wall begin to serve as condensation loci for water vapour in the air. Two separate regions of condensation are critical. At the pointed tip of the spore (the hilum) closest to the supporting basidium, Buller's drop accumulates as a large, almost spherical water droplet.

At the same time, condensation occurs in thin film on the adaxial face of the spore. When these two bodies of water coalesce, the release of surface tension and the sudden change in the centre of mass leads to sudden discharge of the basidiospore. Remarkably, Money (1998) has estimated the initial acceleration of the spore to be about 10,000 g. Successful basidiospore discharge can only occur when there is sufficient water vapour available to condense on the spore.

Evolutionary Loss of Forcible Discharge

Some basidiomycetes lack forcible discharge, although they still form basidiospores. In each of these groups, spore dispersal occurs through other discharge mechanisms. For example, members of the order Phallales (stinkhorns) rely on insect vectors for dispersal; the dry spores of the Lycoperdales (puffballs) and Sclerodermataceae (earth balls and kin) are dispersed when the basidiocarps are disturbed; and species of the Nidulariales (bird's nest fungi) use a splash cup mechanism.

In these cases the basidiospore typically lacks a hilar appendage, and no forcible discharge occurs. Each example is thought to represent an independent evolutionary loss of the forcible discharge mechanism ancestral to all basidiomycetes.

Bolete

A bolete is a type of fungal fruiting body characterised by the presence of a pileus that is clearly differentiated from the stipe, with a spongy surface of pores (rather than gills) on the underside of the pileus.

"Bolete" is also the English common name for fungal species having this kind of morphology.

The boletes are classified in the Boletales, however not all members of that order are boletes; recent discoveries in the micromorphology and molecular phylogeny of this group have established that it also contains a large number of agarics, gasteromycetes, and other fruiting body morphologies.

Similar pore surfaces exist in the polyporales and other orders, but these species usually lack the overall physical structure of boletes, have much firmer (often woody) flesh, or are otherwise differentiated.

Generally the term refers to members of the genus *Boletus*, but as superficially similar fungi have been placed in other genera many of them have retained the common name. These include:

- *Birch bolete:* Leccinum scabrum
- *Bitter bolete:* Tylopilus felleus
- *Manzanita bolete:* Leccinum manzanitae
- *Rhubarb bolete:* Boletellus obscureococcineus

Chlamydospore

A Chlamydospore is the thick-walled big resting spore of several kinds of fungi. It is the life-stage which survives in unfavourable conditions, such as dry or hot seasons.

Chlamydospores are usually dark-coloured, spherical, and have a smooth (non-ornamented) surface. They are multicellular, the cells being connected by pores in septae between cells.

Chlamydospores are a result of asexual (thus being actually conidia called chlamydoconidia) or sexual reproduction(rare). Teliospores are special kind of chlamydospores of rusts and smuts.

Coenocyte

A coenocyte is a multinucleate cell. It can result from multiple nuclear divisions without accompanying cell divisions, or from cellular aggregation followed by dissolution of the cell membranes inside the mass. Coenocytes are found in fungi and some protists, such as algae and slime mould. Some plant structures, such as endosperm, are coenocytic as well.

A type of mycelium in which hyphae lack septa is known as aseptate or coenocytic. Fairy Rings are a prime example of a coenocytic fungi. A large fairy ring may in fact just be one large cell with millions of nuclei.

Conidium

Conidia, sometimes inappropriately termed conidiospores (an intrinsically redundant name), are asexual, non-motile spores of a fungus; they are also called mitospores due to the way they are generated through the cellular process of mitosis. They are haploid cells genetically identical to the haploid parent, can develop into a new organism if conditions are favourable, and serve in biological dispersal.

Asexual reproduction in Ascomycetes (the Phylum

Ascomycota) is by the formation of conidia, which are borne on specialised stalks called conidiophores.

The morphology of these specialised conidiophores is often distinctive of a specific species, and can therefore be used in identification of the species. There are two main types of conidium development:

- *Blastic* conidiogenesis, where the spore is already evident before it separates from the conidiogenic hypha which is giving rise to it, and
- *Thallic* conidiogenesis, where first a cross-wall appears and then the thus created cell develops into a spore.

Health Issues

Conidia are present in the air all the time, but levels fluctuate from day to day and with the seasons. An average person inhales 40 conidia per hour.

Conidia are often the method by which some normally harmless but heat-tolerating (thermotolerant), common fungi establish infection in certain types of severely immunocompromised patients (usually acute leukaemia patients on induction chemotherapy, AIDS patients with superimposed B-cell lymphoma, bone marrow transplantation patients, or major organ transplant patients suffering from graft versus host disease). Their immune system is not strong enough to fight off the fungus, and it may, for example, colonise the lung, resulting in a pulmonary infection.

Cystidium

A cystidium (plural cystidia) is a relatively large cell found on the hymenium of a basidiomycete (for example, on the surface of a mushroom gill), often between clusters of basidia. Since cystidia have highly varied and distinct shapes that are often unique to a particular species or genus, they are a useful characteristic in the microscopic identification of mushrooms and other basidiomycetes.

Cystidia are sometimes referred to as sterile cells because they do not produce spores, however this is not correct. If one were to call them "sterile cells" you would have to call all cells in the fungus except the basidia "sterile cells".

Cystidia may occur on the gill edge (cheilocystidia), on the face of the gill (pleurocystidia), on the surface of the cap (dermatocystidia or pileocystidia) or on the stipe (caulocystidia). Especially the pleurocystidia and cheilocystidia are important for identification within many genera. Sometimes the cheilocystidia give the gill edge a distinct colour which is visible to the naked eye or with a hand lens. Boletes also have cystidia; here, of course, the definitions should refer to pores, not gills.

Chrysocystidia are cystidia whose contents contain a distinct refractive yellow body, that becomes more deeply yellow when exposed to ammonia or other alkaline compounds. Chrysocystidia are characteristic of many (though not all) members of the agaric family Strophariaceae.

In general, the purpose of the cystidia from the point of view of the fungus is a mystery. Cheilocystidia, which occur at the edge of the gills, have an excretory function. The margin of gills with abundant cheilocystidia are often beaded with tiny droplets of liquid, so one function of cystidia may be to exude substances which in this case are visible as droplets.

Ericoid Mycorrhiza

Ericoid mycorrhiza are a symbiotic relationship between fungi and the roots of plants from the order Ericales. Ericoid mycorrhiza are considered crucial for the success of the family Ericaceae in variety of edaphically stressful environments worldwide (Read, 1992). Ericaceous plants commonly co-occur in soils with leguminous or carnivorous plants (Read, 1996), further highlighting the low nutrient status of these soils. Ericoid mycorrhizal fungi, which are predominately ascomycetes, enable their host plant to obtain nutrients in these depauperate soils.

Epacris pulchella, an ericoid mycorrhizal epacrid from eastern Australia.

Many of the fungi which form ericoid mycorrhizas are from the ascomycete order Helotiales. The best studied of these fungi is *Rhizoscyphus ericae* (=Hymenoscyphus ericae). *R. ericae* commonly occurs in symbiosis with a variety of ericaceous species from the northern hemisphere, but is not routinely found associated with Ericaceae (ex Epacridaceae) in the southern hemisphere (Cairney and Ashford, 2002).

Fairy Ring

A fairy ring, also known as fairy circle, Elf circle or pixie ring, is a naturally occurring ring or arc of mushrooms. The rings may grow over ten metres in diameter and become stable over time as the fungus grows and seeks food underground. They are found mainly in forested areas, but also appear in the grasslands and/or rangelands.

Fairy rings are not only detectable by sporocarps in rings or arcs, but also by a necrotic zone (death grass) or a ring of dark green grass. If these manifestations are visible a fairy fungus mycelium is likely present in the ring or arc underneath.

Genesis: There are two theories regarding the process involved in creating fairy rings. One states that fairy ring is begun by a spore from the sporocarpus. Initially the spore lands on suitable ground, and as it starts to grow underground it pushes out mycelium (fungal threads) in all directions.

As the mass of mycelium grows the central part dies off, leaving a disc of mycelium growing at the outer edge. Eventually, when season and weather allow, mushrooms are produced above the mycelium and release spores. Sometimes, several years may go by before mushrooms appear around a fairy ring. The underground presence of the fungus can also cause withering or varying colour or growth of the grass above. Many fair rings are believe to be coenocytic, meaning that an entire ring is just one large cell with multiple nuclei.

The second theory, which is presented in the investigations of Japanese scientists on the *Tricholoma matsutake* species, shows that fairy rings could be established by connecting neighbouring oval genets of these mushrooms. If they make an arc or a ring, they continuously grow about the centre of this object.

It is said that there are about 40-60 mushroom species which can grow in the fairy ring pattern. The best known is the edible Scotch bonnet (*Marasmius oreades*), which is commonly known as a "fairy ring mushroom".

Necrotic or Rapid Growth Zones

One of the three manifestations is a necrotic zone, or area in which grass or other plant life has withered or died. These are caused by the mycelia, which during a very dry year coat the roots of grasses and other herbs in meadows. After some time they are removed by biotic factors from the ground, at which stage we can see a zone on the surface soil.

Patterns other than the basic ring or arc are also possible: circles, doubled arcs, sickle-shaped arcs, and other complicated formations are also formed by this process.

Fungi can deplete the soil of readily available nutrients like nitrogen, causing plant growing within the circle to be stressed which leads to plant discolourisation. Some fungi also produce chemicals that act like hormones called gibberellins, which affects plant growth, causing rapid luxuriant growth.

Types: There are two generally recognised types of fairy ring fungus. Those found in the woods are called *theatered,* due to the fact that they are formed by mycorrhizal fungi living in commensalism with trees.

Meadow fairy rings are called *free,* because they are not connected with other organisms. These mushrooms are saprophytic. The effects on the grass depend on the type of fungus that is growing, when Calvatia cyathiformis is growing in the area grass will grow more abundantly however Clitocybe gigantea will cause the grass to wither.

Fairy Rings in Folktales and Mythology

In English folklore, fairy rings were said to be caused by, elves, fairies or pixies dancing in a circle, wearing down the grass beneath their feet. Toads would then sit on the mushrooms, poisoning them; hence the name *toadstool*. In Sussex they were called, "hag tracks", in Devon people thought that fairies caught young horses and rode them in circles.

In Scandinavian folklore, these circles were attributed to elves or witches and were called *alvdanser,* i.e. *elf dances, alvringar,* or *heksering*.

In German-speaking Europe, fairy rings are known as *Hexenringe*, or "witches rings", stemming from an old medieval belief that the rings represented places where witches would have their gatherings. In Austria people thought that dragons breath burned the land. Similar myths to those in German folklore can be found in Czech, Slovak, Polish and even Russian folk tales. In the Czech language they are called *earodijne kruhy*, as they are thought to have been caused by a dragons having a rest at those places.

Another myth states that fairy rings are doors into the fairies' world, transporting people to other places or making people appear in the same place in a different time.

Young ladies are also warned not to touch dew situated on the grass within the ring, due to the belief that it can cause skin problems.

Germ Pore

A germ pore is a small pore in the outer wall of a mycelial spore through which the germ tube exits upon germination. It can be apical or eccentric in its location, and, on light microscopy, may be visualised as a lighter coloured area on the cell wall.

Apical germ pore is a term applied to mushroom spores which have a pore at one end. Some spores have a hole in the cell wall where the first strand of germinating mycelium emerges. If the cell wall is divided from one end to the other, this is called a germ slit. Commonly the germ pore is at one end of the mushroom spore and is called an apical pore.

Mushroom genera with apical germ pores include *Agrocybe*, *Panaeolus*, *Psilocybe*, and *Pholiota*.

Glabrous

In botany and mycology, glabrous is used to describe a morphological feature as smooth or having no hair or similar growth (see indumentum). In medicine, "glabrous" skin is used to describe the hairless skin on the human body, such as the fingers, palmar surfaces of hands, soles of feet, lips, and penises. *Tinea corporis* is said to target glaborous skin.

Gleba

Gleba is the fleshy spore-bearing inner mass of fungi such as the puffball or stinkhorn.

The gleba is a solid mass of spores, generated within an enclosed area within the sporocarp. The continuous maturity of the sporogenous cells leave the spores behind as a powdery mass that can be easily blown away. The gleba may be sticky or it may be enclosed in a case (peridiole).

It is a tissue usually found in an angiocarpous fruit-body, especially gasteromycetes. Angiocarpous fruit-bodies usually consists of fruit enclosed within a covering that does not form a part of itself; such as the filbert covered by its husk, or the acorn seated in its cupule. The presence of gleba can be found in earthballs and puffballs. The gleba consists of mycelium, and basidia and may also contain capillitium threads.

Hygrophanous

The adjective hygrophanous refers to the colour change of mushroom tissue (especially the pileus surface) as it loses or absorbs water, which causes the pileipellis to become more transparent when wet and opaque when dry.

When identifying hygrophanous species, one needs to be careful when matching colours to photographs or descriptions, as colour can change dramatically soon after picking.

Genera that are characterised by hygrophanous species include *Agrocybe, Psathyrella, Psilocybe, Panaeolus*, and *Galerina*.

Hymenium

The hymenium is the tissue layer on the hymenophore of a fungal fruiting body where the cells develop into basidia or asci, which produce spores.

In some species all of the cells of the hymenium develop into basidia or asci, while in others some cells develop into sterile cells called cystidia (basidiomycetes) or paraphyses (ascomycetes).

Cystidia are often important for microscopic identification. The subhymenium consists of the supportive hyphae from which the cells of the hymenium grow, beneath which is the hymenophoral trama, the hyphae that make up the mass of the hymenophore.

The position of the hymenium is traditionally the first characteristic used in the classification and identification of mushrooms. Below are some examples of the diverse types which exist amongst the macroscopic Basidiomycota and Ascomycota.

- In agarics the hymenium is on the vertical faces of the gills.
- In boletes it is in a spongy mass of downward-pointing tubes.
- In puffballs it is internal.
- In stinkhorns it develops internally and then is exposed in the form of a foul-smelling gel.
- In cup fungi, it is on the concave surface of the cup.
- In teeth fungi, it grows on the outside of tooth-like spines.

Hypha

A hypha (plural hyphae) is a long, branching filamentous cell of a fungus, and also of unrelated Actinobacteria. In fungi, hyphae are the main mode of vegetative growth, and are collectively called a mycelium.

A hypha consists of one or more cells surrounded by a tubular cell wall. In most fungi, hyphae are divided into cells by internal cross-walls called septa (singular septum).

Septa are usually perforated by pores large enough for ribosomes, mitochondria and sometimes nuclei to flow among cells. The structural polymer in fungal cell walls is typically chitin (in contrast plants have cellulosic cell walls, and animal cells lack walls).

Hyphae grow at their tips. During tip growth, cell walls are extended by the external assembly and polymerisation of cell wall components, and the internal production of new cell membrane. The Spitzenkorper is an intracellular organelle associated with tip growth. It is composed of an aggregation of membrane-bound vesicles containing cell wall components.

The vesicles travel to the cell membrane via the cytoskeleton, and dump their contents outside the cell by the process of exocytosis. Vesicle membranes contribute to growth of the cell membrane while their contents form new cell wall. As a hypha extends, septa may be formed behind the growing tip to partition each hypha into individual cells. Hyphae can branch through bifurcation of a growing tip, or by the emergence of a new tip from an established hypha.

Hyphae may be modified in many different ways to serve specific functions. Some parasitic fungi form haustoria that function in absorption within the host cells. The arbuscules of mutualistic mycorrhizal fungi serve a similar function in nutrient exchange, so are important in assisting nutrient and water absorption by plants.

Hyphae are found enveloping the gonidia in lichens,

making up a large part of their structure. In nematode-trapping fungi, hyphae may be modified into trapping structures such as constricting rings and adhesive nets. Cords can be formed to transfer nutrients over larger distances.

Types of Hyphae

- Septate (with septa)
- Aseptate or coenocytic (without septa)

Characteristics of hyphae can be important in fungal classification. In basidiomycete taxonomy, hyphae that comprise the fruiting body can be identified as generative, skeletal, or binding hyphae.

- *Generative* hyphae are undifferentiated and can develop reproductive structures. They are thin-walled, usually have frequent septa, and may or may not have clamp connections.
- *Skeletal* hyphae are thick-walled and very long and straight, with little cell content (they are "dead" once they have formed). They have few septa and lack clamp connections.
- *Binding* hyphae are generally thick-walled and are bendy, with very frequent junctions and ramifications.

Based on these hyphal types, in 1932 E. J. H. Corner applied the terms monomitic, dimitic, and trimitic to hyphal systems, in order to improve the classification of polypores.

- Every fungus must contain generative hyphae. A fungus which only contains this type, as do fleshy mushrooms such as agarics, is referred to as monomitic.
- Skeletal and binding hyphae give leathery and woody fungi such as polypores their tough consistency. If a fungus contains all three types (example: Trametes), it is called trimitic.
- If a fungus contains generative hyphae and just one of the other two types, it is called dimitic. In fact dimitic fungi almost always contain generative and skeletal hyphae; there is just one exceptional genus, *Laetiporus*, which includes only generative and binding hyphae.

Lamella (Mycology)

A gill, or lamella, is one of the papery ribs under the cap of a mushroom, most often but not always an agaric. As fungi are studied in more detail, several other types of fungi exhibit gills while not members of the Agaricales. It is unclear whether this is a case of convergent evolution (i.e. gill-like structures evolved separately) or evolved once.

Other groups of fungi to bear gills include:

- The genera *Russula* and *Lactarius* of the Russulales.
- Several genera in the Boletales, including *Paxillus,* the spike-caps of *Gomphidius* and *Chroogomphus* as well as the velvet roll-rim (*Tapinella atrotomentosa*) and the False chanterelle (*Hygrophoropsis aurantiaca*)

The gills are the hymenophore of the mushroom.

Morphologically, gills are classified into the following types:

- Decurrent
- Subdecurrent
- Emarginate
- Sinuate
- Adnexed
- Adnate
- Free

Mycelial Cord

A linear aggregations of the parallel oriented hyphae. The mature cords are composed of wide, empty "vessel hyphae" surrounded by narrower "sheathing hyphae". Cords may look similar to the plant roots, and also frequently have the similar functions.

Mycelial cords are capable of conducting nutrients over long distances. For instance, they can transfer nutrients to a developing fruitbody, or to enable wood-rotting fungi to grow through soil from an established food base in search of new food sources. For parasitic fungi, they can help to spread infection by growing from established clusters to uninfected parts. The cords of the wood rotting fungi (like *Serpula lacrymans*) may be capable to penetrate masonry.

The mechanism of the cord formation is not yet precisely understood. Mathematical models suggest that some fields, parallel to the cord axis, may be involved.

Mycelium

Mycelium is the vegetative part of a fungus, consisting of a mass of branching, thread-like hyphae. The mass of hyphae is sometimes called shiro, especially within the fairy ring fungi. Fungal colonies composed of mycelia are found in soil and on or in many other substrates.

Mycelia may form fruiting bodies such as mushrooms. A mycelium may be minute, forming a colony that is too small to see, or it may be extensive. The mycelium of a single individual of *Armillaria gallica,* dubbed "the humungous fungus," extended through 37 acres of forest soil and was estimated to weigh at least 9700kg (Smith *et al.* 1992).

It is through the mycelium that a fungus absorbs nutrients from its environment. It does this in a two stage process. Firstly the hyphae secrete enzymes onto the food source, which breaks down polymers into monomers. These monomers are then absorbed into the mycelium by facilitated diffusion and active transport.

Mycelium is vital in terrestrial and aquatic ecosystems for its role in the decomposition of plant material. It contributes to the organic fraction of soil and its growth releases carbon dioxide back into the atmosphere. The mycelium of mycorrhizal fungi increases the efficiency of water and nutrient absorption of most plants and confers resistance to some plant pathogens. Mycelium is an important food source for many soil invertebrates.

Sclerotia are compact or hard masses of mycelium.

Uses

One of the primary roles of fungi in an ecosystem is to decompose organic compounds. Petroleum products and pesticides that can be contaminants of soil, are organic molecules. Fungi therefore should have potential to remove such pollutants from the soil environment, a process known as bioremediation.

Mycelial mats have been suggested as having potential as biological filters, removing chemicals and microorganisms from soil and water. The use of fungal mycelia to accomplish this has been termed "mycofiltration", although there is no reason to suspect that the process is any different from that of bioremediation using fungi.

Oidium

An oidium (plural: oidia) is an asexually produced fungal spore that (in contrast to conidia) is presumed not to constitute the main reproductive preoccupation of the fungus at that time. The hypha breaks up into component cells/ small pieces and develop into spores. Oidia can't survive in unfavourable conditions.

Parenthesome

Within the cells of basidiomycete fungi are found microscopic structures called parenthesomes or septal pore caps. They are shaped like parentheses and found on either side of pores in the dolipore septum which separates cells within a hypha. Their function has not been established, and their composition has not been fully elucidated. The variations in their appearance are useful in distinguishing individual species.

Partial Veil

Partial veil is a mycological term which describes a structure of some mushrooms which protects the developing gills or other spore-producing surface. A partial veil, in contrast to a universal veil, is attached to the stipe (stalk) and edge of the pileus (cap). In some mushrooms, both a partial veil and a universal veil may be present.

The partial veil may be membranous or cobwebby, and may have multiple layers. Some mushrooms have partial veils which are evanescent, disappearing once they rupture, while others may leave a persistent annulus (ring).

The presence, absence, or structure of the partial veil is an aid to identification of mushrooms.

Phialide

Phialide [Gr. phialis, diminuitive of phiale, a broad, flat vessel]

- A flask-shaped projection from the conidiophore of certain fungi.
- Projects from the mycelium without increasing in length without a subsequent increase in the formation of conidia.

Also known as the end cell of a phialophore.

Pileipellis

The pileipellis is the uppermost layer of hyphae in the pileus of a fungal fruiting body. It is more or less synonymous with the cuticle, however, the cuticle generally describes this layer as a macroscopic feature, while pileipellis refers to this structure as a microscopic layer. Pileipellis type is an important character in the identification of fungi. Pileipellis types include the cutis, trichoderm, epithellium, and hymenoderm types.

Pileus (mycology)

The pileus is the technical name for what is commonly known as the "cap" of a fungal fruiting body. It is particularly characteristic of agarics, boletes, and some polypores, tooth fungi, and ascocarps. Other sporocarp types have less clear differentiation between the pileus and the rest of the fruiting body. Sporocarps with a pileus typically have some type of hymenophore, such as lamellae, tubes, or teeth, on the underside of the pileus.

Pilei can be of various shapes, the most familiar being hemispherical or convex, with many of these becoming flat as they mature. Many well-known species have this shape, including the button mushroom, various *Amanita* species and boletes. Some, such as the parasol mushroom, have distinct bosses or umbos and are described as umbonate. Many fungi, such as chanterelles have a funnel- or trumpet-shaped appearance, the pileus is termed infundibuliform.

Pileus Area Index

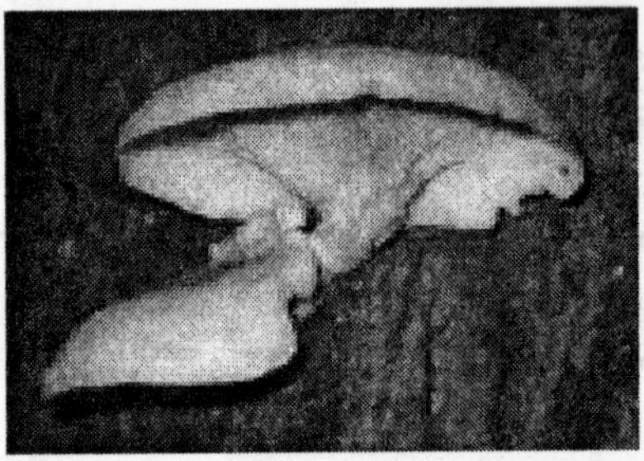

Fruiting body of the Oyster mushroom.

The Pileus Area Index is the technical name used to denote the size of what is commonly known as the "cap" of a fungal fruiting body. It is particularly characteristic of agarics, boletes, and some polypores, tooth fungi, and ascocarps. It was coined by Asher Tarivona Mutsengi a Zimbabwean student at University of Texas at Austin.

Resting Spore

A spore created by fungi which is thickly encysted (has a thick cell wall) in order to survive through stressful times, such as drought.

- Protects the spore from biotic (microbial, fungal viral), as well as abiotic (wind, heat, xeric conditions) factors.

Characteristics

Creates the phenomenon known late potato blight.

- Does this by the virtue that the resting spore can lay dormant within the soil of a field for decades until the right conditions occur (plant host present, rain, fire).

Rhizoid

Rhizoids are a structure in plants and fungi that functions like a root in support or absorption. In fungi, rhizoids are small branching hyphae that grow downwards from the stolons that anchor the fungus.

They release digestive enzymes and absorb digested organic material. In land plants, rhizoids are trichomes that anchor the plant to the ground. In the liverworts, they are absent or unicellular, but multi-celled in mosses. In vascular plants they are often called root hairs, and may be unicellular or multicellular.

Sclerotium

A sclerotium is a compact mass of hardened mycelium (as an ergot) stored with reserve food material that in some higher fungi becomes detached and remains dormant until a favourable opportunity for growth occurs. It is also the name of a genus of fungi which includes *Sclerotium rolfsii*, the causal organism of Southern Blight.

Secotioid

Secotioid mushrooms lack the geotropic orientation of the hymenophore, and basidiospores are not actively shot off, thus they are statismospores. Perhaps an intermediate stage between hymenomycetes and gasteromycetes where gasteromycetation has occurred. A population bottleneck effect may have been involved.

Somatic Hyphae

The lifestage at which a fungus lives, grows, and develops, gathering nutrients and energy.

The fungi uses this stage to proliferate itself through asexually created mitotic spores.

Cycles through somatic hyphae, zoosporangium, zoospores, encystation and germination, and back to somatic hyphae.

Sorus

In ferns, a sorus (pl. *sori*) is a cluster of sporangia on the edge or underside of a fertile frond. In many species, they are protected by an umbrella-like cover called the indusium.

Sori occur on the sporophyte generation, the sporangia within producing haploid meiospores. As the sporongia mature, the indusium shrivels so that spore release is unimpeded. The sporangia then burst and release the spores.

The shape, arrangement, and location of the sori are often valuable clues in the identification of fern taxa. Sori may be circular or linear. They may be arranged in rows, either parallel or oblique to the costa, or randomly. Their location may be marginal or set away from the margin on the frond lamina. The presence or absence of indusium is also used to identify fern taxa.

In fungi and lichens, a sorus is a cluster of sporangia surrounded by an external layer.

Sporangium

A sporangium (pl., sporangia) is a plant or fungal structure producing and containing spores. Sporangia occur on angiosperms, gymnosperms, ferns, fern allies, mosses, algae, and fungi.

Microsporangia are the structures on the stamens of flowers

called anthers, and the pollen-producing structures on the microsporophylls of male conifer cones.

Megasporangia are the comparable "female" structures on these plants, associated with the flower carpel and the megasporangial cone.

On ferns, the mature plant is a sporophyte that develops sporangia, tiny, stalked sacs which contain *meiospores,* on all or just certain leaves (called sporophylls if sporangia are present). The spores are on the underside of the leaves.

In mosses, the little case that rises above the vegetative growth on a thin stalk is a sporangium often called a capsule that, as in ferns, produces meiospores.

This sporophytic (diploid) growth arises out of the gametophytic (haploid) archegonium after the ovum is fertilized. The sporophyte initially has some chlorophyll, but later turns brown and becomes dependent upon the gametophyte for nutrition, which is absorbed through the foot (base of the stalk), embedded in the archegonial tissues.

Categorised based on developmental sequence, *eusporangia* and *leptosporangia* are differentiated in the vascular plants. In a leptosporangium, found only in ferns, development involves a single initial cell that becomes the stalk, wall, and spores within the sporangium.

There are around 64 spores in a leptosporangium. In a eusporangium, characteristic of all other vascular plants and some primitive ferns, the initials are in a layer (i.e., more than one). A eusporangium is larger (hence contain more spores), and its wall is multilayered. Although the wall may be stretched and damaged, resulting in only one cell-layer remaining.

A cluster of sporangia that have become fused in development is called a synangium. This structure is most prominent in *Psilotum* and Marattiaceae such as *Christensenia, Danaea* and *Marattia.*

Spore

In biology, a spore is a reproductive structure that is adapted for dispersion and surviving for extended periods of time in unfavourable conditions. Spores form part of the life cycles of many plants, algae, fungi and some protozoans.

Spores are usually haploid and unicellular and are produced by meiosis in the sporophyte. Once conditions are favourable, the spore can develop into a new organism using mitotic division, producing a multicellular gametophyte, which will eventually go on to produce gametes.

Two gametes fuse to create a new sporophyte. This cycle is known as alternation of generations however "Biological Life Cycle" is a better term as there may be more than one phase, and it cannot therefore be an "alteration" of anything. Haploid spores produced by mitosis (known as mitospores) are used by many fungi for asexual reproduction.

It is useful to contrast spores with gametes: spores are the units of *asexual* reproduction as a single spore develops into a new organism; gametes are the units of *sexual* reproduction as two gametes need to fuse to create a new organism.

The term *spore* may also refer to the dormant stage of some bacteria or archaea; however these are more correctly known as endospores and are not truly spores. The term can also be loosely applied to some animal resting stages. Fungi that produce spores are known as *sporogenous*, and those that do not are *asporogenous*.

Classification of spores

Spores can be classified in several manners.

By function

Diaspores are dispersal units of fungi, mosses, ferns, fern allies, and some other plants. In fungi, chlamydospores are thick-walled resting spores, and zygospores are thick-walled resting spores (hypnozygotes) of zygomycetous fungi which are produced by sexual gametocystogamy and can give rise to a conidiophore ("zygosporangium") with asexual conidiospores.

By Spore-producing Structure

In fungi and fungus-like organisms, spores are often classified by the structure in which meiosis and spore production takes place, such as a telium, ascus, basidium, or oogonium, which produce teliospore, ascospores, basidiospores, and oospores, respectively. Since fungi are often classified according to their spore-producing structures, these spores are often characteristic of a particular taxon of the fungi, such as Ascomycota or Basidiomycota.

By Origin during Life Cycle

Meiospores are the product of meiosis (the critical cytogenetic stage of sexual reproduction), meaning that they are haploid, and will give rise to a haploid daughter cell(s) or a haploid individual. An example is the parent of gametophytes of the higher vascular plants (angiosperms and gymnosperms)—the microspores (give rise to pollen) and megaspores (or macrospores) (give rise to ovules) found in flowers and cones; these plants accomplish dispersal by means of seeds.

A mitospore (conidium, conidiospore) is an asexually produced propagule, the result of mitosis. Most fungi produce mitospores. Mitosporic fungi are also known as anamophic fungi (compare teleomorph or deuteromycetes).

By Motility

Spores can be differentiated by whether they can move or not. Zoospore can move by means of one or more flagellum and can be found in some algae and fungi. Aplanospore cannot move, but may potentially grow flagella. Autospore cannot move and do not have the potential to ever develop any flagella. Ballistospore are actively discharged from the body of a fungal fruit (such as a mushroom). Statismospore are not actively discharged from the fungal fruit body, similarly to a puffball.

Parlance

In common parlance, the difference between "spore" and "gamete" (both together called gonites) is that a spore will germinate and develop into a spiralling, while a gamete needs to combine with another gamete before developing further. However, the terms are somewhat interchangeable when referring to gametes.

A chief difference between spores and seeds as dispersal units is that spores have very little stored food resources compared with seeds, and thus require more favourable conditions in order to successfully germinate. Seeds, therefore, are more resistant to harsh conditions and require less energy to start mitosis. Spores are usually produced in large numbers to increase the chance of a spore surviving.

The endospores of certain bacteria are often incorrectly called spores, as seen in the 2001 anthrax attacks where *anthrax endospores* were incorrectly called *anthrax spores* by the media. Several key differences between bacterial endospores and eukaryotic spores exist: they are primarily a survival mechanism, not a reproductive method, and a bacterium only produces a single endospore.

Diaspores

In the case of spore-shedding vascular plants such as ferns, wind distribution of very light spores provides great capacity for dispersal. Also, spores are less subject to animal predation than seeds because they contain almost no food reserve; however they are more subject to fungal and bacterial predation. Their chief advantage is that, of all forms of progeny, spores require the least energy and materials to produce.

Vascular plant spores are always haploid and vascular plants are either *homosporous* or *heterosporous*. Plants that are *homosporous* produce spores of the same size and type.

Heterosporous plants, such as spikemosses, quillworts, and some aquatic ferns produce spores of two different sizes: the larger spore in effect functioning as a "female" spore and the smaller functioning as a "male".

Under high magnification, spores can be categorised as either monolete spores or trilete spores. In monolete spores, there is a single line on the spore indicating the axis on which the mother spore was split into four along a vertical axis. In trilete spores, all four spores share a common origin and are in contact with each other, so when they separate each spore shows three lines radiating from a centre pole.

Spore Print

The spore print of a mushroom is an important diagnostic character in most handbooks for identifying mushrooms. A spore print is made by placing the spore-producing surface flat on a sheet of dark and white (or just white) paper. The mushroom is left overnight in this manner.

When the mushroom is removed, the colour of the spores should be visible on the paper. Incorporating a glass slide allows for examination of spores under a microscope, or moving the spore print to a darker or lighter surface for good contrast.

Mycologists are reluctant to identify any mushroom from its spore print alone since the spore print is only one characteristic used in determination.

Historically the classification of many families of fungi were based on spore colour, an example being Tricholomataceae being a large family containing many fungi whose common factor was having white spores. However recent molecular research has shown some interesting relationships, with some fungi of disparate spore colours showing close relations.

Sporeling

Historically the classification of many families of fungi were based on spspore, similar to a seedling derived from a germinated seed. They occur in algae, fungi, lichens, bryophytes and seedless vascular plants.

Sporeling Development

Most spores germinate by first producing a germ-rhizoid or holdfast followed by a germ tube emerging from the opposite end. The germ tube develops into the hypha, protonema or thallus of the gametophyte.

In seedless vascular plants such as ferns and lycopodiophyta, the term "sporeling" refers to the young sporophyte growing on the gametophyte. These sporelings develop via an embryo stage from a fertilized egg inside an archegonium and depend on the gametophyte for their early stages of growth before becoming independent sporophytes. Young fern sporelings can often be found with the prothallus gametophyte still attached at the base of their fronds.

Sporocarp (Fungi)

In fungi, the sporocarp (also known as fruiting body) is a multicellular structure on which spore-producing structures, such as basidia or asci, are borne. The fruiting body is part of the sexual phase of a fungal life cycle, with the rest of the life cycle being characterised by vegetative mycelial growth.

The sporocarp of a basidiomycete is known as a basidiocarp, while the fruiting body of an ascomycete is known as an ascocarp. A significant range of different shapes and morphologies is found in both basidiocarps and ascocarps; these features play an important role in the identification and taxonomy of fungi.

Ascocarp of Sarcoscypha Austriaca

Fruiting bodies are termed epigeous if they grow on the ground, as with ordinary mushrooms, while ones which grow underground are hypogeous. Epigeous sporocarps that are visible to the naked eye, especially fruiting bodies of a more or less agaricoid morphology, are often referred to as mushrooms, while hypogeous fungi are usually called truffles or false truffles. Truffles evolutionarily disposed with ability to disperse their spores via air currents, instead opting for animal consumption and subsequent dispersal of their spores.

In amateur mushroom hunting, and to a large degree in academic mycology as well, identification of higher fungi is based on the features of the sporocarp.

Stipe (Mycology)

In mycology a stipe refers to the stem or stalk-like feature supporting the cap of a mushroom. Like all tissues of the mushroom other than the hymenium, the stipe is composed of sterile hyphal tissue. In many instances, however, the fertile hymenium extends down the stipe some distance.

The evolutionary benefit of a stipe is generally considered to be in mediating spore dispersal; an elevated mushroom will more easily release its spores into wind currents or onto passing animals. Nevertheless, many mushrooms do not have stipes, including: cup fungi, puffballs, earthstars, some polypores, jelly fungi, ergots, and smuts.

It is often the case that features of the stipe are required to make a positive identification of a mushroom. Such distinguishing characters include:

1. The texture of the stipe (fibrous, brittle, chalky, leathery, firm, etc.).
2. Whether it has remains of a partial veil (such as an annulus or cortina) or universal veil (volva (mycology)).
3. Whether the stipes of many mushrooms fuse at their base.
4. Its general size and shape.
5. Whether the stipe extends underground in a root-like structure (a rhizome).

When collecting mushrooms for identification it is critical to maintain all these characters intact by digging the mushroom out of the soil, rather than cutting it off mid-stipe.

Sub-sporangial Vesicle

A *Sub-sporangial vesicle* is a vesicle which is below the sporangium on a fungus. Often used in the turgor-building and release to launch the sporangium from the stalk of the fungus using this spore-dispersal method. The Sub-sporangial vesicle in Pilobolus fills with fluid, thereby creating turgor pressure that is released, launching the sporangium out towards the light, with the intent of landing on a plant.

Suspensor

Suspensors are filamentous structural formations employed by Fungi in holding a zygospore between two strains of hyphae.

- Also used by plant zygotes in angiosperms in connecting the endosperm to an embryo. During embryo development in angiosperm seeds, normal development involves asymmetrical division of the unicellular embryo, inducing polarity. The smaller terminal cell divides to become the proembryo while the larger basal cell divides laterally to form the suspensor. The suspensor is analogous to a placental mammalian's umbilical cord.

Teliospore

Teliospore (sometimes called teleutospore) is the thick-walled resting spore of some fungi (rusts and smuts), from which the basidium arises.

Development

They develop in *telia* (or *teliosoruses*).

Morphology

Teliospores are usually dark-coloured, and consist of two dikaryote cells. As the spores germinate, the nuclei undergo karyogamy and thereafter meiosis, giving rise to four-celled basidia with haploid basidiospores. They also have a thick-capped telia as compared to urediniospores.

Trama

In mycology, the trama is a sterile layer of hyphae in the central part of a hymenophore, and is the supportive tissue of the hymenium. By extension, the term *trama* is also used to describe the hyphae composing the inner part of the pileus and stipe, tissue that in macroscopic terms is referred to as context or flesh.

Truffle

Truffle or Truffles may refer to:

- Various kinds of hypogeous fungi, such as:
  - Tuber (genus), the "true" truffles, of the hypogeous ascomycetes genus *Tuber*.
  - Tuberaceae, the hypogeous ascomycete family of that includes *Tuber*.
  - Desert truffles, of the hypogeous ascomycete family Terfeziaceae.
  - False truffles, various taxa of hypogeous basidiomycetes.
- Chocolate truffle, a chocolate confection.
- Psilocybe mexicana, a psychedelic fungus.
- Savoy Truffle (band), a popular, long-time roots-rock band from New Hampshire.

Universal Veil

The Universal veil is a mycological term to describe a structure that envelops all or most of some gilled mushrooms. The young, developing button mushroom, which may resemble a puffball at this point, is protected by this egg-like structure. The veil will eventually become ruptured and dispersed by the growing mushroom, but will usually leave evidence of its former shape with remnants including the volva (mycology), or cup-like structure at the bottom of the stalk stipe and patches or "warts" on the cap pileus.

This macrofeature is very important in wild mushroom identification due to it being an easily observed, taxonomically significant feature which almost always signifies a member of Amanitaceae. This has particular importance due to the disproportionately high number of *deadly* poisonous species contained within that family.

Veil (Botany)

A veil, in botany, is one of several structures in fungi, especially the thin membrane that covers the cap and stalk of an immature mushroom. Veils may be regarded as falling into two categories:

- Partial veil
- Universal veil

Volva (Mycology)

The volva is a mycological term to describe a cup-like structure at the base of a mushroom which is a remnant of the Universal veil. This macrofeature is *very* important in wild mushroom identification due to it being an easily observed, taxonomically significant feature which almost always signifies a member of Amanitaceae. This has particular importance due to the disproportionately high number of deadly poisonous species contained within that family.

Zoospore

A zoospore is a motile asexual spore utilising a flagellum for locomotion. Also called a swarm spore, these spores are created by some algae and fungi to propagate themselves.

Flagella

There are two types of flagellated zoospores, *tinsel* or "decorated", and *whiplash*.

Tinsellated flagella have lateral filaments perpendicular to the main axis which allow for more surface area, and disturbance of the medium, giving it the property of a rudder, that is, the purpose of being used for steering.

- *Whiplash* flagella are straight, to power the zoospore through its medium. There is also the 'default' zoospore, which only has the propelling, 'whiplash' flagella.
- Both *tinsel* and *whiplash* flagella beat in a sinusoidal wave pattern, but when both are present, the tinsel will beat in the opposite direction of the whiplash, to give 2 axes of control of motility.
- There can be many combinations for location of the flagella, such as posterior whiplash; posterior whiplash, anterior tinsel; and anterior whiplash.

The phyla *Chytridiomycota* (Kingdom Fungi), *Oomycota* (Kingdom Chromista), and Hyphochytridiomycota within (Kingdom Chromista), produce zoospores with flagella in the same order as described above (e.g. Hyphochytridiomycota produces anterior whiplash and none else). They are also enumerated in species by 1000+, 580, and 16 respectively.

Zoosporangium

A *zoosporangium* is the sexual structure in which the zoospores develop in a plant, fungi, or protists (such as the oomycota).

Zygospore

A zygospore is a sexual part of a fungus, a chlamydospore that is created by the nuclear fusion of haploid hyphae of different mating types. A zygospore remains dormant while it waits for environmental cues, such as moisture, heat, or chemicals secreted by plants. When the environment is favourable, the zygospore germinates, meiosis occurs, and a sporangium is produced at the end of a sporangiophore. The sporangium sheds spores. A fungus that forms zygospores is called a zygomycete, indicating that the class is characterised by this evolutionary development.

Bibliography

Aberdeen, J. E. C.: *Introduction to the Mushrooms: Toadstools and Larger Fungi of Queensland,* Queensland Naturalists Club, Brisbane, 1979.

Ahmadjian, V. and Paracer, *Symbiosis: An Introduction to Biological Associations,* University Press, New England, 1986.

Ainsworth, G. C.: *The Fungi: An Advanced Treatise,* Academic Press, New York, 1973.

——————— : *Ainsworth and Bisby's Dictionary of the Fungi,* Commonwealth Mycol, Kew, 1971.

——————— : *Introduction to the History of Mycology,* Cambridge University Press, Cambridge, 1976.

Ajello, L. Georg L.K. Kaplan W. and Kaufman L.: *Laboratory Manual for Medical Mycology,* US Publ., New York 1963.

Amaranthus, M. P. and Luoma D. L.: *Diversity of Ectomycorrhizalfungi in Forest Ecosystems,* International Symposium on Microbiology—Santos, Brazil, 1997.

Ames, L. M.: *A Monograph of the Chaetomiaceae,* MacMillan, New York, 1963.

Ammirati, Joseph F. James A.: *Poisonous Mushrooms of the Northern United States and Canada,* University of Minnesota Press, Minneapolis, 1985.

Arnold, Katya and Swope Sam: *Katya's Book of Mushrooms,* Henry Holt and Co., New York, 1997.

Arora, David and Bowers J.: *Mushrooms of the World Coloring Book*, Dover Publications, New York, 1984.

Arora, David: *All that the Rain Promises*, Ten Speed Press, Berkeley, 1991.

————————: *Mushrooms Demystified*, Ten Speed Press, Berkeley, 1986.

Arx, J. A.: *The Genera of Fungi Sporulating in Pure Culture*, J. Cramer, Vaduz, 1981.

Barnes, R. S. K.: *Diversity of Living Organisms*, Blackwell Publishers, New Jersey, 2002.

Barnett, J. A. and R. J. Pankhurst: *A New Key to the Yeasts*, North Holland Publ. Co., Amsterdam, 1974.

Barnett, H. L. and Hunter B. B.: *Illustrated Genera of Imperfect Fungi*, MacMillan, New York, 1987.

Barron, G. L.: *The Genera of Hyphomycetes from Soil*, Williams and Wilkins, Baltimore, 1968.

Barron, George: *Mushrooms of Northeast North America*, Lone Pine Publishing, Renton, 1999.

Bell, A.: *Dung Fungi: An Illustrated Guide to Coprophilous Fungi in New Zealand*, Victoria University Press, New Zealand, 1983.

Beneke, E. S.: *Medical Mycology Laboratory Manual*, Burgess Publ. Co., Minneapolis, 1958.

Benjamin, D. R.: *Mushrooms: Poisons and Panaceas*, WH Freeman and Company, New York, 1995.

Bessette, Alan and Sundberg, Walter J.: *Mushrooms: A Quick Reference Guide to Mushrooms of North America*, MacMillan Publishing Co., New York, 1987.

————————: *Mushrooms of North America in Color: A Field Guide Companion to Seldom-Illustrated Fungi*, Syracuse University Press, Syracuse, 1995.

Bessette, Alan E. Roody, William C.: *North American Boletes*, Syracuse University Press, Syracuse, 2000.

——————— : *North American Boletes: A Color Guide to the Fleshy Pored Mushrooms*, Syracuse University Press, Syracuse, 2000.

Bigelow, Howard E.: *North American Species of Clitocybe*, J. Cramer, Vaduz, 1985.

Bissett, J.: *Mariannaea* and *Paecilomyces*, Canada Agric., Ottawa, 1979.

Boerema, G. H. and Dorenbosch M. M. J.: *The Phoma* and *Ascochyta Species Described by Wollenweber and Hochapfel in Their Study on Fruit-Rotting*, C.B.S. Stud. Mycol, 1973.

Booth, C.: *Methods in Microbiology*, Academic Press, New York, 1971.

Bougher, N. and Syme, K.: *Fungi of Southern Australia*, University of Western Australia Press, Perth, 1998.

Breitenbach, J. and Kranzlin F.: *Entolomataceae, Pluteaceae, Amanitaceae, Agaricaceae, Coprinaceae, Strophariaceae*, Verlag Mykologia, Luzern, 1995.

——————— : *Strobilomycetaceae, Boletaceae, Paxillaceae, Gomphidiaceae, Hygrophoraceae, Tricholomataceae, Polyporaceae (lamellate)*, Verlag Mykologia, Luzern, 1991.

——————— : *Fungi of Switzerland: Non-gilled Fungi*, Verlag Mykologia, Lucerne, 1986.

Brodie, Harold J.: *The Bird's Nest Fungi*, University of Toronto Press, Toronto, 1975.

Brusca, R. C. and Brusca G. J.: *Invertebrates*, Sinauer Associates, Sunderland, 1990.

Buchanan, E. E. and Gibbons N. E.: *Bergey's Manual of Determinative Bacteriology*, Williams and Wilkins, Baltimore, 1974.

Burgess, L. W., Liddell C. M. and Summerell B. A.: *Laboratory Manual for Fusarium Research*, University of Sydney, Sydney, 1988.

Cameron, Eleanor: *Stowaway to the Mushroom Planet*, Little Brown and Company, Boston, 1956.

Carmichael, J. W. and Kendrick W. B.: *Genera of Hyphomycetes*, Alberta Press, Edmonton, 1980.

Chamuris, George P.: *The Non-Stipitate Steriod Fungi in the Northeastern United States and Adjacent*, Berlin, J., Canada, 1988.

Cole, M. Fuhrer B. and Holland A.: *A Field Guide to the Common Genera of Gilled Fungi of Australia*, Inkata Press, Melbourne, 1984.

Cooke, R. C.: *Fungi, Man and His Environment*, Longman, London, 1977.

Corner, E. J. H.: *A Monograph of Cantharelloid Fungi*, Oxford University Press, London, 1996.

——————— : *A Monograph of Clavaria and Allied Genera*, Oxford University Press, London, 1950.

——————— : *Supplement to a Monograph of Clavaria and Allied Genera*, Oxford University Press, London. 1970.

Currah, R. S.: *An Annotated Key to the Genera of the Onygenales*, Systema, Ascomycetum, 1988.

Dashorst, G. R. M. and Jessop J. P.: *Plants of the Adelaide Hills*, Kangaroo Press, Adelaide, 1990.

Delafosse, Claude and Heliadore: *Le Champignon*, Academic Press, France, 1995.

Dennis, R. W. G.: *British Ascomycetes*, J. Cramer, Vaduz, 1981.

Desjardin, Dennis E.: *The Agaricales (Gilled Fungi)*, Mad River Press, California, 1987.

——————— : *Tricholomataceae I: Marasmioid Fungi*, Mad River Press, California, 1987.

Dix, N. J. and Webster, J.: *Fungal Ecology*, Chapman & Hall, London, 1995.

Domsch, K. H. W.: *Compendium of Soil Fungi*, Academic Press, London, 1980.

Dring, D. M.: *Techniques for Microscopic Preparation, Methods in Microbiology,* Academic Press, New York, 1971.

Eberhart, J. L. and Luoma, D. L.: *Truncocolumella Citrina: Pseudotsuga Menziesii,* Victoria B.C., Canadian, 1996.

Ellis, M. B.: *More Dematiaceous Hyphomycetes,* Commonwealth Mycol, Kew, 1976.

Findlay, W. P. K.: *Fungi: Folklore, Fiction and Fact,* Richmond Publishing Company, London, 1982.

Fischer, David W. and Alan E. Bessette: *Edible Wild Mushrooms of North America,* University of Texas Press, Austin, 1992.

Frederick, K: Sparrow and Alfred S.: *A Taxonomic Review with Keys: Ascomycetes and Fungi Imperfecti,* Academic Press, New York, 1973.

Fuhrer, B.: *A Field Companion to Australian Fungi,* Bloomings Books, Melbourne, 2001.

Fuhrer, B. and Robinson R.: *Rainforest Fungi of Tasmania and south-east Australia,* CSIRO, Melbourne, 1992.

Fuller, M. S.: *Lower Fungi in the Laboratory,* Georgia Press, Athens, 1978.

Gilbertson, R. L. and Ryvarden L.: *North American Polypores,* Oslo, Fungiflora, 1987.

—————— : *Fungi That Decay Ponderosa Pine,* University of Arizona Press, Tuscon, 1974.

Gilman, J. C.: *A Manual of Soil Fungi,* Iowa State Coll. Press, Ames, 1957.

Ginsburg, Mirra: *Mushrooms in the Rain,* MacMillan Publishing Co., New York, 1974.

Glick, Phyllis G.: *The Mushroom Trail Guide,* Holt, Rinehart and Winston, 1979.

Gregory, P. H.: *The Microbiology of the Atmosphere,* Leonard Hill, Aylesbury, 1973.

Griffiths, K.: *A Field Guide to the Larger Fungi of the Darling Scarp and South West of Western Australia*, Privately Published, Perth, 1985.

Guba, E. F.: *Monograph of Monochaetia and Pestalotia*, Harvard University Press, Cambridge, 1961.

Halling, Roy E.: *The Genus Collybia (Agaricales) in the Northeastern United States and Adjacent*, Cramer, Canada, 1983.

Hawksworth, D. L. Kirk P. M.: *Ainsworth & Bisby's Dictionary of the Fungi*, CAB International, New York, 1995.

Heller, Ruth: *Plants that Never Bloom*, Grosset & Dunlap, New York, 1984.

Hennebert, G. L.: *Echinobotryum, Wardomyces* and *Mammaria*, Trans. Brit. Mycol, 1968.

Hesler, L. R. and Alexander H. Smith: *North American Species of Crepidotus*, Hafner Publishing Company, New York, 1965.

————————: *North American Species of Hygrophorus*, University of Tennessee Press, Knoxville, 1963.

————————: *North American Species of Lactarius*, University of Michigan Press, Ann Arbor, 1979.

————————: *North American Species of Gymnopilus*, Hafner Publishing Company, New York, 1969.

Hoog, G. S. de and J. Guarro: *Atlas of Clinical Fungi*, C.B.S., Baarn, Netherlands, 1995.

————————: *The Black Yeasts and Allied Hyphomycetes*, C.B.S. Stud. Mycol, 1977.

Hoog, G. S. de, M. TH. Smith and E. Guého: *A revision of Geotrichum*, C.B.S. Stud., Mycol,1986.

Hudler, George: *Magical Mushrooms, Mischievous Molds*, Princeton University Press, Princeton, 1998.

Huffman, Donald M.: *Mushrooms and Other Fungi of the Midcontinental United States*, Iowa State University Press, Iowa, 1989.

Ingold, C. T.: *Dispersal in Fungi,* Oxford University Press, London, 1953.

Inui, T. Takeda Y. and Iizuka H.: *Taxonomical Studies on Genus Rhizopus,* MacMillan, New York, 1965.

Jenkins, David T.: *Amanita of North America,* Mad River Press, Eureka, 1986.

Johnson, Emily: *North American Mushrooms Photo Postcards,* Dover Publications, New York, 1988.

Johnson, Sylvia: *Mushrooms,* Lerner Publishing Company, Minneapolis, 1982.

Jong, S. C. and Davis E. E.: *Contibution to the Knowledge of Stachybotrys and Memnoniella in Culture,* Mycotaxon Press, Sydney, 1976.

Jordan, Michael: *The Encyclopedia of Fungi of Britain and Europe,* David & Charles, Devon, 1995.

Jordan, Peter and Wheeler, Steven: *The Ultimate Mushroom Book,* Smithmark Publishers, New York, 1995.

Katsaros, Peter: *Familiar Mushrooms, North America,* The Audubon Society Pocket Guides, New York, 1990.

—————— : *Illustrated Guide to Common Slime Molds,* Mad River Press, Eureka, 1984.

Kaveler, Lucy: *The Wonders of Fungi,* John Day Co., New York, 1964.

Kaye, T. Liston A.: *Conservation and Management of Native Plants and Fungi,* Native Plant Society of Oregon, Corvallis, 1997.

Kendrick, B.: *The Fifth Kingdom,* Mycologue Publications, Sidney, 2001.

Kendrick, Bryce: *A Young Person's Guide to the Fungi,* Mycologue, Canada, 1986.

Kibby, Geoffrey: *An Illustrated Guide to Mushrooms and Other Fungi of North America,* Dragons World Ltd., London, 1993.

Kirk, P. M. Cannon, P. F.: *Ainsworth and Bisby's Dictionary of the Fungi,* CABI Publishing, Eureka, 2001.

Krahn, Fernando: *Sebastian and the Mushroom,* Delacorte Press, New York,1976.

Kulik, M. M.: *A Compilation of Descriptions of New Penicillium Species,* U.S. Dept., Sydney, 1968.

Kuyper, Thomas W.: *A Revision of the Genus Inocybe in Europe,* Rijksherbarium, Leiden, 1986.

——————— : *Subgenus Insperma and the Smooth—Spored Species of Subgenus Inocybe,* Rijksherbarium, Leiden, 1986.

Laessoe, Thomas and Lincoff, Gary: *Eyewitness Handbooks/ Mushrooms,* D.K. Publishing, New York, 1998.

Laessoe, Thomas: *The Mushroom Book,* D.K. Publishing, New York, 1996.

Largent, David L. and Harry D.: *How to Identify Mushrooms to Genus II: Field Identificaton of Genera,* Mad River Press, Eureka, 1987.

Largent, David L.: *Entolomatoid Fungi of the Western United States and Alaska,* Mad River Press, Eureka, 1994.

——————— : *How to Identify Mushrooms to Genus?: Macroscopic Features,* Mad River Press, Eureka, 1986.

——————— : *The Agaricales (Gilled Fungi) of California: Hygrophoraceae,* Mad River Press, Eureka, 1985.

Larsen, M. J.: *Phellinus (Hymenochaetaceae): A Survey of World Taxa,* Fungiflora, Oslo, 1990.

Larson, D. W.: *Laboratory Response of Net CO_2 Exchange to Moisture, Temperature and Light,* J. Bot, Canadian, 1987.

Lilly, V. G.: *The Chemical Environment for Fungal Growth,* Academic Press, New York, 1965.

Lincoff, Gary H.: *The Audobon Society Field Guide to North American Mushrooms,* Alfred A. Knopf, New York, 1981.

——————— : *Simon and Schuster's Guide to Mushrooms,* Simon and Schuster, New York, 1981.

Lingelbach, Jenipher: *Hands-on Nature,* Vermont Institute of Natural Science, Woshington, 1985.

Lodder, J.: *The Yeasts: A Taxonomic Study,* North Holland Publ. Co., Amsterdam, 1970.

Luoma, D. L. and Trappe J. M.: *Mycorrhizal Fungi in Ecosystems,* Annual Meeting & Symposium, Bend, 1992.

Luoma, D. L. Eberhart J. L.: *Biodiversityof Ectomycorrhizal Types from Southwest Oregon,* Native Plant Society of Oregon, Corvallis, 1997.

—————— : *Communitystructure and Dynamics of Ectomycorrhizal Fungi in Managed Forest Stands,* USDA Forest Service Gen.Tech., Portland, 1996.

Macdonald, R. and Westerman, J.: *A Field Guide to Fungi of South-eastern Australia,* Nelson, Melbourne, 1979.

Margulis, L. and Schwartz K. V.: *Five Kingdoms: An Illustrated Guide to the Phyla of Life on Earth,* W. H. Freeman, New York, 1998.

Marr, Currie D. and Daniel E. Stuntz: *Ramaria of Western Washington (Bibliotheca Mycologica, Band 38),* J. Cramer, Vaduz, 1973.

Martin, G. W.: *Revision of the North Central Tremellales,* State University of Iowa, Iowa, 1952.

Matsushima, T.: *Lcones Fungorum a Matsushima Lectorum,* Kobe, Japan, 1975.

Matthes-Sears, U.: *In Situ Diurnal Field Measurements of Two Sites on a Coast-inland Gradient,* J. Bot, Canadian, 1986.

McCann, I. R.: *Australian Fungi Illustrated,* Macdown Productions, Melbourne, 2003.

McGinnis, M. R.: *Laboratory Handbook of Medical Mycology,* Academic Press, New York, 1980.

McKenny, Margaret and Daniel E.: *The New Savory Wild Mushroom,* University of Washington Press, Seattle, 1987.

—————— : *The New Savory Wild Mushroom*, University of Washington Press, Washington, 1987.

McKnight, Kent H. and McKnight, Vera B.: *Peterson Field Guides, Mushrooms*, Houghton—Mifflin, Boston, 1987.

Methven, Andrew S.: *The Agaricales (Gilled Fungi) of California*, Mad River Press, Eureka, 1997.

—————— : *The Genus Clavariadelphus in North America*, J. Cramer, Berlin, 1990.

Metzler, Susan and Metzler, Van: *Texas Mushrooms: A Field Guide*, University of Texas.Press, Austin, 1992.

Miller, Orson K.: *Mushrooms of North America*, E. P. Dutton, New York, 1980.

Molina R. J.: *Developingan Inventory and Monitoring Protocol for Commercially Harvested Forestmushrooms*, Hillsboro, New Jersey, 1994.

Monsell, Mary Elise: *Crackle Creek*, Atheneum, New York, 1990.

Morquer, R. G. Viala, J.: *Contributiona l'étude Morphogénique du Genre Gliocladium*, Mycol, France, 1963.

Morton, F. J. and Smith G.: *The Genera Scopulariopsis, Microascus* and *Doratomyces*, Mycol, Mycol, 1963.

Moser, Meinhard: *Keys to Agarics and Boleti (Polyporales, Boletales, Agaricales, Russulales)*, Roger Phillips, London, 1983.

Nag, Raj T. R.: *Coelomycetous Anamorphs with Appendage-bearing Conidia*, Mycologue Publications, Canada, 1993.

O'Dell, T. E.: *Diversity and Conservation of Forest Fungi*, USDA Forest Service Gen. Tech., Portland, 1996.

Orr, F. G.: *The Genus Gymnoascus Baranetzky*, Mycopathol, Mycol, 1963.

Orr, Robert T. and Dorthy B. Orr: *Mushrooms of Western North America*, University of California Press, Bereley, 1979.

Orton, P. D.: *British Fungus Flora, Agarics and Boleti, Pluteaceae, Pluteus & Volvariella*, Royal Botanic Garden, Edinburgh, 1986.

Pegler, D. N. Læssøe T. and Spooner B. M.: *British Puffballs, Earthstars and Stinkhorns,* Royal Botanic Gardens, Kew, 1995.

————————: *British Truffles: A Revision of British Hypogeous Fungi,* Royal Botanic Gardens, Kew, 1993.

————————: *The Genus Lentinus: A World Monograph,* HMSO, London, 1983.

Perry, Phyllis J.: *Let's Learn About Mushrooms,* Harvey House, New York, 1974.

Petersen, Ronald H.: *The Genus Clavulinopsis in North America,* Hafner Publishing Co., New York, 1968.

————————: *Ramaria subgenus Echinoramaria,* J. Cramer, Vaduz, 1981.

Phillips, Roger: *Mushrooms of North America,* Brown, & Co., Boston, 1991.

Pitt, J.: *The Genus Penicillium,* Academic Press, New York, 1980.

Ramirez, C.: *Manual and Atlas of the Penicillia,* Elsevier Biomedical Press, New York, 1982.

Raper, K. B. and Fennell D. I.: *The Genus Aspergillus,* Williams and Wilkins Co., Baltimore, 1965.

Rebell, G. and Taplin D.: *Dermatophytes: Their Recognition and Identification,* University Miami Press, Coral Gables, 1970.

Rice, Miriam: *Mushrooms for Color,* Mad River Press, Eureka, 1980.

Rifai, M. A.: *A Revision of the Genus Trichoderma,* Mycol, 1969.

Robinson, R.: *Fungi of the South-West Forests,* Department of Conservation and Land Management, Perth, 2003.

Romanova, Natalia: *Once There Was a Tree,* Dial Books, New York, 1985.

Samson, R. A. Hoekstra E. S.: *Interoduction to Food-borne Fungi,* C.B.S., Baarn, Netherlands, 1995.

—————— : *Revision of the Subsection Fasciulata of Penicillium and Some Allied Species*, C.B.S. Stud, Mycol, 1976.

—————— : *Paecilomyces* and *Some Allied Hyphomycetes*, C.B.S. Stud, Mycol, 1974.

Schaechter, Elio: *In the Company of Mushrooms*, Harvard University Press, Boston, 1997.

Schipper, M. A. A. and Stalpers J. A.: *A Revision of the Genus Rhizopus*, C.B.S. Stud., Mycol, 1984.

—————— : *Zygospore Ornamentation in the Genera Mucor and Zygorhynchus*, Persoonia, New York, 1975.

Selsam, Millicent: *Mushrooms*, William Morrow and Co., London, 1986.

Seth, H. K.: *A Monograph of the Genus Chaetomium*, Beih. Nova Hedw, 1970.

Shanks, Kris M.: *Gilled Fungi*, Mad River Press, Eureka, 1997.

Shepherd, C. J.: *Mushrooms and Toadstools of Australia*, Inkata Press, Sydney, 1988.

Shoemaker, R. A. and Kokko E. G.: *Trichurus Spiralis*, Canada Agric., Ottawa, 1977.

Short, P. S.: *History of Systematic Botany in Australasia*, Australian Systematic Botany Society, Melbourne, 1990.

Shuttleworth, Floyd S.: *A Golden Guide, Non-Flowering Plants*, Golden Press, 1967.

Singer, Rolf: *The Agaricales in Modern Taxonomy*, Koeltz Scientific Books, Koenigstein, 1986.

Sisson, Edith H.: *Nature With Children of All Ages*, Prentice Hall Press, New York, 1982.

Smith, A. H. and Zeller S. M.: *A Preliminary Account of the North American Species of Rhizopogon*, Botanical Gardens, Bronx, New York, 1966.

Smith, A. H. and Thiers H. D.: *The Boletes of Michigan*, University of Michigan Press, Ann Arbor, 1971.

Smith, Alexander H.: *The North American Species of Pholiota,* Hafner Publishing Company, New York, 1968.

——————— : *North American Species of Mycena,* University of Michigan Press, Ann Arbor, 1947.

——————— : *A Field Guide to Western Mushrooms,* University of Michigan Press, Ann Arbor, 1975.

Snell, Walter H. and Esther A. Dick: *A Glossary of Mycology,* Harvard University Press, Cambridge, 1971.

——————— : *The Boleti of Northeastern North America,* J. Cramer, Lehre, 1970.

Stamets, Paul: *Psilocybin Mushrooms of the World,* Ten Speed Press, Berkeley, 1996.

States, Jack S.: *Mushrooms and Truffles of the Southwest,* University of Arizona, Tucson, 1990.

Stevenson, Steven L. and Stempen, Henry: *Myxomycetes, A Handbook of Slime Molds,* Timber Press, Portland, 1994.

St.Germain, G. and R. Summerbell: *Identifying Filamentous Fungi,* Star Publ. Co., Belmont, California, 1996.

Stolk, A. C. and Samson R. A.: *The Ascomycete Genus Eupenicillium and Related Penicillium Anamorphs,* C.B.S. Stud. Mycol, 1983.

Storer, T. I., Usinger R. L.: *General Zoology,* McGraw-Hill Book Company, New York, 1979.

Tesar, Jenny: *Fungi, Our Living World Series,* Blackbirch Press, Woodbridge, 1994.

Thiers, Harry D.: *Gomphidiaceae,* Mad River Press, Eureka, 1985.

Tokumasu, S.: *Notes on Japanese Oidiodendron,* Trans. Mycol, Japan, 1973.

Tylutki, Edmund E.: *Mushrooms of Idaho and the Pacific Northwest, Discomycetes,* University of Idaho Press, Moscow, 1979.

——————— : *Mushrooms of Idaho and the Pacific Northwest: Non-Gilled Hymenomycetes,* University of Idaho Press, Moscow, 1987.

Ulloa, Miguel and Richard T. Hanlin: *Illustrated Dictionary of Mycology*, APS Press, Paul, 2000.

Van Oorschot, C. A. N.: *Taxonomy of the Dactylaria Complex, V. A Review of Arthrobotrys and Related Genera*, C.B.S. Stud., Mycol, 1985.

Waksman, S. A.: *The Actinomycetes*, Williams and Wilkins Co., Baltimore, 1961.

Wang, C. J. K.: *Identification Manual for Fungi from Utility Poles in the Eastern United States*, Type Culture Collection, Rockville, 1990.

Watling, Roy and Evelyn Turnbull: *British Fungus Flora: Agarics and Boleti*, Royal Botanic Garden, Edinburgh, 1998.

Watling, Roy, Norma M. Gregory and Orton P. D.: *British Fungus Flora, Agarics and Boleti*, Royal Botanic Garden, Edinburgh, 1993.

Watling, Roy: *Fungi*, Smithsonian Institution Press, Washington, 2003.

————————: *How to Identify Mushrooms to Genus?: Cultural and Developmental Features*, Mad River Press, Eureka, 1987.

Weber, Nancy Smith: *A Morel Hunter's Companion*, Two Penninsula Press, Lansing, 1988.

Webster, Vera R.: *Plant Experiments*, Childrens Press, Chicago, 1982.

Young, A. M.: *Common Australian Fungi*, New South Wales University Press, Sydney, 2000.

Zambettakis, C.: *Recherches Anatomiqueset Biologiques Sur Sphaeropsidales-Phaeodidymae Des Fungi Imperfecti*, Arch. Mus. Nation, Nat, 1955.

———————— : *Recherches Sur la Systématique Des Sphaeropsidales-Phaeodidymae*, Mycol, France, 1954.

Zycha, H. and Siepmann R.: *Mucorales, Eine Beschreibung Aller Gattungen and Arten Dieser Pilzgruppe*, PHBS Press, Lehre, 1969.

Index

D

E

F

———————— : *North American Boletes: A Color Guide to the Fleshy Pored Mushrooms*, Syracuse University Press, Syracuse, 2000.

Bigelow, Howard E.: *North American Species of Clitocybe*, J. Cramer, Vaduz, 1985.

Bissett, J.: *Mariannaea* and *Paecilomyces*, Canada Agric., Ottawa, 1979.

Boerema, G. H. and Dorenbosch M. M. J.: *The Phoma and Ascochyta Species Described by Wollenweber and Hochapfel in Their Study on Fruit-Rotting*, C.B.S. Stud. Mycol, 1973.

Booth, C.: *Methods in Microbiology*, Academic Press, New York, 1971.

Bougher, N. and Syme, K.: *Fungi of Southern Australia*, University of Western Australia Press, Perth, 1998.

Breitenbach, J. and Kranzlin F.: *Entolomataceae, Pluteaceae, Amanitaceae, Agaricaceae, Coprinaceae, Strophariaceae*, Verlag Mykologia, Luzern, 1995.

———————— : *Strobilomycetaceae, Boletaceae, Paxillaceae, Gomphidiaceae, Hygrophoraceae, Tricholomataceae, Polyporaceae (lamellate)*, Verlag Mykologia, Luzern, 1991.

———————— : *Fungi of Switzerland: Non-gilled Fungi*, Verlag Mykologia, Lucerne, 1986.

Brodie, Harold J.: *The Bird's Nest Fungi*, University of Toronto Press, Toronto, 1975.

Brusca, R. C. and Brusca G. J.: *Invertebrates*, Sinauer Associates, Sunderland, 1990.

Buchanan, E. E. and Gibbons N. E.: *Bergey's Manual of Determinative Bacteriology*, Williams and Wilkins, Baltimore, 1974.

Burgess, L. W., Liddell C. M. and Summerell B. A.: *Laboratory Manual for Fusarium Research*, University of Sydney, Sydney, 1988.

Cameron, Eleanor: *Stowaway to the Mushroom Planet*, Little Brown and Company, Boston, 1956.

Carmichael, J. W. and Kendrick W. B.: *Genera of Hyphomycetes*, Alberta Press, Edmonton, 1980.

Chamuris, George P.: *The Non-Stipitate Steriod Fungi in the Northeastern United States and Adjacent*, Berlin, J., Canada, 1988.

Cole, M. Fuhrer B. and Holland A.: *A Field Guide to the Common Genera of Gilled Fungi of Australia*, Inkata Press, Melbourne, 1984.

Cooke, R. C.: *Fungi, Man and His Environment*, Longman, London, 1977.

Corner, E. J. H.: *A Monograph of Cantharelloid Fungi*, Oxford University Press, London, 1996.

—————— : *A Monograph of Clavaria and Allied Genera*, Oxford University Press, London, 1950.

—————— : *Supplement to a Monograph of Clavaria and Allied Genera*, Oxford University Press, London. 1970.

Currah, R. S.: *An Annotated Key to the Genera of the Onygenales*, Systema, Ascomycetum, 1988.

Dashorst, G. R. M. and Jessop J. P.: *Plants of the Adelaide Hills*, Kangaroo Press, Adelaide, 1990.

Delafosse, Claude and Heliadore: *Le Champignon*, Academic Press, France, 1995.

Dennis, R. W. G.: *British Ascomycetes*, J. Cramer, Vaduz, 1981.

Desjardin, Dennis E.: *The Agaricales (Gilled Fungi)*, Mad River Press, California, 1987.

—————— : *Tricholomataceae I: Marasmioid Fungi*, Mad River Press, California, 1987.

Dix, N. J. and Webster, J.: *Fungal Ecology*, Chapman & Hall, London, 1995.

Domsch, K. H. W.: *Compendium of Soil Fungi*, Academic Press, London, 1980.

Dring, D. M.: *Techniques for Microscopic Preparation, Methods in Microbiology*, Academic Press, New York, 1971.

Eberhart, J. L. and Luoma, D. L.: *Truncocolumella Citrina: Pseudotsuga Menziesii*, Victoria B.C., Canadian, 1996.

Ellis, M. B.: *More Dematiaceous Hyphomycetes*, Commonwealth Mycol, Kew, 1976.

Findlay, W. P. K.: *Fungi: Folklore, Fiction and Fact*, Richmond Publishing Company, London, 1982.

Fischer, David W. and Alan E. Bessette: *Edible Wild Mushrooms of North America*, University of Texas Press, Austin, 1992.

Frederick, K. Sparrow and Alfred S.: *A Taxonomic Review with Keys: Ascomycetes and Fungi Imperfecti*, Academic Press, New York, 1973.

Fuhrer, B.: *A Field Companion to Australian Fungi*, Bloomings Books, Melbourne, 2001.

Fuhrer, B. and Robinson R.: *Rainforest Fungi of Tasmania and south-east Australia*, CSIRO, Melbourne, 1992.

Fuller, M. S.: *Lower Fungi in the Laboratory*, Georgia Press, Athens, 1978.

Gilbertson, R. L. and Ryvarden L.: *North American Polypores*, Oslo, Fungiflora, 1987.

—————— : *Fungi That Decay Ponderosa Pine*, University of Arizona Press, Tuscon, 1974.

Gilman, J. C.: *A Manual of Soil Fungi*, Iowa State Coll. Press, Ames, 1957.

Ginsburg, Mirra: *Mushrooms in the Rain*, MacMillan Publishing Co., New York, 1974.

Glick, Phyllis G.: *The Mushroom Trail Guide*, Holt, Rinehart and Winston, 1979.

Gregory, P. H.: *The Microbiology of the Atmosphere*, Leonard Hill, Aylesbury, 1973.

Griffiths, K.: *A Field Guide to the Larger Fungi of the Darling Scarp and South West of Western Australia*, Privately Published, Perth, 1985.

Guba, E. F.: *Monograph of Monochaetia and Pestalotia,* Harvard University Press, Cambridge, 1961.

Halling, Roy E.: *The Genus Collybia (Agaricales) in the Northeastern United States and Adjacent*, Cramer, Canada, 1983.

Hawksworth, D. L. Kirk P. M.: *Ainsworth & Bisby's Dictionary of the Fungi,* CAB International, New York, 1995.

Heller, Ruth: *Plants that Never Bloom,* Grosset & Dunlap, New York, 1984.

Hennebert, G. L.: *Echinobotryum, Wardomyces* and *Mammaria,* Trans. Brit. Mycol, 1968.

Hesler, L. R. and Alexander H. Smith: *North American Species of Crepidotus,* Hafner Publishing Company, New York, 1965.

—————— : *North American Species of Hygrophorus,* University of Tennessee Press, Knoxville, 1963.

——————: *North American Species of Lactarius,* University of Michigan Press, Ann Arbor, 1979.

——————: *North American Species of Gymnopilus,* Hafner Publishing Company, New York, 1969.

Hoog, G. S. de and J. Guarro: *Atlas of Clinical Fungi,* C.B.S., Baarn, Netherlands, 1995.

——————: *The Black Yeasts and Allied Hyphomycetes,* C.B.S. Stud. Mycol, 1977.

Hoog, G. S. de, M. TH. Smith and E. Guého: *A revision of Geotrichum,* C.B.S. Stud., Mycol,1986.

Hudler, George: *Magical Mushrooms, Mischievous Molds,* Princeton University Press, Princeton, 1998.

Huffman, Donald M.: *Mushrooms and Other Fungi of the Midcontinental United States,* Iowa State University Press, Iowa, 1989.

Ingold, C. T.: *Dispersal in Fungi*, Oxford University Press, London, 1953.

Inui, T. Takeda Y. and Iizuka H.: *Taxonomical Studies on Genus Rhizopus*, MacMillan, New York, 1965.

Jenkins, David T.: *Amanita of North America*, Mad River Press, Eureka, 1986.

Johnson, Emily: *North American Mushrooms Photo Postcards*, Dover Publications, New York, 1988.

Johnson, Sylvia: *Mushrooms*, Lerner Publishing Company, Minneapolis, 1982.

Jong, S. C. and Davis E. E.: *Contibution to the Knowledge of Stachybotrys and Memnoniella in Culture*, Mycotaxon Press, Sydney, 1976.

Jordan, Michael: *The Encyclopedia of Fungi of Britain and Europe*, David & Charles, Devon, 1995.

Jordan, Peter and Wheeler, Steven: *The Ultimate Mushroom Book*, Smithmark Publishers, New York, 1995.

Katsaros, Peter: *Familiar Mushrooms, North America*, The Audubon Society Pocket Guides, New York, 1990.

————————: *Illustrated Guide to Common Slime Molds*, Mad River Press, Eureka, 1984.

Kaveler, Lucy: *The Wonders of Fungi*, John Day Co., New York, 1964.

Kaye, T. Liston A.: *Conservation and Management of Native Plants and Fungi*, Native Plant Society of Oregon, Corvallis, 1997.

Kendrick, B.: *The Fifth Kingdom*, Mycologue Publications, Sidney, 2001.

Kendrick, Bryce: *A Young Person's Guide to the Fungi*, Mycologue, Canada, 1986.

Kibby, Geoffrey: *An Illustrated Guide to Mushrooms and Other Fungi of North America*, Dragons World Ltd., London, 1993.

Kirk, P. M. Cannon, P. F.: *Ainsworth and Bisby's Dictionary of the Fungi*, CABI Publishing, Eureka, 2001.

Krahn, Fernando: *Sebastian and the Mushroom*, Delacorte Press, New York,1976.

Kulik, M. M.: *A Compilation of Descriptions of New Penicillium Species*, U.S. Dept., Sydney, 1968.

Kuyper, Thomas W.: *A Revision of the Genus Inocybe in Europe*, Rijksherbarium, Leiden, 1986.

—————— : *Subgenus Insperma and the Smooth—Spored Species of Subgenus Inocybe*, Rijksherbarium, Leiden, 1986.

Laessoe, Thomas and Lincoff, Gary: *Eyewitness Handbooks/ Mushrooms*, D.K. Publishing, New York, 1998.

Laessoe, Thomas: *The Mushroom Book*, D.K. Publishing, New York, 1996.

Largent, David L. and Harry D.: *How to Identify Mushrooms to Genus II: Field Identificaton of Genera*, Mad River Press, Eureka, 1987.

Largent, David L.: *Entolomatoid Fungi of the Western United States and Alaska*, Mad River Press, Eureka, 1994.

—————— : *How to Identify Mushrooms to Genus?: Macroscopic Features*, Mad River Press, Eureka, 1986.

—————— : *The Agaricales (Gilled Fungi) of California: Hygrophoraceae*, Mad River Press, Eureka, 1985.

Larsen, M. J.: *Phellinus (Hymenochaetaceae): A Survey of World Taxa*, Fungiflora, Oslo, 1990.

Larson, D. W.: *Laboratory Response of Net CO_2 Exchange to Moisture, Temperature and Light*, J. Bot, Canadian, 1987.

Lilly, V. G.: *The Chemical Environment for Fungal Growth*, Academic Press, New York, 1965.

Lincoff, Gary H.: *The Audobon Society Field Guide to North American Mushrooms*, Alfred A. Knopf, New York, 1981.

—————— : *Simon and Schuster's Guide to Mushrooms*, Simon and Schuster, New York, 1981.

Lingelbach, Jenipher: *Hands-on Nature,* Vermont Institute of Natural Science, Woshington, 1985.

Lodder, J.: *The Yeasts: A Taxonomic Study,* North Holland Publ. Co., Amsterdam, 1970.

Luoma, D. L. and Trappe J. M.: *Mycorrhizal Fungi in Ecosystems,* Annual Meeting & Symposium, Bend, 1992.

Luoma, D. L. Eberhart J. L.: *Biodiversityof Ectomycorrhizal Types from Southwest Oregon,* Native Plant Society of Oregon, Corvallis, 1997.

———————— : *Communitystructure and Dynamics of Ectomycorrhizal Fungi in Managed Forest Stands,* USDA Forest Service Gen.Tech., Portland, 1996.

Macdonald, R. and Westerman, J.: *A Field Guide to Fungi of South-eastern Australia,* Nelson, Melbourne, 1979.

Margulis, L. and Schwartz K. V.: *Five Kingdoms: An Illustrated Guide to the Phyla of Life on Earth,* W. H. Freeman, New York, 1998.

Marr, Currie D. and Daniel E. Stuntz: *Ramaria of Western Washington (Bibliotheca Mycologica, Band 38),* J. Cramer, Vaduz, 1973.

Martin, G. W.: *Revision of the North Central Tremellales,* State University of Iowa, Iowa, 1952.

Matsushima, T.: *Lcones Fungorum a Matsushima Lectorum,* Kobe, Japan, 1975.

Matthes-Sears, U.: *In Situ Diurnal Field Measurements of Two Sites on a Coast-inland Gradient,* J. Bot, Canadian, 1986.

McCann, I. R.: *Australian Fungi Illustrated,* Macdown Productions, Melbourne, 2003.

McGinnis, M. R.: *Laboratory Handbook of Medical Mycology,* Academic Press, New York, 1980.

McKenny, Margaret and Daniel E.: *The New Savory Wild Mushroom,* University of Washington Press, Seattle, 1987.

———————— : *The New Savory Wild Mushroom,* University of Washington Press, Washington, 1987.

McKnight, Kent H. and McKnight, Vera B.: *Peterson Field Guides, Mushrooms,* Houghton—Mifflin, Boston, 1987.

Methven, Andrew S.: *The Agaricales (Gilled Fungi) of California,* Mad River Press, Eureka, 1997.

———————— : *The Genus Clavariadelphus in North America,* J. Cramer, Berlin, 1990.

Metzler, Susan and Metzler, Van: *Texas Mushrooms: A Field Guide,* University of Texas Press, Austin, 1992.

Miller, Orson K.: *Mushrooms of North America,* E. P. Dutton, New York, 1980.

Molina R. J.: *Developingan Inventory and Monitoring Protocol for Commercially Harvested Forestmushrooms,* Hillsboro, New Jersey, 1994.

Monsell, Mary Elise: *Crackle Creek,* Atheneum, New York, 1990.

Morquer, R. G. Viala, J.: *Contributiona l'étude Morphogénique du Genre Gliocladium,* Mycol, France, 1963.

Morton, F. J. and Smith G.: *The Genera Scopulariopsis, Microascus* and *Doratomyces,* Mycol, Mycol, 1963.

Moser, Meinhard: *Keys to Agarics and Boleti (Polyporales, Boletales, Agaricales, Russulales),* Roger Phillips, London, 1983.

Nag, Raj T. R.: *Coelomycetous Anamorphs with Appendage-bearing Conidia,* Mycologue Publications, Canada, 1993.

O'Dell, T. E.: *Diversity and Conservation of Forest Fungi,* USDA Forest Service Gen. Tech., Portland, 1996.

Orr, F. G.: *The Genus Gymnoascus Baranetzky,* Mycopathol, Mycol, 1963.

Orr, Robert T. and Dorthy B. Orr: *Mushrooms of Western North America,* University of California Press, Bereley, 1979.

Orton, P. D.: *British Fungus Flora, Agarics and Boleti, Pluteaceae, Pluteus & Volvariella,* Royal Botanic Garden, Edinburgh, 1986.

Pegler, D. N. Læssøe T. and Spooner B. M.: *British Puffballs, Earthstars and Stinkhorns,* Royal Botanic Gardens, Kew, 1995.

———————: *British Truffles: A Revision of British Hypogeous Fungi,* Royal Botanic Gardens, Kew, 1993.

——————— : *The Genus Lentinus: A World Monograph,* HMSO, London, 1983.

Perry, Phyllis J.: *Let's Learn About Mushrooms,* Harvey House, New York, 1974.

Petersen, Ronald H.: *The Genus Clavulinopsis in North America,* Hafner Publishing Co., New York, 1968.

——————— : *Ramaria subgenus Echinoramaria,* J. Cramer, Vaduz, 1981.

Phillips, Roger: *Mushrooms of North America,* Brown, & Co., Boston, 1991.

Pitt, J.: *The Genus Penicillium,* Academic Press, New York, 1980.

Ramirez, C.: *Manual and Atlas of the Penicillia,* Elsevier Biomedical Press, New York, 1982.

Raper, K. B. and Fennell D. I.: *The Genus Aspergillus,* Williams and Wilkins Co., Baltimore, 1965.

Rebell, G. and Taplin D.: *Dermatophytes: Their Recognition and Identification,* University Miami Press, Coral Gables, 1970.

Rice, Miriam: *Mushrooms for Color,* Mad River Press, Eureka, 1980.

Rifai, M. A.: *A Revision of the Genus Trichoderma,* Mycol, 1969.

Robinson, R.: *Fungi of the South-West Forests,* Department of Conservation and Land Management, Perth, 2003.

Romanova, Natalia: *Once There Was a Tree,* Dial Books, New York, 1985.

Samson, R. A. Hoekstra E. S.: *Interoduction to Food-borne Fungi,* C.B.S., Baarn, Netherlands, 1995.

—————— : *Revision of the Subsection Fasciulata of Penicillium and Some Allied Species*, C.B.S. Stud, Mycol, 1976.

—————— : *Paecilomyces* and *Some Allied Hyphomycetes*, C.B.S. Stud, Mycol, 1974.

Schaechter, Elio: *In the Company of Mushrooms*, Harvard University Press, Boston, 1997.

Schipper, M. A. A. and Stalpers J. A.: *A Revision of the Genus Rhizopus*, C.B.S. Stud., Mycol, 1984.

—————— : *Zygospore Ornamentation in the Genera Mucor and Zygorhynchus*, Persoonia, New York, 1975.

Selsam, Millicent: *Mushrooms*, William Morrow and Co., London, 1986.

Seth, H. K.: *A Monograph of the Genus Chaetomium*, Beih. Nova Hedw, 1970.

Shanks, Kris M.: *Gilled Fungi*, Mad River Press, Eureka, 1997.

Shepherd, C. J.: *Mushrooms and Toadstools of Australia*, Inkata Press, Sydney, 1988.

Shoemaker, R. A. and Kokko E. G.: *Trichurus Spiralis*, Canada Agric., Ottawa, 1977.

Short, P. S.: *History of Systematic Botany in Australasia*, Australian Systematic Botany Society, Melbourne, 1990.

Shuttleworth, Floyd S.: *A Golden Guide, Non-Flowering Plants*, Golden Press, 1967.

Singer, Rolf: *The Agaricales in Modern Taxonomy*, Koeltz Scientific Books, Koenigstein, 1986.

Sisson, Edith H.: *Nature With Children of All Ages*, Prentice Hall Press, New York, 1982.

Smith, A. H. and Zeller S. M.: *A Preliminary Account of the North American Species of Rhizopogon*, Botanical Gardens, Bronx, New York, 1966.

Smith, A. H. and Thiers H. D.: *The Boletes of Michigan*, University of Michigan Press, Ann Arbor, 1971.

Smith, Alexander H.: *The North American Species of Pholiota*, Hafner Publishing Company, New York, 1968.

————————: *North American Species of Mycena*, University of Michigan Press, Ann Arbor, 1947.

————————: *A Field Guide to Western Mushrooms*, University of Michigan Press, Ann Arbor, 1975.

Snell, Walter H. and Esther A. Dick: *A Glossary of Mycology*, Harvard University Press, Cambridge, 1971.

————————: *The Boleti of Northeastern North America*, J. Cramer, Lehre, 1970.

Stamets, Paul: *Psilocybin Mushrooms of the World*, Ten Speed Press, Berkeley, 1996.

States, Jack S.: *Mushrooms and Truffles of the Southwest*, University of Arizona, Tucson, 1990.

Stevenson, Steven L. and Stempen, Henry: *Myxomycetes, A Handbook of Slime Molds*, Timber Press, Portland, 1994.

St.Germain, G. and R. Summerbell: *Identifying Filamentous Fungi*, Star Publ. Co., Belmont, California, 1996.

Stolk, A. C. and Samson R. A.: *The Ascomycete Genus Eupenicillium and Related Penicillium Anamorphs*, C.B.S. Stud. Mycol, 1983.

Storer, T. I., Usinger R. L.: *General Zoology*, McGraw-Hill Book Company, New York, 1979.

Tesar, Jenny: *Fungi, Our Living World Series*, Blackbirch Press, Woodbridge, 1994.

Thiers, Harry D.: *Gomphidiaceae*, Mad River Press, Eureka, 1985.

Tokumasu, S.: *Notes on Japanese Oidiodendron*, Trans. Mycol, Japan, 1973.

Tylutki, Edmund E.: *Mushrooms of Idaho and the Pacific Northwest, Discomycetes*, University of Idaho Press, Moscow, 1979.

————————: *Mushrooms of Idaho and the Pacific Northwest: Non-Gilled Hymenomycetes*, University of Idaho Press, Moscow, 1987.

Ulloa, Miguel and Richard T. Hanlin: *Illustrated Dictionary of Mycology*, APS Press, Paul, 2000.

Van Oorschot, C. A. N.: *Taxonomy of the Dactylaria Complex, V. A Review of Arthrobotrys and Related Genera*, C.B.S. Stud., Mycol, 1985.

Waksman, S. A.: *The Actinomycetes*, Williams and Wilkins Co., Baltimore, 1961.

Wang, C. J. K.: *Identification Manual for Fungi from Utility Poles in the Eastern United States*, Type Culture Collection, Rockville, 1990.

Watling, Roy and Evelyn Turnbull: *British Fungus Flora: Agarics and Boleti*, Royal Botanic Garden, Edinburgh, 1998.

Watling, Roy, Norma M. Gregory and Orton P. D.: *British Fungus Flora, Agarics and Boleti*, Royal Botanic Garden, Edinburgh, 1993.

Watling, Roy: *Fungi*, Smithsonian Institution Press, Washington, 2003.

——————: *How to Identify Mushrooms to Genus?: Cultural and Developmental Features*, Mad River Press, Eureka, 1987.

Weber, Nancy Smith: *A Morel Hunter's Companion*, Two Penninsula Press, Lansing, 1988.

Webster, Vera R.: *Plant Experiments*, Childrens Press, Chicago, 1982.

Young, A. M.: *Common Australian Fungi*, New South Wales University Press, Sydney, 2000.

Zambettakis, C.: *Recherches Anatomiqueset Biologiques Sur Sphaeropsidales-Phaeodidymae Des Fungi Imperfecti*, Arch. Mus. Nation, Nat, 1955.

——————: *Recherches Sur la Systématique Des Sphaeropsidales-Phaeodidymae*, Mycol, France, 1954.

Zycha, H. and Siepmann R.: *Mucorales, Eine Beschreibung Aller Gattungen and Arten Dieser Pilzgruppe*, PHBS Press, Lehre, 1969.

Index

C

D

H

I

K

L

M

R

S

T

U

V

❑❑❑

ENCYCLOPAEDIA
OF
FUNGI

ENCYCLOPAEDIA
OF
FUNGI

Vol. 4

Edited by

A.M. Bagulia

ANMOL PUBLICATIONS PVT. LTD.
NEW DELHI - 110 002 (INDIA)

ANMOL PUBLICATIONS PVT. LTD.
H.O.: 4374/4B, Ansari Road, Darya Ganj,
New Delhi-110 002 (India)
Ph.: 23278000, 23261597
B.O.: No. 1015, Ist Main Road, BSK IIIrd Stage
IIIrd Phase, IIIrd Block,
Bangalore - 560 085 (India)
Visit us at: www.anmolpublications.com

Encyclopaedia of Fungi

First Published, 2008
ISBN 978-81-261-3526-4 (Set)

PRINTED IN INDIA

Printed at Mehra Offset Press, Delhi.

Contents

Preface

Fungi are a kingdom of eukaryotic organisms. The fungi are heterotrophic organisms characterised by a chintinous cell wall, and in majority of species, filamentous growth as multicellular hyphae forming a mycelium; while some fungal species grow as single cells. Sexual and asexual reproduction is via pores, often produced on specialised structures or in fruiting bodies. For example, yeasts, moulds, and mushrooms are fungi.

Fungi are found all over the world, and grow in a wide range of habitats, including desert. Most fungi grow in terrestrial environments, but several species occur only in aquatic habitats. Fungi along with bacteria are the primary decomposers of organic matter in almost all terrestrial ecosystems, worldwide. As per estimates there are 1.5 million fungal species of which around 70,000 have been marked. Most fungi grow as thread-like filaments called hyphae, which form mycelium, while others grow as single cells. Human use of fungi for preparation or preservation and other purpose is extensive: yeasts are required for fermentation of beer and bread, some other fungal species are used in the production of soy sauce and tempeh.

Mushroom farming and gathering is a large industry in many countries. Many fungi are producers of antibiotics, including β-lactam antibiotics, such as a pencillin and cephalosporin.

Some of the most well-known types of fungi are edible or poisonous mushrooms. Many species are commercially raised but others are grown in forests. *Agaricus bisporus,* sold as button mushrooms, when small and Protobello mushroom when grownup, are the most commonly eaten species, used in salads, soups, and many other dishes.

Many fungi have important symbiotic relationships with organisms from most, if not all kingdoms. The mycorrhiza symbiosis is most ancient and its existence dates back to 400 million years ago.

This eminent work covers all perfect and realistic features of fungi in an interesting manner. Expectedly, this work would create new waves in the discipline and would prove to be useful for all—students, scholars and general readers.

Editor

1

Control by Fungi

Organic Fungal Control

Plant diseases can be baffling for the gardener to diagnose, even experts can have problems! A variety of microorganisms including fungi, bacteria, viruses and nematodes cause plant disease, and they are known as pathogens.

It is worth remembering that the numbers of beneficial microorganisms far outweigh the plant pathogens. Beneficial microorganisms can live in a symbiotic relationship with plants, improving their fertility and disease resistance. Other microorganisms that live in the soil are predatory, and help to suppress plant diseases; this includes fungi that prey on nematodes. Most significant, though, is the cycling of organic material, within the soil, by the microorganisms, which cause decomposition. The organic gardener relies on this cycling to produce compost and build a healthy soil.

While it is helpful to identify plant disease, most organic solutions are aimed at being preventative, by improving the natural balance within the garden. Plant diseases have complex causes, often related to soil health and plant selection.

In addition to the organic remedies, all of the strategies for disease control will make a difference in the long term. The simplest strategy of all, however, is to always remove a sick looking plant and destroy it. This is putting the overall health of the garden above that of an individual plant's survival.

Diseases Including Fungus, Bacteria and Viruses Plants Affected Description and Damage Organic Remedies General strategies that apply to all diseases are listed below the chart

Fungal Diseases

Anthracnose Rose; fruit trees including avocado, mango, macadamia; wide range of vegetables including bean, watermelon, rockmelon, tomato, cucumber and lettuce In roses, can be confused with Black spot; begins as a black mark with well-defined margins but the centre turns grey and falls out, leaving a black edge. In fruits, circular, sunken spots or craters are formed. Prevalent in cool, humid weather. Spread by gardeners, tools, seed and rain. Select resistant varieties. Avoid overhead irrigation and improve air circulation. Remove and destroy infected plants. Use disease-free seed. On trees, prune all dead wood and twigs and burn prunings (where possible).

Black Spot Rose Black spots with fringed margins on the upper leaf surface, stems and branches. Leaves yellow and fall. Spread by wind, water, gardeners and insects. Worse in warm, humid conditions. Select resistant varieties, remove plants that are constantly infected. Avoid overhead watering and improve air circulation. Remove any damaged leaves as they appear, prune off old flower stems and destroy the prunings. Research has shown oat straw is the best mulch followed by pine bark. A deficiency of potassium makes roses more vulnerable, if needed add 150 g per plant of sulphate of potash. Spray seaweed regularly

Botrytis (Grey mould): Wide variety of plants including strawberry, grape, berry, ornamentals and rose Furry, grey fungus growth covers affected parts. Strawberries become soft, rotten and pale. Worse in cool, humid weather. Improve air circulation, remove lower, older foliage. Spray with Bordeaux when dormant, keep well mulched.

Brown rot Stonefruit Kills blossoms, rots fruit on trees or after harvesting; twig infections show small cankers with gum extruding. Infected fruit has buff coloured powdery spores on soft brown rot.

Stone fruit can be pruned as soon as the fruit is harvested. Clean up dead leaves, remove 'mummified' fruit and prune off diseased wood. Burning prunings (where possible) is the most effective way of destroying spores. Spray with Bordeaux in winter.

Collar Rot Citrus: This is a soil fungus that attacks the tree trunk at ground level and if left untreated can kill the tree. The first signs are splitting, oozing bark and yellowing foliage.

Avoid wetting the trunk when watering. Good air circulation is needed, prevent mulch or long grass building up around the base of the trunk. Avoid using a brushcutter close to fruit trees, as it is a common cause of injury. If the infected area extends fully around the tree trunk, then buy a new tree. If the tree is still viable then treat by cutting affected wood back to clean bark using a sharp knife or chisel. Shape the wound to allow for water drainage.

Damping off Seedlings Seeds rot before germination or the stem of the seedling shrivels at ground level Keep mulch clear of seedling stems in hot, humid weather. Use a good quality seed raising mix. Oxygen is important to the seed's ability to successfully germinate, so avoid overwatering and overcrowding.

Downy mildew Wide range of vegetables, ornamental plants and fruit trees including grape small, angular, light yellow or brownish spots appear on the upper surface of leaves. Underneath the leaf these spots produce white fluffy spores. It causes stunting, wilting, yellowing and the death of seedlings. It is common in cool, wet weather. Spread by wind. Avoid overhead irrigation. Select resistant varieties. Remove leaves as soon as they are affected. Spray vulnerable plants regularly with seaweed based product.

Peach leaf curl Stone fruit particularly peach, nectarine and almond. Thickened pinkish or green blisters on the leaves, tree may become completely defoliated. Spread by water. Collect mummified fruit; destroy by burning in a wood heater or burying in a deep hole. Spray at early bud swell with Bordeaux.

Powdery Mildew: Wide range of vegetables including pumpkin, cucumber, pea and turnip; ornamentals including rose and crepe myrtle; fruit trees including grape, pawpaw, strawberry and apple (particularly Jonathan, Gravenstein and Rome Beauty). Circular white powdery spots on older leaves and stems. These spread, the leaves die, cropping is reduced. Spread by wind. Worse in dry weather with evening dew. Remove leaves as soon as they are affected; thousands of new spores are produced within 4 days. Spray vulnerable plants regularly with seaweed based product. After rose pruning, spray with lime sulphur. For apples, prune out infected shoots in winter. For pawpaws, dust or spray with sulphur, only when below 24°C. Spray grapes when the shoots are first starting, with Bordeaux or wettable sulphur. Milk spray is effective.

Phytophthora spp. (Root rot fungus): Avocado, citrus and macadamia. Foliage starts to die, often on just one side; the tree dies from the top down. Most common in hot, moist conditions, especially with poor drainage. Fungus lives indefinitely in the soil. Beds should be raised to ensure good drainage. Applications of compost have been shown to reduce attack by this fungus.

Rust: Wide range of vegetables including bean, beetroot, silverbeet, spinach; ornamentals including azalea, rose, geranium; stone fruit Orange, brown, powdery pustules underneath the leaves. Leaves yellow and drop early, plant loses vigour. Spores are spread by water and wind.

Shot-hole and Freckle Stonefruit: Particularly Japanese plum and apricot Different symptoms but usually treated the same. Characteristics of shot-hole are brown leaf spots that fall out, and fruit develops scabs and gumming. Characteristics of Freckle are olive green spots on fruit and oval brown spots on twigs. Prune trees back hard. Spray with Bordeaux when dormant.

Sooty Mould: Wide range of plants including citrus and lillypilly. Some pests, including scales and aphids, secrete large amounts of 'honeydew', which sticks to the lower leaves where it is fed on by this fungus. It looks unattractive and interferes with photosynthesis. Honeydew also attracts ants, which feed on it. The ants can 'farm' the pests, protecting them from predators.

The first step is always to control any ants, as without their protection the natural predators will usually be able to keep the pests, and therefore the sooty mould, under control. Keep ants out by banding the trunks with a horticultural glue. Prune any low branches that are touching the ground and make sure tall stems of grass aren't providing an alternative route for the ants.

Verticillium Wilt and Fusarium Wilt: Wide range of plants including rose, tomato and potato. These soil fungi invade roots and cause yellowing, stunting and rapid wilting. Spread by water, seeds and tools. More prevalent in warm, dry weather. Use strategies for soilborne fungus. Remove and destroy infected plants. Ensure good drainage. Long crop rotation (8 years). Select resistant varieties.

Bacterial Diseases

Bacterial Soft Rots: Lettuce, carrot, celery, potato, corn, cabbage family Wet, slimy patches, often foul smelling. No spores present as this is not a fungus. Avoid overwatering and watering during the heat of the day. Use strategies for soilborne disease. Remove and destroy infected plants Crop rotation.

Select resistant varieties; open-hearted lettuces such as Cos and Mignonette lettuce are less vulnerable than crisphead types.

Black rot Cabbage family Discoloured areas on stems, in cross-section black strands can be seen, causes loss of vigour, rotting and death of seedlings. Spread by infected seed, wind, water and insects. Use strategies for soilborne disease. Remove and destroy infected plants. Crop rotation. Select resistant varieties. If Black rot has been a problem previously, soak cabbage seed in hot water at 50°C for 23 minutes, other brassica for 18 minutes, before sowing.

Bacterial canker (gummosis) Stonefruit, cherry Cankers on twigs, dark, splitting bark Prune immediately, disinfecting tools and hands between cuts. Spray with Bordeaux when dormant.

Virus

Mosaic Vegetables including cucurbits, potato, turnip; ornamentals including rose and iris Plants are severely stunted. Leaves are yellow and mottled, sometimes with curled leaf edges Spread by aphids and gardeners. Remove and destroy infected plants. Select resistant varieties. Woody passionfruit virus Passionfruit Fruit has a thickened rind, leaves may be mottled yellow No cure, remove and destroy infected plants. Compost and mulch can help prevent infection.

Nematode

Root Knot Nematode (Eelworms): Wide range of plants, including rose, potato, carrot, tomato, lettuce, zucchini Nematodes burrow into the roots and stimulate the development of galls, or lumps on the roots. Infected plants are stunted and wilt rapidly in hot weather. On potatoes they cause wart-like lesions on the skin. When harvesting, remove as much infected root from the soil as possible and destroy. Use strategies for soilborne disease. Digging fresh chicken manure into a hot, dry soil, something normally to be avoided, has been shown to reduce nematode numbers. Leave the soil undisturbed for at least 3 weeks. Drenching with water and molasses or sugar can also kill nematodes, but will have a negative impact on soil life.

Long Term Strategies to Combat Disease Problems

The overall design of your garden plays an important part in creating and maintaining 'garden health'. Make sure that the soil is well drained, that airflow is good and that only plants that like the shade are planted there. Good plant selection alone can make an enormous difference to the health of the garden.

Plant Selection

Selecting the plants most suited to your area, whether ornamentals, vegetables or fruit trees will reward you with minimal disease problems. At it's most basic, being suitable means the plant evolved in a climatic zone similar to your own. So a plant native to the Mediterranean such as a fig or grape will do best in areas with dry summers. If your area has wet summers than be prepared for problems with Mediterranean plants and do your best to select a planting spot with excellent ventilation and good drainage.

If you live in an area with hot, humid summers then it makes sense to select plants that have developed in these areas; most European vegetables have an Asian equivalent that is likely to be more disease resistant. So try substituting snake beans for French beans, angled luffa for zucchini, Ceylon spinach for silverbeet and so on. Getting the timing right can also reduce disease problems; avoid planting vegetables prone to rust, powdery or downy mildew just prior to periods of expected high humidity.

Be aware that buying certified disease-free planting material, particularly of strawberries and potatoes, will reduce your disease risk. Planting imported garlic, purchased at the greengrocers, also carries a serious viral disease risk.

Soil Health

Plant health is as dependent on the soil, as our own health is on what we eat. Pay attention to feeding the soil and its microfauna. Just as many pests are controlled in an organic garden by predatory insects so can many diseases be controlled by a diverse and abundant soil life. Increase the organic matter by composting, green manuring and mulching. Mulching acts as a barrier and helps restrict the spread of fungal spores from the soil onto the plant. Trace elements are critical to plant health; seaweed is a good way to add these essential nutrients. Avoid high nitrogen fertilizers, especially in summer when fungal problems are at their height, as soft new growth is very vulnerable to disease.

Golden Rule: Stay out of the garden when it is wet.

The strategies below should be seen as a regular part of good gardening practice, they are preventative strategies rather than a response to the actual presence of disease.

Sanitation

Garden and orchard hygiene are basic to reducing disease problems, the old saying 'an ounce of prevention is worth a pound of cure' applies.

Regular clean-ups are part of good gardening practice and help to prevent pathogens over-wintering in the garden. Jobs include: Collection of fallen fruit and 'mummified' fruit still hanging on the tree. Feed it to animals or burn it, as compost is unlikely to be hot enough to break the disease cycle. If chickens or other animals can be allowed to forage under the trees, this job is made easier and the results are more effective.

- Disposal of infected plants is important. The usual recommendation is to burn it; this is the most effective way of destroying spores but is obviously not a very environmental solution. Sealing it in a black plastic bag, and placing it in the sun can provide enough heat to kill spores.

Seasonal collection of dead leaves and plant debris after the harvest.

Pruning of dead wood, including infected twigs that are harbouring disease.

Disinfection of tools during and after cutting diseased wood. Do this by dipping tools into metho or a weak solution of Dettol or tea tree oil.

Keeping on top of weeds, as they can be hosts to both pest insects and diseases. For example, chickweed is believed to be a host to tomato spotted wilt virus.

Crop Rotation

Crop rotation is an essential practice for disease control. To control fungal problems, use long crop rotations of 3 to 5 years. It is important to remember to rotate the plant family, not just the individual vegetable. So if potatoes were planted in an area, a fallowing planting would not include members of the potato family like tomatoes, capsicums and eggplants.

Watering

Keep water off the leaves as much as possible, by watering from below. Fungal spores are often spread by water droplets. Most fungal spores need to be immersed in water for a number of hours in order to germinate, black spot for example needs to be constantly moist for 7 hours to establish. If you have to overhead water, do it early in the day to allow leaves to dry off.

Least Toxic Sprays

Organic sprays are preventative and need to be used before the disease problems are established.

Summer is when fungal problems usually appear. All of the following materials are considered acceptable organic inputs.

Seaweed Spray

Using seaweed as a foliar spray will provide your plants with trace elements they need to resist disease. Seaweed strengthens the plant cell walls and changes the pH of the leaf surface making it less attractive to the fungal spores. Spray every 2 to 3 weeks during peak periods for fungal disease.

Home-made Rose Fungicide

Dissolve 1 rounded tablespoon of baking soda (soda bicarbonate) in 4.5 litres of water. Add $2^1/_2$ tablespoons of Pest Oil. Agitate. Spray weekly as a preventative on roses for black spot and mildew.

Condy's Crystals

Condy's crystals (potassium permanganate) are used to control black spot and powdery mildew. Spray at two week intervals. To make a spray, mix 5g of Condy's crystals with 5 litres of water, use at once.

Lime Sulphur

Lime sulphur is used to control leaf curl, shot hole, freckle and rust on stonefruit; black spot and powdery mildew on apples and ornamentals. Do not use on apricots; if the temperature is over 32°C; or within 10 days of an oil spray. As with all sprays, read the instructions carefully.

Bordeaux

An organic fungicide in use since 1882. The spray can bought pre-mixed. It is useful as a dormant spray for fruit trees.

Milk Spray

Spray with milk as a preventative fungicide for powdery mildew in a ratio of one part milk to 9 parts water. Research is being done to see why this works but the important thing is that it really does seem to work.

Additional Strategies for Soilborne Diseases Solarisation

This is a useful remedy for soilborne diseases; it can also help combat stubborn weeds. To be effective do this in summer and first water the soil well. Then cover the soil with clear 4mm thick plastic. Stretch the plastic over the area, get it as close to the soil as possible. Bury the edges by digging a narrow trench, tucking the plastic in and back-filling. The aim is to raise the temperature to between 45°C and 50°C in the top 10 cms of soil. This is high enough to kill disease pathogens but most beneficial soil organisms will survive. Leave the plastic in place for 4 to 6 weeks and then plant as usual.

Green Manure

Plant a green manure specifically for soil diseases in beds that have had problems over the summer. Use plants such as rapeseed (canola), BQ Mulch, marigolds or Indian mustard that contain high levels of biofumigant compounds. When these plants are dug into the soil, they release chemicals, which control nematodes and fungal pathogens.

Control of Root Disease in Plantation Crops

The amount of disease in the mature plantation will depend upon the previous vegetation, the method of clearing, and the thoroughness with which measures for eradication of root disease have been carried out in the earlier years of the plantation. Once the roots of the plantation trees have expanded fully into their planting sites and made contact with those of their neighbours, the spread of root disease becomes more rapid and more dangerous.

It has then to be decided whether to undertake *ad hoc* control measures to prevent further spread of root disease, and if possible to eradicate it, or to clear and replant. Replanting may be deemed advisable not only on the score of root disease, but also because yield has fallen off for other reasons, or compares unfavourably with that of more recent plantations. Thus Sharples (1936) remarks: "We may justly inquire why, during depressed financial periods, a policy of replanting uneconomic areas should be undertaken. The reply is found in the probable yields of rubber to be obtained by utilising modern methods of planting-up bud-grafted material, *i.e.* trees which have been proved definitely superior and found to possess all the desirable properties. These strains of rubber trees are expected to give greatly increased yields, even to the extent of twice or three times the amount given by the ordinary mixed material planted in past years."

Root Disease Control in Mature Plantations

Let us suppose that a decision has been taken to keep on the standing plantation for a further period. Even if replanting has been postponed indefinitely, root disease control measures should be designed not merely to limit the spread of disease, but to eradicate it wherever possible; in this way, the cost of eventual clearing and replanting will be reduced. The isolation of diseased areas by means of a trench should therefore, be supplemented by removal of dead trees and all buried infected material from the area enclosed by the trench, as originally recommended by Petch (1921). Petch directed that the isolation trench should enclose not only the diseased tree, but also a complete ring of surrounding trees, even although the latter were apparently healthy. The trench had to be dug to a depth of about 2 feet and had to sever all lateral roots; in deeper soils, a greater depth of trench might be necessary.

Further directions were given as follows: "It is now becoming customary to open up and examine the roots of trees which are included within the isolation trench, in order to determine whether they have already been attacked. It is of course impossible to follow up the roots to their extremities, but enough can be done to show whether the tree is seriously affected. The roots should be exposed for a radius of about three feet from the stem, and the tap root to a depth of about two feet. ... If upon examination the tap root is found to be decayed, the tree should be removed. If the fungus is found advancing towards the collar on one or two laterals, these may be cut back into sound tissue, and the diseased laterals dug out. It must be borne in mind that the mycelium usually travels along the underside of the laterals, especially, when the upper side is exposed; this is particularly so in the case of *Fomes lignosus.* In the latter disease, the external strands of mycelium fairly frequently extend over the surface of the root in advance of the penetration of the fungus. The part covered by the youngest mycelium may consequently not have been attacked, or only attacked to a slight depth. In such cases, the mycelium and any underlying decayed bark may be cut away and the wound treated.

This method is especially useful when the tap root only has just been attacked. In Ceylon, it is usual to dress the roots from which mycelium has been scraped off with Brunolinum, or tar, while in the Straits Settlements, Bordeaux paste has been used. It must be remembered that the mycelium travels inside the root as well as on its surface, and consequently, if the decayed parts of the root are not cut away, any treatment is useless. The method is said to have been attended with great success, trees which bore the mycelium of *Fomes lignosus* having been completely cured." It is interesting to see in this early procedure, which Fetch declared to be applicable also to control of the rubber root diseases caused by *F. noxius* and *Ustulina zonata* in Ceylon, the germs of the method later to be developed so successfully by Napper in Malaya for a different purpose, *i.e.* control of root disease and eradication of the causal fungi in *young* plantations.

Fetch's directions for trench isolation and treatment of a diseased area evidently proved fairly satisfactory in practice, for little has been added to them; they were repeated by Sharples and Sanderson (1931) in their recommendations for control of *Ganoderma pseudo ferreum* on old rubber areas in Malaya.

For correct siting of the isolation trench, Sharples (1936) has given the following instructions: "The main requirement of an isolation trench is that it shall be cut outside the advancing limits of the patch. If, therefore, in cutting the trench all the roots severed are healthy, the trench must be fulfilling this requirement. If, however, diseased roots are encountered, the trench is known to be incorrectly placed.... The diseased roots are the indicators for the location of the trenches, and these must be taken further and further away from the diseased patch until all roots severed are healthy."

An improvement in trench maintenance was suggested by Napper (1937), who pointed out that the classical method of keeping the trenches permanently open by clearing out all silt and vegetable debris at regular intervals was completely effective only where a shallow soil overlay a subsoil impervious to roots. In soils of normal depth, new rootlets produced from the ends of the severed laterals found their way down the sides and under the floors of the trenches, and within a year or two of the original digging root contact was again established between diseased and healthy areas.

The trenches thus had to be dug progressively deeper, and the work became more difficult and more expensive at each digging. Napper therefore, suggested that *trenches should be refilled as soon as they had been dug*. All new root contacts would then be established *above* the floors of the trenches, and efficiency could be restored at intervals by re-digging the trenches to their original depth. Such re-filled trenches were usually invaded by small roots within six months of cutting and re-filling, but the spread of disease through this maze of rootlets was slow, and it was unnecessary to dig out the trenches more often than once every 1-2 years. At each re-digging, the root sections had to be examined in the usual way, and any infected roots passing beyond the trench had to be followed up and removed.

For controlling the root disease caused by *Poria hypolateritia* in old tea areas in Ceylon, Gadd (1929a) recommended trenching, with removal and destruction of all bushes within the isolation area since these would sooner or later be killed by the disease if left standing. In a later paper, Gadd (1936a) observed that whereas *Fonics noxius* could usually be eradicated by the simple expedient of uprooting the diseased tea bush and removing the source of infection, this procedure was quite inadequate to check the spread of *Poria hypolateritia*. This was attributed to the fact that whereas *F. noxius* was a quick-killing fungus, and usually killed an infected bush before the roots of neighbouring bushes had become infected, *P. hypolateritia* was a slow-killing fungus, and had usually become well established on the roots of surrounding bushes before the first infected bush was visibly affected by the disease. Later, Gadd (1937) discussed the failure of current control methods to check the spread of *P. hypolateritia* in old tea, which he attributed to the placing of too much reliance upon isolation trenches, at best a passive defence. He declared that control of this fungus could be achieved only by removing a ring of healthy bushes around every diseased one.

Gadd's (1936a) conclusion is supported by Tunstall's (1940) estimate of the root disease situation in the tea plantations of North East India, where the two most important root diseases appear to be those caused by the quick-killing but slow-spreading fungi, *Fonics noxius* and *Ustulina sonata*. Tunstall no longer recommends the digging of isolation trenches, deeming it sufficient to remove the diseased tea bush, together with all dead wood from the soil immediately surrounding it. Healthy bushes need be removed, according to Tunstall, only when their roots occur in very close proximity to the roots of a diseased bush.

Root Disease Control on Replanted Areas

Any decision concerning the extent of measures for root disease eradication to be taken prior to replanting must be dictated by economic considerations. The cost of the work varies from country to country, from district to district, and even from one estate to another, according to soil, situation, and labour available, etc.; more important and more variable, however, is the expectation of root disease in the replanted stand. Few experimental results are available for guidance in making such decisions, so that an experiment carried out by Napper (1938a, 1939-40) on rubber in Malaya, and another by

CRONSHEY and barclay (1939) on rubber in Sumatra will be fully described.

The site of Napper's replanting experiment was a rectangular block of 80 acres in old rubber, where disease patches occupied at least 50 per cent of the total area. The main parasite was *Ganoderma pseudojerreum*, but *Fomes lignosus* and *F. noxius* were also present. Six treatments with 5 replications were arranged in randomised block design, giving a plot unit of 2 per cent acres; the treatments were as follows:

A. Uniform digging to 18 inches, all roots being removed to this depth whether healthy or diseased.

B. As *A*, but digging only to 10 inches.

C. Disease patches dug over to a depth sufficient to ensure complete removal of infected roots. Remainder of the area dug uniformly to 10 inches without collecting roots.

D. Disease patches dealt with as in *C*, but remainder of the area left undug. Apparently healthy trees were felled by jack or grubber to ensure removal of the boles and larger laterals, and to give a further opportunity for root examination.

E. No systematic digging. Diseased trees were dug out individually, all infected roots being followed up and removed. Healthy trees were dealt with as in *D*.

F. No digging or removal of stumps. Trees (both healthy and diseased) were cut off at ground level, the stumps being then poisoned by injection with sodium arsenite.

The different treatments were completed by the end of 1937, and the whole area was then planted with rubber; after planting, the usual periodical tree-to-tree inspection for discovery and eradication of sources of root disease infection, as developed for young plantations on virgin land, was carried out over the whole experimental area. The total incidence of root disease infection in the young rubber up to the end of the first and second years after planting, respectively. Napper also gives numbers of bush cover plants killed by disease under the different treatments, but as these figures run in much the same order as those for infection of the rubber, they need not be quoted here.

Comment upon the results obtained by the end of the first year after planting (1938) is best given in Napper's (1939) own words: In the first place they provide a striking demonstration of the fact, which has gradually become apparent in recent years, that root disease control in connection with replanting is not a single problem, but a combination of two separate and unrelated problems which require separate consideration when planning a policy of control, *viz.*

(a) Control of root disease in areas affected by root parasites in the original stand, and

(b) Control in areas outside the old root disease patches, *i.e.* in areas carrying healthy trees at the time of felling.

The site chosen for the experiment was well adapted to effect a separation and to permit separate study of these two problems, since the parasite mainly responsible for damage in the original stand was *Ganoderma pseudojerreum,* whereas the parasite causing trouble where healthy roots are left to rot in the soil is invariably *Fomes lignosus.*

It was expected, therefore, that the rates of incidence of *Ganoderma pseudoferreum* under the various treatments would provide a measure of the relative efficiencies of the treatments in dealing with problem (a) and would vary inversely with the amounts of digging carried out *within* the boundaries of the root-disease patches, whereas the rates of incidence of *Fomes lignosus* would provide a similar but independent measure of the efficiencies of the treatments in dealing with problem (b) and would vary inversely with the amounts of digging carried out *outside* the boundaries of the root-disease patches. The only inconsistency is the low incidence of *Pomes lignosus* in treatment F, and here an additional factor had been introduced, *viz.* the poisoning of the stumps of the trees after felling by injection with sodium arsenite.

The effect of stump-poisoning in lowering the incidence of *Pomes lignosus* in the *F* plots is a matter of considerable practical interest. The method has often been suggested by non-technical observers as a means of controlling root disease at the time of replanting, but has usually been regarded by its advocates as a method of destroying roots already infected by root parasites, *i.e.* as a method of dealing with problem *(a)* it is, of course, useless for this purpose, a fact which is confirmed in the present experiment by the high incidence of *Ganoderma pseudoferrcum* in treatment *F*. The present results show, however, that as a method of dealing with problem (b) the method merits serious consideration. More green timber was left to rot in the soil in treatment *F* than in any other treatment included in the experiment, yet the incidence of *Fames lignosus* under this treatment was only half the incidence under treatments *D* and *E* in which the chances of infection and spread were greatly reduced by the removal of the stumps and laterals of the healthy trees, and no higher than in treatments *B* and *C* in which all healthy roots were extracted to a depth of 9 inches.....

'The success of the method depends on the fact that poisoning hastens the onset of decay, and therefore, shortens the period between felling and invasion by saprophytic organisms during which the roots are green and abnormally susceptible to attack by *Fames lignosus*. It is during this period that the sources of infection are developed which later give rise to losses in the replanted stand. Shortening the period of susceptibility reduces both the number and size of these sources, and thus provides a useful measure of control. It is still difficult to explain satisfactorily where the *Pomes lignosus* comes from. Some undoubtedly originates from dormant sources which existed in the original stand. Spore infection of the partially exposed and highly susceptible green root-fragments cannot, however, be entirely disregarded.

Napper's conclusions on this experiment may be summarised by saying that 'trouble with *Ganoderma pseudojerreum* in the young plantations occurred only on the sites of root disease patches in the old stand, and could be dealt with by the extraction of all buried infected roots on such sites; trouble with *Pomes lignosus,* on the other hand, was the result of leaving roots of healthy trees in the soil after felling, and constituted, therefore, the same problem as that encountered in young plantations established on jungle sites. Concerning the infections by *Pomes noxius,* Napper stated that nearly 60 per cent of these were directly traceable to contact with diseased rubber logs lying on the surface of the soil; these logs had undoubtedly become infected after felling, most probably by windblown spores of the fungus.

At the end of the second year (1939) of this experiment, the figures for total root infections ran in the same order as those obtained at the end of the first year, and so require no further comment. Napper (1940) was unable to decide whether treatment *F,* stump injection with sodium arsenite, would eventually prove more economical than the methods for dealing with the root systems of healthy trees adopted in treatments *B, C, D* and *E.* He pointed out that where infection by *F. lignosus* did occur under treatment *F,* it was more expensive to deal with, and the higher cost per point treated might outweigh the benefit derived from reduction in number of infections. A full discussion of the costings for this experiment up to the end of the second year after planting has been included in this report by Napper (1940).

Cronshey and Barclay's (1939) replanting experiment, of factorial design, was carried out on the site of a 25-year old rubber plantation on a flat coastal area of clay soil in Sumatra. The old stand had originally been planted at the rate of 121 trees to the acre, had been thinned to 80, and at the time of clearing averaged less than 60 trees to the acre. Collar inspection at clearing showed *15 per cent* of the remaining trees to be diseased, *Fomes lignosus* being the predominant parasite. Details of the experimental treatments are given by Cronshey and Barclay, as follows. Stumps were removed with stump pullers, so that root collars and parts of the tap root and larger laterals were extracted. Roots were removed by "chang-kolling" a changkoll is a native hoe) to a depth of 15-18 inches. In "holing", planting holes 2X2X2 feet were dug, and refilled with top soil free of roots.

These results provide a clear-cut answer to most of the questions that this experiment was designed to elucidate. The presence of cover reduced mortality by about 60 per cent on an average of all comparisons. This confirms the earlier conclusions of Napper (1932a). As might have been expected, "holing" has reduced the incidence of disease. On an average of all comparisons, the removal of stumps and roots has reduced mortality to little more than half that occurring when neither stumps nor roots were removed. An interesting difference is revealed by the comparison of losses following the two treatments "stumps left, roots removed" and "stumps removed, roots left", according to whether the area is clean weeded or under cover. Cronshey and Barclay have not commented upon this curious discrepancy, for which the following explanation is here suggested. In the clean-weeded areas, removal of roots, leaving stumps, has decreased disease losses as might have been expected, on the assumption that roots form the more widely dispersed source of inoculum. But removal of stumps, leaving roots, has actually increased disease mortality above that where stumps and roots were left.

Is it not possible that violent removal of the stumps has increased the susceptibility of the roots left in the soil to infection by latent inoculum of *F. lignosus*, by analogy with the increased incidence of disease on virgin jungle areas burnt after clearing, as compared with unburnt areas (Napper, 1940). In areas under cover, on the other hand, removal of roots but not stumps has not significantly affected the incidence of infection, whereas removal of stumps but not roots has substantially reduced it. To explain this paradox, let us assume that the roots provide more sources of infection, but that these are of smaller volume, whereas the stumps provide fewer but more massive sources of infection. Under clean weeded conditions, where no obstacles are interposed between the roots of the rubber and the inoculum, the roots might be expected to produce the greater number of infections; under cover, however, the "baffling" effect of the cover roots upon the spread of the rhizomorphs, as postulated by NAPPER (1932-34),-might prevent the fungus reaching the rubber from many of the smaller sources of infection provided by the roots, so that the fewer but more massive sources of infection provided by the stumps might ultimately produce the greater number of infections.

CRONSHEY and BARCLAY themselves comment as follows upon their results: Stump-pulling and changkolling are closely interrelated, neither being of much value in the absence of the other. Collectively, however, they have reduced the losses by nearly a half, and it is noticeable that the effect has been larger when other favourable treatments, holing and the establishment of a cover, were also done. In general, the combined effect of the favourable treatments is greater than the sum of their single effects, the percentage reduction being considered as the correct measure of efficiency." They also pointed out that the presence of cover had resulted in a slight reduction in growth of the young rubber, as estimated by increment in diameter of the stem, and suggested that the clean weeding of circles around the trees or of strips along the rows might prove a satisfactory compromise.

On replanted rubber areas in Ceylon, according to MURRAY (1938), the mortality in young rubber from root disease (due almost entirely to *F. lignosus;* occasional outbreaks due to *P. noxius* and *Poria hypobrunnea)* is evidently much lower than that in Malaya or Sumatra. Prior to the replanting of areas occupied by disease patches due to *F. lignosus* in the old stand, MURRAY recommended the entire removal, if possible, of all old roots; on healthy areas, however, laterals might be left to decay *in situ.*

Infection amongst the young rubber trees due to the development of latent inoculum of *F. lignosus* adjacent to or in the laterals of felled stumps in such previously healthy areas could be eliminated, according to MURRAY, by prompt treatment of visibly-affected young trees and removal of the source of infection. MURRAY justified the adoption of this procedure as follows: "This policy could only commend itself if the numbers of plants so lost were small, and if the knots of infection were disclosed before the fungus could spread a great distance. In most of the replanted clearings of which I have knowledge the number of casualties due to root disease amounts so far to considerably less than 1 *per cent* of the total stand. Some of these clearings are quite recent, but others are up to 8 years of age. Even *its per cent* is an insignificant proportion especially as the initial stand is usually much greater than the final stand aimed at. In the first two years or so the casualties can, of course, be replaced with supplies."

MURRAY goes on to suggest that the root disease fungi are likely to spread actively along the laterals of the old stand left in the ground only for the first few months after felling, whilst the roots are still fairly fresh; as for spread along the roots of the young rubber, this should not become dangerously rapid until the root network of the rubber has been fully developed, at an estimated age of 12-15 years. If the existence of sources of infection has not been revealed by this time, such sources may be regarded as dormant, and therefore, comparatively harmless.

In a further communication, MURRAY (1939) reported data derived from replies by 156 estates to a questionnaire on replanting policy. Lateral roots had been removed from the ground prior to replanting on 2465 out of 8124 acres; loss due to root disease fungi (chiefly *F. lignosus* but a little F. *noxius* and *Poria hypobrunnea)* was approximately 0.06 per cent where laterals had been removed, and 0.14 per cent where they had been left in the soil. In the following year, MURRAY (1940) declared the mortality from root disease to reach a maximum in the second or third year from planting, and thereafter to show a marked decline; although this was largely due, no doubt, to the disclosure and eradication of sources of infection by the method just described, there was evidence that fungal activity tended to die down even where infected old roots were left in the ground. No doubt, too, the resistance of the rubber plant to *F. lignosus* increased with age (DE JONG, 1933). MURRAY also stated that several estates with a high incidence of *F. lignosus* in young replantings had been advised to undertake the routine collar inspection treatment developed by NAPPER in Malaya, and that this had proved quite feasible under conditions in Ceylon; it was doubtful, however, whether adoption of this treatment would prove to be economically sound. MURRAY'S policy for replanted areas in Ceylon therefore, seems to have been justified by the low mortality due to *F. lignosus* amongst the young rubber in that country, and by the absence of *Ganoderma pseudoferreum;* a similar conclusion had been reached earlier by DE JONG (1933) in Sumatra, and SHARPLES (1936) has pointed out that a similar policy might be followed in Malaya were it not for the widespread occurrence of the more dangerous root parasite, *Ganoderma pseudojcrreum.*

A suggestion that susceptible bush covers might be employed as "indicators" of root disease in areas of young replanted rubber was made by BERTRAND and MINOR (1937) in Ceylon. They recommended a mixture of three bush covers, so as to reduce incidence of stem and branch diseases, which would spread more rapidly in a pure stand of a single species. MURRAY (1938) criticised this proposal on the score that it would be impossible to grow such bush covers along with creeping covers without the bushes being overwhelmed by the creepers; the bushes alone would not give adequate protection against soil erosion on steep slopes. Moreover, bush covers would not stand being lopped for green manure as did the creepers, and they were likely to "indicate" other root diseases as well as those dangerous to rubber, and thus waste pest gang labour. In a comparison of bush covers and creeping covers for young rubber in Malaya, NAPPER (1940) found the incidence of infection to be higher by a third on the area under creeping cover; from the point of view of root disease treatment, however, creeping covers were more economical, on account of the additional cost of treating the infected bushes. Both as regards disease control and general requirements, therefore, creeping covers appear to be preferable to bush covers.

In North East India, TUNSTALL (1940) recommended that when areas of old tea were being cleared for replanting, patches of vacancies and of infills should be marked off; on these patches, all the roots of both tea bushes and shade trees should be extracted as completely as possible. The remaining bushes might be dug out without taking any special care to remove all the roots from the soil.

Control of Root Disease in Plantation Crops: on Virgin Areas

Any discussion of root disease in plantations must be concerned chiefly with the diseases of tropical crops; root disease is of greater importance in tropical plantations than in those elsewhere. The root diseases of rubber have received the greatest attention, followed by those of tea and cacao; the banana suffers from only one root disease of any importance, *vis.* Panama disease, but that one disease has caused, and is continuing to cause, enormous losses. An outstanding contribution to our knowledge of the subject has been made by the Rubber Research Institute of Malaya, and the principles of root disease control that they have formulated are of general application to plantation crops. No apology need to be made, therefore, for concentrating attention upon the root diseases of tropical crops in general, and upon those of rubber in particular.

Tropical plantations are generally established, in the first instance, upon the site of virgin jungle, which has to be cleared to a greater or less extent before the crop can be planted; various root disease fungi occur throughout the root network of the original jungle, and some of these find congenial hosts in the plants of the young crop. The simplest solution of the root disease problem is to place the cleared area under annual field crops, which are not susceptible to tree root diseases, for a period of years, but this method has never found favour with plantation owners, on account of the long delay involved. Thus NAPPER (1934) commented upon the absence of root disease in rubber planted on land cropped for long periods after clearing with sugar-cane, rice, etc., and also upon the slight amount of root disease to be found on the average smallholding.

Root diseases may develop from the stumps *(i.e.* stumps roots) of the original jungle trees, if these are left behind after felling and clearing. Two additional sources of root disease are provided by the stumps of trees that have to be thinned out because the original plantation was too closely planted, and by the stumps of shade trees that are felled when they have fulfilled their purpose of providing shade for the developing plantation crop. On the subject of thinning out in rubber plantations, FETCH (1921) remarks: "Close planting with the intention of thinning out in later years, was never widely adopted; in the majority of cases estates were planted up at the distances it was considered the trees would remain permanently. It needed, however, only the efflux of time to demonstrate the truth of the prophecies of mycologists who contended that *Hevea* could not possibly exist in a healthy condition when planted two hundred to the acre; and the necessity for thinning out all the earlier-planted Rubber is universally admitted."

Early Methods of Root Disease Control: A good account of the classical methods for root disease control has been given by FETCH (1921). In the first place, clean clearing of the original jungle was favoured wherever practicable, in order to remove all possible sources of infection. Although FETCH admitted the probable presence of infected roots in the jungle stand, it laid much greater emphasis upon another source of inoculum, provided by airborne spore infection of the cut surfaces of the stumps after felling. It supposed that once spore infection of the cut surface had been established, the parasite grew down into the roots of the stump, thence eventually to emerge and infect the roots of the young plantation crop, either through root contact or, with some fungi *(e.g. Fomes lignosus)*, by direct spread through the soil. Diseases considered especially likely to spread in this way were white root disease *(Fomes lignosus)*, brown root disease *(Fonics noxius)*, and charcoal root rot *(Ustulinazonata)*, but other root diseases, such as red root disease (*Ganodermapseudoferreum)*, unknown to FETCH in Ceylon, were not excluded from the same category. This hypothesis was held to explain not only the development of root disease from the stumps of the felled jungle, but also that following the thinning out of shade trees.

It is a remarkable fact that no experiments to prove this hypothesis ever seem to have been reported, although the technical difficulties in the way of protecting the cut surface of the felled stump from spore infection would scarcely seem to have been insuperable; in the mean-time the hypothesis held undisputed sway for many years, and was acted upon in practice, either by extraction of the felled stumps, or by cutting them off below ground level and covering with soil (GADD, 1936b). Indeed, the discrediting of FETCH'S hypothesis has been brought about not by any direct experimental assault, but rather through the gradual accumulation of circumstantial evidence in conflict with the hypothesis, and finally through the establishment by experiment of an alternative explanation.

For the treatment of definite patches of root disease, FETCH recommended the removal of all dead trees, decaying stumps and buried timber in the affected patch, the application of lime to the soil, and the isolation of the area by means of a deep trench. In removing the dead trees, the laterals had to be followed up and dug out; the affected patch had then to be forked over, and all pieces of dead wood collected and burned. The direction to apply lime was apparently based on the assumption that "the majority of fungi prefer an acid medium", an assertion even less valid than most generalisations. This assumption was shown by BRYCE (1922) not to hold for *Fames lignosus,* the most important cause of rubber root disease in Ceylon, since then the application of lime has no longer been advised. The last part of the treatment lay in the digging of an isolation trench, which FETCH declared should enclose not only the diseased tree, but also a complete ring of surrounding, apparently healthy trees. He recommended a depth of 2 feet, but the depth had to be sufficient to ensure the severing of all lateral roots, so that a depth of 3 feet might sometimes be necessary.

The method of trenching to stop the spread of a root disease is an old established one; TUNSTALL (1940) remarks: "More than fifty years ago the foresters in Europe sought to prevent root disease spreading by encircling the diseased trees with shallow trenches. This method was adopted in the case of rubber, tea and coffee." FETCH (1923) evidently viewed the isolation trench as designed primarily to prevent the spread of free mycelium through the soil, and only secondarily to prevent spread of disease by root contact, as shown by the following quotations: *"Ustulina"* usually attacks tea by spreading to its roots from the roots of decaying stumps. The fungus does not spread free through the soil, but only passes from the roots of the stump to those of the tea bush when the two are in contact. Consequently trenching might possibly be dispensed with in the case of this disease, but it is better to err on the safe side and to trench. ... As *Armillaria mellea* produces thick black cords of mycelium (rhizomorphs), which spread freely through the soil, it is essential that the diseased patches should be isolated by trenching.... Since the mycelium of *Femes lignosus* spreads freely underground, affected patches must be isolated by a trench/' FETCH (1921) laid great emphasis on the free mycelial spread of *F. lignosus* through the soil, concerning which he remarks elsewhere: "The majority of fungi only advance within dead wood, but the strands of *Pomes lignosus* can travel for a few feet at least through the soil, unattached to any root or dead wood, except, of course, at their starting point. It is always attached to its base, *i.e.* the stump on which it originated, and it must derive its food from that source until it meets with other dead wood, or a living plant it can attack.... Clear instances of the spread of the mycelium through the soil are not uncommonly met with on estates. In badly-infested areas its presence is often manifested by the appearance of the fructifications on banks by roadsides, or the sides of drains, where no roots or timber is to be seen. In some of these cases the mycelium follows small Rubber roots almost to the surface of the soil, but in others there is no root or wood to be found immediately behind the fructification. In other cases, it is often found running on the under surface of large stones, and in these instances it is readily seen, as a rule, that it is not following a root." The accuracy of FETCH'S field observations has not been questioned by later investigators, though a different interpretation is now put upon them; in the instance quoted above, FETCH failed to discriminate between the spread of the fungus and that of the disease.

The classical view of the behaviour of these tropical root disease fungi, as presented by FETCH, may be summarised as follows: Root disease fungi present on the roots of the original jungle trees persisted on the roots of the stumps after felling; a more dangerous and widespread source of inoculum, however, was furnished by airborne spores of the fungi, alighting on the cut surfaces of the stumps. From the crown, the fungi were assumed to spread down into the roots of the stumps, whence they infected the roots of the young plantation crop either by root contact, or, with *Fomes lignosus, Poria hypolateritia,* and *Armillaria mellea,* by free spread of the mycelium through the soil.

NAPPER's Method of Root Disease Control

The classical theory of root disease control was undermined by the observations of NAPPER (1932-34), upon whose researches at the Rubber Research Institute of Malaya, modern views of the biology and control of rubber root disease fungi are based. In the first place, NAPPER considered that all the root disease inoculum that gave trouble in the young plantation crop was present on the roots of the original jungle, and that spore infection of the felled stumps was of subordinate importance, if, indeed, it occurred at all.

Thus NAPPER (1934) writes: "Being so advantageously equipped for life under forest conditions the root parasites have become generally distributed throughout the jungle. Since, however, they are normal members of the 'jungle association they do no spectacular destruction amongst the jungle trees, but here and there, like any other species, they become locally dominant, each particular parasite forming its own pattern of root disease patches scattered throughout the jungle stand. While the jungle is growing undisturbed, these patterns are perpetually changing in shape and position, but the total incidence of infection remains constant, and the loss of stand never becomes serious."

Whereas FETCH had stressed the difference between those fungi that spread as mycelium through the soil, such as *Pomes lignosus,* and those that spread by root contact, such as *Fonics noxius,* NAPPER, on the other hand, stressed the essential similarity in behaviour of the three important fungi infecting rubber roots in Malaya, *viz. F. lignosus, F. novius,* and *Ganoderma pseudoferreum.* In NAPPER'S view, all three fungi had developed the subterranean organ of propagation, the rhizomorph at the expense of the aerial organ, the fructification. Secondly, NAPPER'S field observations indicated that the importance of free mycelial spread through the soil by *Fomcs lignosus* had been unduly emphasised, and that rhizomorphs could not grow (except under special conditions, and for short distances only) directly through the soil, but required as a vehicle a chain of solid surfaces, preferably those of living host roots.

It was true that rhizomorphs might be found covering the underside of stones and boulders, and growing on dead roots, but they were unable to obtain nourishment from any substrate other than living host roots; the rhizomorphs could not, apparently, invade tissues already colonised by other organisms. NAPPER stressed this effect of microbiological competition in curtailing the spread of the root parasites, whether on the exposed wood surfaces of cut stumps, or on dead roots already occupied by other microorganisms.

From the results of certain experiments originally set up by J. R. WEIR at the Rubber Research Institute of Malaya, on "clean clearing" *versus* no clearing in preparation of a jungle site for rubber, NAPPER derived certain important conclusions as to the behaviour of rhizomorphs, which were to form the basis of a new and outstandingly successful method for root disease control.

Two and a half years after planting, incidence of infection amongst the young rubber was lower where no clearing had been carried out than on the clean-cleared area; it was lowest where secondary jungle ("belukar") had been allowed to grow up as a natural cover. These results were in direct contradiction to the postulate of the classical theory that the amount of disease in the young plantation crop varied directly as the amount of actual and potential inoculum left in the soil after clearing, and were explained by NAPPER as follows:

(i) If "absolute" clean clearing is carried out and even the smallest pieces of buried timber are removed from the soil, *e.g.* by sieving, no actual or potential source of inoculum will be left, and a healthy young plantation will result.

(ii) Under "ordinary" clean clearing, as practised on commercial estates, a small amount of generally distributed timber is left in the soil. Now rhizomorph production does not start until the food material in an infected root section has been almost exhausted by the fungus; rhizomorph production therefore, starts the soonest in the smallest pieces of infected root. It is precisely these small pieces of infected root that are left behind in the soil after "ordinary" clean clearing.

(iii) The larger pieces of infected material must therefore, be considered relatively harmless to the young plantation crop, as they will not arrive at the stage of rhizomorph production for some time; they are not merely harmless, however, but are actually beneficial, as they represent solid surfaces upon which the growth energy of the rhizomorphs must be expended. The greater the area of solid surfaces, or obstacles, to be covered, the slower will be the advance of the rhizomorphs from their source, with the result that some of them may never achieve infection of the roots of the plantation crop.

Upon this last point, NAPPER (1932a) comments: "If the above reasoning is sound, then there is no difficulty in explaining why the presence of a cover crop, even of a woody plant such as *Crotalaria*, which is known to be parasitised by the fungus, tends to reduce the incidence of *Fames lignosus* upon the rubber trees among which it is planted.... There is probably very much more mycelium in the soil under a heavy cover crop than under clean weeded conditions, but this is discounted as indicating the probability of a higher incidence" of attack upon the rubber trees interplanted with a cover because of the larger amount of material in the soil other than the roots of rubber trees upon which the mycelium can feed." NAPPER's conclusions concerning the effect of a cover upon incidence of infection in the young plantation crop have been confirmed by CRONSHEY and BARCLAY (1939) in a well replicated replanting experiment of modern design.

An objection to the policy of planting on the uncleared site was anticipated by NAPPER (19320) in the following words: "Although methods of planting under uncleared conditions, or conditions of secondary jungle, may greatly reduce the loss of stand through root disease during the early years of the clearing, the effect may be one of delayed action only. It has been deduced above that the probability of attack upon the crop plant is greatest when other rottable material in the soil is nearly exhausted, and it may be that, in an estate which was planted under uncleared conditions, or under conditions of secondary jungle, the disease will attack the crop plant in an epidemic fashion in later years when the jungle timber in the soil becomes exhausted, or when the belukar (or other cover crop) is shaded out by the rubber.... Evidence suggests that as the rubber tree grows older it becomes more resistant to the attacks of *Pomes lignosus* on its roots, and further, that the culture of *Pomes lignosus* in the soil of a new clearing stales after a time and is not subsequently so active as during the first few years after clearing. This evidence is based on the observation frequently made that the incidence of the disease in a new clearing tends to rise to a maximum during the period between the second and the fourth years after burning off, and then to drop to a comparatively low figure.

Even where few or no control measures against root disease are carried out, *Fames lignosus* is typically the root disease fungus of young rubber plantations, up to ten years of age, whereas *Ganoderma pseudoferreum* is the root disease fungus characteristic of mature plantations. SHARPLES (1936) explained this by reference to the differences in rhizomorph behaviour of these two fungi. The rhizomorphs of *F. lignosus* grow very rapidly, but soon exhaust their food-bases, whereas those of *G. pseudoferreum* grow much more slowly, and exhaust their food-bases only after a much longer interval.

As a result of these discoveries NAPPER worked out a new method for dealing with root diseases, which is applicable not only to the disease caused by jp. *lignosus,* but also to those caused by *Ganoderma pseudoferreum* and *F. noxius* as well; the method, although primarily designed for young plantations established on jungle sites, can be modified to suit the requirements of replanted areas. The ground, after felling and burning of the jungle, is cleared of the larger stumps, but the young rubber trees are themselves employed for the discovery of residual sources of infection, as follows: Every four to six months, a pest gang of trained labourers makes a tree-to-tree inspection of the whole plantation, in the course of which the upper tap root and laterals of every young tree are examined for rhizomorphs of the root disease fungi. The length of tap root that must be exposed varies from 2 inches at one year after planting to 9 inches at four years, if the soil is a deep one, but the depth of inspection must in any case be sufficient to uncover the undersides of the surface laterals. If rhizomorphs of a root disease fungus are found, the root system is opened up further until the whole extent of the invading mycelium has been exposed, and the source of infection discovered. The source of infection is then removed (and destroyed) and the roots of the jungle stumps are followed up systematically in every direction until all infected material has been disclosed and removed. The root system of the young rubber tree is then pruned of all root sections that have already been penetrated by the parasite.

The method depends for much of its success upon the fact that the rhizomorphs of *F. lignosus,* the most common cause of root disease in young Malayan rubber plantations, extend for 5-15 feet ahead of the infected part of the root. The roots carrying such superficial or epiphytic mycelium can be adequately treated simply by pulling away the larger rhizomorphic strands, and then washing the root surface with a 2 per cent solution of copper sulphate. NAPPER states that this epiphytic habit of the *F. lignosus* rhizomorphs renders it possible, by this periodical treatment, to bring a stand of young rubber through a heavy attack of *F. lignosus* without losing more than 20 per cent of the trees that have been attacked.

The rhizomorphs of the other two members of the Malayan root disease complex, *G. pseudoferreum* and *F. noxius,* are epiphytic only for some 18 inches in advance of penetration, so that a proportionately greater loss of trees infected by these two fungi is inevitable. Owing to the fact that these fungi generally occur sparsely by comparison with *F. lignosus* on jungle sites, however, the efficiency of the control method is not seriously impaired. In this way, root disease can be eliminated before the roots of the trees have expanded into their full planting sites (after some 10 years); once the roots of adjacent trees begin to interlace, spread of infection becomes at once more rapid and more dangerous.

A noteworthy feature of this method is the rejection of trenching, with the realisation that, as spread of these diseases takes place mainly through root contact, trenching will be unnecessary before the root systems of the young trees have expanded fully into their planting sites and made contact with those of their neighbours; by this time, the majority of the buried sources of infection should have been disclosed and removed at the periodical tree-to-tree inspections. Trenching has not, therefore, been advocated by NAPPER (1937), except for control of disease in mature plantations.

Infection of Moribund Stump Roots after Felling

These discoveries of MAPPER'S, whilst throwing doubt upon the hypothesis of airborne spore infection of jungle stumps, provided no explanation to account for the undoubted association between outbreaks of root disease in plantation crops and the operations of thinning-out or felling of shade trees. Gradually, however, a suspicion grew up that an alternative mode of infection was possible, and even probable, *viz.* a direct infection from the soil of the moribund but still living roots of the stump, the resistance of which to infection had been sharply lowered by the act of felling. This implied, of course, the existence of inoculum dormant before the actual felling, either in the form of infected timber lying adjacent to the roots, or as root lesions that had been immobilised by the host defence mechanism.

In the disturbance of the soil incidental to felling and clearing of the virgin jungle, mycelium lying dormant in buried timber might well be roused into activity, especially in low-lying coastal areas where clearing and draining operations resulted in the lowering of a water table too close to the surface for successful planting. The occurrence of latent infections, or localised lesions, of *Fomes lignosus* in the roots of apparently healthy rubber trees was demonstrated by DE JONG (1933) in Sumatra, who states, in the English summary of his paper: "it was observed that *Rigidoporus* decay frequently stops without treatment and of its own accord. In such cases the decayed patches become surrounded by callus and eventually heal over completely. It is striking that the process of recovery ordinarily started at about the time when the wood used for inoculation had itself disappeared through decay. In certain instances, however, it was observed that *Rigidoporus* decay continued to develop after the exhaustion of the original inoculating material. This situation probably results when the decay has progressed to such an extent, before the disappearance of the wood used for inoculation, that the tree becomes seriously weakened."

SHARPLES (1936) obviously appreciated the occurrence and significance of the fall in root resistance to infection after felling, for he remarks:

> "Another point to remember is that a vigorous rubber tree displays a certain amount of resistance to attacks of *G. pseudoferreum* by the production of adventitious roots, but there is no resistance offered by the part of a rubber tree left in the soil after the upper parts have been removed by thinning-out. When the trunk of a tree is removed, the roots do not necessarily die with great rapidity".

However, they become moribund and ultimately death follows. The loss of resistance to the attacks of the fungus may be considered an additional factor which will lead to the rapid spread of root disease after thinning-out. The same point was also taken into account by NAPPER (1937) in a discussion of replanting policy; concerning areas of old rubber that are apparently healthy at the time of felling he remarks: "Of greater interest are the losses which occur in *untreated* soil, *i.e.* in areas of the clearing which were apparently healthy at the time of felling. These losses are invariably caused by contact with infected roots belonging to the original stand, but the existence of such roots in 'healthy' areas is sometimes rather difficult to explain. In general, however, they may be regarded as old infections in which the disease has lain dormant for very many years until brought into activity again by the process of felling. The causes of dormancy are still somewhat obscure. The most common cause in the field appears to be continuous saturation with water. This is shown by the fact that losses of the type under consideration are most abundant in low-lying coastal areas where the water table was originally high, but has been extensively lowered in the process of replanting. The causes of dormancy on inland soils are less important and not so clearly understood. A similar conclusion has been voiced by MURRAY (1938) for rubber areas in Ceylon: "When a stand of old Rubber, in which no case of root disease has ever been recorded, is felled, it was at first supposed that the root fungi were absent, and that there would be no risk of the young plants becoming infected from old roots left in the ground. Recent experience has shown that this is not a safe assumption.... in every investigated case.... it has been possible to trace the source of infection to an old Rubber root bearing the typical rhizomorphs, the fungi concerned being usually *Fomes lignosus* or *Poria hypobrunnea.* This proves that the fungus must have been present before the old trees were felled, and it was probably present in the wood of a root in a condition of what may be termed 'subdued parasitism'.... When the trees are felled the equilibrium is upset."

Proof of this contention, however, was published neither from Malaya, Sumatra nor Ceylon, but came from East Africa, in two contributions by LEACH (1937, 1939), the importance of which it is difficult to exaggerate. In the first paper, LEACH recorded his observation that whereas the percentage of trees killed by *Armillaria mellea* in virgin forest was very small, even in a pure stand of a susceptible species, yet when the forest was cut down, the roots of many species became extensively invaded by *A. mellea;* this implied a breaking-down of root resistance to infection. A similar observation had previously been published by WALLACE (1935) from Tanganyika Territory. In his second paper, LEACH (1939) reported various experiments designed to test the correctness of this conclusion. Twenty-six species of native tree were selected for experiment; in order to simulate the act of felling, lateral roots were sawn through at a distance of 4 feet from the trunk, at the time of inoculation. Two individual trees of each species were tested for resistance to attack by placing pieces of infected *Glyricidia sepium* root in contact with the roots of the tree. Six roots of each tree were inoculated, three being cut through above the point of inoculation, and three being left uncut; the roots were examined 4 months later. Examination of the *uncut roots* showed that roots of 6 species were never invaded by *A. mellea* and appeared to be completely resistant to attack. The roots of all other species were invaded, but, with only two exceptions, were able to confine the parasite within narrow limits, apparently by the formation of gum barriers and callus. Examination of the *cut roots* showed that roots of 24 out of the 26 species were susceptible to invasion; in all 24 of these species, the fungus travelled freely down the roots, though in 5 species *some* of the roots still showed resistance. LEACH compared the resistance shown by the uncut roots in this experiment with that demonstrated in the Myrobolan plum by THOMAS (1934), in his study of infection of susceptible and resistant hosts by *A. mellea.*

As a result of these experiments, LEACH (1939) concluded: "This experiment .... indicates that trees growing near dead stumps in forests may have many localised lesions of *Armillaria* dormant in their roots and that the fungus in these lesions may become active when the forest is felled for a plantation crop. The fungus may spread from these localised lesions even though the roots of the original dead stumps may have been rotted by *Armillaria* beyond the stage in which they would be capable of causing infection." LEACH records two other observations of particular interest: "Some roots on the 'Msopa' tree.... had been invaded at two points close together but on opposite sides. Such roots had lost the power of resistance below the points of infection, *Armillaria* having travelled freely down the root but not up to the crown.... In Nÿasaland, *Eucalyptus* trees are not considered susceptible to *Armillaria*.... Observations showed that a *Eucalyptus* tree forms localised lesions when invaded by *Armillaria* similar to those described for the 'Msopa' tree. If the *Eucalyptus* tree, however, is situated on ground containing a mass of diseased 'Muula' roots it sometimes succumbs to *Armillaria* root disease, its resistance being undoubtedly lowered by the large number of invasions made throughout its root system."

Ring-Barking and Tree Poisoning Methods for Reducing Root Disease Losses after Clearing

In addition to the important observations just described, LEACH (1937) announced another discovery, which suggested a simple and very practical method for avoiding such losses. In a microscopical study of the pathological histology of infected tea seedling roots, LEACH noticed that hyphae of *A. mellea* were virtually absent from the root cortex, but plentiful in the pith; whereas the cortex was devoid of starch, the pith was well supplied. He concluded, therefore, that *A. mellea* required roots with a high carbohydrate content for its development. In a search for *A. mellea* on the roots of felled stumps of "Muula" *(Parinarinin mobola)* trees, which are especially liable to infection by *A. mellea* after felling, LEACH selected for examination 12 trees that had been ringed some time before felling (and in which the roots must have been depleted of carbohydrate before the trees were felled), and 24 non-ringed trees as controls. Twenty out of the 24 stumps or trees not ringed before felling were found to have roots (varying from 6 to 13 per stump) infected by *A. mellea;* the other 4 had roots infected by "dry-rot" fungi of the *Rhizoctonia bataticola* type.

Of the 12 stumps of trees ringed before felling, one stump showed one root that might have been infected by *A. mellea,* whereas all other roots showed a "dry-rot" due to saprophytic fungi. LEACH therefore suggested that if the forest trees were ringed well in advance of felling, their roots would rapidly become devoid of starch, would die early and be invaded by harmless saprophytic fungi instead of by *A. mellea*. In the second paper (1939), he reported experiments demonstrating that once roots or woody prunings had been invaded by saprophytic fungi, they could not be entered by *A. mellea.*

In conclusion, LEACH (1939) suggests an explanation for the association of *A. mellea* with the stumps of particular native trees in Nyasaland: "The results obtained in the experiment on the susceptibility of indigenous trees, showing that nearly all species in a mixed forest are susceptible to *Armillaria* after the roots are severed, raises the critical question why a few only of these species are usually associated with this fungus as a source of infection in tea estates. I suggest that many trees escape infection because their roots die rapidly after felling before *Armillaria* reaches them. Once dead, and invaded by saprophytes, these quick-dying roots cannot be invaded by *Armillaria.* Unless, therefore, roots of one of the quick-dying species are in close contact with a centre of infection at the time of clearing a mixed forest, they are unlikely to become infected before they die.

The species of tree considered the most dangerous. *Armillaria* are those of which the roots die very slowly after felling. In January 1938, a root of 'Muanga' *(Ajroformosia)* was found alive at a depth of 3 ft. in a block of tea planted in 1934. 'Muula' *(Parinarium mobola)* is another such tree.... The ability of the roots of 'Muula' to keep alive so long after felling probably explains why a pure stand forest of 'Muula' trees often gives rise later to such widespread damage in tea gardens, because starting from a few scattered infected stumps, as in the mixed forest, *Armillaria* can spread unimpeded along a network of *moribund,* but still susceptible, roots." TUNSTALL (1940), commenting upon the association between stumps of felled shade trees and outbreaks of root disease of tea due to *Fonics noxius* and *Ustulina sonata* in North East India, remarked that the stumps of hardwood trees, such as Nahor *(Mesua ferrea),* Sal *(Shorea robusta),* and Khair *(Acacia catechu),* were more frequently associated with outbreaks of root disease in the surrounding tea than were those of soft-wooded trees, which decayed more rapidly.

The interest aroused by LEACH'S discoveries has been quickened by NAPPER'S (1939) announcement that stump poisoning of rubber by sodium arsenite produced a result apparently analogous to that of ring-barking, by killing the roots of the stumps and so rendering them liable to invasion by saprophytic fungi rather than by *Fomes lignosus.* This result was obtained in the course of an experiment designed to test out methods for prevention of root disease on a replanted area. At the same time, NAPPER described another experiment designed to compare LEACH'S ring-barking method, arsenite poisoning by injection, (i) of the living trees before felling, (ii) of the stumps after felling, and an untreated control. Although results from this particular experiment in terms of incidence of root disease after replanting are not available, NAPPER'S observations on starch content of the roots in the different series are pertinent to the present discussion. The ring-barking and tree poisoning were carried out early in June, from 4 to 6 months before felling, and the stump poisoning shortly after felling. Comparative estimates of the starch content were made at the end of September, and again early in December; root sections were cut longitudinally, and the exposed surfaces planed, stained with iodine-potassium iodide solution, and compared with similarly treated sections from standing trees. It was found by visual grading that approximate losses of starch were as follows:

(i) In trees ring-barked 4-4 months before felling, the roots had lost 52 *per cent* of their starch content at 3-4 months, and 86 per cent at 6 months after treatment.

(ii) In trees frill-girdled and poisoned with sodium arsenite 4 months before felling, the roots had lost 98 per cent of their starch at $5^1/2$ months after treatment.

(iii) In stumps poisoned by injection with sodium arsenite immediately after felling, the loss was 65 per cent at 2 months after treatment.

(iv) In untreated stumps the loss was only 4 per cent during the first $2^1/_2$ months after felling.

NAPPER observed a fairly close relationship between the rate of loss of starch from the roots and the rate of invasion by saprophytic fungi.

Airborne Spore Infection of Exposed Wood Surfaces

So much, therefore, for the evidence for direct infection of the roots of felled stumps via the soil. Such evidence, although impressive, by no means excludes the possibility that infection of the roots may develop from spore invasion of the exposed cut surfaces of the stumps; the extent and frequency of such hypothetical spore infection can only be gauged by direct experiment.

There is some evidence for the occurrence of such airborne infection by *Ustulina sonata*, which is well known not only as a root parasite of tea and rubber in Malaya and Ceylon, but also as a wound parasite of the stem, gaining entrance through spore infection of pruning snags, and of hoe cuts in the stem just above ground level. *Pomes noxius* can also, apparently, behave in a similar way; NAPPER (1938a) has described an apparent instance of spore infection of felled rubber logs used for terracing, *viz.*: "At about six months after felling numerous cases of Brown Root Disease appeared among bush cover plants (*Tephrosia vogelii*) established below the logs on the banks of the terraces. These infections were thought at first to be caused by contact with infected roots overlooked during the clearing of the original stand. Field investigation showed, however, that this was not the case; in every instance the source of infection was found in a neighbouring log. There was no evidence to suggest that these logs had themselves been infected by contact with the soil or with buried infected roots. It was therefore, concluded that infection had been caused by windblown spores."

The question has been recently revived by GADD (1940), in connection with the aftermath of felling shade trees on tea estates in Ceylon, concerning "an estate which felled about 6,000 *Grevillea* trees which had been grown as shade for tea.

About two years after felling *30 per cent* of the stumps had given rise to Brown Root Disease with the result that 3,000 bushes had to be removed. The source of infection was attributed to wind-carried spores which presumably found lodgment on the cut surfaces of the stumps."

The discussion of this controversial question of stump infection after felling may be concluded with NAPPER'S (1940) account of yet another factor affecting the incidence of root disease in young plantations on cleared jungle sites: "The most striking fact revealed was the reduced incidence of root disease during the early years after planting in areas *cleared without burning*, as compared with areas burnt off in the usual way.... Up to the end of the third year from planting (which is as far as the present records go) it appears normal to expect at least twice as many root infections in burnt as in unburnt areas, and in one experiment a ratio as high as 5 to 1 was observed. No satisfactory explanation of the phenomenon has yet been advanced. It was thought at first to indicate a delayed attack in the unburnt areas, due to the greater bulk of timber left after clearing in these areas as compared with areas burnt off in the normal way. Recent observations suggest, however, that this is not the case, but that burning actually *encourages* the spread of root infection among the root systems of the jungle trees during the susceptible period after felling."

Economics of Root Disease Control: In the formulation of a policy for root disease control on any area, it is not the mycologist but the economist who has the last word. A valuable feature of the experiments on root disease control carried out by the Rubber Research Institute of Malaya has been the recording of costings for all operations. In Sumatra, DE JONG (1933), as a result of extensive observations and experiments on estates where *Fowes lignosus* was the only important root parasite of rubber, concluded that natural recovery of young rubber from infection by *F. lignosus* was commoner than generally supposed, and that conclusions based on a single root examination might exaggerate the importance of this root disease. Much of the superficial or epiphytic mycelium found on the roots at any examination might never achieve infection; even established lesions might become surrounded by callus and heal over completely. DE JONG therefore, criticised NAPPER'S (1932a) earlier conclusions on the score that an estimate of the root disease situation in any area should not be based upon incidence of mycelium on the roots, nor upon a single examination for the presence of root decay, but rather upon repeated examinations to show the progress of decay, or, better still, upon records of mortality due to the disease. He therefore, refused to make any general recommendations for control of *F. lignosus,* preferring to consider individual outbreaks of root disease in relation to local conditions on the area concerned.

NAPPER has justified his policy of periodical tree-to-tree inspection by pointing out that the same treatment will bring not only *F. lignosus* and *F. noxius,* but also *Ganoderma pseudoferreum* (with which DE JONG in Sumatra was not concerned) under control; *G. pseudoferreum* is likely to cause most serious losses in plantations of 10 years old and upwards if preventive measures are not taken earlier, as by this time the root systems of adjacent trees are all in contact. This view has been upheld by SHARPLES (1936): "If DE JONG'S investigations can be confirmed by observations in Malaya, it is obvious that on lightly infected areas, the disease caused by F. *lignosus* will not call for special treatment, and that it may be possible, as certain planters claim, to limit treatment to removing the roots and stems of the trees killed by the fungus and re-supplying immediately. But if this type of treatment is undertaken and no attention is given to eradicate disease knots of *Ganodcnna pseudojerrcum* in the early years, then it can only be expected that more time must be spent and further expense incurred when the trees are about the age of ten years. It cannot be emphasised too often that the recommendations now being made for treatment of *F. lignosus* are largely dependent on the fact that such treatment will also result in the early extermination of disease knots of *G. pseudoferreum".*

Control of Root Disease in Field Crops: Diseases Control Under the Growing Crop

Variation in the incidence of soilborne diseases was analysed as far as possible, and related to variation in the component factors of the soil environment, *viz.* temperature, moisture content, texture, reaction, organic content, and concentration of plant nutrients. Control measures will be discussed under the following heads:

(i) Mechanical methods.

(ii) Amelioration of soil temperature.

(iii) Management of soil moisture content and aeration, and variation in depth of planting.

(iv) Modification of soil reaction.

(v) Use of organic supplements.

(vi) Application of artificial fertilizers.

Mechanical Methods: Mechanical methods have attained their greatest development as measures for control of root disease in plantation crops, and have thus far found few applications in field crop practice. Some use has been made of mechanical barriers for delimiting the spread of *Phymatotrichum omnivorum* into clean areas, by means of trenches filled with mixtures of soil and heavy oils, sulphur, ammonia, carbolic acid or salt, both in Texas (Taubenhaus and Ezekiel, 1931) and in Arizona (Streets, 1937). Quite a different type of barrier was developed in Texas by Taubenhaus and Ezekiel (1931), who found that 4-12 rows of sorghum stopped the spread of P. *omnivorum;* the roots of the sorghum prevented those of the cotton on either side of the barrier from making contact with one another.

A direct mechanical effect of inter-row cultivation of potatoes in breaking root contacts between adjacent rows of the crop was suggested by McKay (1926) as the reason for failure of *Verticillium alboatrum* to spread from one row to another during the growing season, though it spread freely along the rows, as far as the second or third plant away from the one first showing wilt symptoms. McKAY suggested that the practice of planting in check rows, so as to permit of cultivation in either direction, would probably prevent the spread of this disease, and that this procedure might eliminate the necessity for "three-plant rouging" in crops intended for "seed" production.

Amelioration of Soil Temperature: Control of soil temperature cannot be directly secured in the field as it can in the glasshouse, but the range of temperature under which the crop develops can be varied to some extent by alteration in planting date. Information concerning the optimum temperature for the development of particular soilborne diseases and the safest time for planting of the host crops concerned.

An interesting new method of root disease control, apparently achieved through reduction of soil temperature, has been independently reported by Vasudeva and Ashraf (1939) in India, and by Hansford (1940) in Uganda. Vasudeva and Ashraf, investigating the Indian root rot of cotton, which is caused by *Rhizoctonia solani* and *R. bataticold* found the optimum temperature for disease development in Wisconsin-type soil temperature tanks to be rather high; seedling mortality due to *R. solani* was highest at 35°C, and that due to *R. bataticola* at 39°C. In a mixed crop of cotton and sorghum, the deaths of cotton plants from root rot were significantly fewer than those in cotton alone. Vasudeva and Ashraf attributed this reduced mortality to the shading effect of the sorghum in lowering soil temperature; the humidity of the air was higher in the mixed crop, but no consistent increase or decrease in soil moisture content was found. Vasudeva (1941) later reported further experiments confirming and amplifying these results. Whereas sorghum tended to reduce the yield of cotton in mixed cropping even on heavily infected land, the growth of "moth" *(Phaseolus aconitifolius)* in between the cotton rows not only reduced the mortality in the cotton from more than 50 per cent in the pure stand to a negligible percentage in the mixed crop, but also significantly increased the yield of seed cotton. Soil and air temperatures were lower in the mixed crop, and humidity was higher, than in the pure cotton. In a mixed crop of cotton and "swank" *(Panicum colomiin)*, in which the "swank" got a poor start, soil and air temperatures were no lower than those in the pure cotton plots; nevertheless, mortality due to root rot was significantly lower in the mixed crop than in the pure stand of cotton. Vasudeva therefore concluded: "Temperature may be an important factor, but it is likely that some other factor also comes into play which helps in reducing the incidence of the disease, as in the case of *swank* where the temperature was not materially affected in the mixed crop but the incidence of the disease was reduced." One such "other factor" that is likely to operate in reducing mortality due to disease in a mixed crop (where the plant used for inter-cropping is immune to infection) is the interruption of root contacts between susceptible plants by the roots of immune plants.

Hansford (1940) reported a similar result obtained by J. D, Jameson in field experiments on the wilt disease of cotton (more than 90 per cent of which was attributed to *Verticilliuni dahliae*, and the remainder to strains of the Elegans section of the genus *Fusarium)*, carried out at Bukulasa Station in Uganda. The incidence of wilt was diminished both by reducing the distance between the cotton rows, and by interplanting with groundnuts or beans. The effect was attributed to lowering of the soil temperature, which was stated to vary at Bukulasa from 76°-86°F. (24°-30°C). In the Eastern Province of Uganda, on the other hand, where mean soil temperatures were higher than at Bukulasa, the wilt disease tended to be more prevalent under shade.

Management of Soil Moisture Content and Aeration, and Variation in Depth of Planting: The incidence of a root disease favoured by high soil moisture content can often be reduced by attention to field drainage. A reduction in moisture content of the soil around the roots and base of the plant, with beneficial results in diminishing the incidence of a soilborne disease, can sometimes be secured by planting crops "on the ridge", as advocated by Coons *et al.* (1941) for control of black leg (species of *Pythiutn, Rhizoctonia,* and *Aphanomyces)* in sugar-beets, and by Garcia (1933) for control of wilt *(Fusarium annuuwi)* in chilli pepper.

Crops grown under irrigation form an exception to the general rule that soil moisture content under field crops cannot be varied except by indirect methods. More diseases, unfortunately, seem to be favoured by high soil moisture content than are favoured by low soil moisture content, so that the grower more often reduces incidence of disease by abstention rather than by exercise of his advantage over nature. A notable exception is offered by the cereal smuts, the control of which under irrigation has recently been investigated with conspicuous success by Jones and Seif-El-Nasr (1940). As a brief account of cereal cultivation under irrigation in Egypt is necessary for comprehension of these results, that given by Jones and Seif-El-Nasr is quoted: "Wheat and barley under perennial irrigation in Egypt are either planted by the *herati* method of broadcasting seed on moist land and ploughing it in, or else by the *afir* method of broadcasting on dry land, harrowing in the seed with a baulk of wood and irrigating.

Examination of pairs of wheat fields in which all other factors were said to be the same, but in which both methods of planting were used, showed that *herati-sown* plots consistently gave more flag smut than herati-sown plots. Later experiments showed that there was two to three times more flag smut in herati-sown plots, and similar tests with barley proved that the *herati* plots were six times more heavily attacked by covered smut than the *afir* plots, thus explaining satisfactorily the result. Later still, it was found that bunt of wheat and grain smut of millet and broom corn were almost equally sensitive to difference in method of planting. The most obvious difference between the *herati* and *afir* methods of planting is in soil moisture, the *herati* plots being only moist enough for good ploughing, while *afir* plots, being irrigated immediately after sowing, are wet. A second difference lies in the depth at which the seed is planted. In *herati* plots the seed is covered with a plough set at 12-15 cm. depth, so that the seed is buried at any depth between the surface and the plough sole: seeds which happen to be left near the surface fail to germinate owing to lack of moisture, while the most deeply buried seeds germinate poorly, so that the average depth of effective planting is about 8 cm. In the *afir* plots the seeds broadcast on the surface are covered by harrowing with a wooden baulk while the land is dry, and are thus buried only about 4 cm deep before they are irrigated.

Jones and Seif-El-Nasr proceeded to determine the reasons for this difference in incidence of cereal smuts under the *afir* and *herati* methods of sowing, by comparing the incidence of smut in plots sown under moist and wet conditions, and at 4 depths of planting, *viz.* 0.5, 4, 8 and 12 cm. The moisture content of the "moist" soil corresponded to that of the *herati* plots, arid the moisture content of the "wet" soil to that of the *afir* plots. Experiments were carried out with the diseases caused by four smut fungi, *viz.* flag smut *(Urocystis tritici)* and bunt *(Tilletia Icvis)* of wheat, covered smut *(Ustilago hordei)* of barley, and grain smut *(Spacelotheca sorghi)* of millet and broom corn. The effect of higher moisture in decreasing the incidence of smut was consistent throughout the experiments, all diseases behaving in the same way.

The effect of depth of planting was still more pronounced, and almost as consistent; the percentage of smut increased with depth of sowing, presumably (as Jones and Seif-El-Nasr suggested) because the period during which the host was susceptible to infection was thereby prolonged. The only exceptions to this rule were afforded by the incidence of bunt of wheat at 12 cm. depth in wet soil, which was less than that at 8 cm. depth, and by the incidence of grain smut of broom corn, which decreased as depth of sowing increased from 4 through 8 to 12 cm. In explanation of these exceptions, Jones and Seif-El-Nasr suggested that under the conditions of deep planting in wet soils the smut fungi tended to become destructively parasitic, and to kill the infected seedling below ground; the *apparent* percentage of infected plants, as determined by counts at harvest, would in this way be diminished. Nevertheless, infection by all the smut fungi concerned in these experiments is favoured by good soil aeration, as shown by the consistently higher incidence of infection in the moist as compared with the wet soils; it seems possible, therefore, that in the wet and less well aerated soil the advantage offered to the fungus by deeper sowing, through prolongation of the susceptible period of the cereal seedling, might be more than offset by the disadvantage to it of further reduction in oxygen supply.

These exceptions, however, are of minor practical interest; the extensive data collected by Jones and Seif-El-Nasr in these experiments indicate that the advantage of the *afir* over the *herati* method of sowing in smut control is to be attributed both to the shallower sowing and to higher soil moisture content during the pre-emergence period. As Jones and Seif-El-Nasr have pointed out, the logical extension of the *afir* method of planting for better control of smut is "mud sowing" at zero depth and maximum soil moisture; this could be practised on land prepared either as for *herati* or as for *afir* planting.

Concerning the very satisfactory control of flag smut obtained through mud sowing, Jones and Seif-El-Nasr declared that method appeared to promise increased yields and better quality of grain, apart from the control of smut, and that there was no sign of increased tendency towards lodging, as might perhaps have been expected. This method should be particularly valuable for control of flag smut in Egypt, where infection, as in other countries, is soilborne as well as seed-borne.

The method of preparing the seed-bed has also a most important influence upon the occurrence of the take-all *(Ophiobolus graminis)* disease of cereals, which is encouraged by loose, badly compacted soils; control must therefore be sought through preparation of a particularly firm seed-bed for wheat. Such a seed-bed is best secured through early preparation and sufficient working at the correct times, so that the underlayers of the soil become consolidated by the joint action of implements and the weather; the consolidating effect of rain has been particularly stressed. A poor seed-bed, on the other hand, results from too late and hasty a preparation, when a tilth has to be "forced" by the use of implements, instead of maturing with time and action of the weather. In the same way, the ploughing-in of long straw or dry grass, which opens up the soil, or dry ploughing and working of the land, which produces a similar effect, has been found to encourage the disease.

Modification of Soil Reaction: Control of the clubroot disease of crucifers by application of lime is too well-established to require detailed comment. For this purpose, the oxide and hydroxide of lime are much more effective than the carbonate, with which disappointing results have frequently been reported. Thorough incorporation of the lime with the soil down to ploughing depth is obviously essential for complete control of this disease; for this reason, the lime is best applied a year or so before the cruciferous crop is sown. The widespread adoption of lime for control of clubroot owes much to the fact that liming of acid soils greatly increases the yield, and often the quality, of the majority of field crops, apart from control of clubroot.

In contrast to the general employment of lime against clubroot, the adoption of sulphur as a soil-acidifying agent for control of such diseases as Texas root rot of cotton, scab of potatoes, take-all of cereals, and black root rot of tobacco has scarcely loft the experimental stage. The acidification of neutral soils by addition of sulphur is likely to lower the yield of many crops; on highly alkaline soils, such as those of arid regions (and especially soils under irrigation), the application of sulphur may improve crop growth by decreasing soil alkalinity, but the cost of the very heavy applications of sulphur necessary to produce this effect is likely to be too high for commercial farming. Although soil acidification with sulphur cannot as yet be recommended as a practical measure for control of these diseases, useful cautionary advice can be given to farmers on slightly acid or neutral soils, *viz.* to be careful in the use of lime, and preferably to leave the soil with a slight lime requirement.

Use of Organic Supplements: The control of potato scab by green manuring is sometimes extremely successful, especially in private gardens and on small-holdings where grass cuttings, etc., can be incorporated with the soil in amounts that would be quite impracticable in the field. In the field, indeed, green manuring is by no means always successful in controlling scab, as Sanford (1939) has observed. In pot experiments demonstrating the control of scab and other diseases by organic manuring, the green manure has usually been dried and ground, so that incorporation with the soil has been very uniform; in small plot experiments, the green manure has usually been chopped up with a chaff cutter, for the same reason. When green manuring is practised in the field, however, the organic material is usually ploughed into the soil in swathes, so that the resulting mixture of soil and organic material is far from homogeneous; a more uniform and thorough incorporation can be achieved by means of the rotary cultivator. For the cotton crop, King (1937) has ingeniously surmounted this difficulty by planting the cotton rows right on top of the trenches in which the fresh organic material has been buried.

Even the most sanguine investigator should not be surprised, therefore, if the history of his pot experiments fails to repeat itself in the field. The slow progress made by such comparatively new methods as organic manuring for control of root disease in field crops is not surprising; wherever possible, the farmer achieves a satisfactory degree of control through crop rotation, modification of cultural practice, and application of artificial fertilizers.

Application of Artificial Fertilizers: Recommendations to apply artificial fertilizers to assist in root disease control are more likely to be adopted by farmers than most other suggestions made by plant pathologists, because an increase in crop yield can usually be guaranteed even in the absence of disease.

Control of Root Disease in Glasshouse Crops

The control of root diseases of crops under glass presents problems similar in broad outline to those encountered in field crops; the control measures may similarly be classified under the three heads of crop rotation, plant sanitation, and checking of disease development in the growing crop. The relative importance of these three groups of control measures under glass is, however, very different to that under field conditions. Whereas in the field, crop rotation is of paramount importance in root disease control, in the glasshouse its scope is limited by high overhead costs, and consequently restricted choice of suitably remunerative crops. Another factor limiting the choice of crops for rotation is that of glasshouse design; a house ideal for tomato culture is less suitable for the raising of cucumbers, which require a higher temperature for optimum growth.

Under glass, the most important control methods are those comprised within the general term of plant sanitation. The greater value of the crop under glass permits of closer attention to individual plants than is economically possible in the field; where plants are grown in pots, raised beds, or otherwise in individual containers or compartments, the practice of rogueing, with appropriate disposal of infected soil and plant residues, is likely to be more effective than where plants are grown directly in the ground. Moreover, the more rapid growth of crops under glass, together with the greater opportunity for dispersal of parasites during cultural operations, necessitates a stricter attention to the principles of plant sanitation; especially is it important to avoid introduction of parasites either with the seed, or in infected propagating material, such as cuttings, etc. By far the most important control measure to be included under the heading of plant sanitation is that of partial sterilisation of the soil, or "soil sterilisation", as it is commonly though inaccurately called. Methods of partial sterilisation are described in detail below; they form the basis of root disease control in glasshouse crops.

The checking of root disease development by modification of the soil or aerial environment of the growing crop should theoretically prove more feasible under glass than out in the field. Nevertheless, the requirements of the growing crop for optimum vegetative and fruiting development must often conflict with the tentative recommendations of the plant pathologist for root disease control.

Just as the efficiency of crop rotation for control of root disease in field crops has been responsible for the stillbirth of many ingenious but less simple control methods, so the efficiency of partial sterilisation of the soil has had a like effect under glass. Although soil moisture content, for example, is directly controlled by the glasshouse grower, little practical advantage is taken of this possibility for root disease control. Greater use is made by growers of their control over soil and air temperatures. By BEWLEY'S (1922) method of controlling the *Verticil-Hum alboatrum* wilt of tomatoes through raising glasshouse temperature to at least 25°C, a spectacular "cure" of this disease can be effected.

BEWLEY (1923) gives the following directions for control of the disease in this way: "As soon as the wilt appears and it is proved that *Verticillium alboatrum* is the pathogen, the average temperature of the houses should be raised above 25°C by suitably increasing the boiler heat, regulating the ventilation, and by closing down the houses from two to four hours in the middle of the day. A light dressing of whitewash on the glass makes the conditions still more favourable for the plants. As little water as possible should be given to the roots, as heavy watering merely aggravates the wilting, but a light overhead damping helps the wilted plants to recover. The plants should be encouraged to make fresh roots above the original diseased ones by placing fresh soil round the base of the plant." BEWLEY records that, in one nursery where these control measures were carried out, the percentage of wilted plants was reduced from 68 to 10 in a fortnight.

Another advantage enjoyed by the glasshouse grower over the farmer lies in his greater degree of control over soil composition. Glasshouse crops are often grown not in the original soil itself, but in an artificial soil or compost. Even if the original soil is employed, the heavy dressings of organic manures and artificial fertilizers commonly applied in glasshouse practice may rapidly change its physical texture and chemical composition. In the making of composts, a bewildering variety of "recipes" was formerly employed by professional gardeners; the number of composts in use by some of these artists approached the number of plant species under cultivation.

Order and simplicity have recently been introduced into the practice of compost making by LAWRENCE and NEWELL (1936, 1941). The prime ingredients for a compost have almost invariably been loam, leaf-mould or manure, and sand. The leaf-mould or manure supplied both the organic matter for production of the desired texture in the compost, and some at least of the essential plant nutrients. Owing to the wide variation in composition of these organic materials, according to their original constitution and degree of decomposition, it has been impossible to produce a standard compost from such components. LAWRENCE and NEWELL achieved standardisation by discarding these dual-purpose organic materials, and employing peat as the source of humus. The manurial value of peat is relatively low, and does not fluctuate widely within any one type of peat; the physical properties of peat make it a more desirable constituent of compost than leaf-mould or manure.

LAWRENCE and NEWELL found these composts suitable for growth of a wide variety of glasshouse plants, when used in conjunction with a balanced mixture of artificial fertilizers (for which they supplied a formula), together with chalk.

Standardisation of the compost, as recommended by LAWRENCE and NEWELL, may prove useful in the solution of certain root disease problems as well.

DUNLAP (1936) has suggested that control of damping-off in seedling pans might be achieved more easily by raising seedlings in quartz sand moistened with a nutrient solution than in partially sterilised soil, wherein a reintroduced damping-off fungus may spread with especial rapidity. An example of an extreme development in this type of root disease control is afforded by the commercial culture of glasshouse crops in tanks of nutrient solution.

After this brief general account of root disease control in crops under glass, more detailed consideration may now be given to partial sterilisation of the soil, which is as important to the glasshouse grower as is crop rotation to the arable farmer.

Partial Sterilisation of the Soil: This is by no means a new practice. By partial sterilisation is meant any treatment of the soil by heat, chemicals or other agent, whereby the soil microflora and fauna are reduced and simplified. One object of partial sterilisation is to kill all disease-producing micro-organisms and insect pests, whilst leaving alive at least some of the beneficial soil micro-organisms; the spore-forming bacteria usually escape destruction, and the more common non-sporing forms are reintroduced and re-establish themselves with little delay. Partial sterilisation produces not only microbiological but also physical and chemical changes in the soil; a notable effect of heat treatment is the liberation of plant nutrients, especially nitrogen, in an available form through breaking-down of complex organic compounds. Partial sterilisation by heat, in particular, may also improve the physical structure of the soil; according to NEWHALL (1935), the water-holding capacity of the soil may be lowered by steaming, but this is not necessarily a disadvantage.

The history of partial sterilisation, together with its modern applications, has been discussed by WAKSMAN (1931), Newhall *et al.* (1934), NEWHALL (1935), RUSSELL (1937), BEWLEY (1939) and MARTIN (1940). The increase of crop following a fire over the soil was well known to the ancients; such improved growth has usually been attributed to the high potash content of the wood ash, but a liberation of nitrogen through heating of the surface soil must also occur. Concerning pioneer farming in the early days of the U. S. A., NEWHALL *et al.* (1934) write: "In newly-cleared land, places where brush piles were burnt dotted the fields. The early settlers learnt that delicate plants like watermelons, muskmelons, and even beans, could be grown in these spots where they would damp-off otherwise."

It is not surprising to find that authors disagree concerning the credit to be given for the "discovery" of the benefits of partial sterilisation. On this point, MARTIN (1940) remarks: "The first observations of the increase of yield following partial sterilisation were made independently by GIRARD and by OBERLIN in 1894. These workers had employed carbon disulphide as their sterilising agent. Similar results were found after the application of other volatile antiseptics, and in 1907, DARBISHIRE and RUSSELL showed that partial sterilisation by heat produced a like result." Waksman (1931), however, has drawn attention to an earlier paper by Franks (1888), who correlated the increased crop growth observed to follow heating of soil with enhanced solubility of soil mineral constituents and organic matter.

Much was done by Russell and his collaborators in their investigation of the complex causes of soil sickness in English glasshouses; concerned in bringing about this condition were not only parasitic fungi, nematodes, and insect pests, but also a deterioration in the physical and chemical condition of the soil through repeated heavy dressings of artificial fertilizers and continual heavy watering. BEWLEY (1934) has since shown that the last evil can be combatted by incorporating layers of fresh straw with the soil. Whatever the causes of soil sickness, however, the cure advocated by RUSSELL and his co-workers in England proved such a success that partial sterilisation of the soil has become a routine measure in English commercial glasshouses. Although credit has probably been justly assigned to these investigators, in as much as the widespread adoption of partial sterilisation was due largely to their efforts, yet it appears that steaming of the soil was first tried out in the U. S. A. at a still earlier date.

Thus NEWHALL (1935) states: "The first use of steam for commercial greenhouse soil sterilisation seems to have been made in 1883 by W. H. RUDD in Greenwood, Illinois. He laid three steam pipes perforated at 18-in. intervals in the bottom of a wooden bin 4 ft. deep and holding 480 cu. ft. of soil."

Complete sterilisation of the soil, on the other hand, is not beneficial but harmful; in addition to extinction of all beneficial, as well as harmful, soil micro-organisms, there is a loss of soil organic matter, and often a marked deterioration in soil structure or tilth as well. This may happen when dry heat is used as an agent for partial sterilisation as in baking, or in using the "New York type" of electric partial steriliser with soil insufficiently moist. Concerning this possibility, BEWLEY (1939) observes: "When baking a soil, it is possible to overheat it, especially if the soil is on the dry side, and, when this occurs, the soil becomes 'dead' and will not support plant growth. Seedlings planted in it turn blue and hard, and grow very slowly, if at all."

Methods of partial sterilisation at present in use may be classified as follows:

(a) By heat

  (i) Steam,

  (ii) Baking,

  (iii) Electricity,

  (iv) By chemicals.

By Steam: This is by far the most effective agent for partial sterilisation as yet discovered; unlike dry heat, it can be used to treat the soil *in situ* in the beds. It has the advantage over all chemical agents so far tried of being equally lethal to parasitic micro-organisms, including plant-infecting nematodes, to insect pests, and to weed seeds. Various contrivances have been designed for the steaming of relatively small quantities of loose soil, intended for the filling of pots, flats and seedling pans, etc. The chief practical interest centres, however, in methods of steaming soil in the beds, of which a brief description may now be given.

In the *buried tile method,* a system of 3-4-inch diameter clay drainage tiles is laid 13-16 inches deep in rows 18 inches apart, in permanent position in the beds (NEWHALL *et al.* 1934). When steaming is to be carried out, the system is connected up to the boiler; the steam escapes into the soil from the cracks between the tiles. According to NEWHALL *et al.*, such tile systems may be expected to last for at least 12-15 years without resetting. An additional advantage accruing from such a buried tile system is the use to which it can be put either for sub-irrigation, or, in conjunction with flooding, for the leaching of an excess of soluble salts out of the soil. Although apparently popular in the U. S. A., the tile method of steaming is little used in England, and no mention of it is made by BEWLEY (1939).

In the *buried perforated pipe method,* perforated pipes are laid in parallel trenches 6-18 inches deep, and at a distance apart depending on the depth, and covered with soil. One end of each pipe is connected by an elbow to a short upright pipe, through which the live steam is conducted into the buried perforated pipe. In England, where this system is known as the "Hoddesdon", the horizontal perforated pipes are customarily 6-7 feet long (BEWLEY, 1939), but in the U. S. A., they may range from 20 to 120 feet (NEWHALL *et al.*, 1934). Whilst soil steaming is proceeding through one set of parallel pipes, the workmen lay another set in position. A variant of the buried perforated pipe method employed in England is the *small grid method;* horizontal perforated pipes 2-2$^1/_2$ feet long and 10-12 inches apart are connected at their proximal ends to a cross pipe 8-10 feet in length. The apparatus thus resembles a comb, built-up out of tubes.

By the *steam harrow method*, which is sufficiently described by its name, the labour required for steaming, and consequently the cost of the operation, is much reduced. The effect produced is more superficial, however, and the method is not considered suitable for heavy soils. The *inverted steam pan method*, which in England is known as the *tray method*, is similarly characterised by lower labour costs, and by a reduced effective depth of working; it is therefore, especially suitable for the treatment of seed-beds and raised benches, where depth of soil is limited. Various mechanical contrivances for moving the heavier types of pan are in common use. NEWHALL (1939) has recently described an electrically heated inverted steam pan devised at the Cornell Agricultural Experiment Station; steam is generated by electric strap heaters operating in troughs of water placed inside the inverted pan.

Certain precautions are essential to success in the use of all these steaming methods alike. The steaming must, of course, be thorough and must be effective to a sufficient depth in the soil. To conserve heat, it is usual to cover the soil with sacking or other insulating material. The soil should be neither too wet nor too dry, if the best result from steaming is to be obtained; BEWLEY (1939) suggests that "generally speaking, a soil sufficiently dry for potting purposes is in a suitable condition for steaming". With such a suitable moisture content, optimum channels for conduction of the live steam are probably provided by the pore spaces of the soil. If the soil is too dry, the pore spaces are too wide, and the steam escapes before it has completed its work of partial sterilisation; if, on the other hand, the soil is too wet, then insufficient pore space is provided for thorough penetration of the steam to all parts of the soil. Moreover, if the soil is too wet, heat is wasted in raising the temperature of the excess soil moisture to boiling point, and the soil will also take longer to dry-out after steaming.

Further precautions, again, must be observed in the care

and use of soil after partial sterilisation by steam. Steamed soil is a medium more favourable than unsteamed soil for the activity of most root-infecting fungi; nutrients are set free by the steaming, and the competitive effect of other micro-organisms is much reduced. For this reason, HARTLEY (1921) suggested that steamed soil should be reinoculated with saprophytic micro-organisms before being used for the raising of seedlings. Especial care is therefore, necessary to prevent contamination of steamed soil by untreated soil infected with one or more of the root parasites; the feet of workmen, for instance, may so contaminate the freshly steamed soil during the actual progress of steaming.

BEWLEY and BUDDIN (1921) have shown that the glasshouse water supply may constitute an important source of infection. Where the soil beds are not separated from the subsoil, recolonisation of the beds by root-infecting fungi is only a matter of time, since the subsoil serves as a reservoir of infection, which may be tapped by the roots of the first crop grown in the steamed soil. Under such conditions, therefore, periodical soil steaming, like crop rotation, serves not so much to eliminate root parasites as to restrict the damage that they do, and keep it at a commercially unimportant level. Where such spread of a root-infecting fungus from infected subsoil is especially rapid, however, the culture of plants in raised beds or benches, which can be treated *in toto,* may be adopted, as recommended by WICKENS (1935) for control of the *Verticillium cineresccns* wilt of carnation.

A further, though temporary, disadvantage of steamed soil is its toxicity to most seedlings, and to delicate plants; the effect is least noticeable with robust and "strongly feeding" plants like the tomato. Steaming, or other heat treatment, is well known to increase the content of soil solutes very considerably; the concentration of solutes increases with the duration and intensity of the heat" treatment. JOHNSON (1919) demonstrated that the toxicity of steamed soil to seedlings was due chiefly to the ammonia produced through the heating. The toxic effect is greater in rich soils with a high organic content than in poor soils; it is increased by addition of chalk or lime to the soil before steaming (LAWRENCE and NEWELL, 1936). In the past, it has been deemed inadvisable to use steamed soil sooner than 3-6 weeks after steaming; alternatively, ammonium salts could be leached out of the soil by flooding at least once with water, a wasteful procedure. A better alternative is that of applying a sufficient dressing of superphosphate to the steamed soil to balance the excess of soluble nitrogen; this valuable effect of phosphate in combatting the ammonium toxicity in steamed soil seems to have been discovered by LAWRENCE and NEWELL (1936, 1941).

By Baking: Methods of baking soil by dry heat have been described by NEWHALL *et al.* (1934) and, in greater detail, by BEWLEY (1939). They are suitable only for relatively small quantities of soil, and cannot be used to treat soil *in situ* in the beds, as in steaming. If the soil is too dry when treated, there is danger of overheating, with resulting loss of organic matter, and, in the extreme instance, with complete sterilisation.

By Electricity: The use of electricity to effect partial sterilisation of the soil was apparently first reported by ELVEDEN (1921) in England, but further development of this method is due almost entirely to American investigators, whose work has been summarised by NEWHALL and NIXON (1935). Two types of electric partial steriliser have been developed; in the Ohio type, the soil is heated directly by passage of a current through the soil from one electrode to another, whereas in the New York type, the soil is indirectly heated by passage of a current through metal heating elements. In the Ohio type of partial steriliser as well as in the New York type, the soil is partially sterilised by the heating action of the current, and not by the wave effect. Both types of partial steriliser have so far taken the form of containers designed.for treatment of limited quantities of loose soil; no satisfactory method for effecting partial sterilisation of soil in the beds seems so far to have been reported, although various attempts have been made to solve this problem. According to NEWHALL and NIXON, uneconomic lengths of heating element would be required for use of the New York method in this way, whereas use of the Ohio method is complicated by the difficulty of getting satisfactory plate to soil contacts in the uncompressed soil of the beds, by the development of uneven temperatures, and by leakage of power, with attendant personal risk to the operator. If these difficulties can be overcome, electricity may eventually compete with steam as a method for partial sterilisation of soil *in situ*. An interesting new machine for continuous partial sterilisation of loose soil has been described by NEWHALL (1939); approximately 7 cubic feet of soil, or enough to fill 18-20 flats, is turned out in an hour by this machine, the principle of which is similar to that of an electric baker's oven.

For satisfactory partial sterilisation by electricity, the soil must be at a suitable moisture content. If it is too dry in the New York container, it becomes overheated, with consequent loss of fertility, around the heating elements. With the Ohio method, electrical conductivity of the soil is too low for effective heating below a certain limiting moisture content; for this reason, overheating of the soil is impossible, as the current is reduced to zero as soon as the soil starts to dry out. The electrical conductivity of quartz sand, as used for propagating benches, is too low for satisfactory partial sterilisation by the Ohio method, but this difficulty may be overcome, according to NEWHALL and NIXON, by addition of 0.5 per cent potassium nitrate to the sand before treatment. Similar conclusions concerning the relation between electrical conductivity of the soil and its moisture and salt content were expressed by VAN DER MUYZENBERG and VAN RIJN (1936), from their experimental application of these American methods in Holland.

By Chemicals: For partial sterilisation of the soil with especial reference to the need for controlling root-infecting fungi, the performance of formaldehyde is at present outstanding; the consideration of other chemicals for this purpose is at present probably a waste of time. As an agent for general purpose partial sterilisation, formaldehyde, like all other chemicals upon which report has thus far been made, is inferior to heat; it is ineffective against nematodes, insect pests, and most weed seeds, and is less lethal to fungal sclerotia than are the heat treatments, On the other hand, it is a cheap chemical, and has exceptional penetrating power; HUNT *et al.* (1925) demonstrated that its penetration was equal to that of the water carrying it in solution. Concerning rate of application, NEWHALL *et al.* recommended that one gallon of formalin (40 per cent formaldehyde in aqueous solution) be diluted to 50 gallons with water, and that half a gallon of the dilute solution be applied per square foot of soil surface. BEWLEY (1939) recommended a solution of the same dilution, to be applied at either of two rates, which work out at half a gallon per 0.9 square foot and per 1.62 square foot, respectively.

He suggested that the soil should be turned over one spit deep after treatment, on the same day. NEWHALL *et al.*, on the other hand, directed that the soil should be covered for a day or two after treatment in order to retain the fumes of formaldehyde for as long as possible. An interval of 14 days after treatment is usually deemed sufficient for dissipation of the gas; after this, the soil can safely be planted. Effective as formaldehyde is by comparison with other chemicals for partial sterilisation, it cannot compete with steam where this is available. Some interesting crop returns have been quoted by BEWLEY in a comparison of the yield of tomatoes following treatment of the soil by steam; formaldehyde, and cresylic (carbolic) acid; yields of fruit were 51, 45, and 40.25 tons per acre, respectively, as compared with a yield of 37.9 tons on the untreated control soil.

A method was devised by BEACHLEY (1937) for treating soil with a mixture of steam and formaldehyde vapour, by means of the inverted steam pan; the formaldehyde solution was vaporised by pumping it through a heated coil. The formaldehyde vapour was found to penetrate the soil as deeply as the steam; a deeper penetration was thus achieved than when formaldehyde was applied as a liquid drench. By using 1 Ib. of formaldehyde per 100 square feet of soil surface treated by the combined steam and formaldehyde method, the necessary period of steaming could be reduced from 30 to 20 minutes, and the cost cut from 1 dollar to 80 cents per 100 square feet. Moreover, most of the vegetable and flower seeds that were tested germinated and developed satisfactorily when sown in soil treated only 24 hours previously.

Many other chemicals have been tested as agents for partial sterilisation, but none can compete with formaldehyde for cheapness and fungicidal efficiency. The most promising, perhaps, is chloropicrin, which is an efficient insecticide (ROARK, 1934) and has been developed by GODFREY (1935) and his collaborators (GODFREY *et al.*, 1934) as a nemacide. GODFREY (1936) later showed that chloropicrin was a useful fungicide as well; it has since given encouraging results in the hands of other workers, when employed against a variety of soil-borne diseases, *viz.* black leg *(Moniliopsis aderholdi)* of cabbage (EzERS-KAYA, 1937), black root rot *(Thielaviopsis basic ola)* of tobacco (GROOSHEVOY *et al.*, 1940), wilt *(Fusarium bulbigenum* var. *lycopersici)* of tomato (YOUNG, 1940), foot rot *(Rhizoctonia solani)* of annual stocks (DiMOCK, 1941), southern sclerotial rot *(Sclerotium rolfsii)* of sugar-beet (DAVEY and LEACH, 1941), clubroot *(Plasmodiophora brassicae)* of crucifers and other diseases (HOWARD, 1940). Adequate confinement of the chloropicrin vapour is essential for successful use of this fumigant; for this purpose GODFREY (1934) developed a glue-coated paper as a soil cover, with which GODFREY *et al.* (1934) obtained improved nematode control.

Non-persistence in soil is an essential quality for any chemical to be used in wholesale partial sterilisation of glasshouse beds. This requisite eliminates many chemicals of proved fungicidal efficiency, such as the salts of copper and mercury; for the same reason, sulphur and sulphuric acid are unsatisfactory as a substitute for formaldehyde. The only chemicals likely to be adopted by the glasshouse grower for routine partial sterilisation are those that give a substantial temporary increase in soil fertility, in addition to destroying parasites and pests.

Control of Root Disease in Field Crops: Plant Sanitation

The problems arise chiefly during and after harvest of a crop in which a soilborne disease has been prevalent, since they concern the best method for treatment and disposal of infected seeds, propagating sets, and other crop residues. Infected seeds or propagating sets must be either eliminated or so treated that they do not carry infection into the next crop and perhaps onto "clean" land as well. Other infected residues, of no value for propagation, must be destroyed at once, if possible; if not, precautions should be taken to prevent their passive dispersal by the agencies of wind, water, man and his domestic animals. The gradual decline of a parasite when the land is occupied by non-susceptible crops has just been discussed; there remain for consideration those measures that can be taken to accelerate this "normal" disappearance.

Preventing Dispersal of Parasite by Infected Seed: The prevention of disease transmission by infected seed or propagating sets is perhaps the most important part of crop sanitation; loss due to disease in a following crop, though considerable, may be negligible by comparison with the cumulative loss due to infection of clean land. The cereal smuts are the best known seed-transmitted diseases. In some of these, the spores are held mechanically, either adhering to the seed coat, as in bunt *(Tillctia tritici* and *T. levis)* of wheat, or enclosed within the glumes, as in covered smut *(Ustilago hordei)* of barley; these diseases can be controlled by seed treatment with a protective fungicide, such as the organic mercurial compounds now widely used for this purpose. In others, such as loose smut of wheat *(Ustilago tritici)* and of barley *(U. nuda)*, the smut fungus is carried as a latent mycelial infection within the tissues of the seed, and so cannot be reached by surface disinfectants. The "modified hot-water treatment", invented by Jensen (1888), is still used for control of these loose smut diseases of wheat and barley, but is not widely employed on account of the difficulties of operation, which involve skilled supervision, and also because the treatment is liable to reduce percentage germination of the seed. An ingenious application of this principle of control has been worked out by Luthra and Sattar (1934) for Indian conditions (Luthra, 1941). The treatment is carried out in June or July; the infected seed is presoaked in water at indoor temperature for 4 hours from 8 a.m. to 12 noon, and is then taken out and exposed to the sun from 12 noon to 4 p.m. Complete control of loose smut has been obtained in this way, and the reduction in germination of the treated seed is negligible (not exceeding 2 *per cent).*

The method is said to be applicable only to the hotter parts of the Punjab plains; maximum shade and sun temperatures at Lyallpur in June are stated to be 49° and 55°C (120° and 131°R), respectively. Further information concerning seed treatment may be found in the reviews of Leukel (1936), Cunningham (1935), and Martin (1940).

Preventing Dispersal of Parasite by Infected Propagating Sets: The best known soilborne diseases transmitted by vegetative propagation of the host plant are those of the potato. They include wart *(Synchytrium endobioticutn)*, scab *(Actinomyces scabies)*, stem canker and tuber scurf *(Rhizoctonia solani)*, and wilt *(Verticil-Hum alboatrum, Fusarium oxysporum* and *F. solani* var. *eumartii)*. No effective method of tuber treatment for elimination of wart is in use, and the movement of seed tubers from an infected area has been controlled by quarantine legislation in many countries. Tuber treatment by steeping or dusting has been widely advocated for prevention of stem canker, black scurf and scab, but the value of such treatments for this purpose is often questionable. Both *R. solani* and *A. scabies* are widely distributed in most potato-growing soils; the planting of infected "seed" must often fail to increase either the amount of disease in the current season's crop, or the volume of inoculum of either parasite left behind in the soil after lifting of the crop. *R. solani* admittedly has a considerable power of movement through the soil; its mycelium probably grows not only from the seed or parent tuber along the sprouts, perhaps causing stem canker, but may also extend directly through the soil to the later-formed daughter tubers, there to cause the black scurf form of this disease.

Actinomyces scabies, on the other hand, appears to possess little or no power of free movement through the soil, as demonstrated by Sanford (1933) in a masterly experimental analysis of the arguments for and against seed tuber treatment for control of scab.

Sanford's experiments may be summarised as follows: (i) four tuber steeps, *viz.* hot formaldehyde, cold formaldehyde, mercuric chloride and a proprietary compound, failed to reduce incidence of scab on the crop in a field experiment, nor was scab increased by the use of planting sets almost covered with scab lesions, (ii) when sets were coated with a virulent agar culture of *A. scabies*, only a slight increase of scab occurred on the daughter tubers as a result of this, and then only on a few of the tubers that grew very close to the inoculated planting set, (iii) when planting sets were surrounded by a culture of *A. scabies* on sterilised soil, only that part of any daughter tuber which grew actually into the soil inoculum was covered by scab, (iv) in sterilised soil, *A. scabies* grew vigorously almost in contact with potato cores treated with the different protective steeps and dusts. Sanford therefore concluded that tuber treatments, whether or not they reduced the amount of *A. scabies* inoculum persisting in the soil from the original planting set, certainly failed to decrease the incidence of scab in the current season's crop.

An interesting study of the transmission of *Verticillium alboatrum* to the tubers, and of the spread of this parasite under the growing crop, has been made by McKay (1926), who showed that inspection of seed tubers was of little value as an aid to detection of infected ones; separation of tubers with discoloured stem ends, and discarding of stem ends in cut seed tubers alike failed to eliminate carry-over of infection in the seed tubers. McKay therefore concluded that rouging of the parent crop in the field was the only effective way of reducing "seed" transmission of this wilt disease to negligible proportions. Periodic rouging of wilted plants (*i.e.* those showing above-ground symptoms), alone, was insufficient, since by the time such individuals had become obviously diseased, the neighbouring plants on either side of them in the row had usually contracted the infection as well. McKay, therefore, recommended "three-plant rouging", whereby the wilted plant was removed together with the two adjacent plants in the same row, at periodical crop inspections throughout the growing season. In one experiment, where the percentage of visibly wilted plants was 22, single-plant rouging reduced the percentage of infected tubers to 8.4, and three-plant rouging to 3.2. From the data obtained in these experiments, however, McKay concluded that infected tubers could be eliminated from the crop as effectively by marking the plants for elimination, and digging separately at harvest, as by pulling them when first seen; even three-plant rouging failed to reduce spread of the disease to the second plants away from the wilted one in the row.

Destruction of Other Infected Crop Residues: One of the simplest and most effective methods for disposing of infected aerial parts of diseased cereals is to fire the straw after harvest, a procedure that has been followed for many years in Australia for control of the take-all disease; this procedure also eliminates plant material infected by the flag smut organism, *Urocystis tritici*. In the past, Australian wheat has usually been harvested by the "stripper", which removes the heads alone, leaving the long straw standing; the straw never fails to burn well after the hot, dry summer, and the basal infected part of the stubble is usually completely destroyed. Although O. *graminis* continues to survive in the infected root and crown left behind in the soil, this source of infection is generally less massive, and therefore likely to disappear more quickly, than the infected base of the straw. Nevertheless, this practice of firing the straw is to be deprecated except when dealing with extensive infections, since the rapid depletion.

Of the soil humus is likely to lead to serious soil erosion in countries with a climate like that of Australia.

Extraction of infected crop residues from the soil, either by hand labour or by machine, is a measure deserving some consideration. No such expedient seems to have been tried out against any of the important soilborne diseases due to fungi; in the Anglo-Egyptian Sudan, however, the carry-over of the leaf curl virus of cotton from one season to the next has apparently been prevented by hand-pulling (after harvest) of the plants, which are "yanked" out of the ground by an ingenious implement devised by Massey (1934). Treatment of the soil by fungicides might seem to be the obvious method for destroying infected crop residues, but this expedient is rarely economically sound for field crops, although soil sterilisation by heat or by chemicals has become a routine practice in intensive glasshouse culture.

Preventing Dispersal of Parasite by Other Agencies: Dispersal by Wind: Wind dispersal of fungus spores and infected plant debris is probably of less importance than was formerly supposed. Wind-borne fragments of infected wheat straw and dried grass haulm carrying *Ophiobolus graminis* were thought by McAlpine (1904) to be responsible for spread of the take-all disease in Australia, and Samuel (1924) has described an apparent instance of such dispersal; nevertheless, Fellows and Ficke (1939) tried experimentally to simulate such natural dispersal of infected soil by air and water in Kansas, and concluded that spread of *O. graminis* in this way must be slow and uncertain. Infection of roots by spores often fails to occur in the absence of a food-base, though provision of a food-base external to the fungus may be unnecessary in the event of a "mass-infection" by a number of spores simultaneously; the aggregate reserves of a sufficient number of spores presumably provide a food-base adequate to sustain the fungal invasion and establish infection in spite of host resistance.

Now Wind-dispersal must Eliminate Opportunity for Mass-infection by Fungus Spores: The efficiency of spores in transmission of disease with the seed of many species of plant may be attributed in part to the occurrence of mass-infection; thus Heald (1921) demonstrated a relationship between the spore-load of contaminated wheat seed and percentage of bunted heads in the resulting crop. Similarly, Shen (1940) demonstrated that the percentage incidence and the severity of seedling blight due to *Fusarium culmorum* in wheat seedlings depended upon the spore-load of the seed.

The most convincing argument against wind as an important agent for dispersal of soilborne parasites is the efficiency of crop rotation in controlling these parasites; although instances of failure to achieve control of root disease by crop rotation are known, there is little evidence to suggest that such failures are at all frequently due to wind dispersal of the parasite. The best-authenticated example of air-borne contamination of soil is that afforded by bunt of wheat. Infection via the soil is especially important in the Pacific Northwest of the USA, according to Heald (1932), for the following reasons: Winter wheat in the Pacific Northwest is generally sown on summerfallow; crops mature earliest in the western or southwestern part of the area, and the "smut shower" released by threshing of these early-maturing crops is carried by the strong prevailing westerly winds over the summerfallows of the rest of the Pacific Northwest area. The summers are exceptionally dry with little or no rain in July and August to cause germination of the bunt spores that have fallen on the fallows; the autumn temperatures at and shortly after sowing time are favourable for infection of the winter wheat seedlings by the bunt fungus.

Dispersal by Water: The contamination of water supplies by the spores of parasitic fungi was investigated by Bewley and Buddin (1921), who found surface water such as that from ponds and brooks to be much more heavily contaminated than Company's water or that from deep artesian wells. Curzi (1927) found that *Phytophthora hydrophila,* causing root rot of chilli peppers in Italy, was spread by water; a single irrigation was said sometimes to result in the complete destruction of the crop. For the control of a similar disease (here attributed to *P. palmivora)* of chillies in Batavia, Muller (1936) recommended that a network of shallow trenches should be dug, with water pits at intervals to prevent the rain water running off over the surface. If infection occurred in one of the squares isolated by the trenches, the soil and diseased plants were to be watered with 1 per cent solution of copper sulphate, and weeding temporarily discontinued. Crawford (1934) has commented upon the spread by irrigation water of the *Fusarium annuum* wilt of the same crop in New Mexico. Thung (1932, 1938) found *Phytophthora parasitica* var. *nicotianae,* the agent of black shank of tobacco in Java, to be carried by water from old tobacco fields, manure heaps, and curing barns. The movement of water in erosion and drainage from infected upper levels was considered by King *et al.* to result in the infection of lowlands and deltas by *Phymatotrichum omnivorum.*

Peltier (1937) repeatedly observed that when root rot occurred near the headwaters of a stream, it was usually distributed throughout the drainage basin, and that the incidence of the disease increased at the lower levels. Hewitt (1934) has made a similar observation on the spread of *Armillaria mellea.* The dispersal of sclerotia of *Sclerotinia sclerotiorum* by irrigation and flood water has been noted by Brown and Butler (1936) in their study of lettuce "drop" in Arizona.

The importance of the passive spread of pathogenic fungi by water movement has thus been well established. Active spread by free-swimming zoospores of *Phycomycetes* and *Myxoinycetes* is probably of negligible importance as far as actual dispersal is concerned. Thus Haenseler (1926) found that zoospores of *Aphanomyces euteiches* were apparently incapable of migrating even half an inch in the soil, so that practically no spread of the disease in this way occurred in his experiments.

Dispersed by Cultivation Operations: Infected soil may be carried on the feet of farm labourers, and on those of draught animals and stock, as well as on implements. Such methods of spread have been cited, for example, by Gibbs for the spores of *Plasmodiophora brassicae* on New Zealand farms, and by Brown and Butler (1936) for the sclerotia of *Sclerotinia sclerotiorum* in Arizona.

Dispersal by Feeding Diseased Plant Material to Stock: The feeding of diseased roots to stock was shown by Gibbs (1931) to constitute another mode of dispersal of *Plasmodiophora brassicae.* Brown (1937) found *Sclerotinia sclerotiorum* to be spread in Arizona by feeding farm animals on infected lettuce refuse; on the other hand, sheep and other stock could be made to further the ends of plant sanitation by folding them on the infected plants after harvest, and quarantining them for four days, the maximum period for evacuation of living sclerotia. McAlpine (1910) condemned the practice of feeding farm horses on wheat hay infected with flag smut *(Urocystis tritici),* and Clayton (1925) attributed the rapid spread of the "disease in New South Wales to neglect of this advice, having shown that viable spores of *U. tritici* are voided in horse dung. In South Africa, Verwoerd (1929) succeeded in obtaining some *45 per cent* infection of test wheat seedlings not only with freshly voided horse dung, but also with the same dung when tested again the following season; sheep dung was also demonstrated to act as a vehicle for living spores of *U. tritici.*

Dispersal in Compost: The compost heap has often been indicted as a source of infection, whereby root-infecting fungi may be distributed over clean land. The high temperatures developing in a properly made compost heap must, however, be lethal to the mycelium and spores of many plant parasites. Thus Grooshevoy and Levykh (1940) reported that compost heaps prepared for tobacco seed-beds developed internal temperatures of 49°-63°C, which proved fatal both to the chlamydospores of *Thielaviopsis basicola* and to the pseudo-sclerotia of *Rhizoctonia* sp. Moreover, only a very small minority of the root-infecting fungi are sufficiently vigorous saprophytes to spread actively through a compost heap, in face of the competition of obligate saprophytes. An example of such a one has been given by Thompson (1940), who declared that compost heaps in Malaya were sometimes heavily contaminated with *Sclerotium rolfsii,* and that the use of compost carrying this organism had been the cause of loss amongst ornamental plants in gardens.

Accelerating "Natural" Disappearance of Parasite from the Soil: There remain for discussion those procedures whereby the normal rate of disappearance of a parasite from the soil under fallow or a non-susceptible crop can be accelerated by the farmer. Some field crop plants, for instance, do not die after flowering and fruiting, with the result that a root disease fungus may continue to survive and even to spread on the still living roots after harvest. Thus Taubenhaus and Ezekiel (1930fr) demonstrated the survival and spread of *Phy-matotrichum omnivorwn* throughout the winter on living cotton roots, long after the tops of the plants had been killed b} frost. REA (1933) reported that winter cultivations reduced the number of live, overwintering cotton roots, but that reduction in this source of infection alone failed to reduce appreciably the incidence of root rot in the following season.

In general, the period of survival of a root-infecting fungus in the interval between susceptible host crops is likely to be most brief under those conditions most favourable for its vegetative activity, and therefore least favourable for dormancy. Under such conditions, resting spores and sclerotia are likely to germinate more freely, and mycelium in infected tissues more quickly to exhaust the food reserves of the substrate. Thus Bremer (1924) considered that, in the absence of host plants, *Plasmodiophora brassicac* was likely to survive for longest in alkaline soils, which inhibited spore germination; in alkaline soils, moreover, the spores were most resistant to the lethal action of mercuric chloride or high temperatures. Garrett and

Fellows (1941) concluded that the period of survival of *Ophiobolus graminis* in infected host tissues was shortest in warm, well aerated and moist soils, *i.e.* under just those conditions optimum for vegetative activity.

Methods most likely to be of use in hastening the disappearance of a root-infecting fungus from the soil in the absence of host plants may be listed under three heads: (i) cultivation, (ii) organic manuring, (iii) use of selected rotation crops or catch crops. Several experiments on the value of cultivation for eradication of *Phymatotriehum omnivorum* have been carried out; results obtained by Rea (1933) and Ratliffe (1934) were not encouraging, but Streets (1937) and REA (1939) have since quoted more promising results, especially from deep subsoiling. Streets drew attention to survival of P. *omnivorum* below cultivation depths as one of the reasons for the limited success of cultivation in eradicating the fungus, and Rea connected the greater success of modern trials of deep tillage with the more powerful implements employed. Mitchell *et al.* (1941) obtained encouraging results from rotary tillage early in the autumn after cotton harvest, as compared with the usual late autumn ploughing; in the following August, sclerotia were abundant between the 6 and 12 inch soil levels of the ploughed plot, but absent in the upper 12 inches of soil in all the rotary-tilled plots. This reduction in sclerotial population was associated with reduction in the incidence of root rot in the cotton crop; incorporation of farmyard manure or sorghum fodder with the soil by means of the rotary cultivator still further reduced losses from root rot.

Mitchell *et al.,* and Clark (1942), by demonstrating that disappearance of the sclerotia of *P. omnivorum* was greatly stimulated by the addition of organic material to the soil, helped to explain the control of cotton root rot through organic manuring reported by King *et al.* (1934). King's success with this method in Arizona, as compared with its failure in Texas (Rea, 1939), may be attributed in part to the difference between irrigation farming in Arizona and dryland farming in Texas, and in part to the special methods for timing and placement of the application recommended by King (1937): "The method consists of applications of various organic materials in deep furrows during the fall and winter, and in planting the cotton over the buried material. The furrows for the manure should be from 10-14 inches deep, so that the buried material will not interfere with the preparation of a seed-bed. It has been found that as much as 20 tons of horse or cow manure from corals or open stalls per acre may be used without injury to the cotton crop. However, 15 tons per acre probably would be adequate for most conditions. Where green material such as freshly cut alfalfa, clover, field peas, vetch or mustard is available, 30 tons per acre may be used if furrows are made deep enough so that the material may be covered with 5 or 6 inches of soil. Irrigation water should be applied immediately (*i.e.* after covering) so as to compact the soil and encourage the activity of organisms for decomposition. If possible, the organic material should be applied, covered, and irrigated at least a month before planting time, and a second irrigation should be made before planting to ensure adequate moisture for germination of the cotton seed.

Evidence for the view that rotation crops non-susceptible to infection by a particular parasite may yet affect the period of its survival. Germination of the resting spores and sclerotia of root-infecting fungi may sometimes be stimulated by contact with the roots of plants other than their host plants (Noble, 1924); search for such plants seems to offer a promising line of investigation.

Fungicides in Plant Disease Control

Different methods have been adopted for plant disease control. These include: (i) restricting the movement of diseased plant material from one region to another through legislation, (ii) adopting appropriate cultural practices which would either enable a crop to escape from disease or reduce the inoculum potential of pathogen, (iii) growing disease resistant varieties, and (iv) using chemicals to save the crops from diseases. Generally speaking, no one method is exclusively effective in controlling a disease. In the recent past, considerable work has been done in the field of disease control through chemicals. Although chemicals have been used in the control of diseases caused by fungi, bacteria, nematodes, viruses, id nutritional deficiencies, their use in controlling fungus diseases been a good deal more than other diseases.

What is Fungicide?

The word fungicide has originated from two Latin words: viz., *igus* and *caedo.* The word *caedo* means "to kill". Thus literally peaking a fungicide would be any agency which has the ability to kill a fungus. According to this meaning, physical agencies like heat and ultraviolet light should also be called fungicides. However, that is not so. By common usage, the word is restricted to hemicals. Hence the word fungicide should mean a chemical capable of killing fungi.

Some chemicals do not kill fungi. They simply inhibit fungus growth temporarily. If the fungus is freed from such substances, would revive. Such a chemical is called a "fungistat" a fenomenon of temporarily inhibiting the growth. Bordeaux mixture, may inhibit spore production without, affecting the growth of vegetative hyphae. Even though fungistats and anti-sporalants do not "kill" fungi, they are included under the broad term fungicide, because by common usage, the word fungicide has been defined as a chemical substance which has the ability to prevent damage caused by fungi to growing crops and their products. Some workers prefer the use of the term "fungitoxicant" instead of fungicide.

Classification of Fungicides

There are various methods of classifying fungicides. Broadly speaking, classification is done on the basis of their (i) chemical nature, (ii) mode of action against fungi, or (iii) general uses.

On the basis of mode of action, fungicides are classified as given below:

Protectants and therapeutants Fungicide which is effective only if applied prior to fungal infection is called a protectant, e.g. zineb, sulphur, etc. On the other hand, fungicide which is capable of eradicating a fungus after it has caused infection, and thereby "curing" the plant, is called a therapeutant, e.g. 1,4-oxathiin derivatives, 8-quinolinol, antibiotics like Aureofungin, etc.

Horsfall in his book "Principles of fungicidal action" has correctly classified protectant fungicides into two groups, viz., contact and residual. Contact fungicide is the one which is aimed to seek out the fungus at rest, either before or after the fungus has found the host. Fungicides commonly included under "eradicants" can be included under this group. A residual fungicide is the one which is applied as a stationary layer on the outside of the host, there to lie in wait for the mobile fungus, e.g. zineb, maneb, etc.

Protectants and eradicants are those which remove pathogenic fungi from an infection court (area of the host around a propagating unit of a fungus in which infection could possibly occur), e.g., organic mercurials, lime-sulphur, dodine, etc. Since "eradicants" also protect the host from infection. Systemic and non-systemic systemic fungicides are those which, act by entering into the system of the plants. Most of the chemotherapeutants act this way, e.g. 1,4-oxathim derivatives. All other fungicides which do not act systemically are called non-systemic.

Formulation of Commercial Fungicides

Commercial fungicides are formulated in various ways. Most commonly available formulations are:

Water-Dispersible Powders

A large number of fungicides are formulated as water-dispersible (wettable) powders; this is the form most commonly used for spray mixes. Modern water-dispersible powders are easily wetted and disperse well in water. A wetting agent is usually present in most water-dispersible powder formulations, but the addition of a spreader-sticker is sometimes desirable, especially on plants with glossy or waxy leaves. In case of ordinary water-dispersible powders, agitation is generally necessary in the spray tank to keep uniform suspension. However, if the wettable powder is water-dispersible, frequent agitation is not required since it would remain well dispersed for a long time.

A highly developed type of water-dispersible powder is called a colloidal powder. A colloidal powder is a substance which is so finely divided that the individual particles will never sediment out Some copper fungicides have almost colloidal properties.

A typical water-dispersible formulation may be: fungicide (5-50 per cent), non-ionic wetter (1-10 per cent polyethylene oxide condensate), thickening agent (e.g. carboxymethyl cellulose) and a hydrophilic diluent such as bentonite.

Dusts

These formulations generally contain from 4 to 10 per cent active ingredient. As the name indicates, they are applied in dry forms as dusts.

All well-formulated dusts have certain characteristics in common. They are manufactured so that they are light enough to be carried by a slight breeze for a considerable but controlled distance. The particle size and weight must be such that the same amount of active chemical is deposited at five metres from the machine as is deposited at 20 metres from the machine.

One of the chief weaknesses of many old dust formulations was the fact that they were so wasteful in terms of active chemical due to blow off, that large quantities of the dust had to be applied. Modern dust formulations are an improvement over this situation. The better way, which is used wherever possible, is to stick the active ingredient on the carrier particles of carefully selected density and size. Carriers (diluents) more widely used are Attapulgite, kaolin talc, pyrophyllite, bentonite, and diatomaceous earth.

Dust formulations must reach and stick to the foliage. This requires an understanding by the formulator of the micro-climate about the leaves, and also of the sticking properties of the substance employed. Dusts must always be free-flowing so that frequent physical agitation is not needed to keep them in that condition. They should not become sticky under humid conditions or else they will not flow freely.

Sulphur' itself possesses many of the attributes of a carrier. Sulphur dusts, therefore, require only the addition of a small quantity of a conditioning agent—such as talc or magnesium carbonate—which ensures flow ability.

Suspension or Slurries

There are formulations in which a dry form of the active ingredient is mixed with a liquid. Such formulations usually have a high percentage of active ingredient similar to wettable powders. They are mixed with water for final use and require agitation. Fungicide slurries are used in seed processing plants for treating seeds.

Methods of Application

Seed Treatment: The seed treatment with fungicides is essential because a large number of disease causing fungi are carried on and/or in the seed.and when the seed germinates, these fungi also become active and cause either seed and seedling mortality or produce diseases at a later stage. The purpose of seed treatments by the use of fungicides/bactericides is to destroy seed-borne fungi and bacteria that cause seedling blights, seed decays, or other diseases such as smuts. Also such a treatment would protect the germinating seed from the attack of certain soil-inhibiting fungi.

The benefits of fungicidal seed treatment have been well-established through the years. In fact, treatment of seeds of all crops should be done as a routine practice since it is a type of cheap insurance against possible disasters at a later stage.

There are various types of seed treatments. Some do not involve use of fungicides, for example, hot water or solar heat treatment for the control of loose smut of wheat. Fungicidal seed treatment involves the use of more than one fungicide since, there is no one fungicide that may be recommended as a universal preventive for all of the fungi which can attack seeds and seedlings. For example, PCNB seed treatment would be quite effective against *Rhizoetoniaspy.*, whereas captan might fail. On the other hand, PCNB might fail against *Fusarium* spp., whereas thiram might succeed.

Seed treatment fungicides may be applied to seed in small lots using relatively simple seed treaders or they may be applied to large seed lots at seed processing plants. Amongst the simple seed treaders, the most common one available in India is the Rotary Seed Dresser in which dry seed dressing is done. On the other hand, almost all the seed processing plants, owned by the Government as well as private seed producers, have slurry treaders. In these slurry treaders, the requisite quantity of the fungicide slurry is automatically mixed with the specified quantity of the seed before the latter is bagged. The process of seed treatment in a slurry treater is a continuous one and is more efficient than the rotary seed dressers.

Other methods of seed treatment are seed-dip and planter or hopper-box treatment. The seed-dip method involves preparing fungicide suspension in water, often at field rates, and then dipping the seed or propagative material for specified time. In the planter or hopper-box treatment, requisite quantity of the fungicide dust is mixed in the hopper-box of a seed-drill before sowing. There is one advantage in this method. Since the fungicide is mixed at the time of seed sowing, the chances of treating seed quantities in excess of the requirement are eliminated.

As pointed out earlier, seed treatment should be followed as a routine practice in order to ensure good emergence and crop stands. A question is often raised as to whether legume seeds should be treated with fungicides or not. This is because of the apprehension that these seed dressing materials might adversely affect the *Rhizobium* spp. and hence the nodulation. There are diverse opinions on this subject.

However, it is generally accepted that fungicidal seed treatments can be recommended in case of well-adapted and regularly cultivated legumes. It is only in the case of legumes which are being introduced newly in an area that this question assumes importance. One such situation arose when soybean seed from USA was imported at Pantnagar and other places in India. As a result of trials conducted at the G. B. Pant University of Agriculture and Technology, it became clear that fungicides like captan or thiram (0.25 per cent) do not adversely affect nodulation under field conditions. The plants which developed from fungicide-treated and *Rhizobium* inoculated seed had the same number of nodules as those which developed from untreated seed in areas where soybean had not been cultivated earlier (Nene *et al.)*. It would be pertinent to mention here the work of Milthorpe.

He reported that Ceresan, an organomercurial fungicide, prevented all nodulation unless the pea seed was sown immediately after inoculation. This finding might explain why no adverse affect was found in case of soybean at Pantnagar, where it has been observed that the best way of ensuring good nodulation is to coat the seed with the *Rhizobium* just before sowing. In all the fungicidal trials this procedure had been followed.

Soil Treatment: Soil harbours a very large number of plant pathogens. It is well-known that the source of primary inoculum of many plant pathogens happens to be the soil where dead organic matter supports active or dormant stages of these pathogens. Thus one can say that many diseases arise from the soil and that if steps are taken to eliminate the inoculum or reduce the inoculum potential, one can hope to achieve disease control. Sharvelle has broadly classified the soil treatment methods into two types, viz., physical and chemical methods. The physical methods involve the use of heat in the form of steam or hot water or the use of electricity, whereas chemical methods involve the use of both volatile and non-volatile chemicals.

The soil treatment methods involving the use of chemicals are:

(a) *Soil Drenching:* Fungicides are made up with water at about the same concentration as for spraying and applied to the soil surface either before or after plants emerge. Requisite quantity of fungicide suspension is applied with a sprinkler per unit area so that the fungicide in general reaches a depth of at least 10-15 cm. This method is followed for controlling damping-off, root rots, or infections at the ground level.

(b) *Broadcast:* Sometimes non-volatile fungicides mixed with soil or fertilizers are scattered with hand as uniformly as possible over the field and then mixed with the soil with a suitable implement. This method involves consumption of greater quantity of fungicides than other methods and is followed where suitable implement for proper application of fungicide is not available.

(c) *Furrow Application:* In this method fungicides are applied either as dusts or with water to the furrow at the planting time. Such a treatment is for the control of diseases that occur at the base of the plant. Special equipment must be built or purchased for such application since manual application is never uniform. This method requires much less quantity of fungicides per hectare than the broadcast method.

(d) *Fumigation:* Application of certain chemicals to the soil can control fungi and nematodes. Such materials usually produce a gas that distributes itself through the soil. Some impervious cover, like thin polythene sheet, which confines the gas to the soil for a limited period of time is required. This method is usually restricted to small areas and high value crops. Methyl bromide, for example, is applied by this method. Metallic cans containing methyl bromide under pressure are available in some countries. The cans are opened only with a special applicator directly under the polythene cover.

Certain general precautions and suggestions concerning fungi cidal soil treatment are:

(a) *Soil Condition*: (i) The soil must he loose and properly crumbled so that it can be thoroughly penetrated by chemicals, (ii) All lumps, trash, and clods should be broken up, and old plant debris, especially large diseased roots should be removed, (iii) Soil should be in good planting condition when treated, (iv) It should be moist enough to permit seed germination, (v) Soil should not be treated when it is either excessively dry or wet:

(b) *Soil Amendments:* (i) Soil amendments, such as manure, peat, compost, other humus material, and sand must be added before treating, (ii) It is particularly important that the organic matter is fully decomposed, (iii) When using soil fumigants, fertilizers containing ammonia or ammonium salts should not be added to the soil at or near the time of treating.

(c) *Treating Tools:* When using materials like methyl bromide, all the tools to be subsequently used should be sterilised.

(d) *Time Interval between Treatment and Planting*: (i) When chemicals are used, it may take two to four weeks to aerate the soil before it is safe to seed or plant, (ii) Soils rich in organic matter or clay, excessively wet, or treated at low temperatures may retain the chemical at toxic levels for even longer periods of time, (iii) It may be necessary to work the soil at least once to a depth of several inches to allow the gas to escape.

Materials used for soil treatments are subject to many

chemical, physical, as well as biological interactions. Fungicide residues in soils can be depleted by microbial dissimilation, hydrolysis, and chemical reactions with organic matter as well as by purely physical processes such as evaporation and leaching by water. The relative residual toxicity of a fungicide, therefore, would depend upon how effectively it is able to withstand such factors.

Spraying

This is the most commonly adopted method. Spraying of fungicides is done on leaves, stems, and fruits. Wettable powders are most commonly used for preparing sprays. The most common diluent or carrier is water. The dispersion of the spray is usually achieved by its passage under pressure through nozzles of sprayers.

Spraying is of two types, viz., high volume and low volume. These two terms describe the amount of liquid in which a crop protection material is applied to a crop. When sprays involve large quantities of liquid per unit area, they are termed high volume. Six hundred litres and above per hectare would be considered to be of the high volume category. With low volume sprays, it is usually possible to cover one hectare with about 100 litres or less. Machinery for these two types of sprays is different. In conventional high volume sprays applied at rates of 800-3500 litres per hectare, drop size is likely to be of the order of 0.5 to 3 mm. On the other hand, in low volume sprays as little as 15 to 250 litres of liquid per hectare may be used to distribute the same amount of chemical. When air-blast equipment is used, optimum radius of the droplets would be 15-40 microns, while for aircraft application, it is about 35-70 microns.

Dusting

Dusts are applied to leaves, fruits, and stems of plants as an alternative to spraying. Dry powders are used for covering host surface. In these dusts, the active component is diluted with a suitable finely divided "carrier". Generally, dusting is practicable only in calmest weather and a better protective action is obtained if the dust is applied when the plant is wet with dew or rain.

Spraying Versus Dusting

There are advantages and disadvantages both in spraying and dusting. Spraying: (a) gives better coverage, (b) has longer residual effect, (c) can be done under relatively high wind velocities, (d) bothers the operator less than dusting, and (e) materials are less costly on per hectare basis.

On the other hand, in case of dusting: (a) the equipment is lighter and cheaper and can be used under adverse ground conditions, (b) requires no water and therefore has less operational difficulties, and (c) material is ready to use.

Fungicide Pastes or Paints

Fungicides are occasionally mixed with water, alcohol, or other carriers and applied as a paint on a wound surface. For example, Bordeaux mixture can be made up to paint-like consistency by mixing a prepared Bordeaux mixture powder with linseed oil or water for use in painting tree wounds.

Characteristics of a Good Fungicide

A good fungicide should have the following qualities:

High Field Performance

Field performance of fungicides is determined by several factors such as: (a) the inherent fungitoxicity, (b) availability of the active constituent so that it can act against fungi, (c) good coverage of the host surface, (d) initial deposition on host in case of spraying or dusting, and (e) the tenacity of the residue which would determine the residual effect.

Low Phytotoxicity

A fungicide which causes phytotoxicity under varied environmental conditions would naturally be a total failure. However, certain fungicides cause phytotoxicity only under certain environmental conditions, e.g. sulphur dust is toxic to some crops if applied during hot season, but under cooler conditions it does not cause toxicity. In such cases, the fungicide use is restricted to favourable environment.

Stability in Storage

Fungicides have to be stored sometimes for several months or even more than a year. Stability of the active ingredient under a variety of storage conditions is, therefore, an important factor.

Stability after Dilution to Spray Strength

Some chemicals are less effective if diluted. A good fungicide should therefore possess stability after dilution to spray strength, which would be sufficient to ensure the application of' a uniform desired concentration of active constituents.

Low Toxicity to Human Beings and Cattle

A good fungicide should be as least toxic as possible to human beings and cattle. This will eliminate dangers of accidental poisoning and make it safer for an operator to work.

Factors Influencing the Field Performance of Fungicides

It is not enough that a chemical has high fungitoxic activity. Such a chemical may not be of any utility unless it stands out in the tests and gives proof of significant control of diseases under varied field conditions. There are several factors which influence the performance of a fungicide under field condition. These may be either physical or chemical in nature.

Physical Factors: Plant protection fungicides, which are usually finely ground organic or inorganic powders with low solubility, are either sprayed or dusted on the plants. In order to settle on the plants, the particles must have enough momentum to overcome repulsive forces which exist near surfaces. These may be electrostatic in nature or may arise from convection currents caused by temperature differentials between the surfaces and the surrounding atmosphere. Also, the high velocity air streams used to propel dusts and concentrate sprays tend to glide around plant surfaces. Therefore, the particles carried by them must have enough momentum to strike the plant surfaces, instead of being carried away by deflected air currents. According to Burchfield, a particle with a radius of 10 microns and a density of 2 gm per cm^3 would reach a limiting velocity of 2.3 cm per sec and have a momentum of 1.9×10^{-8} gram cm per second. This is about the minimum value that would enable it to penetrate the barriers created by repulsive forces at the plant surfaces. Thus it would seem that dust particles should be about 20 microns in diameter or form aggregates of about this size to obtain satisfactory deposit buildup. However, unfortunately, particles in this size range are easily dislodged by wind and rain, and therefore dusting with most fungicides has not been a very satisfactory method for protecting crops.

Sometimes it is suggested that dust deposition may be improved by the presence of positive electrostatic charges on the individual particles induced by friction in the blowing apparatus. This would cause them to be attracted to negatively charged surfaces of leaves by electrostatic forces. However, according to Gill, the charges induced on dusts during blowing are weak and vary in sign and therefore it is unlikely that they will improve deposition significantly. Also, according to Bowen *et al.* good deposition was obtained when the particles were negatively charged. This, according to Burchfield, shows that the image charge induced on the surfaces of the plants by the dust cloud is positive in sign even though it is generally assumed that leaves are negatively charged.

Disadvantages associated with dust applications are reduced in case of spraying. Here the small primary particles of fungicides are encapsulated in larger liquid droplets that can be given sufficient momentum to penetrate the barrier immediately surrounding the foliage. The droplets spread on the impact, and when the liquid carrier evaporates, the fungicide particles are left on the plant parts. The coverage in this case is rarely uniform, but since it is possible to deposit smaller particles in case of sprays and also because tenacity is more, the results obtained by sprays are better than those by dusting.

The amount of fungicides deposited on plant surfaces should under normal circumstances, be directly proportional to its concentration in the spray suspension. Factors which may disturb this relationship are: (i) variation in the droplet size of the spray, and (ii) deposition after run off. Since most of the fungicides formulations contain surface active agents, the surface tensions of sprays should be somewhat less at high concentrations than at low. Low surface tension is known to cause the formation of smaller drops, which might reduce deposition through greater drift of the spray with wind, Wettability of plant surfaces by spray droplets may contribute more to deposition of spray droplets than electrostatic charge, particularly when the surface to be covered is waxy. In such cases the interfacial tension between the liquid and the surface is high, which leads to high contact angle.

The contact angle is important in determining whether a spray suspension will spread uniformly on the surface of a plant. A contact angle that approaches zero indicates that the adhesive forces between liquid and surface are almost as great as between molecules of liquid, so that spray droplets will form thin films on leaves. On the other hand, a contact angle approaching 180° indicates very poor wettability, and in such cases droplets may roll off the surfaces without wetting them. All types of intermediate situations may exist depending on the plant species, its age, environment, and the physical properties of the fungicide formulation. Plant surfaces which are easy to wet include beans, brinjal, potato, tomato, etc. whereas those which are difficult to wet are banana, most crucifers, colocasia, etc. Fungicide sprays will not wet waxy surfaces satisfactorily unless formulated with adjuvants capable of decreasing contact angles between the impinging droplets and the plant surfaces. For example, Burchfield and Goenaga observed that young banana leaves were sprayed with

Bordeaux mixture, many spray droplets hit the leaf surfaces and bounced off as if they were minute rubber balls. When a non-ionic surfactant was included in the formulations, the leaves were wet uniformly and the amount of copper deposited under these same conditions increased fivefold. On the other hand, no marked improvement was noticed in deposition in case of easily wettable castor leaves even when a surfactant was added.

Particle Size and Redistribution: Deposition of fungicide on the leaf surface alone will not ensure sits to plants include: (i) particle size and shape, (ii) the presence of aggregates, (iii) distribution on the plant surface, (iv) nature of the compounds, and (v) the intensity and duration of weathering.

Sufficient evidence that particle size plays a dominant role in tenacity has been accumulated for copper and sulphur fungicides. There is some evidence supporting the role of particle shape in tenacity. When 10-10-100 Bordeaux mixture is allowed to age, the particles become spherulites and such an aged Bordeaux mixture lacks tenacity as evidenced by exposure to simulated rain (Burchfield and Goenaga).

It has been observed by Burchfield that many fungicides form heterogeneous spray deposits on leaves and when subjected, to rain, large particles as well as aggregates of small particles are very easily washed off, while smaller particles stick on. However, this does not happen in case of Bordeaux mixture and hence it has been stated that these deposits are homogeneous, since the amount of copper removed by each rainfall treatment is directly proportional to the amount present at the beginning of the treatment. Obviously this could happen if all of the spray deposits responded in the same way to weathering (Burchfield).

Distribution or spreading of spray droplets on the plant surface may also influence tenacity (Burchfield and Goenaga). For example, when Bordeaux mixture is applied to banana leaves it collects as discrete droplets at the leaf veins and these are easily washed off with the first rain. Spreading of these droplets can be affected by the addition of non-ionic surfactants to the Bordeaux mixture. Such a deposit has very high tenacity.

Different fungicides may differ in their intrinsic capacity to stick to foliage. It is well-known that Bordeaux mixture has a much better natural tenacity than many other fungicides.

Intensity of weathering is important in relation to adhesion of fungicides to plant surfaces. Heavy monsoon showers in India, for example, may wash off fungicide residues, in spite of the presence of stickers, more quickly than under light showers. Similarly a continuous wet spell may adversely affect tenacity.

Chemical Factors

(a) *Photolysis:* In the presence of sunlight, certain chemicals undergo changes due to absorption of energy sufficient to cause dissociation of the molecule into fragments possessing unpaired electrons.

According to Burchfield, plant protectant chemicals are in a favourable physical condition for photolysis since they are applied as finely divided powders with high specific surfaces. Thus they could easily dissolve in moisture films at foliage surfaces where they were subjected to sunlight. Fortunately, the atmosphere of the earth absorbs much of the harder radiation. Approximately 40 per cent of the total solar energy reaching the ground level is in the 4000-7600 Å (visible) region of the spectrum, while 5 per cent or less is below 4000 Å. Therefore, only those compounds which absorb light in the visible or near ultraviolet regions are likely to be affected significantly.

Quinone fungicides are known to be photosensitive. In aqueous solution p-benzoquinone yields hydro-quinone and another product when exposed to a wave length of less than 5770 Å. The threshold wave length region for the photolysis of quinones decreases with decreasing oxidation potential Thus chloranil with a potential of 0.73 volt and a threshold region in the neighbourhood of 5770 Å naturally decomposes more readily when exposed to light than dichlone with an oxidation potential of 0.43 volt. That is why chloranil is a good seed protectant, but poor in its performance when sprayed on foliage. On the other hand, dichlone is effective both ways.

(b) *Hydrolysis:* Compounds that are amides, esters, or contain reactive halogen atoms or nitrogroups tend to hydrolyse in the presence of water and form compounds which are either inert or have modified activity spectrum. Harry *et al.* reported that glyodin lost its fungi-toxicity when slurried with hot water. Fungicides like captan and folpet which contain active halogen atoms are known to get hydrolysed in aqueous suspensions. It is believed that a halogen is replaced by a hydroxyl ion. For instance, Daines *et al.* found that captan decomposed slowly at pH 7.0 but instantaneously at alkaline pH such as in the presence of sodium hydroxide. PCNB which has an active nitro group gets hydrolysed in the presence of moisture and produces pentachlorophenol (PCP). Both these compounds are good fungicides but PCNB is useful in soil applications, whereas PCP is good for the protection of wood,

(c) *Atmospheric Interactions:* Fungicides are subject to oxidation. Some compounds may lose their fungitoxic properties, while others may get activated during the process of oxidation. For example, nabam gets activated as a result of oxidation. Nabam is a water soluble compound that is fungistatic but not fungicidal. Spores immersed in a 10 per cent solution of nabam are able to germinate after they are removed and washed with water. In spite of being water soluble, nabam shows good residual effect and gives good disease control.

This is because nabam is oxidised on plant surfaces to an insoluble product which has good residual properties.

Carbon dioxide is another component of air which possibly reacts with certain fungicides. According to Wilcoxon and McCallan, CO_2 converts the excess lime of Bordeaux mixture to calcium carbonate within a few hours after spraying. Free found that CO_2 increased the amount of soluble copper in equilibrium with insoluble basic copper carbonate.

(d) *Phytotoxicity:* Often a chemical may be an excellent fungicide but it may be phytotoxic and this would, therefore, limit its use for practical plant disease control. Such phytotoxic chemicals may react with protoplasm of the plant cells or may chelate essential metals. One of the reasons why some chemicals are more phytotoxic is their high solubility in water, because such compounds permeate more readily into the host tissues.

Auxiliary Spray Materials

Auxiliary spray materials, also called adjuvants, are usually relatively inert materials added to a fungicide to improve the physical characteristics of the toxicant and its carrier. Most of these are surface active agents, and therefore cause a variation in either surface tension or interfacial tension. The various adjuvants may be classified as follows:

Wetting Agents

Wetting agents are the materials which are used to ensure that there will be no layer of air between a solid and a liquid. Wetting agents reduce surface tension of particles.

The cuticle of most plants, because of its waxy surface, has little affinity for water. Therefore, when aqueous solutions or suspensions are applied, they fail to wet the cuticle and spread over it, and instead form droplets and roll off the plants. This inability to wet the cuticle is enhanced by hairs, which trap pockets of air between the aqueous liquid and the cuticle, thus preventing wetting. Thus wetting agents when added to aqueous fungicidal preparation help in its deposition on leaves.

Some examples of wetting agents are polyethylene oxide condensate esters of fatty acids, and flour.

Spreaders

Spreaders are materials used to facilitate the spread over a surface so that the area covered is increasingly greater. Wetting and spreading are closely related phenomena. Wetting must precede spreading and in fact, generally speaking, this is the only distinction between wetting and spreading. Spreaders are more commonly used in contact sprays where coverage is more important. Examples are soap and flour.

Stickers (Adhesives)

Materials which increase adhesion of fungicides to host surfaces are called stickers or adhesives. Stickers when added to fungicide preparations increase their tenacity, thus increasing the residual action. Fish oil or linseed oil are used commonly for this purpose. Certain proteinaceous materials like milk casein and gelatine also act as stickers besides acting as spreading and wetting agents. Polyethylene polysulphides have also been suggested as stickers.

Deflocculating Agents

Materials which keep particles away from each other to prevent flocculation, to ensure dispersion, and to retard setting of a solid within a liquid are called deflocculating agents; These retard sedimentation. Examples are gelatine, various plant gums, and milk products.

Common Names of Fungicides

A fungicide frequently is given different names by different commercial firms. It becomes, therefore, very difficult for a person working with fungicides to remember these names.

To avoid unnecessary use of trademark names and provide for better communications between the scientists of different countries, common names for pesticides have been developed. The use of these common names also helps in overcoming the difficulty of writing long chemical names of fungicides. Following is a list of common names of some fungicides.

Common Adopted Names of Some Fungicides

Common name	*Chemical name*	*Some commercial formulations*
Anilazine	4, 6-dichloro-N- (2-chlorophenyl)-1,3,5-triazin-2-amine	Dyrene
Benodanil	2-iodobenzoic acid anilide	BAS-3170
Benomyl	Methyl l-(butylcarbamoyl)-2-benzimidazolecarbarnate	Benlate
Captafol	Cis-N-(1,1,2,2-tetrachloroethyl-thio)-4-cyclohexene-l, 2-dicarboximide	Difolatan
Captan	N-(trichloromethylthio)-4-cyclohexene-1,2-dicarboximide	Orthocide
Carbendazim	Methyl-IH-beazimtdazole-2-yl carbamate	Bavistin Derosal
Chloroneb	l,4-dichloro-2,5-dimethoxybeazene	Demosan
Chlorothalonil	Tetrachloroisophthalonitrile	Daconil 2787
Dichlofluanid	N-dichlorofluoromethylthio-N-N-dimethyl-N-phenylsulphamide	Euparen
Dichlone	2,3-dichloro-l, 4-naplhoquinone	Phygon
Dicloran	2,6-dichloro-4-nitroaniline	Botran, DCNA
Dimethirimol	5-n-butyl-2-dimethyl amino-4-hydroxy-6-methylpyrimidine	Milcurb
Dinocap	A mixture of 2,4-dirtitro-6-octylphenyl crotonate and 2,6-dinitro-4-ocylylphenyl crotonate	Karathane
Ditalimfos	0,0-diethyl phthalimidophos-phonothioate	Plondrel
Dodir	N-dodecylguanidin acetate	Cyprex
Edifenphos	O-ethyl-S,S-diphenyl dithio-phosphate	Hinosan
Ethazol	5-ethoxy-3-tricliloromethyl-l, 2,4-thiadiazole	Terrazole
Ethirimol	5-n-butyl-2- ethylamino-4-hydroxy-6-methylpyrimidine	Milcurb
Fenaminsosulph	Sodium p-(dimethylamino) benzenediazo sulphonate	Dexon
Fentin acetate	Triphenyltin acetate	Brestan, TPTA
Fentin Hydroxide	Triphenyltin hydroxide	Du-Ter, TPTH
Ferbam	Ferric dimethyldithiocarbamate	Hexaferb

Contd...

Common name	*Chemical name*	*Some commercial formulations*
Folpet	N-(trichloromethylthio) phthalimide	Phaltan
Guazatine	9-aza-l, 17-diguanidino-heptadecane salt (either sulphate or acetate)	Panoctine
Mancozeb	Manganese ethylenebis-dithiocarbamate plus zinc	DithaneM-45
Maneb	Manganous ethylenebis-dithiocarbamate	DithaneM-22
Nabam	Disodium ethylenebis-dithiocarbamate plus some metallic sulphates	Dithane A-40
Oxycarboxin	5,6-dihydro-2-metliyl-l 4-oxathiin-3-carboxanilide-4,4-dioxide	PJantvax DCMOD
Oxythioquinox	6-methyl-l, 3-dithiolo (4,5-b) qumoxalin-2-one	Morestan
PCNB	Pentachloronitrobenzene	Brassicbl quintozenc
Piperalin	3-(2-methylpiperidino) propyl-3,4-dichloro-benzoate	Pipron
Propmeb	Zinc propylenebisdithiocarbamate	Antracol
Prothiocarb	S-ethyl-N-(dimethyl-Arninopropyl)	Preview thiol carba-mate hydrochloride
Pyracarbolid	3,4-dihydro-6methyl-n-phenyl-2H-pyran-5-carboximide	Sicarol
Pyroxychlor	2-chloro-6 methoxy-4-(tri-chloromethyl) pyridine Lorvek	Dowco-269
Thiabenda-zole	2-(4-thiazolyl) benzimi-dazole	Mertect, TBZ
Thiophanate	Di ethyl 4,4'-0-phenyleneb is (3-thioalilophanate)	Cercobin Topsin
Thiophanate Methyl	Dimethyl 4,4-0-phenylenebis (3-thioalloplianate)	Cercobin-M Topsin-M
Tridemorph	N-tridecyl-2,6-diraethyl-morpholine	Calixin
Zineb	Zinc ethylenebisdithio-carbamate	Dithane Z-78
Ziram	Zinc dimethyldithiocarbamate	Ziride, Hexazir

Storage and Handling of Fungicides

Most of the fungicides are poisons and therefore require to be stored and handled properly. Any carelessness in doing so might lead to serious accidents. Following are some of the important points which should be borne in mind:

(a) The manufacturer's labels should be studied carefully

and thoroughly. Any good product would always have a label giving all the important information about it.

(b) Fungicides should be stored in a separate room having a lock. This will avoid any accidents involving human or cattle life.

(c) As far as possible, fungicides should be stored separately from other agrochemicals particularly herbicides. Very often herbicides, mistaken for fungicides, have caused serious damage to crops. By the time damage is realised, it is too late.

(d) It is safer to handle fungicides in open air and not in closed room. One should avoid inhalation of any of these materials.

(e) All the spills of spray fluids on the body should be promptly washed off with soap and water. It is a good habit to always wash hands after handling packages of fungicides. Instead of using hands, some sort of stirrer should be used for mixing fungicide in water.

(f) In case of splashing of any fungicide material into the eyes, immediate washing with clean water is absolutely essential, after which a doctor should be consulted.

(g) The empty containers and packages should always be burnt or buried.

(h) It is good to know first-aid measures and the use of antidotes at least of those fungicides which are being used more frequently.

(i) The compatibilities between various fungicides, insecticides, herbicides, and fertilizers should be known before it is decided to mix them for using together.

(j) Seed treatment with fungicides, particularly the dry seed dressing, should be carried out in open or well-ventilated room. It is desirable to use dust masks which are now available in India.

Compatibility of Fungicides and Other Agrochemicals

In plant protection schedules, very often it becomes

necessary to combine applications of different agrochemicals due to several considerations. It is very economical to mix agrochemicals together to save on the cost of material application. There is also the time factor. Sometimes there is little time available for separate sprays, may be because of the emergency in the disease or pest situation or because of short intervals between rains, particularly in the monsoon season. Though combining applications is advantageous, it may create a few problems if one is not careful. Pesticides and fertilizers, being chemicals, tend to react with each other and in this process may form compounds which have no pesticidal value. The fungicide captan, for example, can be mixed with many other fungicides, insecticides, as well as fertilizers, but this is not done if the materials are alkaline in reaction because tinder alkaline conditions captan gets decomposed thereby losing its fungicidal value. Sometimes materials can be mixed for immediate applications, but not for using after a certain time interval. It is, therefore, very essential to know interactions between these materials before these are mixed.

Sometimes mixing with agrochemicals may result in the increase of fungitoxicity of certain fungicides. Several reports on the synergism between fungicides as tested under laboratory conditions have appeared in the past and Scardavi has reviewed these. These are indicated below:

Copper oxychloride with zineb, ziram, captan, quinolinol derivatives, dodine, and sulphur; sulphur with caffaro dust (copper and calcium oxychloride), zineb, ziram, maneb and dodine and copper oxychloride; zineb with thiram, captan, folpet, ferbam and copper oxychloride; captan with barium sulphide and copper oxychloride.

Two things must be borne in mind while considering the reported synergism between fungicides. First, most of the reports are based on laboratory evaluation, and therefore it is not necessary that the synergism would be observed under field conditions. Secondly, in many of these reports, it has been pointed out that mixtures of fungicides may show synergistic effects only when they are mixed in a certain ratios and not otherwise. Obviously, therefore, practicability of utilising the synergistic actions between fungicides is limited.

Fungicidal materials in crude form have been in use since ancient times. People of ancient civilizations, through their experience, had found out effectivity of certain materials against diseases and pests. There are records of the use of fungicidal materials by the Greeks and Romans. People of ancient India were fully aware of plant diseases and pests and their importance. Between 3000 to 2000 BC, an agriculture flourished on the banks of Indus river.

The Indus people left no records of plant diseases, even though archaeological sites, particularly the granaries at mohenjodaro, indicate a highly developed agriculture· By 2000 BC, the Indus civilization was replaced by Sanskrit-speaking Aryans who moved into India. In their *Veda* literature (1500-500 BC), first references to plant protection are found. Later between 500 BC to 500 AD, a brief reference to plant diseases has been found in the *Kallavagga,* typically representative of the Buddhist literature: "The rice crop is attacked by 'mildew'and the same fate meets sugarcane if the diseases called 'blight' attack it." In *Brihit Samhita* (500 AD), it is said that low temperature, dryness, windy conditions, very high temperatures, and excessive rains cause diseases. People of that time had of course no idea of the causal agents of diseases, but they did try to make systematic study and therefore could relate environment with plant diseases. Through experience they also found out a few ways of disease control. For example, in Surpala's *Vrksayurveda* (800 AD), we find references to tree-dressings, plant fumigations, and seed treatments through the use of some materials.

Regarding tree-dressings, for example, it is said: "Usually there are vermins at the root of a tree of which the leaves and flowers wither and the body becomes disfigured without any apparent defect. They should be destroyed with special effort. After their extermination, if the tree is smeared with the paste of cow's urine, *ghee* (butter oil), *Vidanga (Embelia ribis)*, mustard and sesame, and watered with diluted milk as well as fumigated with incense, it recovers very quickly."

Regarding plant fumigation, it is said: "Vegetables like cucumbers become free from disease if fumigated with the fumes of the bones of a cow and a dog and ordure of a cat."

Regarding seed treatment, for instance, it is said: "A seed which has been steeped in milk and rubbed well in cowdung and after drying again, rubbed repeatedly in honey and the powder of *Vidanga (Embelia ribis)* grows without fail."

Again about seed treatment, it is said: "The seeds that have ripened in proper season should be extracted from the fruit, steeped in milk, and dried for five days. Then these should be fumigated with the fumes of *ghee* (butter oil) in which *Vidanga (Embelia ribis)* has been mixed."

Rapid progress in the knowledge of fungicides, however, was made only after the 18th century, mainly because of the advancements in the knowledge of chemistry during the last two centuries. The major advancements in the development of fungicides through centuries are given below in brief.

History in General

Ancient Period: 1000 BC Homer, the Greek poet, made a mention of sulphur with pest controlling properties.

9th Century: Surpala's *Vrksayurveda* mentions tree wound dressing, tree fumigations, and seed treatments for control of diseases and pests.

17th and 18th Century

1637: Remnant mentioned the value of seed treatment with sodium chloride for the control of stinking smuts.

1705: Homberg recommended mercuric chloride as a wood preservative.

1761: Schulthess first suggested the use of copper sulphate on wheat seed against stinking smut

19th Century

1807: Prevost of France showed effectivity of copper sulphate in the control of stinking smut and also showed adverse effect of copper sulphate on spore germination.

1821: Robertson of England stated that sulphur is effective against peach mildew.

1833: Kendrick of the USA proposed lime-sulphur preparation against grape mildew.

1851: Grison in Italy introduced a mixture prepared by boiling 52 equal parts of rimstone and freshly slacked lime together in water for ten minutes. This was known at that time and was used extensively for the prevention of powdery mildews. 1882 Millardet of France made his first report on the control of downy mildew of grapes with Bordeaux mixture.

1885: Ozanne was the first to try copper sulphate in India. At Poona, he used copper sulphate and common salt to pickle sorghum seed against smut before sowing and obtained fairly satisfactory results.

1887: Mason of France introduced Burgundy mixture in which quicklime of Bordeaux was substituted by sodium carbonate.

1888: Trillat reported the fungicidal properties of Formaldehyde.

1889: Weed of the USA, for the first time, used fungicides in combination with insecticides.

1893: Nageli's work on the fungicidal action of copper sulphate was published posthumously.

1894: Saunders, Bedford, and MacKay introduced seed protection by chemicals.

1897: Bo Uey was the first to use formaldehyde for wheat smut control.

20th Century

1904: Lawrence used Bordeaux mixture for the first time in India against leaf spot of groundnut *(Cercospora* spp.).

1906: Sulphur was used for the first time in India (Bombay) on grapes.

1913: Reihm working for the I.G. Farben Industries in Germany introduced the organic mercuries for seed treatment of wheat for smut control.

1914: A mercurial compound was used fin India for the first time by Burns for steeping seed potato to prevent *Rhizoctonia* attack.

1917: Darnell-Smith, an Australian worker, introduced copper carbonate as a dust seed treatment for wheat.

1925: Hilson used organomercurials for the first time in India for the control of sorghum smut through seed treatment.

1932: Low soluble or fixed coppers were accepted as substitutes for Bordeaux mixture.

1934: Tisdale and Williams reported the fungitoxic activity of dithiocarbamates.

1935: Bliss enunciated the principle of probits and their significance for the bioassay of fungicides and other toxicants.

1940: Cunningham and Sharvelle introduced chloranil as a practical organic seed protectant. Burlingham and Reddish proposed the "Zone of Inhibition" technique for fungicidal bioassays.

1942: Goldsworthy and his co-workers published the first field results on the fungicidal value of ferbam.

1943: Dimond and his co-workers introduced the ethylene-bisdi-ftiocarbamates as fungicides.

The American Phytopathological Society proposed methods for fungicidal bioassay.

1947: Wellman and McCallan confirmed the fungicidal properties of 2-heptadecyl-2-imidazoline (glyodin).

1952: Kittleson introduced captan as a fungicide.

1957: Fungicidal properties of N-dodecylguanidine acetate (dodine) were reported.

1966: Von Schmeling and Marshal Kulka reported systemic fungicidal activity of two oxathin derivatives. This has been followed by the production of large number of systemic fungicides.

1968: Benomyl was developed by Du Pont.

History of Fungicides in India

In an excellent review, Bhagwagar and Patel have brought together the fungicide work done in India up to 1940. Leaving out the ancient period, the work on fungicides was initiated by some British workers in late 19th century and later taken up by a few Indian workers in different States. Earlier work seems to have been done mostly in South India on crops like palms, tea, coffee, rubber, betelvine, sorghum, and some other crops. As pointed out earlier, Ozanne (1885) was the first person to use a fungicide in India for the control of a crop disease when he used copper sulphate for the control of sorghum smut. Most of the earlier work in India confined to the use of copper fungicides like Bordeaux and Burgundy mixtures. Inorganic sulphur and mercurial fungicides represented other groups. The use of organic sulphur fungicides and others was made in India mostly after 1950.

1885: Ozanne was the first to try copper sulphate for the control of sorghum smut.

1886: Ozanne was probably the first person to use carbolic acid against sorghum smut. The formula consisted of 1 ounce carbolic acid in 1 lb water to cover 60 lbs of seed.

1904: Lawrence at Poona used Bordeaux mixture for the first time against leafspot of groundnut (*Cercospora* spp.).

1906: Sulphur was used on grapes in Bombay.

Butler recommended Bordeaux mixture brushing for the control of Koleroga of arecanut.

1909: McRae claimed control of blister blight of tea (*Exobasldium vexans)* with 6-4-50 Bordeaux mixture in Darjeeling.

1910: Goleman claimed control of *Phytophthora omnivora* var. *arecae* on arecanut with Bordeaux mixture in which resin was added as a sticker.

1911: McRae advised to spray potato plants with Bordeaux mixture against early and late blights.

1914: For the first time, a mercurial compound (mercuric chloride) was used by Burns for steeping seed potato to prevent *Rhizoctonia* attack.

1923: Hector stated that a weak solution of formalin 0.25 oz in 2 gals when applied in soil completely checked damping-off of tobacco seedlings.

Thomas treated paddy seed for foot rot control with 2 per cent copper sulphate solution for 30 minutes and obtained good results.

1925: Hilson used organomercurials for the first time in India for the control of sorghum smut through seed treatment.

1926: Venkata Rao reported successful control of powdery mildew of betelvine by spraying Bordeaux mixture.

1927: An stead claimed protection of rubber from *Phytophthora* with Bordeaux mixture.

Kamat found Bordeaux paste quite satisfactory against gummosis of citrus.

1929: Uppal reported the control of Brown spot of rice in upper Sind by seed treatment with organomercurials.

1930: Narsimhan suggested the use of linseed oil in Bordeaux mixture for improving coverage and tenacity.

1931: Dastur claimed success in controlling foot rot of betelvine by soil drenching with 2-2-50 Bordeaux mixture.

Mayne obtained control of coffee rust by Bordeaux sprays.

Murray controlled powdery mildew of rubber by sulphur dusting.

Uppal *et al.* obtained control of powdery mildew of vine by sulphur dustings.

Uppal could control fig rust by dusting 300 mesh sulphur applied at monthly intervals.

Uppal and Desai reported control of grain smut of sorghum by seed dressing with sulphur.

1933: Rama Murty stated that foot rot of paddy was successfully controlled by dry seed dressing with organomercurials.

1934: Cotton anthracnose was controlled by Dastur by seed dressing with an organomercurial fungicide after delinting the seed with sulphuric acid.

1935: Mitra obtained partial control of wheat bunts by seed treatment with copper carbonate.

Uppal tried colloidal copper for the control of fig rust and obtained good results,

Dastur stated that citrus canker could be controlled by Bordeaux.

Uppal *et al.* showed that one application of sulphur at the time of flowering checks the powdery mildew of peas.

1938: Uppal stated that one application of sulphur could completely control powdery mildew of betelvine.

1951: Mehta and Singh reported use of zineb in controlling potato blights.

1958: Grewal and Dharam Vir used captan in laboratory tests and reported it to be effective against the jute stem rot pathogen.

1959: Grewal and Dharam Vir for the first time in India

reported control of wheat black and brown rusts through four sprays of nabam + zinc sulphate.

1964: Thirumalachar and his co-workers- reported the development of Aureofungin, a new anti-fungal antibiotic, for use in plant disease control.

1969: Chatrath and his co-workers, for the first time in India, reported successful control of loose smut of wheat through seed dressing with carboxin.

Sulphur Fungicides

Of the different types of sulphur dust formulation, the "ground sulphur" is the most common form available in India. One of the common methods by which ground elemental sulphur is produced is the "air-blast separation process" in which mechanically ground sulphur is fractionated in an air-blast, the heavier particles falling down, and the smaller particles being carried over in the air-blast. Mechanically ground sulphurs have a tendency to form small aggregates which may be overcome by the addition of small amounts of inert material such as Kaolin or lead arsenate. The fungicidal efficiency of a sulphur dust depends on the fineness of its particles; a high proportion should pass a 200 or 300-mesh sieve (particles not exceeding 74 and 47 microns, respectively, in diameter), and still finer division is preferable.

In recent years wettable sulphurs have become more popular. They form uniform suspension in water and are meant for use as sprays. Though this can be in two forms, viz. paste or powder, the latter is the one which is most commonly available in India. Any form of sulphur, such as ground sulphur, can be made into wettable sulphur by grinding with protective colloidal materials such as sulphite lye, casein, bentonite clay, etc. There are many proprietary products consisting of wettable sulphur which are available in India.

There are different types of wettable sulphurs. The one more commonly available in India is the "micronised sulphur". It is prepared by the use of a microniser mill, in which a shallow circular grinding chamber is fitted with a series of high pressure jets. Air pressure of 100 lb per square inch is fed through these jets into the grinding chamber resulting in rapid rotation of the contents of the grinding chamber. Coarse sulphur is fed into the grinding chamber through feed injectors situated on the upper side of the mill. The grinding action is accomplished by the impact of particles against each other. Micronised sulphurs can be easily recognised by the ovoid shape of their particles.

Lime-Sulphur

Lime-sulphur, or a mixture of calcium polysulphides and calcium thiosulphate is the product obtained by the combination of lime with sulphur. It has been used widely for more than 50 years and is still used for some of the common fruit diseases. Lime-sulphur is used mainly as a dormant spray.

Varied proportions of lime and sulphur are mixed in water to get the lime-sulphur. One of such formula is:

Rock lime	20 lbs
Sulphur	15 lbs
Water	50 gallons

Required amount of lime is placed in an open metal container and to this is added about five gallons of water. Required amount of sulphur is then added. Heat from an open fire is then applied to boil the mixture for one hour. After boiling, the mixture is allowed to settle for several hours and the clear supernatent is then filtered off. Such a lime-sulphur preparation has comparatively low polysulphide content (7-10 per cent) and a relatively high thiosulphate content (2-3 per cent).

A standard liquid lime-sulphur should have a Baume density of 31-33 degrees at 60°F and should contain not less than 20 per cent calcium polysulphides. It is true that home-made lime-sulphur has a low polysulphide content as mentioned above than the standard one. However, standardised commercial formulations of lime-sulphur, available in some other countries, are not available in India.

Mechanism of Fungicidal Action

The mechanism of action of elemental sulphur has been a subject of interest for many years. In the past, various theories were advanced and later refuted. One of the earliest theories was the "oxidised sulphur theory". Mach and Portelle first suggested that sulphur dioxide accounted for the fungicidal activity of sulphur. Marcelle suggested that the sulphur oxidised even further in the air to SO_3. Young in 1922 said that pentathionic acid, an oxidised derivative of sulphur, was important in the action. The theory lost its ground when Roach and Glynne and Wilcoxon and McCallan showed that pentathionic acid, sulphur dioxide, and sulphur trioxide were no more toxic than could be accounted for by the hydrogen ion concentration of their solutions in water.

Another theory was the "hydrogen sulphide theory". As early as 1875, Pollacci first suggested that HgS was the active compound. Marsh found that hydrogen sulphide was toxic to spores. This finding was confirmed and supported by various workers. In 1953, Miller *et al.* finally refuted this theory. They found that when dosage was accurately controlled, colloidal sulphur was 5-50 times, depending on the fungus, more fungitoxic on a molar basis than hydrogen sulphide. Their experiments clearly indicated that the formation of hydrogen sulphide from elemental sulphur was not detrimental to the germination of the spores concerned until relatively large quantities were evolved on a spore weight basis.

Another theory is the "direct action" theory. According to this, sulphur acts as a hydrogen acceptor in metabolic systems, in doing so, disturbs the normal hydrogenation and dehydrogenation reactions in the cell. This theory is based on the presumption that sulphur, being next to oxygen in the periodic table, acts much like oxygen and since fungus cells are accustomed to being permeated by oxygen, they should be permeable to sulphur too. Sciarini and Nord pointed out that *Fusarium* 'confuses' oxygen and sulphur and forms BUS instead of H_2O in the fermentation reaction. It would appear, then, that sulphur becomes all mixed up in the compounds of the Krebs cycle, causing thereby a metabolic disturbance.

The process by which fungus tissues reduce sulphur to hydrogen sulphide is not fully understood. Practically all plant tissues reduce small amounts of sulphur to hydrogen sulphide which may be a direct chemical reaction between sulphur and the ubiquitous glutathione to form the disulphide of the latter. The reaction in fungus spores takes place much more rapidly when sulphur and glutathione are brought together. Moreover, spores of some species will reduce quantities of sulphur equivalent to several per cent of their weight in a few hours. It was shown by McCallen and Miller and confirmed by Tweedy, that extra quantities of carbon dioxide were produced at the same time. The work of Tweedy indicates that the cytochrome system is involved in the reduction of sulphur, Owens advocated an interesting theory that poly sulphide free radicals, produced in metabolic breakdown of 8-membered rings of elemental sulphur to H_2S, might be responsible for toxicity. He suggested that the reactive polysulphides free radicals might form stable cross linkages with protein and other cellular components.

Thus it would appear that the basis of toxic action of sulphur in the cell is still not clear, and it presents one of lie most in solved problems in fungicidal mechanisms.

Sulphur fungicides emit sufficient vapour to prevent the growth of fungus spores at a distance of several mm from deposits on leaves. This is an advantage in sulphur fungicides as compared to many others because vapour phase activity ensures protection of gaps on leaf surfaces resulting from weathering or expansion of leaf areas.

Phytotoxicity of Inorganic Sulphur Fungicides

Acute injury by sulphur is rare in temperate climates, but

in warmer climates (above 80°F), severe burning is sometimes caused on cucurbits when sulphur is used for the powdery mildew control. Apples treated with sulphur in semi-arid areas may develop lesions on the sun-exposed side of the fruit, an injury attributed to "sulphur sun scald".

Some of the fruit varieties are sulphur-sensitive. Among sulphur-sensitive apple varieties are Stirling Castle and Lane's Prince Albert.

Toxicity by sulphur containing fungicides has also been noticed, when they are applied during the blossoming period. McDaniels and Furr have shown that sulphur placed on the stigma of apple blossoms inhibits pollen germination and hence reduces the setting of the fruit.

It has been observed that lime-sulphur when applied on green leaves considerably reduces the photosynthesis by these leaves (Kozlowski and Keller).

The mechanism of sulphur phytotoxicity has not been identified but a reduction in CO_2 assimilation appears to be a fundamental factor underlying the phytotoxic effects.

Disease Control by Inorganic Sulphur Fungicides

Both elemental sulphur and lime-sulphur have been widely used as fungicides for the control of different types of diseases, particularly powdery mildews. Whereas elemental sulphur has been used as a residual fungicide, lime-sulphur has been mainly used as a contact fungicide.

Organic Sulphur (Carbamate) Fungicides

The carbamate fungicides form a very important group among fungicides. Most of these are foliage fungicides, while some are used for soil and seed treatments. Tisdale first demonstrated the fungicidal possibilities of the carbamates in 1931 in the laboratories of E.I. DuPont Company, USA, but the commercial production started about a decade later.

All the carbamate fungicides at present available

commercially are derivatives of dithiocarbamic acid. This organic acid does not occur in the free state and was synthesised in the 1920's to accelerate the action of sulphur in vulcanising rubber.

Mechanism of Action

The mode of action of dithiocarbamates has been discussed in detail by Thome and Ludwig and Rich and Horsfall. Dithiocarbamates are divided into two groups primarily on the basis of evidence for two different mechanisms of action.

The first (I) group, the dialkyldithiocarbamates, includes compounds like sodium dimethyldithiocarbamate and tetramethyl-thiuram disulphide (thiram) which are derived from dialkyl amines, whereas the second (II) group, the monoalkyldithiocarbamates, includes compounds like sodium-N-methyldithiocarbamate (vapam) and sodium ethylenebis-dithiocarbamate (nabam) which are derived from monoalkyl amines.

Dialkyldilhiocarbamates

Dialkyldithiocarbamates are strong chelating agents and it was thought for some time that they acted by depriving the cell of needed metals. However, it is indicated that a heavy metal ion is required for high toxicity of these fungicides.

For example, upon interaction of copper ions and dimethyldithiocarbamate (DDC) ions, two complexes are formed, viz. the "unsaturated" positively charged 1: 1 complex and the "saturated" uncharged 1: 2 complex.

According to Goksyr, the 1: 1 complex is responsible for toxicity of dialkyldithiocarbamates at low concentrations. This 1,1 $(DDCCu)^+$ complex presumably attaches to vital components of the cell and thus prevents growth.

Owens suggested that free radicals rather than dithiocarbamate ions may be the active forms of ferbam and thiram. It has been suggested that these are toxic by virtue of their reaction with cellular thiols, which, in turn, results in the interference in the activities of sulphydryl requiring enzymes, Lowe and Phillips observed that diethyldithiocarbamate and other toxic chelating agents catalyse incorporation of copper into the diethyl ester of mesoporphyrin, and they suggested that the mode of action of these compounds might be associated with the incorporation of copper into a porphyrin precursor of an essential haem-type pigment.

Bates and Tweedy studied effect of thiram upon glucose catabolism by *Fusarium oxysporum* and reported that thiram had no effect upon glycolysis but did inhibit the activity of the pentose pathway and acetate metabolism.

According to Sisler the toxicity of dialkyldithiocarbamates in the cell might involve one or more of the following:

(a) Chelation of required heavy metals.

(b) Attachment of a 1:1 complex of metal and dithiocarbamate ions to enzymes.

(c) Attachment of dithiocarbamate ions to metals bound to proteins.

(d) Reaction of free radical intermediates with cellular components.

(e) 'Lethal catalysis' as suggested by Lowe and Phillips. Thus the mechanism of action of dialkyldithiocarbamates in fungus cells still remains obscure.

Monoalkyldithiocarharnates

A direct consequence of introducing hydrogen at the nitrogen atom in a dithiocarbamate structure is a considerably reduced chemical stability. According to Kaars Sipestein and van der Kerk the dithiocarbamates bearing free hydrogen at the nitrogen can split off hydrogen sulphide with the formation of isothiocyanate groups.

Especially in the presence of the ion of heavy metals these equilibria are forced to the right because of the formation of insoluble metal sulphides.

Ziram

Ziram is the coined name of zinc dimethyldithiocarbamate. It was developed in 1930 by E.I. Du Pont de Nemours and Co. and Bayer, A.G. as an accelerator in rubber vulcanisation. Some of the common trade names under which ziram is sold are Corozate, Cuman, Fuklasin, Hexazir, Karbam white, Milbam, Niagara Z-C spray, Vancide 51Z, Zerlate, Ziram, and Zirberk. More commonly it is used as a spray.

The molecular weight of ziram is 305.8. It is soluble to the extent of 65 ppm in water, completely soluble in chloroform, and essentially insoluble in alcohol and ether. Amongst the dithiocarbamate fungicides, ziram is the most stable one. It is a white substance.

Ziram is compatible with most other pesticides except those containing either heavy metals like copper, iron, mercury, or lime.

As far as the toxicity of ziram is concerned, it is toxic to mammals if consumed orally. It may cause irritation to nose, throat, and skin.

Ziram has been widely used in many countries including India as a protective spray against a number of diseases of vegetables and ornamentals. Between late and early blights of potato, ziram has been found to be more useful against the latter. Like any other protective fungicide, ziram is applied 3-7 days before the expected time of disease appearance. The rate generally used for applying ziram as a spray varies from 0.15 to 0.25 per cent. In addition it is known to repel Japanese and cucumber beetles;

Ferbani

The coined name ferbam stands for ferric dimethyldithiocarbamate. It was developed in 1931. In US it was originally developed by Du Pont de Nemours and Co.

Ferbam is sold in the market under the names "Coromet",

Ferbam, Ferberk, Fermate, Fermate-D, Fermocide, Ferradow, Hexaferb, Karbamblack, etc.

The molecular weight of ferbam is 416.5. It is somewhat unstable to heat, light and moisture, and may decompose if stored long. The colour of pure ferbam is black. It is soluble in water to the extent of 120 ppm at room temperature while in acetone, chloroform; or pyridine, it is completely soluble.

Ferbam is compatible with most pesticides, but its use with copper, lime, or mercury compounds tends to reduce the fungicidal efficacy. Since ferbam is somewhat unstable in the presence of moisture, or under excessive heat, it should be stored in a cool and moisture-proof place. Like ziram, ferbam is toxic to mammals if consumed orally.

Ferbam has been successfully used as a protectant against a wide variety of fungus diseases of fruits, vegetables, and ornamentals. It is one of the important fungicides included in the spray-schedule for the control of a number of diseases of apples and other fruits in the temperate regions. Essentially ferbam is a spray or dust fungicide, but sometimes, it has been used for seed and soil treatments. As in the case of other carbamate fungicides, the rate of application of ferbam varies between 0.1 to 0.3 per cent. Like ziram it acts as a repellent towards Japanese beetle.

Prasad and Pranier reported mutagenic activity of ferbam when it was applied to growing *Allium cepa* roots. They observed unusually high chromosomal aberrations in ferbam-treated roots. This indicates the possibility of an injurious effect on certain host plants.

Some of the common trade names under which zineb is sold are Dithane Z-78, Hexathane, Lonacol, Parzate C, and Du Pont Fungicide A. Zineb has a molecular weight of 275.5. It is soluble in pyridine, but its solubility in water is only a little over 10 ppm. It is unstable in the presence of light, heat, or moisture. Pure Zineb is buff or light coloured, but the formulated products, like Dithane Z-78, are yellowish white.

Zineb is compatible with most of the common fungicides and insecticides. However, it should not be used with any fungicide containing lime, as the presence of calcium may change the residual fungicidal action to eradicant action with a loss in residual activity.

Zineb is toxic to mammals if consumed orally. It may cause irritation to skin, nose, and throat. Zineb is phytotoxic in case of zinc-sensitive plants. Sometimes it is known to cause defoliation in vines.

The residual effect of zineb depends to a great extent upon the weather and the season. As mentioned earlier, since zineb is unstable in the presence of light, heat, or moisture, its residual action is reduced under conditions of excessive heat or rains. It is reported that zineb, in comparison to Bordeaux mixture gives better control of late blight of potato in the plains of India than in the hills, whereas the reverse is true in the hills. One reason could be that in the hills, during the summers when potatoes are grown, there are frequent rains, and therefore the residual action of zineb is reduced. It is well-known that Bordeaux mixture has a tenacity of its own and it may be due to this that it gives better control of late blight in the hills in comparison to zineb.

There is one advantage in using zineb. Some plants like tomato and paddy are able to absorb zinc from zineb. Therefore, use of zineb should be of great advantage in areas where zinc is deficient in the soil.

Zineb is mainly used for foliar sprays. However, soil treatments with zineb have also been attempted. Solel and Minz found furrow application of zineb to give good control of the post-emergence damping-off of cotton seedlings caused by *Rhizoctonia solani* and others. Similarly Shahare and Asthana could obtain good control of Rhizome rot of ginger by applying gallons of 0.15 per cent zineb (65 per cent) per 8' × 5' plot before planting and then giving three similar applications after planting at 3-week intervals. When applied to the soil, zineb shows nematicidal properties against *Meloidogyne* spp.

Zineb (65 per cent) when applied at the rate of 1.88 gm/1 to the root system reduces the relative numbers of rhizosphere bacteria (Halleck and Cochrane). A report on nitrogen deficiency disease of sugarcane induced by large number of applications of zineb or maneb to the preceding tomato crop was made by Dubey.

Proprietary mixed formulation containing zineb and other fungicides like copper oxychloride were marketed by some companies. For example, Miltox by Sandoz contains zineb and copper oxychloride.

Nabam

Nabam is the coined name for disodium ethylenbisdithiocarbamate. Nabam happens to be the first ethylenebisdithiocarbamate fungicide. It was discovered by W.F. Hester of the Du Pont Company, USA, in 1935. Dimond *et* al. reported its fungicidal properties in 1943. Nabam, however, was found to be rather unstable and also too soluble for use in the field. Heuberger and Manns in 1943 reported that nabam could be stabilised with the addition of zinc sulphate-lime mixture (2.5 lb nabam +1 lb $ZnSO_4$+1 lb lime in 100 gallons of water). The stabilisation results from the formation of zinebas a result of weathering of nabam-zinc sulphate-lime mixture on the leaves (Barratt and Horsfall). In fact it was this work of Barratt and Horsfall that led to the production of zineb commercially.

Nabam is sold under different trade names such as Chembam, Dithane A-40, Dithane D-1.4, Parzate liquid, etc. It is in the form of a liquid.

The molecular weight of nabam is 256.3. In the powder form, it is relatively unstable to heat, light, and moisture. As pointed out earlier, it is relatively soluble in water.

For field use nabam is always mixed with zinc sulphate alone or zinc sulphate-lime mixture. Two to three lbs of nabam are mixed with 1 lb zinc sulphate and to 1 lb of lime. Lime can be avoided if it is certain that the amount of zinc sulphate used would not be toxic. Nabarn-zinc sulphate mixture has been used for the control of a number of those diseases against which zineb has been successfully used.

Though nabam has been primarily used for foliar applications, successful soil application has been attempted. For example, seedling disease of cotton *(Rhizoctonia solani, Fusarium* spp., and *Pythium* spp.) can be controlled when nabam is applied to the soil at the rate of 5-6 lbs per acre.

SMDC (Vapam)

SMDC was introduced by the Stauffer Chemical Co. of the USA in 1954. Besides having fungicidal properties, it also has nematicidal as well as herbicidal properties. Some insecticidal activity has also been reported. Chemically it is sodium methyl-dithiocarbamate.

Vapam is the trade name given by the Stauffer Chemical Co. Other names under which this chemical is sold are VPlVt, Chem-Vape, 4-S, etc.

SMDC is soluble in water and only moderately soluble in alcohol. It has a boiling point of 119°C.

It is available as a concentrated (32.7 per cent) solution of Sodrum methyldithioppbamate under the name Vapam. Wheti applied to the soil, it releases the futnigant methylisothiocyante (CH3N=C=S). Temperature and moisture influence SMDC decomposition. Increasing the soil temperature from 10 to 40°C, decreases the time required for maximum SMDC decomposition from 7 to 1.5 hours. Also decreasing the moisture content from 20 to 6 per cent decreases the time required from 7 to 2.5 hours.

The mode of action of SMDC is more like ethylene bisdi-thiocarbamates though structurally it is similar to dialkyldithiocarbamates.

SMDC can be applied with watering can or sprinklers. Plant bed has to be prepared one week in advance of treatment when soil moisture is good. Bed area should be cultivated to a depth of 15 to 20 cm prior to treatment and all clods should be broken up or raked off the bed. The dosage normally varies between 1.5 to 2.5 litres of vapam per 10 square metres. After the application of vapam, it is necessary to apply a water seal to upper 2.5 cm of soil. After the top treated soil has dried sufficiently, it should be cultivated to a depth of 2.5 to 5 cm. Planting should not be done for at least three weeks, SMDC has been successfully used in the control of several soilborne diseases. Some of the examples are: Cotton wilt *(Fusarium oxysporum f. gossypii),* damping-off of papaya *(Pythium debaryanum),* damping-off of peas *(Pythium uhimum),* sugarbeet root rot *(Sclerotium rolfsii),* root disease of tea *(Poria hypolatenym),* seedling diseases of tobacco (species of *Pythium, Fusarium, Rhizoctonia),* and *Sclerotium rolfsii* on tomato.

As a nematicide, SMDC has been found quite effective in the control of several diseases caused by nematodes. Some of the examples are: *Tylenchulusr semi-penetrans* on citrus, *Heterodera rostochinensis* on potato, root-knot nematodes, *Haplolaimus uniformis* on carrots.

Copper Fungicides

Although the fungicidal activity of copper sulphate was first recognised in 1807 by Prevost, it was not until 1885 that this fungicidal property was utilised for the protection of foliage from pathogenic fungi. Later copper fungicides predominated the field of fungicidal plant disease control for more than 50 years until synthetic organic fungicides invaded the market. However, even today copper fungicides are used widely in many countries including India. The copper fungicides have been used for the control of many vegetable, fruit and flowering plant diseases. They have also been commonly used in plantation crops like tea, rubber, coffee, etc.

Bordeaux Bordeaux Burgundy Cheshunt Chaubattia mixture paste mixture compound paste of copper sulphate against the grape downy mildew caused by *Plasmopara viticola.* He noticed that grape vines along the roadside at Medoc in the Gironde (France) retained their leaves at the end of October when all the others had been defoliated by mildew. His companion McDavid, steward of Mr. Johnston's estate in the neighbourhood, explained that it was the custom to sprinkle the vines near the road with copper sulphate and lime to discourage pilfering. The following year Millardet and David confirmed the utility of this practice in checking the disease. Further tests under the supervision of Millardet and Gayon, who were professors at the University of Bordeaux, were carried out, and in October 1885, they published illustrations of treated and untreated vines with analytical data. By 1887, great success was reported in the control of the disease.

This mixture of copper sulphate and lime received wide recognition under the name "Bouillie Bordelaise" (Bordeaux mixture). In 1885 its use for controlling potato late blight was suggested and in 1888 its efficacy against the blight was established. Bordeaux mixture in the beginning was in the form of a paste to be applied with a brush. Millardet had proposed the admixture of 100 litre of water in which 8 kg of bluestone was dissolved with 30 litres of a lime suspension prepared from 15 kg lime.

This formula (8-15-100) was soon found to be too strong and apt to cause foliar injury. In 1887, Millardet and Gayori tested a series of formulae ranging from approximately 25-50-50 up to 4-1-50 and found that the mixture with the lower concentrations of copper sulphate and lime was as effective in preventing the germination of the spores of *Plasmopara viticola* (downy mildew) as those with higher concentrations. Since then a wide range of formulae have been tried and the one used relatively more often at the present time is 4-4-50. The method commonly followed in India for preparing Bordeaux mixture is by mixing solutions of copper sulphate and either quick lime or hydrated lime. The latter is commercially available in a more finely divided form of greater purity than quick lime and has advantages in case of preparation of the mixture. Air-slaked quick lime is unsuitable. For preparing the mixture, copper sulphate and lime are separately added in sufficient quantities of water. In case of quick lime, slaking is done slowly. These two solutions are then mixed simultaneously into a third container, while agitating thoroughly, and the Bordeaux mixture is ready. The containers used should be such which do not react with copper sulphate; e.g. copper, wooden, or earthen vessels.

In some European countries Bordeaux mixture is available commercially which saves the botheration of preparing the mixture. These preparations are sold in dry or paste forms.

After preparing the Bordeaux mixture, it is desirable to test the finished mixture for free copper. Free copper in Bordeaux mixture can be phytotoxic and to avoid this more lime would have to be added. To test Bordeaux mixture for free copper, a few drops of a 10 per cent solution of potassium ferrocyanide are added to a small quantity of the prepared Bordeaux mixture. If a reddish brown deposit is formed immediately, more lime should be added to the mixture. If potassium ferrocyanide is not available, a clean iron blade could be used by dipping it for some time in the mixture and observe if copper deposits on it.

Bordeaux mixture should be sprayed while fresh. On standing it losses its fungicidal property. However, the mixture may be stabilised by adding sugar or jaggary at the rate of 1 lb in 50 gallons of the mixture. The mixture stabilised in this manner will not deteriorate for several days under normal conditions.

Bordeaux mixture has both merits and demerits. The merits are: (a) its natural adhesiveness or tenacity, (b) its relative cheapness, (c) its utility in controlling a wide variety of diseases, and (d) its safety to handle. Among the demerits are: (a) its phytotoxicity on certain crops, particularly fruits like apples, peaches, etc. (b) its tendency to delay ripening, (c) the botheration of preparing, and (d) its corroding action on metallic containers of spraying equipment.

Besides being effective against fungal diseases, the Bordeaux mixture has also been found effective in reducing the incidence of certain bacterial diseases like citrus canker.

Solubility of Bordeaux Mixture: The inherent solubility of Bordeaux is quite low. According to Goldsworthy and Green, Bordeaux mixture in equilibrium with water yields a solution containing about 4 ppm of copper. However, McCallan and Wilcoxon found that a well-washed 4-4-50 Bordeaux mixture was soluble only to the extent of furnishing 1 ppm of copper. After this material was thoroughly dried, as in a spray film, the solubility decreased to 0.2 to 0.3 ppm. McCallan and Wilcoxon have shown a comparison between the amounts of Bordeaux mixture and copper sulphate required to inhibit the germination of 90 per cent of the spores of a few species. The Relative Efficacy of Bordeaux Mixture and Copper Sulphate in Inhibiting Spore Germination.

Cu. mgllitre for LD 90

Species *Bordeaux Mixture*	*Copper*	*Sulphate*
Uromyces caryophyllinus	180	1.74
Sclero tinia fructicola	120	1.20
Botrytis paeoniae	390	2.23
Glomerella cingulatd	500	1.40
Alternaria solani	2,400	6.72

If Bordeaux mixture or other copper spray or dust of low solubility provides less than 1 ppm of copper to the solutions with which it is in equilibrium, it is evident that the concentration of copper is too low for any substantial toxicity. A few possible mechanisms have been suggested to explain the manner in which copper is brought in soluble form in quantities enough to be toxic to spores. These are: (i) action of the atmospheric carbon dioxide and ammonium salts dissolved in rain water or dew, (ii) action of the secretions from the fungus, and (iii) secretions from the healthy or wounded surface of the host plant.

Millardet and Gayon observed the formation from Bordeaux deposits, after carbonation of the excess lime of soluble copper by the action of the carbon dioxide and slight traces of ammonium salts present in rain, water and dew. This hypothesis did not, however, find much support because of stronger evidence in support of the other two agencies, particularly the spores themselves.

The suggestion that spore execrations are involved in the fungicidal action of Bordeaux mixture was first made by Barth and Swingle. McCallan and Wilcoxon demonstrated the presence of malic acid and of amino acids in washings from spores of *Neurospom sitophila.* As the spore execrations are almost neutral, these workers considered that the salts of hydroxy-acids such as malate and of amino acids present in spore execrations, act on the Bordeaux deposit to form soluble toxic copper hydroxy and copper amino salts. Wain and Wilkinson confirmed the solution of copper from dried Bordeaux deposit by the lysate of spores of *Neurospora sitophila* and studied the solvent action of a range of compounds likely to be present in the lysate. Active solvents included amino-, hydroxy-, and certain dicarboxylic acids and their salts.

A role of the host plant in bringing copper into soluble form has also been suggested. Barth first suggested that host plant secretions might render the copper of the spray deposit soluble or available. Barker and Gimingham showed that substances exuded from mechanical injuries to the leaf are capable of dissolving copper from Bordeaux precipitates.

The "solubilisation" hypothesis as a whole has, however, been questioned, Somers provided evidence to show that formation of copper complexes, from the reaction between leaf or spore exudates with a Bordeaux deposit, reduces the uptake of copper by the spores of *Neurospora crassa.* This suggests that these water soluble complexes are less effective in penetrating the spore than is the free cupric ion. According to Somers simple accumulation of ionic copper is the mechanism by which copper from Bordeaux mixture becomes toxic to spores.

Reaction between the Constituents of Bordeaux Mixture: Millardet and Gayon thought that this reaction proceeded according to the equation: Thus basic copper sulphate is formed and this remains in suspension.

Bordeaux Paste

Bordeaux paste consists of the same ingredients as that of Bordeaux mixture, but it is in the form of a paste. It is primarily used for tree wound dressings to prevent fungal attacks, e.g. in the control of stem bleeding disease of coconut. It is normally prepared by adding one lb each of copper sulphate and lime in one gallon of water.

Burgundy Mixture

Burgundy mixture was introduced by Mason in 1887 in Burgundy (France) in place of Bordeaux mixture. It is prepared in the same way as the Bordeaux mixture, except that lime is substituted by sodium carbonate crystals. This preparation was developed primarily because at that time good quick lime was not easily available in Europe. A 4-4-50 formula has been reported to be satisfactory. Now that good quality lime is easily available, Burgundy mixture is a thing of past. Besides, it is slightly less effective than Bordeaux mixture.

Cheshunt Compound

The compound contains 60g of copper sulphate and 11 parts of ammonium carbonate. These two are well-powdered and thoroughly mixed and the dry mixture is stored in an airtight receptacle for 24 hours before using. If left exposed, it gradually loses ammonia and becomes less effective. One ounce of this mixture is dissolved in a little hot water and the solution is made up to two gallons with cold water. This solution is then ready for spraying.

The Cheshunt compound is used to drench nursery beds for the control of "damping-off Copper carbonate preparations.

Chaubattia Paste

This paste was developed at the Government Fruit Research Station, Chaubattia, in the district of Almora. This was developed as a wound dressing fungicide be applied to pruned parts of apple, pear, and peach trees for the control of diseases like stem-black *(Coniotheciwn chomatosporum)*, stem-brown *(Botryosphaeria ribis*\ pink disease *Corticium salmoni-colour)*, and stem canker *(Monochaetia mail)*, of apple and pear, and collar rot *(Rosellinia* sp.) of apple, peaches, apricots, and plums. The paste is prepared by mixing copper carbonate (800 gm) and red lead (800 gm) in 1 litre of lanolin or raw linseed oil. These are mixed in some glass or chinaware pot. The paste has the advantage that it is not washed off easily by rain. No better substitute to this paste has so far been found for use in the Kumaon hills (Almora and Nainital districts) to control these diseases. A readymade paste is now sold commercially.

Copper Oxychloride Preparations

Though Bordeaux mixture was used as a major copper fungicide for many years, in view of its toxicity to certain plants, attempts were made to find out some suitable substitutes. Starting in approximately the second decade of the present century, a series of copper fungicides termed as "insoluble", "low-soluble", or "fixed coppers" were found out. In all these the copper is stabilised and rendered less soluble, which considerably reduces the phytotoxic properties of copper. These low-soluble copper fungicides are much less injurious, but they have not been found to be as effective as Bordeaux mixture, because the latter has more natural tenacity. However, in view of the convenience in their use, these low-soluble copper fungicides are frequently used in place of Bordeaux mixture. The low-soluble copper available in India is copper oxychloride. It has been formulated both as wettable powder and as a ready to use dust. Copper oxychloride ($CuCl_2$. 3Cu $(OH)_2$) is readily formed by the action of air on cupric chloride solutions or scrap copper. Formulations containing 4 to 50 per cent metallic copper in the form of copper oxychloride are available in India. Copper oxychloride, containing 4-12 per cent metallic copper is used for dusting and 50 per cent metallic copper for spraying. Some of the commercial preparations are: Blimix' 4 per cent; Blitox-50; Cupramar; Micop D-06; Micop W-50; Fytolan; Vaijfru Tamraghbl; and Blue copper 50. Cuprous oxide preparations Horsfali suggested the use of red oxide of copper as a seed treatment fungicide in 1932, It was used for treatment of vegetable seeds, such as peas, and was also used to a limited extent for spraying leaves.

In 1939, Heuberger and Horsfall reported that the fungicidal and protective values of the cuprous oxides vary inversely as the particle size. These values were found to increase as the colour of the cuprous oxides changes from red through orange to yellow. These workers explained that the increased fungicidal and protective values of the cuprous oxides are due to an increase in the area of chemically reactive surface per unit of weight and a corresponding increase in the rate at which soluble copper is released. This research led to the release of yellow cuprous oxide as a seed and foliage protectant.

One of the commercial preparations, Fungimar, consists of 50 per cent metallic copper in the form of cuprous oxide. Perenox is another example of cuprous oxide preparations.

The low soluble coppers are generally recommended for spraying at the rate of 3-4 lb of material containing 50 per cent copper in 100 gallons of water. If a copper dust containing 4-6 per cent metallic copper is to be used, generally speaking 25-35 lbs of the dust would be required to get a good coverage.

Mechanism of Fungicidal Action

Those workers who believe in "solubilisation" hypothesis presume that the fungal cells accumulate the water soluble copper complex (with amino-, hydroxy-, or dicarboxylic acids in the spore exudates) which later on dissociates to give free cupricion. On the other hand some workers suggest with evidence, that fungal cells accumulate free ionic copper from Bordeaux mixture and that no "solubilisation" is needed. Same behaviour can be expected from the low soluble copper fungicides.

What does then the free ionic copper do? The common explanation of the toxic action of the heavy metals is based upon the properties of these ions of precipitating or denaturing proteins. Enzymes are proteins, and it would therefore be expected that the heavy metals would inactivate these catalysts. However, not all enzymes are equally inactivated by low concentrations of heavy metal ions. The enzymes which require free sulph-hydryl groups for activity appear to be especially susceptible to inactivation by ions of heavy metals. It is, therefore, quite possible that copper kills spores by combining with the sulph-hydryl groups of certain enzymes.

Phytotoxicity

Copper in Bordeaux mixture and other fungicides sometimes causes injury to plants. The injurious effect is characterised by chlorosis followed by brown or purple flecks on the leaves or fruits. In case of leaves, the leaf-veins develop purpling and the leaves fall off. On fruits, such as apple, a russetting may be seen. Lime is added to Bordeaux mixture to reduce phytotoxicity, but sometimes the lime itself may be injurious.

In some of the high yielding varieties of paddy (IR-8) and maize (e.g. Ganga Hybrid Makka No. 3), copper toxicity is characterised by the appearance of chlorotic patches somewhere in the middle of the leaf, particularly where the upper leaf bends. This is followed by brown necrosis. Widespread phytotoxicity due to wrongly recommended copper sprays has been observed in Uttar Pradesh.

An interesting relationship between copper sprays and increased frost damage has been reported by Wilson in case of potatoes and tomatoes. The effect of frost damage in these crops when sprayed with various copper fungicides such as Bordeaux mixture was more severe as compared to those plants which were sprayed with non-copper fungicides like zineb.

Dispersal by Insects

It is one of the most striking facts of botany that transport of pollen grains in angiosperms from flower to flower is largely the work of insects, and this reliance on insects has apparently been a major factor in the evolution of floral structures. Although there are a number of interesting examples of spore dispersal by insects, fungi have tended largely to be anemophilous and only in one small family—Phallaceae—is the sporophore as intimately related to insect visitors as is the coloured, scented, nectar-producing flower so common in angiosperms.

The most familiar British species in Phallaceae is the cosmopolitan stink-horn *(Phallus impudicus)* so common in all kinds of woods from June to October. The fruit-body nearing maturity is an egg-shaped structure half-buried in the soil and leaf-litter of the forest floor. Then, usually in the evening, the papery skin (exoperidium) ruptures and, early the following morning, the compressed spongy tissue of the stalk rapidly elongates to a height of 15-20 cm., bearing at its apex a conical cap thickly covered with a greenish-black, sugary slime containing millions of very minute basidiospores. At first no smell is emitted and no insects appear on the scene, but after an hour or, so a characteristic fetid odour develops which can be detected many yards away. Flies, especially blow-flies and green-bottles, are attracted from far and wide, and it is a common sight to see the cap of a stink-horn crowded to capacity with jostling insects. In two or three hours the spore-slime is completely removed, leaving the bare white, alveolate cap. Spores are carried on the feet and bodies of the insects and also pass rapidly and apparently unharmed through the alimentary canal. However, neither in *Phallus impudicus* nor in any other member of the family has germination of the spores been observed. If the spores have lost their power of germination, effective dispersal of the fungus must be brought about not by the dissemination of basidiospores but by some other means. However, it is more likely that dispersal of this very abundant species is due to transport of basidiospores by insects, but that in the laboratory conditions for spore germination have yet to be realised. In *Mutinus caninus,* very like the stink-horn but smaller and with the cap fused throughout its length to the orange stalk, the odour of the spore-slime is comparatively slight, but *Clathrus cancellatus,* with its beautiful red, spherical network coated with spore-slime, is even more fetid than the stink-horn.

A contemplation of British species of Phallaceae, however, gives but a poor impression of the family, which reaches its richest development in tropical and subtropical regions. There are, for example, the beautiful crinoline fungi *(Dictyophora* spp.) agreeing in general form with *Phallus* but having a skirt-like network, often coloured, displayed below the cap around the stalk like a crinoline. Perhaps the most interesting genus from the point of view of dispersal is the Australian *Aseroe rubra.* In this species, a sporophore is developed which forms a remarkable parallel with the entomophilous flower. There is a stipe (peduncle) 2-4 inches long, a central fertile region at its apex where the odoriferous spore-slime occurs, and around this a magnificent 'perianth' of red sterile lobes (petals).

Another example of specialised entomophily is the conidial stage of *Claviceps purpurea.* This fungus produces its sclerotia (ergots) in rye *(Secale cereale),* rye-grass *(Lolium perenne),* purple moor-grass *(Molinia caeruled),* and, more rarely, in other members of Gramineae. Primary infection of the rye apparently is due to filiform ascospores violently discharged during the late spring and borne by the wind to flowers of the host. The fungus invades the ovary, which becomes replaced by a dense mass of fungal tissue which is highly convoluted and covered by a hymenium of simple conidiophores. These produce vast numbers of minute elongated conidia which become involved in a copious, stinking 'honey-dew in which the sugar concentration may be over 2 molar. The flowers of grasses are not insect-pollinated, but some insects regularly visit the spikelets of rye for the sake of pollen. Others are attracted by the 'honey-dew' secreted by the leaf-louse, an insect found abundantly on the fertile spikes of rye. Insects, especially *Melanostoma mellina,* which feeds both On pollen shading are actually bright and on 'honey-dew' whether of red. The slimy spores are insect or fungal origin, seem to be normal vectors of the ergot disease. The spores may be carried on the surface of the insect or they may be sucked in with the spore-containing honey-dew, to be excreted or regurgitated later, perhaps on a healthy spikelet.

It should be mentioned that, although insects would appear to be the principal agents of the spread of the conidial *(Sphacelia)* stage of ergot, other methods of dispersal may sometimes obtain. In the unnatural conditions of field cultivation conidia may easily be spread by a diseased head knocking against a healthy neighbouring spike, and probably rain-splash as well plays some part in dispersal.

The importance of insects in the dispersal story of rust fungi, though envisaged by Rathay as early as 1883, was brought into prominence by the brilliant work of Craigie who demonstrated that *Puccinia helianthi* and *P. graminis* are heterothallic (self-incompatible). This condition has since been shown to occur in many more rusts belonging to a wide range of genera *(Cronartium, Melampsora, Gymno-sporangium, Phragmidium, Uwmyces).* Indeed, self-incompatibility seems to be the normal state of affairs and, according to Buller, it probably obtains in all rusts with fully developed pycnidia. Homothallic (self-compatible) rusts seem few in number, and in them the pycnidia are rudimentary or absent.

On germination of the teleutospore of *Puccinia graminis* uninucleate basidiospores (sporidia) are produced. These are violently discharged and, being further dispersed by wind, normally fall singly on the leaves of the host. If a basidiospore infects a leaf of barberry, a localised intercellular mycelium is produced on which pycnidia are formed, mainly at the upper surface, and also spherical proto-aecidia embedded in the somewhat hypertrophied leaf tissue near the lower surface.

Each pycnidium is more or less flask-shaped, with its wall of interwoven hyphae within the host tissue but with its ostiole, surrounded by a fringe of stiff red, projecting periphyses, opening to the outside. Within the pycnidium minute pycnidiospores are abstricted and ooze out through the ostiole together with nectar, so that each pycnidium is capped by a little sugary drop having an osmotic pressure of between 12 and 24 atmospheres. Often at a later stage the drops from neighbouring pycnidia flow together to form a single larger drop. In addition to the coloured periphyses there project into the nectar-drop through the ostiole a few rather long flexuous hyphae which are simple or sparingly branched and non-septate.

Apparently in most rusts there is a sweet scent associated with the pycnidial stage. This may most easily be detected in the common thistle rust *(Puccinia suaveolens)* which causes a systemic infection of *Cirsium arvense.* Early in the growing season the under surfaces of the leaves of infected shoots are densely covered with pycnidia. The diseased leaves have a scent, almost like that of violets, which disappears when the succeeding uredospore stage develops and the pycnidia disappear. In such a rust as *Puccinia graminis* where only small local infection spots are involved, the number of pycnidia crowded together is very much less and the scent can be detected only by those with a keen sense of smell.

In Craigie's experiments with *P. graminis* and *P. helianthi* infection of a leaf by a basidiospore led to the development of an infected spot bearing pycnidia at the upper surface and spherical proto-aecidia near the lower. These did not normally develop into aecidia if precautions were taken to exclude insect visitors. When, however, the pycnidial nectar from one infection spot was added to the nectar around the ostioles in another single-spore infection spot, mature aecidia developed in a few days, provided the two infections were of opposite mating types.

A proto-aecidium borne on mycelium must receive nuclei

of the opposite sign if it is to develop into an aecidium. The pycnidiospores would seem to act as diploidising, or more correctly dikaryotising, agents although the cytological details of the process are still very obscure. When pycnidiospores are added to the drop around the ostiole of a pycnidium of opposite sign, the flexuous hyphae in the drop fuse with the introduced spores. Presumably the nuclei of these spores pass into the flexuous hyphae, thence into the intercellular mycelium, and so to the proto-aecidium. Here dikaryotisation of certain cells occurs and it is these that give the chains of aecidiospores.

Graigie has shown that under experimental conditions flies will mix nectar from different monosporidial cultures and so lead to the development of mature aecidia. Very much earlier Rathay studied the insect visitors to pycnidial nectar and recorded a large number of insects that were attracted, including flies, beetles, and ants.

An example somewhat similar to rusts is *Coprinus cinereus*. This coprophilous agaric has the tetra-polar self-incompatibility type of heterothallism that is a feature of most of the larger Basidiomycetes. Each basidiospore gives rise on germination to a primary mycelium of uninucleate cells, and from this mycelium conidia (oidia) are produced on short conidiophores in slimy heads. The primary mycelium can develop into a secondary one bearing fruit-bodies only if it receives nuclei of the appropriate mating type, and one of the ways this can happen is by contact with suitable oidia. It has been shown experimentally in cultures that flies may transport oidia and effect diploidisation of a primary mycelium. No doubt this kind of transport is also effective in nature, although diploidisation must frequently, perhaps usually, occur following the intermingling of primary mycelia derived from basidiospores of compatible mating types germinating in the same dung ball.

In some pathogenic fungi, spore production is in the host flower and dispersal from infected to healthy plants is brought about by insects in the course of pollination. The most familiar example is *Ustilago violacea,* the common anther-smut of campions and other members of Caryophyllaceae. Infection of the host is systemic or semi-systemic but spore production is restricted to the anthers. Diseased plants appear almost normal until flowering, when it is found that blackish-purple smut spores are produced in place of pollen grains. In *Melandrium album,* there are separate staminate and pistillate plants. When the staminate plant is infected by *Ustilago violacea* the floral structure is modified only to the extent that smut spores are formed in place of pollen. When, however, the pistillate plant is infected the staminal rudiments are stimulated to develop into stamens, but from their anthers brand spores, not pollen grains, are set free.

Further, the ovary is somewhat reduced and although ovules are present they do not develop into seeds. The smutted flowers still give out their scent and are visited by night-flying moths which distribute the spores to flowers of healthy plants. There is, conflicting evidence about how infection of the host takes place, but it seems clear that when brand spores are transferred to a flower on a healthy plant they germinate to produce a mycelium that grows back through the flower stalk into the shoot, so that flowers which develop subsequently on that plant may be smutted. The old view that contamination of a flower by smut spores led to the production of infected but viable seed, as in the loose smut of wheat and barley *(Ustilago nuda),* is not borne out by more recent experiments. Again, infection through the flower is probably not the only method by which the fungus is handed on to healthy plants. There is little doubt that seedling infection also occurs.

Another anther-smut in which the spores are probably dispersed by insects is *Ustilago succisae,* found frequently on devil's bit scabious *(Succisa pratensis).* Here spore production is again limited to anthers. However, the brand spores are white or pale cream in the mass—not black, as in most smuts. Infected flower-heads are easily spotted since the projecting anthers are whitish in contrast to their purple colour in normal plants.

The anther mould of red clover caused by the hyphomycete *Botrytis antkophila* shows a remarkable parallelism with the anther smut of campions. The disease is systemic, but spore production is limited to the flowers, the conidiophores of the fungus being produced on the surface of the anthers. The conidia may be carried by a pollinating bee to the stigma of a healthy flower, where they germinate and the germ-tubes penetrate the tissue of the style after the fashion of pollen-tubes. The developing seeds are infected and within these a dormant mycelium is formed.

These infected seeds are viable, but in the following year they give diseased plants. A further example of dispersal of a fungal pathogen by pollinating insects is to be found in endosepsis of figs. The causal organism *(Fusarium moniliforme* var. *fici)* may be introduced into the edible fig by its obligate pollinator, the fig-wasp *(Blastophaga psenes).* Again, nectar-fermenting yeasts are often carried from flower to flower by bees. For example, *Candida reukaufitf* grows in the nectar of a wide range of flowers. The fungus can be isolated by trapping a bee after it has visited an affected flower, rendering it unconscious with ether and finally cutting off its proboscis and planting it on nutrient agar. It is thought that the yeast may spend the winter in the alimentary tract of bumble bees and be introduced into nectaries in the spring.

A number of fungi fructify in the bore-holes and brood galleries of wood- and bark-beetles, which not only disperse the spores but also introduce the fungi into the inner tissues of the host. The best-known example is the 'Dutch elm disease', which has killed so many elms in the past thirty years. In this disease bark-beetles (chiefly *Scolytus* spp.) are responsible for the transport of the spores of the pathogen, *Ophiostoma ulmi.* The insects form characteristic 'engravings' (brood galleries) in the dead or dying elm trunks at the interface of wood and bark. The eggs are laid in the main gallery and the insects develop from egg to pupa in the lateral galleries, on the walls of which the fungus produces its heads of sticky conidia.

Finally the young beetles bore their way directly outwards through the bark from the widened ends of the lateral channels, leaving little round holes in the bark so that from the outside it seems to have been peppered by shot. The emerging insects carry with them, both internally and externally, viable spores of the fungus. *Scolytus* can produce its brood galleries only in the larger limbs and trunks of dead and dying trees, but before doing so it feeds on the bark in the crutches of very small branches of healthy trees. This introduces the pathogenic fungus to the wood of vigorous living trees which, as they succumb in the course of years, become suitable sites for brood galleries.

The fungus *Ceratostomella ips,* which causes a 'blue stain' in the wood of conifers, has a similar dispersal story, being spread by bark-beetles *(Ips* spp.). However, only trees already dead or very much weakened are affected because the beetles do not, as in Dutch elm disease, prepare their future victims by feeding on the younger, healthy parts before boring into the dead trunks for breeding purposes. However, the 'blue-stain' fungus, though so closely similar to the elm parasite, occurs mainly in the perithecial stage, but this is biologically just like the conidial stage of *Ophiostoma ulmi,* for a mass of slimy ascospores is borne at the end of a little stalk, but the stalk is the long neck of a perithecium the base of which is buried in the wood of the brood gallery.

The story of dispersal in the fungi (especially *Lepto-graphium* spp.) 'cultivated' by the ambrosia beetles is somewhat similar. The beetles bore deeply into the wood of dead and weakened trees, producing branching galleries. The fungi grow in the wood immediately surrounding the galleries and cause a characteristic darkening of the tissue. Sporulation occurs on the gallery walls and the spores are used by the grubs as food. When beetles migrate to another tree, they carry viable spores of the fungus with them and so a continual supply of 'ambrosia' is assured.

A somewhat similar association of insect and fungus is that of the fungus-growing ants. These ants cultivate certain fungi in their nests in a state of more or less pure culture. When a queen establishes a new colony she carries an inoculum of the fungus with her in her infrabucal pouch.

Again, wood wasps may enter into special relations with certain lignicolous fungi. An interesting example is the great wood wasp, *Sirex gigas,* which attacks coniferous trunks. The female has a long ovipositor. Near the fixed end of this in the abdomen of the insect are two little pouches in which *Stereum sanguinolentum* is to be found in the oidal condition. As an egg enters the ovipositor from the main body of the insect an oidial inoculum is exuded on to it from the pouches so that as each egg is laid in a coniferous trunk the fungus is introduced at the same time. The advantage to the insect of this regular association is far from clear. Perhaps the fungus predigests the wood for the larva, or perhaps the larva is, in fact, mycophagous. So far as the fungus is concerned this means of dispersal, though very efficient, is merely additional to its regular spread by airborne basidiospores.

In connection with insect dispersal of fungi growing on tree trunks, *Cryptoporus volvatus* is of considerable interest. It is a smallish bracket polypore, found in North America and Northern Asia on dead and dying trunks of conifers severely attacked by bark-boring beetles. When a fruit-body is about to be formed the mycelium accumulates in an old abandoned bore-hole and where this emerges the sporophore is formed. Both in general appearance and in texture it is rather like a small specimen *of Polyporus betulinus,* but the tube layer is covered in below by a firm volva. At first this is continuous, but later a small round hole is formed in it. Spores rain down from the hymenial tubes as in any other polypore, and a very small proportion of these fall through the hole and thus escape into the outer air. The vast majority, however, are deposited on the inner surface of the volva. Through the hole in the volva, no doubt attracted by the dimness within, many insects crawl. These are mainly beetles, though not bark-beetles.

These become contaminated with numerous spores and on

emerging may go on to explore abandoned bore-holes made by bark-beetles in the trunk of another tree. So, perhaps, the fungus is dispersed, though no doubt the spores that escape directly through the hole in the volva may infect nearby trunks. The dispersal story is in need of closer study, but in view of the unique structure of the fruit-body the fungus is more likely to be entomophilous than anemophilous.

Hemipterous insects with their piercing sucking mouth-parts are of great importance in the transmission of virus diseases of plants. They may also pick up the spores of certain pathogenic fungi during feeding and later introduce them into healthy plants. This type of disease is particularly found in the fruits of a wide range of species and is referred to as stigmatomycosis. The fruits are affected internally but remain sound on the outside. An example of considerable economic importance that has been much studied is the internal boll disease of cotton. This is caused by *Ashbya gossypii* and allied fungi and leads to staining of the lint in immature, unopened bolls. The fungus grows and sporulates in the lint, but the pathogen makes no appearance at the surface of the boll. The chief vectors of the disease are bugs of the genus *Dysdercus.* When a bug feeds on a diseased boll the tip of its fine sucking apparatus punctures the ovary wall.

This apparatus, throughout the greater part of its length, is sheathed in the split-tube formed by the labium, the tip of which rests on the surface of the fruit. The sucking apparatus is composed of four stylets (actually a pair of maxillae and a pair of mandibles) tightly pressed together, and contains two longitudinal tubes: a narrower one down which saliva is pumped and a slightly wider one up which the food solution is sucked. The long, narrow ascospores, each with a whip-like tail, are sucked up with the liquid food and may be stored unharmed in the stylet pouches that occur in the head. Others may pass down the alimentary canal but these, apparently, soon cease to be viable.

In a rather similar but much more casual manner, wasps

may spread *Sclerotinia fructigena* amongst ripe fruit during the late summer. This 'brown-rot' organism is abundant on apples, pears, and plums, both on fruit attached to the tree and on windfalls. The rot rapidly spreads throughout an infected fruit and later, on its surface, pustules of powdery spores are formed in characteristic concentric rings. Entry is possible only through a wound and this is often provided by the bite of a wasp, which, if it has already visited a rotten fruit bearing conidia, may introduce the pathogen at the same time. This is probably the normal method by which the fungus spreads, but the spores are dry and powdery and no doubt wind dispersal occurs to some extent.

This discussion of entomophilous dispersal is by no means exhaustive and many examples have been omitted. Two points, however, stand out clearly: first, the types of insect concerned are very various (flies, beetles, moths, plant bugs, ants, bees, and wasps); second, the fungi involved are equally diverse (stink-horns, a bracket poly-pore, flask-fungi, smuts, rusts, yeasts, and imperfect fungi). No doubt with further work instances of insect dispersal will be multiplied, but the general statement that ehtomophily is rare in the fungi is likely to remain unassailed.

It should be observed that other small invertebrate animals may be responsible for the spread of fungi. Slugs, for example, frequently feed on the larger agarics and, apparently, the spores pass uninjured through the alimentary canal and then geminate freely. Indeed, it has been claimed that slugs play a very important part in the dispersal of fleshy toadstools, but this seems unlikely. Again, mites may be effective agents in the spread of moulds. Sometimes in a mycological laboratory they become established in a few tube-cultures of moulds and then, with alarming speed, contaminants appear in nearby pure cultures. Even eel-worms may disperse fungal spores. Thus in the *Dilophospora* disease of cereals Atanasoff claims that it is nematodes that normally carry spores to the susceptible growing-point of the shoot.

Dispersal by Larger Animals

The larger animals may have an important part in the spread of certain fungi. Thus the dung fungi are not only dispersed by herbivorous animals, but find an ideal natural cultural medium in the faeces; some woodland subterranean fungi—truffles and their like—are, apparently, dispersed by rodents; water fowl almost certainly play an essential part in the long-distance dispersal of freshwater aquatic fungi; and finally, it is, perhaps, logical to include the activities of civilized man, who has unwittingly been responsible for the spread of many plant pathogenic fungi even from one continent to another.

When the freshly deposited dung of herbivorous animals (e.g. horse, cow, rabbit) is kept moist under a bell-jar in the light, a rich flora of fungi, mainly minute, develops and in a fairly definite succession. Under natural conditions, in a field for example, the same flora develops but usually not in such a regular fashion, since weather changes may interfere in a drastic manner with the potential succession. Amongst Phycomycetes it is the Mucorales that provide the chief genera (*Mucor, Pilaria,* and *Pilobolus*) often attacked by parasites of the same order (e.g. *Piptocephalis Syncephalis,* and *Chaetocladium*), The coprophilous Ascomycetes are mostly very minute forms including Discomycetes (e.g. *Ascobolus, Dasyobolus, Coprobia, Ascophanus, Saccobolus, Rhyparobius, Thelebolus*) and Pyrenomycetes (*Sordaria, Delitschia, Philocopra, Sporormia,* and *Ckaetomium*). In Basidiomycetes the most characteristic coprophilous genus is *Coprinus,* many species being restricted to dung. Most of these fimicolous *Coprinus* spp. are small and in the latest stages in the development of the dung flora, there is often a forest of tiny fruit-bodies. Further, conidial fungi are commonly to be seen. These may be 'imperfect stages of Ascomycetes or Basidiomycetes, but a number of Fungi Imperfecti also occur. One of the commonest of these is *Arthobotrys oligospora* which catches and consumes the eel-worms that may occur in vast numbers in the dung.

In the development of this coprophilous flora there tends to be a succession. Phycomycetes come first, followed by Ascomycetes, leading to a final stage dominated by Basidiomycetes (*Coprinus* spp.). The phases, indeed, overlap, but the story of succession is usually fairly definite. However, there has been no careful study of the causes of this succession. Very probably the early appearance of the mucoraceous fungi is due to the rapidity with which they grow and sporulate, and the delay in development of agarics to relatively slow growth of the mycelium and the late stage at which sporophores are differentiated.

The general story of dispersal in most of the dung fungi would seem to be the same. The spores produced by the sporophores on the dung reach the surrounding grass, which may later be eaten by a horse, cow, or rabbit. The spores not only pass uninjured through the alimentary canal of the animal, but encounter there conditions stimulating germination, which takes place usually in the deposited dung. The spores of some coprophilous species will germinate only if treated to some of the conditions encountered in the intestine, such as a relatively high temperature, or an alkaline reaction, or both. However, some species can germinate without any such treatment. Spores of the dung *Coprini* need no special conditions, and, although the, germination of *Pilobolus* spores is greatly stimulated by a high temperature, a small percentage do germinate under ordinary conditions. Since the spores of some fimicolous fungi can germinate in nature without passage through an animal, it is clear that some of the species that make their appearance on dung may develop from air-borne or insect-borne spores, but there is little doubt that the normal channel for dispersal is through the animal.

So far as their apparatus of dispersal is concerned, some of the dung fungi seem little, if at all, modified in connection with their peculiar habitat. Little or no specialisation is to be seen in the coprophilous species of *Mucor, Coprobia, Chaetomium,* or *Coprinus,* but, on the other hand, species such as *Pilobolus* spp., *Dasyobolus immersus, Saccobolus* spp., and *Sordaria* spp. show, in the organisation of their spore discharge, beautiful and parallel adaptation. This specialisation is associated with the first stage of dispersal—the passage of spores from the dung to the grass. In these fungi the spores are discharged to such a distance (10-200 cm.) that they can reach the grass without the aid of wind. The distance *(d)* to which a microscopic spherical projectile is shot when discharged with a given initial velocity is given by: $d = Kr^2$ where K is a constant depending on the physical characters of the air and r is the radius of the projectile.

The long range of the spore-guns of specialised coprophilous fungi is due rather to the comparatively large size of the projectile than to a greater muzzle velocity. For such minute bodies as spores the resistance of the air is the principal factor limiting the height or distance of discharge, and the smaller the projectile the more serious is this factor, In some of the specialised coprophilous fungi the individual spores are small, but they are bound together by mucilage to form projectiles of larger size.

In some other species, notably *Dasyobolus immersus* and *Sordaria fimiseda,* the eight spores of the ascus are not only stuck together by mucilage but also are relatively enormous. In some of the more specialised fungi growing on the dung of herbivores the parallel evolution of spore-discharge equipment is outstanding. This may be best emphasised by considering the three common types: *Pilobolus, Dasyobolus immersus,* and *Sordaria* sp. In all the spore-projectile is relatively large, the spores being held together by mucilage. In all the distance of discharge is considerable: up to 200 cm. in *Pilobolus,* about 60 cm. in *Dasyobolus immersus* and some 25 cm. in *Sordaria curvula.* These distances of discharge are sufficient to carry the spores on to the surrounding grass without the aid of wind. In all discharge is by day and the spore gun is orientated by the phototropism of the sporangiophore in *Pilobolus,* the individual projecting asci in *Dasyobolus,* and the perithecium neck in *Sordaria.* This positive phototropism helps to ensure that the spore projectile shall be thrown clear of the substratum. In all the projectile is strongly adhesive, the mucilage cementing it firmly to the stems and leaves of the surrounding herbage. There is some evidence that light has an injurious effect on fungal spores, and it is therefore of interest to notice in these three fungi that in the spore-masses discharged on to the grass the protoplasm is shaded by the pigment of the spore-wall in *Dasyobolus* and *Sordaria* and by the black part of the sporangial wall in *Pilobolus* which effectively covers up the discharged mass of almost colourless spores cemented to a blade of grass.

A strong case can be made for adding *Sphaerobolus stellatus* to the list of highly specialised coprophilous fungi, although in Britain this common fungus is much more familiar as a lignicolous species. There is no doubt, however, that it may at times be coprophilous, and then its dispersal story is exactly like that *of Pilobolus.* It has nearly all the features of the specialised coprophilous species, with its large, adhesive spore-mass (peridiolum) projected to a considerable distance (5 metres) and aimed towards the incident light by the phototropism of the sporophore. Discharge is diurnal and the spores and gemmae in the projected peridiolum are screened from light by a thin, brown membrane. Another gasteromycete that belongs to the coprophilous flora is *Cyathus stercoreus,* fairly common on old manure in North America. The blackish, egg-like peridiola within the 'splash-cup' of the open fruit-body are scattered by rain on to the surrounding grass, where each sticks firmly by means of its gelatinous funiculus. No doubt the peridiola are eaten with the grass and the same cycle occurs as in most coprophilous species.

It must again be emphasised that by no means all the dung-fungi show a high degree of structural specialisation in relation to their peculiar environment. In species of *Ascophanus* and *Coprobia* the spores are colourless and are not glued together in the ascus to form a single mass, but are discharged so that they separate in the air, as in most non-coprophilous

Ascomycetes. In consequence the distance of discharge is only a few centimetres and they rely on wind to carry the spores to the nearby grass. *Thelebolusnanus* is an interesting example of a partially specialised form. This normally produces only one ascus at a time (probably daily for a few days) but it contains approximately 256 spores. These are bound together in a single spore-mass and are discharged to a height of about 7 cm. However, the spores are quite hyaline and unprotected from strong light.

Again, in the coprophilous agarics specialisation, if it may be said to exist, is very slight. The spores are liberated as in other gill-bearing fungi and wind is responsible for dispersal to the surrounding grass. It is, however, interesting to notice that coprophilous agarics have black *(Coprinus, Anellaria, Panaeolus)* or dark-brown *(Strophariasemi-globata)* spores, so that again the protoplasm is protected from light during the sojourn of the spores on the grass before it is eaten. Although black-spored agarics are by no means limited to the dung habitat, it seems unlikely that the regularity with which black or near-black spores are associated with coprophilous species is a matter of pure chance.

Just how the spores of the coprophilous Mucorales (especially *Mucor* spp. and *Pilaria anomala)* reach the grass from the dung is not clear. It has already been noted that most species of *Mucor* form sporangial drops. A projectile of *Pilobolus* might, immediately after discharge, brush against such a drop, picking up some spores which might then be carried as passengers to the surrounding herbage. That this does happen can easily be demonstrated by catching *Pilobolus* sporangia in mid-air on the surface of sterile nutrient agar, when a growth of *Mucor* frequently develops from the projectile, but this seems improbable as the main method of primary dispersal; perhaps splashing rain or insects play a part.

On the excreta of frogs a minute phycomycete, *Basidiobolus ranarum* occurs with great regularity. It seems to be on its own and no coprophilous *flora* develops as on the dung of herbivores. Interest centres on the fact that the fungus is so like a tiny *Pilobolus* although, systematically, there is a considerable gulf between them.

In its intestine the conidium, which is really an immature sporangium, completes its development and gives rise to eight sporangiospores. These are voided with the excreta and germinate to give mycelia there.

The hypogeal fungi form an interesting biological group. The fruit-bodies, varying from a few millimetres to several centimetres in diameter, are buried in the soil beneath the leaf carpet in woods. In most of them the story of dispersal would seem to be the same, but observations are few and experimental work is apparently completely lacking. In a number of these fungi the fruit-body gives off a smell which attracts rodents. These grub up the fruit-bodies and eat them. The spores no doubt pass through the alimentary canal and are distributed in the droppings. But these fungi are not coprophilous. They are soil fungi but some, perhaps all, enter into mycorrhizal relations with forest trees.

In hypogeal fungi convergent evolution has clearly occurred, the same type of fruit-body having evolved in Endogonaceae (a family of Mucorales), in Hymeno-gastraceae (a family of Basidiomycetes) and in two very distinct families of Ascomycetes (Elaphomycetaceae and Tuberaceae). Even in Tuberaceae there does not appear to have been a single line of evolution. Apparently the hypogeal types included in this family have arisen in three rather distinct ways from epigeal forerunners.

Perhaps the commonest British hypogeal fungus is *Elaphornyces granulatus* (hart's truffle) occurring both in deciduous and coniferous woods. Often the little excavations where rodents have grubbed for the fruit-bodies are to be seen and sometimes an abandoned half-nibbled specimen may be discovered left behind, no doubt, when the animal was disturbed. It is well known that the French, much more inclined to mycophagy than the British, train dogs and pigs to locate truffles, but which animals are mainly responsible for dispersal in nature does not seem to have been studied. Studies on the germination of the spores of these fungi might yield interesting results.

Hypogeal fungi are usually regarded as rather rare, but this is only because few have taken the trouble to look for them. In Britain many records were made last century largely due to the activities of Broome who, armed with a hand-rake, searched methodically for these fungi. In very recent years this practice has been revived by Dr. L. E. Hawker and her students. On the foray of the British Mycological Society at Bangor, North Wales, in 1950 an hour's raking by half a dozen members under Dr. Hawker's leadership brought to light under a single tree six different species of hypogeal fungi divided amongst Phycomycetes, Ascomycetes, and Basidiomycetes.

There is probably a certain amount of casual dispersal of resistant fungal spores which are able to pass uninjured through the alimentary canal of larger animals. Sometimes dispersal of this kind may be of some importance. An example is *Plasmodiophora brassicae,* causing club-root of crucifers. If diseased turnips are fed to cattle, the minute resting spores get into the dung. If this is used as manure viable spores are added to the soil and may cause infection in a cruciferous crop. For this reason many gardeners prefer to use artificial fertilizers rather than run the risk of introducing this fungus by the use of natural manure.

Birds have occasionally been suspected of transporting fungal spores and there is some evidence that they may have played a part in the disastrous spread of the bark disease of American chestnut in the second decade of this century. Plant pathologists in the United States made a thorough study of the dispersal of the causal organism *Endothia parasitica.* The local spread of the disease from tree to tree in a single wood was clearly due to airborne ascospores discharged from perithecia during wet weather, but it was thought that the occasional long-distance dispersal to remote plantations might be due to ascospores or pycnidiospores carried by birds, especially woodpeckers and tree creepers, on their breasts, tail feathers, feet, and beaks. Birds were shot in the neighbourhood of infected trees and a 'poured plate' technique was used in estimating the spore load on each bird. Two downy woodpeckers were each found to be carrying over half a million and a brown tree creeper over a quarter of a million viable spores. The general conclusion was reached that many of the local centres of infection isolated from the general area of chestnut bark disease might well have originated from spores carried by these birds.

No doubt water fowl are concerned in the dispersal of freshwater aquatic fungi. In spite of the fact that these fungi have no air-borne spores, the distribution of individual species is just as wide, if not wider, than that of terrestrial species. This raises the problem of the dispersal of these fungi from one isolated freshwater system to another, and there can be little doubt, in spite of the absence of direct evidence, that, as with other aquatic plants, water birds play an essential part in their long-distance dispersal.

It would be long and rather tedious to discuss in detail the influence of Man on the dispersal of fungi. Occasionally he deliberately aids in dispersal such as when he distributes mushroom 'spawn' or cultures of moulds for the manufacture of cheese, but usually dispersal is an incidental concomitant of his global activities from which he may suffer considerably. It is in the intercontinental spread of pathogenic fungi that Man's influence has been most marked.

Such vast expanses as the Atlantic and Pacific Oceans appear to offer formidable or perhaps insuperable barriers to the spread of fungal spores. But where Nature has failed Man has unwittingly succeeded. The story of plant pathology in Britain is punctuated by invasions from the Western Hemisphere, e.g. potato blight (introduced about 1840) and American gooseberry mildew (about 1900). And the North American Continent has received its share of pathogens from Europe such as the blister rust *(Cronartium ribicola)* of white pine (introduced about 1906) and Dutch elm disease (about 1930). Further at least one serious disease in North America, the bark disease of chestnut, appears to have originated in China or Japan.

The spread of fungal diseases from one continent to another has, apparently, been due largely to the introduction of infected material (e.g. nursery stock) of the host plant rather than to the accidental transport of individual fungal spores. Through bitter experience Man has now learnt that great care must be taken to prevent the spread of disease-producing fungi to new areas, and in most countries has taken steps, often fairly effectively, to guard against the introduction of further pathogenic species from abroad.

Seed-borne Fungi

A number of fungi are dispersed in an association of some kind with the dispersal units of the host. These units are usually seeds, or seed-like fruits such as the caryopses of cereals or the mericarps of umbellifers. Although seed-borne fungi have but a small place in the general picture of fungal dispersal, they are of great significance to the plant pathologist since many important plant diseases are seed-borne.

There are a number of endophytic fungi, causing systemic infections of their hosts, that are normally seed-dispersed and, at least under natural conditions, do not seem to be dispersed in any other way. Thus in members of Cistaceae an endophytic fungus is regularly present. The story has been studied in some detail in *Helianthemum chamaedstus.* The fungus is found in root and shoot, and when flowers are produced enters the ovary wall. The outer gelatinous coat of the testa is infected, but the fungus is prevented by a barrier of tannin-filled cells from passing into the embryo and endosperm. Infection of the seedling from the dormant mycelium in the testa occurs at germination and under natural conditions for continued growth such infection seem to be necessary. Much earlier work by Rayner on the endophytic mycorrhiza of *Calluna vulgaris* and other ericads gave a picture essentially like that in *Helianthemum.* However, workers on ericaceous mycorrhiza have reached conflicting conclusions. There is no certainty that infection of *Calluna* is systemic nor that the *Phoma* isolated from the seeds is responsible for the mycorrhizal roots. All is obscure and the whole subject is in need of critical reinvestigation.

Another fungus that apparently relies entirely on seed dispersal is the well-known endophyte of *Loliwn* spp. which does not seem to be a symbiont but merely a harmless parasite. The fungus forms an intercellular mycelium throughout the shoot system of the host. In due course, the ovary is infected and the caryopsis carries a dormant mycelium in the testa. With some difficulty the fungus has been isolated in pure culture, but no spores were produced so, as with the endophyte of *Helianthemum* its systematic position is unknown. From *Lolium* a second endophyte can sometimes be isolated which grows well in culture and produces microconidia. Although most specimens of *Lolium* are infected some are free from either endophyte. On making reciprocal crosses between infected and uninfected plants it has been shown that the fungus is 'inherited' only through the ovule.

Another interesting fungus that is sometimes seed-borne is *Epichloe typhina,* the choke of grasses. This attacks a number of species belonging to several genera causing systemic infection. In *Dactylis glomerata* flowering is prevented in affected plants and the disease cannot, therefore, be seed-borne. However, when *Festuca rubra* is attacked it still produces seed and these may carry the fungus as a resting mycelium in the embryo and endosperm. Further, plants of *F. rubra* may show no external manifestations of disease, no stroma being formed. Where this type of latent infection occurs the fungus is dispersed solely with the seed of the host and a condition of affairs very like that of *Lolium* and its endophyte is attained.

Amongst important plant diseases that are normally seed-borne, the smuts of cereals are outstanding. Most smuts are systemic with the fine intercellular mycelium extending throughout the host plant. Spore production, however, is usually localised in and about the reproductive organs. In cereal smuts, the ovary is the chief centre of spore production and no good grains are formed by infected plants, but instead of the grain there is formed a covered or uncovered mass of blackish brand spores. In stinking smut (bunt) of wheat caused by *Tilletia caries,* the wall of the caryopsis is unaffected but the interior is filled with black spores smelling of decaying fish due to the production of trimethylamine. During harvesting, these grains are broken and the brand spores get dusted over healthy grains where they remain dormant until sowing time.

The grains of wheat and the brand spores germinate at the same time and so infection of the seedlings is assured. Here the fungus is carried merely on the surface, so that the disease is easily controlled by shaking the grain in a drum with an organo-mercurial dust. It is of interest that, although this fungus is normally seed-borne, it is not necessarily so. Thus where an infected crop is threshed on the spot, spores may settle on the nearby soil, remain dormant but viable, and infect clean seed sown in the following season.

In *Ustilago nuda,* the loose smut of wheat and barley, the story is rather different. Spikelets of diseased plants are reduced to an exposed mass of brownish-black brand spores at a time when the uninfected plants are in flower. The dry powdery brand spores are wind borne and may reach the stigmas of healthy plants, but instead of being resting-spores they germinate at once to produce a mycelium which penetrates the ovary and enters the developing seed. The seeds, however, are not destroyed but simply contain the fungus in a resting condition, so that at germination the seedlings are already infected. In such smuts surface sterilisation of the grain is useless and control depends on the fact that the fungus is more sensitive to heat than the embryo of the seed, so that steeping in hot (50-52° C.) water for a short time (about fifteen minutes) kills the parasite without greatly affecting the viability of the grain.

Another group of cereal diseases that are typically seed-borne are the 'stripe diseases' caused by *Helminthosporium* spp. *H. avenae* is responsible for leaf stripe and seedling blight of oats. Except for a very local spread of the disease due to airborne conidia during the growth of the crop, dispersal is dependent on infected grains which have a Turner and Millard (1931).

Seed-borne fungi are specially prominent in the Gramineae, but they are also common in other families of flowering plants. Some of the diseases of pulses are seed-borne. A well-known example is anthracnose of french and runner beans due to *Colletotrichum lindemuthianum.* On all parts of the shoot this fungus produces small circular lesions in association with which slimy conidia are formed and are dispersed locally by rain-splash. Where the pods are infected the parasite may spread to the developing seed which ripen with the fungus deeply situated in the tissue of the cotyledons. The appearance of disease in the following season is due almost entirely to the use of infected seed. The fungus is too far embedded in the embryo for seed sterilisation to be effective, and control consists largely in the use of clean seed. The story of leaf, stem, and pod spot of peas caused by *Ascochyta pisi* is very similar.

Flax is of special interest because its principal fungal diseases are seed-borne. Thus seedling blight caused by *Colletotnchum linicola* is carried over from one season to the next as a resting mycelium in the outer layers of the seed-coat. Stem break and browning due to *Polyspora lini* is transmitted in much the same manner, and this is also true of the grey mould *(Botrytis cinerea)* which may hibernate as a resting mycelium in the outer layers of the testa. However, *B. cinerea* is a very unspecialised and common parasite occurring on a large range of hosts and also as a saprophyte on dead plant material, but the early infection of the flax seedlings is commonly from the seed-borne mycelium, and only in connection with later infections are other sources of inoculum important. Other flax diseases such as foot rot, due to *Phoma* spp., and wilt, caused by *Fusarium lini,* may also be seed-borne, but infection of the new crop is often directly from the soil. In Northern Ireland, a great flax-growing centre, the examination of flax seed for the presence of seed-borne fungi has become an important aspect of seed-testing. In the method used (Ulster method) a random selection of seeds from a sample is placed on sterile malt agar in petri dishes. These are incubated at 22° C. and examined five days later. If a seed carries one of the important pathogens that are capable of rapid growth on *Colletotnchum linicola, Polyspora lini, Phoma* sp., *mtrytis cinerea* and *Fusarium lini,* a characteristic colony develops on the medium around the Muskctt and Colhoun (1947).

Seedling (Plate VIII): Although the seeds may carry surface spores of saprophytic moulds (especially *Penicillium* spp.), their growth does not, in practice, interfere with this simple and effective test, which can be carried out on a large scale and gives a clear picture of the degree to which the seed is contaminated and of the nature of the pathogens present. In Holland also examination of seed samples of crop plants for the presence of parasitic fungi has become an important part of seed-testing.

So far examples have been considered where the fungus is seed-borne as a resting mycelium or as spores or both, but sometimes a whole fruicpig structure is involved. This is so in *Septoria apii,* which causes leaf-spot of celery. All parts of the shoot may be affected, causing Doyer (1948).

The developing fruits are often infected and on the mature mericarps ('seeds') pycnidia are frequently formed. These remain alive but dormant when the 'seeds' are dry, but when they are soaked and germination occurs, spores ooze out from the pycnidia and infect the seedlings.

Apart from those diseases in which the fungus is actually carried in some way or other on or in the viable seed, there are a number of fungal diseases where the pathogen, though not actually attached to the living seed, tends to be dispersed as a contaminant of the seed sample. This type of dispersal is somewhat unnatural, but may be of considerable economic importance. For example, a sample of rye seed may contain sclerotia of ergot *(Claviceps purpurea)* and at sowing-time these are scattered with the grain; Again, in the blind-seed disease of *Loliumspp.* caused by *Phialea temulenta* infected seeds are not viable, but in a commercial sample may be mixed with healthy ones.

Both types are sown together and the 'blind' seeds may give rise to apothecia from which ascospores are discharged, infecting the ovaries of rye-grass and darnel and thus leading to the formation of more 'blind' seeds. Even some rust fungi may be 'seed-borne' by being included in a commercial seed sample. When seed from a rusted flax crop is harvested teleutospores *of Melampsora lini* attached to small fragments of fruit-walls and leaves get mixed with the seed. In next season, these fragments are scattered with the seed and in due course the teleutospores germinate to produce basidiospores which are discharged and infect the young flax plants.

The various types of seed-borne fungi have been illustrated by reference to a few scattered examples, but there are very many more, and plant pathologists are becoming increasingly aware of the importance of this type of dispersal in agricultural practice. How far seed dispersal is of importance for fungal pathogens attacking plants under natural conditions is not yet known. Harley reports the examination of seeds of a number of wild and garden plants removed with due precautions from unopened capsules and left overnight in sterile water. He says: 'Excluding Cistaceae and Ericiceae, in all of which fungi occurred, eleven out of twenty-five species were seen to bear hyphae on some of the seeds.'

A consideration of seed-borne fungi shows all types of association, from the most casual one in which the spores simply adhere to the surface of the grain (e.g. *Tilletid caries)* to the most highly organised where the fungus has ceased to sporulate and relies entirely for its dispersal on a seed-borne mycelium (e.g. the endophyte of *Lolium).*

From the point of view of disease prevention the location of the parasite in the seed is of great importance. Surface infection, with the spores or resting mycelium on or near the surface of the seed or seed-like fruit, is the commonest type and can usually be tackled successfully by seed treatment with a suitable fungicide. However, where deeper tissues are affected disinfection cannot be achieved in this way. Luckily, deep-seated infection is rarer. In *Ustilago nuda,* as already mentioned, seed sterilisation can be effected by hot-water treatment, but this is not a method of general application. In anthracnose of beans, for example, the only efficient method of control is to avoid the use of seed from a diseased crop.

The types of fungi that are normally seed-borne, in the sense that the fungus is actually in or on the seed, are rather limited. Amongst Basidiomycetes they are smuts. Otherwise, they are mostly Fungi Imperfecti or Ascomycetes in the 'imperfect' stage. The main groups of truly obligate parasites (i.e. Peronosporaceae, Erysiphales, and Uredinales) are not usually seed-borne. It is interesting to note that in all these there is not only a well-developed airborne stage, but also an efficient hibernating stage. Seed-borne infection as often as not serves rather as a method of hibernation than of dispersal.

The plant pathologist often extends the conception of seed-borne fungi to include those carried by vegetative reproductive units or 'planting units' such as tubers, bulbs, and runners. But although dispersal of this nature may be of great practical significance, it has little importance In nature, and comes under the heading of 'Man as an agent in the spread of fungi'.

Water and Dispersal of Spores

It is necessary to distinguish between the part played by water in the dispersal of terrestrial fungi and the state of affairs encountered in the true aquatic fungi, where water is the medium in which the spores are dispersed. In the development of all land fungi water is a factor of utmost importance, and without an adequate supply growth and sporulation are impossible. Again, in fungi in which the spores are violently discharged, spores liberation can occur only whilst the cells of the hymenium remain turgid, and the supply of water to maintain this turgidity is vital for spore liberation.

However, in land fungi water may play a more direct part in dispersal and the kinetic energy of moving water may be used in scattering the spores. This is probably so in a large number of slime-spore fungi. For example, *Colletotrichum lindemuthianum,* the cause of anthracnose in dwarf and runner beans, produces its spores on the pods in pink, slimy masses. Heavy rain falling on the diseased pods may bespatter the neighbouring healthy pods with spores of the pathogen and so spread infection. Another example, to which reference has already been made, is the conidial *(Fusicladium)* stage of apple scab (*Venturia inaequalis).*

Although they are 'dry-spores', the conidia are too firmly attached to their conidiophores to be liberated by the wind alone, but separate easily when wetted and are apparently normally dispersed by wind-driven rain. Again, in the puff-balls *(Lycoperdon* spp. and *Geaster* spp.) heavy rain, or large drops of water dripping from the foliage of trees, may bring about the liberation of visible puffs of spores. A large drop striking the dry, papery, unwettable peridium, locally and momentarily depresses it, with the result that the fungus behaves like bellows and there issues a puff of spore-laden air—a puff bearing millions of spores.

In all these examples the relation of rain-splash to dispersal is somewhat casual and it can hardly be said that the fungal structure has become modified in connection with splash dispersal. However, in the bird's-nest fungi *(Cyathus* spp. and *Crucibulum vulgare)* there is a strong suggestion that the open fruit-body with its contained peridiola has a special survival value as a 'splash-cup'. Drops of water falling from a height into the open fruit-body are broken up and reflected from the interior of the splash-cup, often carrying peridiola with them to a distance of several yards. This method of dispersal can easily be demonstrated by allowing drops of water to fall from a height into an open fruit-body of *Crucibulum vulgare* or *Cyathus* spp. A peridiolum striking an object becomes firmly attached because of the sticky nature of its funiculus.

In the field, around fruit-bodies of *Cyathus striatus,* the bird's-nest fungus with the most specialised splash-cup, dispersed peridiola can often be seen hanging by the funiculus from nearby leaves. Splash-cups are not limited to these fungi, for the same mechanism has arisen in the liverwort *Marchantia polymorpha* where the gemma-cups are essentially similar, and perhaps also the soredia-lined podetia of the lichen *Cladonia pyxidata* may be interpreted in the same way.

Trickling rainwater may also play a small part in the spread of some fungi. In the American chestnut bark disease caused by *Endothia parasitica,* spread of infection from tree to tree is usually due to ascospores violently discharged into the air. However, in addition, the fungus produces pycnidiospores which ooze from the pycnidia in slimy masses and are spread by rain trickling over the surface of the trunk, perhaps causing further infections in the same tree. Again, rain percolating through the soil may be of importance in the dispersal of some species. In *Phytophthora, infestans* the conidia produced on the leaves of diseased plants may fall on to the soil below and be washed down by rain to infect the potatoes below. As the water percolates downwards, spores will tend to be stranded on soil particles and on organic debris. Consequently tuber infection can be much reduced by thorough earthing up of the plants, so that spores are more likely to be filtered out of suspension before the rainwater reaches the tubers.

The downward movement of fungal spores in a sandy soil has been studied experimentally and it has been found that whilst wettable slime-spores such as those of *Zygorrhyncus vuillemini* and *Gliomastix convoluta* are readily washed down through the soil, the spores of *Penicillium* sp., which have non-wettable coats, remain at the surface and are not carried down by percolating water. In conformity with this result it appears that the fungi isolated from the lower layers of soil tend to have slime-spores, whilst species isolated at or near the soil surface tend to be dry-spored.

Apart from its actual surface the soil is a more or less aquatic environment, and it therefore causes no surprise to find some true water moulds (e.g. *Achlya* spp., *Saprolegnia* spp., and certain chytrids) as normal members of its fungus flora.

There are a number of fungi which, while themselves essentially terrestrial, grow on waterside or reed-swamp species of higher plants. Reproductive units of these fungi may be dispersed on the surface of the water. For example, in Sclerotinia a parasite of *Scirpus lacustris,* the sclerotia formed in the old stalks are liberated into the water when these break up, and are later to be seen floating on the surface and stranded on the shores of lakes. Floating and stranded sclerotia produce aerial apothecia in the Spring, from which discharged spores are airborne to their hosts.

In the completely aquatic fungi the spores are produced, liberated, and normally dispersed below water. Fungi are predominantly land organisms with between 50,000 and 100,000 'good' species so far described, but of these only about 2 per cent, are aquatic. In the majority of aquatic fungi, the whole evolutionary story seems to have been in water and they are branded as primitive aquatics by the possession of zoospores. On the other hand, there are a number of submerged aquatic Ascomycetes and Fungi Imperfecti which appear to have migrated from land to water. No basidiomycete is known to have taken this seemingly retrograde step.

Amongst the primitive water fungi chytrids are the most numerous, but the true water-moulds (Saprolegniales) are the most familiar. In all these Phycomycetes the zoospore is the normal unit of dispersal, and quite clearly such a spore is suitable only for the aquatic environment.

It is interesting to consider the biological value of the zoospore as contrasted with a non-motile spore. So far as actual transport is concerned the motility of the zoospore has little advantage over the non-motile spore, since the actual movement of water in a pond, lake, or stream is probably great in comparison with the feeble efforts of a zoid. The value of the zoospore would seem to reside in its power to 'select' the substratum on which it settles by virtue of its taxisms.

Amongst the aquatic fungi derived apparently from terrestrial ancestors are Ascomycetes (Discomycetes, Pyrenomycetes, and some lichens) and Fungi Imperfecti, especially Hyphomycetes.

One striking and abundant flora of aquatic Hyphomycetes is regularly to be found on submerged, decaying leaves of deciduous trees and shrubs in well-aerated water, particularly in small streams that are not too sluggish. They occur especially on the petioles and larger veins, the branched septate mycelium ramifying in the decaying leaf tissue and the conidiophores projecting at right angles into the water. The whole story of dispersal is strictly aquatic, the spores being produced, liberated, and dispersed below water. Some twenty-five species of these fungi have been described and most of them are very common.

In aquatic Hyphomycetes the spores are hyaline and mostly of a somewhat unusual form. Most species have branched spores and in these the commonest type is a spore consisting of four straight arms diverging from a common point. A study of spore development shows clearly that the mode of formation of the tetraradiate conidium differs markedly in different genera, suggesting strongly that parallel evolution has occurred and that this type of spore has some significance—some survival value—in the aquatic environment. This diversity of development of the tetra-radiate spore may be illustrated by considering certain examples.

In *Clavariopsis aquatica*, the spore-primordium is a club-shaped cell separated terminally from its conidiophore by a cross-wall. Then from the obtuse apex of this cell there appear simultaneously three little outgrowths which develop into three narrow arms, the original primordium forming the fourth, the shortest and fattest arm of the spore. As soon as the conidium is fully mature it separates from its conidiophore by disarticulation at the separating cross-wall.

In *Tetrachaetum elegans,* the spore-primordium, as in *Clavariopsisy* is a single straight cell terminal on a simple conidiophore. The primordium increases by apical growth and, at a certain stage, two little protuberances appear side by side just behind the growing-point. The appearance of these causes the growing apex of the primordium to be deflected into a new direction, an angle of about 120° being formed between the original and the later directions of growth. The two protuberances and the original apical growing-point grow at the same rate until the mature four-armed spore is produced with all its arms of much the same length and width. By the time, the spore is fully grown a small separating cell has been cut off at the end of the conidiophore, and it is by the breakdown of this that the spore is liberated.

In the genus *Articulospora* two species have been described. *A. tetracladia* is very common but *A. infiata* is rarer. Both are characterised by the same type of development in which the four arms of the spore make their appearance in strict succession. In *A. inflata* thc spore usually occurs singly at the end of a simple conidiophore. As in *Tetrachaetum* the spore-primordium is a straight cell with apical growth marked off by a wall from its conidiophore. When it reaches a certain length a cross-wall is formed delimiting a lower cell, which is the first-formed arm of the spore, from an upper which after further growth forms the second arm. Then, just below the wall separating these two arms, a small protuberance develops and begins to grow into the third arm. Much as in *Tetrachaetum* the development of this protuberance has the effect of bending the original straight hypha formed by the first and second arm. Still later a further outgrowth occurs below the septum dividing the first arm from the second and develops into the fourth arm. When all four arms are of approximately the same length the spore separates from its conidiophore as in *Clavariopsis.*

In *Clavariopsis, Tetrachaetum* and *Articulospora* the mature spore is attached to the conidiophore by the tip of one of its four arms, but in two genera *(Lemonniera* and *Triscelophorus)* the attachment is at or near the point of divergence of the arms. Thus in the very common species *Lemonniera aquatica* the long conidiophore that juts out into the water is usually sparingly branched, terminating in a small group of spore-producing cells or phialides. The spore starts as a spherical primordium blown at the end of a phialide, but soon becomes tetrahedral, each angle of the tetrahedron being a growing-point. Growth at these points takes place simultaneously and at the same rate until the arms are fully developed, when the spore separates, but only at the last moment is there formed a cross wall between the spore and its phialide. As soon as one conidium is liberated another begins to form from the phialide. In *Lemonniera,* the conidium is a phialospore whilst in *Clavariopsis, Tetrachaetum* and *Articulospora* it is a terminal thallospore. However, the liberated spores of *Lemonniera aquatica, Tetrachaetum elegans,* and *Articulospora infiata* are so much alike that it is only after some experience that they can be told apart.

Triscelophorus monosporus agrees with *Lemonniera* in having its spore attached to its conidiophore near the point of divergence of its four arms, but differs in having a spore that is a terminal thallospore and in the fact that the arms arise in succession. The conidiophore is short and usually unbranched. The tip swells to form a spherical primordium delimited by a cross-wall. In this primordium apical growth is established, it becomes obclavate, and divides into an upper and a lower cell. By continued growth the upper cell develops into one long, straight arm of the spore whilst the others, arising in succession, bud out from the lower cell. When mature the tetraradiate terminal thallospore separates from its conidiophore.

These five examples of the different ways in which tetraradiate spores may arise suffice to support the view that this unusual kind of spore has evolved along a number of different lines of evolution. But other examples could be given. Thus in *Heliscus longibrachiatus* spore development is like that of *Clavariopsis*, and *Alatospora acuminate* like *Tetrachaetum*, but in both species the spore is of the phialospore type, not a terminal thallospore.

It is of considerable interest that *Orbimyces spectabilis*, a marine species, has branched conidia rather like those of the freshwater aquatic Hyphomycetes. Further, in *Sphacelaria*, a genus of brown seaweeds, certain species reproduce largely in a vegetative manner by propagules, which are branches of limited growth and definite form that are shed when mature. In *Sphacelaria fuscl*, these are tetraradiate and show an extraordinary agreement with the spores of such Hyphomycetes as *Tetrachaetum*, *Articulospora*, and *Lemonniera*.

The tetraradiate reproductive unit is not limited to aquatic organisms. It occurs, for instance, in *Tetraploa aristata* whose spores, as recorded long ago by Cunningham, may be airborne. However, in aquatic Hyphomycetes the tetraradiate spore is of such frequent occurrence that it is natural to suppose it has some biological significance in the aquatic environment. What this significance may be is pure speculation.

Perhaps a spore of this kind settles relatively slowly in water and so stands a good chance of adequate dispersal; perhaps, on the other hand, it acts as an anchor and readily.

Another type of spore common amongst the Hyphomycetes of submerged decaying leaves is the long filamentous or vermiform spore, twisted so that its curvature lies in more than one plane. Two species are very common: *Anguillospora longissima* where the conidium is a terminal thallospore, and *Flagellospora curvula* where it is a phialospore. Very probably the long filamentous spore has the same biological significance as the tetraradiate type, and it is of interest that other aquatic fungi, for example the freshwater discomycetes *Apostemidium guernisaci* and *Vibrissea truncorum*, and the marine pyrenomycete *Halophiobelus salinus* also have spores of this shape. However, it must be borne in mind that many terrestrial fungi have long, thread-like spores (e.g. *Cordyceps militaris* and *Rytisma acerinuni*). However, in this connection it is a suggestive fact that in *Zprtera*, the only British genus of flowering plants with pollination occurring below water, the pollen grains are long and filamentous.

There is another widespread flora of aquatic Hyphomycetes which occurs on submerged leaves decaying under conditions that appear to be almost anaerobic. Leaves of deciduous trees dredged from the bottom of stagnant ponds or lakes have a characteristic black appearance.

The fungi concerned belong to such genera as *Helicodendron, Helicoon,* and *Clathrosphaerina.* Apparently, in the submerged leaves the mycelium grows, though somewhat slowly, in spite of the anaerobic or nearly anaerobic conditions. However, sporulation can occur only in the presence of a gas phase, but the gas is not necessarily air, since, at least in some species, growth and sporulation can take place in pure cultures on malt agar grown in an atmosphere of nitrogen. These particular fungi of submerged decayed leaves are not confined to the aquatic habitat, though that appears to be their characteristic environment. They have also been found under sub-aerial conditions on damp, rotten wood.

The most striking feature of the spores is that they are not readily wettable. They float on water, helped by the fact that the spores are built in the form of a helix or spherical network in which air is enmeshed and not easily displaced. In nature it is probable that sporulation occurs when blackened leaves containing the mycelium of these fungi are exposed, as, for example, around the edge of a drying pond. When the water-level again rises the leaves are submerged but the spores are left floating. Some species (e.g. *Helicodendron triglitziensis)* can sporulate even when the leaves are submerged so long as the water is only a few centimetres deep. Under these conditions fine hyphae grow up to the surface and produce spores there. Although wind dispersal of the spores of these fungi can obviously occur, it seems likely that they are usually dispersed on the surface of the water. Falling on to the water and floating there, leaves become infected, and when they become waterlogged and sink growth of the fungi continues.

In addition to Hyphomycetes, quite a large number of minute Ascomycetes are submerged aquatic fungi. These are so widely scattered in any 'natural' scheme of classification that it seems clear they are not primitive aquatic fungi, but are almost certainly derived from terrestrial-ancestors.

In the freshwater habitat minute Discomycetes and Pyrenomycetes are common on submerged dead stalks of reed-swamp plants (e.g. *Phalaris arundinacea, Phragmites communis, Scirpus lacustris, Equisetum fluviatile)* and on submerged, waterlogged twigs and branches of trees. These fungi are little different from those occurring on land, but the spores tend to have certain features that seem to stamp them as belonging to aquatic organisms. Thus the ascospores may have well-developed mucilage sheaths or appendages. They are often considerably elongated and are almost always densely crowded with minute oil globules. Perhaps here again spores with these features may settle comparatively slowly, and perhaps the mucilage may be important in connection with anchorage.

It is to be noted that asci, whether arranged in an apothecium or a perithecium, can function quite well under water. On the other hand, the basidium is an apparatus suited only to sub-aerial conditions, and it is a striking fact that submerged aquatic Basidiomycetes are unknown. On the whole, spore liberation is similar in aquatic and land Ascomycetes. Thus, in the submerged aquatic *Ophiobolus typhae*, a saprophyte on the leaf-bases of *Typha latifolia*, spore liberation from the perithecium is of the detached-ascus type already described for terrestrial Pyrenomycetes.

Again, in *Pleospora scirpicola*, in which spore discharge was illustrated nearly a hundred years ago by Pringsheim, the mechanism is of the jack-in-the-box type. Although Pringsheim gave such a clear and beautiful account of successive spore discharge from the ascus in this common aquatic fungus, he failed to note the extraordinary development of the mucilage sheath around the spore.

This can best be seen if freshly liberated ascospores are mounted in Indian ink. When the spores are first freed from the ascus the sheath is comparatively small and is bounded by a firm membrane, but following discharge the mucilage rapidly absorbs water so that a long, worm-like strand of fluid mucilage, about half a millimetre long, is formed containing the large spore. Again, in the minute pyrenomycete *Loramyces juncicola,* occurring on submerged dead stalks of *Juncus spp.* and *Scirpus lacustris,* the two-celled ascospore is surrounded by a thick mucilaginous sheath and, in addition, the lower cell is prolonged into a long, straight tail, so that at a glance the spore looks like a large chytrid zoospore.

In these two fungi and in others like them the mucilage is of a very fluid nature and after a short time the sheaths disappear completely, but in *Ceriospora caudaesuis* a saprophyte on submerged sticks of ash, there is a coiled, tendril-like appendage of persistent mucilage at each pole of the spore. Turning to the marine Ascomycetes, there are two distinct biological types. One type belong those species that grow as saprophytes on submerged wood and cordage: species of the other type are parasites of the larger seaweeds. Knowledge of the saprophytic species comes mainly from the work of Barghoorn and Linder.

The saprophytic lignicolous forms show a striking parallelism with the freshwater Ascomycetes. In some species *(Halophiobolus* spp.) the spores are long and thread-like and in addition are furnished with adhesive tips. In others, although the spores are not thread-like, they have various types of mucilaginous appendage.

Linder suggests that these modifications 'appear to serve both for the attachment of spores to the substratum, as well as to assist in suspending the spores hi water and facilitating their dispersal. He notes how successfully the sticky extremities attack the spores of *Halophiobolus* spp. to an object, and because of this it is very difficult to discharge spores from a glass pipette. A feature noted by Linder in all the saprophytic marine Pyrenomycetes was the breakdown of the ascus wall at maturity. Apparently in these marine species the original method of discharge has been lost and the spores ooze out in a mucilaginous mass. This type of spore liberation may occur in the freshwater Pyrenomycetes, but it is by no means general.

Knowledge of the Pyrenomycetes that attack the larger seaweeds is due largely to the work of Cotton and of Sutherland. Some of these fungi also have long, worm-like spores, for example *Ophiobolus laminariae*, a parasite found on the stipe of *Laminaria digitata* and *Trailia ascophylli*, which attacks *Ascophyllum*. One of the most interesting of these marine species is *Mycosphaerella ascophylli*. This causes a systemic infection of *Ascophyllum* and apparently it is invariably present in the host plant. However, the production of perithecia is limited to the receptacles. Ascospores of the parasite and eggs of the seaweed are liberated at the same time, and no doubt infection of the young seaweed occurs at a very early stage in its career, although the story of infection has yet to be told.

Much remains to be discovered concerning the dispersal of water fungi. Further information is needed about the mechanism of spore liberation from the perithecia of aquatic Ascomycetes, especially those in which deliquescence of the asci occurs. Observations on the rate of fall of the spores in water might yield interesting information, and a study of how readily spores adhere to underwater objects would be of special interest.

Special Problems

In the development of methods for control of root disease in plantation crops, primary emphasis has been laid upon eradication of the sources of infection. In the earlier days of plantation pathology, the complete removal of all buried timber from the plantation site before planting up was sometimes advocated, but this was generally admitted to be a counsel of perfection. A revolutionary change in methods of root disease control was brought about by NAPPER (1932-34) at the Rubber Research Institute of Malaya when he proposed that eradication of root disease fungi should be carried out under the young plantation crop, instead of before planting; a great economy in labour was effected by this new routine treatment, inasmuch as excavation was limited to actual instead of potential sources of root infection. A further advance was made by LEACH (1937) in Nyasaland, when he demonstrated that the development of many potentially dangerous sources of infection after felling of the jungle could be prevented by ring-barking the jungle trees well in advance of felling. LEACH'S discovery was followed by NAPPER'S (1939-40) development of the tree and stump poisoning methods in Malaya for the same purpose.

The amount of space thus far devoted to a description of these sanitation methods reflects their relative importance to the plantation grower. The usefulness of soil fungicides for root disease control in plantations is strictly limited, though such fungicides may sometimes be used in conjunction with other control methods, as in the employment of copper sulphate solution by NAPPER as described above.

Soil fungicides have also been used, though not with conspicuous success, in various attempts to eradicate root disease fungi introduced into "clean" areas. On the whole, however, phytocidal chemicals, such as the sodium arsenite used by NAPPER (1940) for stump poisoning in Malaya, and the heavy gas oil employed by SMITH (1932) for destruction of banana stumps in Jamaica, are probably of more importance to the plantation grower than are fungicidal chemicals. TUNSTALL (1926) has emphasised the difficulties of eradicating root-infecting fungi ensconced within bulky pieces of timber buried at some depth in the soil by any method except that of excavation. Nevertheless, soil fumigation with carbon disulphide, applied through an injector, has given promising results in attempts at eradication of *Armillaria mellea* from the soil of Californian citrus orchards (THOMAS, 1938; THOMAS and LAWYER, 1939; GRIFFITH, 1941).

When choosing a site for establishment of a plantation crop, it is well to remember that some soils are more favourable than others for the development of a particular root disease. Some advantage of this fact has been taken in banana growing, following REINKING'S (1935) discovery that Panama disease *(Fusarium oxysporum cubense)* is more prevalent on soils of light texture; according to WARDLAW (1941), heavy alkaline soils are least favourable to the disease.

Low-lying sites, unless properly drained before planting, will predispose a plantation crop to attack by such a parasite as *Sphaerostilbe repens,* the cause of violet root rot. Concerning the relation between incidence of root disease and soil conditions, less evidence is available for plantation crops than for field crops. Developments in the manuring of field crops have outstripped those in the manuring of plantation crops.

Observations and experiments on the relation between manuring and incidence of root disease have therefore, been made almost exclusively on field crops; only few and casual reported observations, such as that of DE JONG (1933) concerning the slightly greater incidence of *Fomes lignosus* on rubber in manured plots, can be found on the relation between manuring and occurrence of root disease in plantation crops.

Certain root diseases, to which further space is devoted below, require special precautions or treatment for their prevention or cure. Such are the root diseases due to species of *Rosellinia,* which are favoured by a humid atmosphere and accumulation of humus, diseases due to fungi that are primarily wound parasites, such as *Ustulina sonata,* diseases in which root infection, by fungi like *Sphaerostilbe repens,* cannot occur in the absence of some such predisposing factor as waterlogging, and diseases, such as internal root rot of tea, formerly attributed to *Botryodiplodia thcobromac,* in which fungi are purely secondary. These diseases, and the problem presented by the introduction of a root disease fungus into a "clean" area, are more fully considered below.

Root Diseases Requiring Special Treatment

Root diseases due to Rosellinia spp. The growth habit and parasitic activity of these fungi have already been described (pp. 50 and 61); it remains to discuss the special control measures appropriate to such a fungal growth habit. Little has been added to the recommendations made 'by NOWELL (1923) in the West Indies and by FETCH (1923) in Ceylon. Accumulations of dead leaves and other vegetable litter, which favour the saprophytic spread of these fungi, must be cleared away, leaving the ground, in FETCH'S phrase, "as bare as a tennis court." Ventilation should be improved by severe pruning of bushes or trees surrounding the diseased individual or group. As WATERSTON (1941) has pointed out, however, parasitic activity of *Rosellinia* is not confined to those situations in which the fungus can make a rapid saprophytic surface spread amongst vegetable litter under dense shade and high rainfall; *Rosellinia* may occur as a parasite under conditions of lighter shade and lower rainfall, with little or no accumulation of surface litter.

In the latter type of situation, spread of the fungus is more or less confined to movement along the roots of its host plants; such subterranean spread as a parasite is necessarily slower than the rapid surface spread as a saprophyte. WATERSTON investigated the occurrence in such situations of a root disease due to *R. pepo* on cacao estates in Trinidad and Grenada, but was unable to correlate its incidence with soil reaction, or with organic matter, nitrogen, phosphate, or potash content of the soil. The disease occurred chiefly on soils of light texture a limitation attributed by WATERSTON to the aeration requirements of the fungus.

Charcoal Root rot Due to Ustulina Sonata

The fungus causing this disease, which is well known on tea and rubber estates in Ceylon, and on rubber plantations in Malaya (FETCH, 1921, 1923; SHARPLES, 1936), spreads from plant to plant underground by root contact, in the same way as *Fomes lignosus, F. noxius* and *Ganoderma pseudoferreum. U. sonata* is also a well known stem parasite, gaining entrance through spore infection of wounds, so that certain special precautions must be observed when dealing with it. SHARPLES (1936) remarks that all authorities consider the collar to be the commonest seat of infection. Whereas in Java emphasis has been laid on infection through collar wounds made during changkolling (hoeing), in Malaya, it is thought that accumulation of scrap rubber at the base of the tree provides suitable conditions for infection by *U. sonata,* through suffocation of the cortical tissue beneath such airtight pads. Control measures should include prompt removal of felled rubber logs on which the fungus may develop as a saprophyte and later produce fructifications, periodical removal of scrap rubber accumulating at the base of the stem, and avoidance or adequate protection of wounds.

Violet Root rot Due to Sphaerostilbe Repens

Infection by this fungus can apparently occur only under certain soil conditions that lower the resistance of the roots; in Malaya, according to SHARPLES (1936), the disease is most frequently seen on rubber areas that have been flooded. TUNSTALL (1940) remarks that violet root rot is usually associated with badly aerated, stiff soils in the tea plantations of North East India. In exceptionally sticky soils, single lines of good plants may sometimes be seen along the edges of drains, with the intervening bushes poor and most of them severely attacked by *S. repens.* TUNSTALL also observes that excessive defoliation of the tea bushes through plucking and pruning may lead to such a lowering of the vitality of the roots that *S. repens* can effect entry; on the lighter soils, the place of *S. repens* is taken by *Botryodiplodia theobromae.*

In the British West Indies, *S. repens* is commonly associated with a condition known as red root disease of limes; BAKER (1938), who has reviewed the history of this disease in the British West Indies, has concluded that the status of 5 *repens* as a primary root parasite of limes is still unproven. He has recommended the budding of limes on sour orange root-stocks, which are resistant not only to the red root disease, but also to gummosis.

Internal Root Rot of Tea, Formerly Attributed to Botryodiplodia Theobromae

Ideas concerning this "disease" of the tea bush were revolutionised by GADD (1928-1929), who demonstrated that infection by *B. theobromae* was a secondary phenomenon, and that the 'disease' was really caused by cultural practices. Internal root rot was stated by GADD to occur in Ceylon chiefly at middle and low elevations, where the plucking rounds were most frequent, and the pruning cycle the shortest; it was manifested as a failure to make fresh growth after pruning, and by subsequent death of the bush. GADD showed that failure to recover after pruning appeared invariably to be associated with absence of starch reserves in the roots, and that this condition actually *preceded* infection by *B. theobromae.*

Lack of starch reserves thus appeared to be the cause, and the subsequent general infection of the weakened roots by *B. theobromae* merely an effect of the disease. In tea grown at the higher elevations, where assimilation could better keep pace with growth, the disease was almost unknown. GADD also observed that whereas roots infected by *B. theobromae* were completely deficient in reserve starch, those infected by *Poria hypo-lateritia*, *Rosellinia* spp., and *Ustulina sonata* showed a normal starch content.

This rather constant association of infection by *B. theobromae* with low starch content of the roots is a phenomenon of considerable mycological interest, especially in connection with the findings of LEACH (1937) in Nyasaland. TUNSTALL (1940) in North East India agreed that susceptibility to internal root rot was almost always associated with lack of starch reserves, which might also be occasioned through defoliation due to drought or attacks of Red Spider. Of especial interest is his observation, cited above, that the roots of trees weakened through defoliation were infected by *B. theobromae* in light-textured soils, but in the heavier soils by *Sphaerostilbe repens.*

Eradication or Immobilisation of an Introduced Root Disease Fungus: No better commentary can be given upon the problems involved in eradication or immobilisation of an introduced root parasite in a plantation crop than by recounting the story of Panama disease of bananas in Jamaica, full particulars of which have been given by COUSINS and SUTHERLAND (1930) and by SMITH (1932a). The disease was identified on the Island early in 1912, and two more outbreaks, in widely separated localities were discovered soon afterwards.

These outbreaks, considered to be of independent origin, were attributed to plantation labourers returning from the diseased banana lands of Central America, and bringing back infected tools and planting material, in spite of strict watch for such practices on the part of customs officers and plant quarantine inspectors. The Jamaican Department of Agriculture suspected from the first that Panama disease spread from one plant to another by root contact rather than directly through the soil; COUSINS and SUTHERLAND very aptly described this mode of spread as "chain action".

A drastic Quarantine Regulation was therefore, drafted in 1912, whereby all plants within one-chain radius of the original diseased plant had to be removed and destroyed, making a quarantine area of four square chains around the site of each diseased plant; the quarantine area was not to be replanted with bananas. The selection of a one-chain (22 yards) radius was based on the finding that the longest roots of a banana plant did not extend for a radius of more than half a chain (11 yards).

In proof of the correctness of the original assumption that the disease spread by "chain action" along the roots of the plants, COUSINS and SUTHERLAND cited as an example plots rogued by this method in 1915, where in 1930 healthy bananas still surrounded the original quarantined areas. This was considered to disprove the assertion that the causal fungus, *Fusarium oxysporum cubense,* could travel independently through the soil in the absence of the banana plant. The loss of all the healthy plants on an area of four square chains was a heavy one to the planter, in spite of compensation, and with increase in incidence of the disease in any locality, a compromise was eventually made between economic considerations and the requirements of the rogueing treatment.

Under a Panama Disease Order of 1925, a graded reduction in the quarantine area was permitted, as follows. For the first three cases on any property, an area of four square chains was to be enforced, for the fourth and fifth cases, one square chain, and after the fifth case, an area of half a square chain was to be quarantined. Thereafter, with approval of the Director of Agriculture, the quarantine area might be reduced to that of the "nine root system", whereby only the eight healthy plants surrounding each diseased one had to be removed and destroyed. COUSINS and SUTHERLAND remarked that experience had shown the rate of spread of the disease to increase with each relaxation of the quarantine restriction. SMITH (19320) confirmed this conclusion, and stated that the quarantine regulations had met with success at least in checking the rate of spread of Panama disease. The "one root system", whereby the diseased plant alone was removed and destroyed, was to be deprecated as very dangerous, and permissible only when the property was officially gazetted as infected, and adjacent lands were not endangered.

At about this time, a considerable improvement in the method of disposing of the plants to be destroyed was introduced by SMITH (1932&). According to the method originally adopted, the diseased plant was dug up and burnt immediately; the surrounding healthy plants were dug up, chopped into small pieces and heavily treated with lime.

The area so treated was then fenced in, and maintained in quarantine. Apart from the difficulties and expense involved in destruction of banana plants by this method, there was the danger attendant upon moving soil, especially on river banks and steep slopes liable to wash. SMITH therefore, sought for a method of killing the stumps of the felled plants *in situ;* many chemicals were tried, of which *heavy gas oil* (the fraction from the "topping" process of refining) proved most satisfactory in all respects, including that of cost.

Only 2-3 pints are required per stump; after clearing away surface trash, the oil is applied around the stump, rather than on the cut surface. Oil so applied penetrates to the underground buds and kills the "eyes", from which new growth would otherwise start; large applications of oil to the cut surface of the stump were found to be quite unnecessary. An indirect benefit resulting from this method, by which sucker growth from the stumps is eliminated once and for all, has been pointed out by SMITH: "It has been found in the past that the adventitious suckers growing in a quarantined area have an extra-ordinary fascination for the small settler, who appears to prefer obtaining his planting material from such dangerous areas."

Unfortunately for the strenuous and well-directed efforts of the Jamaican Department of Agriculture, however, rogueing could prevent only the active spread of the wilt fungus, and could not stop other modes of dispersal. From the beginning, according to SMITH (1932a), the effectiveness of the quarantine operations was vitiated by non-cooperation on the part of some of the planters, more particularly small-holders, who not only failed to report infections, but carried out the unauthorised one root treatment on their own.

In efforts to discover a "cure" for the disease, certain individual planters carried out all sorts of secret experiments on their own land, generally without taking proper precautions against the spread of infection. Diseased material was sometimes burnt, but might even be thrown into an adjacent stream, to avoid detection! Small settlers, again, had been known to plant up a new area with diseased banana suckers obtained from adventitious growth on quarantined land. One of the most effective agencies for passive spread of *Fusarium oxysporum cubense* had been that of flood and river water.

SMITH considered that dissemination of the fungus by surface wash from local flooding after heavy rains was of particular importance; another dangerous practice was that of planting bananas right up to the banks of rivers bordering plantations. Finally, after commenting upon the loss of banana cultivation in large areas of the best alluvial soils in the river valleys and level lands, due to flood dispersal of the wilt fungus, SMITH (1936) reported the decision of the Department of Agriculture, taken in 1935, to permit one-root treatment of the disease throughout the Island.

An apparently introduced root disease that has recently caused considerable concern in the English hop industry is the wilt due to *Verticillium alboatrum,* which has been investigated by KEYWORTH (1942). Whatever the origin of this disease may have been, its subsequent devastating spread in a number of hop plantations was not properly anticipated, with the result that insufficiently stringent measures were taken for eradication of the parasite. KEYWORTH has now given the following directions for eradication of a small outbreak of this disease in a new plantation: "Cut down and burn all the bines of diseased plants and *treat every adjacent plant similarly.* Then excavate a hole 1 yd. square and 1 yd. deep at each place where a treated plant has been and remove the soil from the hopfield. Pour 8 gal. of 2 per cent formalin into and upon the walls of this hole and then fill with uninfected soil. Then treat the whole area where plants have been removed with 8 gal./sq.yd. of 2 per cent formalin, retaining this with a bank if necessary. The spaces may be replanted after 3-4 weeks, but the young plants and surrounding old plants should be observed for wilt during the next season, and the procedure should be repeated if any wilt is seen."

This wilt disease of hops certainly presents unusual obstacles to the successful application of the usual routine measures of plant sanitation. KEYWORTH found that the fungus eventually reached all parts of badly diseased bines except the bracts and bracteoles of the hop cones. Such diseased bines were found to be dragged all over the hop plantations by the tines of the cultivators; this fact was considered by KEYWORTH to explain differences in spread of the disease in "square-plant" hop gardens with cross cultivation, and "Worcester-plant" gardens with one-way cultivation. KEYWORTH also experimentally demonstrated that diseased leaves could act as inoculum, and might thus further the spread of the fungus to other parts of the same field, to other fields on the same farm, and even, perhaps, to other farms.

2

Various Facets

Chromoblastomycosis and Etiologic Fungi

Clinical Aspects

Definition: Chromoblastomycosis is a chronic, infectious, apparently non-contagious skin disease confined most frequently to one of the lower extremities and characterised clinically by the formation of nodular, verrucous, or cauliflower-like lesions. The infection may be caused by several species of dematiaceous fungi.

Synonymy: The term chromoblastomycosis, first used by TERRA *et al.* in 1922, is not only a long word, but a misnomer falsely ascribing the disease to a *Blastomyces* and conveying the erroneous impression of an extraordinary colour element in the clinical picture. Notwithstanding this, we believe that, for the present, this name should be preserved. It has been extensively used in the literature and it expresses the frequent resemblance of the disease to true blastomycosis. Furthermore, the prefix "chromo," though misleading in regard to the clinical picture, is descriptive of the etiologic fungus.

The following synonyms have been used or suggested for this disease: *"blastomycose negro,"* (Pedroso, 1911, cited by Fonseca and Leao, 1930); *"figueira"* (Rudolph, 1914); *"dcrmatite verrucosa por Phialophora verrucosa"* (Pedroso and Gomes, 1920); *"formigueiro"* (Gomes, 1921); chromomycosis (Moore and Almeida, 1935); *"dermatite verrucosa cromomicosica"* (Redaelli, 1935); *"dermatitis verrucosa blastomicotica"* (Bogging, 1937); Pedroso's disease, Fon-seca's disease or Gomes' disease (Weidman and Rosenthal, 1941); Pedroso and Carrion's disease (Barros-Barreto, 1943). Some of the synonyms just mentioned are almost as long or longer than chromoblastomycosis and are still misnomers; others might be criticised for bearing personal names. If the term chromoblastomycosis were to be changed at all, it should be done at an International Congress where the subject could be thoroughly discussed and settled by general agreement.

History: It is generally accepted that chromoblastomycosis was discovered in Brazil. In 1911, Pedroso, of Sao Paulo, noted the presence of large, dark-brown to yellowish, spherical bodies in a biopsy from a patient with nodular and ulcerated skin lesions of the foot and leg (Pedroso and Gomes, 1920). He suspected a mycotic infection and was able to isolate from the lesions a dark coloured fungus. The disease became known as *"blastomycose negra"* (black blastomycosis) in Pedroso's laboratory (Pedroso, quoted by Fonseca and Leao, 1930), but the study of the presumptive causative agent was postponed and the discovery was not reported until 1920 (Pedroso and Gomes).

In 1914, Rudolph published his observations on a skin disease popularly known as *"figueira"* in Minas Geraes and Goyaz. The excellent clinical description and the mycologic findings given by Rudolph in his report clearly indicate that he was dealing with the same disease observed by Pedroso in Sao Paulo three years before.

In 1915, Medlar and Lane, in the United States, published two elaborate communications on "A Cutaneous Disease

Caused by a New Fungus" in a patient from Boston. The infection in this patient had lasted for only a year and the clinical picture was not particularly specific; but the pathologist discovered a blastomycetic type of lesion with numerous, spherical, pigmented fungus cells, and the cultures revealed the presence of an organism producing a dark mycelium. The fungus was carefully studied and was named *Phialophora verrucosa.* The term *Phialophora,* representing a new genus (Thaxter, cited by Medlar, 1915) means cup-bearer and is descriptive of the conidiophores produced by the fungus in culture, while *verrucosa* indicates the type of lesion induced by the parasite on the human skin.

From the descriptions of Medlar and Lane, it was apparent that the infection in the Boston patient was identical in nature with that previously observed in Brazilian patients by Pedroso and by Rudolph. This fact was recognised by Pedroso and Gomes in 1920 when they published a report on four cases of the disease in Brazil, including the original case found by Pedroso in 1911. In their report, Pedroso and Gomes referred to the malady as a form of *"dermatite verrucosa."* It was not until two years later that the term Chromoblastomycosis was first used to designate the infection (Terra *et al.,* 1922).

In 1922, Brumpt established that the organism responsible for Pedroso's original case was not *Phialophora verrucosa* but a new species which he named *Hormodendrum pedrosoi: Hormodendrum,* because some of the conidia observed in culture were produced in branching chains; *Pedrosoi,* to honour the investigator who discovered the disease. In the following year Fonseca and Leao conducted careful morphologic observations on eight fungus isolates from Brazilian cases and arrived at the conclusion, like Brumpt, that the Brazilian organism was not a *Phialophora;* but they classed it in the genus *Acrotheca* because they found that the outstanding feature of the species was not the *Hormodendrum* sporulation already observed by Brumpt, but the production of terminal spore clusters of characteristic *Acrotheca* type. Consequently, they renamed the fungus

Acrotheca Pedrosoi (Fonseca and Leao, 1923). In 1936, Negroni, who made a mycologic study of another case of Chromoblastomycosis occurring in Argentina, confirmed the presence of both the *Hormodendrum* and *Acrotheca* types of sporulation in his patient's isolate. Considering that neither of the two genera above-mentioned could legitimately admit a fungus showing the sporulating habits of the other, Negroni created the genus *Fonsecaea* to include organisms possessing the double method of sporulation characteristic of *Pedrosoi* and the name of the species was changed, accordingly, to *Fonsecaea Pedrosoi* (Brumpt, 1922) Negroni, 1936, *comb. nov.* Finally, in September 1935 and later in 1936 (Emmons and Carrion, p. 703), it was demonstrated that *Fonsecaea Pedrosoi* possesses, in addition, a semi-endogenous type of sporulation by which conidia are produced in phialides. This observation, not only added a new and fundamental character to the species, but established its relationship with a third generic group, namely *Phialophora* Medlar, 1915.

In June, 1935, and in the following year, one of us described in a Puerto Rican case, a third etiologic agent for chromoblastomycosis. The Puerto Rican fungus is generally a *Fonsecaea* with its characteristic triple method of sporulation, but it possesses sufficiently distinct qualities to warrant its registration as a new species which bears the name *Fonsecaea compactum*.

It is convenient to state at this moment that, after the publications of Medlar and Lane in North America and of Pedroso and Gomes in Brazil, interest in chromoblastomycosis became increasingly progressive; year after year new cases of the disease were discovered, not only in Brazil and the United States, but in many other regions throughout the world; and by the end of 1945, at least 151 legitimate cases had been recognised. With the publication of these cases many supposedly new etiologic species were described and a long and confusing list of specific names was added to the mycology of chromoblastomycosis. Careful comparative studies have demonstrated that the large majority of these names are synonymous with *Fonsecaea Pedrosoi;* but there is no doubt that some of the newly described species, which have not been generally recognised as yet, are worthy of consideration and will be included in our discussion on "Mycologic Aspects".

Incidence and Geographic Distribution: During the thirty-five years that have elapsed since the discovery of the first Brazilian case, the existence of chromoblastomycosis has been recognised in other parts of South America, in North and Central America, the West Indies, Europe, Africa, the East Indies, Japan, and Australia, leaving Continental Asia as the only part of the world as yet without reported cases. A careful review of the literature for this period has revealed a large number of infections classed as chromoblastomycosis, but the information given for some of these infections is not considered sufficient to establish their authenticity beyond doubt.

It is only in 151 of the reported cases that the diagnosis appears to have been made on a sound basis and this number may be accepted as an approximate index of the recognised authentic cases of the disease during the last 35 years. The geographic distribution of this group of cases is as follows: Cuba, 43; Brazil, 41, South Africa, 12; United States, 9; Venezuela, 9; Puerto Rico, 7; Russia, 5; Costa Rica, 4; Dutch East Indies, 4; Japan, 3; Algiers, 2; Argentina, 2; Australia, 2; and Rhodesia, Dominican Republic, Guatemala, Canal Zone, Mexico, Canada, Paraguay, and Uruguay, 1 case each. By plotting these cases on the map, it will be found that 119, or 79 per cent, occurred in tropical or subtropical regions, and only 32, or 21 per cent, in the temperate zones. According to these observations, Chromoblastomycosis might be considered as a cosmopolitan disease, but its distribution would be predominantly tropical. This was to be expected from the fact that we are dealing with a fungus infection and it is well known that fungi generally thrive better in warm, moist climates.

Symptomatology: The disease usually affects one of the lower extremities. It begins as a small papule or warty growth which may develop anywhere on the extremity, but is located, as a rule, on some part of the foot, whence the infection spreads upward through the gradual development of satellite lesions. The course of the pathologic process is extremely slow and the clinical history often reveals that the infection had existed for ten or more years at the time of examination.

In a typical, well advanced case of Chromoblastomycosis the foot and leg are generally swollen and somewhat elephantiasic in appearance. The lesions occur in great numbers, especially towards the distal portion of the extremity, and they are conspicuously varied in morphology. For the purpose of description it is convenient to class them into five different types, namely, the nodular, the tumorous, the verrucous, the plaque, and the cicatricial types.

The nodular type includes the youngest and smallest elements in the clinical picture and consists of moderately elevated, fairly soft, dull-pink to violaceous growths, the surface of which may be smooth, verrucous or scaly.

Through further development many nodules are gradually transformed into lesions of the second, or tumorous type. This type is represented by much larger and prominent, distinctly papillomatous, sometimes lobulated, tumour-like masses, partly or wholly covered with dirty gray epidermal debris, crusts, and horny particles. On the foot and lower leg, where the pathologic process tends to be most exuberant, the tumour masses often reach enormous dimensions, taking on the characteristic appearance of cauliflowers.

In the third, or verrucous type of lesion, hyperkeratosis is the outstanding feature; the efflorescences are warty in appearance and may resemble *Verruca vulgaris.* Growths of the verrucous type are frequently encountered along the borders of the foot.

The plaque type is the least common of the lesions of chromoblastomycosis. It consists of fairly flat, slightly elevated, variously sized and shaped areas of infiltration. They are reddish to violaceous in colour, superficially scaly, and some of them show exaggeration of the lines of cleavage. The development of small, papillomatous vegetations, or larger nodules, within a plaque is sometimes observed and may lead to great variations in the morphology. When present in the clinical picture, the plaques are generally found on the higher portions of the extremity and never on the lower leg or foot.

Finally, the cicatricial type of lesion is represented by growths that enlarge by peripheral extension while healing takes place at the centre with the production of sclerotic or atrophic scarring. Cicatricial lesions may cover more or less extensive areas and are usually annular, arciform or serpiginous in contour.

The lesions of chromoblastomycosis develop slowly but progressively and, in the course of time, a few or many of them may coalesce to form extensive and often bizarre aggregates. The infected tissues are easily traumatised and bleed readily. When pressed with the fingers many of the lesions discharge a whitish, caseous material and sometimes fluid pus at one or more points. Secondary bacterial infections and ulceration frequently complicate the clinical picture and are mostly responsible for the foul smelling character of the eruption, a feature that is usually perceptible at a distance from the patient. Subjectively, pruritus may be an important symptom and some patients complain of pain. In advanced cases there is partial or total incapacity for work.

The deeper tissues are not usually involved. The lymphatic glands draining the diseased focus may participate in the process, but this is not the rule. However, adenitis due to bacterial complications is not infrequent. Metastases through the blood stream appear to be extremely rare, but there is no question that they can be produced. Finally, no systemic symptoms have yet been recorded from the infection.

It should be emphasised that the dermatologic picture just given is a general representation of the symptomatology in well developed, typical cases of chromoblastomycosis of the lower extremities. Although location in one of the lower extremities is the rule, the disease may affect the skin in almost any other part of the body, either as aberrant lesions produced during the course of a limb infection, or as the original site of attack. Involvement of the upper extremities ranks second in order of frequency, the initial lesion being usually located somewhere on the hand or wrist, although it may appear in any other segment of the limb. Infection of the face has been recorded in three instances, while the neck and trunk have been involved only once each and in different cases.

In chromoblastomycosis, as in any other chronic, infectious dermatosis, the clinical type of the infection in individual patients may vary in accordance with the duration of illness, the degree of virulence of the infecting parasite, the nature of the host reaction, and the location of the pathologic process. The possible correlation of clinical types with one or another of the specific parasites producing the disease has been suggested by various workers, but our present knowledge is too limited to warrant final conclusions on this phase of the subject.

Histopathology: Chromoblastomycosis falls in the group of infectious granulomas. The pathologic process affects both the cutis and the epidermis, and it develops towards the surface with little or no tendency to involve the deeper structures. Infection is apparently extended to adjacent regions by autoinoculation through the epidermis, or by way of the superficial lymphatics. Metastases through the blood stream are very rare but may be produced.

When the lesions are examined microscopically, the epidermis is generally thickened and often folded to fit the underlying papillomatous elevations. The thickening is due largely to hyperplasia of the *stratum Malpighii* which not only is broader than normal but shows irregular growths penetrating more or less deeply into the cutis. These growths may be sufficiently pronounced to resemble an epithelioma of the prickle cell type. The *stratum corneum* shows marked hyperkeratosis and is often distorted. Polymorphonuclear leucocytes are sometimes noted infiltrating the epidermal layers and, not infrequently, they form miliary, abscess-like accumulations.

In the cutis, the pathologic reaction is essentially granulomatous with a varied cellular infiltrate consisting of lymphocytes, plasma cells, large mononuclear leucocytes, polymorphonuclear leucocytes, eosinophiles, epithelioid cells, and occasional giant cells of both the Langhans' and foreign body types. The infiltrate may be focal or diffuse. In many places it is distinctly tuberculoid. Miliary abscesses are frequently observed, but widespread necrosis, softening and free suppuration are not so common. The lesions of chromoblastomycosis show a constant tendency towards fibrosis as a natural defence mechanism to wall off the infective process. This tendency becomes especially noticeable in the older foci of infection where the major part of the lesions often consists of fibrous tissue.

The most significant character in the pathologic picture is undoubtedly the presence of the infecting fungus. This may be seen as rounded, occasionally crescent-shaped, often septate bodies measuring about ten microns in diameter, with fairly thick and dark cell walls and a coarsely granular protoplasm possessing a pigment which has been variously described as ochre, olivaceous, yellowish-green and dark chestnut. The parasitic cells may occur singly or in groups. In the cutis they are variously located within giant cells, free in the tissues and, not infrequently, in the centre of microabscesses; but they may be found also within epithelial pearls and microabscesses in the epidermal layers. Evidence of germination is often noted in the *stratum corneum.*

Etiologic Factors: Chromoblastomycosis occurs most frequently during the period of active adult life. Among 109 authentic cases of the disease in which the age was recorded, there were 2 patients who contracted the infection during the first decade of life; 8 were infected during the second decade; 23 during the third; 29 during the fourth; 25 during the fifth; 13 in the sixth; 6 in the seventh, and 3 in the eighth. According to these records it is evident that the highest incidence of the disease lies between the ages of 20 and 50 years (ca. 71 per cent of the cases). It may be added that the extreme ages at which infection took place were 3 and 76 years respectively.

Chromoblastomycosis is decidedly more common among males. A review of 138 authentic cases in which the sex was registered, showed that 132 were males, a proportion of 96 per cent.

There seems to be no race immunity. However, in a collection of 124 clinical histories which contained data regarding the race of the patients, it was found that 74 cases, or 60 per cent, were Caucasians and 38 cases, or 30 per cent, were Negroes. Other races reported in the remaining 10 per cent included: the Mongolian (6 cases); the Malayan (2 cases); the Hindu (2 cases); a Mexican "mestizo" and a Jamaican.

A large majority of the victims of Chromoblastomycosis have been farm labourers working barefooted in the fields at the time infection was contracted. In a study of 73 patients whose histories included the occupation, it was found that 60, or 82 per cent, were engaged in this type of work, while three others were farmers who were also engaged in other occupations (carpentry, mining, nursing). Among the remaining patients, there were three day labourers, two animal tenders, one cobbler, one merchant, one printer, one literary worker, and one reported as a "city dweller."

Treatment: Chromoblastomycosis has been subjected to many different methods of treatment with various degrees of success. When the lesions are small and discrete, as is usual in incipient cases, it may be possible to completely eradicate the infection by surgical excision or electrocoagulation. On the contrary, when the infection is advanced and the lesions numerous, extensive and confluent, these measures would be impracticable and treatment becomes a more difficult problem, the method of attack depending on the individual circumstances.

In late, lower-limb infections the leg and foot are liable to be elephantiasic and heavy; secondary bacterial complications and ulceration are not uncommon and may be disturbing to the general health; the lesions are foul-smelling and repulsive in appearance; the patient becomes physically disabled and mentally demoralised. Under such circumstances surgical amputation and an artificial leg might solve the therapeutic problem, provided that the patient is otherwise healthy, not aged, and is willing to undergo the operation. This procedure would not only eliminate the physical and mental sufferings inherent to the disease but would bring the victim back to an active, more productive and happier life.

When the patient is old or unwilling to accept radical surgery, or when the infection affects a region, other than the leg and foot, not so liable to serious complications, treatment should be conducted on more conservative principles and should consist of both systemic and local measures. Internally, many chemotherapeutic agents—including arsenic, mercury, bismuth, gold, antimony, methylene blue, and the iodides of potassium and sodium—have been used to combat the infection. Among these, only the iodides would seem to be partially effective. They should be administered both by mouth and intravenously when possible, and the daily dose should be gradually raised to, and maintained at, the limit of tolerance over long periods.

By this method of therapy it has often been possible to reduce the fibrosis and the size of the lesions, some of which may even disappear, promoting a general feeling of optimism in the patient. However, careful follow-up observations have not offered convincing evidence of the curative action of these compounds. Recent experiments would seem to indicate that sodium sulphamerazine possesses distinct inhibiting properties against *Fonsecaea Pedrosoi in vitro*. On the basis of this knowledge the drug is now being administered to Puerto Rican patients, but it is too early yet to evaluate its true merits as a therapeutic agent in Chromoblastomycosis.

Local treatment should aim, in the first place, at the elimination of all incipient efflorescences and as many of the older ones as can be effectively removed surgically or by electrocoagulation. Since the disease is slow in spreading and the possibilities of metastases are so meagre, the fulfilment of that aim should check further progress of the infection. In the second place, lesions of larger size should be destroyed as far as possible by similar methods, and local treatment then conducted on general principles, giving preference to sodium sulphamerazine preparations. Iontophoresis with copper sulphate, which was employed with good results in a patient with lesions on the hand and forearm (Martin *et al.*, 1936) is worthy of further trial. Treatment with Roentgen rays has often promoted the improvement or complete clinical disappearance of some lesions. However, follow-up observations have not been satisfactory and recurrences on the radiated areas have been observed. This method of therapy might, perhaps, be effective on lesions of limited extent, but further investigative work is required for the development of an adequate technique.

Mycologic Aspects

Chromoblastomycosis has been ascribed to several species of dematiaceous fungi, the biology of which is only partially known. These fungi may be classed in two different groups. The first group consists of organisms that are mouldy in appearance, habitually producing an abundance of healthy, dry, aerial hyphae in laboratory cultures. In this group are included *Fonsecaea Pedrosoi* (Brumpt, 1922) Negroni, 1936 *comb, nov.; Fonsecaea compactum; Phialophoraverrucosa,* Medlar, 1915; *Hormodendrum species* (Simson *et al.*, 1943); *Hor-modendrum species;* and *Torula poikilospora,* Taka-Hashi, 1937. The etiologic importance of *Fonsecaea Pedrosoi, Fonsecaea compactum,* and *Phialophora verrucosa* has been generally accepted. The other three species mentioned are not so generally known, and some of them have not been extensively studied, but there are good reasons to suspect their possible etiologic relation to the disease, and it has been thought convenient to include them in this discussion.

The second group of fungi associated with chromoblastomycosis consists of four organisms which produce primarily soft, dark, moist colonies on the usual laboratory media, undergoing what appears to be a yeast-like phase in the course of their development. As the cultures age, a good part of the thallus becomes filamentous, but the growth as a whole retains its moist appearance. Only two of the organisms of this group have been published in the literature, and they were described as new species, namely, *Hormiscium dermatitidis*, Kano, 1937; and a "black *Candida-like...* species" (Berger *et al.*, 1945). Of the two remaining fungi, one was isolated by Bonne from a patient with chromoblastomycosis in the East Indies, and the other was obtained by us from a Puerto Rican case. The descriptions of these two fungi are not yet available. From a preliminary study of the four organisms included in this group, we have been led to suspect the existence of close interrelations among them. Further comparative studies are now being conducted in our laboratories with a view towards elucidating their true taxonomic position and relationship.

Geographic Distribution: *Fonsecaea Pedrosoi* has never been found in nature outside of the human body, but the widespread distribution of its victims would seem to indicate its ubiquity. *F. Pedrosoi* is by far the most common etiologic agent of chromoblastomycosis. Among 90 legitimately classified organisms isolated from the disease throughout the world 76, or 84 per cent, belong to this species. The geographic distribution of the fungus is conspicuously higher in the warmer climates. This is evident from the fact that 63, or 83 per cent, of the 76 isolates above mentioned, were obtained from patients who contracted the infection in tropical or subtropical regions, while only 13, or 17 per cent, were found in temperate zones. It might be of interest to note here that these proportions appear to be reversed in the case of another important etiologic species of chromoblastomycosis, namely, *Phialophora verrucosa*, of which there are 6 known isolates and 5 of them, or 83 per cent, correspond to the colder climates. However, the fact that most of the isolates of *P. verrucosa* have been found in temperate regions is no indication of predominance, for even in such regions the incidence of *F. Pedrosoi* has been found to be higher than that of *P. verrucosa* in a proportion of 13 to 5.

The continental distribution of *F. Pedrosoi* has been found to be as follows: North America, 5 isolates, including 4 from the United States and 1 from Mexico; Central America, 4 isolates, including 3 from the Canal Zone and 1 from Guatemala; South America, 25 isolates, including 19 from Brazil, 4 from Venezuela and 2 from Argentina; West Indies, 33 isolates, including 18 from Cuba, 11 from Puerto Rico and 2 from the Dominican Republic; Europe, 4 isolates, all of them from Russia; Africa, 4 isolates, including 1 from Algiers and 3 from the Union of South Africa; Asia, 1 isolate from Japan; and the East Indies, 2 isolates, including 1 from Java and 1 from Sumatra.

Morphology in Pathologic Tissue: Since chromoblastomycosis in human beings is essentially a skin disease, the infecting fungus is to be found almost exclusively in cutaneous lesions. Here it may be recognised in both the dermis and epidermis as characteristic, usually spherical, occasionally crescent-shaped, brownish-yellow bodies, measuring about 10 microns in diameter.

These bodies may occur either singly or in clumps, rarely in short chains. They are found within giant cells, or free in the tissues, or in the centre of microabscesses, or, finally, enclosed within epithelial pearls in the Malpighian and horny layers of the epidermis.

The fungus elements possess a dark, fairly thick cell wall, part of which is occasionally swollen, bulging towards the interior of the cell. The wall is sometimes covered with a crusty layer of refractile material. In histopathologic preparations stained with hematoxylin and eosin, this layer takes the eosin stain, but it may be seen just as well in fresh unstained preparations of the infected epidermal scales. Not infrequently, the cell wall appears thin and degenerate due, undoubtedly, to the defence reaction of the body tissues. The protoplasm is granular and usually contains variously-sized, mostly coarse, spherical or irregular refractile inclusions, and it possesses a conspicuous natural pigment which has been described as ochre, olivaceous, yellowish-green, and chestnut brown. The nucleus is not apparent in fresh or stained preparations. Germination of the parasitic cells is frequently observed in the *stratum corneum* of the epidermis, the filaments being comparatively short, irregular, septate and branched, and showing the same pigmentation and structure noted in the mother elements. The authors have never seen typical filamentous structures in the dermis, but some investigators have reported their occurrence in this location.

In the infected tissue, multiplication of the fungus takes place by fission, different stages of the process being noticeable in both the dermis and epidermis. The first stage is represented by the development of an internal septum which divides the fungus cell into two hemispherical compartments. The parasitic body is thus transformed into a bicellular structure. Upon further growth, each of the two compartments tends to reproduce the spherical shape of the mother cell and a constriction becomes apparent at their plane of contact. Finally, the two daughter elements may separate from each other and become free cells, but they often remain united and, on maturity, they start again the process of fission. The repetition of cell division without separation and the fact that the dividing process may take place in more than one plane often lead to the formation of irregular conglomerations of fungus cells not unlike the sclerotia sometimes noted in old fungus cultures. It is thus that the parasitic cells seen in tissues have been frequently described as "sclerotic cells."

Fonsecaea Pedrosoi, as present in the host tissues, has often been called a "black yeast," or a "*Blastomyces*" or a yeast-like fungus and some investigators have mentioned the production of conidia in the lesions by budding. Indeed, histopathologic sections showing spherical fungus cells with bud-like outgrowths have been occasionally observed. However, neither the descriptions, nor the illustrations published, nor the personal observations of the authors have yet offered convincing evidence that this fungus may multiply in the tissues by a budding process similar to that noted in North American blastomycosis which is produced by a different fungus, namely *Blastomyces dermatitidis*. The bud-like process noted in the infected tissues, might possibly represent germinating tips, or might result, perhaps, from a difference in the rate of growth of the two constituent elements of a septate fungus cell.

In animals experimentally infected, the parasite may be found in lesions produced, not only on the skin, but also in one or more of the internal organs, according to the route of inoculation. The morphology of the fungus in the pathologic tissues of such animals is essentially similar to that already noted in human lesions. Published descriptions mention the presence of spherical or oval, often septate, pigmented bodies, and also of filamentous structures, some of which are, undoubtedly, remnants of the inoculum.

In 1945, Area-Leao *et al.* observed the presence of chestnut-coloured "grains," of the type produced in maduromycosis, in the testicular lesions of a rat experimentally infected with the fungus of a Brazilian patient. The "grains" consisted of one or more sclerotic cells or mycelial filaments surrounded by numerous, closely set, but conspicuous, refractile, club-shaped bodies. Although the specific fungus was not named, it is presumable that these investigators were working with a specimen of *Fonsecaea Pedrosoi*, which is the parasite usually encountered in Brazilian chromoblastomycosis. In their report, they do not state, either, if the fungus was recovered in culture from the "grains."

Gross Morphology in Culture: On Sabouraud's "*milieu*

d'eprewve" after the fourth week and at room temperature, cultures resemble a flattened cone measuring about 5.5 cm in diameter; centre of culture elevated about 8 mm. above medium, sometimes forming mamillary prominence; rest of culture gently sloping, often showing radial folding; border circular, or festooned; the aerial mycelium forming dark, gray, greenish, olivaceous-gray or dark-brownish, felt-like network of short, delicate, ascending hyphae arising from a more consistent, somewhat elastic, black, membranous mat; substrate mycelium light gray and moderately abundant. On 4 per cent dextrose agar, cultures usually less regular on the surface and in contour; aerial hyphae less intensely pigmented, more profuse and less uniformly developed. On Czapek's agar, colonies poorly developed; mycelium mostly submerged, gray to olivaceous, forming, at border, arborescent outgrowths; central zone showing shallow layer of aerial hyphae, consisting mostly of conidiophores.

Microscopic Morphology in Culture: Microscopic characters usually more conspicuous after third week of growth. Vegetative hyphae long, straight or undulated, 1.25 to 3 microns in diameter, septate, branching, cell walls thick and dark; protoplasm olivaceous, granular, with refractile droplets.

Sporulation of three different types occasionally combined in same spore head: *(a) Hormodendrum* type more or less abundant, generally more conspicuous on Czapek's agar cultures, sometimes predominating, sometimes obsolete, according to individual isolate; fertile branches erect or ascending, often arborescent, with terminal cell, or conidiophore, sometimes darker, having at tip several tiny, truncate, conical prominences to which spores are attached. Conidia borne in chains, usually short, which tend to branch by multiple budding at distal pole of each successive conidium, resulting in the formation of complicated spore heads; conidia unicellular, ovoid, often elongated, those at the base of the chain frequently shield-shaped and occasionally bicellular; measuring 3 to 5 by 1.5 to 3 microns (basal elements 2.5 to 3.5

by 7 to 10 microns); olivaceous in colour, walls smooth, thick and dark; disjunctors moderately to poorly developed. *(b) Fonsecaea* type more or less abundant, usually more conspicuous on corn meal agar cultures, sometimes predominating, sometimes obsolete according to individual isolate; conidiophores short or long, straight or irregular, exceptionally branched, consisting of one, sometimes more, articles disposed terminally, laterally, or intercalarily, often derived from spore element in *Hormodendrum* head; surface of conidiophore wholly or partly verrucous due to the presence of tiny, truncate, conical prominences to which spores are attached; pigmentation of conidiophore often darker than vegetative mycelium; conidia sometimes single (not catenate) forming clusters which may be indistinguishable, when terminal, from those of *Acrotheca,* some clusters showing tendency to chain formation as in *Hormodendrum;* morphology of conidia resembling terminal and subterminal elements of *Hormodendrum* heads already described, *(c) Phialophora* type, usually scant to moderate in abundance, rarely predominating; morphology similar to that of *"Phialophora verrucosa".*

Chlamydospores of the "sclerotic cell" type already noted in tissue, sparsely produced in culture, the elements being spherical, 8 to 12 microns in diameter, with coarse, dark walls, protoplasm yellowish-brown, granular, usually containing variously sized, refractile inclusions; some elements showing cell division by internal septation in different planes, with production of bi-, tri-, or tetracellular structures.

Morphologic Varieties of Fonsecaea Pedrosoi: According to the various proportions in which the three methods of conidial formation above described may be represented in different isolates of *Pedrosoi,* this species has been subdivided into the following varieties:

1. *Fonsecaea Pedrosoi* var. *typicus,* corresponding with *Pedrosoi* as originally presented by Brumpt (1922); not a frequent variety; *Fonsecaea* sporulation most highly

developed and in overwhelming predominance, especially on corn meal agar; *Hormodendrum* sporulation scant, or depauperate or abnormal in morphology; *Phialophora* sporulation usually rare.

2. *Fonsecaea Pedrosoi var. Cladosporioides* corresponding to fungi of the type described as *Hormodendrum algeriensis* (Montpellier and Catanei, 1927); *Hormodendrum* sporulation most highly developed and in overwhelming predominance; *Fonsecaea* sporulation scant; *Phialophora* sporulation usually rare.
3. *Fonsecaea Pedrosoi* var. *Phialophorica* Carribn, 1942, represented by a single isolate described as *Phialophora macrospora*; *Phialophora* sporulation in overwhelming predominance; *Fonsecaea* sporulation scant; *Hormodendrum* sporulation obsolete.
4. *Fonsecaea Pedrosoi* var. *communis* the most common variety encountered; sporulation of *Hormodendrum* and *Fonsecaea* types always abundant, although in various relative proportions; *Phialophora* sporulation sometimes as abundant as the other two types, but usually scant.

In recent years, many isolates of *F. Pedrosoi* have been described as new and independent species on the basis of predominance of one or another of the methods of sporulation characteristic of this fungus and a great deal of time-consuming and painstaking work was required to establish the identity of such species with *Fonsecaea Pedrosoi.* The subdivision into varieties, as defined above, offers a broader concept of the species, it will promote adequate orientation for the classification of unknown isolates, and it will help to avoid the prevailing tendency to create new and illegitimate species.

Behaviour in Artificial Culture Media: *Fonsecaea Pedrosoi* has been grown in a large variety of solid and liquid culture media. According to most investigators the formulae most appropriate for optimum development of gross cultural characters are those containing agar, 2 per cent, with 1 per cent of peptone, and 4 per cent of either maltose or dextrose as nutrients. In these media the fungus grows faster and more luxuriantly and the aerial mycelium is more abundant and forms more regular patterns, sometimes showing concentric zonation. These characters are specially noticeable in cultures produced on Sabouraud's *milieu d'epreuve*" (Sabouraud, 1910). The fungus has also been found to thrive luxuriantly on wort agar, tubers and root vegetables (potato, carrot, red beet, turnip, radish, gourd, manioc, sweet potato and others).

Cultures on the media above mentioned may be used, in addition, for the study of microscopic morphology. However, due to the fact that *Fonsecaea Pedrosoi* possesses three distinct types of sporulation, all of which are not always readily produced in such media, it is frequently necessary to resort to other formulae that are more stimulating to the process of sporulation in order to bring out the morphologic features of the fungus. Experience has shown that corn-meal infusion agar (white corn-meal, 4 per cent; agar, 2 per cent) and Czapek's synthetic medium modified by Dox (1910), and by Thom and Church (1926), not only stimulate the production of conidia, but each of them tends to induce more selectively a different type of sporulation.

Thus, the *Hormodendrum* sporulation is more conspicuous, as a rule, on Czapek's agar; whereas the *Fonsecaea* type is usually better developed on corn-meal agar. The use of these two media, therefore, may be essentially important in the diagnosis of certain isolates of the species F. *Pedrosoi* and its varieties. It has been claimed that the *Phialophora* method of sporulation becomes more evident on Czapek's agar. From personal experience we have been led to suspect that this claim may be correct, but further observations will be required before definite conclusions can be established on this point.

Hydrogen-ion Concentration, Temperature,. and Oxygen Requirements: It is generally admitted that *Fonsecaea Pedrosoi* requires an acid medium with a pH ranging between 5.5 and

6.5, and it has been claimed that, when the pH is above 7, growth is delayed and may even be inhibited.

The optimum temperature for this fungus has not been precisely determined, However, its mycelium develops profusely at room temperature and some of the most important studies on its morphology have been conducted on cultures produced under ordinary laboratory conditions. It has been reported that artificial incubation at 37° or 37.5° C. promotes better growth; but there has been no definite claim, so far, of any substantial advantage derived from cultivation at such temperatures, except, perhaps, that primary sowings with infected material may disclose positive results a few days earlier when artificially incubated, a fact that might be of value in cases requiring a prompt diagnosis and which show no fungus elements in the epidermal scrapings.

Fonsecaea Pedrosoi grows best under aerobic conditions. This is evident from the abundance of healthy aerial mycelium and superficial sporulation usually produced under normal laboratory conditions on solid culture media. There is no doubt, however, that the fungus may grow to a certain extent at reduced oxygen tensions. Close examination of agar cultures will invariably show a certain amount of mycelium growing in the substrate, a feature which is particularly conspicuous on Czapek's synthetic medium. Anaerobic growth has also been observed in stab cultures on glucose agar (Weidman and Rosenthal, 1941) and on various liquid media. It should be noted, however, that mycelium produced under such conditions is usually depauperate and devoid of sporulation.

Chromatic Affinities: Although many attempts have been made to determine the staining reactions of *Fonsecaea Pedrosoi,* no essential information has been obtained, by this means, about the internal structure of the fungus cells. Tschernjawski (1929) has stated that the parasitic elements stain dark blue with azure II; dark brown with methyl green pyronine; green with polychrome blue, and dark violet with the Gram-Weigert stain. According to Pozojewa (1930) the fungus would take a faint blue colour with methylene blue. In preparations stained with Giemsa, its colour has been described as "dark blue" (Nuuck, 1931) and "dark green-blue" (Binford *et al.*, 1944). The fungus structures have shown no apparent reaction to Gram's method of staining (Pozojewa, 1930; Nauck, 1931), and, with the Ziehl-Neelsen technique, they take on a darker and somewhat reddish colour (Nauck, 1931). Finally, Weidman and Rosen-Thal (1941) have noted that osmic acid would intensify the dark pigmentation of the conidiophores produced in culture.

Biochemical Reactions: The biochemical activities of *Fonsecaea Pedrosoi* do not appear to be important. In 1930, Meriin tested the fermentative action of a Russian isolate, named *Hormodendrum rossicum (Fonsecaea Pedrosoi),* against saccharose, mannitol, glycerol, dulcitol, galactose, maltose, dextrose, and starch, with negative results. His fungus liquefied gelatine but would not coagulate milk. Takahashi made similar tests with another isolate, the so-called *Hormodendrum japonicum,* using dextrose, galactose, mannose, levulose, arabinose, xylose, rhamnose, saccharose, maltose, lactose, raffinose, dextrin, inulin, and mannitol, also with negative results.

In the following year, a Venezuelan specimen of *F. Pedrosoi* failed to ferment levulose, maltose, saccharose, galactose, inulin, raffinose, or glycerol under aerobic and anaerobic conditions and, in the same year, Pereira (1938) published his observation on a Brazilian isolate *(Hormodendrum Negroni),* reporting no liquefaction of gelatin; coagulation and acid production on milk; and no fermentation of rafftnose, xylose, mannitol, adonitol, saccharose, arabinose, dextrin, glycerol, inulin, galactose, levulose, glucose, lactose, or dulcitol. Mazza and Nino (1939) have stated that an isolate of *F. Pedrosoi* from Argentina (the so-called *Hormodendrum chaquense)* liquefied gelatin slowly, peptonised milk with mild acidification of the medium, did not liquefy agar, showed no action on starch, liquefied coagulated serum with brownish discolouration of the medium, did not ferment glucose, and did not alter the pH of glucose-containing media. Finally, Castro-Palomino *et al.* (1941), were unable to produce fermentation of maltose or lactose with two Cuban isolates.

Immunologic Reactions—Intracutaneous Tests: Meriin (1932 and 1938) claims to have induced specifically positive intracutaneous tests in patients with chromoblastomycosis using, as antigens, elaborately prepared culture extracts of *Fonsecaea pedrosoi (Hormodendrum rossicum)* isolated from two different patients.

Complement Fixation and Agglutination Tests: It was also Meriin (1930 and 1932), who first obtained a positive complement fixation reaction in this disease, but his results were found to be non-specific.

Balina *et al.* (1932) observed an intensely positive complement fixation test in a patient with chromoblastomycosis in Argentina. The antigen consisted of a mixture of cultures of *Fonsecaea pedrosoi (Acrotheca Pedrosoi)* and *Phialophora verrucosa,* the tests being controlled with the sera of other patients not suffering from chromoblastomycosis.

In 1936 Martin *et al.* reported on a series of strongly positive and specific complement fixation reactions with the serum of a patient infected with *Fonsecaea pedrosoi* in North Carolina (U. S. A.). The tests were performed with 12 antigens, each of them consisting of a culture extract from a different isolate of *pedrosoi,* including the one from North Carolina. This experiment was controlled with two additional series of tests. In the first control series, 16 other fungi, some of them pathogenic, were used as antigens against the patient's serum. In the second control series, the serum of 28 presumptive syphilitics, four of them positive, were tested, respectively, against three isolates of *F. pedrosoi.* All the controls were negative.

In 1937, Conant and Martin immunised each of two groups of rabbits, respectively, against a different isolate of *Fonsecaea Pedrosoi* by intracutaneous inoculation of the animals with a suspension of the respective fungus. Subsequent immunologic studies carried out with the sera of these animals showed that complement fixing antibodies were present in both sera, and in high titre, for both of the *F. Pedrosoi* isolates used in the experiment. It was also shown that the serum of the immunised animals would induce positive cross reactions, in an equally high titre, against *Fonsecaea compactum*, and, in a lower titre, against *Phialophora verrucosa* (Uruguay); thus confirming, immunologically, the morphologic relationship previously established among the species *F. Pedrosoi*, *F. compactum* and *P. verrucosa*.

A few attempts to agglutinate *Fonsecaea Pedrosoi* with the serum of patients suffering from chromoblastomycosis have been unsuccessful (Montpellier and Catanei, 1927; Meriin, 1930).

Response to Antibiotic, Chemical, and Therapeutic Agents:

1. *Alcohol:* The observations of Weidman and Rosenthal (1941) and of Binford *et al.* (1944) would tend to establish that ethyl alcohol (95 per cent) is lethal to the fungus. From long experience in our laboratory, where parts of the specimens received for culture are routinely treated with ethyl alcohol (80 and 95 per cent) prior to inoculation, we have been led to believe that this chemical may actually kill the parasite, but its effects are conditioned by the factors of time and concentration. Further investigations are now in progress with the purpose of gaining more definite knowledge on this point.
2. *Formaldehyde:* Weidman And Rosenthal (1941) have reported what appeared to be growth of the fungus in a mixture of equal parts of formaldehyde (10 per cent) and dextrose bouillon. It was not stated if the organism could be recovered in culture from this mixture.
3. *Sulphonamides:* In a series of carefully conducted experiments to determine the *in vitro* action of sulphonamides on pathogenic fungi, Keeney *et al.* were impressed by the inhibiting effects of sodium sulphamerazine against *Fonsecaea Pedrosoi.* On the basis

of these experiments, the clinical trial of this drug in patients with chromoblastomycosis, both by mouth and locally on the skin lesions, should be encouraged. As already stated, some of our patients are now receiving treatment with this sulphonamide, but the period of observation has been too short to warrant conclusions as to final results.

4. *Fatty acids:* Keeney *et al.* have also studied the effects of fatty acids on the same group of fungi *in vitro,* and have found that certain salts of these acids, and particularly sodium caprate and sodium undecylenate, possess striking fungistatic and fungicidal action against *Fonsecaea pedrosoi.* In view of the toxicity of these salts for the albino mouse, they do not recommend the internal administration of such salts in the treatment of the deep mycoses of man, until further animal experiments warrant their use. There is no reason, however, why these new drugs should not be tried locally on the cutaneous lesions of chromoblastomycosis.
5. *Penicillin:* The sensitivity of *Fonsecaea Pedrosoi* to penicillin has been tested by Keeney *et al.,* as well as by Fleming and Queen (1946), with negative results in both cases.
6. *Other Therapeutic Agents:* The effects of other chemical preparations used in the treatment of chromoblastomycosis have been already discussed.

Host Range: Man is the only well known natural host susceptible to infection with *Fonsecaea pedrosoi.* It is possible that a spontaneous infection noted by Carini (1910) in the lungs and kidneys of a batracian *(Leptodactylus pentadactylus),* was produced by this or a closely related fungus. Rudolph (1914) observed in Brazilian cattle another disease similar to human chromoblastomycosis, but no information is given regarding the true nature of the pathologic process.

Many investigators have endeavoured to produce artificial

infection with *Fonsecaea pedrosoi.* Takahashi reported on the reproduction of typical lesions of chromoblastomycosis in the healthy skin of a patient who was suffering from the disease and also in a volunteer not previously infected, by inoculating a suspension of *Fonsecaea pedrosoi (Hormodendrum japonicum)* isolated from his patient. The induced lesions were consistent histopathologically with chromoblastomycosis and the etiologic fungus could be recovered in culture.

Attempts to produce artificial infection have been carried out on eight different species of experimental animals, namely, rats (Ru-Dolph, 1914; Pedroso and Gomes, 1920; Tschernjawski, 1929; Meriin, 1930; Carrion and Koppisch, 1933; Takahashi, 1937, P. 53; Pereira, 1938; Briceno-Ragorri, 1938; Mazza and Nino, 1939; O'daly, 1943; and Azulay, 1945); Mice (Montpellier and Catanei, 1927; Meriin, 1932; Takahashi, 1937, P. 53; and Levy and Black-schaffer, 1945); Rabbits (Pedroso and Gomes, 1920; Montpellier and Catanei, 1927; Gomes and Pessoa, 1929; Meriin, 1932; Carrion and Koppisch, 1933; Takahashi, 1937, P. 53; Pereira, 1938; and Mazza and Nino, 1939); Dogs (Gomes and Pessoa, 1929); Monkeys (Rudolph, 1914; Gomes and Pessoa, 1929; Carrion and Koppisch, 1933; and Simson *et al.*, 1943); Guinea-pigs (Pedroso and Gomes, 1920; Meriin, 1932; Balina *et al.*, 1932; Takahashi, 1937, P. 53; Briceno-Ragorri, 1938; Pereira, 1938; Mazza and Nino, 1939; Simson *et al.*, 1943; and Azulay, 1945); Pigeons (Montpellier and Catanei, 1927); Frogs (Almeida, 1934). Successful infections have been claimed in all but the last two species enumerated above. The animals most frequently used in the tests were the rat (11 investigators), the guinea-pig (9 investigators), the rabbit (8 investigators), the monkey (4 investigators), and the mouse (4 investigators); but the higher proportion of positives occurred in the rat and the mouse in the order given. In most of the induced infections the inoculum consisted of suspensions of the fungus grown on laboratory media. Only a few investigators used tissue emulsions and purulent exudate from lesions to infect their animals. The routes of inoculation have varied with different investigators and with the animals used. The intraperitoneal route was successfully employed in rats by Meriin (1930), Carrion and Koppisch (1933), Takahashi, and by Mazza and Nino (1939); and in mice, by Meriin (1932), Takahashi and Levy and Black-Schaffer (1945). The subcutaneous route was used with positive results in rats by Tschernjawski (1929), Meriin (1930), Takahashi and Pereira (1938); and, in mice, by Meriin (1932), and Takahashi. Intracutaneous inoculation has been effective in rats, and possibly in rabbits (Gomes and Pessoa, 1929, and Mazza And Nino, 1939). The intratesticular route was used by Mazza And Nino (1939) and, with remarkable success, by Azulay (1945) in both rats and guinea-pigs; and finally, Levy And Black-schaffer (1945) produced systemic infections in mice by the intravenous route. According to the above observations it would seem that, among the different animals tested, the rat and the mouse are the species most susceptible to infection with *Fonsecaea pedrosoi;* that infection may be effected by more than one route of inoculation; and, finally, that the intraperitoneal, the subcutaneous, and the intratesticular are the most adequate routes to produce infection.

Laboratory animals artificially infected with *Fonsecaea pedrosoi* have not shown the chronic, slowly progressive, incurable, clinical manifestations inherent to the human disease, but the lesions produced have been sufficiently characteristic to confirm the pathogenic properties of the fungus. Infection in laboratory animals may be more or less limited in extent or it may be systemic, and even fatal, according to the method of inoculation. The lesions are nodular; variously sized, usually small; sometimes suppurating; single, few, or numerous; cutaneous or internal (liver, lungs, kidneys, peritoneum, testicle, etc.); when cutaneous, sometimes verrucous. The histopathology represents a granulomatous type of tissue reaction similar to that noted in human lesions, with the infecting parasite constituting a prominent part of the picture.

Taxonomy: Up to the present time the species *F. pedrosoi* has shown no evidence of sexual reproduction and, consequently, it has its place among the Fungi Imperfecti. In view of the general morphology, sporulating habits, and dark colour of the parasite, it has been included in the order *Moniliales,* family *Dematiaceae.* Finally, the simultaneous occurrence, in this organism, of three types of sporulation, which correspond to the genera *Hormodendrum, Phialophora* and *Acrotheca,* respectively, has led to the classification of the species in the new genus *Fonsecaepa* (Negroni, 1936) Carrion *emend.* (1940 and 1942). *Fonsecaea pedrosoi* has been subdivided into the four varieties *typicus, Cladosporioides, Phialophorica,* and *communis,* as already defined.

The generic position of the *"pedrosoi"* group has been a much debated subject. Differences of opinion have arisen mainly from (a) the triple sporulating ability of the fungus; (b) the outstanding and misleading predominance of one of the sporulating types in each of the varieties *typicus, Cladosporioides,* and *Phialophorica;* (c) failure to recognise the specific identity of these three varieties as established by the existence of intergrading forms (variety *communis)* which link them together and preclude any subdivision of the group beyond the ranks of varieties. Fortunately, many names under which different members of this species were described in the literature, have fallen into synonymy.

At the present time, some respectable authorities refer to *"Pedrosoi"* as a *Hormodendrum,* while others believe that the parasite should be included in the genus *Phialophora. Hormodendrum* was the name used by Brumpt in his original description of the species, and therefore, has the advantage or priority. It is further supported by the fact that most members of the *"Pedrosoi"* group present the *Hormodendrum* sporulation as an outstanding morphologic feature. On the contrary, *Hormodendrum* is objectionable because, by definition, this genus includes fungi where conidia are produced only in branching chains, and would not admit the varieties *"Pedrosoi typicus"* and *"Pedrosoi Phialophorica"* where the characteristic *Hormodendrum* chains are considerably reduced, abnormal or obsolete, while other methods of sporulation, representative of other genera *(Phialophora* or *Acrothecd)* constitute the outstanding morphologic features. The genus *Hormodendrum* could be emended to legitimately admit the species *Pedrosoi,* but this would introduce a serious source of confusion in the taxonomy of the group. Indeed, the history of *Pedrosoi* has clearly shown that objections to the generic name *Hormodendrum* have contributed more than any other cause to the long and confusing synonymy of the species. According to the International Rules of Botanical Nomenclature and as a matter of principle, it is objectionable to "use... names which may cause error or ambiguity, or throw science into confusion."

Thom (1940) has proposed to include *Pedrosoi* in the genus *Phialophora,* based on his opinion that the strains of that species and those of *verrucosa* are too homogeneous to warrant classification in different genera. Our own observations and those of others would not lend support to THOM'S opinion in this respect. The occurrence of phialides in *Pedrosoi* is truly indicative of some relationship with *Phialophora,* but there is no doubt whatsoever that, morphologically, the *Pedrosoi* group is infinitely closer to *Hormodendrum* than it is to *Phialophora.*

There is only one isolate of *Pedrosoi,* the so-called *"macrospora",* in which the *Phialophora* sporulation is a predominant character; in all the other isolates the *Hormodendrum* method, or the *Fonsecaea* method—which Thom (1940) himself has interpreted as a reduced form of *Hormodendrum*—constitute the predominating feature. Moreover, Emmons and Carrion (1936) have shown that the *Phialophora* conidiophores in *Fonsecaea pedrosoi* may be produced by transformation of one or more *Hormodendrum* spores in an otherwise normal *Hormodendrum* spore head, indicating that the *Hormodendrum* sporulation is the primary biologic character of the species. If there were no other alternative but to place *Pedrosoi* in either *Phialophora* or *Hormodendrum,* the latter should be considered the more legitimate (1) because the morphology of *Pedrosoi* is *by nature* closer to that genus, and, (2) because the species was originally described under that generic name which, thus, has the right of priority.

The occurrence of certain hotnogeneal characters among species does not always imply actual cogeneric relationships. This fact has been repeatedly recognised. The morphology and parasitic habits of the Dermatophytes, for instance, point to close interrelations among the members of that group; yet they have shown sufficient differences to warrant their classification into different genera. "The rusts, to take another conspicuous example, are caused by many undoubtedly related fungi which have evolved along divergent lines and are classified in different genera". Finally, it is quite evident that the *Phialophora* type of conidiophore is not an exclusive character of the genus *Phialophora,* since similar morphologic structures are to be found also in other fungi, such as *Pleurage anserina* (B. O. Dodge, 1936), *Podospora cwnwla* (Satina, 1917, quoted by B. O. Dodge, 1936), *Cylindrocarpon radicicola* which are classed in widely separated generic groups.

Binford *et al.* (1944) have accepted that a change in the generic name of *Pedrosoi* is necessary and, like Thom, they have stated their belief that this species should be transferred to the genus *Phialophora.* Their statement is made on the basis that "the *Phialophora*... sporulation was demonstrated *sparsely in several* strains of *Pedrosoi*" and *predominantly in one:* the so-called *"macrospora"*. These observations are evidently in support of the fact that the *Phialophora* sporulation is not the most important morphologic character in *Pedrosoi* and, consequently, cannot be considered as a sound basis for transferring the species. It should be noted that the isolate called *"macrospora"* is not closer to *Phialophora* than the isolate called *"algeriensis"* is to *Hormodendrum.* Indeed, it is much easier to find sporulation of the *Acrotheca* type in *"macrospora"* than to find phialides in *"algeriensis"* If the generic position of *Pedrosoi* were to be solved by emending a preexisting genus, as suggested by Binford and his associates, the rule of priority would point, not to *Phialophora,* but to *Hormodendrum,* which is part of the original binomial used by Brumpt in his description of the species.

The classification of *Pedrosoi* as a *Fonsecaea* has been accepted on the following basis: (1) *Fonsecaea* (Negroni, 1936) Carrion, 1942, *emend*, is a legitimately created generic group including, at least two, well differentiated species, namely, *Pedrosoi* and *compactum;* (2) it admits, *a priori*, the simultaneous occurrence of the *Hormodendrum, Acrotheca*, and *Phialophora* methods of sporulation as noted in its constituent species; (3) it covers without strain the different varieties of *Pedrosoi;* (4) it leaves out fungi that are exclusively *Hormodendrum*, or *Acrotheca*, or *Phialophora* in type (5) it recognises the existence of a characteristic *Fonsecaea* method of sporulation which is present in all the known strains of the group, reaching its highest degree of development in *Fonsecaea Pedrosoi typicus;* (6) as a generic name, it is neither misleading nor confusing, and, finally, (7) *Fonsecaea* has already become extensively popular, as noted in a review of recent publications on chromoblastomycosis.

In the preceding paragraphs we have presented the different trends of thought regarding the generic position of the species *Pedrosoi*. In view of the fact that this is an imperfect fungus possessing exceptional morphologic characters which link it closely to different form genera, it seems probable that differences of opinion will continue to exist in the future. The authors believe, however, that, with the broad generic concept of *Fonsecaea*, and the subdivision of *F. pedrosoi* into varieties, students in medical mycology will be able to work more effectively and lose less time in the classification of new or unknown isolates of this specific group. Further research towards determining the perfect phase of *F. pedrosoi* in either its natural habitat—most probably the soil or vegetable matter therein—or in artificial media should be encouraged as the only way leading to a scientific solution of this taxonomic problem.

Phialophora-verrucosa-medlar

Synonymy: *Cadophora americana* (Melin and Nannfeldt, 1934).

Geographic Distribution: *Phialophora verrucosa* has been encountered in only six chromoblastomycotic infections. The majority of these infections by *Phialophora verrucosa* occurred in continental United States and were distributed as follows: Boston, 1 case; Texas, 1 case; Missouri, 2 cases. The two remaining cases came from infections registered in Uruguay (South America) and Algiers (North Africa), respectively. Although the small number of *Phialophora* infections does not warrant final conclusions regarding the geographic distribution of the fungus as yet, it is of interest to note that five of the isolations were made in temperate climates while only one, the North African strain, corresponds to a tropical climate.

Natural Habitat: The isolation of a pathogenic fungus from its normal habitat in nature has been accomplished in only a few instances. In 1925, Kress *et al.* (quoted by Martin, 1938), working with material from the wood pulp industry in the United States and Canada, reported the isolation of several dark coloured fungi, some of which were later classified by Melin and Nannfeldt (1934, quoted by Martin, 1938) as species of the genus *Cadophora* Lagerberg, 1927. Subsequent studies have established beyond doubt that one of these species, the so-called *Cadophora americana* is identical (Conant, 1937; Martin, 1938) with *Phialophora verrucosa.* Thus, *Phialophora verrucosa* has become an additional member of the small group of pathogenic fungi so far discovered in their natural environment.

Morphology in Pathologic Tissue: In parasitised cutaneous tissue the fungus is indistinguishable from *F. pedrosoi.* It may be observed in both the dermis and epidermis as characteristic, usually spherical, sometimes polyhedric bodies measuring 8 to 15 microns in diameter. These bodies may occur singly or in clusters, rarely in chains. They are found within giant cells, free in the tissues, and in the centre of microabscesses.

The fungus elements possess a dark and coarse cell wall, which may vary in thickness, even in an individual cell. The protoplasm is finely granular, yellowish-brown in colour, and frequently contains variously-sized, mostly coarse, spherical or irregular refractile inclusions. There is no distinct nucleus discernible. Germination of the parasitic cells has been observed in old lesions or necrotic tissue, the filaments being short, thin-walled and septate, and showing the same pigmentation and structure noted in the mother elements (Moore and Mapother, 1940). In host tissues, the fungus evidently multiplies by internal septation, the process being identical with that already described in *Fonsecaea pedrosoi.* Multiplication by budding, endogenous sporulation and the production of conidia in tissue have been variously reported in the literature. However, as in the case of *F. pedrósoi,* neither the descriptions nor the illustrations published offer convincing evidence that *Phialophora verrucosa* may multiply in tissue by processes other than direct cell division.

Gross Morphology in Culture: Healthy colonies are produced on various laboratory media, but with great variation in the rate of growth according to individual isolates, cultures three weeks old on dextrose agar ranging in diameter between 0.5 cm. (Boston isolate) and 3.5 cm. (St. Louis isolate). Gross cultural characters generally more evident between the sixth and eighth weeks of growth; cultures roughly conical in shape, measuring 3.5 to 6 cm. in diameter on the eighth week, with summit elevation of 5 to 10 mm.; surface irregular due sometimes to radial folding, sometimes to the development of extensive, prominent, lobular, central outgrowths, and sometimes to both types of alterations; border festooned or indented; aerial hyphae profuse, short, variously described as grey, dark olivaceous and dark brown, forming velvety or felt-like growth arising from a more membranous and darker mycelial mat which is attached to the substrate by a fairly thick growth of submerged hyphae.

Microscopic Morphology in Culture: Microscopic characters usually more conspicuous between the fourth and sixth weeks of growth. Vegetative hyphae straight, cylindrical, branching, septate, with articles 8 to 25 microns long by 2 to 6 microns in diameter, septation often closer forming oidioid or moniliform mycelium; cell wall coarse, dark brown, with cell membrane thin and hyaline; protoplasm yellowish-brown, finely granular, and containing refractile droplets of various sizes; nucleus not apparent in fresh preparations, but visible with hematoxylin stain. Conidiophores either terminal or lateral, unicellular, measuring 5 to 12 by 2 to 3.8 microns, each conidiophore consisting of three parts: the base, largest of all, containing protoplasm and nucleus; a constricted portion or neck; and a cup-shaped outlet ranging in diameter, at the lip, from 1.1 to 4.7 microns, usually from 2 to 3; as a rule, neck and cup forming the terminal portion of the conidiophore which thus becomes flask-shaped; otherwise cup and neck arising laterally from fertile hyphal article. Conidia semi-endogenous, suggesting spermatia of Ascomycetes, produced singly and successively by budding through neck into cup, often numerously, overflowing cup and agglutinating to form spherical spore masses loosely adherent to cup; conidial elements oval, sometimes elongated, surface smooth, thin-walled, hyaline, measuring 0.8 to 2.8 by 1.4 to 7.8 microns, mostly 2 to 4 by 1 to 2.

Chlamydospores resembling "sclerotic cells" noted in tissue, sparsely produced in artificial culture media. Elements large, spherical; with coarse, dark walls; protoplasm yellowish-brown, granular, usually containing variously sized, refractile inclusions; some elements showing cell division by internal septation in different planes, with production of bi-, tri-, or tetracellular structures.

Behaviour in Artificial Culture Media: *Phialophora verrucosa* will grow in most of the laboratory media usually employed in medical mycology, but the morphology of the cultures varies according to the composition of the substrate. In media containing carbohydrates, and particularly, in dextrose agar, the colonies develop faster and aerial mycelium is produced in greater abundance (Medlar, 1915; Wilson *et at.*, 1933; Mackinnon, 1936; Montpellier and Catanei, 1944). These media are useful, therefore, for the study of gross cultural characters.

Czapek's synthetic medium appears to stimulate the development of mycelium in the substrate, partially inhibiting aerial growth (Mackinnon, 1936; Moore and Mapother, 1940). In the isolate studied by Moore and Mapother, the rate of growth was found to be faster in Czapek's agar. According to Medlar, Loeffler's blood serum and hydrocele agar would tend to produce moist colonies in which the aerial mycelium would lose its usual velvety appearance.

Some strains of *Phialophora verrucosa* have the property of producing a diffusible, dark, sepia-brown pigment, the formation of which appears to be related to the composition of the medium. The Boston isolate exhibits this property equally well on hydrocele agar and dextrose agar (Medlar, 1915); the Uruguayan (Mackinnon, 1936) and the St. Louis (Moore and Mapother, 1940) isolates produce the pigment on dextrose agar only; while the Texas isolate (Wilson *et al.* 1933) failed to stain the agar in culture.

The influence of ingredients of the medium on the microscopic morphology of the fungus has been studied by various workers. Moore and Mapother (1940) noted that the vegetative hyphae of *Phialophora verrucosa,* when grown on Czapek's and malt extract agars, have a smaller diameter than those developed in media containing peptone. The production of moniliform hyphae has been reported to be more prominent on media containing sugar, but this type of mycelium has also been produced on other laboratory media. Medlar (1915) observed that hydrocele agar and Loeffler's blood serum would induce the formation of chlamydospores of sclerotic type, which were not produced in any of the other media used in his studies. Moore *et al.* (1943) confirmed the occurrence of sclerotic cells on the last named medium. However, the production of these morphologic structures does not appear to be limited to the two media above mentioned. Sclerotic cells have been found on Czapek's agar and it is possible that they can be produced on other substrates, though perhaps in more scant numbers.

According to some investigators, Czapek's agar would be the most adequate substrate to stimulate sporulation in this species. It has been stated that, in this medium, phialides are produced in greater numbers and in the largest variety of forms.

Hydrogen-ion Concentration, Temperature, and Oxygen Requirements: *Phialophora verrucosa* thrives best on an acid substrate. The hydrogen-ion concentration of the media used in cultural studies of this fungus has ranged between 4.3 and 7 (Moore and Mapother, 1940). Its optimum temperature has not been precisely determined, but the fungus is known to develop well at 37° C. (Medlar, 1915) and at 28° C. (Mackinnon, 1934). In our own laboratories, satisfactory cultures have been obtained at room temperature (24° to 33° C.). Although growth is usually more luxuriant under aerobic conditions, the development of mycelium and sporulation in the depth of the substrate (Medlar, 1915; Mackinnon, 1936), especially in Czapek's agar, indicates that the organism may grow to a certain extent at reduced oxygen tension.

Chromatic Affinities: The chromatic affinities of *Phialophora verrucosa* were studied by Medlar in 1915 and, since then, no essential addition has been made to his observations. In smears from cultures, the fungus elements would stain well, though diffusely, with all the routine bacteriologic stains, structural details being more evident after staining with dilute methylene blue for ten minutes or longer, followed by thorough washing in water. In fresh preparations of the mycelium stained with hematoxylin, practically all the cells reveal the presence of a nucleus and, in the case of conidia, the nucleus also becomes visible with eosine and methylene blue. The protoplasmic droplets noted in vegetative elements and in conidia, stain bright red with scharlach and with carbol fuchsin, showing that they are fatty in nature. It has been found also that young fungus cells are Gram-positive, while the older structures take the blue irregularly when stained by Gram's method.

According to Medlar, culture sections fixed in Zenker's

fluid should be stained with eosine and methylene blue, or with Mallory's iron hematoxylin, or with phosphotungstic acid hematoxylin, the latter stain being specially valuable for its decolourising effect on the dark cell wall of the fungus, which permits more definition in structural detail. For culture sections fixed in alcohol, Mallory's hematoxylin or alum hematoxylin are equally effective. Medlar states further that the fungus in tissue may stain satisfactorily with the usual eosine-methylene-blue reagents.

Biochemical Reactions: Medlar reported the following biochemical reactions for *Phialophora verrucosa*. In litmus milk, no pellicle is formed, the milk is not coagulated or peptonised, and is gradually changed to a strongly alkaline reaction. In Dunham's peptone solution, no indol is produced, and the medium is coloured a dark brown to chocolate brown in old cultures. The fungus does not ferment lactose, maltose, saccharose, mannitol, inulin, or levulose. Slight acid production occurs in dextrose and dextrin.

Virulence: From the clinical picture noted in human infections, the slow course of the disease and the histopathologic reactions noted in the infected tissue, it is evident that *Phialophora verrucosa* is an organism of comparatively low virulence. In 1933 Wilson *et al.* made experimental inoculations to determine the variation in virulence produced by animal passage, noting that "Passage increased the virulence to a slight extent as judged by the greater emaciation of the animals and the somewhat larger size of the lesions."

Moore (1942) claims to have succeeded in producing infection of the chorio-allantois of chick embryos with the Boston and St. Louis strains of *Phialophora verrucosa*. The Boston strain, which had been kept in culture in the laboratory for 27 years, induced a definite, but mild and superficial reaction; while the St. Louis strain, of more recent isolation (four years), led to the development of more intensive and deeper lesions, indicating a higher degree of pathogenicity. As a result of his experiments, Moore was led to the conclusion that experimental infection of the chorio-allantois of chick embryos may be considered a more valuable and quicker method for determining comparative virulence among fungi, than the ordinary inoculation of other laboratory animals.

Immunologic Reactions: In 1936 Martin *et al.* demonstrated the presence of complement-fixing antibodies for *Phialophora verrucosa* in the serum of a patient infected with *Fonsecaea pedrosoi*. In a subsequent series of well controlled experiments, Conant and Martin (1937) showed that the serum of a rabbit artificially immunised against *P. verrucosa* (Uruguayan strain) possessed complement-fixing antibodies, in high titre, for this species. The complement fixation test was also positive, although in a low titre, for *Fonsecaea pedrosoi*, and it was negative for *Fonsecaea compactum*. It was further shown that *znti-pedrosoi* rabbit serum, which produced an intensely positive reaction with a *pedrosoi* antigen, would also react with *P. verrucosa*, although in a comparatively lower titre. Finally, anti-compactum rabbit serum was found to induce positive reactions of practically equal intensities, not only with *F. compactum*, but also with *F. pedrosoi* and *P. verrucosa*. The cross reactions noted by these investigators would point to the existence of both specific and group fractions in the antigenic structures of these three fungi.

A saprophytic strain of *Phialophora verrucosa (Cadophora americana)* isolated from wood pulp by Kress *et* al. (1925, quoted by Martin, 1938), has been shown to be antigenically similar to other pathogenic strains of the species (Martin, 1938).

Host Range: Man is the only known natural host susceptible to infection with *Phialophora verrucosa*. Artificial infection has been attempted in four species of experimental animals, namely, the rat (Medlar, 1915; Wilson *et al.* 1933; Mackinnon, 1934); the mouse (Medlar, 1915); the chick embryo (Moore, 1942) and the guinea-pig (Medlar, 1915). Success has been claimed in all except the last named species. In every instance the inoculum consisted of a suspension of the fungus grown on laboratory media. The routes of inoculation varied with different investigators and with the animals used. In rats and mice, infection could be elicited by subcutaneous, intraperitoneal or intratesticular inoculations; in chick embryos, the inoculation and infection were effected on the chorio-allantoic membrane. Although the attempts to produce infection with *Phialophora verrucosa* in experimental animals have not been numerous, the results obtained appear to be generally consistent with those noted in similar experiments with *Fonsecaea pedrosoi.*

Taxonomy: The species *P. verrucosa* has not shown a sexual phase of reproduction in laboratory cultures and should be classed, therefore, among the Fungi Imperfecti. It is remarkable that the semi-endogenous spores produced by this species resemble the spermatia noted in certain types of Ascomycetes. However, attempts to induce ascus formation by the pairing of different strains in laboratory cultures have not been successful (N. F. CONANT, personal communication to the authors).

The production of conidiophores from superficial hyphae at any point on the surface of the thallus places this fungus in the third order of Saccardo's classification, namely, the *Hyphomycetales (Moniliales);* while the dark colour of the cultures sets it among the *Dematiaceae.* Since the conidia are dark and unicellular, the position of the fungus lies in the first division of this family, with the *Phaeosporae.* According to the method of conidial formation, it falls in the subdivision *Chalareae.* The species was placed in the new genus *Phialophora* MEDLAR at the suggestion of Thaxter (Medlar, 1915) due to the agglutination of the conidia into sporangium-like masses at the mouths of the phialides. Finally, the specific name *verrucosa* was selected because the lesions produced by the fungus resemble those of *tuberculosis verrucosa cutis.*

Fonsecaea Compactum

Geographic Distribution: Up to the present time, there are only two recognised representatives of the species *Fonsecaea compactum:* one of them is the fungus originally discovered in a case of chromo-blastomycosis in Puerto Rico; the other was isolated more recently from the skin lesions of a patient in the Armed Forces in the State of Tennessee (USA).

Morphology in Pathologic Tissue: The fungus is present in the cutaneous lesions and may be readily recognised in both the dermis and epidermis. It is represented by characteristic, usually spherical, sometimes hemispherical or ovoid, occasionally crescent-shaped bodies, measuring from 7 to 10.5 microns in diameter. These bodies occur either singly or in clumps. They may be found within giant cells, free in the tissue, in the centre of microabscesses, and in the Malpighian and horny layers of the epidermis.

The fungus elements possess a dark and fairly thick cell wall. The protoplasm is granular, it is naturally coloured yellowish-brown to olivaceous, and contains, as a rule, variously sized, mostly coarse, spherical or irregular, refractile inclusions. The nucleus is not apparent either in fresh or stained preparations. Germination of the parasitic cells is frequently observed in the *stratum corneum* of the epidermis, the filaments being comparatively short, irregular, septate and branched, and showing the same pigmentation and structure noted in the mother elements. Filamentous structures have not been observed in the dermis. In the infected tissue, multiplication of the fungus takes place by internal septation, the process being identical to that already described in *Fonsecaea pedrosoi.*

Gross Morphology in Culture: On Sabouraud's growth slow; cultures six weeks old developed at room temperature, roughly conical in shape, measuring 2.5 cm. in diameter, with summit elevation of approximately 6 mm.; centre of culture forming irregular mamillary prominence, about 10 mm. wide; rest of culture moderately sloping and uneven; marginal zone shallow, slightly depressed in places, showing faint tendency to radial furrowing; border irregular and indented; aerial hyphae profuse and plushlike, coarse, erect, brownish-black, arranged in characteristic tufts, arising from a more compact and brittle mycelial mat which is attached to substrate by a thin growth of submerged hyphae.

On 4 per cent dextrose agar, cultures essentially similar to the above, but not as plushlike or bushy; aerial growth smoother and more velvety in appearance; hyphae finer; peripheral radial furrowing more pronounced; substrate mycelium extending slightly beyond border of culture, forming marginal rim, less than 1 mm. wide.

On Czapek's solution agar, cultures characteristic, though poorly developed, measuring 10 to 16 mm at the end of sixth week; mycelial growth chiefly in substrate, forming olive-black layer limited by an irregular, finely dentate border; centre of colony showing scant aerial growth, powdery and dusty in appearance, brownish-black in colour, reaching a thickness of about 1 mm. at point of inoculation.

Microscopic Morphology in Culture: Microscopic characters usually more conspicuous after third week of growth. Vegetative hyphae long; coarse (2.5 to 5.2 microns); septate; branching, sometimes dichotomously; borders usually irregular; cell walls thick and dark; protoplasm olivaceous, granular, with refractile droplets; hyphal fusions frequent; nucleus not apparent in fresh preparations. Sporulation of three different types, occasionally combined in same spore head: (a) *Hormodendrmn* type predominating. Fertile branches erect or ascending; often arborescent; with terminal cell (conidiophore) swollen, often flask-shaped and darker, having at tip broad facets to which spores are united. Conidia produced in short and profusely branching chains, some of the branches growing basipetally; conidial elements unicellular, spherical to subspherical, mostly barrel-shaped; cell wall smooth, coarse and dark; protoplasm olivaceous and granular; disjunctors absent, each conidium being compactly linked to its immediate neighbours in the chain by broad articulating facets; size of conidia 2.5 to 4.8 by 2.5 to 3.8 microns, basal elements 3.8 to 6 by 3 to 4.5 microns, (b) *Fonsecaea* type not conspicuously abundant. Conidiophores consisting of an article, usually short, straight or irregular; more darkly pigmented than vegetative mycelium; disposed terminally, laterally or intercalarily;

sometimes produced by transformation of spore element in *Hormodendrum* spore head; surface of conidiophore irregular due to presence of broad facets to which spores are attached. Conidia individually similar in morphology to terminal and subterminal elements of *Hormodendrum* chains already described in type (a), but produced acropleurogenously, and singly (not catenate), forming clusters; some clusters, however, showing tendency to chain formation as in *Hormodendrum*. (c) *Phialophora* type scant, morphology similar to that of *Phialophora verrucosa*.

Chlamydospores resembling "sclerotic cells" noted in tissue, sparsely produced in culture. Elements spherical, 8 to 12 microns in diameter; with coarse dark walls; protoplasm yellowish-brown, granular, usually containing variously sized, refractile inclusions; some elements showing cell division by internal septation in different planes with production of bi-, tri-, or tetracellular structures.

Behaviour in Artificial Culture Media: *Fonsecaea compactum* will grow in most of the laboratory media usually employed in medical mycology, but the morphology of the cultures varies according to the composition of the substrate. The formula most appropriate for optimum development of gross cultural characters is one containing agar, 2 per cent, with 1 per cent of peptone and 4 per cent of either maltose or dextrose as nutrients. In media prepared according to this general formula, and particularly in Sabouraud's *"milieu d'epreuve"*, the fungus grows more luxuriantly, the aerial mycelium is produced in greater abundance, and the colonies form more regular patterns. The above-mentioned media may be useful also for the study of microscopic morphology. However, corn meal agar and Çzapek's solution agar appear to be more stimulating to the process of sporulation and are generally employed in morphologic studies of the fungus.

Hydrogen-ion Concentration, Temperature, and Oxygen Requirements: *Fonsecaea compactum* requires an acid medium with a pH ranging between 5.5 and 6.5. Its optimum temperature has not been precisely determined, but the fungus develops well at both room (25°C) and incubator (37°C) temperatures. Although growth is usually more abundant under aerobic conditions, the development of mycelia in the depth of the agar, especially in Czapek's medium, indicates that this organism may grow at reduced oxygen tensions. It should be noted, however, that mycelia produced under such conditions are usually depauperate and devoid of sporulation.

Immunologic Reactions: In 1936 Martin *et al.* were able to demonstrate the presence of complement-fixing antibodies for *Fonsecaea compactum* in the serum of a patient infected with *Fonsccaea pedrosoi.* In a subsequent series of well-controlled experiments, Conant and Martin (1937) showed that the serum of a rabbit artificially immunised against *F. compactum* possessed complement-fixing antibodies, in high titre, not only for the immunising species, but also for *Fonsecaea pedrosoi.* Conversely, the serum of another rabbit immunised against *F. pedrosoi* produced equally strong positive reactions with culture antigens of both *Pedrosoi* and *compactum.* These observations are indicative of the existence of a close antigenic relationship between the two *Fonsecaeas: Pedrosoi* and *compactum.* The immune sera of the two animals also induced positive reactions, although in a lower titre, with a culture antigen prepared with *Phialophora verrucosa.* Finally, *Fonsecaea compactum* has been found to be antigencally related (Martin, 1938) to the so-called *"Cadophora americana"* (Melin and Nannfeldt, 1934), a saprophytic strain of *Phialophora verrucosa* isolated from wood pulp.

Viability: In a desiccated glucose agar culture of *Fonsecaea compactum* which was kept in the laboratory under ordinary conditions, the fungus was found to be still viable after a period of over two years. At the end of this period, inoculations of material from the dry culture into fresh media gave rise to characteristic cultures in which the specific morphology was typically reproduced.

Taxonomy: The species *compactum* has failed to reveal a sexual phase of reproduction in laboratory cultures, and should be classed, at least for the present, among the Fungi Imperfecti. The production of conidiophores from superficial hyphae at any point on the surface of the thallus places it in the third order of Saccardo's classification, namely, the *Hyphomycetales (Moniliales).* Its familial position, as determined by the dark colour of the mycelium and conidia, is among the *Dematiaceae.* It has been included in the genus *Fonsecaea* because it possesses the three types of sporulation characteristic of that genus, namely, the *Hormodendrum,* the *Fonsecaea,* and the *Phialophora.* The specific name *compactum* was given in consideration of the compact articulation of the spores in the *Hormodendrum* chains.

Hormodendrum Species Source: This fungus has been isolated from chromoblastomycotic lesions in only one instance, the infection taking place in a patient from the Cape Province, Union of South Africa. Original descriptions of the clinical picture in this patient and of the infecting fungus were published by Simson *et al.* in 1943.

Morphology in Pathologic Tissue: In the infected tissues the parasite was described (Simson *et al.*, 1943) as: "Brownish, round fungus bodies... lying free among the pus cells in the centres of the follicles, or included within the cytoplasm of the giant cells. These bodies measured 6 to 8 microns in diameter."

Gross Morphology in Culture: The fungus grows very slowly, cultures five weeks old incubated at room temperature on 2 per cent glucose agar measuring only 2 cm. in diameter. "The surface is rather irregular in cross-section, and, when the organism was first isolated, aerial hyphae" were not very abundant. "On repeated subculture, they increased in number until the surface became plushlike in character."

Microscopic Morphology in Culture: Vegetative hyphae long, septate, branching, measuring about 2.5 microns in diameter. Sporulation of *Hormodendrum* type only: spores adjacent to conidiophore measuring 8 to 10 by 2.5 to 4 microns, their size gradually decreasing towards the distal end of the chain, "average intermediate spore being about 3-5 microns."

Animal Inoculations: Heavy suspensions of the spores and hyphae of this fungus were injected subcutaneously into guinea-pigs and a *rhesus* monkey. Although the animals were examined repeatedly over a period of more than six months, no lesion developed.

Hormodendrum Species Source: This *Hormodendrum* was obtained from one of the lesions of a patient with chromoblastomycosis in Venezuela, the fungus being cultured from exudate aseptically aspirated from the lesion. Descriptions of the clinical picture and of the infecting parasite were published by O'daly in 1943.

Morphology in Pathologic Tissue: In purulent exudate from the cutaneous nodules the parasite is described as double-contoured, dark coloured, sometimes septate bodies, occurring either singly or in clusters. These bodies are also represented in photomicrographs of the histopathology.

Gross Morphology in Culture: Cultures six weeks old developed at room temperature on 4 per cent dextrose agar are irregularly circular in contour, and are superficially uneven. They measure 35 to 40 mm. in diameter, and, centrally, they are elevated about 6 mm. above the surface of the medium. The central portion of the culture is dome shaped, and is surrounded by a gently sloping, variously folded intermediate zone which is, in turn, surrounded by a smooth peripheral zone somewhat depressed below the surface of the medium. The border is festooned. Superficially the colonies are covered with aerial mycelium. In the centre this mycelium forms a brownish-green, velvety network of short, delicate, ascending hyphae; over the rest of the colony the growth is finer and more powdery. The cultures are attached to the agar by a moderate growth of substrate hyphae which extend peripherally beyond the border of the aerial mycelium, forming a glistening-black marginal rim.

On 2 per cent dextrose agar the colonies are generally more filamentous, and, on honey agar, the central prominence appears more elevated and is covered with shorter, finer, and more glossy aerial mycelium.

At the time of isolation, the cultures produced a dark diffusible pigment, but this property was not apparent in subcultures.

Microscopic Morphology in Culture: Microscopically this fungus has the general morphology of a *Hormodendrum.* The fungus has a fine mycelium, the spore chains are fairly long, and the spores are uniform in size. *No* further details are given in the original description.

Experimental Inoculations: By inoculating spores of this *Hormodendrum* on the tail of white rats, O'daly claims to have produced lesions indistinguishable from those produced by other etiologic fungi of chromoblastomycosis in the same animals.

Torula Poikilospora Source: This fungus was repeatedly isolated from purulent exudate and fragments of lesions in a case of chromoblastomycosis observed in Tokio, Japan. A report of the case and a description of the fungus were published by Takahashi in 1937.

Morphology in Pathologic Tissue: The parasite, as observed in the exudate, consisted of "round or oval, double-contoured, light to dark brown bodies, mostly unicellular, seldom bicellular, measuring 5.5 to 9 microns in diameter." In tissue sections these bodies usually occurred "within microabscesses and giant cells, seldom in the lumen of newly formed blood vessels." In shape and colour, they were similar to those noted in the exudate, but their diameter was given as 5.5 to 10 microns for the round forms, and 6-9 by 5.5-7.5 microns for the oval forms. The cell wall consisted of a coarse, dark brown, outer layer, and a thinner, light brown, inner membrane.

Lesions induced on experimental animals revealed the fungus structures in two different forms: the "yeast-like" and the mycelial. The "yeast-like" form corresponded closely with the round or oval bodies already noted in human lesions. The mycelial form could be best observed in the necrotic tissue of mouse lesions, and was represented by a network of olive-green, sparsely branched hyphae, 1.5 to 3.5 microns in width, divided into rather short articles by transverse septa. According to Takahashi, the mycelial structures were less abundantly produced in rabbit and rat lesions, and they were never observed in guinea-pig infections.

Gross Morphology in Culture: On Sabouraud's glucose agar, growth slow, colonies one month old measuring 3 to 3.5 cm. in diameter; shape of culture roughly conical, showing central hemispherical prominence, surrounded by a more extensive, gently sloping, radially furrowed zone; border circular, filamentous, finely arborescent; aerial hyphae short, downy, brownish-gray, arising from a soot-black, moist, granular mycelial mat.

On Sabouraud's maltose agar and on "Pollacci's medium," cultures essentially similar to the above. On plain agar and ascitic agar, growth slower, and furrowing less conspicuous. On potato and carrot plugs, growth slow, colonies hemispherical, reaching a diameter of 2 to 2.5 cm. after two months; covered by brownish-gray, short, and downy mycelium. On liquid media (glucose broth, maltose broth, and "potato water"), growth noted at the surface and bottom, the liquid remaining clear.

Microscopic Morphology in Culture: Vegetative hyphae straight, undulated, or bent at acute angles; cylindrical; profusely branched, some filaments showing multiple terminal branching; septate, with articles 3 to 25 microns long by 1.5 to 4.5 microns wide; protoplasm olive-green, older hyphae more deeply coloured and containing variously shaped deposits of a black pigment. Fertile hyphae undifferentiated. Arthrospores developed at tip and sides of filaments, forming long, branching chains, the acrogenous chains being longer and more branched than the pleurogenous; spores firmly linked to immediate neighbours in chain by broad articulations; spores olive-green, the older containing black pigment deposits; some spores spherical (2 to 9 microns in diameter) or oval (2.5 to 14 by 1.4 to 8 microns); most spores, however, irregular in shape, measuring 2 to 2.5 by 1.5 to 16 microns, the irregularity resulting from development of knob-like or short filiform appendages on the tip or on the side, never on the base, of spore cell, the appendages sometimes producing a secondary element and often separated from each other and from mother spore by septation.

Chlamydospores developed in old cultures, sometimes in chains; elements double-contoured, light brown to brownish-black; terminal or intercalary.

Temperature Requirements: The optimum growth temperature for *Torula poikilospora* is given as 37°C.

Biochemical Reactions: The fungus does not ferment glucose, galactose, mannose, levulose, arabinose, xylose, rhamnose, saccharose, maltose, lactose, raffinose, dextrin, inulin and mannite.

Experimental Infections: Takahashi has reported the reproduction of typical lesions of chromoblastomycosis on the healthy skin of his patient, following local inoculations with suspensions of *Torula poikilospora*. The histopathology of the induced lesions was similar to that of the natural foci of infection noted in the clinical picture, and the etiologic fungus could be recovered in culture.

Experimental infections were also produced in rats, mice, rabbits, and guinea-pigs, susceptibility being more pronounced on the first three of the species mentioned. The inoculum used in these experiments consisted of saline suspensions of the fungus grown for three months on Sabouraud's glucose agar. The inoculations were made by the subcutaneous route in 4 mice and 2 white rats; the intraperitoneal route was used in 12 mice, 7 white rats, and 4 guinea-pigs; and the intravenous route, in 3 rabbits.

The subcutaneous inoculations were followed by the development of ulcerating nodules locally. The intraperitoneal inoculations produced progressive granulomatous nodules on all the abdominal organs of the rats and mice, and on the omentum of the guinea-pigs; finally, the intravenous inoculations induced a granulomatous reaction in the lungs of all the rabbits tested. Microscopically, the lesions were of granulomatous type and resembled those noted in human beings, with the infecting parasite constituting a prominent part of the picture.

Taxonomy: *Torula poikilospora* is an imperfect fungus corresponding to the family *Dematiaceae*. Following the classification of Saccardolindau, Takahashi has included this species among the *Toruleae*, and has placed it provisionally in the genus *Torula* Persoon. The specific name *poikilospora* was given in accordance with the polymorphic character of the spores produced by the fungus.

Hormiscium Dermatitidis Source: *Hormiscium dermatitidis* was originally isolated from the lesions of a patient with hromoblastomycosis in Japan and the fungus was described as a new pathogenic species by Kano in 1937.

Morphology in Pathologic Tissue: The parasite could be observed in the cutaneous lesions of the patient and it was present in both the dermis and epidermis. In the infected epidermal scrapings, as well as in tissue sections, it was described as brownish, round or oval bodies, measuring 2 to 10 microns in diameter, with a dark brown, double-contoured membrane, and a brownish, granular protoplasm. These bodies occurred either singly or in small groups, sometimes in chains of 2 to 5 elements. They were found within giant cells, or free in the tissues, or in the centre of microabscesses.

Gross Morphology in Culture: On 4 per cent maltose agar, growth slow, cultures six weeks old at room temperature measuring 8 mm. in diameter; cultures hemispherical in shape; coal-black in colour; surface finely granular and wrinkled; substrate stained by intensely black, diffusible pigment which extends beyond limits of colony.

On 4 per cent grape sugar agar, on mulberry, and on

potato plugs, growth essentially similar to that described on maltose agar, cultures most exuberant on potato plugs. On peptone agar, growth poor. In glucose and maltose peptone water, loose floccular colonies developed at bottom of tube, with rest of the liquid clear.

Microscopic Morphology in Culture: Microscopic characters relatively simple and constant in a large number of laboratory media, both solid and liquid, employed in the study of the fungus. Cultures four days old on "glucose-peptone agar" developed at a temperature of 30° C, showed "short chains of spherical, oval, rod-shaped, clavate, or swollen cells," the chains usually consisting of ten or more elements; branching definite, though rudimentary, dichotomous, sometimes trichotomous, and taking place at the tip or side of the cell, often close to an intercellular septum. Cellular elements of different size, the spherical forms measuring about 10 microns in diameter, the others ranging from 3 to 5 microns in width by 5 to 20 in length; cell membrane thick and dark, becoming double-contoured and more deeply pigmented with age; some cells showing internal septum perpendicular to longitudinal axis.

Large chlamydospores, of "sclerotic" type, present in old cultures; some of these chlamydospores showing internal septation in one or more planes; some showing constriction at the line of contact of the cell wall with the internal septum; some producing daughter cells by lateral sprouting of the mother element.

On carrot-plug cultures, rudimentary mycelium produced in moderate abundance, each mycelial structure consisting of about ten relatively long and thin articles, measuring 2 to 3 microns in diameter, and 6 to 8 in length.

Optimum Temperature: In cultures on maltose agar, optimum growth was produced between 20° and 30° C. At 37° C. the development was very poor, and at 43° C. it was inhibited.

Sugar Fermentation Reactions: Fermentation activities were tested with maltose, dextrin, levulose, lactose, glucose, arabinose, galactose, saccharose and xylose with negative results in every instance.

Experimental Infections: Kano has reported the reproduction of a typical nodule of chrōmoblastomycosis on the healthy skin of his patient by inoculating, locally, a suspension of *Hormiscium dermatitidis.* The histopathology of the nodule was similar to that of other lesions naturally developed on the patient's skin.

Experimental infections were also produced on three mice, three guinea-pigs, and three rabbits, the inoculum consisting of a suspension of the fungus grown in artificial cultures. The subcutaneous route of inoculation was used in all the animals. However, additional and simultaneous inoculations were made intraperitoneally to the mice and guinea-pigs, and intratesticularly to the rabbits. The subcutaneous inoculations were negative for the mice, but, in the guinea-pigs and rabbits, they were followed by discrete and progressive local indurations. Histopathologically, these indurations were of the type of an infectious granuloma with microabscesses. The intraperitoneal inoculations produced progressive granulomatous nodules in the abdominal organs of all the animals, the lesions taking a tuberculoid character in the mice and guinea-pigs. The testicular lesions developed in the rabbits consisted of a diffuse induration of the organ. Microscopically, the histopathologic reaction was of granulomatous and tuberculoid type, resembling that noted in human lesions, with the infecting parasite constituting a prominent part of the picture.

Taxonomy: This fungus has not shown a sexual phase of reproduction in artificial cultures and, consequently, has been classed among the Fungi Imperfecti. Following the classification of Saccardolindau, Kano has included this species in the family *Dematiaceae,* Fries subdivision *Phaeosporeae,* Saccardo; genus *Hormiscium* Kunze; and, in accordance with OTA'S advice, the fungus has been given the provisional name *Hormiscium dermatitidis.*

Candida-like Species Source: This blade, *Candida-like,* fungus was repeatedly isolated by Berger *et al.* (1945) from the cutaneous lesions of a Canadian patient who had been suffering from papillomatous growths over a period of fourteen years. From the original focus of infection, located on the left hand, the disease gradually spread to the opposite hand, the left forearm, the left foot, the right ear, the right buttock, and both cheeks. Histopathologically, the lesions presented the classical picture of Chromoblastomycosis, although the infiltrate was sometimes less nodular, and the fungus smaller, more delicate, and distinctly budding.

Morphology in Pathologic Tissue: In the host tissues, the parasite appeared as yeast-like bodies, the morphology of which was somewhat different in two series of biopsies done at different periods. In the earlier biopsies, these bodies were very abundant, sometimes outnumbering the inflammatory cells, especially in the papillary layer of the dermis. The fungus cells were rather small and delicate, and had a fine membrane, a few of them showing distinct budding, while others would form short, bead-like chains. In biopsies done at a later period, the parasitic cells were not so abundant, and were located in the middle and deeper layers of the dermis. They were coarser, thick-walled, irregular, devoid of budding, generally clustered, often engulfed by giant cells and frequently degenerated.

Gross Morphology in Culture: In his description of the fungus, as originally isolated, Berger reports: "Colonies on corn meal and hay infusion agar, Loeffler's serum and blood agar, remain small (3 to 4 mm.), and die soon on the two latter media, but are larger on glycerolated potato and carrot, and on Sabouraud's agar (4 to 7 mm.), and still larger on Lowenstein's and Petragnani's media (up to 15 mm.). On Sabouraud's agar, they are dark brown from the beginning, and half-spherical; on the other media, they are coal-black; they are always moist, of creamy consistence and glossy..." "In potato water (Langeron and Talice), small, fluffy colonies rest on the bottom, and the liquid contains a great number of single or budding yeast cells; there is no veil."

"Cultures made from a biopsy in 1944 yielded brown to black colonies which grew much more slowly, and were rough and dry, and presented microscopical features differing from those of colonies obtained before."

Microscopic Morphology in Culture: Differences in microscopic morphology were likewise noted between the early and late isolations of the fungus. Berger states that cultures of the early isolations "at pH 7 are predominantly formed by yeast-like cells and of only a scant pseudomycelium; at lower pH's, pseudomycelial branching filaments extend on the surface and into the underlying medium, and bear clusters of pseudoconidial blastospores; there are no aerial hyphae." In cultures obtained from a later biopsy (1944), the fundamental characters are still yeast-like, but the cells are larger and thicker, often double-walled; outlines often irregular, and many elements show internal septations in different planes. The resulting multicellular structures become very complicated, and are made up of intricately intertwined, irregularly curved, and anastomosing, rather thick threads of clustered elements which grow in all directions. These... cells may contain one or more round lipoid inclusions.... They may be considered as sclerotic cells and are obviously a variant of the initially yeast-like pseudomycelial fungus.

In hanging drop and slide cultures, many budding cells may be seen which remain attached to the mother cell, and form, by successive buddings, moniliform strands of various lengths; the latter become pseudomycelial through elongation of the individual elements. Apical blastospores appear and give in turn rise to pseudomycelial threads, but the branches are few and the resulting arborisation is, therefore, rather simple. Some blastospores show intensive budding of many small conidia-like blastospores, which form soon more or less large clusters of easily detached cells. There is no trace whatsoever of conidiophores or of any other specialised apparatus of sporulation, although many colonies were kept alive for over two years.

Effects of Temperature, Anaerobiosis, and pH on Growth: Berger's organism grows equally well at room and incubator temperatures; it does not grow in the depth of agar stabs; at pH 7 its morphology resembles that of the yeast, and at lower pH's, the cultures become filamentous.

Biochemical Reactions: The fungus does not liquefy gelatin and has no action on milk. It acidifies agar media containing glucose, fructose, mannose, galactose, cellobiose, and xylose respectively, but there is no gas formation in the corresponding liquid sugar media.

3

Biological Features

Biology of Coccidioides

Coccidioides immitis is perhaps the best known of the fungi which cause generalised mycoses of man, yet it challenges the student with many unanswered questions. For several years after its discovery it was classified among the Protozoa. Ophuls and Moffitt (1900) convincingly demonstrated its fungus nature, yet its anomalous life cycle has led modern investigators to assign it to positions as far apart in taxonomic mycology as the Ascomycetes and Phycomycetes. Its relationships in the latter group (where it undoubtedly belongs) are still undetermined.

The disease which it causes is not transmitted from man to man, and it seems certain from epidemiological and case history studies that most infections follow inhalation of spore-laden dust, yet it is not proved whether the fungus normally grows as a saprophyte in soil or whether it is primarily a disease of rodents by which the soil is contaminated. The disease is geographically limited to relatively small endemic areas in which it may be very common, yet we do not know the ecological factor which imposes this limitation unless it be the required presence of certain species of rodents which, unlike more widely distributed species, are properly balanced as to susceptibility and resistance as to provide a suitable reservoir. Besides its generalised and highly fatal form the disease occurs as a benign, self-limited, respiratory infection, as shown by Dickson (1937, 1938), Gifford (1936), Smith (1940, 1943), and others; yet the very mild, subclinical form of infection by which persons become sensitised and apparently immunised passes unrecognised in most infant residents of an endemic area.

History: The mycosis caused by *C. immitis* was first reported from Argentina by Posadas and Wernicke in 1892. They described the disease as protozoan. The second and third recorded cases were studied in southern California by Rixford and Gilchrist (1896) who, seeing a resemblance to *Coccidium,* described the disease as coccidioidal pseudotuberculosis, and named the organism *Coccidioides immitis.* Ophuls and Moffitt (1900) reported the third North American case and, having isolated the organism in culture, described its various growth phases and proved that it was a fungus. Ophuls (1905) supplemented the description of the parasitic growth phase of the fungus given by Rixford and Gilchrist by a description, complete in essential details, of the growth of the fungus in tissue, its production of spores by the formation of cleavage furrows, the transformation of this growth phase to that of a mould when the fungus is planted on agar, and the production of chlamydospores in culture. Many subsequent investigators have confirmed Ophuls' observations and corrected and supplemented them in details.

Most of the early observations of the mycosis were made in southern California and especially in the San Joaquin Valley. Many of the reported cases from South America were incorrectly diagnosed and were, in fact, paracoccidioidal granuloma or South American blastomycosis, as later South American investigators pointed out. Only a few authentic South

American cases are recorded. As the disease became better known and was searched for it was found in southern Arizona and western Texas. The extensive use of coccidioidin in skin-testing by C. E. Smith has to a considerable extent determined the limits of the disease in this general region. The observations of Aronson, Saylor and Parr (1942) have indicated that there is a high incidence of the mycosis in southern Arizona. The few reported cases from other parts of the United States are in some cases in persons who in all probability were infected while travelling or residing within the known endemic areas of the disease or who may have been infected from packing materials, food products or similar material originating in those areas. A possible endemic focus may exist in Italy. It is generally recognised, however, that the important endemic area is the arid region of southwestern United States.

Primary Coccidioidomycosis

Until 1937 the mycosis, generally known as coccidioidal granuloma, was recognised as a generalised infection having a mortality rate of 40 to 50 per cent. At that time Gifford (1936) and Dickson (1937), investigating a relatively mild respiratory disease common in the San Joaquin Valley and known to the layman as Valley fever, desert rheumatism, etc. proved that this was the primary form and by far the commonest manifestation of infection by *C. immitis*. Dickson proposed that this be called *primary Coccidioidomycosis*. Infection follows inhalation of wind-blown spores and occurs following dust storms and in people exposed in such dusty occupations as fence or road building and picking dusty grapes or cotton. Many infections, as revealed by coccidioidin skin tests are subclinical. This appears to be usual in the case of infants or young children who are native to the endemic area. Aronson, Saylor and Parr (1942) showed that in southern Arizona as many as 97 per cent of Indian children may react to coccidioidin, yet the disease remains unrecognised in this age group and population. On the other hand, in adults entering the endemic area the primary infection is often severe and may be characterised by cough (usually non-productive), pleural pain, malaise and loss of appetite, fever, chills, nightsweats, headache, and backache. There may be a transient rash. Some or all of these symptoms may appear 10 to 14 days after exposure. From 2 to 20 days after the acute respiratory stage the lesions of erythema nodosum and sometimes of erythema multiforme may appear. This constitutes the classical picture of "Valley fever" (primary coccidioidomycosis).

Roentgenograms may show a widening of the hilar shadows, pneumonic infiltration, nodular lesions in the lung parenchyma, small pleural effusions, and occasionally thin-walled cavities, which close spontaneously. Despite the dramatic onset and severity of the symptoms in some cases, the prognosis in primary coccidioidomycosis is very good. Cavities, when present, heal much more quickly than in tuberculosis and the pulmonary lesions may completely disappear, or calcification may follow healing. In most cases, there are no sequelae and the only evidence of a past infection is the persistence of skin sensitivity as revealed by the coccidioidin intradermal test.

Progressive Coccidioidomycosis: In perhaps 0.2 per cent of persons infected with *C. immitis, secondary* or *progressive coccidioidomycosis* develops, usually within a few weeks, but occasionally after a lapse of several years. This is the generalised form of the disease which was first recognised (coccidioidal granuloma). It presents many clinical variations, involving as it does almost any organ with the usual exception of the gastrointestinal tract. The infection extends from the lungs to involve the other visceral organs, brain, meninges, bones, joints, subcutaneous tissues, and skin. Fistulas draining deep-seated lesions develop; there are often enormous sacculate subcutaneous abscesses, and ulcers of the skin are frequent. In rare cases primary skin lesions develop. In the exudate from lesions the fungus can be found, often in great numbers. The mortality rate is high but spontaneous arrest of the infection occurs in some cases as shown by Cox and Smith (1939).

Laboratory Diagnosis: Diagnosis is made by the laboratory-demonstration of the fungus and its isolation in culture. Pus or sputum can be examined by mounting a small drop (with or without an equal amount of 10 per cent NaOH) on a slide under a cover slip and searching it for the presence of the large spherical cells of the fungus. There may be some doubt about the identity of the organism unless mature cells containing endospores can be found. The fungus grows rapidly on a wide variety of media and isolation in culture is therefore relatively easy if the material is not heavily contaminated by secondary invaders. It is necessary only to streak the pus out on the surface of such a medium as Sabouraud's agar and incubate the culture so made at a temperature between that of the room and 37°C.

If the pus is contaminated or if the isolation is to be made from sputum the presence of bacteria may interfere with the ready isolation of the fungus. C. E. Smith recommends the use of a selective medium containing 1 per cent ammonium chloride, 1 per cent sodium acetate, 0.8 per cent tribasic potassium phosphate, and 2 per cent agar. This is autoclaved for 10 minutes at IS Ibs. pressure and 0.04 per cent cupric sulphate is added just before the agar is poured. *C. immitis* makes a slow scanty growth on this medium, but most other organisms are inhibited. Any doubt regarding the identity of the fungus isolated can be resolved by injecting its spores intraperitoneally into white mice.

Therapy: Treatment of coccidioidomycosis is not specific and consists principally in supportive therapy. This is usually satisfactory in the case of the primary disease. In progressive coccidioidomycosis there is poor response to therapy. Jacobson has recommended the use of vaccines and colloidal copper. Penicillin has been tried without success. When lesions are localised, excision or amputation may be indicated.

Coccidioidin: Coccidioidin has been widely used to determine the existence of coccidioidomycosis in an area and to detect a previous exposure or infection in individuals. This sterile, skin-testing antigen is prepared by Smith (1943) by growing ten strains of C. *immitis* on Bureau of Animal Industry Synthetic Medium (the asparagine broth medium used in making tuberculin) for one to two months. After testing a sample for potency the broth is sterilised by Berke-feld filtration, tested for sterility, and merthiolate to make a final concentration of 1:10,000 is added. Potency and specificity tests are run on persons who have recovered from the mycosis and whose degree of skin sensitivity is known from previous tests, as well as by testing persons who are known to lack sensitisation to the fungus.

Lots of Coccidioidin which are not potent or are not specific are discarded. The undiluted Coccidioidin keeps several years, and when diluted 1:100 and kept sterile and refrigerated retains its potency for several months. However it is not recommended that it be used for more than one month after the dilution is made. In testing, 0.1 ml of a 1:100 dilution should be injected intracutaneously. This may cause severe reactions in occasional individuals, and if the person to be tested has had erythema nodosum he should be tested only with a dilution of 1:1000 or 1:10,000.

The test is read in 24 and 48 hours, preferably the latter if only one reading is to be made. Edema only is measured and areas of edema of 5 mm. or more in diameter are considered positive. According to Smith, there is no danger of reactivating arrested lesions, even when the reaction is severe. Persons who have had primary coccidioidomycosis, and particularly those who have had lesions of erythema nodosum or erythema multiforme accompanying the infection, give the most severe reactions. In persons having progressive coccidioidomycbsis the reactions are less severe, and in the terminal phases sensitisation may entirely disappear.

Coccidioidin, when properly prepared and standardised is generally considered to have a good degree of specificity. The studies of Aronson *et al.* show that there is no cross reaction with tuberculin. There is cross reaction with *Haplosporangium parvum*, an apparently related fungus which was isolated from a pulmonary mycosis in rodents, but which has not yet been isolated from man. Perhaps 10 per cent of persons living in certain areas of the mid-West and East who have had no known exposure to *Coccidioides* react to coccidioidin. The significance of these non-specific reactions is not known. A skin reaction to coccidioidin usually means past or present infection with *Coccidioides*, but it is possible that there are other fungi which also sensitise to this fungus and whose presence in other areas explains the non-specific reactions. Skin sensitivity once acquired persists for many years, and this must be kept in mind if coccidioidin is being used as a diagnostic agent.

Hassid, Baker and Mccready (1943) isolated an immunologically active polysaccharide from coccidioidin prepared in the standard way on synthetic asparagine medium. This polysaccharide consisted of galacturonic acid, glucose and some unidentified sugar in the approximate ratio of 1:6.3. A non-protein nitrogen compound was associated with it. Skin reactions in sensitised individuals were elicited by the polysaccharide but not by a polysaccharide regenerated from an acetyl derivative. Both these sugars were satisfactory for demonstrating precipitins, however.

Serology: Cooke in 1915 demonstrated the presence of precipitins and found no agglutinins in coccidioidal granuloma. Davis (1924) also failed to demonstrate agglutinins. He described a complement fixation test, using a suspension of a culture as antigen.

Coccidioidin has been used as an antigen in precipitin and complement fixation tests in coccidioidomycosis, although these serologic tests are not widely used because of the somewhat erratic results obtained. Smith reported that the tests are rarely positive in mild or inapparent infections. Precipitins are usually present in the more severe primary infections, but not in progressive coccidioidomycosis. In the latter complement fixation may be complete in 1:256 serum dilution. A fall in precipitins and an increase in complement fixing antibodies may therefore herald dissemination of the infection. The sedimentation rate is increased in coccidioidomycosis and this test may be of value, therefore, in helping to interpret a positive coccidioidin test.

Cultures: When pus or sputum containing *C. immitis* is planted on agar and incubated at 30°C growth of the fungus is apparent in three or four days, contrasting with the relatively slow development of most other pathogenic fungi. The appearance is that of a gray or white mould. Although actually the hyphae are rather delicate and of small diameter their tufted and tangled arrangement gives the colony the aspect of a coarse mould. In a colony of a typical strain ten days old there is a central tuft of hyphae surrounded by a narrow zone in which surface growth is sparse. Outside this is a wider zone of cottony tangled mycelium. The colour is white at first, becoming grey, and in many strains becoming brown. Some strains develop a lemon yellow pigment. There is little colour in the reverse side of an agar culture, although there is an indefinite dark area near the centre of the colony in many strains, and an occasional strain produces marked darkening of peptone-containing media. Variation among strains can be readily demonstrated if many strains are compared. Some strains are almost entirely confined to the substratum with very scant development of aerial mycelium. Baker, Mrak and Smith (1943) have described some of the variations commonly found.

In many strains the first spores to form in culture appear on specialised branches of the mycelium. These may be simple side branches or they may be more or less elaborately branched. The stalk of one of these branches is of the same diameter as the vegetative hyphae, but it abruptly widens to approximately twice that diameter in its fertile portion. This is particularly apparent in the larger, more complex structures. These specialised fertile hyphae were first described by Emmons (1942) in many strains of the fungus, and it is probable that they are normally present. They can be demonstrated only in young cultures soon after the time spores begin to form by carefully teasing out the mycelium in a microscopic mount. When older cultures are examined these specialised hyphae have lost their identity or have disintegrated with the formation of spores, and additional hyphae, poorly or not at all differentiated morphologically have become fertile. One finds then only a tangled mass of vegetative hyphae and fragments of spore chains.

The manner of spore formation is distinctive, although occasional fungi are found as contaminants in which the manner of sporulation is so similar as to be confusing when one is attempting to identify a culture. The fertile hyphae which are destined for spore production are large and engorged, and become divided by frequent septa into barrel-shaped segments. Within these cells the protoplasm condenses so that it occupies one end or, usually, the middle portion of the cell. This results in a chain of spores each separated from the next by a space lacking protoplasm.

The original septa dividing the fertile hypha can usually be seen midway between the chlamydospores so formed. When condensation of the protoplasm in two adjacent cells occurs by chance next to the common wall separating them the resulting chlamydospores are of course adjacent to each other, and this breaks the symmetry of the usual arrangement of a chain of spores separated from each other by spaces.

The chlamydospores dissociate rather easily when old, but younger spores may remain attached in pairs or short chains. This persistent association of some spores is an important consideration in the interpretation of the structures observed in animals experimentally infected from cultures, as will be noted in a later paragraph. The large number of spores produced in a culture, and the ease with which they are blown about makes the handling of old cultures a dangerous procedure. C. *immitis* should not be grown in plate cultures unless these are destroyed before cultures become old.

Transition to Parasitic Growth Phase

When the chlamydospores of *C. immitis* are injected into an animal they are transformed directly into the parasitic growth phase of the fungus. Each spore rounds up and increases in size until it becomes a spherical cell. When the cell becomes mature (in some cases before it reaches the indicated size, in other cases after becoming much larger) its protoplasm is divided by the formation of cleavage planes into an indeterminate number of endospores (sporangiospores). If the suspension of chlamydospores injected contained short chains of spores these may remain connected during the transformation to the parasitic growth phase and be found as chains of large cells (sporangia) filled with sporangiospores. It is not uncommon to see pairs of such transformed chlamydospores attached to each other by a short tube which represents an empty or dead cell in the original chain of chlamydospores. These bear a superficial resemblance to conjugating pairs of cells but should not be so interpreted. With the rupture of the sporangium and the subsequent escape of the spores the parasitic growth phase of the fungus properly begins.

Parasitic Growth Phase: Sporangiospores vary considerably in size but in most cases where subdivision of the protoplasm has progressed normally the liberated spores are 2-3/x in diameter. These may be taken up by any phagocytic cell, perhaps most often by giant cells," or they may remain intercellular. In either case, unless destroyed by the protective mechanism of the host, they proceed to grow. In some tissues, and dependent in part perhaps on strain differences, the growing cell early develops a central vacuole which soon occupies most of the cell and limits the stainable protoplasm to a relatively thin peripheral layer. This layer of protoplasm increases in thickness as the cell matures, and in some cases the central vacuole may be lacking during most of the development of the cell. In the material illustrated, taken from experimentally infected mice, the vacuole was a conspicuous feature except in the mature sporulating cells. As the cell matures small vacuoles appear in the peripheral cytoplasm and these apparently determine the location and direction of the cleavage furrows which initiate sporulation. The first cleavage cuts out large multinucleate masses of protoplasm, and these are progressively subdivided by further cleavage into the ultimate sporangiospores. This process of sporulation seems to be in all essential respects like that found in the Phycomycetes.

When the sporangium becomes filled with spores its outer wall is ruptured, apparently by internal pressures, and the spores escape into the surrounding tissues where they repeat the cycle just outlined. Hyphal formation and budding do not occur during the parasitic growth phase. When material is removed from animal tissue to agar the fungus immediately reverts by the formation of hyphae to the mould or saprophytic growth phase. Spores still within the sporangium, newly liberated spores, partly grown vegetative cells and fully grown sporangia before the initiation of sporulation are capable of this transformation by the production of germination tubes.

The development of sporangia in cultures was demonstrated by Lack (1938) who incubated chlamydospores in glucose broth and partially coagulated egg albumen under reduced oxygen tension. Baker and Mrak (1941) reported that sporangia ("culture spherules") were produced in ordinary agar cultures of some strains. The nature of these spherule-like bodies found in old cultures of some strains needs further investigation.

Nuclei were demonstrated in material from experimentally infected mice following fixation with Wilson's modification of Bouin's solution, careful dehydration and infiltration, and staining with iron alum hematoxylin. The nuclei are small, with a delicate nuclear membrane, a small amount of chromatin distributed peripherally and a single nucleolus. Unless the process of sporulation is disturbed the ultimate sporangiospores are uninucleate but nuclear division quickly follows their growth.

Variation: During the early studies of coccidioidomycosis two species of *Coccidioides* were described, based on differences in the size of sporangiospores. Subsequent studies showed that spore size is variable and depends upon whether sporulation proceeds normally or whether it is interrupted at some point. The size of the sporangium varies widely also, depending in part upon strain differences, but more importantly upon the host or the tissue in which it is growing. In material from man the cells of *C. immitis* may be uniformly filled with protoplasm. In infected mice most strains produce cells with a conspicuous central vacuole and the latter may be so large that only a thin peripheral layer of cytoplasm is seen. It seems probable that large size and the presence of a large vacuole are both indicative of rapid growth in tissue which presents little opposition to the growth of the cell either because of natural susceptibility or because the normal immunological response has been prevented by an overwhelming invasion of the infecting fungus.

C. immitis is equally variable in culture. The appearance of a typical strain has already been described. In some strains, the typical tufted and tangled aerial mycelium is lacking and the mycelial growth is confined to the substratum where it forms a moist tough layer at and below the surface. Some aerial hyphae may be present near the centre, aggregated into strands which resemble coremia. The usual colour of the fungus is gray, becoming brownish with age and the formation of spores, but some strains remain pure white while others are yellow. The virulence of strains varies, but is not correlated with any particular colony type. An apparent lessening of virulence is sometimes due to the sparse production of spores with a consequent difficulty in getting a heavy spore suspension for animal inoculation.

Transmission: Although the sporangiospores of *C. immitis* are infectious and it is possible to transmit the disease experimentally with them they do not seem to be important factors in the direct transmission of the disease. Direct transmission from man to man is not known and probably it is very infrequent. On the contrary the chlamydospores of the fungus are the important disseminating elements and infection occurs usually by the inhalation of these spores. Rarely they are introduced into a wound or abrasion and primary skin lesions result. The spores may be present in windblown dust and it has been generally assumed that the fungus grows in soil during and following the rainy season and that during the dry season the spores mature and are disseminated.

Coccidioides in Soil: *C. immitis* can be isolated from soil by making a suspension of soil in salt solution (30 per cent) in a large cylinder. This is stirred and the coarse material which immediately rises to the top is skimmed off and discarded. The suspension is allowed to stand for an hour, the upper layer is withdrawn, diluted to make approximately a physiological salt solution, centrifuged, and the sediment is planted on agar or injected intraperitoneally into guinea pigs. If cultures are made the selective medium previously mentioned can be used to decrease the growth of other organisms or it can be plated out and suspected colonies can be picked when they appear and transplanted to agar slants until identified. Inoculation into guinea pigs has proved more productive in the author's experience, although undoubtedly many soil samples containing spores are not detected because the amount of inoculum is too small to initiate an infection in the animal.

There are three records of the isolation of *C. immitis* from soil. Stewart and Meyer (1932) isolated it from soil near the bunkhouse of a ranch at Delano, California, where there were men with coccidioi-dal granuloma. Davis and others (1942) isolated it repeatedly in Pan-oche Valley, California, from soil taken in the immediate vicinity of a burrow from which a rattlesnake had been dug by a group of University students who subsequently developed primary coccidioidomycosis. Emmons (1943) isolated the fungus from five soil samples taken at random on the desert in the area around San Carlos, Arizona. The occurrence of the fungus in soil is therefore definitely established and supports the concept of airborne transmission of spores from soil to man. The ecological relationships of this fungus in soil are not known however and there is some doubt whether it normally grows as a saprophyte in soil, as has been usually assumed. It has not been explained, for example, why it is limited in the United States to a comparatively small desert area although it is not fastidious as to mineral nutrients, moisture and temperature in artificial culture in the laboratory. If it had an obligatory association with some desert plant or animal its failure to spread beyond these areas would be more comprehensible. No parasitic or commensal association with a plant has yet been found.

Animal Reservoir

There is some reason to suspect a necessary association with animals. A wide variety of animals are susceptible to experimental inoculation and coccidioidomycosis occurs spontaneously in cattle and sheep grazing in endemic areas. The disease in these animals, however, is usually restricted to the mediastinal nodes. Presumably these animals, as in the case of man are accidentally infected and do not constitute a necessary link in the chain. Several cases of coccidioidomycosis have been reported in the dog but here too the disease is not recognised as common and the dog can hardly be considered as a reservoir.

In studies conducted in Arizona, Emmons (1942, 1943) observed what may prove to be a true animal reservoir. When small desert rodents were trapped it was found that 80 per cent of the pocket mice *(Perognathus spp.)* examined had a pulmonary mycosis. This was due to one or both of two fungi, *C. immitis* and *Haplosporangium parvum.* The latter is a fungus which appears on immunological and morphological grounds to be related to *C. immitis.* In 124 specimens of *Perognathus, C. immitis* was found in 19 and *H. parvum* in 95. The kangaroo rat *(Dipodomys merriami)* and the Harris ground squirrel *(Citellus harrisi)* were also frequently infected. On the contrary, the white-footed mouse *(Peromyscus eremicus),* which was even more abundant than *Perognathus,* and the grasshopper mouse *(Onychomys spp.)* were rarely infected although they were living in burrows adjacent to pocket mice, and the exposure to a fungus living in soil must have been similar. *Peromyscus* is susceptible to a rapidly fatal infection under conditions of experimental infection in the laboratory, but apparently is relatively free of the infection in the field.

In the animals trapped the observed infection was confined to the lungs, but in some cases the destruction of lung tissue was extensive. It was assumed that the animals eventually succumb but this was not actually proved. Predators and scavengers are common on the desert, and it is assumed that sick and dead rodents are so quickly removed that they were not observed. The infected animals trapped were in many cases pregnant or lactating females and it was apparent that the disease was chronic and progressed so slowly that it did not greatly reduce the rodent population.

It should be noted that the high incidence of infection in rodents awaits confirmation by other observers in other areas, and that proof that the fungus is primarily a parasite of rodents rather than a saprophyte in soil is not yet complete. However, the occurrence of pulmonary coccidioidomycosis in a high percentage of certain species of rodents and its infrequency in other closely associated species which are nevertheless susceptible to laboratory infection, the chronic and slowly progressive form which the mycosis takes in *Perognathus,* and the coincidence between the distribution of these rodents and the endemic area of coccidioidomycosis was interpreted as evidence for a true animal reservoir of coccidioidomycosis. It is suggested that the fungus is present in soil which has been contaminated by infected rodents, and that the distribution of appropriate species limits the distribution of the mycosis.

Biology of Pityrosporum Ovale

Pityrosporum ovale is a yeast-like organism, oval or elongate in form, which is found consistently in dandruff scales. It may easily be demonstrated in methylene blue preparations of the scales. In such preparations the organism appears as small blue staining bodies, some dense, some vacuolated, of the characteristic round or elongated bottle shape. To Rivolta (1873) goes the credit for first describing this fungus, in his Parasiti Vegetali. He described it, under the name *Cryptococcus psoriasis*, as a rounded cell with a double contoured membrane which does not produce a nucleus. He had found it in a case of psoriasis vulgaris.

The next year Malassez (1874) gave a more precise description of the organism which he found in scales from alopecia areata; he considered the organism to be pathogenic. His so-called "spores" budded and were ovoid, rarely spherical, showing at all times some differences as to form and size. He found three types of cells, (*1*) which budded and was large, in diameter, (2) which also budded but was smaller, in diameter, (*3*) did not bud and the diameter was less than second type. The last mentioned he regarded as daughter cells. Malassez found similar "spores" in simple pityriasis, but these were always spherical and more numerous, these he also considered pathogenic. Klaman (1884) without referring to the work of Malassez came to the same conclusion concerning these "spores" and their relation to pityriasis. Logie found the "spores" in the epidermal scales and likewise concluded that they were pathogenic.

A number of the early students of these conditions held to the view that this organism was the cause of the lesion in which it was found. Among these may be listed Courrges, Hardy (1876) and Hodara (1894). Others, namely Horand (1874, 75) and Nystrom (1875), held the view that it was a harmless saprophyte.

Bizzozero (1884) found small spherical cells, 2.5-5μ, in diameter, in scales from regions of the body covered with hair: the scalp, chin, lip, and pubis. To these he gave the name *Saccharoinyces sphaericus*. He also found smaller ovoid cells which he considered identical with the "spores" of Malassez and which he called *Saccharoinyces ovalis*. Since these cells were found to occur so abundantly on so many people, he did not consider them to possess any pathogenic significance. Boeck (1886) noted these "spores" both in pityriasis capitis and eczema marginatum and considered them to be merely saprophytes. Unna (1891), observing what apparently were the same structures in all cases of pityriasis capitis, thought they were a bacillus and introduced the name "flaschen bacillus" or bottle bacillus, a name which clings to this day and is made familiar in many advertisements offering a cure for "infectious dandruff" so called. Hodara (1894) agreed with Unna, and in addition, claimed to have cultured the bacillus, but reported that it would not live long on agar. Sabouraud (1902, '04) reported he always found the "spore" of Malassez in pityriasis and in the scales of exfoliated epidermis; he considered them to be the etiological agent in these conditions. Jacquet and Rondeau (1905) and Darier (1907) were opposed to the microbic etiology of pityriasis as was Tieche (1908) who claimed that seborrhea and excessive desquamation are seen in all cutaneous infections.

This controversy on the etiological significance of *Pityrosporum* in seborrheic dermatitis and pityriasis capitis emphasised the need for obtaining cultures of these tiny yeast-like organisms; many attempts were made to do so. The task proved to be more difficult than would be anticipated from the frequency with which the organism is found on the scalp.

Attempts to Grow P. Ovale in Culture: Hodara (1894) stated that Unna (1891) was the first to culture *Pityrosporum*, that he was able to obtain subcultures on agar, but that they did not live long as the medium was not favourable. Hodara says that he himself obtained primary cultures but was never successful in subculturing.

Van Hoorn (1895) claimed that Pekeliiaring, a Dutch worker, preceded UNNA in culturing from hairs in cases of simple pityriasis the "spore" of Malassez and of

Bizzozero. Oudemans, a botanist, studied this fungus and classed it in the genus *Saccharomyces*, naming it *Saccharomyces capillitii*. He also designated under the names S. *sphaericus* and S. *ovali* yeasts which he considered identical to each other and to those of Pekelharing. The descriptions he gave are vague and much doubt exists as to whether the organisms were actually species of *Pityrosporum*. Sabouraud (1904) held this opinion and announced that the "spore" of Malassez had not been cultured on artificial medium stating that he himself had tried it on many media without success. He claimed that the organism belongs to the lower cryptograms and, because of its habit of budding, with the blastomycetes.

Later reports of the successful cultivation of this fungus included those of Marzinowski and Bogrow (1907) and Bold (1910). The description by these authors, however, leaves one in doubt as to the identity of their cultures. Meirowsky (1911) stated that the bottle bacillus of Unna is not a bacillus, but a yeast and that it may easily be cultured on a medium containing lanolin, but gave no details concerning the cultures. KRAUS (1913) also obtained a culture on a medium containing lanolin. His description of the organism stated that, as seen in culture, it appears identical with that in the scales. He did, however, mention a mycelium and this at once makes one doubtful as to the identity of his culture.

To come to the more recent studies, Castellani (1925) isolated from a case of stomatitis an organism which he named *Cryptococcus graciloides*. Ota and Huang (1933) felt that Castellani in actuality had isolated *Pityrosporum*. Templeton (1926) seems also to have cultured *Pityrosporum* from scales of pityriasis simplex, on media containing oleic acid. Acton and Panja (1927) obtained what seems to be a valid culture on a medium containing glycerine. The culture of Macleod and Dowling (1928) was obviously not *Pityrosporum*, this was so stated by Dowling in a later publication (1939). The yeast isolated by Benedek (1930) was very different from the *Pityrosporum*, although he thought it identical, and tried to explain the discrepancies by saying the organism was different in culture from that found in the scales. Moore's (1935) strain has been shown to be a *Cryptococcus (Torulopsis)* belonging to BENIIAM'S group III. It grows very readily and abundantly on the usual media, quite the opposite of *Pityrosporum.*

Necessity of Fatty Substances in Culture Media: It would seem that it was not until the important paper by Ota and Huang (1933) that the question of the culture of the organism, seen so abundantly in dandruff scales, was cleared up. The authors studied the culture of Acton and Panja (1927) and the *Cryptococcus graciloides* of Castellani and found that these were identical with their own cultures. From their descriptions, and the difficulty with which the cultures were obtained, there is no doubt as to the identity of the cultures with *Pityrosporum ovale.* They found it necessary to add butter or lecithin to the medium. On untreated media, they obtained cultures of ordinary yellow or red yeasts. Templeton (1926) probably had the true organism but failed to carry on subcultures.

Benham (1939) was able to isolate eight strains of *P. ovale* and to maintain five of them in subculture on a medium to which was added fats and fatty substances. These cultures appeared identical with the organism seen in the scales and with the description of the cultures given by Ota and Huang. The fats were added to the medium in the form of ether extracts or solutions pipetted over the surface of the agar (wort) in tubes. The ether evaporated leaving a thin film of the fat on the surface of the agar. Ether extracts of crude lanolin and butter were found to be most effective. Emmons (1940) corroborated these findings by successfully culturing and maintaining in subculture a number of strains which corresponded in every way to the description of *P. ovale* and which appeared identical with Benham's strains. Emmons isolated his strains by planting the scales in acid dextrose broth containing 23 to 44 per cent glycerine, and incubating at 30 to 37° C. His subcultures were maintained on media treated with ether extracts of lanolin or oleic acid. It will be noticed that all who were successful in culturing P. *ovale* did so only when some fatty substance was added to the medium. Benham (1941), using a synthetic medium, proved that this organism will not grow unless fats are added, thus indicating for the first time a case in which a microorganism requires a fatty substance for continued growth.

Classification and Nomenclature: The fact that P. *ovale* was first described by Rivolta (1873) under the name *Cryptococcus psoriasis* indicated that the organism was yeast-like. However, Malassez (1874) described it merely as a spore, the "spore" of Malassez. Unna (1891) added further complications by terming it a bacillus and the organism is still commonly called the bottle bacillus. It is to Sabouraud that we owe the name *Pityrosporum,* from the Greek *Pityron* meaning scale. He gave it the specific name *Malassezi* in honour of Malassez. Brumpt (1936) accepts the name *Pityrosporum,* but in a new combination, using as the specific name *ovale,* originating from Bizzozero (1884). This is the commonly accepted name for the organism.

Since the organism is yeast-like, reproducing by budding, and does not produce a mycelium or ascospores, it would of necessity be placed in the *Torulopsidoideae* near the genus *Cryptococcus (Torulopsis).* Since it differs from this in size and manner of vegetative reproduction it seems better to keep it as a distinct entity under the now familiar name *Pityrosporum.* It reproduces by a process between budding and septation much as found with the genus *Saccharomycodes* and therefore may be an imperfect form of this genus.

Cultural Characteristics: Wort agar has been found to be a suitable medium for the isolation of *Pityrosporum ovale* from dandruff scales. If fats are added as described above, this same medium will serve for subcultures. Media which have begun to dry out are not satisfactory for supporting growth of this microorganism. In the original cultures growth is first noted about the scales, cream to tan or orange coloured and dry in appearance. If a suspension of this growth is pipetted over the surface of wort agar to which oleic acid has been added growth will appear in four or five days. It appears first as tiny cream-coloured colonies which later fuse, becoming dry and wrinkled with the colour varying from cream to tan to orange. In a synthetic liquid medium (described later in this paper) in flasks, *Pityrosporum* grows both in granules which settle to the bottom of the flask and as a film or scum covering the surface and spreading up the sides of the flask above the level of the medium. The earliest growth appears as fine particles in or attached to globules of oleic acid which float on the surface.

Microscopic Characteristics: *Pityrosporum ovale* characteristically has several forms in culture though usually it appears as an elongate structure. It frequently appears two-celled, one of the cells being smaller suggesting a bud. In some instances the bud has elongated forming a neck-like projection. This is the so-called "bottle form." Often the cells appear joined by a septum, indicating that the reproduction is intermediary between budding and septation, as described for the genus *Saccharomycodes*. These yeasts form buds which become separated from the parent cell by the formation of a septum at the base rather than by constriction of the base. This is a rare group and only one species is well known, *Saccharomycodes Ludwigii.* This species forms ascospores and is much larger than *Pityrosporum ovale.*

Sometimes the cells lie in parallel pairs. The daughter cell, after separating from the mother cell, continues to lie near it. There is no indication of mycelial development and no multiple budding. Large aggregate masses form by the clumping of cells. Individual cells may contain small vacuoles. Some spherical thick walled cells are seen. These may be in the nature of resistant cells and are characteristic of most yeasts.

Fats and Fatty Acids in the Nutrition of P. Ovale: *Pityrosporum ovale* has been shown to require a fatty substance for continued growth. Of a number of fats tried, crude lanolin and butter seem to be the most effective. Of the pure fatty acids tested, oleic gave the best results. Oleic acid was used by Fleming (1909) for cultivation of the acne bacillus and has since been used in a number of bacteriological mediums. Cohen and Mueller (1940) have shown its importance for *C. diphtheriae*. However, *Pityrosporum* is probably the first microorganism reported to need fats or fatty acids as an indispensable factor for growth.

Culture of P. Ovale in a Synthetic Medium: If oleic acid is supplied, *Pityrosporum ovale* will sometimes grow feebly on an inorganic solution similar to Currie's medium containing NH_4Cl, KH_2PO_4, $MgSO_4$, and dextrose (Benham, 1945). Substitution of asparagine as a nitrogen source caused increased growth. If thiamin is added, some growth will be obtained regularly even with NH_4Cl as the nitrogen source, and with asparagine there will be an increased yield. Dextrose regularly increases growth when asparagine is used as the nitrogen source, and with thiamin gives yields practically equal to the best obtained. When dextrose is added to the series with NH_4Cl as a nitrogen source, with or without thiamin, there is no increase compared with parallel flasks without dextrose.

The addition of ethyl oxaloacetate causes a striking acceleration of growth. With oxaloacetate, and asparagine as a nitrogen source, the yields, determined as dry weight of the fungus, are equal to or slightly higher than those with asparagine plus thiamin. With NH_4Cl, the yields are about as good as those with asparagine. The addition of both thiamin and oxaloacetate give no greater average yield than oxaloacetate alone. The addition of dextrose increases the yield both with asparagine and NH_4Cl when oxaloacetate is present.

To prevent the carrying over of 'any thiamin with the inoculum, the suspensions used for seeding must be dilute and the cells carefully washed. Only two to three drops of such a suspension should be used to each flask containing 50 cc. of medium.

These facts point to *Pityrosporum ovale* as an interesting organism from the standpoint of accessory food substances. As it has definite requirements of its own, it may help in an understanding of the function of some of these substances. For example, it would seem that NH_4Cl and ethyl oxaloacetate can be substituted for asparagine and thiamin in the cultivation of *Pityrosporum ovale.* It appears that if a suitable keto acid is supplied, neither thiamin nor an amino acid is necessary. On the other hand, the asparagine may furnish both the carbon and nitrogen source needed by the organism, yielding ammonia and a carbon skeleton, probably oxaloacetate. Thiamin increases growth when added to the asparagine by hastening the production of the carbon skeleton in some way. When this carbon skeleton is supplied as oxaloacetate, thiamin is no longer necessary. It will be interesting to continue the study of the nutrition of *P. ovale,* especially using fatty acids other than oleic, and to investigate the role played by the various vitamins.

4

Medicinal Utility

Uses of Fungi in Medicine

The fungi have only a very restricted application in modern *materia medica,* but in bygone days they had a much greater vogue. The reasons for their gradual desuetude are exactly parallel with those obtaining in the case of the medicinal herbs, formerly so extensively used in "simples," but nowadays replaced by the purer and stronger drugs extracted from the plants, or, as is more frequently the case, prepared synthetically by a chemical process.

Nor must this former larger use in medicine be considered as contradictory to the undoubtedly great popular repugnance to the fungi in nearly all their aspects, for those who employed them in this manner were the physicians and astrologers, who were not only accustomed to surround themselves with skulls, toads, serpents and suchlike objects of horror and loathing, as emblems of their craft, but some of which they used in the preparation of potions of supposed medicinal value. The ingredients of the "hell-broth" that the three witches were concocting in *Macbeth* may be referred to in this connection.

The various medicinal purposes for which fungi have at different times been employed are legion.

Counterirritants: The earliest mention of the employment of a fungus in this manner is found, in the fifth century BC, in the writings of Hippocrates, who recommends that cauterisation should be made, by means of a fungus, in order to cure certain complaints:

> "One should cauterise the osseous and nervous parts with *fungi (mukes)*"; "Quickly cauterise in eight places so as to intercept the extremities of the spleen"; and again, "When the liver has attained its greatest volume, one should cauterise with fungi."

The fungi here indicated are probably some of those used in the manufacture of *amadou,* or German tinder, and were possibly either *Polyporus officinalis,* the "Female Agarick" of the old Pharmacopoeias, or *Fomes fomentarius,* which was later used on a very large scale for the production of tinder; while the treatment, after the fungus had been softened, immersed in a solution of saltpetre, and dried, so as to increase its inflammability, would simply involve lighting it, and then, as it smouldered, applying it to the affected part. In the early practice of surgery, while frequently a really remarkable knowledge of anatomy was displayed, the methods of treatment were of a rudimentary and drastic nature; and these, among which may be numbered bleeding and cauterisation, find only a very restricted use in modern practice.

It is of interest to note that cauterisation by means of fungi appears to have survived among uncivilized races almost to the present day, and it is quite possible that it may still be employed. According to Rees, "the Laplanders have a way of using funguses, or common toadstools, as we call them, as the Chinese and Japanese do the moxa, to cure pains. They collect the large funguses which they find on the bark of beech and other large trees, and dry them for use. Whenever they have pains in their limbs, they bruise some of this dried matter, and pulling it to pieces with their fingers, they lay a small heap of it on the part nearest to where the pain is situated, and set it on fire. In burning away, it blisters up the part, and the water discharged by this means generally carries off the pain. It is a coarse and rough method, but generally a very successful one, especially when the patient has prudence enough to apply it in time, and resolution enough to bear the burning to the necessary degree."

Hooker found the same use being made of pufballs in Sikkim in 1850, and relates:

> "My servant having severely sprained his wrist by a fall, the Lepchas wanted to apply a moxa, which they do by lighting a piece of puffball, or Nepal paper that burns like tinder, laying it on the skin, and blowing it till a large open sore is produced; they shook their heads at my treatment, which consisted in transferring some of the leeches from our persons to the inflamed part."

It is difficult to see how such cauterisation can be of any real efficacy, and when successful appears only to demonstrate that the complaint was amenable to a kind of faith-cure. It would, however, have much to recommend it in the eyes of the primitive medical practitioner, for its failure must naturally always be due entirely to a lack, on the part of the patient, of the necessary degree of fortitude.

Universal Medicines

Such a use, however, is not that for which *Polyporus officinalis* was once held in such high repute, for, under the name "agarick," it was taken internally as a sort of universal remedy for all complaints and disorders. The belief in a universal remedy was held to some extent by the Greeks and Romans; while, in a later day, it allured many of the alchemists, and stimulated them in their assiduous search for the Elixir of Life. To Dioscorides, who flourished about 200 AD, *agaricum* was certainly little removed from being the universal remedy, and in his book *De Medicina* he expatiates upon its virtues as follows:

> "Its properties are styptic and heat-producing, efficacious against colic *(strophous)* and sores, fractured limbs, and bruises from falls: the dose is two obols weight with wine and honey to those who have no fever; in fever cases with honeyed water; it is given in liver complaints, asthma, jaundice, dysentery, kidney diseases where there is difficulty in passing water, in cases of hysteria, and to those of a sallow complexion, in doses of one drachma; in cases of phthisis it is administered in raisin-wine, in affections of the spleen with honey and vinegar. By persons troubled with pains in the stomach and by those who suffer from acid eructations, the root is chewed and swallowed by itself without any liquid; it stops bleeding when taken with water in three-obol doses; it is good for pains in the loins and joints, in epilepsy when taken with an equal quantity of honey and vinegar. It prevents rigour if taken before the attack; in one- and two-drachma doses, it acts as a purgative when taken with honeyed water; it is an antidote in poisons in one-drachma doses with dilute wine. In three-obol doses with wine it is a relief in cases of bites and wounds caused by serpents. On the whole it is serviceable in all internal complaints when taken according to the age and strength of the patient; some should take it with water, others with wine, and others with vinegar and honey, or with water and honey."

Another fungus, not so famous as "agarick," was regarded by the Romans, if not as a universal remedy, certainly as a cure for a number of different complaints. This was *Boletus edulis,* which by all authorities is considered to be the *suillus* (or hog-fungus) of Pliny and Martial, of which the former states: "Suilli are dried and hung up, being transfixed with a rush, as in those which come from Bithynia. These are good as a remedy in fluxes from the bowels which are called *rhcumatismi,* and for fleshy excrescences of the anus, which they diminish and in time remove; they remove freckles *(lentiginem)* and blemishes on women's faces; a healing lotion is also made of them, as of lead, for sore eyes; soaked in water they are applied as a salve to foul ulcers and eruptions of the head and to bites inflicted by dogs."

The attitude of the ancients towards nature was somewhat careless and entirely empirical, and even the careful observer of natural objects preferred rather to indulge in wild speculation than to carry out a simple experiment that might have demonstrated the falsity of his hypothesis. It, therefore, followed that many opinions were based on very slight evidence or were simply pure conjecture. Yet such an opinion, when expressed by a high authority, became a creed which it was heresy to doubt.

The dogmas, many erroneous, of the classical writers undoubtedly held ground for a very lengthy period, and, in fact, found no opposition until the alchemistic period ended with the coming of Paracelsus in the sixteenth century. Thus the herbalists of the Middle Ages used the axioms and doctrines of such Arabian physicians as Geber and his pupils Rhazes, Avicenna, and others; and the latter prepared their medicines, which were almost entirely of vegetable origin, according to the original receipts of Galen, Andromachus and others, these having been transmitted to them by the Nestorians. The apothecaries' shops before the advent of Paracelsus were, therefore, nothing more than stores for roots, herbs, and syrups and confections of all kinds, the latter being prepared exclusively by the pharmacists.

The extent to which these still relied, for their knowledge of materia medica, upon classical sources, is quite evident from a study of the famous herbal of Gerard, which was first published in the year 1597. To Gerard, "agaricke" was still a universal remedy, and he thus describes its applications:

"Agaricke is hot in the first degree and dry in the

second, according to the old writers. It cutteth, maketh thin, clenseth, taketh away obstructions or stoppings of the intrailes, and purgeth also by stoole."

Agaricke cureth the yellow jaundise proceeding of obstructions, and is a sure remedy for cold shakings, which are caused by thick and cold humours.

The same being inwardly taken and outwardly applied, is good for those that are bit of venomous beasts which hurt with their cold poison.

It provoketh urine and bringeth downe the menses: it maketh the body well coloured, driveth forth wormes, cureth agues, especially quotidians and wandering feavers, and others that are of long continuance, if it be mixed with fit things that serve for the disease: and these things it performes by drawing forth and purging away grosse, cold and flegmaticke humours, which cause the diseases.

From a dram weight, or a dram and a halfe, to two, it is given at once in substance or in powder: the weight of it in an infusion or decoction is from two drams to five.

But it purgeth slowly, and doth somewhat trouble the stomacke; and therefore it is appointed that Ginger should be mixed with it, or wilde Carrot seed, or Lovage seed, or Sal gem, in Latine, *Salfossilts.*

Galen, as *Mesue* reporteth, gave it with wine wherein Ginger was infused: some use to give it with Oxymel, otherwise called syrrup of vineger, which is the safest way of all.

Agaricke is good against the paines and swimming in the head, or the falling Evill, being taken with syrrup of vineger.

It is good against the shortnesse of breath, called Asthma, the inveterate cough of the lungs, the ptysicke, consumption, and those that spet bloud: it comforteth the weake and feeble stomacke, causeth good digestion, and is good against wormes."

Many of these virtues are probably quite imaginary. Even when Agarick was not prescribed as a universal remedy in itself, it was still an essential constituent of any of these panaceas, and, in particular, of perhaps the most famous of them, that known as "Mithridate." This preparation, which was formerly one of the capital medicines of the apothecaries' shops, was named after Mithridates VI., or Eupator, who was king of Pontus from about 120 to 63 BC, and who was much skilled in poisons, for which he was in the habit of concocting antidotes, by the aid of which he could take the poisons with impunity. His reputation was, therefore, commemorated by the term "Mithridate," which was composed of a very large number of ingredients in the form of an electuary, regarded as a universal antidote or preservative against poisons or infectious diseases, being also considered a cordial, opiate, sudorific and alexipharmic. One recipe for it reads as follows:

> "Take of cinnamon, fourteen drachms; of myrrh, eleven drachms; *agarick*, spikenard, ginger, saffron, seeds of treacle-mustard, frankincense, Chio turpentine, of each ten drachms; camel's hay, costus, Indian leaf, French lavender, long pepper, seeds of hartwort, juice of the rape of cistus, strained storax, opopanax, strained galbanum, balsam of Gilead, or in its stead, expressed oil of nutmegs, Russian castor, of each an ounce; poly-mountain, water germander, the fruit of the balsam tree, seeds of the carrot of Crete, bdellium strained, of each seven drachms; Celtic nard, gentian root, leaves of dittany of Crete, red roses, seed of Macedonian parsley, the lesser Cardanum seeds freed from their husks, sweet fennel seeds, gum Arabic, opium strained, of each five drachms; root of the sweet flag, root of wild valerian, anise-seed, sagapenum strained, of each three drachms; spignel, St. John's wort, juice of acacia, the bellies of scinks, of each two drachms and a half; of clarified honey, thrice the weight of all the rest: dissolve the opium first in a little wine, and then mix it with the honey made hot. In the meantime, melt together, in another vessel, the galbanum, storax, turpentine, and the balsam of Gilead, or the expressed

oil of nutmeg (I have no doubt that one will do quite as well as the other; and this must be highly satisfactory for sufferers to know), continually stirring them round, that they may not burn, and, as soon as these are melted, add to them the hot honey, first by spoonsful, and afterwards more freely. Lastly, when this mixture is nearly cold, add by degrees the rest of the spices reduced to powder *and,* as the French quack used to say of his specific for the toothache, if it does you no good, it will certainly do you no harm. For my own part, I think the remedy worse than the disease; but a gentleman just poisoned may be of another opinion; and I can only say, that if with prussic acid knocking at his pylorus, he has leisure to wait till the above prescription is made up for him—till the bellies of scinks and the camel's hay are procured, and till the ingredients are amalgamated by degrees—he will, if he survive the poison, the waiting, and the remedy, have deserved to be calle the patient."

From the end of the sixteenth century, after Paracelsus had enunciated the doctrine that the object of chemistry (e.g., of alchemy) was not to make gold, but to prepare medicines, officinal preparations other than those of vegetable origin, i.e., metallic salts and the like, came more and more into use, and there gradually developed the more enlightened practice of medicine based on actual physiological tests on the human organism rather than on conjecture. With the advance of scientific pharmacy, the empirical remedies in general, and the vegetable simples in particular, gradually lost favour, although still surviving in some degree even to the present day. Examples of this survival are the herbalist shops still occasionally to be met with, and the herbal remedies still employed in rural households, and by some of the less civilized races of mankind. Some of these remedies naturally possess some therapeutic virtue, while others are of doubtful or of no efficacy. We are naturally only concerned with those vegetable recipes in which fungi are employed.

Cathartics: From the nature of some of the complaints for which *agarick* was regarded as a specific remedy, it is obvious that it must have some definite physiological effects, and if there is one instance where it found proper application, it is for the purpose in question, where, however, it has also been superseded in modern times by other less empirical drugs, the effects of measured doses of which are subject to no such variability as probably characterised those of the more primitive remedy. When, however, its use was in favour, it was considered that "agaric . . . purgeth phlegm, and opens obstructions in the Liver." This use has to a large extent and needs no further elaboration here. It may, however, be mentioned that various other fungi, not possessing definitely poisonous principles, have an emetic or purgative value. Among these may be mentioned *Hypholo mafasciculare, H, elaeodes, H. sublateritium* and *Hirneola auricula-judae.*

The employment of agaric has now been almost entirely abandoned in surgery, as indeed was inevitable, when its qualities are contrasted with those of the perfectly aseptic and antiseptic materials with which the resources of mo0dern science has furnished us. It is interesting to note, however, that as recently as 1910 a reference is made by Whitla to *Ly coper don giganteum* (= *L. Eovistd),* as forming "a soft and comfortable surgical dressing. The dusty powder is a powerful haemostat." Locally, however, puffballs are apparently still used as a household remedy, and Swanton states that "*Calvatia gigantea (=Lycoperdon Bovista), C. caelata (=L. caelatum), Lycoperdon gemmatum* and apparently any other species of Puffball, though the larger kinds were always given preference, were at one time kept in the farmhouses and cottages of West Sussex for use as styptics. The custom still lingers but is rapidly dying out." There is, however, one styptic use of a fungus still remaining, for *amadou,* of course in its natural or unimpregnated condition, but after the usual preparation, is sometimes used by dentists as an absorbent for drying cavities in teeth before filling them.

Among fungi whose employment was based entirely upon superstition, and which use has naturally been abandoned, may be mentioned the Jew's Ear, *Hirneola auricula-judae*, which, according to Berkeley, "owed its reputation in throat cases, probably to the fancied resemblance of its hymenial surface to the fauces." Gerard says it "is much used against the inflammations and all other sorenesses of the throat, being boiled in milk, steeped in beere, vineger, or any other convenient liquor." It was doubtless more efficacious when the second most "convenient liquor "was employed. In a similar manner, *Elaphomyces granulatus* was formerly regarded as an aphrodisiac, being known as *Fungi cervini* and used in the preparation of love potions. A curious household specific, which again is of entirely superstitious origin, is mentioned by Swanton, who says:

> "*Daldinia concentrica* was carried until quite recently by old men in the fold 'district of West Surrey and Sussex as a charm against cramp. At Haslemere the little spherical tumours so frequently seen on beech and oak trunks were carried instead and were considered equally efficacious. Specimens of these cramp-balls,' as they were called, given me by old villagers, have been placed in the Haslemere museum."

Their efficacy would appear to be entirely imaginary. The same writer mentions another interesting surviving application of a fungus in this locality. It is that of *Fomes pomaceus* which has a great reputation as a dressing for a swollen face, the fungus being ground down on a nutmeg grater and heated in an oven before application.

Considering now some of the cases in which the use of a particular fungus has been advocated, we find in comparatively recent times Badham speaking of the value of *Trametes suaveolens* in the cure of consumption; while as recently as 1884, Gautier deprecates the employment of *Lactarius piperatus* and *L. torminosus*, which had recently been recommended for the same purpose, but which gave no results of any value. The same writer mentions other medicinal applications. Several species of the same genus have been considered as diuretic against gravel, and their juice has also been vaunted as of great efficacy in the removal of warts, but would appear to be less successful than that of the wall celandine, well known for this use. *Amanita muscaria*, the Fly Agaric, has been strongly recommended as a remedy for epilepsy and other nervous maladies, scrofula, etc. The lotion made from it has been applied externally, and even internally, against ringworm and skin eruptions. It has also been proposed as a narcotic, but there are other more active and more easily administered agents. This fungus yields the uncrystallisable alkaloid, muscarine, the nitrate of which was formerly used in medicine for purposes similar to those of pilocarpine, but which appears not now to be in general use. Another *Amanita* is referred to by Gautier, who says: "Fairly recently Dr. Curtis has recommended a tincture of *A. phalloides* in use against cholera, Bright's disease and intermittent fevers, conduct which can only be described as an 'haute fantaisie therapeutique."

At the present day, a number of fungi are used by less civilized people. One of the most celebrated Chinese medicines, "Tong-chong-ha-cho," consists of a fungus, *Cordyceps sinensis*, which is parasitic upon a species of caterpillar, these being finally killed by its effects. The dead caterpillars, with the fungus attached to them, are made up into bundles and sold and consumed as an invaluable specific, the caterpillar being thus an integral part of the dose. The approved method of administration is to place a bundle of the necessary dimensions in the stomach of a duck, which is roasted, and consumed by the patient. Analogous methods of disguising the "medicine" will doubtless be recalled by the reader. The remedy in question has probably been in use in China for thousands of years, and is much praised by Du Halde, but probably on somewhat scanty evidence, and Berkeley describes it as "a drug of doubtful virtues." However, to this day, it holds, together with "Toad-spittle cakes" and "Powdered tiger's bones," a worthy place in the ranks of that heterogeneous and mysterious medley which comprises Chinese materia medica. *Lysurus Mokusin* is another fungus used by the Chinese, in the treatment of gangrenous ulcers.

In India, the common mushroom is officinal in the Punjab. The young dried specimens are known as "Mokshai" and sold in the bazaars, being regarded as alternatives. In the same bazaars are found on sale the spores of a puffball, probably *Lycoperdon gemmatum*, which are used for "expelling cold and bilious humours." In Burma, *Polyporus anthelminticus* grows at the roots of old bamboos, and is employed as an anthelmintic. In the Malay States, *Polyporus sacer*, under the name "Susu Rimau" or "Tiger's Milk," is used for treating consumption and colds, the whole plant being utilised; while *P. sanguineus*, known as "Chendawan Merah" or "Red Fungus," is employed in cases of dysentery.

Fungi in Modern Pharmacology

In modern pharmacology, which may be considered to date from the beginning of the nineteenth century, to include any still more recent uses, and being not necessarily restricted rigidly to present-day practice, only four fungi are admitted to the ranks of drugs possessing undoubted therapeutic value. These are *Claviceps purpurea, Ustilago Maydis, Saccharomyces cerevisiae* and *Polyporus officinalis* which are dealt with separately below, and of which only the first appears in the British Pharmacopoeia at the present day.

Claviceps Purpurea: This, under the name of Ergot of Rye, has long been used as a drug of the first importance for obstetric purposes. It is not mentioned by classical writers as being so employed, although it was presumably as widespread in its distribution as it is today. The earliest specific reference to this application is made towards the middle of the sixteenth century by Adam Loricer of Frankfort, who describes the appearance of the ears of rye on which the ergot parasitically occurs, and relates that it is considered by women to be of sovereign efficacy. Johannes Kalius in 1588 also describes its appearance and effects. "Rathlaw, a Dutch accoucheur, employed ergot in 1747. Thirty years later, Desgranges of Lyons prescribed it with success, but its peculiar and important properties were hardly allowed until the commencement of the present century, when Dr. Stearns of New York succeeded in gaining for them fuller recognition. Ergot of rye was not, however, admitted into the London Pharmacopoeia until 1836." Ergot is the sclerotium of this fungus, and originates in the ovary of rye *(Secale cereale]* by the absorption of the ovarian tissues; rye which has been so attacked, being sometimes known as "spurred" rye. The fungus is found all over Europe and gives risc commercially to Spanish, Russian and German ergots, of which the first is the boldest and is usually considered the best; but physiological experiments have demonstrated that the Russian variety is generally more active. Sclerotium in section seems roughly cylindrical or somewhat triangular, longitudinally furrowed on each side, and curved and tapering towards the ends; dark or violet-black externally and pinkish white inside, with a characteristic and disagreeable smell and taste. It is used in cases where active contraction of the muscles of the uterus is necessary, the dose varying between 15 and 60 grains.

Ustilago Maydis: This fungus, parasitic upon maize in the United States, is known as Corn Smut, Corn Ergot or Maize Ergot. It occurs on the Continent and is also recorded for this country, and is used in parturition instead of rye ergot.

It is found in commerce as an irregular globose mass, sometimes six inches thick, consisting of a blackish membrane, enclosing a large number of brownish black nodules, and has an unpleasant smell and taste. As with rye ergot, it should be kept whole and dry, and must not be stored for longer than one year. It is stated to increase the force without increasing the duration of uterine contractions. The dose consists of from 15 to 60 grains.

Saccharomyces Cerevisiae

The ferment produced by brewing beer is known medicinally as Cerevisiae Fermentum, or Faex Medicinalis. In character it is viscid, frothy, flocculent and semi-fluid, and in the compressed pasty form shows under microscopic examination numerous rounded or oval cells, either separate or arranged in clusters, sometimes in branching chains. It has a characteristic and peculiar smell and a bitter taste.

It is used as an antiseptic in poultices for application to ulcers. Internally it is used for various purposes, in cases of boils; as a remedy for diabetes, enabling the patient to take more carbohydrates; for septic endocarditis; while it is also said to increase the opsonic power of the blood to an invading organism. In acne, fresh yeast, in doses of from half a teaspoonful to a tablespoonful, may be taken with a little water at meals.

Whole, not in a powdered condition, and dry, while the official requirement is that it shall not be used if more than a year old. It is a complex substance, containing many different constituents, and published analyses differ considerably; but there is no doubt that its chief alkaloidal principles are ergotoxine and ergotinine, of which the former is definitely most active. Its B.P. preparations are:

Extractum Ergotae, used in the preparation of Irjectio Ergotae Hypodermica, one part in three.

Extractum Ergotae Liquidum, I part in i. Infusum Ergotae, one part in 20. Tinctura Ergotae Ammoniota, one part in 4 of a mixture of 60 per cent, alcohol and ammonia solution.

Martindale and Westcott quote "one authority" as recommending the use of ergot in all diseases in which a drug inducing muscular contractions is required, e.g., in disturbances of the circulatory system, skin affections, in nervous complaints (caused, for example, by excessive smoking), and in cases of surgical shock. It has also been recommended by some of the faculty in typhoid and intermittent fevers. The application that has been made of ergot for various forms of internal haemorrhage is entirely incorrect, for, by increasing the blood pressure, it may cause the bleeding of a wound which had stopped, to start again, thus acting in a manner precisely contrary to that for which it had been prescribed.

Ergots from other graminaceous plants are also used instead of rye ergot. Among these may be mentioned *ergot of wheat,* which is shorter and thicker than the official kind; *ergot of oat,* which is much more slender; and *ergot of diss (Arundo ampelodesmus* a North African grass), which is said to be twice as active as rye ergot.

Yeast dried at 30° C. is a light grey powder, which is used in half-gramme doses for constipation, being injected per rectum to break up the faeces. It has recently been employed in tuberculous affections and in dysentery. By extracting fresh yeast with alcohol is obtained Extractum Cerevisiae Fermenti, or Faexin Extract, which has also been employed for all the purposes for which fresh yeast has been used, viz., in acne, erysipelas, furunculosis, folliculitis, leucorrhoea, diabetes, typhoid, rheumatoid arthritis and other complaints. It is interesting to note that yeast enters into the composition of several present-day patent medicines.

Polyporus Officinalis: This, the most famous fungus used in medicine, is known as *Agaricus albus,* White Agaric or Purging Agaric. Its earlier reputation and applications have already been dealt with, and, although fallen from its high estate, it is still found in the pharmacopoeias. For officinal purposes, the outer skin is removed, only the inner portions being utilised. These are found in commerce in irregular pieces, about six inches broad, and hoof-shaped or conical. They are white, light, somewhat fibrous, spongy and friable; having a faint smell and a sweetish taste, with a bitter pungent aftertaste.

The active principle is agaricic acid, known alternatively as agaric or laricic acid, which is obtained in minute silvery plates or prisms by the extraction of the fungus, either with dry ether or with 90 per cent, alcohol. The alcoholic extraction is sometimes known as *agaricin.*

Hippocrates and Euripides refer to cases of fungus poisoning. From the writings of Theophrastus, about 300 BC, it is clear, not only that they were eaten on a fairly large scale, but that some of them, particularly truffles, were articles of commerce and export. There is no doubt, however, that in the later years of the Roman Republic and under the regime of the Empire, their consumption had much increased, some of them being regarded as very great luxuries. When the sterner virtues of the Romans had decayed, and they had acquired many luxurious tastes, one of the most significant was that for the pleasures of the table, in which it is to be regretted that the fungi played no small part.

Sumptuary Laws of Romans

To check these self-indulgent habits, at different times a number of sumptuary laws were enacted, each framed so as to check some definite and growing evil. Most of them were apparently very little observed and quickly fell into disuse. They were naturally very difficult to enforce, and even when the letter of any particular one of them was observed, it led to a fresh excess in other directions.

The first laws to prescribe the nature of the foods and liquors consumed at private dinners were the Lex Fannia in 161 BC and the Lex Didia in 143 BC; while the Lex Licinia, promulgated about 103 BC, limited the amounts of meat and fish that were to be consumed on ordinary occasions, and encouraged the consumption of vegetables.

Increased use of Fungi by the Romans

Among the vegetable foods which thus came into greater favour were the fungi, in regard to which Cicero says: "While these elegant eaters wish to bring into high repute the products of the soil which are not included in the Act, they prepare their fungi, helvellae, and all vegetables with such highly-seasoned condiments that it is impossible to conceive anything more delicious." In effect, they were so treated as to resemble animal foods, while the same extravagance continued. The most celebrated mushroom that was so employed was the imperial mushroom, which the Romans of the Empire, past-masters in gluttony, called the food of the gods, *Cibus deorum* the agaric of the Caesars. This is undoubtedly z*Amanita caesarea* of modern mycologists, known to the French today as the *oronge,* and termed by the Romans *boletus.* Pliny (23-79AD) paints a vivid picture of the manner in which boleti and some other fungi were regarded, and in his remarks on the cooking of fungi, he says that these were "the only food which dainty voluptuaries themselves prepare with their own hands, and thus, as it were, by anticipation feed on them, using amber knives and silver service." Juvenal, who flourished towards the close of the first century AD, regretfully says: "Nor will that youth allow any relative to hope better of him who has learnt to peel truffles and to pickle boleti."

Fungus Luxuries of the Romans: In the cooking of boleti, special vessels, called *boletaria,* were used, and these were not to be employed for more ignoble purposes.

Martial (43-104 AD), in an Epigram, depicts one of these vessels as bewailing its fallen state in the functions of a Roman kitchen: "Although boleti have given me so noble a name, I am now used, I am ashamed to say, for Brussels sprouts," In another Epigram of Martial, there appears a certain host, Caecilianus, who invites various friends to dinner and then eats all the boleti himself, upon which one of the guests protests as follows:

The reference is to the death of the emperor Claudius through eating a dish of poisoned mushrooms. In the opinion of Martial, boleti were very great delicacies, which, if sent as a gift to a friend, should never be entrusted to a slave, for he would most certainly devour them on the way, although gold and silver might safely be committed to his care:

"*Argentum atqueaurum* facile est, laenamque togamque
Mittere: boletos mittere est." Juvenal refers specifically

only to the boleti and describes all other fungi as "ancipites."

"Vilibus ancipites fungi ponentus amicis, Boletus domino."

"Doubtful fungi shall be served to his clients, the boletus to the lordly patron."

As further evidence of the madness which had befallen the Romans, Suetonius relates that the Emperor Tiberius (14-37 AD) awarded the sum of 200,000 sesterces (about £2,000) to Asellius Sabinus for a Dialogue he had written, in which boleti, beccaficos, oysters and thrushes were depicted as contending for the honour of being considered the choicest fare.

Divided Opinions of Romans as to the Food Value of the Fungi. It would appear, however, that the Romans were by no means unanimous as to the virtues of the fungi, and Seneca thus delivers himself: "Good gods ! how many men does one belly engage ! What! Do you think that those boleti—a pleasant poison—albeit they hurt not now, conceal within them no hidden mischief, and again: "For they are not food, they serve only to tickle the appetite, constraining those that are full to eat more; a very gratifying amusement to such persons as stuff themselves with such things as readily go down and as readily return." Cicero (106-143 AD) had previously contributed some evidence in agreement with that of Senéca, and relates in a letter to Gallus how he had attended a banquet given by the augurs at the house of Lentulus, where, being unable to indulge in such delicacies as oysters and eels, because of the sumptuary law forbidding them, he had instead eaten various dishes of highly-seasoned fungi, with the result that he was seized with a violent attack of diarrhoea, so as to be hardly able to stand up for about ten days. He says plaintively: "I who willingly abstained from delicate fishes, found myself taken in by vulgar herbs. He finishes his letter by saying that in future, he will be" more careful, which recalls the old proverb, "The devil was ill," etc.

Deprecation of their Use: Galen (130-218 AD), whose fame in medicine lasted for many centuries, also threw the weight of his authority against the fungi. He says: "Of fungi, the boletus, when well boiled, must be counted among insipid things; it is generally eaten with various kinds of spices, as is done with other insipid food. These fungi, after being eaten in large quantities, yield cold, clammy, noxious juices as their nourishing qualities; the boleti are the most harmless, and after them the amanitae; as for the rest, it is far safer to have nothing to do with them," and elsewhere; "Of all such kinds of food, fungi have the coldest, most viscid, and thickest juice; however, among them the boleti have never been known to cause anyone's death; still, to some persons, even they cause cholera and indigestion." As would, however, be expected, the populace were not to be deterred by the efforts of the "reformers," from indulging in their favourite fungi, and they continued, in general, to eat heartily, and occasionally, to suffer dreadfully.

Fungi used by the Romans: The reasons for the identification of the boletus with the modern *Amanita gaesarea* applies to the other species in which the Romans indulged. It should be remarked that the *amanitae* of Galen were almost certainly the Common Mushroom, *Psalliota campestris,* and not representatives of the modern genus of *Amanita.* It is a matter of history that the emperor Claudius was poisoned in 54 AD by a dish of boleti, of which he was very fond, but it is quite clear, from the writings of Tacitus, Suetonius, and Pliny, that the fungi were only the means, and that a poison was added to the dish, either by, or at the instigation of Agrippina, the wife of Claudius.

Besides boleti, Roman writers also frequently mention *fungi* and *suilli* ("hog-fungi"), so called because swine were fond of them. The latter have been identified as *Boletus edulis,* a fungus belonging to a genus of an entirely different character, whose members are characterised by having the hymenium covering tubes or pores, and not gills, as with the genus *Amanita.* Suilli were often contrasted unfavourably with boleti by the Romans, and Martial in an Epigram complains to Ponticus that when he was invited to a banquet, Pònticus had boleti, but *he* had to be content with suilli.

"Sunt tibi boleti: fungos ego sumo suillos."

The Romans also used as foods, several kinds of truffles, puffballs, morels, the Common Mushroom, and a variety of other fungi, whose identifications with modern species have been more or less surely arrived at.

Use by the Greeks: Not only were the Romans addicted to luxury and extravagance, but also the Greeks, and Plutarch, the Greek writer contemporary with the Roman Juvenal, speaks of the folly of indulging in costly foods and liquors, merely because they are so, and therefore accessible only to the rich. Among such he mentions Italian mushrooms, which were evidently an article of import into Greece, the latter country being notoriously poor in mushrooms to the present day.

Classical Recipes for Preparation of Fungi

A large number of recipes for the preparation of fungi in different ways for the table are given in the work De Re Coquinaria, ascribed to Caelius Apicius, and written in the third century AD Apicius is the name of no fewer than three famous Roman epicures, of whom the best known was M. Gavius Apicius, who lived under Tiberius, and who is said to have written on cookery, but the work in question appears to be based on earlier Greek writings. Apicius refers to four kinds of fungi in his recipes:

> Boleti, tubera (truffles), fungi farnei (thought to be *Boletus edulis)* and Sfonduli, Funguli or Spongioli (believed to be the same as the *spongiole* of Porta, 1592, and the modern Italian *spugnola,* the morel). Houghton has translated these recipes, of which those for the preparation of truffles may be given.
>
> "For *Tubera:* (1) Slice, boil, sprinkle with salt, and transfix with a twig (surculo infigis); partly roast, and place in

a cooking vessel with oil, liquor, sweet boiled wine (caroenum), unmixed wine, pepper and honey; while boiling, beat up with fine flour, take out the twigs and serve. (2) Another recipe: Boil and sprinkle salt, transfix with twigs, partly roast, place in a cooking vessel with liquor, oil, greens, sweet boiled wine, a small quantity of unmixed wine, pepper, and a little honey, and let it boil; while boiling beat up with fine flour; prick the tubers that they may absorb, take out the twigs and serve. If you like you may surround the tubers with the omentum of a pig, then roast and serve."

Other recipes are given for cooking truffles and the other fungi in question, and among other ingredients occur green coriander and pounded coriander seeds, cummin, echinus eggs (uvam), leeks, lovage, mint, parsley, rue and seseli. It is probable that the recipes of Apicius closely resemble those in use by the Romans.

Medieval use as Foods: In times more recent, one may find many references by our own writers to the use of fungi as foods, with, at times, warnings as to the attendant dangers, and a description of the methods of discrimination between the edible and poisonous kinds. It is of interest to learn, from the few monastic diet rolls that have come down to us, that the consumption of mushrooms, both in summer and winter, was certainly enormous; while, on the other hand, there are no records relating to vegetables. Many religious houses certainly had herb-gardens, in which were cultivated the plants used in simples, and also, presumably, potherbs for domestic use; but to what extent they cultivated the ordinary vegetables remains a matter of conjecture.

The culinary use of fungi is mentioned by King Richard's chief cook, whose name is forgotten, but whose initials, C.S.S., survive, for over them he wrote one of the first Cookery Books in English, "On the Forme of Cury," which was published about the year 1390. Here one is told to take funges and pare hem clene, and dyce [=dice] hem. What the "funges" were, we do not know, but they were most likely the Common Mushroom. Horman, however, have had something quite different in mind, for the Common Mushroom never grows on trees, although quite a large number of edible kinds are found so growing.

Half a century or so later, Maplet emphasises the need for care in collecting them for food, and says: "The Mushroom or Toadstoole.. hath two sundrie kinds, ... for the one may be eaten: the other is not to be eaten."

Gerard, in his Herbal, thought even less of the fungi from the gastronomic standpoint, for he divides them into two classes "whereof some are very venomous and full of poyson, others not so noisome; and neither of them very wholesome meate." He quotes Galen to the effect that they "ingender a clammy, pituitous and cold nutriment if they be eaten, and he clearly thought that a liking for them was strongly to be deprecated, as indicating a depraved taste. He concludes by saying: "Therefore I give my advice unto those that love such strange and new-fangled meates, to beware of licking honey among thornes, lest the sweetnesse of the one do not countervaile the sharpness and pricking of the other." The epicures of that day paid doubtless as little attention to his strictures as did the Romans the earlier condemnation of the fungi by Seneca and other writers.

At the beginning of the next century, Topsell speaks of a curious natural phenomenon as follows: "also writeth this of the Lycurium, that it groweth in a certaine stone, and that it is a kind of mushroom, or padstoole, which is cut off yearly, and that another groweth in the room of it, a parte of the roote or foot being left in the stone, groweth as hard as a flint, and thus doth the stone encrease with a naturall fecundity; which admirable thing (saith he) I could never be brought to beleeve, untill I did eate thereof in myne owne house."

This irresistibly reminds us of the story told by that entertaining voyager, Sir John Mandeville, who, writing of course much earlier, relates that "in that vale [of Ebron or Hebron] is the felde where men draw out of the earth a thinge the which men in that countrey call Chambell and they eate that thinge in the stede of spyce, and they beare it to sell, and men may not grave [i.e., dig] there so deepe ne so wyde, but it is at the yeeres ende full againe to the sydes through the grace of God." But whether there is any etymological connection between "chambell" and "champignon" we will not venture to say. However, it certainly appears not impossible, in view of the collateral evidence of Topsell, that this "thinge" was a fungus, and there was of course often some substratum of truth in Mandeville's stories.

A little later than Topsell, Francis Bacon, in his "Naturall Historic," written about the year 1624, but not published until 1627, the year after Bacon's death, writes:

> "The mushrooms have two strange properties; the one that they yield so delicious a meat; the other, that they come up so hastily, as in a night, and yet they are unsown."

Truffles and Morels

These historical references to the uses of fungi as foods would hardly be complete without some mention of those justly-celebrated delicacies, the truffles and morels, and an early reference to the former is made by Evelyn, who relates in his diary for the year 1644 "We got to Vienne in Dauphine... here we supped and lay, having amongst other dainties, a dish of Trufles, an earth-nut found out by an hogg trained to it, and for which those animals are sold at a great price." This statement is of some interest, for Evelyn moved in "the best circles" and would not have so phrased his description, if truffles had been used as a luxury in England in his day. They do occur in this country, but were not commonly found or eaten until later. Evelyn writes elsewhere concerning Morilles and Truffs: (the first whereof is a certain delicate red mushroom.

A few years later, Sorbiere says, in relation to the Londoners, "as that for Champignons, and Moriglio's, they were as great strangers to them as if they had been bred in Japan and we are not so sure that he would have much better to say about the Londoners of today. Sorbiere betrays, however, a lamentable ignorance of the use, centuries old, made of fungi as foods by the Japanese."

More Recent Uses as Foods: Turning from the historical side of the matter to consider the modern utilisation of the different fungi as foods, it is of interest to note that the savage and more primitive races of the present day make a much greater use of them than do their more civilized brothers. This is not surprising, in that the regular cultivation of staple foods renders unnecessary the casual gathering of a very variable crop of nature's own providing, and one, moreover, of decidedly inferior food value.

Uses by Savage Races: The staple edible of the Fuegians is a *Cyttaria*, which grows freely on living twigs of the evergreen beech during a considerable portion of the year. Under the name of "Thunder dirt," the Maories of New Zealand used formerly, because of the lack of better food, the gelatinous volva of *Ileodictyon*, which Berkeley describes as "an execrable article of diet." The "native bread" of the Australians is also a fungus, *Polyporus mylittae*, said to be considerably better than the last, and "when dry in some conditions looks like hard compacted lumps of sago. "According to Guppy, fungi are often cooked and eaten by the natives of the Solomons, being known as "magu," but the author cannot give the species.

Recourse is had to fungi as food in Central Africa, and H.M. Stanley, on his third journey to the Nyanza in 1888, in describing the privations endured by his column on the Ihuru river, relates that "our Nyanza people were provident and eked out their stores with mushrooms and wild fruit." Apparently, however, not all of the natives knew how to distinguish poisonous kinds, perhaps in their desperation being impelled to try some whose properties were unknown to them, for on returning to camp on Nov. 12th, Stanley found that "six people had succumbed, a Madi from a poisonous fungus."

Uses among other Civilized Races

Passing on to the more civilized races of the world, one is immediately struck by the varying degree in which edible fungi enter into their dietary; a variability which, indeed, is almost national. We have seen that fungi are extremely cosmopolitan in their distribution, and thus the same or a similar group of esculent species can be found in almost every temperate country. Yet it does not follow that they are used to the same extent in all these countries. For example, our own countrymen exhibit a preference for the Common Mushroom, and, save for a small band of enthusiastic mycophagists, regard with indifference or contempt a number of other species of equal or superior esculence; species which the French seek with eagerness and devour with avidity. It would, therefore, be of interest to consider briefly the chief countries of the world in which edible fungi are consumed to any appreciable degree, and to trace the extent to which the best-known esculents are used.

France: In the markets of Paris, before about the year 1876, only the Common Mushroom and certain truffles were allowed to be exposed for sale, although doubtless many other kinds were collected and eaten privately. Even the morel was at this time forbidden. At Nantes, however, the right of entry was not so restricted, and Genevier, writing in 1876 of this market, calculated the quantity of wild mushrooms sold each year as about 30,000 kilos (or, roughly, 66,000 Ibs), which he considered might approximately be subdivided into three parts of about 10,000 kilos each; one third being *Psalliota campestris;* another third composed of three edible kinds very common locally, namely, *Boletus edulis, B. aereus* and *Lepiota procera;* while the third part consisted of some ten or more common kinds, among which were several species of *Lepiota, Morchella* and *Amanita.* Probably much the same obtains today, when, however, a very different state of affairs is found in Paris than that existing before 1876, for the market has now been extended so as to include commonly some thirty kinds, while others would be permitted, if they were forthcoming. In addition to the truffle and the Common Mushroom, the two species which are dried or preserved in other ways are *Boletus edulis* (la cepe) and *Amanita Caesarea* (Poronge), the annual value of all kinds preserved being some 250,000 francs. The inspection of the edible fungi entering the markets of St. Etienne in the Department of the Loire began in 1897, in which year All the above occur in this region, but those most abundant on the market, in order of decreasing importance, are as follows:

Cantharellus Cibarius.

- Boletus edulis.
- Hydnum repandum.
- Hydnum imbricatum.
- Lepiota procera.
- Tricboloma equestre.
- Tricholoma portentosum.

In France there is scarcely any city near to a woody region which has not become an important market for many kinds of wild mushrooms in their season, the chief species sold varying with the locality. Thus, in Lyons, in 1907, about 26,000 Ibs. of different kinds were so disposed of, the greater part of which consisted of the chanterelle *(Cantharellus cibarius).* with *Tricholoma terreum* second in amount; while in other markets *Boletus edulis* and *B. granulatus* are often the species sold in greatest quantity. Besides these "wild" mushrooms, there appear in almost all markets the products respectively of the great mushroom-growing industry of Paris and of the truffle industry of Southern France, both of which are of such magnitude as to require separate consideration.

Switzerland: In Switzerland, both Lausanne and Geneva are notable markets. Thus, in the former town, no fewer than seventy-eight different kinds are allowed to be sold, to the amount of some 60,000 Ibs annually; while in the latter, a particular portion of the market is allocated solely for their sale, and during the season constitutes a veritable fungus exhibition. These are all very carefully inspected, in order that they may be passed as wholesome before being sold, and even those who gather wild fungi for their own use are invited to submit their spoils to be thus certified.

Italy: With regard to Italy, it is of interest to note that, as early as the year 1820, the Government of Milan published a list of about half a dozen species as being edible, but provided for the varying kinds to be found in different localities, by decreeing, that each district should similarly issue its own local list of wholesome fungi. These were even then of some economic importance. The varieties sold in Italian markets today are naturally the lineal descendants of those known to the Romans and some account of them is given later in this connection.

Germany: According to Duggar, who visited Berlin and Leipzig in 1899, these markets were at that time not well organised to encourage the sale of any large number of different species, and only about a dozen common kinds were represented, this being, possibly, partly the result of a Prussian law passed early in 1812, restricting the sale of wild mushrooms to the following species:

- Morchella esculenta
- M. conica
- Psalliota campestris
- Lactarius deliciosus
- Agaricus cebaceus (Hygropborus ceraceus)
- Merulius cantharellus (Cantharellus cibarius)
- Boletus edulis

All other species were to be rejected, some as being poisonous themselves, and others as being liable to be confused with deadly kinds. In Bavaria, however, possibly because the moist woods of the southern portion of this state are climatically very favourable for their growth, a very different state of affairs is found, and Duggar says that the city of Munich is the largest market in the world for wild mushrooms. According to Prof. Giesenhagen, the quantity sold in the provision market of that city in the summer and autumn of 1901 was about 850,000 kilos.

The sale of thirty-one species, therefore, occurred on this market. Doubtless this number would be somewhat increased by the fact that several species may be included under *Psalliota campestris,* and the same may be true of some other forms. Aside from this, it is to be noted that this list is a local one and that many fairly common edible fungi are not included.

India: In Asia, as in Europe, the Common Mushroom does not belie its name, and it is abundant in many parts of India, being found particularly in cattle-fields in many parts of the Central Punjab after the rains, in the desert tracts of Central and Southern Punjab, and also in Afghanistan. It is consumed extensively by the natives, either fresh or after drying in the sun. It is described by Europeans as excellent, and equal in quality to the English variety.

The morel *(Morchella esculenta),* which also closely resembles the European form, is found in abundance locally, and, according to Troup, is exported in large quantities from Kashmir and elsewhere in the Himalayas to the plains. They "are collected in April and strung up on a string to dry in the sun previous to export." It is of interest to note that the Mahommedans will only eat morels, as they consider other kinds to be impure food. On the other hand, the Hindoos will eat any mushroom having an agreeable taste and smell. Besides the two species in question, the other chief edible fungi of India are *Helvella crispa* and *Hydnum coralloides.*

"Edible truffles, known in Burma as kaingu,' are obtained in the rainy season from the roots of the kaing (elephant-grass); while another form of edible truffle grows on the roots of *Pinus khasya.*" Stewart mentions an "underground mushroom," which may possibly be a truffle, and which is found in cultivated ground near Multan and known in the vernacular as "boenphal." This, he says, is also eaten by the natives. Balfour refers to an underground fungus, *Mylitta,* found in the Nilgiri hills, which he considers to be closely allied to the so-called "native bread" of the Tasmanians, *Polyporus Mylittae.* He does not say, however, whether the Indian species is eaten by the natives.

Another doubtful species mentioned by Stewart is known as "shirian" in the Jhelam, and "bat-bakri" in the Kair valley. Stewart describes this as "a thin flat ragged-looking fungus, yellow above and with white gills below, which is found on dead trees in various parts of the Punjab Himalaya at 8,000 to 8,500 feet. The natives slice and cook it either fresh or dry, and eat it as a relish with bread. I have tried this species in stews, etc. but found it leathery and flavourless." A *Lepiota,* of which the species is not stated, is referred to by Gibbon "as being found in the nests of white ants and eaten with relish by the natives," while Berkeley states that "*Hypoxylon vernicosum* is consumed by the Bhoteans in the Himalayas."

Tibet: In the pinewoods of Tibet, Hooker records that a large mushroom, known by the natives as "Onglau," is abundant, and constitutes a favourite article of food. This species, which was afterwards named and described by Berkeley as *Cortinarius emodensis,* is also called by the Tibetans "Yungla tchamo," the latter word signifying a toadstool. Doubtless many other fungi are also eaten by the natives of different parts of India, but the literature of the subject is scanty and a complete list cannot be given.

Japan: This country has been called "the land of mushrooms With its humid climate and large extent of forest, it certainly possesses great natural advantages for their production, and the total annual crop exceeds 5,000 tons. Of the dozen or more species in general use, the "shii-take" *(Armillaria shiitake)* and the "matsu-take" *(A. edodes)* are by far the most important, and their huge production may well be compared with the mushroom and truffle industries of France, with which they have received detailed consideration.

The chanterelle *(Cantharellus cibarius)* held in high repute in our own country, is also known in Japan, as "shiba-take," for Rein tells us that "under that name, persons were offering for sale by the basketful our well-known egg-mushroom, in September, 1874, in the village at the foot of Fugi-san. I saw it in other places, too, but cannot find it anywhere mentioned as growing in Japan." In the forests of Fugi-san there occurs also another British species, *Glavaria botrytis,* known as "nedzumi-take," with an ally, *C. flava* both being mentioned by Rein as sold in the neighbouring villages.

In sandy localities, where the pines grow, is found in spring the "shoro," the name being derived from "sho," for "matsu," a pine; and "ro-tsuyu," dew. The name is applied equally to three allied species, *Rhizopogon aestivus, R. rubescens* and *R. virens,* of which the second occurs also in this country. The members of this genus, although quite distinct, resemble truffles, and the species in question are highly prized in Japan as ingredients in soups and also as separate dishes. They are said to be very tender and of a delicate flavour, and are often dried for future use. The same method of preserving is followed in the case of the "kikurage" *(Hirneola polytricha),* which grows on the mulberry and elm, and the "kawa-take" *(Hydnum olidum),* found in the dense parts of forests, and having a pleasant smell and exquisite taste. According to an official Japanese publication, both of these are "very nice food."

Among mushrooms also eaten by the Japanese may be mentioned the "hara-take" *(Psalliota campestris),* the "hira-take" *(Pleurotus ostreatus),* the "shiba-take" *(Boletus subtomentosus),* the "amigasa-take" *(Morchella conica),* and the "hatsudake" *(Lactarius hatsudake),* the last-named growing in the forests and having two varieties, one brown and the other green. There are also the "ko-take," shimeji-take and tsuga-take, the botanical names of which we have not yet been able to trace; while probably still other fungi are used on a smaller scale. At the present day, the rapid "Westernisation" of the Japanese, with its concomitants in the form of new habits of eating, does not appear to have led to any falling-away in this notable industry, which does not fear comparison with that carried on by any other country.

China: The Chinese mushroom trade is remarkable by reason of the scale on which the so-called "Jew's Ear (which, however, is not *Hirneola auricula-judae*, but its ally, *H. polytricha*, is cultivated. This subject, however, is dealt with later. In addition, the Chinese use as food a number of other kinds, those gathered on the largest scale being *Cantharellus cibarius, Lactarius deliciosus, Psalliota campestris* and *Tricholoma gambosum.*

Australasia: *Hirneola polytricha*, which we have just referred to, is of decided economic importance in New Zealand and the South Sea Islands, particularly Tonga; from both of which localities it is exported in a dry state to China. The New Zealand trade does not seem so flourishing as formerly, but the fungus is still regularly collected in the northern forests, three to sixpence a pound being given for the dried material. The trade apparently began in 1871, and increased rapidly.

United States: Fungi are plentiful in the United States, but those sold in the local markets are restricted to a few species. Thus, of the thirty or so good-sized edible species that are to be found in the District of Columbia, only four kinds are sold in any abundance in the Washington markets. These, which will be seen on almost any market day in autumn and sometimes under favourable conditions in spring and summer, are the Common Mushroom *(Psalliota campestris)*, the Horse Mushroom *(Psalliota arvensis)*, which is not always distinguished by the market people from the first; the Shaggy-cap *(Coprinus comatus)*, incorrectly called "French mushroom" by the market women; and the puffball *(Lycoperdon Eovistd)*. Some of the other edible species of this district are occasionally brought to market, generally to fill some particular order, but they are slow to find a buyer in the open market, since even those who bring them are a little doubtful as to their wholesomeness, and progress towards their coming into general use is very tardy.

Great Britain: Returning to our country, we find that the market is almost entirely confined to wild and cultivated forms of the Common Mushroom *(Psalliota campestris)* or of its ally, the Horse Mushroom *(P. arvenis)*. The trade in such wild mushrooms as the blewitt *(Tricholoma personatum)*, the Common Parasol *(Lepiota procera)*, the so-called "Summer Truffle" *(Tuber aestivum) Craterellus cornucopioides* and others, which was formerly a feature of Covent Garden and of some provincial markets, has fallen away or entirely disappeared. This is doubtless primarily due to the more extensive supplies of the Common Mushroom which are now available; but a further factor, that the collection and marketing of these other edible fungi never received official encouragement in the form of the appointment of referees, who could, as in some other countries, vouch for the authenticity of species offered for sale, should not be overlooked. This economic neglect of practically all species but the Common Mushroom is doubtless also due in a measure to the occasional cases of poisoning which have from time to time occurred through the eating of kinds other than this species. The ill-repute into which the fungi have thus fallen has seriously detracted from their more general employment in this capacity.

The need for a greater encouragement of their use as Foods. In this connection, we would most strongly urge that some serious attempt should be made to encourage the use of the other edible fungi on a scale comparable with that of France and Germany, for most of the species which are employed so extensively in these countries are available in Britain, and a livelier interest in them would 7 per cent not only afford supplies of cheap and agreeable food, but would also tend to the development of an industry among our country folk which would stand them in good stead at a time when farmers' work is none too plentiful.

We need not tabulate the many species which might be

used. It should be sufficient to say that for those enthusiastic enough and sufficiently well-versed in their identification, some edible species are furnished by most months of the year.

Discrimination between Edible and Poisonous Fungi. The eating of fungi, apart from the well-known cultivated form, has always been attended by the danger of using poisonous kinds, for there are no general rules by which to tell one from the other. Various empirical methods have at different times received popular support, but all of these are utterly untrustworthy, for not only would a reliance on any one of them rule out many edible species, but it would not exclude many poisonous ones. Some of these beliefs are first recorded by Greek and Roman writers. Thus Horace (65-8 BC) says that "fungi which grow in meadows are the best; it is not well to trust others." Many growing in the open, however, are deadly, while some of those found on or under trees are quite harmless and also most excellent. Like Horace, Pliny (23-70 AD) refers to the method of distinguishing harmful kinds by their situation, but also judges from their general appearance. He records that "some persons have discriminated the kinds of fungi from the kinds of trees on which they grow, such as those found on the fig, the fennel (ferula) and gummiferous trees. We have already spoken of those which are produced on the beech, the oak, or the cypress. All the poisonous fungi have a livid colour."

He says also that "those kinds which remain hard after cooking are injurious, while those which admit of being thoroughly well cooked, when eaten with saltpetre are harmless," while in Holland's edition of his works, we read that "the nearer that a Mushrome or Toadstoole commeth to the colour of a fig hanging upon the tree, the lesse presumption there is that it is venomous."

Dioscorides, a century or so later, tells us that poisonous fungi "have a thick coating of mucus, and when laid by after being gathered, quickly become putrid, but others not of this kind impart a sweet taste to sauces." A few years later, Athenaeus, in his *Deipnosophistae* ("Banquet of the Learned") quotes from a much earlier writer, Diphilus, a Greek physician, who wrote a book on *Diets Suitable for Persons in Good and Bad Healthy* the following passage: The wholesome kinds appear to be those which are easily peeled, are smooth and readily broken, such as grow on elms and pines; the unwholesome kinds are black, livid and hard, and such as remain hard after boiling; such when eaten produce deadly effects." The general appearance is of course no guide, for a toadstool of vivid hue, red or blue in colour, may be not only innocuous but also pleasant to the palate; while in another, snow-white and of fair outward seeming, may lie hid the promise of an unhappy fate for the unwary gatherer.

The empirical tests which have been advanced from time to time, are well tabulated by Ramsbottom, as follows:

(1) "Poisonous fungi have either a repugnant appearance; are viscid; do not 'peel'; are green, red or white in colour; have a foetid odour; have a sharp bitter taste; exude a milky fluid when broken; have their flesh rapidly turning colour when broken. Every one of these characters occurs in both poisonous and edible species."

(2) "A fungus which has been nibbled by rabbits, squirrels, etc., or which is slug-eaten, is edible. *Slugs thrive on Amanita phattoides, the most poisonous fungus known.* (It may be added that the experimental evidence acquired by feeding domestic animals on doubtful species is not absolutely trustworthy).

(3) "Fungi which blacken a silver coin or silver spoon when cooking; which coagulate milk; which turn an onion bluish or brown, or which make parsley leaves yellow, are poisonous. This is not true: and, further, fungi which do not act in any of these ways are not necessarily edible."

(4) "Fungi which grow on highly manured places are poisonous. Such fungi may be either poisonous or edible, e.g. the common mushroom".

With regard to the second, a widely-held presumption, there is perhaps a little to be said for it, for, although rabbits and squirrels are less discriminating than man in their selection of species on which to feed, they appear, on the whole, to avoid those kinds which he considers poisonous. Slugs, snails and insects, however, show a much greater catholicity of taste, and in regard to the relation of the last to the belief in question, the great French naturalist, J. H. Fabre, says:

> "The insect which feeds on one sort of mushroom and refuses others, cannot tell us anything about the kinds that are good or bad for us. Its stomach is not ours. It pronounces excellent what we find poisonous; it pronounces poisonous what we think excellent. That being so, when we are lacking in the botanical knowledge which most of us have neither time nor inclination to acquire, what course are we to take. The course is extremely simple."

There is, of course, only one sure method of discrimination. It is to know one kind from another, as surely as one can distinguish between a carrot, a parsnip and a beetroot; to restrict one's choice rigidly to those recognised with certainty, and not to add any to the number without independent and unchallengeable authority. Furthermore, it is advisable, in one's first essays, to exclude certain edible ones, because others, which are most toxic in their effects, resemble them, more or less closely, according to the variability of the species and the knowledge of the gatherer. Thus the Blusher *(Amanita rubescens)* should be rigidly avoided on principle, not that the expert would be misled for a moment, but the deadly *A. pantherina* is often mistaken for it by the novice.

Thus, under the name "Golmotte," large quantities of *A. rubescens* are annually eaten in Lorraine and the Ardennes, where every year *A. santherina* the "Fausse Golmotte," is frequently collected instead, with unhappy consequences. In the same country there is a similar and most deplorable confusion between the delicious *A. caesarea*, which is common in the departments of the Midi and the Centre, and the poisonous Fly Agaric, *A. muscaria,* known as the "Fausse Oronge." It is, therefore, better to avoid those esculents in which such possibilities of confusion occur. It is of course not the student who suffers, but the imitator.

It is, naturally, inadvisable to eat any which have begun to change colour and decay, for they are then just as apt to produce toxins as is meat or fish in a similar condition, and this has led on various occasions to the improper condemnation of quite harmless varieties. It is also advisable to make sure that the fungi are not already tenanted by any of the various grubs which may infest them. These often seem to appreciate the fungi quite as well as does man, and when he comes second, the flavour may be the worse. Their presence, however, can be ascertained by first examining the top of the cap to see that there are no holes in it, and then by cutting cleanly across the stalk, near to the gills, when, if again no holes are found, it is clear that no grubs have penetrated into the cap from the ground. The stalk is usually rejected, being tougher, but can again be cut across, and used.

We must not metamorphose into a Cookery Book, having no space for the multifarious recipes which have been concocted for the preparation of fungi. As with other foods, however, a good cook can serve up one of indifferent flavour as an appetising dish; while, contrariwise, an excellent variety may be quite spoilt by unsuitable cooking. In concluding the consideration of the wild fungi as foods, it may be remarked that their greater use has been retarded, not only by the difficulty, to the layman, of identifying them, with the consequent danger of confusing edible and poisonous kinds, but also to the fact that the wild fungi are much more frequently grub and rabbit-eaten, and generally to the precarious and uncertain nature of the "crop."

Bibliography

Aberdeen, J. E. C.: *Introduction to the Mushrooms: Toadstools and Larger Fungi of Queensland,* Queensland Naturalists Club, Brisbane, 1979.

Ahmadjian, V. and Paracer, *Symbiosis: An Introduction to Biological Associations,* University Press, New England, 1986.

Ainsworth, G. C.: *The Fungi: An Advanced Treatise,* Academic Press, New York, 1973.

——————— : *Ainsworth and Bisby's Dictionary of the Fungi,* Commonwealth Mycol, Kew, 1971.

——————— : *Introduction to the History of Mycology,* Cambridge University Press, Cambridge, 1976.

Ajello, L. Georg L.K. Kaplan W. and Kaufman L.: *Laboratory Manual for Medical Mycology,* US Publ., New York 1963.

Amaranthus, M. P. and Luoma D. L.: *Diversity of Ectomycorrhizalfungi in Forest Ecosystems,* International Symposium on Microbiology—Santos, Brazil, 1997.

Ames, L. M.: *A Monograph of the Chaetomiaceae,* MacMillan, New York, 1963.

Ammirati, Joseph F. James A.: *Poisonous Mushrooms of the Northern United States and Canada,* University of Minnesota Press, Minneapolis, 1985.

Arnold, Katya and Swope Sam: *Katya's Book of Mushrooms,* Henry Holt and Co., New York, 1997.

Arora, David and Bowers J.: *Mushrooms of the World Coloring Book,* Dover Publications, New York, 1984.

Arora, David: *All that the Rain Promises,* Ten Speed Press, Berkeley, 1991.

———————— : *Mushrooms Demystified,* Ten Speed Press, Berkeley, 1986.

Arx, J. A.: *The Genera of Fungi Sporulating in Pure Culture,* J. Cramer, Vaduz, 1981.

Barnes, R. S. K.: *Diversity of Living Organisms,* Blackwell Publishers, New Jersey, 2002.

Barnett, J. A. and R. J. Pankhurst: *A New Key to the Yeasts,* North Holland Publ. Co., Amsterdam, 1974.

Barnett, H. L. and Hunter B. B.: *Illustrated Genera of Imperfect Fungi,* MacMillan, New York, 1987.

Barron, G. L.: *The Genera of Hyphomycetes from Soil,* Williams and Wilkins, Baltimore, 1968.

Barron, George: *Mushrooms of Northeast North America,* Lone Pine Publishing, Renton, 1999.

Bell, A.: *Dung Fungi: An Illustrated Guide to Coprophilous Fungi in New Zealand,* Victoria University Press, New Zealand, 1983.

Beneke, E. S.: *Medical Mycology Laboratory Manual,* Burgess Publ. Co., Minneapolis, 1958.

Benjamin, D. R.: *Mushrooms: Poisons and Panaceas,* WH Freeman and Company, New York, 1995.

Bessette, Alan and Sundberg, Walter J.: *Mushrooms: A Quick Reference Guide to Mushrooms of North America,* MacMillan Publishing Co., New York, 1987.

———————— : *Mushrooms of North America in Color: A Field Guide Companion to Seldom-Illustrated Fungi,* Syracuse University Press, Syracuse, 1995.

Bessette, Alan E. Roody, William C.: *North American Boletes,* Syracuse University Press, Syracuse, 2000.

——————— : *North American Boletes: A Color Guide to the Fleshy Pored Mushrooms,* Syracuse University Press, Syracuse, 2000.

Bigelow, Howard E.: *North American Species of Clitocybe,* J. Cramer, Vaduz, 1985.

Bissett, J.: *Mariannaea* and *Paecilomyces,* Canada Agric., Ottawa, 1979.

Boerema, G. H. and Dorenbosch M. M. J.: *The Phoma* and *Ascochyta Species Described by Wollenweber and Hochapfel in Their Study on Fruit-Rotting,* C.B.S. Stud. Mycol, 1973.

Booth, C.: *Methods in Microbiology,* Academic Press, New York, 1971.

Bougher, N. and Syme, K.: *Fungi of Southern Australia,* University of Western Australia Press, Perth, 1998.

Breitenbach, J. and Kranzlin F.: *Entolomataceae, Pluteaceae, Amanitaceae, Agaricaceae, Coprinaceae, Strophariaceae,* Verlag Mykologia, Luzern, 1995.

——————— : *Strobilomycetaceae, Boletaceae, Paxillaceae, Gomphidiaceae, Hygrophoraceae, Tricholomataceae, Polyporaceae (lamellate),* Verlag Mykologia, Luzern, 1991.

——————— : *Fungi of Switzerland: Non-gilled Fungi,* Verlag Mykologia, Lucerne, 1986.

Brodie, Harold J.: *The Bird's Nest Fungi,* University of Toronto Press, Toronto, 1975.

Brusca, R. C. and Brusca G. J.: *Invertebrates,* Sinauer Associates, Sunderland, 1990.

Buchanan, E. E. and Gibbons N. E.: *Bergey's Manual of Determinative Bacteriology,* Williams and Wilkins, Baltimore, 1974.

Burgess, L. W., Liddell C. M. and Summerell B. A.: *Laboratory Manual for Fusarium Research,* University of Sydney, Sydney, 1988.

Cameron, Eleanor: *Stowaway to the Mushroom Planet*, Little Brown and Company, Boston, 1956.

Carmichael, J. W. and Kendrick W. B.: *Genera of Hyphomycetes*, Alberta Press, Edmonton, 1980.

Chamuris, George P.: *The Non-Stipitate Steriod Fungi in the Northeastern United States and Adjacent*, Berlin, J., Canada, 1988.

Cole, M. Fuhrer B. and Holland A.: *A Field Guide to the Common Genera of Gilled Fungi of Australia*, Inkata Press, Melbourne, 1984.

Cooke, R. C.: *Fungi, Man and His Environment*, Longman, London, 1977.

Corner, E. J. H.: *A Monograph of Cantharelloid Fungi*, Oxford University Press, London, 1996.

———————— : *A Monograph of Clavaria and Allied Genera*, Oxford University Press, London, 1950.

———————— : *Supplement to a Monograph of Clavaria and Allied Genera*, Oxford University Press, London. 1970.

Currah, R. S.: *An Annotated Key to the Genera of the Onygenales*, Systema, Ascomycetum, 1988.

Dashorst, G. R. M. and Jessop J. P.: *Plants of the Adelaide Hills*, Kangaroo Press, Adelaide, 1990.

Delafosse, Claude and Heliadore: *Le Champignon*, Academic Press, France, 1995.

Dennis, R. W. G.: *British Ascomycetes*, J. Cramer, Vaduz, 1981.

Desjardin, Dennis E.: *The Agaricales (Gilled Fungi)*, Mad River Press, California, 1987.

———————— : *Tricholomataceae I: Marasmioid Fungi*, Mad River Press, California, 1987.

Dix, N. J. and Webster, J.: *Fungal Ecology*, Chapman & Hall, London, 1995.

Domsch, K. H. W.: *Compendium of Soil Fungi*, Academic Press, London, 1980.

Dring, D. M.: *Techniques for Microscopic Preparation, Methods in Microbiology,* Academic Press, New York, 1971.

Eberhart, J. L. and Luoma, D. L.: *Truncocolumella Citrina: Pseudotsuga Menziesii,* Victoria B.C., Canadian, 1996.

Ellis, M. B.: *More Dematiaceous Hyphomycetes,* Commonwealth Mycol, Kew, 1976.

Findlay, W. P. K.: *Fungi: Folklore, Fiction and Fact,* Richmond Publishing Company, London, 1982.

Fischer, David W. and Alan E. Bessette: *Edible Wild Mushrooms of North America,* University of Texas Press, Austin, 1992.

Frederick, K. Sparrow and Alfred S.: *A Taxonomic Review with Keys: Ascomycetes and Fungi Imperfecti,* Academic Press, New York, 1973.

Fuhrer, B.: *A Field Companion to Australian Fungi,* Bloomings Books, Melbourne, 2001.

Fuhrer, B. and Robinson R.: *Rainforest Fungi of Tasmania and south-east Australia,* CSIRO, Melbourne, 1992.

Fuller, M. S.: *Lower Fungi in the Laboratory,* Georgia Press, Athens, 1978.

Gilbertson, R. L. and Ryvarden L.: *North American Polypores,* Oslo, Fungiflora, 1987.

————————: *Fungi That Decay Ponderosa Pine,* University of Arizona Press, Tuscon, 1974.

Gilman, J. C.: *A Manual of Soil Fungi,* Iowa State Coll. Press, Ames, 1957.

Ginsburg, Mirra: *Mushrooms in the Rain,* MacMillan Publishing Co., New York, 1974.

Glick, Phyllis G.: *The Mushroom Trail Guide,* Holt, Rinehart and Winston, 1979.

Gregory, P. H.: *The Microbiology of the Atmosphere,* Leonard Hill, Aylesbury, 1973.

Griffiths, K.: *A Field Guide to the Larger Fungi of the Darling Scarp and South West of Western Australia*, Privately Published, Perth, 1985.

Guba, E. F.: *Monograph of Monochaetia and Pestalotia*, Harvard University Press, Cambridge, 1961.

Halling, Roy E.: *The Genus Collybia (Agaricales) in the Northeastern United States and Adjacent*, Cramer, Canada, 1983.

Hawksworth, D. L. Kirk P. M.: *Ainsworth & Bisby's Dictionary of the Fungi*, CAB International, New York, 1995.

Heller, Ruth: *Plants that Never Bloom*, Grosset & Dunlap, New York, 1984.

Hennebert, G. L.: *Echinobotryum, Wardomyces* and *Mammaria*, Trans. Brit. Mycol, 1968.

Hesler, L. R. and Alexander H. Smith: *North American Species of Crepidotus*, Hafner Publishing Company, New York, 1965.

——————— : *North American Species of Hygrophorus*, University of Tennessee Press, Knoxville, 1963.

——————— : *North American Species of Lactarius*, University of Michigan Press, Ann Arbor, 1979.

——————— : *North American Species of Gymnopilus*, Hafner Publishing Company, New York, 1969.

Hoog, G. S. de and J. Guarro: *Atlas of Clinical Fungi*, C.B.S., Baarn, Netherlands, 1995.

——————— : *The Black Yeasts and Allied Hyphomycetes*, C.B.S. Stud. Mycol, 1977.

Hoog, G. S. de, M. TH. Smith and E. Guého: *A revision of Geotrichum*, C.B.S. Stud., Mycol,1986.

Hudler, George: *Magical Mushrooms, Mischievous Molds*, Princeton University Press, Princeton, 1998.

Huffman, Donald M.: *Mushrooms and Other Fungi of the Midcontinental United States*, Iowa State University Press, Iowa, 1989.

Ingold, C. T.: *Dispersal in Fungi*, Oxford University Press, London, 1953.

Inui, T. Takeda Y. and Iizuka H.: *Taxonomical Studies on Genus Rhizopus*, MacMillan, New York, 1965.

Jenkins, David T.: *Amanita of North America*, Mad River Press, Eureka, 1986.

Johnson, Emily: *North American Mushrooms Photo Postcards*, Dover Publications, New York, 1988.

Johnson, Sylvia: *Mushrooms*, Lerner Publishing Company, Minneapolis, 1982.

Jong, S. C. and Davis E. E.: *Contibution to the Knowledge of Stachybotrys and Memnoniella in Culture*, Mycotaxon Press, Sydney, 1976.

Jordan, Michael: *The Encyclopedia of Fungi of Britain and Europe*, David & Charles, Devon, 1995.

Jordan, Peter and Wheeler, Steven: *The Ultimate Mushroom Book*, Smithmark Publishers, New York, 1995.

Katsaros, Peter: *Familiar Mushrooms, North America*, The Audubon Society Pocket Guides, New York, 1990.

——————— : *Illustrated Guide to Common Slime Molds*, Mad River Press, Eureka, 1984.

Kaveler, Lucy: *The Wonders of Fungi*, John Day Co., New York, 1964.

Kaye, T. Liston A.: *Conservation and Management of Native Plants and Fungi*, Native Plant Society of Oregon, Corvallis, 1997.

Kendrick, B.: *The Fifth Kingdom*, Mycologue Publications, Sidney, 2001.

Kendrick, Bryce: *A Young Person's Guide to the Fungi*, Mycologue, Canada, 1986.

Kibby, Geoffrey: *An Illustrated Guide to Mushrooms and Other Fungi of North America*, Dragons World Ltd., London, 1993.

Kirk, P. M. Cannon, P. F.: *Ainsworth and Bisby's Dictionary of the Fungi,* CABI Publishing, Eureka, 2001.

Krahn, Fernando: *Sebastian and the Mushroom,* Delacorte Press, New York,1976.

Kulik, M. M.: *A Compilation of Descriptions of New Penicillium Species,* U.S. Dept., Sydney, 1968.

Kuyper, Thomas W.: *A Revision of the Genus Inocybe in Europe,* Rijksherbarium, Leiden, 1986.

——————— : *Subgenus Insperma and the Smooth—Spored Species of Subgenus Inocybe,* Rijksherbarium, Leiden, 1986.

Laessoe, Thomas and Lincoff, Gary: *Eyewitness Handbooks/ Mushrooms,* D.K. Publishing, New York, 1998.

Laessoe, Thomas: *The Mushroom Book,* D.K. Publishing, New York, 1996.

Largent, David L. and Harry D.: *How to Identify Mushrooms to Genus II: Field Identificaton of Genera,* Mad River Press, Eureka, 1987.

Largent, David L.: *Entolomatoid Fungi of the Western United States and Alaska,* Mad River Press, Eureka, 1994.

——————— : *How to Identify Mushrooms to Genus?: Macroscopic Features,* Mad River Press, Eureka, 1986.

——————— : *The Agaricales (Gilled Fungi) of California: Hygrophoraceae,* Mad River Press, Eureka, 1985.

Larsen, M. J.: *Phellinus (Hymenochaetaceae): A Survey of World Taxa,* Fungiflora, Oslo, 1990.

Larson, D. W.: *Laboratory Response of Net CO_2 Exchange to Moisture, Temperature and Light,* J. Bot, Canadian, 1987.

Lilly, V. G.: *The Chemical Environment for Fungal Growth,* Academic Press, New York, 1965.

Lincoff, Gary H.: *The Audobon Society Field Guide to North American Mushrooms,* Alfred A. Knopf, New York, 1981.

——————— : *Simon and Schuster's Guide to Mushrooms,* Simon and Schuster, New York, 1981.

Lingelbach, Jenipher: *Hands-on Nature,* Vermont Institute of Natural Science, Woshington, 1985.

Lodder, J.: *The Yeasts: A Taxonomic Study,* North Holland Publ. Co., Amsterdam, 1970.

Luoma, D. L. and Trappe J. M.: *Mycorrhizal Fungi in Ecosystems,* Annual Meeting & Symposium, Bend, 1992.

Luoma, D. L. Eberhart J. L.: *Biodiversityof Ectomycorrhizal Types from Southwest Oregon,* Native Plant Society of Oregon, Corvallis, 1997.

————————— : *Communitystructure and Dynamics of Ectomycorrhizal Fungi in Managed Forest Stands,* USDA Forest Service Gen.Tech., Portland, 1996.

Macdonald, R. and Westerman, J.: *A Field Guide to Fungi of South-eastern Australia,* Nelson, Melbourne, 1979.

Margulis, L. and Schwartz K. V.: *Five Kingdoms: An Illustrated Guide to the Phyla of Life on Earth,* W. H. Freeman, New York, 1998.

Marr, Currie D. and Daniel E. Stuntz: *Ramaria of Western Washington (Bibliotheca Mycologica, Band 38),* J. Cramer, Vaduz, 1973.

Martin, G. W.: *Revision of the North Central Tremellales,* State University of Iowa, Iowa, 1952.

Matsushima, T.: *Lcones Fungorum a Matsushima Lectorum,* Kobe, Japan, 1975.

Matthes-Sears, U.: *In Situ Diurnal Field Measurements of Two Sites on a Coast-inland Gradient,* J. Bot, Canadian, 1986.

McCann, I. R.: *Australian Fungi Illustrated,* Macdown Productions, Melbourne, 2003.

McGinnis, M. R.: *Laboratory Handbook of Medical Mycology,* Academic Press, New York, 1980.

McKenny, Margaret and Daniel E.: *The New Savory Wild Mushroom,* University of Washington Press, Seattle, 1987.

———————: *The New Savory Wild Mushroom*, University of Washington Press, Washington, 1987.

McKnight, Kent H. and McKnight, Vera B.: *Peterson Field Guides, Mushrooms*, Houghton—Mifflin, Boston, 1987.

Methven, Andrew S.: *The Agaricales (Gilled Fungi) of California*, Mad River Press, Eureka, 1997.

———————: *The Genus Clavariadelphus in North America*, J. Cramer, Berlin, 1990.

Metzler, Susan and Metzler, Van: *Texas Mushrooms: A Field Guide*, University of Texas Press, Austin, 1992.

Miller, Orson K.: *Mushrooms of North America*, E. P. Dutton, New York, 1980.

Molina R. J.: *Developingan Inventory and Monitoring Protocol for Commercially Harvested Forestmushrooms*, Hillsboro, New Jersey, 1994.

Monsell, Mary Elise: *Crackle Creek*, Atheneum, New York, 1990.

Morquer, R. G. Viala, J.: *Contributiona l'étude Morphogénique du Genre Gliocladium*, Mycol, France, 1963.

Morton, F. J. and Smith G.: *The Genera Scopulariopsis, Microascus* and *Doratomyces*, Mycol, Mycol, 1963.

Moser, Meinhard: *Keys to Agarics and Boleti (Polyporales, Boletales, Agaricales, Russulales)*, Roger Phillips, London, 1983.

Nag, Raj T. R.: *Coelomycetous Anamorphs with Appendage-bearing Conidia*, Mycologue Publications, Canada, 1993.

O'Dell, T. E.: *Diversity and Conservation of Forest Fungi*, USDA Forest Service Gen. Tech., Portland, 1996.

Orr, F. G.: *The Genus Gymnoascus Baranetzky*, Mycopathol, Mycol, 1963.

Orr, Robert T. and Dorthy B. Orr: *Mushrooms of Western North America*, University of California Press, Bereley, 1979.

Orton, P. D.: *British Fungus Flora, Agarics and Boleti, Pluteaceae, Pluteus & Volvariella*, Royal Botanic Garden, Edinburgh, 1986.

Pegler, D. N. Læssøe T. and Spooner B. M.: *British Puffballs, Earthstars and Stinkhorns,* Royal Botanic Gardens, Kew, 1995.

——————— : *British Truffles: A Revision of British Hypogeous Fungi,* Royal Botanic Gardens, Kew, 1993.

——————— : *The Genus Lentinus: A World Monograph,* HMSO, London, 1983.

Perry, Phyllis J.: *Let's Learn About Mushrooms,* Harvey House, New York, 1974.

Petersen, Ronald H.: *The Genus Clavulinopsis in North America,* Hafner Publishing Co., New York, 1968.

——————— : *Ramaria subgenus Echinoramaria,* J. Cramer, Vaduz, 1981.

Phillips, Roger: *Mushrooms of North America,* Brown, & Co., Boston, 1991.

Pitt, J.: *The Genus Penicillium,* Academic Press, New York, 1980.

Ramirez, C.: *Manual and Atlas of the Penicillia,* Elsevier Biomedical Press, New York, 1982.

Raper, K. B. and Fennell D. I.: *The Genus Aspergillus,* Williams and Wilkins Co., Baltimore, 1965.

Rebell, G. and Taplin D.: *Dermatophytes: Their Recognition and Identification,* University Miami Press, Coral Gables, 1970.

Rice, Miriam: *Mushrooms for Color,* Mad River Press, Eureka, 1980.

Rifai, M. A.: *A Revision of the Genus Trichoderma,* Mycol, 1969.

Robinson, R.: *Fungi of the South-West Forests,* Department of Conservation and Land Management, Perth, 2003.

Romanova, Natalia: *Once There Was a Tree,* Dial Books, New York, 1985.

Samson, R. A. Hoekstra E. S.: *Interoduction to Food-borne Fungi,* C.B.S., Baarn, Netherlands, 1995.

———————— : *Revision of the Subsection Fasciulata of Penicillium and Some Allied Species,* C.B.S. Stud, Mycol, 1976.

———————— : *Paecilomyces* and *Some Allied Hyphomycetes,* C.B.S. Stud, Mycol, 1974.

Schaechter, Elio: *In the Company of Mushrooms,* Harvard University Press, Boston, 1997.

Schipper, M. A. A. and Stalpers J. A.: *A Revision of the Genus Rhizopus,* C.B.S. Stud., Mycol, 1984.

———————— : *Zygospore Ornamentation in the Genera Mucor and Zygorhynchus,* Persoonia, New York, 1975.

Selsam, Millicent: *Mushrooms,* William Morrow and Co., London, 1986.

Seth, H. K.: *A Monograph of the Genus Chaetomium,* Beih. Nova Hedw, 1970.

Shanks, Kris M.: *Gilled Fungi,* Mad River Press, Eureka, 1997.

Shepherd, C. J.: *Mushrooms and Toadstools of Australia,* Inkata Press, Sydney, 1988.

Shoemaker, R. A. and Kokko E. G.: *Trichurus Spiralis,* Canada Agric., Ottawa, 1977.

Short, P. S.: *History of Systematic Botany in Australasia,* Australian Systematic Botany Society, Melbourne, 1990.

Shuttleworth, Floyd S.: *A Golden Guide, Non-Flowering Plants,* Golden Press, 1967.

Singer, Rolf: *The Agaricales in Modern Taxonomy,* Koeltz Scientific Books, Koenigstein, 1986.

Sisson, Edith H.: *Nature With Children of All Ages,* Prentice Hall Press, New York, 1982.

Smith, A. H. and Zeller S. M.: *A Preliminary Account of the North American Species of Rhizopogon,* Botanical Gardens, Bronx, New York, 1966.

Smith, A. H. and Thiers H. D.: *The Boletes of Michigan,* University of Michigan Press, Ann Arbor, 1971.

Smith, Alexander H.: *The North American Species of Pholiota,* Hafner Publishing Company, New York, 1968.

——————— : *North American Species of Mycena,* University of Michigan Press, Ann Arbor, 1947.

——————— : *A Field Guide to Western Mushrooms,* University of Michigan Press, Ann Arbor, 1975.

Snell, Walter H. and Esther A. Dick: *A Glossary of Mycology,* Harvard University Press, Cambridge, 1971.

——————— : *The Boleti of Northeastern North America,* J. Cramer, Lehre, 1970.

Stamets, Paul: *Psilocybin Mushrooms of the World,* Ten Speed Press, Berkeley, 1996.

States, Jack S.: *Mushrooms and Truffles of the Southwest,* University of Arizona, Tucson, 1990.

Stevenson, Steven L. and Stempen, Henry: *Myxomycetes, A Handbook of Slime Molds,* Timber Press, Portland, 1994.

St.Germain, G. and R. Summerbell: *Identifying Filamentous Fungi,* Star Publ. Co., Belmont, California, 1996.

Stolk, A. C. and Samson R. A.: *The Ascomycete Genus Eupenicillium and Related Penicillium Anamorphs,* C.B.S. Stud. Mycol, 1983.

Storer, T. I., Usinger R. L.: *General Zoology,* McGraw-Hill Book Company, New York, 1979.

Tesar, Jenny: *Fungi, Our Living World Series,* Blackbirch Press, Woodbridge, 1994.

Thiers, Harry D.: *Gomphidiaceae,* Mad River Press, Eureka, 1985.

Tokumasu, S.: *Notes on Japanese Oidiodendron,* Trans. Mycol, Japan, 1973.

Tylutki, Edmund E.: *Mushrooms of Idaho and the Pacific Northwest, Discomycetes,* University of Idaho Press, Moscow, 1979.

——————— : *Mushrooms of Idaho and the Pacific Northwest: Non-Gilled Hymenomycetes,* University of Idaho Press, Moscow, 1987.

Ulloa, Miguel and Richard T. Hanlin: *Illustrated Dictionary of Mycology*, APS Press, Paul, 2000.

Van Oorschot, C. A. N.: *Taxonomy of the Dactylaria Complex, V. A Review of Arthrobotrys and Related Genera,* C.B.S. Stud., Mycol, 1985.

Waksman, S. A.: *The Actinomycetes,* Williams and Wilkins Co., Baltimore, 1961.

Wang, C. J. K.: *Identification Manual for Fungi from Utility Poles in the Eastern United States,* Type Culture Collection, Rockville, 1990.

Watling, Roy and Evelyn Turnbull: *British Fungus Flora: Agarics and Boleti,* Royal Botanic Garden, Edinburgh, 1998.

Watling, Roy, Norma M. Gregory and Orton P. D.: *British Fungus Flora, Agarics and Boleti,* Royal Botanic Garden, Edinburgh, 1993.

Watling, Roy: *Fungi,* Smithsonian Institution Press, Washington, 2003.

——————: *How to Identify Mushrooms to Genus?: Cultural and Developmental Features,* Mad River Press, Eureka, 1987.

Weber, Nancy Smith: *A Morel Hunter's Companion,* Two Penninsula Press, Lansing, 1988.

Webster, Vera R.: *Plant Experiments,* Childrens Press, Chicago, 1982.

Young, A. M.: *Common Australian Fungi,* New South Wales University Press, Sydney, 2000.

Zambettakis, C.: *Recherches Anatomiqueset Biologiques Sur Sphaeropsidales-Phaeodidymae Des Fungi Imperfecti,* Arch. Mus. Nation, Nat, 1955.

——————: *Recherches Sur la Systématique Des Sphaeropsidales-Phaeodidymae,* Mycol, France, 1954.

Zycha, H. and Siepmann R.: *Mucorales, Eine Beschreibung Aller Gattungen and Arten Dieser Pilzgruppe,* PHBS Press, Lehre, 1969.

Index

D

E

F

G

H

I

U

V

W

X

Z

❑❑❑